CERI Resource Center

STRATIGRAPHICAL ATLAS OF FOSSIL FORAMINIFERA
Second Edition

ELLIS HORWOOD SERIES IN GEOLOGY

Editors: D. T. DONOVAN, Professor of Geology, University College London, and J. W. MURRAY, Professor of Geology, University of Exeter

This series aims to build up a library of books on geology which will include student texts and also more advanced works of interest to professional geologists and to industry. The series will include translation of important books recently published in Europe, and also books specially commissioned.

RADIOACTIVITY IN GEOLOGY: Principles and Applications
E. M. DURRANCE, Department of Geology, University of Exeter
FAULT AND FOLD TECTONICS
W. JAROSZEWSKI, Faculty of Geology, University of Warsaw
A GUIDE TO CLASSIFICATION IN GEOLOGY
J. W. MURRAY, Department of Geology, University of Exeter
THE CENOZOIC ERA: Tertiary and Quaternary
C. POMEROL, University of Paris VI
Translated by D. W. HUMPHRIES, Department of Geology, University of Sheffield, and E. E. HUMPHRIES
Edited by D. CURRY and D. T. DONOVAN, Department of Geology, University College London

BRITISH MICROPALAEONTOLOGICAL SOCIETY SERIES

This series, published by Ellis Horwood Limited for the British Micropalaeontological Society, aims to gather together knowledge of a particular faunal group for specialist and non-specialist geologists alike. The original series of Stratigraphic Atlas or Index volumes ultimately will cover all groups and will describe and illustrate the common elements of the microfauna through time (whether index or long-ranging species) thus enabling the reader to identify characteristic species in addition to those of restricted stratigraphic range. The series has now been enlarged to include the reports of conferences, organized by the Society, and collected essays on specialist themes.
The synthesis of knowledge presented in the series will reveal its strengths and prove its usefulness to the practising micropalaeontologist, and to those teaching and learning the subject. By identifying some of the gaps in this knowledge, the series will, it is believed, promote and stimulate further active research and investigation.

PALAEOBIOLOGY OF CONODONTS
Editor: R. J. ALDRIDGE, Department of Geology, University of Nottingham
CONODONTS: Investigative Techniques and Applications
Editor: R. L. AUSTIN, Department of Geology, University of Southampton
FOSSIL AND RECENT OSTRACODS
Editors: R. H. BATE, British Museum Natural History, London, and Stratigraphic Services International, Guildford, E. ROBINSON, Department of Geology, University College London, and L. SHEPPARD, British Museum Natural History, London, and Stratigraphic Services International, Guildford
NANNOFOSSILS AND THEIR APPLICATIONS
Editor: J. A. CRUX, British Petroleum Research Centre, Sunbury-on-Thames. and S. E. VAN HECK, Shell Internationale Petroleum, Mattschappij
A STRATIGRAPHICAL INDEX OF THE PALAEOZOIC ACRITARCHS AND OTHER MARINE MICROFLORA
Editors: K. J. DORNING, Pallab Research, Sheffield, and S. G. MOLYNEUX, British Geological Survey, Nottingham
MICROPALAEONTOLOGY OF CARBONATE ENVIRONMENTS
Editor: M. B. HART, Department of Geological Studies, Plymouth Polytechnic **A STRATIGRAPHICAL INDEX OF CONODONTS**
Editors: A. C. HIGGINS, Geological Survey of Canada, Calgary, and R. L. AUSTIN, Department of Geology, University of Southampton
STRATIGRAPHICAL ATLAS OF FOSSIL FORAMINIFERA, 2nd Edition
Editors: D. G. JENKINS, The Open University, and J. W. MURRAY, Department of Geology, University of Exeter
A STRATIGRAPHICAL INDEX OF CALCAREOUS NANNOFOSSILS
Editor: A. R. LORD, Department of Geology, University College London
MICROFOSSILS FROM RECENT AND FOSSIL SHELF SEAS
Editors: J. W. NEALE, and M. D. BRASIER, Department of Geology, University of Hull
THE STRATIGRAPHIC DISTRIBUTION OF DINOFLAGELLATE CYSTS
Editor: A. J. POWELL, British Petroleum Research Centre, Sunbury-on-Thames
OSTRACODA
Editors: R. C. WHATLEY and C. MAYBURY, Department of Geology, University College of Wales

ELLIS HORWOOD SERIES IN APPLIED GEOLOGY

Published by Ellis Horwood Limited for the Institution of Geologists, Berlington House, Picadilly London W1V 9AG
The books listed below are motivated by the up-to-date applications of geology to a wide range of industrial and environmental factors; they are practical, for use by the professional and practising geologist or engineer in the field, for study, and for reference.

A GUIDE TO PUMPING TESTS
F. C. BRASSINGTON, North West Water Authority
QUATERNARY GEOLOGY: Processes and Products
JOHN A. CATT, Rothamsted Experimental Station, Harpenden
TUNNELLING GEOLOGY AND GEOTECHNICS
Editors: M. C. KNIGHTS and T. W. MELLORS, Consulting Engineers, W. S. Atkins & Partners
PRACTICAL PEDOLOGY: Manual of Soil Formation, Description and Mapping
S. G. McRAE, Department of Environmental Studies and Countryside Planning, Wye College (University of London)
MODERN STANDARDS FOR AGGREGATES
D. C. PIKE, Consultant in Aggregates, Reading
LASER HOLOGRAPHY IN GEOPHYSICS
S. TAKEMOTO, Disaster Prevention Research Unit, Kyoto University

STRATIGRAPHICAL ATLAS OF FOSSIL FORAMINIFERA

Second Edition

Editors:

D. G. JENKINS, B.Sc., Ph.D., D.Sc.
Department of Earth Sciences
The Open University, Milton Keynes

and

J. W. MURRAY, B.Sc., A.R.C.S., Ph.D., D.I.C., F.G.S.
Department of Geology
University of Exeter

QE
772
STR
1989

Published by
ELLIS HORWOOD LIMITED
Publishers · Chichester

for
THE BRITISH MICROPALAEONTOLOGICAL SOCIETY

Halsted Press: a division of
JOHN WILEY & SONS
New York · Chichester · Brisbane · Toronto

First published in 1989
ELLIS HORWOOD LIMITED
Market Cross House, Cooper Street,
Chichester, West Sussex, PO19 1EB, England
*The publisher's colophon is reproduced from James
Gillison's drawing of the ancient Market Cross, Chichester.*

Distributors:
Australia and New Zealand:
JACARANDA WILEY LIMITED
GPO Box 859, Brisbane, Queensland 4001, Australia

Canada:
JOHN WILEY & SONS CANADA LIMITED
22 Worcester Road, Rexdale, Ontario, Canada

Europe and Africa:
JOHN WILEY & SONS LIMITED
Baffins Lane, Chichester, West Sussex, England

North and South America and the rest of the world:
Halsted Press: a division of
JOHN WILEY & SONS
605 Third Avenue, New York, NY 10158, USA

South-East Asia
JOHN WILEY & SONS (SEA) PTE LIMITED
37 Jalan Pemimpin # 05–04
Block B, Union Industrial Building, Singapore 2057

Indian Subcontinent
WILEY EASTERN LIMITED
4835/24 Ansari Road
Daryaganj, New Delhi 110002, India

© **1989 D.G. Jenkins and J.W. Murray/Ellis Horwood Limited**

British Library Cataloguing in Publication Data
Stratigraphical atlas of fossil formaminifera. — 2nd ed.
1. Fossil Feraminiferida
I. Jenkins D.G. (David Graham), *1933–*
II. Murray, John W. (John William) *1937–*
III. Series
563'.12

Library of Congress Card No. 88–13890

ISBN 0–7458–0153–6 (Ellis Horwood Limited)
ISBN 0–470–21226–8 (Halsted Press)

Printed in Great Britian by
The Camelot Press, Southampton

COPYRIGHT NOTICE
All Rights Reserved. No part of this publication may be reproduced, stored in a retrieval system, or transmitted, in any form or by any means, electronic, mechanical, photo-copying, recording or otherwise, without the permission of Ellis Horwood Limited, Market Cross House, Cooper Street, Chichester, West Sussex, England.

Table of Contents

Preface to the first edition 9
Preface to the second edition 11

Chapter 1   **Some Early Students of Foraminifera in Britain** 13
J. W. Murray, Department of Geology, University of Exeter, Exeter EX4 4QE (from October 1989, University of Southampton, Southampton SO9 5NH)

Chapter 2   **Cambrian to Devonian** 20
J. M. Kircher, Paleoservices Ltd., Unit 15, Paramount Industrial Estate, Sandown Road, Watford, WD2 4XA
M. D. Brasier, Department of Earth Science, University of Oxford, Parks Road, Oxford OX1 3PR

Chapter 3   **Carboniferous** 32
M. D. Fewtrell, The British Petroleum Company plc, BP Research Centre, Chertsey Road, Sunbury-on-Thames, Middlesex TW16 7LN
W. H. C. Ramsbottom, Brow Cottage, Kirby Malzeard, Ripon, North Yorksire HG4 3KY
A. R. E. Strank, The British Petroleum Company plc, BP Research Centre, Chertsey Road, Sunbury-on-Thames, Middlesex TW16 7LN

Chapter 4   **Permian** 87
*J. Pattison, British Geological Survey, Keyworth, Nottingham NG12 5GG

Chapter 5   **Triassic** 97
P. Copestake, BP Exploration, 301 St. Vincent Street, Glasgow G2 5DD

Chapter 6   **Jurassic** 125
        6.1 Introduction 125
           P. Copestake, BP Exploration, 301 Vincent Street, Glasgow G2 5DD
        6.2 The Hettangian to Toarcian (Lower Jurassic) 129
           P. Copestake, BP Exploration, 301 Vincent Street, Glasgow G2 5DD

Table of Contents

        B. Johnson, Gearhart Geoconsultants, Howe Moss Place, Kirkhill Industrial Estate, Dyce, Aberdeen AB2 0GL

  6.3 **The Aalenian to Callovian (Middle Jurassic)** 189
        P. H. Morris, GeoStrat Ltd., Unit 23A, Motherwell Business Centre, Dalziel Street, Motherwell ML1 1PJ
        B. E. Coleman, 59 Malvern Close, Mitcham CR4 1EH

  6.4 **The Oxfordian to Portlandian (Upper Jurassic)** 237
        D. J. Shipp, Robertson Research International Ltd., Ty'n-y-Coed, Llanrhos, Llandudno, Gwynedd LL30 1SA

Chapter 7   **Cretaceous** 273
        M. B. Hart, Department of Geological Sciences, Polytechnic South West, Drake Circus, Plymouth PL4 8AA
        H. W. Bailey, Paleoservices Ltd., Unit 15, Paramount Industrial Estate, Sandown Road, Watford WD2 4XA
        S. Crittenden, Gearhart Geoconsultants Ltd., Howe Moss Close, Kirkhill Industrial Estate, Dyce, Aberdeen AB2 0GL
        *B. N. Fletcher, British Geological Survey, Keyworth, Nottingham NG12 5GG
        R. Price, Amoco Canada Petroleum Company Ltd., Calgary, Alberta T2P 072
        A. Swiecicki, BP Thailand, Sitthivorakit Building, 5, Soi Tpat, Siolm Road, Bangkok, Thailand

Chapter 8   **Cretaceous of the North Sea** 372
        C. King, H. W. Bailey, C. Burton and A. D. King, all of Paleoservices Ltd., Unit 15, Paramount Industrial Estate, Sandown Road, Watford WD2 4XA

Chapter 9   **Cenozoic of the North Sea** 418
        C. King, Paleoservices Ltd., Unit 15, Paramount Industrial Estate, Sandown Road, Watford WD2 4XA

Chapter 10   **Palaeogene** 490
        J. W. Murray, Department of Geology, University of Exeter, Exeter EX4 4QE (from October 1989, University of Southampton, Southampton SO9 5NH)
        D. Curry, Mallard Creek, Spinney Lane, Itchenor, Chichester, PO20 7DJ
        J. R. Haynes, Institute of Earth Studies, University College of Wales, Aberystwyth SY23 3DB
        C. King, Paleoservices Ltd., Unit 15, Paramount Industrial Estate, Sandown Road, Watford WD2 4XA

Table of Contents

Chapter 11 **Neogene** 537
D. G. Jenkins, Department of Earth Sciences, The Open University, Walton Hall, Milton Keynes MK7 6AA
C. King, Paleoservices Ltd., Unit 15, Paramount Industrial Estate, Sandown Road, Watford WD2 4XA
*M. Hughes, British Geological Survey, Keyworth, Nottingham NG12 5GG

Chapter 12 **Quaternary** 563
B. M. Funnell, School of Environmental Sciences, University of East Anglia, Norwich NR4 7TJ

Chapter 13 **An Outline of Faunal Changes through the Phanerozoic** 570
J. W. Murray, Department of Geology, University of Exeter, Exeter EX4 4QE (from October 1989, University of Southampton, Southampton SO9 5NH)

General Index 574

Index of genera and species 578

*These authors publish with permission of the Director, British Geological Survey.

Preface to the first edition

The British Micropalaeontological Society may be said to have come of age when it published *A Stratigraphical Index of British Ostracoda* in 1978. This was the first of a series of special publications aimed at bringing together knowledge for a particular faunal group in order to make it more useful to specialists and non-specialists alike. The present volume has been written with the same objectives in mind.

Once the idea had been mooted, a group of prospective contributors met at the Institute of Geological Sciences in Leeds during the course of a Foraminifera Group weekend conference. General agreement was reached on the format and content which to a large extent follows that of the ostracod precursor. However, one major difference of approach is that we agreed to link our foraminiferal data to lithostratigraphical rather than biostratigraphical successions. The reasons for this are that it is easier to re-sample accurately with reference to a lithological succession at a specific locality; also biostratigraphical divisions (zones, subzones) may have their boundaries revised in the light of new evidence. Where possible we have correlated our lithostratigraphical stratigraphy with zones and stages.

Our original intention was to restrict the volume to include only species of short time range, i.e. true index species. However, this proved to be impractical for two reasons. First, most of our faunas are benthic, tied to facies, and therefore not truly index forms. Second, strict application of this approach led to some fossiliferous parts of the succession appearing to be barren because their faunas are made up of long-ranging species. It was therefore decided to broaden the scope to include the common elements of the fauna whether index or long-ranging species. In this way we hope to convey a general impression of the fauna, allowing the user to identify the common species as well as those of restricted stratigraphical range. For economy of space the original name is given for each species so that readers may have ready access to the type description of the Catalog of Foraminifera (Ellis and Messina, 1940 et seq.).

There has been a general policy not to include too much data from offshore because many users have no access to such material and also because problems of confidentiality limit the detail which can be given. Nevertheless, in the chapters on the Mesozoic, similarities and differences between onshore and offshore faunas are discussed and for the Cenozoic we have a chapter summarizing the principal features of the North Sea succession. The chapter on the Neogene depends heavily on offshore material as the successions on land are incomplete.

Inevitably a synthesis of knowledge such as we present here reveals not only its strength but also its weaknesses. We hope it will prove to be useful to the practising micropalaeontologist and also to those who are being trained in the subject. We believe that by identifying some of the gaps in our knowledge we shall promote further active research.

All the contributors to this volume volunteered their services. The editors are indebted to them for their help and their patience during the two year period of gestation.

The editors wish to record their gratitude for the interest, help and encouragement given by Dr. C. G. Adams during the preparation of this work.

D. Graham Jenkins

October, 1980   John W. Murray

Preface to the second edition

The authors and editors were gratified by the enthusiastic reception afforded to the publication of the first edition in 1981. The preparation of a second edition was prompted firstly by the wealth of new information which has become available (especially from the offshore areas) and secondly by the continued demand for the volume even after it went out of print.

A couple of the first edition reviewers commented on the absence of 'British' from the title and implied that this was misleading. We do not believe this to be so for either edition: this publication is part of the British Micropalaeontological Series; also there is extensive cross-referencing to distributions outside Britain and the ranges of many of the pre-Cenozoic taxa are applicable to Europe and in some cases to parts of North America.

Our objectives have been much the same as in the first edition. However, we have provided range charts based on 'tops' to assist industrial colleagues. There is significant new information for the Cambrian to Devonian, Triassic, Jurassic and for the Cretaceous and Cenozoic of the North Sea. New zonal schemes are given for the Jurassic and for the Cretaceous.

We would like to thank all the contributors for their willingness to participate in this team venture. Professor John Haynes read most of the typescript and offered valuable advice. We are indebted to him for his help and encouragement. Mrs. Judith Jenkins gave invaluable help in the editing and compilation of the index.

D. Graham Jenkins

July, 1988 John W. Murray

1

Some early students of foraminifera in Britain

J. W. Murray

The scientific discoveries of one generation are built upon the foundations laid by earlier workers. The synthesis of data presented in this book may appear to rest largely on the labours of micropalaeontologists working in the twentieth century, but we should not overlook the contributions made by earlier workers, many of whom earned their living in business or in other fields of science or medicine. These early micropalaeontologists made contributions to science which were fundamental and their value is recognised throughout the world. Fortunately many of the collections on which these contributions were based are still available for examination (Adams et al., 1980).

It is perhaps natural that many of these earlier studies should have been on Recent material from the shores of the British Isles, from the continental shelf and from the ocean basins. There were many naturalists who took part in scientific voyages of discovery during the nineteenth century and who were prepared to exchange material with colleagues. Early publications reveal a great deal about this collaboration.

As can be seen from the brief survey presented below, the contribution of British workers to the study of Recent foraminifera was large. However, the fossil forms were somewhat neglected especially in comparison with the labours of European workers in the same time period. By the end of the nineteenth century something was known of Carboniferous, Permian, Lower Cretaceous, Palaeocene and Plio-Pleistocene foraminifera of Britain but very little was known of the Mesozoic in general, or of the Eocene and Oligocene. Indeed, it is really only since the 1940s that much of the British stratigraphical record has received serious attention.

In each chapter there is a historical survey of past work and here it is intended only to emphasise the role of prominent researchers in the period up to the 1930s.

The study of foraminifera started as a hobby for eighteenth century gentlemen who took an interest in microscopy. The earliest of these was William Boys of Sandwich in Kent, a surgeon and ardent naturalist who first studied foraminifera from shore sand. Together with Walker he published descriptions of various small animals (Walker and Boys, 1784) to which binomial names were applied by Walker and Jacob (1794). Colonel George Montagu (1803, 1808) described foraminifera from the south coast. However, the really major contributions were made by W. C. Williamson, W. B. Carpenter, and H. B. Brady.

William Benjamin Carpenter, 1813–1835, (Plate 1.1, Fig. 1) was born in Exeter, but spent most of his childhood in Bristol. In 1828 he commenced studies at the Bristol Medical School, continued at University College, London, in 1834 and Edinburgh in 1835, graduating MD in 1839. He won an essay prize of £30 which he used to purchase a microscope. He commenced a lectureship in Medical Jurisprudence in the Bristol Medical School in 1837 and developed a keen interest in physiology. He was Fullerian Professor of Physiology at the Royal Institution, London, from 1844 to 1847, then Swineyan Lecturer in Geology at the British Museum, 1847, Professor of Medical Jurisprudence at University College, London, 1849, and finally Registrar of the University of London from 1856 until 1879.

It was probably in the 1840s that Carpenter commenced his interest in foraminifera and he published a series of papers from 1849 to 1883 (see Murray and Taplin, 1984, for a complete bibliography), including his classic work *Introduction to the study of Foraminifera* (1862). He was particularly interested in the detailed structure and morphology of larger foraminifera and he worked on material collected on early oceanographic cruises, such as those of the 'Porcupine', 'Lightning', 'Valorus' and 'Challenger'. He also published widely on medical matters, microscopy in general, and on echinoderms.

In 1816, William Crawford Williamson (Plate 1.1, Fig. 2) was born in Scarborough. His father was a gardener with a keen interest in geology, entomology and conchology, and he served as curator of Scarborough Museum for 27 years. This interest in natural history had a strong influence on his son. In 1832, W. C. Williamson started medical studies, but these were interrupted in 1835 when, at the age of 19, he became curator of the Manchester Museum. In 1838 he resumed his medical studies and during 1840 was a student at University College, London.

On 1 January 1841, Williamson started a medical practice in Manchester and found that he had plenty of time to indulge his interest in natural history. In 1842 he read an abstract of Ehrenberg's work on chalk and was keen to see the microscopic objects described. A friend, Mr Sidebotham, received a sample of marine mud from the Levant, i.e. the eastern Mediterranean, supplied by Mr Reckitt of Boston. This was examined under the microscope and Williamson saw his first foraminiferid: '. . . my eye fell upon an object which made me spring out of my chair'. (Williamson, in Watson and Thomas, 1985, p. 105). His enthusiasm was such that he purchased a new microscope, obtained his own supply of Levant mud from Reckitt, and corresponded with a wide circle of scientists who supplied more material: Bailey, Darwin, Carpenter, Mantell. His studies led to the 1848b paper on *Lagena* which was well received and stimulated Williamson to undertake further work. One technique that Williamson pioneered was to attempt to separate foraminifera from sediment by flotation using dried sample sprinkled into water.

Plate 1.1
Fig. 1. – W. B. Carpenter.
Fig. 2. – W. C. Williams.
Fig. 3. – H. B. Brady.
Reproduced with permission of The Royal Society (1), the Vice Chancellor of Manchester University (2), and Dr C. G. Adams, BMNH (3).

Plate 1.1

In 1848 Williamson and Carpenter agreed to produce '... some elaborate work on the foraminifera'. Williamson was to work on the Recent forms and Carpenter on a general study, and both were published by the Ray Society (in 1858 and 1862 respectively).

After completion of this work, Williamson carried out no further research on foraminifera. He published two further notes: one on the anatomy and physiology of foraminifera (1865) and another of corrections to the Levant mud paper of 1872 (1876). Because of his major contribution to early foraminiferal studies, a complete list of his publications is given in the references.

Williamson was a distinguished surgeon, was Professor of Natural History at Owen's College from 1851 to 1888, during which he taught botany, zoology and geology, and published numerous papers on a great variety of topics, but especially on fossil plants. He died in 1895.

William Bowman Brady, 1835–1891, (Plate 1.1, Fig. 3) is best known for his major contributions to the Recent foraminifera collected on the Challenger Expedition (Brady, 1884). He was born in Gateshead, the son of a medical practitioner and surgeon. In 1855 he qualified as a pharmacist and set himself up as a wholesale and retail pharmacist in Newcastle-upon-Tyne. His business flourished and he took a prominent role in the affairs of the British Pharmaceutical Conference and the Pharmaceutical Society. He was also a very keen naturalist. At the age of 41 (in 1876) he had already published 20 papers on foraminifera and he retired from pharmacy to devote himself to their study. The Challenger Report (1884) has been described by Adams (1978) as setting '... a new standard for British Students of Foraminifera and ensured that the name of its author would be as familiar today as those of Alcide d'Orbigny and Joseph A. Cushman'.

Brady showed an interest in ecology and biogeography. Adams (1978) believes that Brady must have been one of the first researchers to stress the importance of preserving Recent material in alcohol so that living and dead tests could be distinguished.

Brady also studied fossil foraminifera (1876) and he collaborated with Carpenter in studies of *Parkeria* (which is now known not to be foraminiferal) and *Loftusia* (Carpenter and Brady, 1870).

William Kitchen Parker, 1823–1890, at the age of fifteen was apprenticed to a pharmacist and three years later to a medical practitioner. During this time he collected and named plants, studied dissections and prepared skeletons of various animals. In 1844 he went to London where he trained to become a Licentiate of the Society of Apothecaries and then practised in Pimlico. He made significant contributions to the study of foraminifera. Together with Thomas Rupert Jones, he published during the period 1857 to 1871 numerous papers on the taxonomy of Recent species, especially those erected by Linnaeus, Gmelin, Walker, von Fichtel and von Moll, Lamarck, Denys de Montfort, de Blainville, Defrance, d'Orbigny, Ehrenberg and Batsch. He also assisted W. B. Carpenter in the preparation of *Introduction to the study of the Foraminifera*. Apart from this he published widely on the skeletal anatomy of birds, both modern and fossil, and various other vertebrates. He was elected Fellow of the Royal Society in 1865 and had numerous other honours bestowed upon him. In 1873 he became Hunterian Professor of Comparative Anatomy and Physiology at Kings College, London, (Jones, 1891).

Fortescue William Millett, 1833–1915, became interested in foraminifera in 1883 when he retired to south-west England. In later years he became a recluse (Sherborn, 1915). He is best known for his work on the Malay Archipelago (1899–1904) and for his studies of the St Erth Clays. For the latter he was awarded the Bolitho Medal by the Royal Geological Society of Cornwall (Anon. 1915(a)). On his death his collection of 10 000 slides was bought by Heron-Allen. (Anon. 1915(b)).

Edward Heron-Allen and Arthur Earland published numerous joint works. Heron-Allen, 1861–1943, trained as a solicitor and succeeded his father as head of the family firm of solicitors. He retired at 50 and devoted himself to his hobby, microscopy. In 1907 he contacted Earland to seek help in identifying foraminifera from the shore sands of Bognor. This partnership continued for twenty-five years and led to many publications on Recent foraminifera from our own coasts (e.g. 1913, 1930) and from various other parts of the world including the Kerimba Archipelago, the Falkland Islands, and Egypt. Much of this work was carried out in the British Museum (Natural History) but, in Earland's words '... conditions were not favourable, there were too many other distractions at the Museum for rapid progress, and they resulted in friction between us.' (Earland, 1943). In 1932 the partnership broke up and Heron-Allen withdrew from scientific study.

Arthur Earland, 1866–1958, entered the Civil Service in 1885 as a 'boy clerk' and retired in 1926 (due to ill health) as an assistant controller. Like Heron-Allen his heart was not in his work and in 1887 he took up microscopy as a hobby. Hawkyard encouraged him to study foraminifera and later he received help from Millett, Wright, Lister and d'Arcy Thompson (Hedley, 1958). Macfadyen (1959) described Earland as 'the last representative of the Brady era in this country'.

Apart from their joint publications, Heron-Allen and Earland gathered together valuable reference collections and an extensive library, all of which now form part of the collections in the Protozoa Department of the British Museum (Natural History).

The first major account of British fossil foraminifera is that by Brady (1876) on Carboniferous and Permian forms. This monograph is beautifully illustrated showing both complete specimens and thin sections. The material studied came not only from Britain but also from Ireland, Belgium, Russia, United States, Canada and Germany.

The monograph on the Foraminifera of the Crag (Jones et al., 1866–1897) was a major collaborative exercise. Part 1, published in 1866, was prepared by T. R. Jones, W. K. Parker and H. B. Brady. After its completion the authors were busily engaged in other matters (e.g. Brady prepared the 'Challenger' report) and so the preparation of subsequent parts was delayed. Then both Brady and Parker died. Jones retired from his post as Professor at the Military College, Sandhurst, in 1881 and moved to Chelsea (Robinson, in Bate and Robinson, 1972). There he spent his retirement preparing the later parts of the monograph. Because of failing eyesight he took on Charles Davies Sherborn as a young assistant, (Elliot, 1978). An incomplete manuscript for Part II had been left by Brady and this was completed by Jones and Sherborn with assistance from H. W. Burrows, F. W. Millett, R. Holland and F. W. Chapman. All these authors contributed to Parts III and IV, these being published in 1895, 1896 and 1897 respectively.

The Foraminifera of the Crag is the largest single work on fossil foraminifera in Britain. Apart from the systematic description and illustration of the species, their geographical and stratigraphical distributions are discussed and brief comments are made on the use of the foraminifera in determining the environments of accumulation of the Crags.

Burrows and Holland also made a significant contribution to the study of the Thanet Beds at Pegwell Bay. They systematically sampled the succession and recorded the stratigraphical distribution of the foraminifera (Burrows and Holland, 1897).

Sherborn, 1861–1942, published three short papers with Chapman on foraminifera from the London Clay from drainage works in Piccadilly (1886, 1889) and from the cliffs at Sheppey (Chapman and Sherborn, 1889).

Frederick Chapman, during the period 1891–1896 produced a ten part illustrated account of the Foraminifera of the Gault. He also described faunas from the Rhaetic, the Bargate Beds and the Cambridge Greensand. Chapman left England in 1902 to take up a post at the Museum in Melbourne and from there he published on a wide variety of Recent and fossil material (Bate and Robinson, 1978).

William Archibald Macfayden, 1893–1985, trained first as a chemist and later as a geologist. He had a varied geological career in the Middle East and Africa. From 1932–1942 he published a series of papers on late Pliocene, Pleistocene and 'post-glacial' foraminifera from East Anglia and the borders of the Bristol Channel–Severn Estuary. After a break of 13 years he published his final paper on this theme on Somerset Levels material in 1955. He also made contributions to Mesozoic studies in Egypt and Dorset (1941).

This early research was carried out by men with scientific curiosity. Their objectives were to describe, record, and to define principles where appropriate. Thus, by the 1920s this pure science was ready to be applied to the practical problems of stratigraphical correlation and palaeoecology in the newly established oil exploration industry.

REFERENCES

Anon. 1915(a). Fortescue William Millett, *J. Quekett microsc. Club*, (2) **12**, 559–560.

Anon. 1915(b). *Nature*, **95**, 180.

Adams, C. G. 1978. Great names in Micropalaeontology. 3. Henry Bowman Brady, 1835–1891. *Foraminifera*, **3**, 275–280. Academic Press.

Adams, C. G., Harrison, C. A. and Hodgkinson, R. L. 1980. Some primary type specimens of foraminifera in the British Museum (Natural History). *Micropaleontology*, **26**, 1–16.

Bate, R. H. and Robinson, E. 1978. A stratigraphical index of British Ostracoda. *British Micropalaeontological Society, Special Publ.*, **1**, 538 pp., Seel House Press, Liverpool.

Brady, H. B. 1863. Report on the Foraminifera. In Mennell, H. T. (ed.) Report on the dredging expedition to the Dogger Bank and the coasts of Northumberland. *Trans. Tyneside Nat. Fld. Cl.*, **5**, 291–294.

Brady, H. B. 1876. A monograph of Carboniferous and Permian Foraminifera (the genus *Fusulina* excluded). *Palaeontogr. Soc. Monogr.*, London, 1–66, pls 1–12.

Brady, H. B. 1884. Report on the Foraminifera dredged by H.M.S. Challenger during the years 1873–76. *Rep. Scient. Results Challenger Exped. Zoology*, 9, 1–800, 115 pls.

Burrows, H. and Holland, R. 1897. The Foraminifera of the Thanet Beds of Pegwell Bay. *Proc. Geol. Assoc.*, **15**, 19–52.

Carpenter, W. B. 1862. Introduction to the study of the Foraminifera. Ray Soc. London, 319 pp.

Carpenter, W. C. 1883. Report on the specimens of the genus *Orbitolites* collected by H.M.S. Challenger during the years 1873–1876. *Rep. Scient. Results Challenger Exped. Zoology*, 7, (21), art. 4, 1–47. pls 1–8.

Carpenter, W. B. and Brady, H. B. 1870. Description of *Parkeria* and *Loftusia* two gigantic types of arenaceous Foraminifera. *Phil. Trans. R. Soc.*, **159**, 721–754.

Chapman, F. 1891–1898. The Foraminifera of the Gault of Folkestone. *Jl. R. microsc. Soc.* 10 parts. *Jl. R. microsc. Soc.* (Part 1) 1891, 565–575; (Part 2) 1892, 319–330; (Part 3) 1892, 749–758; (Part 4) 1893, 579–595; (Part 5) 1893, 153–163; (Part 6) 1894, 421–427; (Part 7) 1894, 645–654; (Part 8) 1895, 1–14; (Part 9) 1896, 581–591; (Part 10) 1898, 1–49. Also available as a bound reprint by Junk (1970)

Chapman, F. and Sherborn, C. D. 1889. The Foraminifera from the London Clay of Sheppey. *Geol. Mag.* Dec. 111, **6**, 497–499.

Elliott, G. F. 1978. Charles Davies Sherborn, 1861–1942: an appreciation. *Foraminifera*, **3**, 267–273. Academic Press.

Earland, A. 1943. Edward Heron-Allen, F.R.S. *Jl. R. microsc. Soc.*, **63**(3), 48–50.

Heron-Allen, E. and Earland, A. 1913. Clare Island Survey Foraminifera. *Proc. R. Ir. Acad.*, **31**(64), 1–188.

Heron-Allen, E, and Earland, A. 1930. The Foraminifera of the Plymouth District. *Jl. R. microsc. Soc.*, **50**, 46–84, 161–191.

Hedley, R. H. 1958. Mr. Arthur Earland. *Nature*, **181**, 1440–1441.

Jones, T. R. 1891. William Kitchen Parker. *Proc. Roy. Soc.*, **48**, XV–XX.

Jones, T. R., Parker, W. K. and Brady, H. B. 1866–1897. A monograph of the Foraminifera of the Crag. *Palaeontogr. Soc. Monogr.*, 1–402.

Macfadyen, W. A. 1941. Foraminifera from the Green Ammonite Beds, Lower Lias of Dorset. *Phil. Trans. R. Soc.*, **B231**, 1–173.

Macfadyen, W. A. 1959. Arthur Earland. *Jl. R. microsc. Soc.*, **77**, 146–150.

Millett, F. W. 1898–1904. Report on the Recent Foraminifera of the Malay Archipelago. *Jl. R. microsc. Soc.*, (17 parts). (Part 1) 1898, 258–269; (Part 2) 1898, 499–513; (Part 3) 1989, 607–614; (Part 4) 1899, 249–255; (Part 5) 1899, 357–365; (Part 6)

1899, 557–564; (Part 7) 1900, 6–13; (Part 8) 1900, 273–281; (Part 9) 1900, 539–549; (Part 10) 1901, 1–11; (Part 11) 1901, 485–597; (Part 12) 1901, 619–628; (Part 13) 1902, 509–528; (Part 14) 1903, 253–275; (Part 15) 1903, 685–704; (Part 16) 1904, 489–506; (Part 17) 1904, 597–609.

Montagu, G. 1803. *Testacea Britannica, or natural history of British shells marine, land and freshwater, including the most minute.* pp. 1–606, pls 1–16. Romsey.

Montagu, G. 1808. *Testacea Britannica: Supplement*, pp. 1–183, pls 1–30, Exeter.

Murray, J. W. 1971. The W. B. Carpenter Collection. *Micropaleontology*, **17**, 105–106.

Murray, J. W. and Taplin, C. M. 1984. The W. B. Carpenter Collection of Foraminifera: a catalogue. *J. Micropalaeontol.*, **3**, 55–58.

Sherborn, C. D. 1915. Fortescue William Millett. *Geol. Mag.* (6), **11**, 288.

Sherborn, C. D. and Chapman, F. 1886. On some microzoa from the London Clay exposed in the drainage works, Piccadilly, London, 1885. *Jl. R. microsc. Soc.*, **6**, 737–767.

Sherborn, C. D. and Chapman, F. 1889. Additional note on the Foraminifera of the London Clay exposed in the drainage works, Piccadilly, London, 1885. *Jl. R. microsc. Soc.*, 483–488.

Walker, G. and Boys, W. 1784. *Testacea minuta rariora*, London.

Walker, G. and Jacob, E. 1794, In Kanmacher, *Adam's essays on the microscope.* Ed. 2, London.

Watson, J. and Thomas, B. A. 1985. Facsimile edition of Williamson, W. C., *Reminiscences of a Yorkshire Naturalist*, 288 pp.

Williamson, W. C. 1847. On some microscopical objects found in the mud of the Levant and other deposits, with remarks on the mode of formation of calcareous and infusorial siliceous rocks. *Manchr. Lit. Phil. Soc. Mem.*, **8**, 1-128.

Williamson, W. C. 1848a. On the structure of the shell and soft animal of the *Polystomella crispa*. With some remarks on the zoological position of the Foraminifera. *Micr. Soc. Trans.*, **2**, 159–180.

Williamson, W. C. 1848b. On the Recent species of the genus *Lagena*. *Ann. Mag. nat. Hist.*, ser. 2, **1**, 1–20.

Williamson, W. C. 1850. On the minute structure of the calcareous shells of some recent species of Foraminifera. *Trans. Microsc. Soc.*, **3**, 105–128.

Williamson, W. C. 1853. On the minute structure of a species of *Faujasina*. *Quart. J. Microsc. Sci.*, **1**, 87–92.

Williamson, W. C. 1858. *On the Recent Foraminfera of Great Britan.* Ray. Soc., pp. 1–107, pls 1–7.

Williamson, W. C. 1865. On the anatomy and physiology of the Foraminifera. *Popular Science Review*, **4**, 171–179.

Williamson, W. C. 1876. Corrections of the nomenclature of the objects figured in a memoir 'On some of the minute objects found in the mud of the Levant'. *Manchr. Lit. Phil. Soc. Mem.*, **25**, 131–136.

2

Cambrian to Devonian

J. M. Kircher and M. D. Brasier

2.1 INTRODUCTION

British pre-Carboniferous foraminifera have previously received little attention. Most of our current knowledge has been gained from agglutinated forms extracted by acid residue techniques. Mabillard and Aldridge (1982) reported an assemblage of Silurian agglutinated foraminifera from the Llandovery–Wenlock boundary and this was followed by extensive studies into Silurian agglutinated foraminifera by J. M. Kircher. This latest work has added much new information on the stratigraphic and geographic distribution, palaeobiology and palaeoecology of these simple forms. Together with Ordovician and Devonian occurrences reported here for the first time, it suggests there is much more to be discovered about British Pre-Carboniferous foraminifera.

The importance of Cambrian and Ordovician foraminifera for stratigraphic correlation is currently difficult to assess owing to the paucity of material. In the Silurian, both genera and species are long ranging and appear to be facies controlled, with limited biostratigraphic potential. By Frasnian and Fammenian times, however, the group begins to approach the better known Carboniferous assemblages in terms of biostratigraphic utility, and Devonian foraminifera have been widely employed for stratigraphy in North America and the Soviet Union (Toomey and Mamet, 1979; Poyarkov, 1979).

2.2 LOCATION OF IMPORTANT COLLECTIONS

The British Museum (Natural History) houses the material described by H. B. Brady (1888), Mabillard and Aldridge (1982) and the Silurian material of J. M. Kircher illustrated in this work.

The British Geological Survey has the material described from the basal Cambrian by Rushton (1978).

The Earth Sciences Department of the University of Cambridge houses some of the unpublished material of M. Orchard from the Ashgillian of northern England.

The Earth Science Department of the University of Oxford currently has material from the Cambrian of Nuneaton described by

Brasier (1986) and the material from the Upper Devonian of the Torquay area collected by C. Castle.

2.3 STRATIGRAPHIC DIVISIONS

British Cambrian to Devonian stratigraphy and correlations have been reviewed in four Special Reports of the Geological Society of London (Cocks et al., 1971; Williams et al., 1972; Cowie et al., 1972, House et al., 1977). These, and the standardisation of Silurian chronostratigraphy outlined by Holland (1985), are followed in this report.

2.4 FACIES AND FAUNAS

Too little is yet known about Cambrian and Ordovician faunas to form a clear impression of their facies distribution, but those remains attributed to foraminifera have generally been found in shallow to mid-shelf carbonates or mudstones with abundant shelly benthos.

Studies on the palaeoecology of Silurian foraminifera in England and Gotland by J. M. Kircher provide a clearer picture, with two main biofacies. Large stromatoporoid and coral bioherms bear a relatively diverse *Ammodiscus* assemblage, with *Ammodiscus, Tolypammina, Hyperammina, Hemisphaerammina, Webbinelloidea, Thurammina, Colonammina, Psammosphaera, Saccammina, Storthosphaera, Stomasphaera, Atelikamara, Lagenammina, Stegnammina, Tholosina,* and *Rhabdammina*. Level-bottom carbonates and mudstones yield a lower diversity *Hemisphaerammina* assemblage, dominantly composed of the latter and *Tholosina*. Smaller bioherms support an intermediate fauna. Deeper shelf sediments, graptolitic muds and turbidites appear to be barren of foraminifera at this time.

The distribution of foraminiferid remains in the marine Devonian of southwest England is more poorly known. Adherent agglutinated forms were reported by Tucker and Van Straaten (1970) from the Upper Devonian limestones of Chudleigh, Devon. Shallower platform carbonates such as those in South Devon contain primitive calcareous forms, with various *Parathurammina* spp., calcareous septate *Paratikhinella* and *Eonodosaria*, and coiled *Nanicella*. Late Frasnian and Fammenian carbonate assemblages have not yet been described from Britain.

2.5 OCCURRENCES

2.5.1 Cambrian [M. D. B.]

The often flattened, siliceous agglutinated tubes of *Platysolenites antiquissimus* Eichwald are regarded by Loeblich and Tappan (1964), Glaessner (1978) and Føyn and Glaessner (1979) as the earliest foraminiferid tests because they closely resemble the younger *Bathysiphon*. They differ from modern forms in being comprised of mineral grains and silica spherules rather than sponge spicules. These *Platysolenites* tubes are also much larger than the Silurian *Bathysiphon* reported here, so that the relationship may be regarded as requiring further investigation. *Platysolenites* first appears near the base of the basal Cambrian Dolwen Grits of the Harlech Dome, Wales (Rushton, 1978) marking the start of the Non-Trilobite Zone of Cowie et al. (1972). It is now known from a similar level in the Withycombe Farm borehole near Banbury in Oxfordshire (Dr A. W. A. Rushton, pers. comm. 1987). The taxon reappears in higher beds of this zone at Nuneaton, Warwickshire, and is last seen there at the base of the Olenellid Zone (Brasier, 1986). The top of this zone has yielded a single ?*Rhabdammina* sp. in the Ac3 beds of the Comley Limestone at Comley, Salop. Foraminifera have not yet been identified from higher beds in Britain.

Platysolenites antiquissimus provides an index for the lowest Cambrian zone across the East European Platform and the Baltic Shield (Føyn and Glaessner, 1979; Urbanek and Rozanov, 1983) and is also proven in the lowest Cambrian of southeastern Newfoundland and California (unpublished data).

2.5.2 Ordovician [J. M. K., M. D. B.]

Limited information of Ordovician foraminiferids indicates that assemblages were mainly of primitive agglutinated forms with long stratigraphic ranges. Silicified remains from ostracod residues have been obtained by Dr C. Jones (formerly of Leicester University) from the Llandeilo Kincoed and Cwm Agol Quarries, and the basal Caradoc Velfry Quarry of Lampeter, Wales; these include foraminifera we refer to ?*Hemisphaerammina* sp. An assemblage from the Ashgillian Cautley Mudstone of the Howgill Fells, Yorkshire has been kindly shown to us by Dr M. Orchard (following written communication, 1986). This contains a richer assemblage of *Amphitremoida*, *Bathysiphon*?, *Hippocrepina*?, *Hyperammina*?, *Psammosphaera*, *Rhabdammina*?, *Tolypammina* and *Turritelella*, to which we would add *Hemisphaerammina*. This faunal succession compares with assemblages from North America reviewed by Conkin and Conkin (1982) but awaits fuller taxonomic treatment.

The calcareous ?parathuramminacean *Saccamminopsis* cf. *fusulinaformis* (M'Coy), sometimes known as *Saccammina carteri* Brady, is recorded from the Upper Llandeilo Stinchar Limestone and the Caradoc Craighead Limestone of the Girvan district of SW Scotland (Nicholson and Etheridge, 1879; Lapworth, 1882; Pringle, 1948; Cummings, 1952). Other Ordovician occurrences of this type were reviewed by Schallreuter (1983) who described calcareous, uniserial, septate remains from some Middle and Upper Ordovician erratic boulders of northern Germany. These were placed by him in *Saccamminopsis*? sp. cf. *syltensis* and *S*.? sp. cf. *teschenhagensis*. The foraminiferid affinities of these Ordovician *Saccamminopsis* require further investigation.

2.5.3 Silurian [J. M. K.]

Agglutinated remains attributed to the foraminifera in Silurian rocks are morphologically simple. They are placed here in the suborder Textulariina, represented by members of the superfamilies Ammodiscacea, Hyperamminacea, Astrorhizacea and Saccamminacea. The latter is elevated herein from family to superfamily rank.

Several early records of British Silurian foraminifera such as those of *Dentalina* (Blake, 1876; Keeping, 1882) were shown by Wood (1949) to be inorganic mineral growths. Others were misattributed to extant taxa so that '*Lagena*' from the Silurian of Shropshire (Smith, 1881; Brady, 1888) was later placed in *Saccamminopsis* cf. *fusulinaformis* by Cummings (1952).

Tectinous remains tentatively assigned to foraminifera were discovered in palynological residues (Aldridge et al., 1979, and Dorning, pers. comm. 1986). Remains referrable to *Archaeochitosa lobosa* Eisenack were illustrated from the Much Wenlock Limestone Formation by Dorning (1978).

The first authenticated reports of agglutinated foraminifera were two abstracts by Ireland (1958, 1967) who recovered material from the type Wenlock and Ludlow areas, noting eight genera and 19 species. Although Aldridge et al. (1979) observed that agglutinated foraminifera were sporadically abundant in the Silurian of Britain and Ireland, they noted their distribution had not been investigated. Aldridge et al. (1981) later recorded the group in the Wenlock Series of the Welsh Basin. Mabillard and Aldridge (1982) then described and illustrated a foraminiferal fauna from the Llandovery to Wenlock strata of Shropshire. More recently, an extensive investigation and systematic revision of Wenlock to Ludlow foraminifera from the English West Midlands and Welsh Borderland by J. M. Kircher has further extended our knowledge of these early forms. Some 19 genera and 38 species were recognized and two assemblages identified. These are a diverse *Ammodiscus* assemblage characterized by numerous specimens and genera, and a second *Hemisphaerammina* assemblage which is less diverse.

British Silurian foraminiferal faunas appear to have been ecologically rather than stratigraphically controlled, with a distinct preference for carbonate-rich environments. This is demonstrated in the Much Wenlock Limestone Formation at Wenlock Edge, Shropshire in which the high diversity *Ammodiscus* assemblage occurs. The fauna includes *Ammodiscus, Atelikamara, Ceratammina, Hemisphaerammina thola* (Moreman), *Hyperammina, Lagenammina, Psammosphaera cava* Moreman, *Rhabdammina, Saccamina, Sorosphaerella, Stegnammina, Stomasphaera brassfieldensis* Mound, *Storthosphaera, Tholosina, Thurammina, Tolypammina* and *Webbinelloidea*.

The second assemblage is dominantly composed of *Hemisphaerammina* and *Tholosina*. It is typically recovered from Ludlow sections, but is generally more widespread in time and space than the *Ammodiscus* assemblage, being found widely in both Wenlock and Ludlow strata across the Welsh Basin. It appears to have been tolerant of varying environmental conditions. Deeper shelf graptolitic muds and turbidites appear to be barren of foraminifera at this time.

2.5.4 Devonian [M. D. B.]

Although the Devonian was an important period of evolution for the foraminifera, studies in Britain lag behind those in the U.S.S.R. (e.g. Poyarkov, 1979) and North America (e.g. Toomey and Mamet, 1979) with no illustrated true records to date. New data presented here, however, indicate that some stratigraphically useful assemblages occur in the Hercynian terrain of South Devon.

The only previous record of adherent agglutinated forms is from the Upper Devonian of Chudleigh (Tucker and Van Straaten, 1970). Scarce agglutinated saccaminaceans have since been identified in conodont residues of Upper Devonian limestones of the Torquay region by Dr C. Castle (formerly of Hull University) who also found septate planispiral *Nanicella* cf. *gallowayi* casts in the Frasnian limestones of Oddicombe Road and Castle Road, Torquay. A more diverse calcareous assemblage has now been identified by M. D. Brasier from Barton Limestone material (lower *Polygnathus asymmetricus* Zone of the Frasnian) at Babbacombe Cliff, Torquay. Thin sections contain (in comparison with the nomenclature of Poyarkov, 1979): *Archaelagena* cf. *borealia*, *Bisphaera* cf. *elegans*, *Diplosphaerina* cf. *isphaeramensis*, *Eonodosaria* cf. *evlanensis*, *Eovolutina* cf. *magna*, *Irregularina* cf. *angulata*, *Nanicella* cf. *gallowayi*, *Paratikhinella* cf. *cannula*, *Parathurammina* (*Parathuramminites*) cf. *paracushmani*, *P.* (*P.*) cf. *cushmani*, *P.* (*Salpingothurammina*) cf. *cordata*, and *Uslonia* cf. *permira*. These occur with numerous radiosphaerid and non-radiosphaerid calcispheres. *Wetheredella* sp. has been noted from Daddyhole, east of Torquay by Dr C. T. Scrutton (pers. comm. 1986). The systematics of these assemblages await analysis. They compare with Frasnian assemblages from the platforms of the Soviet Union and Eastern Europe (Poyarkov, 1979) and of North America (Toomey, 1972; Toomey and Mamet, 1979), notably in the presence of *Eonodosaria* cf. *evlanensis*, *Nanicella* cf. *gallowayi* and *Parathurammina* (*Parathuramminites*) cf. *paracushmani*.

SUBORDER TEXTULARIINA

Ammodiscus exsertus Cushman
Plate 2.1, Fig. 1 (× 125), Much Wenlock Limestone Formation, Wenlock Edge. Description: test free, proloculus followed by non-septate tubular chamber, close-coiled for several whorls then uncoiling at right angles for varying lengths in same plane; aperture terminal at open ends of tube. Distribution: Wenlock Series, Welsh Borderland.

Ammodiscus sp.
Plate 2.1, Fig. 2 (× 125), Much Wenlock Limestone Formation, Wenlock Edge. Description: test free, proloculus followed by non-septate tubular chamber, close-coiled for several whorls without uncoiling; aperture terminal at open end of tube. Distribution: Wenlock Series, Welsh Borderland.

Atelikamara incomposita McClellan
Plate 2.1, Fig. 3 (× 50), Much Wenlock Limestone Formation, Wenlock Edge. Description: test free, irregular in outline, conical to irregularly inflated on dorsal surface, ventral surface usually planar, maybe angular; incomplete subdivision of interior, exterior may appear nodular; no visible aperture. Distribution: Wenlock and Ludlow Series, Welsh Borderland.

Bathysiphon sp.
Plate 2.1, Fig. 4 (× 50), Much Wenlock Limestone Formation, May Hill. Description: test free, tubular, straight cylindrical; apertures at open ends of tube. Distribution: Wenlock Series, West Midlands and Wenlock and Ludlow Series, Welsh Borderland.

Ceratammina cornucopia Ireland
Plate 2.1, Fig. 5 (× 50), Much Wenlock Limestone Formation, Wenlock Edge. Description: test free, tubular, horn-shaped; no aperture visible. Distribution: Wenlock Series, Welsh Borderland.

Colonammina bituba Dunn
Plate 2.1, Fig. 6 (× 25), Much Wenlock Limestone Formation, Wenlock Edge. Description: test attached, planoconvex, circular in outline; two projections from convex surface each bear a terminal aperture. Distribution: Wenlock Series, Welsh Borderland.

Colonammina conea Moreman
Plate 2.1, Fig. 7 (× 50), Much Wenlock Limestone Formation, Wenlock Edge. Description: test attached on two sides producing a wedge-shaped appearance; upper surface convex; terminal circular aperture on broad short neck near middle of convex surface. Distribution: Wenlock Series, Welsh Borderland.

Hemisphaerammina thola (Moreman)
Plate 2.1, Figs 8, 9 (× 50), Bringewood Formation, Aymestrey; Figs 10, 11 (× 50), Much Wenlock Limestone Formation, May Hill. Description: test attached, of one to many hemispherical chambers which may have a bordering flange; chambers loosely to closely joined in various configurations; no visible apertures. Distribution; Wenlock Series of West Midlands and Welsh Borderland, and Ludlow Series of Welsh Borderland.

Hyperammina curva (Moreman)
Plate 2.1, Fig. 12 (× 25), Much Wenlock Limestone Formation, Wenlock Edge. Description: test free, long and tubular, with gentle curvature in one plane, distinctly tapering towards minute proloculus (usually missing); aperture at open end of tube. Distribution: Wenlock Series, West Midlands and Welsh Borderland.

Lagenammina stilla Moreman
Plate 2.1, Fig. 13 (× 125), Much Wenlock Limestone Formation, Woolhope Inlier. Description; test free, bottle shaped, that part of the chamber below the neck being roughly ellipsoidal in outline; terminal aperture at the open end of a short neck. Distribution: Wenlock Series, Welsh Borderland.

Psammosphaera cava Moreman
Plate 2.1, Fig. 14 (× 50), Much Wenlock Limestone Formation, Wenlock Edge; Fig. 15 (× 50), Much Wenlock Limestone Formation, May Hill Inlier. Description: test free, one to many subspherical to globular chambers, variously arranged, loosely to firmly joined to one another; no visible apertures. Distribution: Wenlock Series of West Midlands and Welsh Borderland, and Ludlow Series of Welsh Borderland.

Rhabdammina bifurcata Browne and Scott
Plate 2.1, Fig. 16 (× 125), Much Wenlock Limestone Formation, Wenlock Edge. Description: test free, of three nearly straight tubular arms which unite to form a Y-shaped test; longest arm is double the length of two shorter arms which are subequal in length; aperture terminal at open ends of tubes. Distribution: Wenlock Series, Welsh Borderland.

Rhabdammina irregularis Carpenter
Plate 2.1, Fig. 17 (× 50), Much Wenlock Limestone Formation, Wenlock Edge. Description: test free, tubular, dichotomously branching; aperture terminal at open ends of tubes. Distribution: Wenlock Series, Welsh Borderland.

Saccammina moremani Ireland
Plate 2.1, Fig. 18 (× 50), Much Wenlock Limestone Formation, Wenlock Edge. Description: test free, single spherical chamber with distinct short neck bearing a terminal circular aperture. Distribution: Wenlock Series, Welsh Borderland.

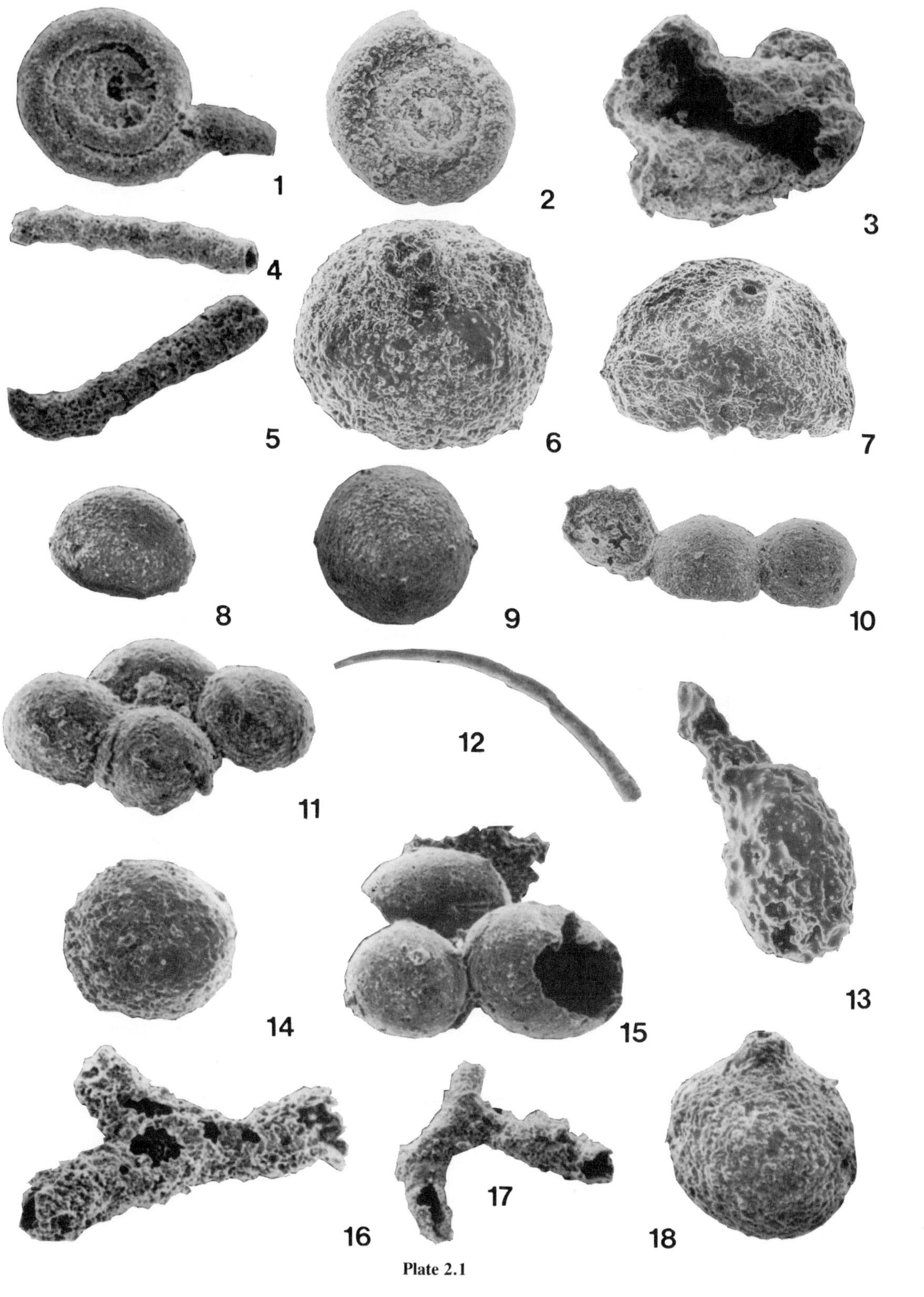

Pl. 2.1]   **Cambrian to Devonian**   25

Plate 2.1

Sorosphaerella cooperensis Conkin, Conkin and Canis
Plate 2.2, Fig. 1 (× 50), Much Wenlock Limestone Formation, Wenlock Edge. Description: test attached, single globular subspherical chamber bearing an attachment groove; no visible aperture. Distribution: Wenlock Series of West Midlands and Welsh borderland, and Ludlow Series of Welsh Borderland.

Stegnammina elongata Moreman
Plate 2.2, Fig. 2 (× 50), Much Wenlock Limestone Formation, Wenlock Edge. Description: test free, cylindrical with slightly rounded ends; no visible apertures. Distribution: Wenlock Series, West Midlands and Welsh Borderland.

Stegnammina hebesta Moreman
Plate 2.2, Fig. 3 (× 50), Much Wenlock Limestone Formation, Wenlock Edge. Description: test free, box-like to cylindrical, no visible apertures. Distribution: Wenlock Series, Welsh Borderland.

Stomasphaera brassfieldensis Mound
Plate 2.2, Figs 4 (× 125), 5 (× 50), Much Wenlock Limestone Formation, Wenlock Edge. Description: test free, one to many subangular to spherical chambers; aperture single, flush, round to oval. Distribution: Wenlock Series of West Midlands and Welsh Borderland, and Ludlow Series of Welsh Borderland.

Storthosphaera malloryi McClellan
Plate 2.2, Fig. 6 (× 50), Upper Bringewood Formation, Aymestrey. Description: test free, irregular globular chamber, appears lumpy on exterior; no visible aperture. Distribution: Wenlock Series, West Midlands and Welsh Borderland.

Tholosina sedentata Ireland
Plate 2.2, Fig. 7 (× 50), Much Wenlock Limestone Formation, Dudley, Warwickshire. Description: test attached, hemispherical and slightly oval; aperture terminal in a short tube at one end of the test, close to the plane of attachment. Distribution: Wenlock Series of West Midlands and Welsh Borderland, and Ludlow Series of Welsh Borderland.

Tholosina sp.
Plate 2.2, Fig. 8 (× 50), Much Wenlock Limestone Formation, Wenlock Edge; Fig. 9 (× 50), Much Wenlock Limestone Formation, Tortworth Inlier. Description: test attached, planoconvex hemispherical chamber; aperture single, flush with test surface, located at or near the basal margin. Distribution: Wenlock Series of West Midlands and Welsh Borderland, and Ludlow Series of Welsh Borderland.

Thurammina papillata Brady
Plate 2.2, Fig. 10 (× 50), Much Wenlock Limestone Formation, Wenlock Edge. Description: test free, spherical chamber; apertures terminal on the ends of nipple-like projections, distributed evenly over the surface of the test. Distribution: Wenlock Series, Welsh Borderland.

Thurammina tubulata Moreman
Plate 2.2, Fig. 11 (× 50), Much Wenlock Limestone Formation, May Hill Inlier. Description: test free, spherical chamber with numerous well-formed tube-like projections randomly scattered over surface: apertures terminal at ends of projections. Distribution: Wenlock Series, West Midlands and Welsh Borderland.

Tolypammina sp.
Plate 2.2, Figs 12, 13 (× 50), Much Wenlock Limestone Formation, Wenlock Edge. Description: test attached, proloculus followed by a second tubular chamber which winds irregularly over the surface of attachment; aperture terminal, at open end of tube. Distribution: Wenlock Series, West Midlands and Welsh Borderland.

Webbinelloidea similis Stewart and Lampe
Plate 2.2, Figs 14–17 (× 50), Much Wenlock Limestone Formation, Wenlock Edge. Description: test attached, one to many chambers, variously attached and arranged; chambers planoconvex; aperture single, dorsal, flush with surface, subcentral in position. Distribution: Wenlock Series of West Midlands and Welsh Borderland, and Ludlow Series of Welsh Borderland.

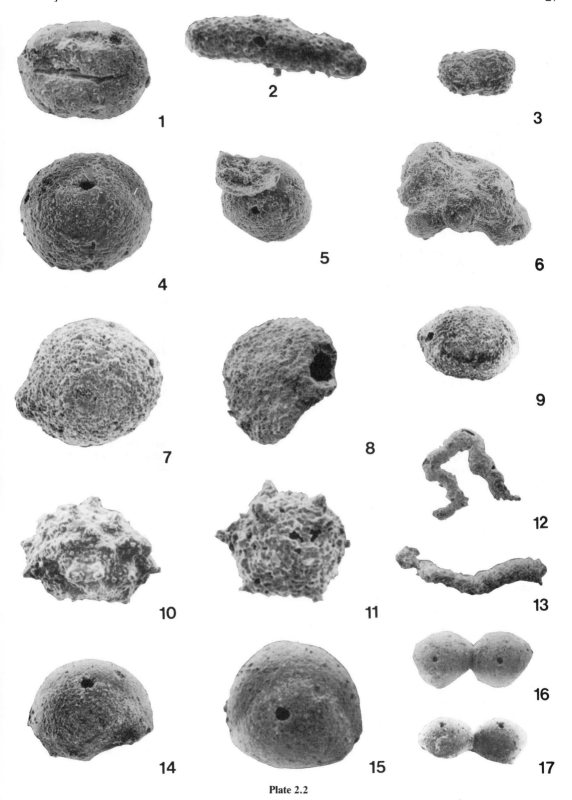

Plate 2.2

SUBORDER FUSULININA

Archaelagena cf. *borealia* Pronina
Plate 2.3, Fig. 1 (× 101), Barton Limestone, Babbacombe Cliff, South Devon. Description: test free, globular, unilocular, with dark, calcareous microgranular wall, plus small carbonate grains; single aperture on a moderately extended tubular neck. Distribution: Frasnian Stage, South Devon.

Bisphaera cf. *elegans* Vissarionova
Plate 2.3, Figs 2 (× 108), 3 (× 61), Barton Limestone, Babbacombe Cliff, South Devon. Description: test free, globular, unilocular, with dark, calcareous microgranular wall; shape elongate with one or more minor insections of test; no apertures visible. Distribution: Frasnian Stage, South Devon.

Diplosphaerina cf. *isphaeramensis* Poyarkov
Plate 2.3, Fig. 4 (× 103), Barton Limestone, Babbacombe Cliff, South Devon. Description: test free, globular, bilocular, with dark, calcareous microgranular wall; small subspherical proloculus is partly enveloped by much larger, nearly spherical second chamber; no apertures visible. Distribution: Frasnian Stage, South Devon.

Eonodosaria cf. *evlanensis* Lipina
Plate 2.3, Fig. 5 (× 122), Barton Limestone, Babbacombe Cliff, South Devon. Description: test free, multilocular, rectilinear uniserial, with up to ten short (brevithalamous) chambers; wall calcareous, two-layered, with dark inner layer and very pale outer layer; aperture single, central and terminal. Distribution: Frasnian Stage, South Devon.

Eovolutina cf. *magna* Poyarkov
Plate 2.3, Fig. 6 (× 115), Barton Limestone, Babbacombe Cliff, South Devon. Description: test free, globular, bilocular, with one ovate chamber enclosing a nearly spheroidal chamber; no apertures visible; wall dark, calcareous microgranular. Distribution: Frasnian Stage, South Devon.

Irregularina cf. *angulata* Poyarkov
Plate 2.3, Figs 7 (× 80), 8 (× 112), Barton Limestone, Babbacombe Cliff, South Devon. Description: test free, globular, unilocular, with elongate chamber bearing several irregular tubular extensions; chamber wall dark, calcareous, microgranular. Distribution: Frasnian Stage, South Devon.

Nanicella cf. *gallowayi* (Thomas)
Plate 2.3, Figs 9 (× 87), 10 (× 91), Barton Limestone, Babbacombe Cliff, South Devon. Description: test free, multilocular, planispiral evolute; profile slightly biconvex; globular proloculus followed by several tight whorls of evenly formed brevithalamous chambers; wall pale, calcareous and hyaline; aperture single, basal equatorial slit. Distribution: Frasnian Stage, South Devon.

Paratikhinella cf. *cannula* (Bykova)
Plate 2.3, Figs 11 (× 59), 12 (× 51), Barton Limestone, Babbacombe Cliff, South Devon. Description: test free, multilocular, rectilinear to curved uniserial; chambers elongate along axis and rather irregularly separated by weak septa; wall pale, calcareous and ?hyaline; aperture single, central, terminal. Distribution: Frasnian Stage, South Devon.

Parathurammina (Parathuramminites) cf. *paracushmani* Reitlinger
Plate 2.3, Fig. 13, (× 126), Barton Limestone, Babbacombe Cliff, South Devon. Description: test free, globular unilocular, with three or four unevenly scattered apertures appearing as elongate tubes extending away from the chamber, penetrating through the thickened wall of the test; wall dark, calcareous, microgranular. Distribution: Frasnian Stage, South Devon.

Parathurammina (Parathuramminites) cf. *cushmani* Suleimanov
Plate 2.3, Fig. 14, (× 119), Barton Limestone, Babbacombe Cliff, South Devon. Description: test free, subrounded globular, unilocular, with relatively thick wall; several unevenly scattered flush apertures connected to chamber by radial tubular pores; wall dark, calcareous, microgranular. Distribution: Frasnian Stage, South Devon,

Parathurammina (Salpingothurammina) cf. *cordata* Pronina
Plate 2.3, Figs 15 (× 125), 16 (× 108), Barton Limestone, Babbacombe Cliff, South Devon. Description: test free, globular unilocular, with apertures appearing as four or five long tubules extending radially from a central globular chamber; wall dark, calcareous, microgranular. Distribution: Frasnian Stage, South Devon.

Uslonia cf. *permira* Antropov
Plate 2.3, Fig. 17 (× 63), Barton Limestone, Babbacombe Cliff, South Devon. Description: test free, unilocular, elongate irregular with angular extremities; wall dark, calcareous microgranular; aperture not seen. Distribution: Frasnian Stage, South Devon.

Cambrian to Devonian

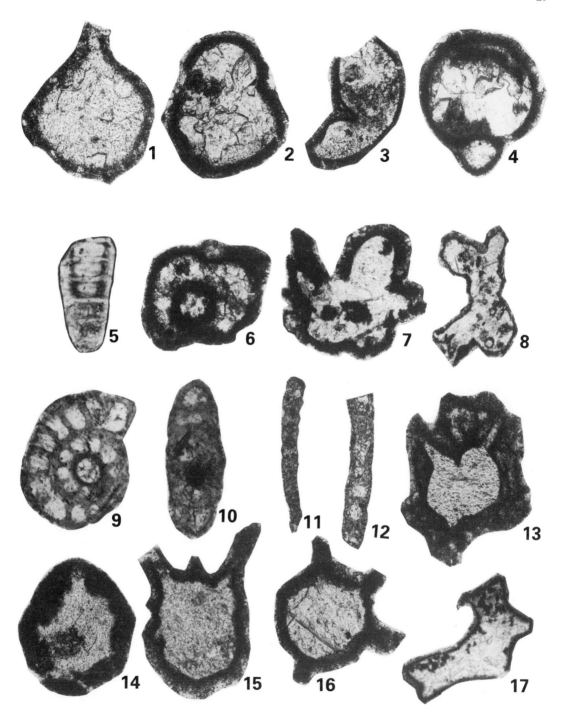

Plate 2.3

ACKNOWLEDGEMENTS

We thank all those who have encouraged the progress of this work, especially Dr R. J. Aldridge, Professor J. E. Conkin and Dr David J. Siveter for access to their Silurian and Ordovician material; Dr A. W. A. Rushton for access to some Cambrian material; Dr M. J. Orchard for information concerning his Ordovician assemblages; and Dr C. Castle and Dr C. T. Scrutton for information on the Devonian assemblages. The doctoral research on Silurian foraminifera by J. M. Kircher was undertaken whilst in receipt of a NERC studentship at Hull University under the supervision of Dr M. D. Brasier and Dr Derek J. Siveter and their support is gratefully acknowledged.

References

Aldridge, R. J., Dorning, K. J., Hill, P. J., Richardson, J. B. and Siveter, D. J. 1979. Microfossil distribution in the Silurian of Britain and Ireland. In Harris, A. L., Holland, C. H. and Leake, B. L. (eds), *The Caledonides of the British Isles – reviewed*, pp. 433–438. Scottish Academic Press, Edinburgh, for the Geological Society of London.

Aldridge, R. W., Dorning, K. W. and Siveter, D. J. 1981. Distribution of microfossil groups across the Wenlock Shelf of the Welsh Basin. In Neale, J. W. and Brasier, M. D. (eds) *Microfossils from Recent and fossil shelf seas*, pp. 13–30. British Micropalaeontological Society Series, Ellis Horwood, Chichester, 380pp.

Blake, J. F. 1876. Lower Silurian Foraminifera. *Geol. Mag.*, n.s., **3**, 134–135.

Brady, H. B. 1888. Notes on some Silurian Lagenae. *Geol. Mag.*, n.s., **5**, 481–484.

Brasier, M. D. 1986. The succession of small shelly fossils (especially conoidal microfossils) from English Precambrian–Cambrian boundary beds. *Geol. Mag.*, **123**, 237–256.

Cocks, L. R. M., Holland, C. H., Rickards, R. B. and Strachan, I. 1971. A correlation of Silurian rocks in the British Isles. *Spec. Rept Geol. Soc. Lond.*, No. 1.

Conkin, J. E. and Conkin, B. M. 1982. North American Paleozoic agglutinated Foraminifera. In Broadhead, T. W. (ed.) *Foraminifera, Notes for a Short Course*, (Studies in Geology 6), pp. 177–191. University of Tennessee Department of Geological Sciences, 192pp.

Cowie, J. W., Rushton, A. W. A. and Stubblefield, C. J. 1972. A correlation of Cambrian rocks in the British Isles. *Spec. Rept Geol. Soc. Lond.*, No. 2.

Cummings, R. H. 1952. *Saccamminopsis* from the Silurian. *Proc. Geol. Ass.*, **63**, 220–226.

Dorning, K. J. 1978. Foraminifer *Archaeochitosa lobosa* Eisenack 1959, Wenlock Limestone, Silurian, Dudley, England. *J. Univ. Sheffield Geol. Soc.*, **7**(3), preface.

Føyn, S. and Glaessner, M. F. 1979. *Platysolenites*, other animal fossils, and the Precambrian–Cambrian transition in Norway. *Norsk geol. Tidsskrift*, **59**, 25–46.

Glaessner, M. F. 1978. The oldest foraminifera. In Belford, D. J. and Scheibnerova, V. (eds), The Crespin Volume: essays in honour of Irene Crespin. *Bur. Min. Res. Bulletin*, **192**, 61–65.

Holland, C. H. 1985. Series and Stages of the Silurian System. *Episodes*, **8**, 101–103.

House, M. R., Richardson, J. B., Chaloner, W. G., Allen, J. R. L., Holland, C. H. and Westoll, T. S. 1977. A correlation of Devonian rocks of the British Isles. *Spec. Rept. Geol. Soc. Lond.*, No. 8.

Ireland, H. A. 1958. Microfauna of Wenlockian and Ludlovian Silurian beds in Western England. *Bull. Geol. Soc. Amer.*, **69**, 1592 (abstract).

Ireland, H. A. 1967. Microfossils from the Silurian of England. *Bull. Am. Assoc. petrol. Geol.*, **51**, 471 (abstract).

Keeping, W. 1882. On some remains of plants, Foraminifera and Annelida, in the Silurian rocks of central Wales. *Geol. Mag.*, **9**, 485–491.

Lapworth, C. 1882. The Girvan succession. *Quart. Jl Geol. Soc. Lond.*, **38**, 537–666.

Loeblich Jr, A. R. and Tappan, H. 1964. Sarcodina, chiefly 'thecamoebians' and Forminiferida. In Moore, R. C. (ed.), *Treatise on invertebrate paleontology*, part C, pp. 1–2. Geological Society of America and Kansas Press, 900pp.

Mabillard, J. E. and Aldridge, R. J. 1982. Arenaceous foraminifera from the Llandovery/Wenlock Boundary Beds of the Wenlock Edge area, Shropshire. *J. micropalaeontol.*, **1**, 129–136.

Nicholson, H. A. and Etheridge, R. 1879. *A monograph of the Silurian fossils of the Girvan District in Ayrshire*, **1**, 334pp. Blackwood, Edinburgh.

Poyarkov, B. V. 1979. *Evolution and distribution of Devonian foraminifera*. Isdateldstvo "Nauka", Moscow. (In Russian). 172pp.

Pringle, J. 1948. The south of Scotland, 2nd edition. *Br. reg. Geol.* HMSO, London, 87pp.

Rushton, A. W. A. 1978. Appendix 3. Description of the macrofossils from the Dolwen Formation. *Bull. geol. Surv. G. B.*, **61**, 46–48.

Schallreuter, R. E. L. 1983. Calcareous foraminifera from the Ordovician of Baltoscandia. *J. micropalaeontol.*, **2**, 1–6.

Smith, J. 1881. Notes on a collection of bivalved Entomostraca and other Microzoa from the Upper Silurian strata of the Shropshire District. *Geol. Mag.*, **8**, 70–75.

Toomey, D. F. 1972. Distribution and palaeoecology of upper Devonian (Frasnian) algae and foraminifera from selected areas in Western Canada and the northern United States. *24th IGC*, 1972, Montreal, Section 7, 621–630.

References

Toomey, D. F. and Mamet, B. L. 1979. Devonian Protozoa. In House, M. R., Scrutton, C. T. and Bassett, M. G. (eds), The Devonian System. *Spec. Paps Palaeont.*, **23** 189–192.

Tucker, M. E. and Van Straaten, P. 1970. Conodonts and facies on the Chudleigh Schwelle. *Proc. Ussher Soc.*, **2**, 160–170.

Urbanek, A. and Rozanov, A. Yu. (eds) 1983. *Upper Precambrian and Cambrian Palaeontology of the East European Platform*. Wydawnictwa Geologiczne, Warsaw 158pp.

Williams, A. Strachan, I., Bassett, D. A., Dean, W. T., Ingham, J. K., Wright, A. D., and Whittington, H. B. 1972. A correlation of Ordovician rocks in the British Isles. *Spec. Rept. Geol. Soc. Lond.*, No. 3.

Wood, A. 1949. The supposed Silurian foraminifera from Cardiganshire. *Proc. Geol. Ass.*, **60**, 226–229.

3

Carboniferous

M. D. Fewtrell, W. H. C. Ramsbottom,
A. R. E. Strank

3.1 INTRODUCTION

A British Carboniferous *Endothyra*, illustrated by John Phillips (1846) was the first fossil ever to be figured from a thin section. However, no extensive study of the group was made until H. B. Brady produced his *Monograph on Carboniferous and Permian foraminifera*, published by the Palaeontographical Society in 1876. This book, which forms a foundation for most subsequent work on late Palaeozoic foraminifera, is remarkable for its time in that there was generally a close stratigraphical control and fairly full details of collecting localities. Mostly this is because a large proportion of the material studied by Brady came from the Yoredale facies rocks of Northumberland and Scotland, where the stratigraphy was well known.

Brady was supplied with specimens by several of the numerous amateur collectors of the time – notably James Bennie in Scotland (later to join the Geological Survey there) and by Walter Howchin in Northumberland. Howchin's story is remarkable. A Primitive Methodist minister, he travelled around Northumberland on foot collecting the weathered debris from the numerous small working quarries then available, later sieving and washing the foraminifera. In 1881 he was advised to go to Australia for the sake of his health, which was very poor; actually, he lived to 92 and only died in 1937. He took his collection with him to Australia but much of it has recently been returned and is now in the British Museum (Natural History). Although he published only one paper on Carboniferous foraminifera (1888), he left some uncompleted manuscripts. The achievements of these early pioneers are all the more remarkable in that they usually worked with only a hand-lens and did not always investigate the internal structure of the specimens.

Virtually no more work was done in Britain until the 1940s, and this postdated the revolution in Carboniferous foraminiferal studies introduced by Russian authors who used only thin sections for their systematics. In Britain, A. G. Davies did some work, mostly on boreholes (1945, 1951) and R. H. Cummings published several papers (1955–61), mainly on Palaeotextulariidae, in which he began to tackle the problem of using randomly orientated thin sections. He devised and used his own zonal scheme, the details of which were never published but which is quoted (in the form F.Z.1, F.Z.2, etc.) by other authors (e.g. Sheridan, 1972). In Belgium, pioneer studies by R. Conil and M. Lys and their collaborators have used foraminifera in Dinantian stratigraphy with considerable success. Detailed studies are being made in the British Isles and enough has been done to show that foraminifera have very considerable potential here too. The ultimate aim of integrating the results as they continue to improve has implications for the refinement of Dinantian stratigraphy in the British Isles. Recent papers on the British faunas include those of Hallett (1970), Conil and George (1974), Marchant (1974), Fewtrell and Smith (1978), and Conil, Longerstaey and Ramsbottom (1980). New studies on British Carboniferous foraminifera are now in progress, and the present account is only an interim report.

The authors have contributed principally as follows: Courceyan to Arundian, M. D. F.; Holkerian to Brigantian, A. R. E. S.; Namurian, W. H. C. R.

3.2 LOCATION OF IMPORTANT COLLECTIONS

The majority of the specimens figured by Brady (1876) are in the British Museum (Natural History). The whereabouts of the specimens figured in Howchin's paper of 1888 is unknown, but the residue of Howchin's collections is also in the British Museum (Natural History), as are the specimens figured by Hallett (1970). Cummings' specimens are mostly in the Hunterian Museum, Glasgow. The largest collection of Carboniferous foraminiferal thin sections in Britain is in the British Geological Survey, Keyworth, from which many of the specimens figured here came. Others are mainly from the Sedgwick Museum, Cambridge.

Specimens figured by Conil and Longerstaey (1980) are mostly at the Geological Institute, Louvain la Neuve, Belgium – some are from the collections of the British Geological Survey, Edinburgh.

3.3 STRATIGRAPHIC DIVISIONS

British Carboniferous stratigraphy and correlations have been recently reviewed in two Special Reports of the Geological Society of London (George *et al.* 1976; Ramsbottom *et al.* 1978) and the limits and divisions used here are those proposed and used in these reports (see Fig. 3.1).

Although foraminifera were used in choosing the local boundaries of some of the stages of the Dinantian proposed by George *et al.* (1976), no foraminiferal zones have yet been proposed in Britain. The foraminiferal zones erected by Conil *et al.* (1978) for Belgium (Fig. 3.1) are, in effect, assemblage zones. They are not directly adopted here because several discrepancies exist and our studies have not yet established whether this zonation is the most suitable that could be devised for Britain. At present, foraminiferal faunas are assigned to a chronostratigraphical stage framework, though usually it is possible to assign a fauna to the early, middle or late part of a stage.

3.4 OCCURRENCE

Calcareous foraminifera are more or less restricted to the shallow water shelf areas where limestone or highly calcareous mudstones were being deposited. Generally, foraminifera are found in the bioclastic limestones which contain a limited proportion of finer crinoid debris. They may also be common and well-preserved in oolitic and pellet limestones. They are usually rarer in micritic or arenaceous limestones, although some families seem to have been well adapted to the

34 Carboniferous [Ch. 3

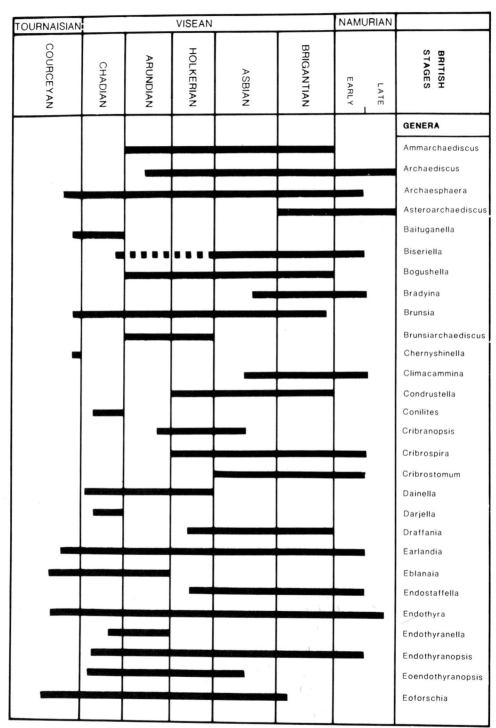

Fig. 3.1 – Range of chart of Carboniferous genera.

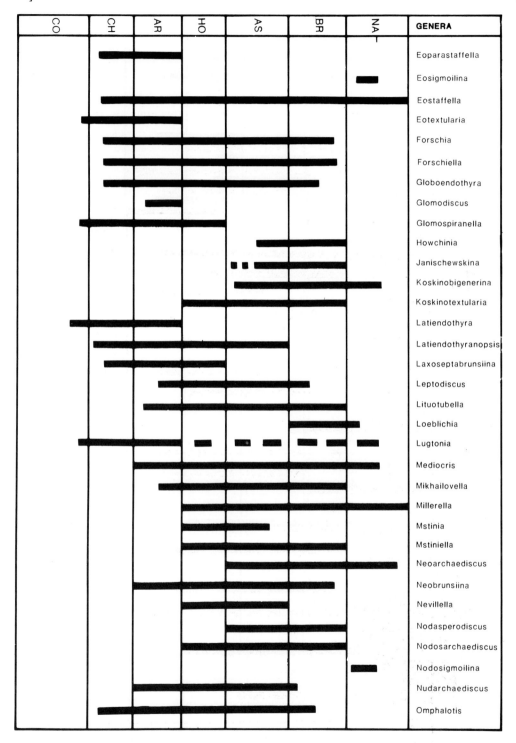

Fig. 3.1 – *Continued*.

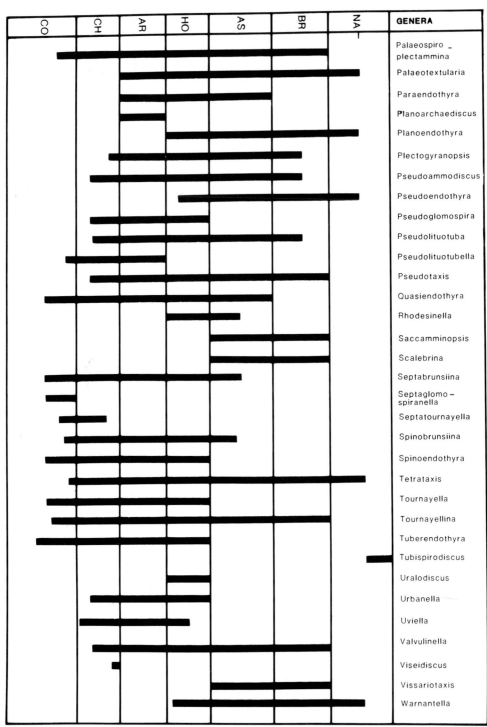

Fig. 3.1 – *Continued*.

environment, e.g. small Archaediscidae. It has been observed that an abundance of bryozoa in a limestone means that there will be few foraminifera.

Nearly all work is now done on thin sections and, indeed, this is the only way that the wall structure, much used in classification, can be seen. Thus, although most work has been done with the foraminifera found in limestones, and complete foraminiferal tests may readily be washed from calcareous shales, the determination of whole specimens is difficult or impossible. It is known, however, that the foraminifera present in limestones are not necessarily identical with those found in adjacent shales, and that each fauna contains ecologically restricted elements.

The arenaceous forms found commonly in the Westphalian marine bands are closely indicative of a particular facies (Calver, 1969), but since they have not proved to be stratigraphically diagnostic they will not be mentioned further here.

3.4.1 Courceyan

Courceyan strata are widespread in the British Isles but are often in non-foraminiferal facies. This is particularly true in the lower part, and the base-Courceyan boundary stratotype in Co. Cork lacks foraminifera altogether. The main areas of study so far are Ireland (Midlands and Dublin Basin), and the Craven Basin of northern England.

The lowest known fauna recorded in the British Isles has been found in the Irish Midlands (Marchant, pers. comm. 1980). The fauna is dominated by *Tuberendothyra* sp. and *Eoforschia* sp. which are also present in later Courceyan rocks. The restricted fauna is not related to any obvious facies control. Later faunas found elsewhere in the British Isles also tend to be sparse and low in diversity. Forms present in the Swinden No. 1 borehole (the lowest horizon reached in the Craven Basin) include *Septabrunsiina* sp., *Septatournayella*, *Latiendothyra* cf. *parakosvensis*, *Endothyra* sp., *Tournayella discoidea*, *Spinoendothyra* sp., and *Palaeospiroplectammina tchernyshinensis*. Similar forms are present in the stratigraphically higher Haw Bank Limestones in the Craven Basin but with the addition of rare *Chernyshinella glomiformis*, and *Septaglomospiranella primaeva*; other more common forms include *Brunsia spirillinoides*, *B. pseudopulchra*, *Lugtonia concinna*, *Pseudolituotubella* sp., *Earlandia vulgaris*, *Archaesphaera*, *Palaeospiroplectammina mellina*, *Spinoendothyra recta*, *S. mitchelli*, *S. costifera*, *Spinobrunsiina ramsbottomi*, *Latiendothyra* cf. *latispiralis*, *Endothyra* including *E. bowmani*, *E. laxa*, *E. prokirgisana*, *E. prisca*, *E.* cf. *freyri*, *E.* cf. *nebulosa*, *E.* cf. *tumida*, and several previously undescribed forms are also present. The genera *Endothyra*, and *Septabrunsiina* become the dominant elements of the fauna in the Craven Basin.

Higher beds in the Haw Bank Limestones contain *Eblanaia michoti* which is a useful marker. Other species of *Eblanaia* were recorded by Conil (1976) at a similar level in the Dublin Basin with *Tournayella kisella* (in Marchant, 1974). Both *Tetrataxis* sp. and *Eotextularia* sp. have been recorded from the late Courceyan, the former from the south west Province and Dublin Basin, the latter from north of Dublin (George *et al.* 1976). Conil *et al.* (1980) have illustrated fauna from the Black Rock Limestone of the Bristol area which contain *Eblanaia*, *Septabrunsiina*, and poorly preserved endothyrids which they referred to *Granuliferella*. The Black Rock Limestone is of later Courceyan age.

3.4.2 Chadian

Limestones of Chadian age are more widely developed in the British Isles and foraminifera are recorded from a somewhat greater number of localities. Published records however are from similar areas to those for the Courceyan, i.e. the south west Province (Mendips and Gower); the Dublin Basin and elsewhere in Ireland; the Craven Basin and

Ravenstonedale in northern England.

Early Chadian faunas have much in common with those of the high Courceyan particularly in the Craven Basin where the transition is best developed in a thick sequence. The boundary, which has its stratotype in this area, is much less clearly marked in terms of foraminifera than it is in Belgium where there are probably substantial non-sequences at the Tournaisian–Viséan boundary. The Bankfield East beds which immediately overlie the base of the Chadian in the Chatburn road cutting near Clitheroe, Lancashire, contain the following forms; *Tournayella discoidea, T. kisella, Septabrunsiina* sp., *Brunsia*, including *B. pseudopulchra* and *B. spirillinoides, Glomospiranella* spp., *Spinoendothyra costifera, S. recta* and *S mitcheli*. Many of the endothyrids mentioned in the Haw Bank Limestones are present, in particular *E. bowmani, E. freyri, E. laxa, E. danica*, and several undescribed forms. *L.* cf. *parakosvensis, Septabrunsiina* sp., *Spinobrunsiina ramsbottomi* and *Eblanaia* sp. are also characteristic forms in this type of section. *Palaeospiroplectammina* sp. and *Eotextularia diversa* are represented. In addition to these forms however, probable *Dainella*, a Chadian marker is present in the lower part of the section.

It has not been possible to confirm a report (Conil pers. comm.) of *Eoparastaffella* in these beds; a thorough study of the Clitheroe sequence is needed and has now been initiated. *Dainella* and *Eoparastaffella* enter together at the base of the Belgian Viséan, along with many of the elements of the above fauna. A 'basal' Chadian fauna with both *Dainella* and *Eoparastaffella* was reported by Conil and George (1973) from the Caninia (now the Gully) Oolite in Gower, S. Wales, but *Dainella* appears without *Eoparastaffella* in several sequences in the Craven Basin, including Ravenstonedale (Stone Gill Beds, with *Endothyra similis*; Ramsbottom in Holliday, Neves and Owens, 1979) as well as in the base-Chadian stratotype. Conil *et al.* (1980) have illustrated further material from the Stone Gill Beds including *Spinoendothyra mitchelli, Brunsia spirillinoides*, and cf. *Eblanaia*. There is a good possibility that the sequence in northern England is more complete than in Belgium (see Fewtrell and Smith (1981) for further discussion of this problem). The implication by Conil *et al.* (1980, caption to Plate 3.2) that the topmost Courceyan at Chatburn is basal Viséan is not supported by the available evidence.

In the eastern Craven Basin, the Thornton Limestones (Clitheroe Formation) are particularly noteworthy, although the junction with Courceyan strata is not seen. The beds contain a rich, diverse and well preserved fauna, which characteristically includes the following: *Earlandia vulgaris, E. minor, Lugtonia, Darjella, Tetrataxis, Pseudotaxis, Brunsia pseudopulchra* and *B. spirillinoides, Forschia, Forschiella prisca, Septatournayella lebedevae, Septabrunsiina* sp., *Pseudolituotubella, Conilites dinanti, Eotextularia diversa, Palaeospiroplectammina mellina, Spinoendothyra recta, Latiendothyra latispiralis, Latiendothyra parakosvensis, Endothyra danica, E. lensi, E. costifera, Omphalotis* cf. *chariessa, Endothyranopsis compressa, E. crassa, Eoendothyranopsis, Tuberendothyra tuberculata magna, Dainella cognata, D. fleronensis* and *Urbanella fragilis. Eoparastaffella* is only doubtfully present. Similar faunas have been described from Ireland at this level, e.g. Conil and Lees (1974), Austin *et al.* (1973), Marchant (1974). *Eoparastaffella* appears in abundance, with *Eostaffella*, in several localities in the Craven area in which Chadian beds are immediately overlain by Arundian (e.g. Embsay Limestone, Haw Crag, Hetton Beck Limestones; Fewtrell and Smith, 1978). These faunas also include taxa appearing for the first time, such as *Plectogyranopsis exelikta, E. apposita, E. campinei, E. convexa, E. delepinei, Biseriella bristolensis*, and *Pseudolituotuba gravata. Biseriella bristolensis* is also found, along with *Brunsia* spp., *Dainella* spp., *Eoparastaffella* sp. and endothyrids in the Gully Oolite in

Gower, South Wales (Conil and George, 1973). Further material of Chadian age has been illustrated by Conil et al. (1980) from Furness, Derbyshire, and the south west Province.

3.4.3 Arundian

Arundian faunas hae been studied in both shelf and basin facies in the south west Province, northern England and in Ireland. The base of the Arundian Stage is defined at the first lithological change below the entry of archaediscid foraminifera in the boundary stratotype section at Hobbyhorse Bay, Pembrokeshire (George et al., 1976, p9). The entry of this particularly distinctive and biostratigraphically important group is therefore a useful marker in correlating this boundary in sections elsewhere in the British Isles. Above the base, the evolution of the Archaediscidae continues to provide biostratigraphical data upwards into the Namurian.

At Hobbyhorse Bay, the earliest archaediscids to appear are *Brunsiarchaediscus* and *Ammarchaediscus*. *Uralodiscus* and *Archaediscus* enter rather higher in the sequence. The fauna also includes *Eotextularia*, *Forschia*, *Omphalotis*, cf. *minima*, *Latiendothyranopsis menneri* and *Mediocris* including *M. breviscula* and *M.* cf. *carinata*. In the higher part of the Arundian in the same area, the fauna is characterised by distinctively small forms of *Archaediscus* at the '*involutus*' stage of evolution (Pirlet and Conil, 1978), *L. menneri* also being present.

The earliest Archaediscidae in the Craven Basin are found in the higher part of the Embsay Limestone (Clitheroe Formation), in the limestones in the core of the Lothersdale anticline and in the calcareous shales overlying the reef limestones of Haw Crag (cf. Fewtrell and Smith 1978, p. 267); these units are thereby dated as Arundian. *Brunsiarchaediscus* and *Ammarchaediscus* are among the first archaediscids to appear, as at the boundary stratotype, followed by large *Uralodiscus* (*Permodiscus* of Fewtrell and Smith 1978, George et al., 1976). *Glomodiscus* sp. and *Ammarchaediscus* are also present as are *Omphalotis*, *Urbanella fragilis*, *Endothyra* spp. including *E. prisca*, *Pseudolituobella*, *Pseudotaxis*, *Tetrataxis*, *Pseudoammodiscus*, *Eostaffella*, *L. menneri* and *Mediocris*, common in the south west Province are very rare in the Craven Basin. *Valvulinella* has been found in the Arundian only in Ireland (Conil and Lees, 1974, Conil, 1976).

The Arundian of the Bristol area generally lacks the earliest archaediscids because of widespread non-sequence at this level. A fauna from the upper Burrington Oolite in the Mendips (Austin et al., 1973) includes *Brunsia*, *Eotextularia*, *Endothyra bowmani*, *Dainella* spp., *Eostaffella* sp., *Archaediscus stilus* and *Nodosarchaediscus* (i.e. *Neoarchaediscus*). A 'nodose archaediscid' was also reported at a similar (high Arundian) stratigraphic horizon in Gower by Conil and George (1973); nodose forms are normally associated with Asbian and later faunas (see below). Marchant (1974), Conil and Lees (1974) and Conil (1976) have all recorded Arundian faunas from Ireland, with characteristics similar to those described already. Reference was made by George et al. (1976) to Arundian faunas from the south Askrigg Block and from localities in Ireland.

Additional taxa recorded from the Arundian by Conil et al. (1980) include the following. From the Bristol area: *Eoparastaffella*, *Uviella*, *Spinoendothyra recta*, *Endothyra* spp., and *Glomodiscus*. From northern England, *Rectoseptaglomospiranella*, *Globoendothyra*, *Pseudoammodiscus*, *Uralodiscus* (as *Rectodiscus*), *Plectogyranopsis* spp., *Glomodiscus*, *Uralodiscus* (as *Tubispirodiscus*) *settlensis*, *Archaediscus varsanofievae*, and *Quasiendothyra* sp.

3.4.4 Holkerian

The more primitive foraminifera of the two earlier Viséan stages are rapidly replaced by a new assemblage brought in with the Holkerian transgression. Not only is this a period of rapid development and colonisation by certain taxa but also a period when characteristically restricted foraminiferal assemblages existed. A comparable occurrence of restricted faunas among the macrofossils has been taken to indicate abnormal salinity, probably hypersalinity.

Several new genera make their appearance in the Holkerian including *Koskinotextularia*, *Mikhailovella* and *Archaediscus* at the concavus stage. Many others use this stage as a period of development and diversification, including: *Omphalotis*, *Lituotubella*, *Nevillella*, *Bogushella*, *Globoendothyra*, *Endothyranopsis*, *Plectogyranopsis*, *Rhodesinella*, *Mstiniella*, and *Nodosarchaediscus*. By way of contrast, *Endostaffella*, *Millerella* and *Pseudoendothyra* make only an occasional appearance in the Holkerian but are all much more common and stratigraphically useful in the younger Viséan stages.

All the genera so far mentioned can be common in the Holkerian as a whole but they are frequently totally absent in the restricted assemblages. Essentially these consist of *Eostaffella parastruvei*, *E. mosquensis*, *Endothyra*, *Quasiendothyra nibelis*, *Dainella holkeriana*, *Rhodesinella avonensis*, *Septabrunsiina* (*Spinobrunsiina*) and the alga *Koninckopora*. Individually these forms may occur in older and/or younger rocks, but this particularly characteristic association is very typical of the Holkerian. Here, *Eostaffella* and *Koninckopora* are very abundant but *Q. nibelis* in particular is never better developed or present in greater numbers than in this stage.

3.4.5 Asbian

This stage can be divided into two distinct mesothems, the early and later Asbian. Of the two, it would appear that the early Asbian transgression was much less widespread as it is often missing in British stratigraphical successions.

The early Asbian can be distinguished from the underlying Holkerian by the appearance of several forms. Amongst these the genus *Vissariotaxis* is the most striking but unfortunately it is not very common and its usefulness is therefore restricted. Further new genera appear in the form of primitive *Nodasperodiscus* and *Neoarchaediscus* but these are also relatively uncommon, often poorly defined, and not well preserved at this level. An additional introduction to the rapidly expanding Archaediscidae is the development of *Archaediscus* at the angulatus stage. This form is not only widely distributed but also fairly abundant and easy to identify. *Nodosarchaediscus* also becomes very widespread at this level. Several genera become less common in the Asbian compared with a relative abundance in the underlying Holkerian. Of these *Quasiendothyra nibelis* is the most notable example.

In contrast to the rest of north west Europe the early Asbian in Britain contains double-walled Palaeotextulariidae in the form of primitive *Cribrostromum*, a genus which was not introduced until the late Asbian transgression in Belgian stratigraphy.

A typical early Asbian assemblage in the British Isles might consist of *Eostaffella mosquensis*, *Omphalotis minima*, *Endothyranopsis crassa*, *E. compressa*, *Forschiella prisca*, *Forschia*, *Globoendothyra globula*, *Endostaffella fucoides*, *Pseudoendothyra sublimis*, *Valvulinella latissima*, *Palaeotextularia*, *Koskinotextularia*, *Lituotubella*, *Bogushella ziganensis*, *Mstinia*, *Palaeospiroplectammina*, *Archaediscus* at the angulatus stage, *Neobrunsiina*, *Nevillella*, *Mikhailovella*, *Scalebrina*, *Millerella*, *Endothyra spira* and other species of the above genera.

The late Asbian introduces a rich variety of fauna including several new genera. The most striking change is in the Palaeotextulariidae where plentiful large *Koskinobigenerina* and *Climacammina* appear for the first time; *Cribrostomum* now becomes more abundant and well preserved. The new forms *Howchi-*

nia and *Bradyina* have unique wall or septal structures which can be readily identified even when only small fragments are available and are thus very useful stratigraphically.

A marked increase in the number of small stellate *Neoarchaediscus* and *Nodasperodiscus* is evident at this level together with *Archaediscus* at the angulatus stage. *Pseudoendothyra*, *Biseriella* and *Endostaffella* tend to be present in increasing quantities.

A typical late Asbian assemblage may include elements of the above mentioned genera together with species of Forschiinae, *Lituotubella*, *Cribranopsis*, *Bogushella*, *Koskinotextularia*, *Palaeotextularia*, *Nodosarchaediscus*, *Omphalotis*, *Vissariotaxis*, *Mikhailovella*, *Plectogyranopsis*, *Millerella*, *Palaeospiroplectammina*, *Cribrospira*, *Valvulinella*, and another new genus *Sacamminopsis*.

3.4.6 Brigantian

This stage yields a foraminiferal fauna which is in many ways similar to that of the Late Asbian, but the introduction of widely distributed new forms makes it easily distinguishable. During this stage there is a marked variation in foraminiferal faunas from province to province. In most areas of the British Isles the Brigantian may be divided into two mesothems (D6a and D6b of Ramsbottom, 1979) which yield characteristically different faunas. Brigantian foraminifera tend to be rich and varied and to occur in abundance, and are frequently excellently preserved.

Stratigraphically the most useful development is that of the small stellate *Nodosarchaediscus* (*Asteroarchaediscus*), which is widely distributed and often very abundant. These tend to be well preserved even in oolitic, dolomitic or siliceous limestones where all other genera are frequently destroyed or absent. Thus they are particularly useful in this stage where transitions to clastic beds are common. Another new form, slightly less useful due to its relative scarcity, is the very distinctive *Janischewskina*. Its characteristically complex septal areas render it easily identifiable when only small fragments of the test are preserved. The numerous subquadrate chambers of *Loeblichia*, another new genus introduced in the Brigantian, are very distinctive in thin section.

Associated Brigantian foraminifera may include giant tetrataxids, *Bradyina*, *Howchinia bradyana*, *Koskinobigenerina*, *Climacammina*, *Cribrostomum*, *Warnantella*, *Valvulinella*, large *Archaediscus* at the angulatus stage, *Loeblichia paraammonoides*, *Saccamminopsis*, *Biseriella*, *Endothyranopsis*, *Millerella*, *Lituotubella*, *Pseudoendothyra* and *Endostaffella*. Also frequently present as an aid to distinguishing these assemblages from those of the late Asbian is the alga *Calcifolium*.

In contrast to these flourishing genera a general and rapid decline of the Tournayellidae, Endothyridae, Calcisphaeridae, *Earlandia*, *Brunsia* and *Mediocris* began in the Brigantian.

3.4.7 Namurian

Foraminifera are abundant in the lower Namurian limestones of northern England and Scotland, but very little has been published about them. Foraminifera have also been found recently in the few thin limestones of upper Namurian age in northern England.

The lower Namurian (Pendleian and Arnsbergian stages – Early Namurian of the range chart) contain many of the genera and species which are also present in the Brigantian, and there is no really sharp change in the foraminiferal fauna at the base of the Namurian. Particularly common are forms such as *Endothyranopsis*, *Bradyina*, *Eostaffella*, *Asteroarchaediscus* and *Neoarchaediscus*. A totally new element in the fauna of the early part of the Namurian is *Eosigmoilina*, used by Conil *et al.* (1976) as the zonal index of their zone Cf7. In England, however, this fossil has not been found except in the E_{2a}

Zone, and the same appears to be true in Scotland. The closely related *Nodosigmoilina* has been found in some of the E_{2b} limestones in the Alston Block area. Compared with faunas of the same age in the U.S.S.R. the early Namurian fauna is rather restricted in the number and variety of taxa present.

Late Namurian faunas are recorded here for the first time in Britain, but are restricted to beds of R_1, R_2 and G_1 zone age (Kinderscoutian to Yeadonian stages). No foraminifera have been found as yet in the Chokerian and Alportian stages, and the few limestones known of these ages are of an unsuitable facies. In the Kinderscoutian a few archaediscids and endothyrids together with *Millerella* s.s. are known in the N7 mesothem in the Throckley Borehole in Northumberland, and foraminifera are also known in the N6 and N8 mesothems, though they are poor and fragmentary. The highest foraminiferal fauna yet found is in the *Gastrioceras cumbriense* Marine band in Lincolnshire; small archaediscids occur here, almost to the exclusion of other foraminifera.

3.5 DESCRIPTION OF GENERA

Although species are illustrated in Plates 3.1 to 3.12, at the present state of knowledge of British Carboniferous foraminifera the most practicable taxa for biostratigraphical studies are genera. For most taxa the ranges are given on Fig. 3.1. Two genera recorded by Conil *et al.* (*Bessiella* and *Florenella*) are not yet validly published and are not included below. The classification followed is that of Loeblich and Tappan (1964).

3.5.1 Suborder Fusulinina Wedekind, 1937
Test calcareous, microgranular to granular, advanced forms with two or more layers in the wall. Within this suborder, Dinantian foraminifera are divided into the superfamilies Parathuramminacea, Endothyracea, and Fusulinacea. In the following generic description the test is free unless otherwise stated.

Ammarchaediscus Conil and Pirlet, 1978. Type species *A. bozorgniae* Conil and Pirlet, 1978. Test lenticular, with planispiral coiling throughout. Wall double-layered, internal dark microgranular layer always present; external fibrous layer developed in central axial region only, or weakly throughout the test. Aperture simple, terminal.

Archaediscus Brady, 1873, emend. Conil and Pirlet, 1978. Type species *A. karreri* Brady, 1873. Test lenticular. Proloculus followed by nonseptate tubular chamber coiled in various planes. Evolute. Wall composed of clear finely fibrous outer layer and thin microgranular inner layer. Aperture a simple opening at end of tube.

Archaesphaera Suleimanov, 1945, emend. Conil and Lys, 1976. Type species *Diplosphaera inaequalis* Derville, 1931. Very simple microgranular test, varying from hemispherical or spherical to diplospherical (with two intersecting spheres of different size). Not certainly a foraminiferid, but conventionally regarded as such in much of the literature.

Asteroarchaediscus Miklukho-Maklai, 1956. Type species *Archaediscus baschkiricus* Krestovnikov and Theodorovich, 1936. Test lenticular or discoidal, often with an uneven serrated surface. Proloculus followed by irregularly coiled tubular chamber which is almost completely occluded by pseudofibrous nodosities throughout ('stellate' condition). Wall with inner dark microgranular layer weakly developed and outer fibrous layer well developed throughout.

Baituganella Lipina, 1955. Type species *B. chernyshinensis* Lipina, 1955. Test highly irregular, consisting of a number of irregular chambers with angular constrictions and pseudosepta. Lacks a proloculus. Wall microgranular to granular, secreted, including agglutinated material. Aperture simple, terminal.

Biseriella Mamet, 1974. Type species *Globivalvulina parva* Chernysheva 1948. Test subglobular. Proloculus followed by biserially arranged chambers with marked expansion of coiling. Initial biseriamminid coil followed by open helical coil. Wall microgranular with tectum. Aperture simple, lobate.

Bogushella Conil and Lys, 1978. Type species *Mstinia ziganensis* Grozdilova and Lebedeva, 1960. Test irregularly coiled with raised terminal whorl. Pseudochambers well developed. Wall generally well-differentiated into two layers; external layer irregularly granular and agglutinated; internal layer thin, dark, microgranular. Aperture terminal, cribrate.

Bradyina Moeller, 1878. Type species *Nonionina rotula* Eichwald, 1860. Test robust, planispiral and nautiloid, involute. Not more than four whorls, with 3–9 chambers in last whorl. Chamberlets or canals formed by converging septal lamellae or infolding of outer wall to form septa. Wall thick, microgranular, coarsely perforate with distinct radial lamellae. Aperture interiomarginal, with additional large areal pores forming a terminal cribrate aperture; sutural apertures well developed.

Brunsia Mikhailov, 1939. Type species *Spirillina irregularis* Moeller, 1879. Test discoidal, consisting of proloculus followed by coiled non-septate tubular chamber. Early stages irregularly coiled, later portion planispiral, or with slight oscillations. Wall thin, microgranular. Aperture simple, terminal.

Brunsiarchaediscus Conil and Pirlet, 1978. Type species *Propermodiscus contiguus* Omara and Conil, 1965. Test *Brunsia*-like. Internal dark layer clearly developed, external fibrous layer forming umbilical lateral fillings. Coiling involute, becoming evolute in final whorls. Aperture simple, terminal.

Brunsiina Lipina, 1953. Type species *B. uralica* Lipina, 1953. Test small, discoidal, consisting of proloculus followed by poorly septate tubular chamber. Early portion irregularly coiled, later whorls planispiral. Slow expansion of the whorl with coiling. Wall thin, microgranular. Aperture simple, terminal. The attempt by Conil and Lys (1978) to synonymise *Brunsiina* with *Glomospiranella* Lipina, 1953 confused it with *Brunsia*. Although *Brunsiina* is probably therefore still available, *Glomospiranella* is used herein.

Chernobaculites Conil and Lys, 1978. Type species *C. beschevensis* (Brazhnikova, 1965 al. *Ammobaculites sarbaicus* (Malakhova, 1956) var. *beschevensis* Brazhnikova (1965). Test initially coiled, later part straight, uniserial; uncoiled portion predominates. Wall thick, differentiated; outer layer thick, agglutinated; inner layer thin, microgranular. Chambers of coiled portion chernyshinellid in form; chambers of straight portion drop-shaped, with protruding septal necks. Aperture simple, terminal. Recorded from the low Chadian Bankfield East beds, Clitheroe. Conil *et al.*, 1980.

Chernyshinella Lipina, 1955. Type species *Endothyra glomiformis* Lipina, 1948. Small to medium test, with 1–4 simple to highly skewed whorls. Septa short, periphery lobate. Chambers typically 'droplet' shaped or 'chernyshinellid'. Later chambers inflated. Wall microgranular.

Climacammina Brady, 1873. Type species *Textularia antiqua* Brady, 1871. Test large, initial portion biserial, terminal portion uniserial and cylindrical. Chambers increase gradually in size through the biserial portion. Sutures depressed. Wall consists of inner fibrous radiate layer and outer granular layer with agglutinated particles. Aperture slit-like in biserial portion and cribrate in uniserial part.

Condrustella Conil and Longerstaey, 1980. Type species *Mstinia modavensis*, Conil and Lys, 1980. Coiling irregular. Chambers chernyshinellid, well defined and few in number. 2–3 whorls. 4–5 chambers per whorl. Wall thick, coarsely agglutinated with thin microgranular internal layer. Aperture low, simple.

Conilites Vdovenko, 1970. Type species *Ammobaculites dinantii* Conil and Lys, 1964. Initial part of test planispirally coiled for $3\frac{1}{2}$–4 whorls, followed by an uncoiled straight uniserial portion. Pseudoseptate throughout, with 8–9 chambers in last spiral whorl. Wall thick, internal layer microgranular, external layer agglutinated. Aperture cribrate, terminal.

Cribranopsis Conil and Longerstaey, 1980. Type species *Cribrospira fossa* Conil and Naum, 1977. Test involute, planispiral or with slight initial oscillations. Wall granular and finely perforate, initially thin, becoming thicker. Septa short, truncated, thick and massive. Chambers subquadrate, numerous, 8–13 in final whorl, with tendency to uncoil in terminal chambers. Aperture cribrate.

Cribrospira Moeller, 1878. Type species *C. panderi* Moeller, 1878. Test nearly involute, coiling somewhat irregular. Chambers increase rapidly in size; whorls few in number. Septa short. Wall microgranular to granular with tectum. Aperture cribrate, consisting of large pores on apertural face. Intercameral opening large.

Cribrostomum Moeller, 1897. Type species *C. textulariforme* Moeller, 1879. Test large. Chambers broad and low, biserially arranged, increasing rapidly in size; sutures depressed. Wall calcareous, with an inner radiating fibrous layer, and an outer granular layer with agglutinated particles. Cribrate aperture in last chambers.

Dainella Brazhnikova, 1962. Type species *Endothyra chomatica* Dain, 1940. Test globular, irregularly coiled with rapid changes in coiling axis. Early whorls in particular have numerous closely packed chambers; outer whorls with larger chambers, often with high angled changes in coiling direction between whorls. Chomata well developed. Wall microgranular, single layered.

Darjella Malakhova, 1963. Type species, *D. monolis* Malakhova, 1963. Test uniserial, numbers of chambers variable, usually 2–4. Chambers spheroidal, increasing in size. Each chamber envelops the preceding one at the connecting neck. Wall agglutinated, although the amount of agglutinated material in the cement is variable.

Draffania Cummings, 1957. Type species *D. biloba* Cummings, 1957. Test flask-shaped or pyriform. Central thin tube around which are arranged two hemispherical chambers. Tube extends into long neck. Periphery rounded; surface smooth and with minute pore openings. Wall calcareous, perforate, thick, layered. Aperture terminal, circular.

Earlandia Plummer, 1930. Type species *E. perparva* Plummer, 1930. Spherical proloculus, followed by cylindrical or flaring tube. Wall microgranular. Aperture simple, terminal.

Eblanaia Conil and Marchant, 1976. Type species *Plectogyra michoti* Conil and Lys, 1964. Test large, discoidal, with large and well marked umbilici. Evolute. Coiling planispiral to oscillating initially, outer whorls usually planispiral. Supplementary deposits in the form of basal nodes, becoming spines, projecting forwards, in the final chambers; corner fillings may also be present. Chambers arched, periphery lobate, septa endothyroid; early stages chernyshinellid with pseudochambers. Wall poorly to clearly differentiated, with a tectum, a medial thick granular layer, and an inner darker microgranular layer.

Endochernella Conil and Lys, 1978. Nomen nudum (no type species designated, although Conil and Lys apparently intended it to be *Endothyra quaestita* Ganelina, 1966). Test entirely enrolled, the inner part of *Chernyshinella*-like form, outer part *Endothyra*-like. Wall tends to show differentiation into three layers; no basal deposits; corner fillings accentuate the chernyshinellid form of the chambers. Recorded from the *Michelina grandis* beds (Arundian), Ravenstonedale and from the Chadian at the Eyam borehole, Conil et al., 1980.

Endostaffella Rozovskaya, 1961. Type species *Endothyra parva* Moeller, 1879. Test lenticular or discoidal, concave or bilaterally symmetrical. Coiling involute, final whorls may be evolute. Initial whorls coiled at large angle to terminal portion. Chomata or pseudochomata in later whorls. Aperture simple, basal. Wall thin, microgranular.

Endothyra Phillips, 1846 emend. Brady, 1876. Type species *Endothyra bowmani* Phillips, 1846, emend. Brady, 1976. Test septate, involute, coiling irregular to planispiral, normally of 3–4 whorls, usually less than ten chambers in last whorl, periphery lobate. A variety of secondary deposits may be present (e.g. spines, tubercules). Wall dark, microgranular, with thin tectum.

Endothyranella Galloway and Harlton, 1930. Type species *Ammobaculites powersi* Harlton, 1927. Early portion coiled and endothyroid; younger portion uncoiled, uniserial, and much larger. Wall microgranular. Aperture simple, basal and crescentiform in spiral part; terminal, circular to oval in rectilinear part. Septate throughout.

Endothyranopsis Cummings, 1955. Type species *Involutina crassa* Brady, 1869. Test globular, involute. Coiling planispiral. Septa thick and long; chambers large, subquadrate or slightly rounded. Periphery smooth; sutures poorly developed. Wall thick, coarsely granular, agglutinated and finely perforate. Aperture simple, lunate.

Eoendothyranopsis Reitlinger and Rostovceva, 1966. Type species *Parastaffella pressa* Grozdilova, 1954. Test discoidal, laterally compressed, markedly biumbilicate. Coiling involute, planispiral, may have feeble initial oscillations. Periphery rounded to subrounded. Chambers subrounded to subquadrate with well defined sutures. Wall granular, with inclusions, two-layered. Terminal forward-projecting spine well developed. Aperture simple, basal.

Eoforschia Mamet, 1970. Type species *Tournayella moelleri* Malakhova, 1953. Test large, thick-walled, predominantly planispirally coiled, with pseudosepta. Wall two-layered; inner layer may have agglutinated grains. Aperture a simple basal opening.

Eoparastaffella Vdovenko, 1954. Type species *Pseudoendothyra simplex* Vdovenko, 1954. Test involute, lenticular to discoidal, planispirally coiled, may have slight initial oscillations. Secondary deposits present in the form of simple paired chomata. Wall microgranular.

Eosigmoilina Ganelina, 1956. Type species *E. explicata* Ganelina, 1956. Test small, lenticular. Proloculus followed by coiled, non-septate chamber; coiling sigmoidal. Wall porcellanous, lacks fibrous layer.

Eostaffella Rauser-Chernoussova, 1948. Type species *Staffella parastruvei* Rauser-Chernoussova, 1948. Test lenticular to nautiloid, laterally compressed, biumbilicate, involute, planispiral. Periphery often keeled in early whorls. Chambers subquadrate, numerous, 12–20 in last whorl. Septa nearly straight and at right angles to the wall. Wall dense, granular to microgranular, with pseudochomata or chomata throughout. Aperture simple, basal.

Eotextularia Mamet, 1970. Type species *Palaeotextularia diversa* Chernysheva, 1948. Initial portion chernyshinellid in character with very few chambers. Terminal portion uncoiled and biserial; chambers increase rapidly in size. Wall thick with coarse agglutinated outer layer and thin, dark, microgranular inner layer. Aperture simple, basal.

Forschia Mikhailov, 1939. Type species *Spirillina subangulata* Moeller, 1879. Test planispiral. Tubular chamber without distinct septation; pseudosepta only. Wall differentiated; outer layer granular, may have agglutinated grains; inner layer dark, microgranular. Aperture terminal, cribrate where tube flares terminally.

Forschiella Mikhailov, 1935. Type species *F. prisca* Mikhailov, 1935. Coiling initially planispiral and *Forschia*-like. Later stage is uncoiled, uniserial and partially septate. Aperture cribrate.

Globoendothyra Reitlinger, 1959. Type species *Nonionina globulus* Eichwald, 1860. Test nautiloid. Coiling oscillating or aligned; involute or with last whorl evolute. Wall thick with outer tectum; thick granular layer. Septa inclined. Spine-like projection in last chamber. Corner and lateral fillings with low nodosities in some cases. 3–5 whorls, 8–10 chambers in last whorl. Aperture simple.

Glomodiscus Malakhova, 1973. Type species *G. biarmicus* Malakhova, 1973. Test discoidal to lenticular. Proloculus followed by an open tubular chamber, involutely coiled, with oscillating coiling axis as in *Archaediscus*. Wall consisting of two layers; inner dark microgranular layer well developed and thickened laterally; outer fibrous radiate layer also well developed and completely enveloping each whorl. Aperture simple, terminal.

Glomospiranella Lipina, 1953. Type series *G. asiatica* Lipina, 1953. Proloculus followed by irregularly coiled tubular chamber, initially undivided, then divided by pseudosepta. Wall dark, microgranular, single-layered. Aperture simple, terminal.

Granuliferella E. J. Zeller, 1957. Type species *Endothyra rjausakensis* Chernysheva, of which the original type species *Granuliferella granulosa* (Zeller, 1957) is generally regarded as a junior synonym. *E. rjausakensis* is alternatively regarded as belonging to *Latiendothyra* (e.g. Vachard, 1978)*. Test endothyroid. Wall tending to be differentiated into a thin internal dark microgranular layer, and an outer, more or less coarsely granular layer; an outer tectum may be present. Septate throughout. Recorded as *Granuliferella avonensis* sp. nov. from the Black Rock Limestone of the Avon Gorge, and the Bankfield East beds Clitheroe (Chadian) by Conil *et al*., 1980.

Haplophragmella Rauser-Chernoussova, 1936. Type species *Endothyra panderi* Moeller, 1879. Early portion of test irregularly coiled, later portion uncoiled and uniserial. Coiled portion of about two whorls; uncoiled portion of about seven chambers. Wall thick, coarse, agglutinated, with dark inner layer. Aperture simple and interiomarginal in coiled stage; terminal and cribrate in uncoiled stage.

Howchinia Cushman, 1927. Type series

Patellina bradyana Howchin, 1888. Test trochoid, consisting of non-septate tube coiled in a high spire around a slightly depressed umbilicus filled with supplementary deposits of microcrystalline calcite. Vertical tubular structures developed in the umbilicus of some species. Spiral suture depressed, bridged by many extensions of shell matter, with small pits between. Wall with dark microgranular inner layer, and outer clear radiate layer. Aperture terminal.

Janischewskina Mikhailov, 1935. Type species *J. typica* Mikhailov, 1935. Test planispiral. Coiling involute with few chambers and whorls. Wall dark, microgranular, thin, and finely perforated. Septal chamberlets formed by infolding of outer wall. Cribrate aperture in apertural shield. Secondary sutural openings.

Koskinobigenerina Eickhoff, 1968. Type species *K. breviseptata* Eickhoff, 1968. Test biserial in early portion, uniserial in terminal portion. Wall single-layered, dark, granular with some agglutinated grains. Cribrate aperture in uniserial chambers.

Koskinotextularia Eickhoff, 1968. Type species *K. cribriformis* Eickhoff, 1968. Test biserial with cribrate aperture. Wall single-layered, granular, with some agglutinated grains. Septa with inflated ends.

Latiendothyra Lipina, 1963. Type species *Endothyra latispiralis* Lipina, 1955. Small to medium endothyroids with skew coiling in first few whorls and planispiral outer whorls, increasing rapidly in size. Wall microgranular, single layered (unlike other endothyrids).

Latiendothyranopsis Lipina, 1977. Type species *Endothyra latispiralis grandis* Lipina, 1955. Large involute endothyranopsinae with oscillating coiling. Wall coarsely agglutinated with dark internal layer and traces of a tectum. Chambers subquadrate, usually more than 8 per whorl, septa thick. Sutures poorly developed. Aperture simple.

Laxoseptabrunsiina Vachard, 1978. Type species, *L. valuzierensis* Vachard, 1978. Test irregularly coiled, with rapid expansion of

---

*This would leave *Granuliferella* as a *nomen nudum*.

whorl height. Tubular juvenarium leading to pseudochambers and true chambers in later portions of the test. Wall simple microgranular to granular. Aperture simple, basal.

Leptodiscus Conil and Pirlet, 1978. Type species *Permodiscus umbogmaensis* Omara and Conil, 1965. Test planispiral and lenticular. Internal dark microgranular layer clearly to weakly developed. External radiate layer forms a thick umbilical, lateral filling and thin peripheral layer around test. Chamber has convex floor.

Lituotubella Rauser-Chernoussova, 1948. Type species *L. glomospiroides* Rauser-Chernoussova, 1948. Test non-septate and irregularly coiled in early stages; uniserial, uncoiled, with regular constrictions of the wall forming pseudochambers in later stages. No true septa. Wall differentiated; outer layer granular with or without agglutinated particles; inner layer dark and microgranular.

Loeblichia Cummings, 1955. Type species *Endothyra ammonoides* Brady, 1873. Test discoidal, planispiral and evolute with numerous whorls. Wall finely granular. Chambers small, numerous (13–20 in final whorl) and quadrate. Sutures distinct, radial. Pseudochomata weakly developed. Septa straight, inclined towards the aperture. Aperture simple, basal, and crescent shaped.

Lugtonia Cummings, 1955. Type species *Nodosinella concinna* Brady, 1876. Uniserial test; proloculus followed by several chambers divided by convex septa and expanding in size. Wall microgranular; aperture simple.

Mediocris Rozovskaya, 1961. Type species *Eostaffella mediocris* Vissarionova, 1948. Test lenticular, discoidal or oval to subspherical. Periphery usually rounded. Coiling planispiral; may oscillate slightly in early whorls. Involute, rarely evolute in outer part. Wall microgranular, undifferentiated. Extensive lateral fillings. Aperture basal.

Mikhailovella Ganelina, 1956. Type species *Endothyrina gracilis* Rauser-Chernoussova, 1948. Initial part of test endothyroid, involute. Later part uncoiled, uniserial, subcylindrical, with few chambers. Wall microgranular to granular. Aperture simple in early stages, cribrate in last chambers of spiral portion and in uniserial portion.

Millerella Thompson, 1942. Type species *Millerella marblensis* Thompson, 1942. Test lenticular, nautiloid, laterally compressed, biumbilicate, planispiral. Evolute in ultimate and sometimes penultimate whorl. Periphery usually rounded. Septa arcuate, forward-pointing, regular and numerous. 16–20 chambers in last whorl. Chomata present. Wall microgranular. Aperture simple.

Mstinia Mikhailov, 1939. Type species *Mstinia bulloides* Mikhailov, 1939. Test irregularly coiled with small number of whorls and well developed chambers. Involute. Subglobular with lobulate periphery. Thick wall with coarse agglutinated outer layer and dark microgranular inner layer. Chambers chernyshinellid. Septation in early stages not always developed. Aperture terminal, cribrate.

Mstiniella Conil and Lys, 1978. Type species *Mstinia fursenkoi* Mikhailov *in* Dain, 1953. Test irregularly coiled. Initial tubular portion may be divided into pseudochambers, developing into later endothyroid chambered portion. Wall thick, well differentiated with external agglutinated layer and compact microgranular layer. Aperture basal, cribrate.

Neoarchaediscus Miklukho-Maklai, 1956. Type species *Archaediscus incertus* Grozdilova and Lebedeva, 1954. Coiling irregular. Wall double layered; inner dark microgranular layer very thin, outer radiating fibrous layer thick. Central whorls occluded, resulting in confused stellate flaring. Final 1–2 whorls well defined and without nodosities. Aperture simple, terminal.

Neobrunsiina Lipina, 1965. Type species *Glomospiranella finitima* Grozdilova and Lebedeva, 1954. Test large, globular to discoidal and umbilicate. Initial portion irregularly coiled with occasional terminal alignment of whorls. 4–7 whorls. Pseudochambers developed in terminal portion of test. Wall thick, outer layer granular with agglutinated grains; inner layer dark, microgranular. Aperture simple, basal.

Nevillella Conil, in press. (= *Nevillea* Conil and Longerstaey, 1980, = *Georgella* Conil and Lys, 1978) Type species *Haplophragmella dytica* Conil and Lys, 1977. Initial coiling chernyshinellid, terminal portion uncoiled, uniserial. Wall thick, clearly differentiated into a thin, dark, microgranular layer and an external coarsely granular agglutinated layer. Aperture cribrate.

Nodasperodiscus Conil and Pirlet, 1978. Type species *Archaediscus saleei* var. *saleei* Conil and Lys, 1964. Test similar to that of *Nodosarchaediscus* but having a stellate central region, caused by occlusion of the central whorls and almost total disappearance of the dark internal layer. Final whorls with thick fibrous layers and thin, feeble microgranular layers. Aperture simple, terminal.

Nodosarchaediscus Conil and Pirlet, 1978. Type species *Archaediscus maximus* Grozdilova and Lebedeva, 1954. Test lenticular to globular. Wall bilayered: internal dark microgranular layer very thin; outer radiating fibrous layer well developed. Nodosities present in part or all of tubular chamber, but no central stellate flaring developed. Aperture simple, terminal.

Nodosigmoilina nom. nud. Conil *in* Conil *et al.*, 1980. Type species apparently intended to be *Eosigmoilina rugosus* Brazhnikova, 1964, but not designate. Like *Eosigmoilina* but chambers more or less occluded by nodosities.

Nudarchaediscus Conil and Pirlet, 1978. Type species *Planoarchaediscus concinnus* Conil and Lys, 1964. Test irregularly coiled. Wall initially consisting of a thick, dark, microgranular layer and an outer fibrous layer. The latter is absent in the final whorls, where the microgranular layer is thick and well developed. Convex floor to tubular chamber. Aperture simple, terminal.

Omphalotis Schlykova, 1969. Type species *Endothyra omphalota* Rauser-Chernoussova and Reitlinger, 1936. Large endothyrid, involute; small number of whorls, large chambers. Distinctive wall structure, with external thin dark layer (tectorium), thick microcrystalline tectum, and continuous secondary deposits lining the chambers and projecting in places in the form of spines or tubercules.

Palaeospiroplectammina Lipina, 1965. Type species *Spiroplectammina tchernyshinensis* Lipina, 1948. Test elongate. Initial coiled portion, small, followed by larger uncoiled biserial portion. Wall microgranular to granular, undifferentiated. Aperture simple, basal.

Palaeotextularia Schubert 1921. Type species *P. schellwieni* Galloway and Rynicker, 1930 (by subsequent designation). Test biserial. 5–12 pairs of biserially arranged chambers; septa straight or gently arched, with lunate ends. Double layered wall; inner layer fibrous, radiate, outer layer granular. Aperture simple, lunate, on interiomarginal arch.

Paraendothyra Chernysheva, 1940. Type species *P. nalivkini* Chernysheva, 1940. Test coiled throughout; predominantly planispiral, initially irregular. Wall structure distinctive, thick, granular, with inclusions. Chambers inflated, periphery lobate, septa long and curved. Prominent basal deposits increasing in height and becoming hook-shaped towards the aperture, which is raised.

Planoarchaediscus Miklukho-Maklai, 1956. Type species *Archaediscus spirillinoides* Rauser-Chernoussova, 1948. Test lenticular. Proloculus followed by coiled tubular chamber, irregular at first, then planispiral. Wall dark, microgranular, with a fibrous layer restricted to the early, irregularly coiled whorls. Aperture simple, terminal.

Planoendothyra Reitlinger, 1959. Type species *Endothyra aljutovicha* Reitlinger, 1950. Test discoidal, biconcave. Initial irregular whorls followed with sharp change in coiling axis by later, slightly oscillating whorls. Initial coils involute, later evolute. 8–11 chambers in last whorl. Wall microgranular. Pseudochomata, lateral fillings and floor linings present. Aperture a simple slit.

Plectogyranopsis Vachard, 1978. Type species *Plectogyra convexa* Rauser-Chernoussova, 1948. Test nautiloid, septate. Coiling irregular, may be aligned or oscillating. Septa thick and inflated. Involute. Wall microgranular to granular, thin in initial whorls, thicker in final whorl; some species finely perforated; wall may be coarse and agglutinated. No basal layer.

Pseudoammodiscus Conil and Lys, 1970. Type species *Ammodiscus priscus* Rauser-Chernoussova, 1948. Test planispiral, umbilicate. Proloculus followed by non-septate tubular coiled chamber with only very slight increase in whorl height. Wall thin, microgranular and compact. Aperture simple, terminal.

Pseudoendothyra Mikhailov, 1939. Type species *Fusulinella struvei* von Moeller, 1879. Test lenticular, laterally compressed, biumbilicate, involute, mainly planispiral; early whorls may deviate slightly. Wall with four layers (including diaphanotheca). Periphery often keeled. Chambers regular, subquadrate, very numerous in last whorl. Septa straight. Pseudochomata or low chomata present. Aperture simple, low crescentiform.

Pseudoglomospira Bykova, 1955. Type species *P. devonica* Bykova, 1955. Test consisting of globular proloculus and tubular undivided chamber which is streptospirally coiled. Wall dark, thin, microgranular. Aperture simple, terminal.

Pseudolituotuba Vdovenko, 1971. Type series *Haplophragmella gravata* Conil and Lys, 1965. Test fixed or encrusting, tubular, partly coiled. Septa absent or in the form of rudimentary internal protuberances. Wall thick, and clearly agglutinated, with a dark internal microgranular layer.

Pseudolituotubella Vdovenko, 1967. Type species *P. multicamerata* Vdovenko, 1967. Test large. Early portion irregularly coiled with a few septa; later portion uniserial, uncoiled, and with septa throughout.

Wall thick, outer layer granular or agglutinated; inner layer dark and microgranular. Aperture terminal, cribrate

Pseudotaxis Mamet, 1974. Type species *Tetrataxis eominima* Rauser-Chernoussova, 1948. Test trochospirally coiled, with concave base. Low chambers arranged in helical spiral; 3–5 chambers per whorl visible around broad umbilicus in transverse section. Wall microgranular to granular with tectum but no fibrous layer. Aperture terminal.

Quasiendothyra Rauser-Chernoussova, 1948. Type series *Endothyra kobeitusana* Rauser-Chernoussova, 1948. Test discoidal or nautiloid. Coiling involute and glomospiral in early stage, evolute and aligned or irregular in later stages. Septa commonly inflated, curved. Wall microgranular to granular, chomata well developed. Aperture simple.

Rectoseptaglomospiranella Reitlinger, 1961 (subgenus of *Septaglomospiranella*). Type species *S. (R.) asiatica* Reitlinger, 1961. The emendation of Conil and Lys (1978), by which *Rectoseptaglomospiranella* is removed from *Septaglomospiranella* which itself becomes a synonym of *Septabrunsiina*, is not accepted here. Test coiled initially, then straight uniserial. Initial portion irregularly coiled; septate throughout. Aperture simple, terminal. The form recorded as *Rectoseptaglomospiranella* sp., by Conil et al. (1980, pl. IX, Fig. 1) from the *Michelinia grandis* beds, Ravenstonedale (Arundian) is poorly septate and has a cribrate aperture and this conforms neither to the original diagnosis nor to the emended one.

Rhodesinella Conil, et al., 1980 (= *Rhodesina* Conil and Longerstaey, 1980). Type species *Cribrospira pansa* Conil and Lys, 1965. Coiling planispiral, often with slight initial oscillations. Wall thick, granular to coarsely granular with agglutinated elements. Chamber size increases rapidly towards the final whorl. Terminal chamber raised into a short, almost uncoiled section. Aperture cribrate.

Saccamminopsis Sollas, 1921. Type species *Saccammina carteri* Brady, 1973, = *Nodosaria fusulinaformis* McCoy, 1849. Test uniserial, with globular to ovate chambers and strongly contricted sutures. Wall thin, microgranular. Aperture terminal, simple, rounded.

Scalebrina Conil and Longerstaey, 1980. Type species *S. compacta* Conil and Longerstaey, 1980. Test fixed for encrusting. Wall moderately thick, microgranular. Tubular chamber irregularly coiled. Septa absent or rudimentary. *Scalebrina* is distinguished from *Pseudolituotuba* Vdovenko, 1971 only by the wall structure and smaller size.

Septabrunsiina Lipina, 1955. Type species *Endothyra krainica* Lipina, 1948. Test coiled. Poorly septate or non-septate in early portion; later planispiral portion with well developed septa. Wall microgranular with basal supplementary deposits; terminal projections and nodosities. Aperture a simple opening at the end of tube.

Septaglomospiranella Lipina, 1955. Type species *Endothyra primaeva* Rauser-Chernoussova, 1948. Test small to medium, with 1–4 skew-coiled whorls. Periphery smooth to slightly lobate. Pseudosepta or septa present. Wall microgranular, single-layered. Aperture a simple opening. Conil and Lys (1978) suggested synonymising this genus with *Septabrunsiina*; this is not followed here although further consideration of this question is needed.

Septatournayella Lipina, 1955. Test species *Tournayella segmentata* Dain, 1953. Test planispiral, with pseudosepta or septa which increase in importance towards the outer whorls. Wall microgranular. Lacks secondary deposits. Aperture simple.

Spinobrunsiina Conil and Longerstaey, 1980 (subgenus of *Septabrusiina*). Type species *S. ramsbottomi* Conil and Longerstaey, 1980. Test coiled, non-planispiral. Wall microgranular, non-differentiated. Inner whorls have pseudochambers; outer whorls have true chambers. Supplementary deposits in the form of nodosites, arches, basal projections, basal and lateral thickenings. Aperture simple.

Spinochernella Conil and Lys, 1978. Type species *S. brencklei* Conil and Lys, 1978 (probable junior synonym of *Pseudochernyshinella* Brazhnikova, 1974, see Conil *et al.*, 1980, p. 58). Test coiling nearly planispiral. Wall more or less granular, tending to show differentiation into an inner, more compact layer. Septa and chambers chernyshinellid in inner whorls, becoming endothyrid in outer whorls. Supplementary deposits in the form of persistent thin basal projections. Distinguished from *Eblanaia* only by lack of a particularly prominent projection in the outermost chamber. (Recorded as *S.* cf. *brenklei* Conil and Lys by Conil *et al.* (1980) from the Chadian of the Chatburn bypass section.)

Spinoendothyra Lipina, 1963, emend. Mamet, 1976. Type species *Endothyra costifera* Lipina, 1955. Endothyrid with near planispiral or variable coiling. Wall microgranular. The distinctive feature is the presence of secondary deposits in the form of a spine at the base of each chamber. Aperture simple.

Tetrataxis Ehrenberg, 1854. Type species *T. conica* Ehrenberg, 1854. Test free or attached, trochospirally coiled, with a concave base, consisting of low to flattened chambers arranged in a helical spiral; 4 chambers per whorl. Umbilical cavity broad. Wall differentiated into two layers; outer layer dark, microgranular; inner layer fibrous, radiate. Simple umbilical aperture.

Tournayella Dain, 1953. Type species *T. discoidea* Dain, 1953. Test discoidal, planispirally coiled, lacking septa but sometimes having small pseudosepta. Evolute. Wall microgranular. Aperture simple.

Tournayellina Lipina, 1955. Type species *T. vulgaris* Lipina, 1955. Small number of whorls of which the terminal whorl is markedly expanded. Few chambers per whorl, usually 3–5, globular (chernyshinellid) in shape. Wall dark, granular. Aperture low, simple, terminal.

Tuberendothyra Skipp, 1969, emend. Mamet, 1976. Type species *Endothyra tuberculata* Lipina, 1948. Endothyroid with irregular coiling and a highly lobate periphery. Each chamber has a large rounded tubercle of secondary material at its base. Aperture simple.

Tubispirodiscus Browne and Pohl, 1973. Type species *T. simplissimus* Browne and Pohl, 1973. Test evolute. Coiling planispiral. Tubular chambers without nodosities. Dark microgranular layer of wall very thin or absent. Very gradual increase in size of tubular chamber towards simple aperture.

Uralodiscus Malakhova, 1973. Type species *U. librovichi* Malakhova, 1973. (= *Rectodiscus* Conil and Pirlet, 1978). Test discoidal to lenticular. Proloculus followed by a simple tubular chamber more or less planispirally coiled. Involute. Wall consisting of two layers; inner dark microgranular layer well developed, with lateral thickenings; outer fibrous radiate layer also well developed, and completely enveloping each whorl. Aperture simple.

Urbanella Malakhova, 1963. Type species *Quasiendothyra urbana* Malakhova, 1954. Test small, essentially planispiral. Numerous chambers, rather square in cross-section; periphery slightly lobate. Wall dark, microgranular. Differs from *Loeblichia* only in flatter axial section; *Urbanella* may be a junior synonym of *Loeblichia*.

Uviella Ganelina, 1966. Type species *U. aborigena* Ganelina, 1966. Test with $3\frac{1}{2}$–$6\frac{1}{2}$ whorls, initial whorls irregularly coiled and skew to remainder, which are evolute and planispiral. Slight constriction in inner whorls, developing into rudimentary septa in outer whorls. Wall thick and commonly contained agglutinated grains. Aperture a narrow opening.

Valvulinella Schubert, 1907. Type species *Valvulina youngi* Brady, 1876. Test conical. Chambers trochospirally arranged with only 2–3 chambers per whorl. Chambers subdivided into additional chamberlets by horizontal and vertical partitions. Chamberlets vary from rudimentary, to distinct and regular. Wall microgranular, single-layered. Aperture simple, umbilical.

Vissariotaxis Cummings, 1966. Type species *Monotaxis exilis* Vissarionova, 1948. Test trochoid, consisting of non-septate tube coiled in an elevated spire around a depressed umbilical region. Wall microgranular. Aperture simple, terminal.

Viseidiscus Mamet 1975. Type species *Permidiscus* (?) *primaevus* Pronina, 1963. (= *Parapermodiscus transitus* Reitlinger, 1969). Proloculus followed by semi-cylindrical planispiral tube, deviating slightly only in inner two whorls. Evolute. 5–7 whorls in all. Outer pseudofibrous wall layer of archaediscid type developed only in umbilical region. Aperture simple, terminal. N.B. *Ammarchaediscus* Conil and Pirlet, not published until 1978, is a probable junior synonym in part at least.

Warnantella Conil and Lys, 1978. Type species *Glomospira tortuosa* Conil and Lys, 1964. Test irregularly coiled with zig-zag convolutions. Wall microgranular, compact and very dark. Tubular non-septate chamber enlarges gradually in diameter. Periphery irregular, angular, subquadrate. Aperture terminal.

PLATE 3.1—All figures × 75. Specimen numbers refer to collection of M. D. Fewtrell of which representative rock samples are held in the Sedgwick Museum, Cambridge, unless otherwise stated.

Earlandia vulgaris (Rauser-Chernoussova and Reitlinger, 1937)
Plate 3.1, Fig. 1, HOW 43 Haw Bank Limestone, Chatburn Formation, Skipton, Yorkshire. Description: large *Earlandia*, up to 2.5 mm long, with simple cylindrical chamber. Wall thick, about 70 μm, dark microgranular.
Range: late Courceyan–early Namurian.

Archaesphaera inaequalis (Derville, 1931)
Plate 3.1, Fig. 2, HOW 124 Haw Bank Limestones, Chatburn Formation, Skipton, Yorkshire. Description: diplosphaerid form of *Archaesphaera*; overall size up to 250 μm; smaller sphere up to 100 μm, located within the wall of the larger sphere; outer wall dark microgranular.
Range: late Courceyan–Brigantian.

Archaesphaera reitlingerae (M. Maklai, 1958)
Plate 3.1, Fig. 3, FS 514 Haw Bank Limestones, Chatburn Formation, Skipton, Yorkshire. Description: test consists of a flattened sphere attached to a basal disc which may be flat or convex. Overall size, up to 250 μm, wall dark microgranular.
Remarks: considered to be growth stage of *A. inaequalis* by Conil, Groessens and Lys, 1976. Range: late Courceyan–Brigantian.

Archaesphaera firmata (Conil and Lys, 1964)
Plate 3.1, Fig. 4, FS 514 Haw Bank Limestones, Chatburn Formation, Skipton, Yorkshire. Description: test consists of single chamber semicircular to lenticular in cross-section. Overall size, up to 500 μm, wall dark microgranular.
Remarks: differs from *A. reitlingerae* in the overall shape, and relative thickness of the test. Range: late Courceyan–Asbian.

Tournayella discoidea (Dain, 1953)
Plate 3.1, Fig. 5, FS 186 Haw Bank Limestone, Chatburn Formation, Skipton, Yorkshire. Description: planispiral tubular chamber with slight pseudoseptation. Microgranular wall of medium thickness.
Range: late Courceyan–early Chadian.

Baituganella sp.
Plate 3.1, Fig. 6, HOW 8, Haw Bank Limestone, Chatburn Formation, Skipton, Yorkshire. Description: large, irregular sac-like form, thick-walled test of microgranular calcite; irregular constrictions or pseudosepta.
Range: late Courceyan–late Chadian.

Palaeospiroplectammina tschernyshinensis (Lipina, 1948)
Plate 3.1, Fig. 7, B.G.S. Coll. Swinden Borehole (Chatburn Formation), Yorkshire. Description: test elongate, small initial coiled portion followed by biserial portion of greater breadth and up to nine arched chambers separated by curved and interlocking septa. Wall microgranular.
Range: late Courceyan–late Chadian.

Septaglomospiranella primaeva (Rauser-Chernoussova, 1948)
Plate 3.1, Fig. 8, CA 72 Haw Bank Limestones, Chatburn Formation, Skipton, Yorkshire. Description: small test of about three irregularly coiled whorls. Thick short septa poorly developed. Lobate periphery. Wall dark, granular.
Range: late Courceyan.

Chernyshinella glomoformis (Lipina, 1948)
Plate 3.1, Fig. 9–10, FS 263 Haw Bank Limestones, Chatburn Formation, Skipton, Yorkshire. Description: small; $2-2\frac{1}{2}$ whorls, with rectangular changes in coiling axis; short, forward-pointing septa; highly lobate periphery. Wall calcareous, more or less granular.
Range: late Courceyan.

Brunsia spirillinoides (Grozdilova and Glebovskaia, 1948)
Plate 3.1, Fig. 11, HOW 51, Haw Bank Limestones, Chatburn Formation, Skipton, Yorkshire. Description: discoidal without central hump, 7–8 whorls, evolute irregularly coiled for first 4–5 whorls, then planisprial. Pseudosepta may be present in outer whorl. Wall dark, microgranular.
Range: Courceyan–Chadian.

Endothyra sp.
Plate 3.1, Fig. 12, HOW 4, Haw Bank Limestones, Chatburn Formation, Skipton, Yorkshire. Description: small endothyrid of three whorls, nearly planispiral except initially. Wall rather thick, dark microgranular. Chambers numerous (about 12 in outer whorl), squarish. Septa at right angles to wall, sutures slightly depressed.
Remarks: wall thicker and sutures clearer than in *E. prisca* (Plate 3.3, Fig. 13). Range: late Courceyan.

Endothyra nebulosa (Malakhova, 1956)
Plate 3.1, Fig. 13, HOW 81B Haw Bank Limestones, Chatburn Formation, Skipton, Yorkshire. Description: test of $2\frac{1}{2}-3$ whorls irregularly coiled; $6\frac{1}{2}-7\frac{1}{2}$ chambers in outer whorl. Periphery slightly lobate. No secondary deposits.
Range: late Courceyan–early Chadian.

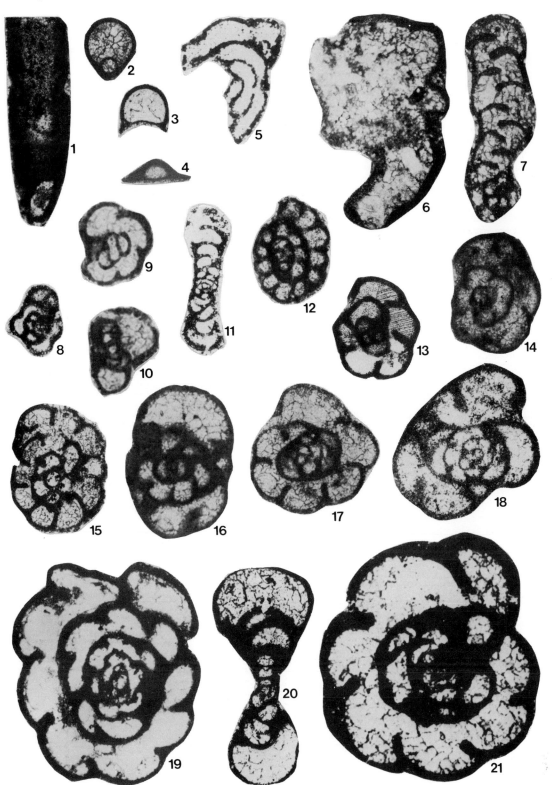

Plate 3.1

Endothyra tumida (Zeller, 1957)
Plate 3.1, Fig. 14, HOW 23, Haw Bank Limestones, Chatburn Formation, Skipton, Yorkshire. Description: small to medium endothyrid, tightly and irregularly coiled at first, then looser and approximately planispiral. Septa short. Periphery lobate. Septa may be secondarily thickened posteriorly.
Remarks: resembles *E. laxa* (which is also present in this fauna) but differs in being essentially planispiral. Range: late Courceyan–early Chadian.

Spinoendothyra recta (Lipina, 1955)
Plate 3.1, Fig. 15, SK 30, Haw Bank Limestones, Chatburn Formation, Skipton, Yorkshire. Description: test more or less planispirally coiled, 4–5 whorls, about ten chambers in last whorl. Periphery nearly smooth. Septa straight. Basal deposits in the form of nodes, tending to become spines or hooks in ultimate chambers.
Range: late Courceyan–early Chadian.

Endothyra cf. *freyri* (Conil and Lys, 1964)
Plate 3.1, Fig. 16, HOW 55, Haw Bank Limestones, Chatburn Formation, Skipton, Yorkshire. Description: small endothyrid of $3-3\frac{1}{2}$ whorls of which the last $\frac{1}{2}-1$ whorl is at right angles to the earlier, planispiral part. Corner fillings may be present, otherwise secondary deposits are absent.
Range: late Courceyan–early Chadian.

Endothyra laxa (Conil and Lys, 1964)
Plate 3.1, Fig. 17, HOW 14, Haw Bank Limestones, Chatburn Formation, Skipton, Yorkshire. Description: endothyrid of $3\frac{1}{2}-4$ whorls irregularly coiled initially, planispiral in outer $1\frac{1}{2}-2$ whorls. Six chambers in last whorl. Chambers large, separated by short, inclined septa. No secondary deposits. Wall thin, microgranular.
Range: late Courceyan–Arundian.

Eblanaia michoti (Conil and Lys, 1964)
Plate 3.1, Fig. 19–20, HOW 143, Haw Bank Limestones, Chatburn Formation, Skipton, Yorkshire. Description: large, evolute planispiral with occasional deviations. Up to five whorls; 7–9 chambers in outer whorl. Pseudosepta becoming fully developed septa in outer whorl and pointing strongly forward. Lobate periphery.
Remarks: Axial section (e.g. Fig. 20) shows deep umbilici and evolute, nearly planispiral coiling. Range: late Courceyan–early Chadian.

Septabrunsiina sp.
Plate 3.1, Fig. 18 and 21, FS 263, Haw Bank Limestones, Chatburn Formation, Skipton, Yorkshire. Description: test large, initial whorls skew-coiled outer whorls nearly planispiral. Septa short and forward pointing; $5\frac{1}{2}-8$ chambers in outer whorl.
Remarks: basal deposits to chambers subdued as compared to spinose development in specimens 1 and 2, Plate 3.2. Large specimens of *Latiendothyra parakosvensis* are distinguished from forms such as this by the absence of spines, slightly thinner walls, and endothyrid septa; having said this the identification cannot always be made with certainty.
Range: late Courceyan–mid Chadian.

PLATE 3.2—All figures × 75. Specimen numbers refer to collection of M. D. Fewtrell, of which representative rock samples are held in the Sedgwick Museum, Cambridge, unless otherwise stated.

Septabrunsiina krainica (Lipina, 1948)
Plate 3.2, Fig. 1, B.G.S. Coll. SAD 812A, Chatburn bypass, Lancashire; Fig. 2, TLS 48, Lower Thornton Limestone, Clitheroe Fm, Broughton, Yorkshire. Description: test large, up to four whorls; inner whorls skew coiled; outer planispiral. Septa short, very poorly developed in initial chambers, anteriorly directed. Spines well developed in outer chambers.
Remarks: Conil *et al.* (1980 pl. V, Figs 1–2) illustrate a rather similar form as *Endothyra* cf. *kosvensis* Lipina. Range: late Courceyan–early Chadian.

Endothyra danica (Michelsen, 1971)
Plate 3.2, Fig. 3, TLS 46, Lower Thornton Limestone, Clitheroe Formation, Broughton, Yorkshire. Description: small endothyrid, up to four whorls, of which the outer two are planispiral. Septa pointed and directed forwards. Chambers increase moderately in size. Periphery slightly lobate.
Range: late Courceyan–mid Chadian.

Septatournayella cf. *lobedevae* (Poyarkov, 1961)
Plate 3.2, Fig. 4, SQB 80, Thornton Limestone, Clitheroe Formation, Broughton, Yorkshire. Description: small *Septatournayella* with up to three planispiral whorls following a rather large proloculus. Simple forward-pointing septa throughout; chamber size increases only slowly.
Remarks: the initial whorls of *Eblanaia* have a similar appearance. Range: late Courceyan–mid Chadian.

Carboniferous

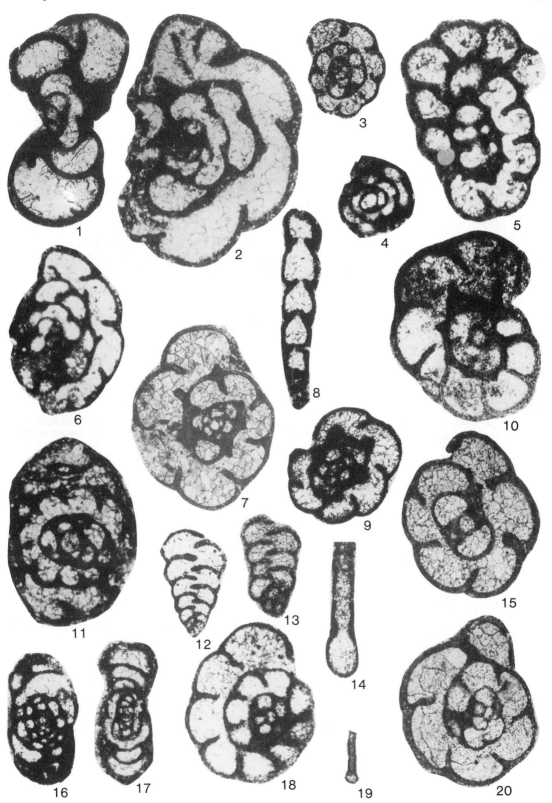

Plate 3.2

Spinoendothyra cf. *praeclara* Conil and Longerstaey, 1980
Plate 3.2, Fig. 5, B.G.S. Coll. LL2033, Stone Gill Limestone, Ravenstonedale. Description: irregularly coiled endothyrid of moderate size. Lobate periphery. Chambers numerous, up to 12 in outer whorl. Septa strong, straight. Secondary deposits in form of basal spines, one to each chamber in outer whorl. Thick dark microgranular wall. Range: Early Chadian.

Spinoendothyra cf. *mitchelli* Conil and Longerstaey, 1980
Plate 3.2, Fig. 6, B.G.S. Coll. SAD806, Chatburn bypass, Clitheroe, Lancs. Description: test of 4–5 whorls streptospirally coiled; $8\frac{1}{2}$–9 chambers in last whorl. Periphery slightly lobate; septa forward pointing and often thickened terminally. Secondary deposits in the form of ridges of which those in the last chambers may be hook-shaped in cross-section. Wall rather thin compared to Fig. 5.
Range: late Courceyan–early Chadian.

Endothyra bowmani (Phillips, 1846) emend. Brady, 1876
Plate 3.2, Fig. 7, TLS 89, Lower Thornton Limestone, Clitheroe Formation, Broughton, Yorkshire. Description: irregularly coiled endothyrid becoming planispiral in outer whorls. Septa tend to point forward, separating subglobular chambers. Periphery lobate. Secondary deposits various, including nodes in last few chambers.
Range: late Courceyan–Arundian.

Lugtonia concinna (Brady, 1876)
Plate 3.2, Fig. 8, SQB 1, Lower Thornton Limestone, Clitheroe Formation, Broughton, Yorkshire. Description: slightly flaring cylindrical test with constrictions at the septa. Uniserial; chambers pear-shaped, increasing moderately in size. Septa pointing markedly towards aperture.
Remarks: not previously recorded earlier than Namurian. Range: late Courceyan–Chadian–Arundian.

Tuberendothyra tuberculata (Lipina, 1948) subsp. *magna* Lipina and Safonova, 1967.
Plate 3.2, Fig. 9, OQB 34, Lower Thornton Limestone, Clitheroe Formation, Broughton, Yorkshire. Description: small endothyrid with about four whorls, coiling highly skewed throughout, periphery moderately lobate. Septa convex; may be secondarily thickened. Internal elements in form of pronounced rounded tubercules at base of each chamber.
Range: mid-Chadian.

Endothyra delepinei (Conil and Lys, 1964)
Plate 3.2, Fig. 10, HC 31, Haw Crag, Clitheroe Formation, Eshton anticline, Yorkshire. Description: test of about $2\frac{1}{2}$ whorls of moderate size, 7–8 chambers in outer whorl, periphery smooth. Secondary deposits in the form of floor deposits and nodes which become produced into large spines in last few chambers.
Range: mid-late Chadian.

?*Dainella fleronensis* (Conil and Lys, 1964)
Plate 3.2, Fig. 11, B.G.S. Coll. LL 1906T$_2$, Chatburn bypass, Clitheroe, Lancs. Description: test $3\frac{1}{2}$–$4\frac{1}{2}$ whorls, irregularly coiled with rectangular change in coiling direction between last two whorls. Twelve chambers in last whorl. Apparently lacking in chomata.
Range: Chadian.

Plate 3.2

Palaeospiroplectammina mellina (Malakhova, 1965)
Plate 3.2, Fig. 12–13. 12, HED21B, Embsay Limestone, Clitheroe Formation, Skipton. 13, TLS48, Lower Thornton Limestone, Clitheroe Formation, Broughton, Yorkshire. Description: small biserial test with initial coiled portion of about five chambers; coiled portion small and usually indistinct.

Earlandia minor (Rauser-Chernoussova, 1948)
Plate 3.2, Fig. 14, SQB 80. Lower Thornton Limestone, Clitheroe Formation, Broughton, Yorkshire. Description: small *Earlandia*; wall of medium thickness (about 20 μm) but thickened around junction between proloculus and tubular chamber.
Range: mid-Chadian–Brigantian.

Endothyra sp.
Plate 3.2, Fig. 15. TLS 89, Lower Thornton Limestone, Clitheroe Formation, Broughton, Yorkshire. Description: moderately large endothyrid, initially irregularly coiled but planispiral outer $1-1\frac{1}{2}$ whorls. Axial section tends to have oblong outline. 8–10 chambers in last whorl. Wall rather thick, dark microgranular. Septa blunt and forward-pointing; may be secondarily thickened.
Range: late Courceyan–early Chadian.

Dainella cf. *elegantula* (Brazhnikova, 1962)
Plate 3.2, Fig. 16, OQB 30, Lower Thornton Limestone, Clitheroe Formation, Broughton, Yorkshire. Description: involute, skew-coiled subglobular test with many chambers. Secondary deposits in the form of chomata. Microgranular, single-layered wall.
Remarks: sections through *Dainella* are very variable and specific identification is not always possible. Range: low to mid-Chadian.

Glomospiranella cf. *barsae* (Conil and Lys, 1968)
Plate 3.2, Fig. 17, B.G.S. Coll. LL 1898, Chatburn bypass, Clitheroe, Lancs. Description: test of many whorls, irregularly coiled initially, becoming essentially planispiral in outer whorls. Tubular chamber of low profile, pseudoseptate in outer whorls. Wall thin, dark, microgranular.
Range: late Courceyan–early Chadian.

Endothyra sp.
Plate 3.2, Fig. 18, TLS 89, Thornton Limestone, Clitheroe Formation, Broughton, Yorkshire. Description: endothyrid of about $2\frac{1}{2}$ whorls, planispiral outer whorl following sharp change in coiling axis from inner whorls. 5–6 chambers in outer whorl, periphery slightly lobate. Septa blunt, pointing slightly forwards. Secondary deposits lacking except for possible septal thickenings.
Remarks: simple endothyrids of this character are common in the lower part of the Craven Basin succession; they are not readily divided into species. Range: late Courceyan–Chadian.

Earlandia elegans (Rauser-Chernoussova and Reitlinger, 1937)
Plate 3.2, Fig. 19, HE D21 Embsay Limestone, Clitheroe Formation, Skipton, Yorkshire. Description: spherical proloculus followed by rectilinear tubular chamber, frequently with an internal constriction at the junction of the two. Maximum dimensions 330 μm × 120 μm.
Range: mid-Chadian–Holkerian.

Endothyra cf. *prokirgisana* Rauser-Chernoussova, 1948
Plate 3.2, Fig. 20, FS263, Haw Bank Limestone, Chatburn Formation, Skipton, Yorkshire. Description: endothyrid of moderate size. Coiling irregular to nearly planispiral. Six to eight chambers in last whorl. Septa slightly anteriorly directed. Chambers subquadrate; periphery slightly lobate.
Remarks: morphologically close to *Latiendothyra* of the group *latispiralis* (Lipina, 1954). Assignment to *Latiendothyra* depends on recognition of wall structure, which is not a reliable character in the available material which is subject to diagenetic alteration. Range: late Courceyan–early Chadian.

PLATE 3.3—All figures × 75. Specimen numbers refer to Fewtrell Collection of which representative rock samples are held in the Sedgwick Museum, Cambridge, unless otherwise stated.

Conilites dinantii (Conil and Lys, 1964)
Plate 3.3, Fig. 1, TLS 55, Lower Thornton Limestone, Clitheroe Formation, Broughton, Yorkshire. Description: initial part of test nearly planispiral, evolute, of $3\frac{1}{2}$–4 whorls, divided into chambers by pseudosepta becoming more clearly defined in last whorl. Later part of test uncoiled, straight, septate. Chambers increasing in size throughout. Periphery lobate. Wall thick, aperture cribrate.
Range: mid- to late-Chadian.

Biseriella bristolensis (Reichel, 1946)
Plate 3.3, Fig. 2, HEF 140, Embsay Limestone, Clitheroe Formation, Skipton, Yorkshire. (= *Globivalvulina bristolensis* Reichel, 1946. *Eclog. Geol. Helv.* 38 (2), 524–560, pl. 19) Description: small *Biseriella*, rarely more than 300 μm high. Axial section resembles an orange segment; biserial character is readily apparent only in tangential sections. Periphery smooth, non-lobate. Chambers increase in height rapidly between tightly coiled inner and loosely coiled outer portions of test.
Range: late Chadian.

Darjella monilis (Malakhova, 1963)
Plate 3.3, Fig. 3, TLS 25, Lower Thornton Limestone, Clitheroe Formation, Broughton, Yorkshire. Description: test uniserial. Chambers large, flask-shaped, with a septal neck protruding into succeeding chamber. Wall thick agglutinated, cement dark, microgranular.
Remarks: differs from *Lugtonia* in characteristically having few chambers, which are larger and more globular in outline. Range: mid-late Chadian.

Brunsia pseudopulchra (Lipina, 1955)
Plate 3.3, Fig. 4, RDC Lower Thornton Limestone, Clitheroe Formation, Broughton, Yorkshire. Description: discoidal, evolute, 6–9 whorls, planispiral in outer $1\frac{1}{2}$–5 whorls, with central hump where the coiling is irregular. Slight constrictions may be present in the outer whorl. Wall microgranular.
Range: late Courceyan–mid-Chadian.

Viseidiscus sp.
Plate 3.3, Fig. 5, QDR 18 Dogber Rock Quarry, Clitheroe Formation, Yorkshire. Description: primitive archaediscid with essentially planispiral coiling, deviating only in first two whorls. Evolute throughout. Microgranular wall layer well developed throughout; fibrous layer developed only in umbilical region.
Remarks: this genus, transitional to those with fully developed fibrous layer, is the only archaediscid believed to appear below the Arundian.
Range: late Chadian–early Arundian.

Tournayella kisella (Malakhova, 1965)
Plate 3.3, Fig. 6, HED 21, Embsay Limestone (late Chadian), Clitheroe Formation, Skipton, Yorkshire. Description: planispiral, about four whorls, moderate to small in size. Wall microgranular, of moderate thickness. There might be a slight suggestion of wall constrictions in some specimens.
Range: Courceyan–Chadian.

Pseudoglomospira curiosa (Malakhova, 1956)
Plate 3.3, Fig. 7, SQB 1, Upper Thornton Limestone, Clitheroe Formation, Broughton, Yorkshire. Description: test comprising a proloculus and an irregularly coiled non-septate tubular chamber. About six whorls, increasing very gently in height. Wall dark, microgranular. Overall size up to 500 μm.
Range: Chadian–Holkerian.

Plectogyranopsis exelikta (Conil and Lys, 1964)
Plate 3.3, Fig. 8, QDR 18, Dogber Rock Quarry, Clitheroe Formation, Coniston Cold, Yorkshire. (= *Plectogyra exelikta* Conil and Lys, 1964, p. 185, Pl. XXVII, Fig. 555–563.) Description: test about $2\frac{1}{2}$ whorls, planispiral except initially. Periphery lobate. Five to $5\frac{1}{2}$ chambers in outer whorl; chambers globular. Septa well developed, straight or slightly curved. Wall rather thick, microgranular to agglutinated: no supplementary deposits.
Remarks: the wall structure and the form of the septa and chambers require transfer of this species to *Plectogyranopsis* Vachard.
Range: late Chadian.

Palaeospiroplectammina mellina (Malakhova) subsp.
Plate 3.3, Fig. 9, RCD 42, Lower Thornton Limestone, Clint Rocks Quarry, Clitheroe Formation, Broughton, Yorkshire; Fig. 10, OQB 34, Lower Thornton Limestone, Clitheroe Formation, Old Quarry, Broughton, Yorkshire. 1976 cf. *Palaeospiroplectammina mellina* (Malakhova) – Conil, pl. II, Fig. 15. Description: subspecies of *P. mellina* in which the biserial portion is twisted half way along its length. Initial coiled portion succeeded by wider and gently flaring biserial portion of up to about 10 chambers. About half way along the biserial portion, there is a slight but distinct twist in the biserial plane. The chambers in the biserial portion are roundly crescentic; the periphery is slightly lobate. Max. length about 300–500 μm, breadth 150–250 μm.
Remarks: Conil's specimen from the Lane Limestone, Co. Dublin appears very similar to subspecies. T. R. Marchant (pers. comm.) has also reported a similar form from the Dublin Basin at a similar stratigraphic level.
Range: mid-Chadian–Arundian.

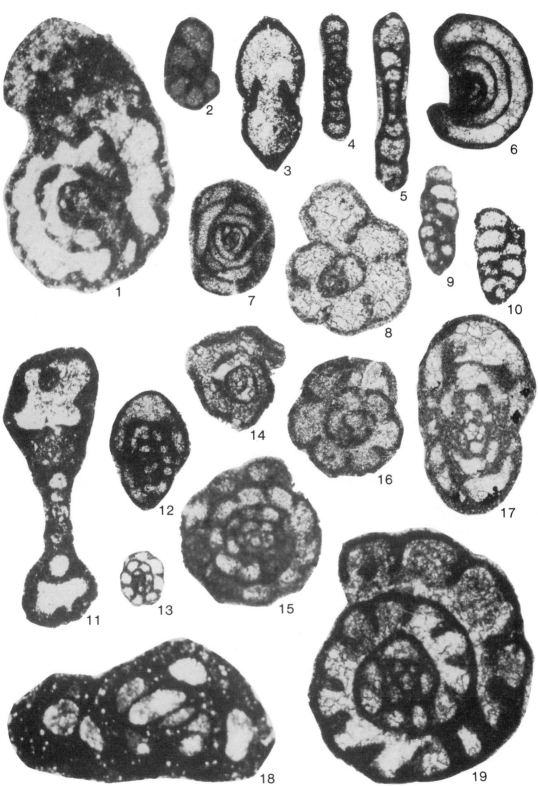

Plate 3.3

Forschiella prisca (Mikhailov, 1935)
Plate 3.3, Fig. 11, SQB 1, Upper Thornton Limestone, Clitheroe Formation, Broughton, Yorkshire. Description: test discoidal, planispiral, evolute, of 4–5 whorls of rapidly increasing cross-sectional diameter. Last whorl partly uncoiled. Aperture cribrate.
Range: mid-Chadian–Asbian.

Eoparastaffella simplex (Vdovenko, 1954)
Plate 3.3, Fig. 12, HED 14B, Embsay Limestone, Clitheroe Formation, Skipton, Yorkshire. Description: lenticular to ovoid test, involute planispiral coiling. Numerous chambers in each whorl. Secondary deposits forming simple paired chomata.
Range: Chadian–Arundian.

Endothyra prisca (Rauser-Chernoussova and Reitlinger, 1936)
Plate 3.3, Fig. 13, HEH 47, Embsay Limestone, Clitheroe Formation, Skipton, Yorkshire. Description: small, mainly planispiral endothyrid of rather variable form. 2–4 whorls, 6–13 chambers in last whorl. Outer 2–3 whorls planispiral, initial part less regularly coiled. Septa straight or forward-pointing. Chambers low, periphery more or less smooth. Wall thin, dark, microgranular, secondary deposits absent.
Remarks: the variability of this species has led to the proposal of a new genus, *Priscella* by Mamet (1974), but this is not used here.
Range: Courceyan–Brigantian.

Forschia cf. *parvula* (Rauser-Chernoussova, 1948)
Plate 3.3, Fig. 14, OQB 34, Thornton Limestone, Clitheroe Formation, Broughton, Yorkshire. Description: test discoidal, planispirally coiled, about three whorls. In the later part of the outer whorl there are several constrictions. Aperture slightly flared and apparently cribrate. Wall thick, probably agglutinated. Size of test about 500 μm.
Remarks: not certainly identified with *F. parvula*, which is an Upper Viséan form.
Range: mid-late Chadian.

Spinobrunsiina cf. *ramsbottomi* (Conil and Longerstaey, 1980)
Plate 3.3, Fig. 15, LL 1930, Clitheroe Formation, Chatburn bypass, Lancs. Description: small initial 'pellet' followed by 2–3 aligned evolute whorls; $5\frac{1}{2}$–7 whorls altogether, $8\frac{1}{2}$–10 chambers in outer whorl, pseudochambers in initial part. Secondary nodes developed, particularly in last half whorl. Diameter 400–700 μm. Undifferentiated microgranular wall.
Remarks: it should be noted that this form also closely resembles *Eoendothyranopsis spiroides* (Zeller, 1957).
Range: late Courceyan to Arundian.

Endothyra convexa (Rauser-Chernoussova, 1948)
Plate 3.3, Fig. 16, HEG 20, Embsay Limestone, Clitheroe Formation, Skipton, Yorkshire. Description: endothyrid with two skew-coiled whorls followed by a more or less planispiral outer whorl. Chambers increase rapidly in size, being globular in outer whorl: lobate periphery. Septa may be thickened; secondary deposits otherwise absent.
Remarks: this form bears some resemblance to the inner whorls of *Plectogyranopsis settlensis* Conil and Longerstaey, 1980.
Range: late Chadian.

Eostaffella sp. cf. *E. parastruvei* (Rauser-Chernoussova, 1948)
Plate 3.3, Fig. 17, HED 21A, Embsay Limestone, Clitheroe Formation, Skipton, Yorkshire. Description: test ovoid in cross-section, keels well-rounded, flanks slightly umbilicate. 4–5 whorls, involute throughout. Chomata small but distinct, paired.
Remarks: differs from *E. parastruvei* in its more rounded keel and less sharply defined umbilicus.
Range: mid-Chadian–Arundian.

Pseudolituotuba gravata (Conil and Lys, 1965)
Plate 3.3, Fig. 18, QDR 18, Dogber Rock Quarry, Coniston Cold, Yorkshire. Description: large highly irregular, probably fixed, test. Size up to 2.5 mm. Test comprises an irregularly coiled non-septate tube with numerous whorls. Wall very thick (often 200 μm) and agglutinated, with large included grains.
Range: Chadian–Holkerian.

Endothyranopsis crassa (Brady, 1876)
Plate 3.3, Fig. 19, B.G.S. coll. HR 2553, Peach Quarry Limestone, Clitheroe Formation, Clitheroe, Lancs. Description: test large. $2\frac{1}{2}$ to $3\frac{1}{2}$ whorls, 10–12 chambers in last whorl, subspherical. Involute, whorl height increases steadily throughout the test. Coiling planispiral; may be slightly irregular initially. Septa massive, often pointing towards the aperture. Sutures depressed only in last half whorl. Wall thick and with agglutinated grains.
Range: Chadian–Brigantian.

PLATE 3.4—All figures × 75. Specimen numbers refer to Fewtrell Collection of which representative samples are held in the Sedgwick Museum, Cambridge, unless otherwise stated.

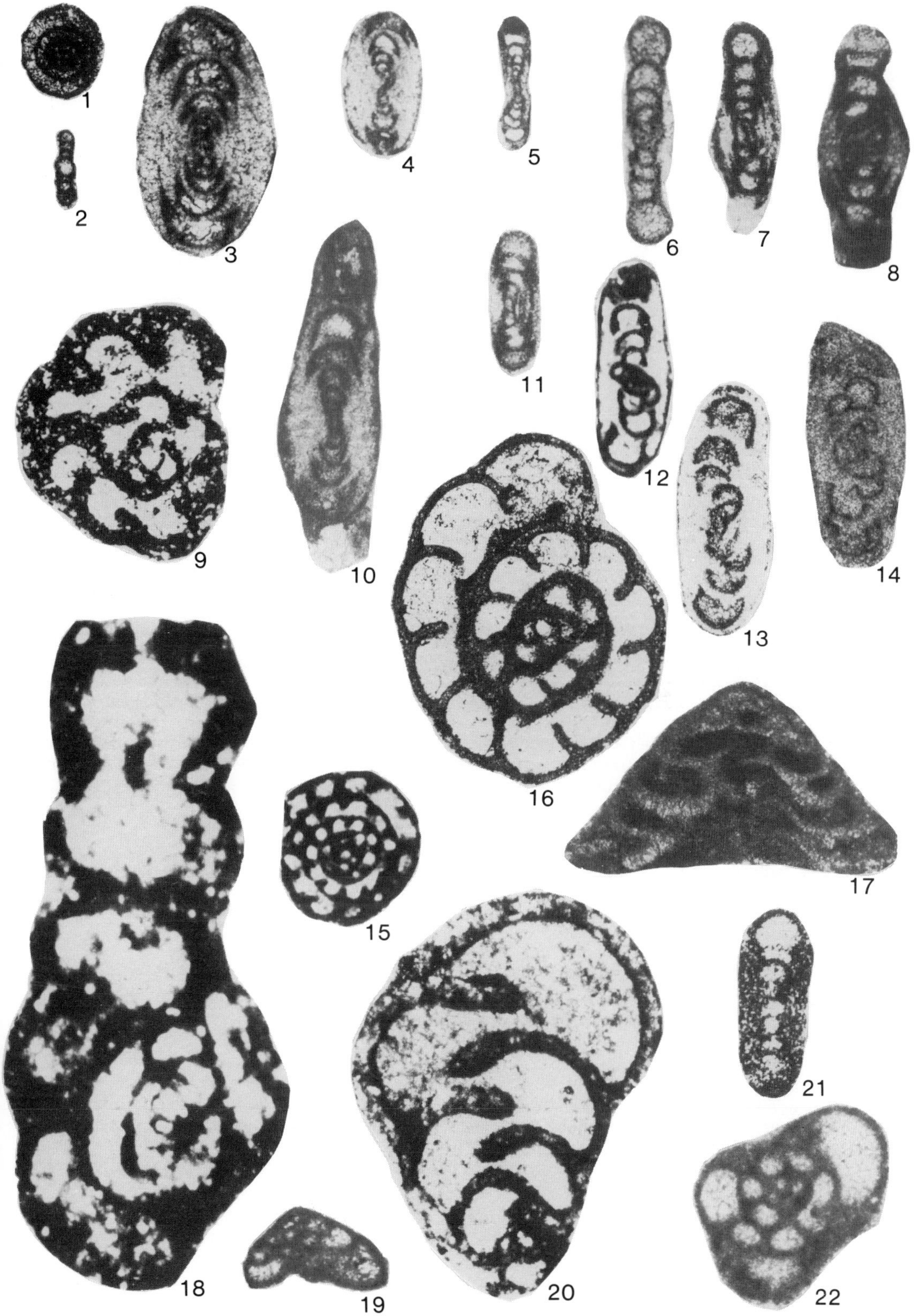

Carboniferous

Plate 3.4

Pseudoammodiscus sp.
Plate 3.4, Fig. 1–2, Embsay Limestone, Clitheroe Formation, Skipton, Yorkshire. Fig. 1, HEH 17, equatorial section; Fig 2, axial section, ?megalospheric form, HEF 69. Description: test discoidal, comprising a proloculus followed by a planispirally coiled tubular chamber, increasing very little in height, 3–4 whorls. Overall diameter small, between 100 μm and 200 μm.
Range: mid-Chadian–Arundian.

Uralodiscus rotundus (Chernysheva, 1948)
Plate 3.4, Fig. 3. LOT LEQ, Clitheroe Formation, Lothersdale, Yorkshire. Description: test involute, planispirally coiled, roundly lenticular. Up to six whorls. Wall two-layered: inner dark microgranular layer moderately well developed, outer fibrous layer enveloping each whorl. Inner layer thickened on either side of the base of the aseptate chamber, which has a convex floor. Maximum diameter 300–400 μm.
Range: Arundian.

Glomodiscus sp.
Plate 3.4, Fig. 4, LOT 100, Clitheroe Formation, Skipton, Yorkshire; Fig. 10, LOT MEQ, Clitheroe Formation, Lothersdale, Yorkshire. Description: skew-coiled initially, more nearly planispiral in outer whorls, roundly lenticular in overall shape. Wall two-layered, inner dark microgranular layer well developed and thickened laterally; outer fibrous layer envelopes each whorl. Chamber floor is convex. Differs from *Uralodiscus* spp. in its non-planispiral mode of coiling and in the better development of the inner dark layer. Size 250–350 μm.
Remarks: most closely resembles *Propermodiscus oblongus* Conil and Lys, 1964, pl. XX, Fig. 406, not here transferred to *Glomodiscus*, as *Propermodiscus* appears to remain a valid taxon.
Range: Arundian.

Ammarchaediscus spirillinoides (Rauser-Chernoussova, 1948)
Plate 3.4, Fig. 5, HEH 39 Embsay Limestone, Clitheroe Formation, Skipton, Yorkshire; Fig. 11, HE H17, Embsay Limestone, Clitheroe Formation, Yorkshire. Description: test discoidal, small. Planispirally coiled, deviating slightly only in initial 1–2 whorls; 5–6 whorls altogether. Fibrous layer developed only in central portion of test, not covering outer 1–2 whorls, and forming parallel to concave flanks to the test. Microgranular layer well developed, sometimes developing corner fillings.
Remarks: Pirlet and Conil (1978) did not formally transfer this species to *Ammarchaediscus* although they indicated that they regarded it as belonging here. The name is therefore proposed here as *Ammarchaediscus spirillinoides* (Rauser-Chernoussova, 1948) Fewtrell comb. nov.; basionym *Archaediscus spirillinoides* Rauser-Chernoussova, 1948, pl. 2, Figs. 7–8.
Range: Arundian.

Brunsiarchaediscus sp.
Plate 3.4, Fig. 6, HEH17, Clitheroe Formation, Embsay Limestone, Skipton, Yorkshire. Description: test discoidal with a slight axial swelling. Proloculus followed by coiled aseptate tubular chamber, irregularly coiled for first few whorls, then planispirally coiled for about three whorls. Wall microgranular, with a fibrous layer partially enveloping the inner, irregularly coiled whorls only. Test moderately large, about 400 μm.
Range: Arundian.

Planoarchaediscus cf. *concinnus* (Conil and Lys, 1964)
Plate 3.4, Figs. 7–8, LOT MEQ, Clitheroe Formation, Lothersdale, Yorkshire. Description: test discoidal with a marked axial swelling. Coiling initially irregular or oscillating, then planispiral in outer 3–4 whorls, becoming progressively more evolute. Fibrous layer developed mainly in axial region. Microgranular layer well developed throughout.
Remarks: Differs from *P. concinnus* in having an axial swelling, and in its earlier range.

Latiendothyranopsis menneri (Bogush and Juferev, 1962)
Plate 3.4, Fig. 9, ING 15, Horton Limestone, Thornton Force, Yorkshire. Description: 2–3 whorls, increasing progressively in height; 7–10$\frac{1}{2}$ chambers in last whorl. Chambers rather arched, sutures weak to clear. Septa massive, may be swollen at tips, usually at right angles to wall. Corner and septal thickenings frequent. Coiling more or less planispiral.
Range: Arundian.

Archaediscus krestovnikovi (Rauser-Chernoussova, 1948), ('*involutus* stage' of Pirlet and Conil, 1978)
Plate 3.4, Fig. 12, HR 2869; Fig. 13, KR 3343; Fig. 14, HR 2885; IGS coll, Pen-y-Holt Formation, Hobbyhorse Bay, Pembrokeshire. Description: *Archaediscus* with dark microgranular inner layer moderately well developed and chamber floors flat to convex. Coiling oscillates more or less markedly.
Range: Arundian–Brigantian (*involutus* stage ranges from Arundian to Asbian).

Urbanella fragilis (Lipina, 1951)
Plate 3.4, Fig. 15, HEH18, Embsay Limestone, Clitheroe Formation, Skipton, Yorkshire. Description: test small. Planispiral with very low rate of whorl expansion. Numerous squarish chambers, blunt septa. Wall thin, dark, microgranular.
Range: mid-Chadian–Arundian.

Omphalotis minima (Rauser and Reitlinger, 1936)
Plate 3.4, Fig. 16, KR 3351, B.G.S. coll. Pen-y-holt Formation, Hobbyhorse Bay, Pembrokeshire. Description: endothyrid with skew-coiled initial whorls, becoming planispiral in outer 2–3 whorls. Periphery lobate; septa curved. Some secondary internal thickenings.
Range: mid-Chadian–Arundian.

Tetrataxis conica s.l. Ehrenberg, 1854.
Plate 3.4, Fig. 17, LOT MEQ, Lower Thornton Limestone, Clitheroe Formation, Broughton, Yorkshire. Description: test conical with apical angle between 80° and 100°. Flanks generally smooth, wall with well developed fibrous layer. Umbilical area pronounced. Chambers increase gradually in size.
Range: late Courceyan–Arundian–Brigantian.

Pseudolituobella sp.
Plate 3.4, Fig. 18, HEH17 Embsay Limestone, Clitheroe Formation, Skipton, Yorkshire. Description: Large. Coiled portion of 2–3 whorls, straight uniserial portion with several chambers. Septate throughout. Cribrate aperture.
Range: Courceyan–Arundian.

Pseudotaxis sp.
Plate 3.4, Fig. 19, LOT 16, Clitheroe Formation, Lothersdale, Yorksire. Description: test conical, trochospiral, with at least five whorls. Apical angle obtuse, flanks smooth; small, wall microgranular, single-layered.
Remarks: the form shown here most closely resembles *Tetrataxis pusillus* Conil and Lys, 1964, the name of which is preoccupied by *T. pusillus* Golubsov, 1954.
Range: mid-Chadian-Arundian.

Eotextularia diversa (Chernysheva, 1948)
Plate 3.4, Fig. 20, KR 3354/1, B.G.S. coll. Pen-y-Holt Formation, Hobbyhorse Bay, Pembrokeshire. Description: large test: biserial with a coiled initial part of a single whorl and about five chambers; biserial part of 5–10 chambers, increasing rapidly in breadth. Periphery rounded, slightly lobate. Apical angle rather variable. Septa thick, curved, and terminally thickened.
Range: ?Courceyan–Arundian.

Mediocris breviscula (Ganelina, 1956)
Plate 3.4, Fig. 21, ING 21 Horton Limestone, Thornton Force, Yorkshire. Description: test small, discoidal with flattened flanks and rounded keel, 3–4 whorls. Secondary deposits well developed in the form of lateral fillings.
Range: Arundian–Asbian.

Endothyra sp.
Plate 3.4, Fig. 22, LOT 33, Clitheroe Formation, Lothersdale, Yorkshire. Description: test small, irregularly coiled, slightly lobate periphery. Septa gently curved to the anterior, of moderate length. Secondary deposits not developed.
Remarks: small endothyrids of this type are numerous at this level in the Craven Basin.
Range: High Courceyan–Arundian.

PLATE 3.5—All the figured specimens are in the B.G.S. collection at Keyworth and are × 75, except where stated otherwise.

Dainella holkeriana (Conil and Longerstaey, 1980)
Plate 3.5, Fig. 1, AL 1364, Holkerian, Stackpole Limestone, South Bay, Tenby, S. Wales, 74 m below top of section. Description: test very irregularly coiled, involute. Whorl height initially low; two final whorls increase rapidly in size and their division into chambers much less numerous than in earlier whorls. Septa well inclined towards aperture.
Range: Holkerian.

Millerella excavata (Conil and Lys, 1974)
Plate 3.5, Fig. 2. ARE 923, Holkerian, Garsdale Limestone, R. Clough, near Sedbergh, Yorkshire. Description: test deeply and widely umbilical: periphery well rounded. Coiling often slightly distorted at origin, becoming planispiral. Four whorls. Final whorl very large, evolute. Chomata or pseudochomata very well developed.
Range: Holkerian–Asbian.

Nevillella tetraloculi (Rauser-Chernoussova)
Plate 3.5, Fig. 3, AL 1363 (× 38); Fig. 4, AL 1353 (× 38); Holkerian, Stackpole Limestone, South Bay, Tenby, S. Wales. = *Haplophragmella tetraloculi* Rauser-Chernoussova, 1948. Description: test initially irregularly coiled, laterally compressed. Later portion straight cylindrical or expanding slightly. Coiling irregular, 2–3 whorls, four chambers in final whorl. Chambers slightly convex in spiral portion, sub-rectangular in straight portion. 3–4 chambers in uniserial section. Chamber width is twice its height. Diam. spiral portion 0.62–0.85 mm: width of straight portion 0.67–0.80 mm, length of straight portion up to 1.6 mm. Wall up to 75–80 μm thick in uniserial section. Aperture low, slit-like initially, cribrate in final two chambers of spiral portion and all of uncoiled portion. Convex or almost flat apertural face.
Range: Arundian–Asbian.

Nodosarchaediscus sp.
Plate 3.5, Fig. 5, AL 1367 (× 140), Holkerian, South Bay, Tenby. Description: test laterally compressed, lenticular, irregularly coiled. Well developed nodosities. Well rounded hemispherical roof to chamber. 6–7 whorls. Dark microgranular wall well developed.

Endothyra ex gr. *phrissa* (D. Zeller)
Plate 3.5, Fig. 6, AL 1368, Holkerian, Stackpole Limestone, South Bay, Tenby, S. Wales. = *Plectogyra phyrissa* Zeller, 1953. Description: test discoidal, umbilicate. Proloculus large. Chambers large and strongly inflated. Secondary deposits well developed, a hook being present in the final chamber and nodes in the preceding chambers with connecting basal coverings. Total rotational distortion very high. Septa moderately long and strongly arcuate, with some evidence of light secondary deposits on posterior surfaces.

Glomospiranella aff. *barsae* (Conil and Lys)
Plate 3.5, Fig. 7 = *Brunsiina barsae* Conil and Lys, 1968, ARE 571, Holkerian, Tunstead Quarry, Derbyshire. Description: coiling initially irregular, final whorls oscillate slightly from one stabilised plane. Final $2-2\frac{1}{2}$ whorls almost planispiral. 7–8 whorls in total. Tubular chamber initially thin, slowly widening in course of coiling. Pseudochambers formed by inflexions of the wall well developed in final two whorls. Initial whorls free of all divisions. 12–15 pseudochomata in total. Wall microgranular – may enclose some larger grains.

Draffinia biloba Cummings, 1957
Plate 3.5, Fig. 8, ARE 554, Early Asbian, Timpony Limestone, near Stump Cross Cavern, Grassington, Yorkshire.
Range: Holkerian–Brigantian.

Rhodesinella avonensis (Conil and Longerstaey)
Plate 3.5, Fig. 9 (see also Plate 3.7, Fig. 5), = *Rhodesina avonensis* Conil and Longerstaey, 1980, ARE 923, Holkerian Tunstead Quarry, Derbyshire. Description: evolute $2\frac{1}{2}$–3 whorls; chamber height increases gradually towards aperture, chambers swollen, sutures well defined: septa short, thick, wedge-shaped, inclined towards aperture. $7\frac{1}{2}$ chambers in final whorl.
Range: Holkerian–Early Asbian.

Quasiendothyra nibelis (Durkina, 1959)
Plate 3.5, Fig. 10 (see also Plate 3.6, Fig. 1; Plate 3.8, Fig. 10), ARE, Holkerian, Garsdale Limestone, R. Clough, near Sedbergh, Yorkshire.
Range: Holkerian–Asbian.

Brunsia sp.
Plate 3.5, Fig. 11, ARE 571, Holkerian, Tunstead Quarry, Derbyshire.

Tetrataxis sp.
Plate 3.5, Fig. 12, ARE 552, Early Asbian, Timpony Limestone, near Stump Cross Cavern, Grassington, Yorkshire.

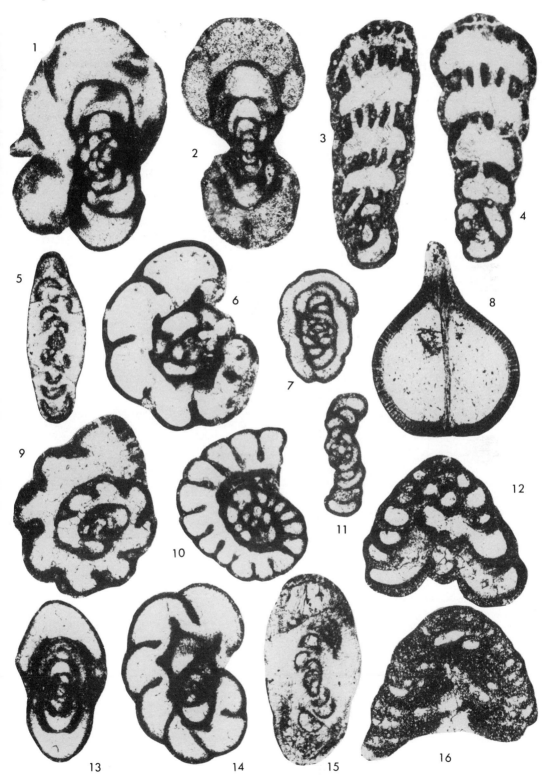

Plate 3.5

Eostaffella parastruvei (Rauser-Chernoussova)
Plate 3.5, Fig. 13, AL 1368, Holkerian, Stackpole Limestone, South Bay, Tenby, S. Wales. = *Staffella parastruvei* Rauser-Chernoussova, 1948. Description: test slightly compressed laterally with moderate umbilici. Periphery carinate or slightly rounded, never pointed. Whorls increase gradually in size towards the final significantly larger revolution. 4–5 whorls. Pseudochomata in the form of bands. Chomata very rare. Diameter 450–700 μm. Width 250–350 μm. Width/diameter ratio 0.43–0.58.
Range: Holkerian–Brigantian.

Endothyra maxima (D. Zeller)
Plate 3.5, Fig. 14, ARE 549, Early Asbian, Timpony Limestone, near Stump Cross, Grassington, Yorkshire. = *Plectogyra maxima* Zeller, 1953. Description: test large, elongate. Wall thin in relation to size of shell. Chambers large, swollen. Proloculus small. Supplementary deposits well developed in form of basal coverings and large forward-curving hooks directly behind each septum.
Remarks: distinguished by its extreme size, the development of large hooks and its high angular distortion.
Range: Asbian–Namurian.

Archaediscus sp.
Plate 3.5, Fig. 15, AL 1354 ($\times$ 140). Holkerian, Stackpole Limestone, South Bay, Tenby, S. Wales. Description: test laterally compressed. Coiling irregular. Tubular chamber increases slowly in size until the final whorl, where a massive expansion occurs, producing a chamber approximately twice the size of the penultimate whorl. 5–6 whorls.

Valvulinella cf. *conciliata* (Ganelina)
Plate 3.5, Fig. 16 (see also Plate 3.10, Fig. 15), = *Tetrataxis conciliatus* Ganelina, 1956, ARE 551, Early Asbian, Timpony Limestone, near Stump Cross, Grassington, Yorkshire.
Remarks: *V. conciliata* is distinguished from other species by its thin wall and 90° apical angle.

PLATE 3.6—All the figured specimens are in the B.G.S. collections at Keyworth, and are $\times$ 75, unless stated otherwise.

Quasiendothyra nibelis (Durkina, 1959)
Plate 3.6, Fig. 1 (see Plate 3.5, Fig. 10, and Plate 3.8, Fig. 10), ARE 341, Early Asbian, Tandinas Quarry, Anglesey.
Range: Holkerian–Asbian.

Cribrospira mira (Rauser-Chernoussova)
Plate 3.6, Fig. 2, BLE 4697 ($\times$ 40). Late Asbian, Danny Bridge Limestone, Raydale Borehole, Askrigg, depth 13 m. Description: test involute, increasing rapidly in height and less rapidly in width, giving an elongate, ovoid section. Two whorls, regularly coiled, expanding markedly in final chamber. Total number of chambers 11–17 with 7–8 in final whorl. Septa long, hook-like, situated at almost equal distances from each other. Sutures distinct. Aperture cribrate in final chamber, high and slit-like in other chambers.
Remarks: differs from *C. panderi* Moeller in having a compressed test along shell axis, in its lower whorl and lower final chambers, in having the septa evenly spaced and fewer apertural openings.
Range: Holkerian–Brigantian.

Endothyranopsis sphaerica (Rauser-Chernoussova and Reitlinger)
Plate 3.6, Fig. 3 (= *Endothyra sphaerica* Rauser-Chernoussova and Reitlinger, 1937) (see also Plate 3.9, Figs. 2, 3), ARE 1422, Early Asbian, Strandhal, Isle of Man.
Range: Asbian–Brigantian.

Mediocris mediocris (Vissarionova)
Plate 3.6, Fig. 4 (see also Plate 3.8, Fig. 5 and Plate 3.7, Fig. 4), ARE 1439, Early Asbian, Strandhall, Isle of Man. = *Eostaffella mediocris* Vissarionova, 1948.
Range: Tournaisian–Namurian.

Nevillella dytica (Conil and Lys)
Plate 3.6, Fig. 5, AL 1350, Holkerian, Stackpole Limestone, South Bay, Tenby, S. Wales, 8 m below top of section. = *Haplophragmella dytica* Conil and Lys, 1977. Description: test massive, uncoiled part subcylindrical with a diameter close to that of coiled part. $2-3\frac{1}{2}$ whorls, $4-4\frac{1}{2}$ chambers per whorl. Chambers initially chernyshinellid then swollen in uncoiled section. Wall relatively thin at the origin, becoming rapidly thicker. Sutures well defined. Aperture cribrate in last 1–2 chambers of uniserial section.
Range: Holkerian–Brigantian.

Palaeotextularia aff. *longiseptata* (Lipina, 1948)
Plate 3.6, Fig. 6, ARE 719 ($\times$ 40). Asbian, Urswick Limestone, 1.5 m below Woodbine Shale, Stainton Quarry, Dalton-in-Furness, Cumbria. Description: chambers slightly convex. Septa long, straight and slightly thickened at ends. Wall bilaminate, thickness in final chambers 36–50 μm. 6–9 chambers each side of test. Apertural face flat or slightly curved.
Remarks: distinguished by its long, straight septa.
Range: Late Asbian–earliest Namurian.

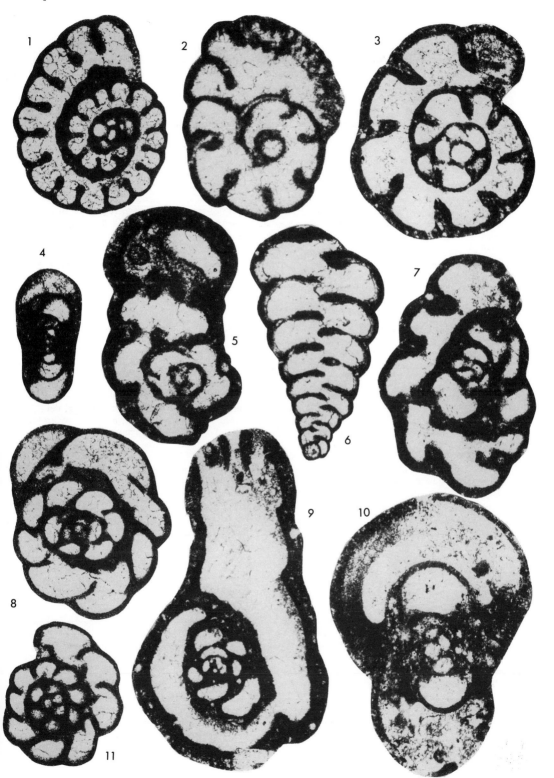

Plate 3.6

Mstiniella sp.
Plate 3.6, Fig. 7, AL 1319, Holkerian, Stackpole Limestone, South Bay, Tenby, S. Wales, 64 m below top of section. Description: wall initially fairly thin, increasing rapidly in thickness towards the final whorls. Sutures well defined. Septa fairly short, thick, blunt and forward pointing. Chambers swollen and well rounded. Eight chambers in final whorl. 4–5 whorls. Aperture low, cribrate in final chamber.

Endothyra excellens (D. Zeller)
Plate 3.6, Fig. 8, ARE 1156, Early Asbian, Potts Beck Limestone, Little Asby Scar, Cumbria. = *Plectogyra excellens* Zeller, 1953. Description: test large, discoidal with rounded periphery. Wall thin, finely granular with tectum. Proloculus small. Chambers large. Secondary deposits massive, continuous, very well developed. Well defined hook present in last chamber. Septa short, directed slightly anteriorly. Aperture high.
Remarks: differs from *E. pandorae* (Zeller) in its strong supplementary deposits, more swollen chambers and thinner walls.
Range: Asbian–Lower Namurian.

Lituotubella magna (Rauser-Chernoussova)
Plate 3.6, Fig. 9 (see also Plate 3.9, Fig. 5), ARE 1244, Asbian, Oxwich Head Limestone, near Swansea. = *L. glomospiroides magna* Rauser-Chernoussova, 1948. Description: width of straight portion increases slightly. Glomospiral portion laterally compressed. Septa and septal sutures initially almost imperceptible. $3\frac{1}{2}$–5 whorls in glomospiral portion. Up to eight chambers in uniserial section, septa more distinct.
Range: Asbian–Brigantian.

Endothyranopsis pechorica (Rauser-Chernoussova)
Plate 3.6, Fig. 10, ARE 1432, Early Asbian, Strandhall, Isle of Man. = *Endothyra crassa pechorica* Rauser-Chernoussova, 1936. Description: deeply umbilicate. Chamber flat, sutures faint, $2\frac{1}{2}$–4 whorls, 7–10 chambers in final whorl. W/d = 0.7–0.8. Heavy secondary infilling in axial region.
Range: Asbian–Brigantian.

Endothyra ex. gr. *spira* (Conil and Lys)
Plate 3.6, Fig. 11, ARE 643, Asbian, Trowbarrow quarry, Silverdale, Cumbria. Description: coiling planispiral; slight initial oscillations. Whorls increase fairly rapidly in height; $2\frac{1}{2}$–3 whorls. Chambers swollen, becoming globular in final whorl. Sutures distinct, especially between terminal chambers; eight in final whorl. Supplementary deposits well developed. Nodosities increase in importance until the final chamber where they are replaced by a spinal projection inclined towards aperture.
Range: Asbian–Brigantian.

PLATE 3.7—Unless otherwise stated, the figured specimens are in the B.G.S. collections at Keyworth, and are × 75.

Endothyranopsis crassa (Brady) emend. Cummings, 1955.
Plate 3.7, Fig. 1, ARE 973, late Asbian, Danny Bridge Limestone, R. Clough, Sedbergh, Cumbria. = *Endothyra crassa* Brady, 1876. Description: test free, large, nautiloid, subglobular and slightly asymmetrical. Three whorls present increasing moderately in height with complete embracement throughout. Approx. ten chambers in final whorl. Sutures moderately well defined. Periphery broadly rounded with lobulation.
Range: Holkerian–Namurian.

Endothyra sp. ex gr. *phrissa* (D. Zeller, 1953)
Plate 3.7, Fig. 2 (see also Plate 3.5, Fig. 6), ARE 1434. Early Asbian, Strandhall shore, Isle of Man.

Nodosarchaediscus sp.
Plate 3.7, Fig. 3, ARE 539 (× 140), early Asbian, Stump Cross Limestone, near Stump Cross, Grassington, Yorkshire. Description: test compressed laterally but slightly swollen. Periphery wall rounded and fairly even. Coiling initially oscillating, becoming more aligned in final whorls. $6\frac{1}{2}$ whorls. Gradual increase in size of tubular chamber towards last two whorls, where dimensions are similar.

Carboniferous

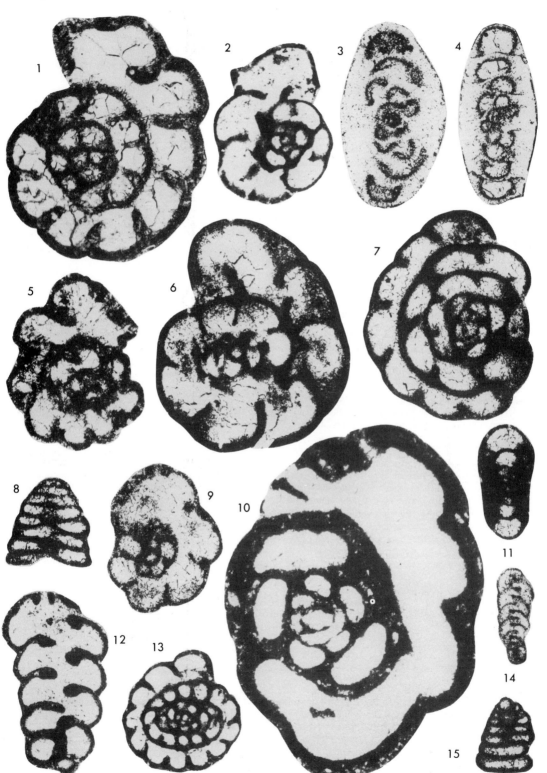

Plate 3.7

Archaediscus reditus (Conil and Lys)
Plate 3.7, Fig. 4, ARE 1194 (× 140), early Asbian, Groups Hollows, near Little Asby Scar, Cumbria. = *A. krestovnikovi reditus* Conil and Lys, 1964. Description: test lenticular, slightly flattened. Coiling initially compact and oscillating, tending to become aligned and more spacious in final whorls. 5–7 whorls. L/d 0.42–0.55. Proloculus small, 25–45 μm diameter.
Remarks: Differs from *A. koktubensis* in mode of coiling and less compressed test.
Range: Holkerian–late Asbian.

Rhodesinella avonensis (Conil and Longerstaey)
Plate 3.7, Fig. 5 (see also Plate 3.5, Fig. 9), AL 1344, Holkerian, South Bay, Tenby, 58 m below top section. = *Rhodesina avonensis* Conil and Longerstaey, 1980.
Range: Holkerian–early Asbian.

Omphalotis cf. *volynica* (Brazhnikova, 1956)
Plate 3.7, Fig. 6, ARE 18, Late Asbian, Trowbarrow Quarry, Cumbria. Description: test irregularly coiled. Whorl height and chamber size increase rapidly from the third whorl towards the aperture. Second whorl perpendicular to the third. $3\frac{1}{2}$–4 whorls. Chambers well rounded, convex, and divided by curved, almost hooked septa. Seven chambers in final whorl. Sutures deep and well defined. Supplementary deposits in the form of large nodosities grading into well-developed spines in the last two chambers.

Planoendothyra sp.
Plate 3.7, Fig. 7, ARE 1010, late Asbian, Oxwich Limestone, Pwll Ddu Head, Gower, S. Wales. Description: test discoidal, compressed laterally. Proloculus followed by an initially irregularly coiled involute spire. Final whorls more regularly coiled, oscillating and evolute. Chambers rounded with forward pointing septa of same thickness as the wall. Septa generally short. Probably 10–13 chambers in final whorl.

Vissariotaxis exilis (Vissarionova)
Plate 3.7, Fig. 8, ARE 719, early Asbian, Urswick Limestone, 1–5 m below Woodbine Shale, Stainton Quarry, Dalton in Furness, Cumbria. = *Monotaxis exilis* Vissarionova, 1948. Description: test small, conical with wide umbilicus. 7–8 whorls, rarely nine. Chambers distinct. Aperture wide, opening into umbilical cavity. Apical angle 73°–81°.
Remarks: differs from *V. compressa* by smaller number of whorls, large apical angle and less compressed form of test.
Range: Asbian.

Cribrospira pansa (Conil and Lys, 1965)
Plate 3.7, Fig. 9, ARE 1329, late Asbian, Fifth Limestone, Yeathouse Quarry, Frizington, W. Cumbria. Description: test initially coiled and compact, followed by a final whorl which grows very rapidly in height. 2–$2\frac{1}{2}$ whorls. 5–$5\frac{1}{2}$ chambers in the final whorl. Chambers moderately swollen and separated by straight septa which tend to be slightly

inclined towards the aperture. Sutures well defined. Supplementary deposits absent. Aperture cribrate over large area of final chambers.
Remarks: differs from *C. rara* by tighter, more compact chambers, more regular coiling and a cribrate aperture found only in final whorl (present in last three in *C. rara*).
Range: late Arundian–Asbian.

Bogushella ziganensis (Grozdilova and Lebedeva)
Plate 3.7, Fig. 10, ARE 10 (× 40), late Asbian, Trowbarrow Quarry, Cumbria. = *Mstinia ziganensis* Grozdilova and Lebedeva, 1960. Description: involute, initial whorls undivided, later part with short pseudosepta. Six whorls, six chambers in final whorl. Cribrate aperture.
Range: Holkerian–Asbian.

Mediocris mediocris (Vissarionova)
Plate 3.7, Fig. 11, (see also Plate 3.6, Fig. 4 and Plate 3.8, Fig. 5), ARE 1190, early Asbian, Groups Hollows, opposite Little Asby Scar, Cumbria. = *Eostaffella mediocris* Vissarionova, 1948.
Range: Tournaisian–Namurian.

Koskinotextularia sp.
Plate 3.7, Fig. 12, LL 2142 (× 48), early Asbian, Little Asby Scar, Cumbria. Description: chambers (nine in total) arranged biserially throughout. Initial spherical proloculus. Chambers swollen and well rounded. Sutures well defined. Septa with swollen and rounded extremities. Septa initially overstep central axial plane of shell but become more widely separated later and do not overlap the central position. Aperture cribrate in the last two chambers, where the apertural faces significantly overstep the central axis plane.

Loeblichiidae
Plate 3.7, Fig. 13, (in B.G.S. Edinburgh), PS 2197 (EA 3183), late Asbian, Archerbeck Beds, Archerbeck Borehole, 2028 ft.

Palaeospiroplectammina syzranica (Rauser-Chernoussova)
Plate 3.7, Fig. 14, AL 1319, Holkerian, South Bay, Tenby, 10 m below top of section. = *Spiroplectammina? syzranica* Rauser-Chernoussova, 1948. Description: shell very small and delicate, elongate with subparallel walls. 8–16 chambers in uncoiled part of shell. Chambers well-rounded with long septa overlapping central axial plane. Septa same thickness as wall and uniform throughout. Length of shell up to 360 μm.
Range: Holkerian–Brigantian.

Vissariotaxis compressa (Brazhnikova)
Plate 3.7, Fig. 15, ARE 5, late Asbian, Trowbarrow Quarry, Cumbria. = *Monotaxis exilis compressa* Brazhnikova, 1956. Description: test tall, narrow and conical. 8–10 whorls. Umbilicus almost cylindrical, straight and very deep. Flanks very regular and sutures feeble.
Range: Asbian.

PLATE 3.8—Unless otherwise stated, the figured specimens are in the B.G.S. collections at Keyworth and are × 75.

cf. *Omphalotis* sp.
Plate 3.8, Fig. 1, ARE 695, late Asbian, Knipe Scar Limestone, Little Asby Scar, Cumbria. Description: wall of constant thickness throughout. Coiling planispiral. 3–3¼ whorls. Chambers smoothly rounded with septa curved towards the aperture. Final chamber very well rounded and curved inwardly. 9½ chambers in final whorl. Basal supplementary deposits and poor nodosities in last few chambers. Spine-like projection in final chamber. Aperture low, simple.

Mikhailovella gracilis caledoniae (Conil and Longerstaey, 1980)
Plate 3.8, Fig. 2, (in I.G.S. Edinburgh), late Asbian, Cornet Limestone, Archerbeck Borehole, depth 1652 ft. Description: test initially coiled – aligned or oscillating, later uncoiled. Whorl initially low, growing regularly towards the final chamber. Three chambers in uniserial section, six in last whorl or coiled section. Corner fillings may sometimes be present. Aperture cribrate in last few chambers.
Range: late Asbian.

Pseudoendothyra sublimis (Schlykova, 1951)
Plate 3.8, Fig. 3, ARE 702, late Asbian, Knipe Scar Limestone, Little Asby Scar, Cumbria. Description: test laterally compressed: diaphanotheca and chomata very well developed. Whorl increases gradually in size, becoming slightly more keeled than rounded in the final revolutions. 4–6 whorls. w/d = 0.47–0.55.
Range: Asbian–Brigantian.

Eostaffella mosquensis (Vissarionova, 1948)
Plate 3.8, Fig. 4, ARE 598, late Asbian, Knipe Scar Limestone, Little Asby Scar, Cumbria. Description: test small, lenticular, laterally compressed with small shallow umbilici. Inner whorls with rounded periphery; outer whorls with rounded acute periphery. w/d = 0.5–0.56. Spire expands fairly rapidly throughout the 4–5 whorls. Pseudochomata well developed.
Remarks: Differs from *E. ikensis* Vissarionova by its notable lateral compression, the small umbilici and discontinuity of pseudochomata.
Range: Holkerian–Asbian.

Mediocris mediocris (Vissarionova)
Plate 3.8, Fig. 5 (see also Plate 3.6, Fig. 4; Plate 3.7, Fig. 11), ARE 1409, early Asbian, Strandhall shore, Isle of Man. = *Eostaffella mediocris* Vissarionova, 1948.
Range: Tournaisian–Namurian.

Pseudolituotuba wilsoni (Conil and Longerstaey, 1980)
Plate 3.8, Fig. 6, (in B.G.S. Edinburgh) PS, late Asbian, Archerbeck Beds, Archerbeck Borehole, depth 1863.5 ft. Description: test irregularly coiled on fine support around a central fixed point. Internal divisions of tubular chamber not observed. Max. diameter of internal tube 80 μm. Wall thickness up to 65 μm in known specimens.
Remarks: Differs from *P. gravata* Conil and Lys by smaller dimensions and thinner, less coarsely agglutinated walls, and from *P. berwicki* Conil and Longerstaey by larger dimensions and a thicker, coarser wall, giving it a more robust appearance. Range: late Asbian–Brigantian.

Cribrostomum lecomptei (Conil and Lys, 1964)
Plate 3.8, Fig. 7, (in B.G.S. Edinburgh) PS 2159 (× 50), late Asbian, Archerbeck Beds, Archerbeck Borehole, depth 1912 ft. Description: 15 biserial chambers. Septa overlap axial line of shell initially and spread apart gradually later. Septa swollen and club-shaped. Last two chambers very large and significantly overlap axis of test. Sutures often deep and well defined. Apical angle 30°.
Range: late Asbian–Brigantian.

Endothyra sp.
Plate 3.8, Fig. 8, ARE 2, late Asbian, Trowbarrow Quarry, Cumbria. Description: test irregularly coiled, approximately four whorls. Whorl height increases gradually towards the final chamber. Septa long, projected slightly towards the aperture and of similar thickness to the wall. Chambers rounded but not swollen. Ten chambers in final whorl. Supplementary deposits very well developed in the form of thick basal layers, nodosities in final chambers and a terminal spine which is projected towards the aperture. Aperture low and simple.

Quasiendothyra nibelis (Durkina, 1959)
Plate 3.8, Fig. 9 (see also Plate 3.5, Fig. 10; Plate 3.6, Fig. 1), ARE 596, late Asbian, Knipe Scar Limestone, Knipe Scar, Cumbria. Description: coiling glomospiral initially followed by regular aligned terminal whorls. Whorl height increases gradually towards aperture. Chambers numerous, 15–17 in final whorl. Septa long, straight, perpendicular to the walls and slightly swollen at the ends. Sutures well defined between the bulbous chambers. Basal supplementary deposits thick, increasing towards the aperture. Strong chomata. Three whorls. Small proloculus. Aperture simple, terminal.
Range: Holkerian–Asbian.

Carboniferous

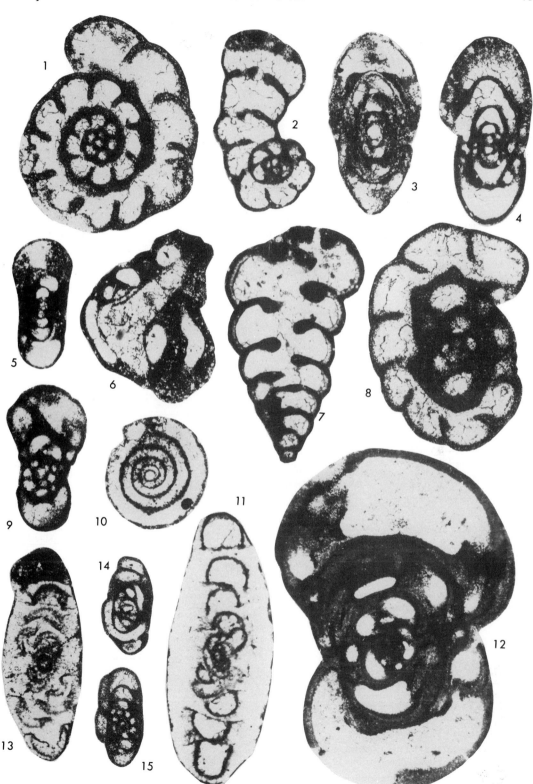

Plate 3.8

Pseudoammodiscus aff. *volgensis* (Rauser-Chernoussova)
Plate 3.8, Fig. 10, (in B.G.S. Edinburgh) PS 2170, late Asbian, Archerbeck Beds, Archerbeck Borehole, depth 1922 ft. = *Ammodiscus volgensis* Rauser-Chernoussova, 1948. Description: $4\frac{1}{2}$ whorls from central proloculus. Tubular chamber grows gradually in height towards the final chamber, where it is approximately four times its original size.

Archaediscus stilus (Grozdilova and Lebedeva, 1954)
Plate 3.8, Fig. 11, (in B.G.S. Edinburgh) PS 2027 (× 140), Archerbeck Borehole, depth 2928 ft. Description: test free, lenticular flattened or slightly swollen. Wall bi-layered. Coiling aligned but not planispiral. $4\frac{1}{2}$–6 whorls. W/d 0.44. Diameter usually in the region of 170–380 μm.
Remarks: Differs from *A. stilus eurus* by its smaller size and smaller number of whorls.
Range: Asbian–Namurian.

Omphalotis samarica (Rauser-Chernoussova)
Plate 3.8, Fig. 12, (in B.G.S. Edinburgh) PS 2140, late Asbian, Archerbeck Beds, Archerbeck Borehole, $1863\frac{1}{2}$ ft. = *Endothyra samarica* Rauser-Chernoussova, 1948. Description: test coiled, umbilical. Last two planispiral whorls grow rapidly in height so that shell seems flattened. Initial whorls irregularly coiled. Septa almost perpendicular to wall, slightly inclined towards aperture. Wall thickness up to 40 μm. Aperture slit-shaped.
Range: Holkerian–Asbian.

Nodosarchaediscus cornua (Conil and Lys)
Plate 3.8, Fig. 13, ARE 1211 (× 140), early Brigantian, Ravensholme Limestone, Ravensholme Quarry, Downham, Lancashire. = *Archaediscus cornua* Conil and Lys, 1964. Description: test lenticular, compressed. Coiling oscillating. 6–7 whorls. Large well-rounded nodosities. Tubular chamber very small at origin, grows regularly in size towards aperture. Small proloculus.
Range: Holkerian–Brigantian.

Brunsia jactata (Conil and Lys)
Plate 3.8, Fig. 14, ARE 602, late Asbian, Knipe Scar Limestone, Knipe Scar, Cumbria. = *Glomospira jactata* Conil and Lys, 1964. Description: test discoidal with very irregular coiling. 6–7 whorls. Wall of moderate thickness. Diameter 275–360 μm, width 125–200 μm.
Range: Holkerian–Asbian.

Endostaffella sp.
Plate 3.8, Fig. 15, ARE 702, late Asbian, Knipe Scar Limestone, Little Asby Scar, Cumbria.

PLATE 3.9—Unless otherwise stated, all the figured specimens are in the B.G.S. collections at Keyworth and are × 75.

Koskinobigenerina sp.
Plate 3.9, Fig. 1, HR 3454, Brigantian, Eelwell Limestone, Spittal Shore, Northumberland. Description: chambers biserially arranged from proloculus. Initial eight chambers separated by septa overstepping the central axial plane of test. Septa well rounded and increasing gradually in thickness throughout the shell. Chambers rounded and swollen with well defined sutures. Last two chambers distinguished by cribrate apertural faces reaching over whole width of shell.
Range: Late Asbian–earliest Namurian.

Endothyranopsis sphaerica (Rauser-Chernoussova and Reitlinger)
Plate 3.9, Fig. 2, (see also Plate 3.6, Fig. 3), PS 1879, late Brigantian, Buccleuch Limestone, Archerbeck Borehole, depth 952 ft; Fig. 3, ARE 486, early Brigantian, Lower Bath House Wood Limestone, River North Tyne. = *Endothyra sphaerica* Rauser-Chernoussova and Reitlinger, 1937. Description: w/d ratio 0.9–1.0 mm. Umbilicus poorly defined. Chambers separated by poorly defined thin septal furrows. 10–14 chambers in outer whorl. $3-3\frac{1}{2}$ whorls slowly increasing in height. Max. diam. 1.33 mm. Aperture low, wide, semilunar at base of apertural face. Septa inflated at ends.
Range: Asbian–Brigantian.

Biseriella parva (Chernysheva)
Plate 3.9, Fig. 4, R. Conil Coll. (in B.G.S. Edinburgh) PS 1882 (× 38). Late Brigantian, Buccleuch Limestone, Archerbeck Borehole, depth 955 ft. = *Globivalvulina parva* Chernysheva, 1948. Description: test small, almost hemispherical with slightly concave apertural face and faintly lobulate peripheral margin. Initial portion of test coiled in trochoid spire; final whorl completely embraces earlier portion. Sutures well defined. External surface exhibits biserial arrangement of chambers. Aperture simple, opening into a curved apertural depression in each chamber.
Remarks: Distinctive features include small test and arrangement, convexity, and number of chambers.
Range: late Asbian–lower Namurian.

Lituotubella aff. *magna* (Rauser-Chernoussova, 1948)
Plate 3.9, Fig. 5 (see also Plate 3.6, Fig. 9), PS 1882 in B.G.S. Edinburgh, late Brigantian, Buccleuch Limestone, Archerbeck Borehole.

Carboniferous

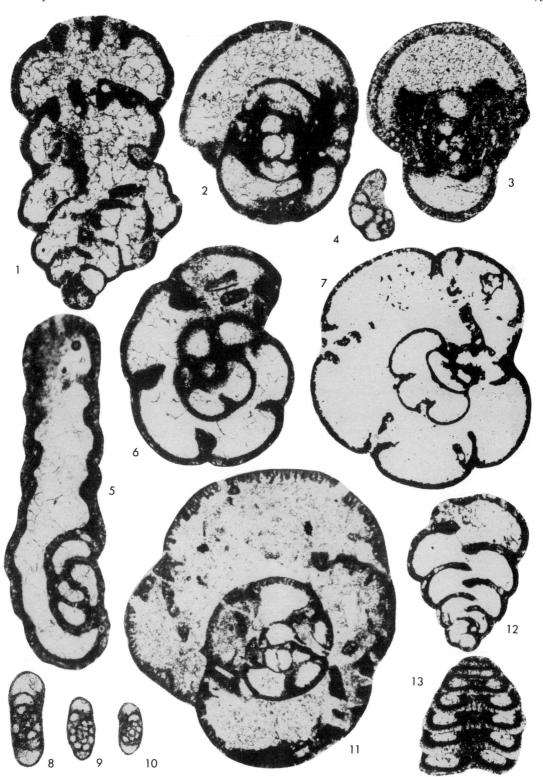

Plate 3.9

Plectogyranopsis ampla (Conil and Lys)
Plate 3.9, Fig. 6, Bk 3381, early Brigantian, Rownham Hill Coral Bed, Ashton Park Borehole, Bristol, depth 1638 ft. = *Plectogyra exelikta ampla* Conil and Lys, 1964. Description: test large. Coiling irregular, becoming planispiral. $2-2\frac{1}{2}$ whorls. Terminal whorl raised and divided into $5-5\frac{1}{2}$ large swollen chambers. Sutures well defined. Septa thick, feebly inclined towards aperture. Wall granular, increasing in thickness with growth to $40-50\ \mu m$ in final chambers. Supplementary deposits absent.
Range: Arundian–Brigantian.

Janischewskina operculata (Rauser-Chernoussova and Reitlinger)
Plate 3.9, Fig. 7, RC 0000, ($\times 50$), Petershill Quarry, Bathgate. = *Samarina operculata* Rauser-Chernoussova and Reitlinger, 1937. Description: test subspherical, slightly compressed laterally. Wall perforate, thin, becoming progressively thicker (up to $25-55\ \mu m$) in later whorls. 2–3 whorls, 5–6 chambers per whorl. Chambers slightly inflated, with double septal sutures. Septa initially short and curved but later fused in pairs. Umbilicus closed. Apertural face covered with 3–4 rows of convex apertural plates. Supplementary septal apertures well developed. Diameter often up to 2.15 mm. Remarks: differs from *J. miniscularia* Ganelina 1956 by the shape of the test, larger size and greater number of whorls. Differs from *J. calceus* Ganelina, 1965, by the latter's asymmetrical position in the first whorl.
Range: Brigantian.

Endostaffella fucoides (Rozovskaya, 1963)
Plate 3.9, Fig. 8, ARE 519, late Brigantian, Cockleshell Limestone, Whitfield Gill, Askrigg, Figs 9, 10, HR 3442. Early Brigantian, Oxford Limestone, Spittal Shore, Northumberland. Description: test free, last whorls evolute, periphery rounded, w/d = 0.35–0.45. Axial region tends to be swollen. Last $1\frac{1}{2}$ whorls aligned.
Range: Asbian–Brigantian.

Bradyina rotula (Eichwald)
Plate 3.9, Fig. 11, PS 1882 ($\times 28$), late Brigantian, Buccleuch Limestone, Archerbeck Borehole, depth 955 ft. = *Nonionina rotula* Eichwald, 1878. Description: test large; whorl low at the origin, increasing rapidly in height towards aperture; chambers large, slightly swollen and separated by thick septa inclined towards aperture. Wall reaches $120\ \mu$ in thickness; $2\frac{1}{4}$ whorls; six chambers; diameter up to $1200\ \mu$.
Remarks: differs from *B. cribrostomata* by larger number of chambers in final whorl.
Range: late Asbian–Brigantian.

Palaeotextularia aff. *lipinae* (Conil and Lys, 1964)
Plate 3.9, Fig. 12, ARE 1358, Brigantian, Middle Limestone, Garth Gill, Askrigg Block. Description: test conical. Septa well rounded and slightly overstepping the axial plane of shell leaving only small spaces between each other. Septa tend to be swollen at their extremities. Sutures well defined due to swollen nature of chambers. 8–11 chambers. Apical angle 30–45°.

Howchinia bradyana (Howchin)
Plate 3.9, Fig. 13 (see also Plate 3.10, Figs. 9, 10), ARE 430 ($\times 80$), late Brigantian, Acre Limestone, Beadnell, Northumberland. = *Patellina bradyana* Howchin, 1888. Description: test free, conical, trochoid. Sutures well defined. Shell consists of spiral tube wound round central umbilical core of microcrystalline calcite. 5–12 whorls. Aperture slit-like from periphery to concave umbilicus.
Range: Late Asbian–Brigantian.

PLATE 3.10—All the figured specimens (except Figs 9–12) are in the B.G.S. collection at Keyworth and are $\times 75$ unless stated otherwise.

Nodasperodiscus stellatus (Bozorgnia)
Plate 3.10, Fig. 1, HR 3695 ($\times 105$), late Brigantian, Scar Limestone, Trout Beck, Cumbria. = *Rugosoarchaediscus stellatus* Bozorgnia, 1973. Description: test lenticular, swollen, rounded periphery. Coiling irregular. Chamber size increases regularly towards very large final whorl. Inner whorls fully occluded forming a stellate flare. Outer 1–2 whorls without occluding nodosities.
Range: Asbian–Namurian.

Archaediscus karreri Brady, 1973 (at angulatus stage)
Plate 3.10, Fig. 2, 3, ARE 465 ($\times 105$), Late Brigantian, Beadnell, Northumberland. Description: test large, lenticular swollen. Irregularly coiled. $5-6\frac{1}{2}$ whorls.
Range: Brigantian.

Forschia sp.
Plate 3.10, Fig. 4, ARE 1453, late Asbian, Poyll Ritchie, near Strandhall, Isle of Man. Description: whorl increases in size gradually towards aperture. $3\frac{1}{2}-4$ whorls.

Carboniferous

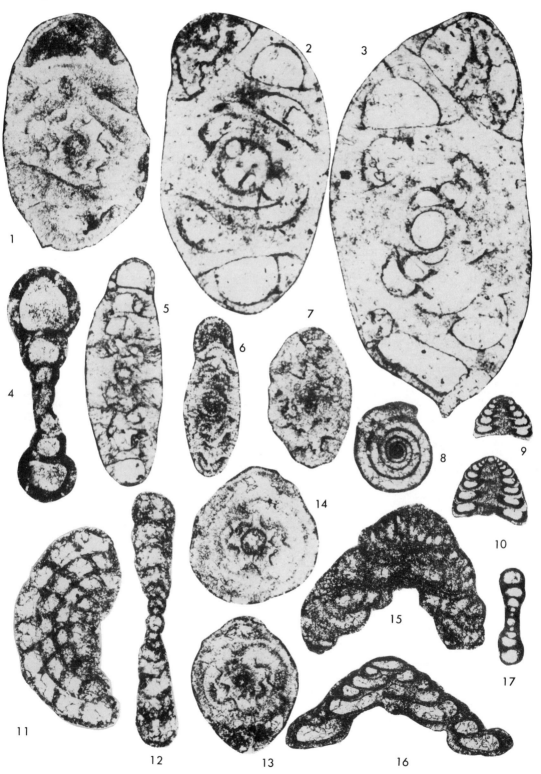

Plate 3.10

Archaediscus stilus eurus (Conil and Longerstaey, 1980)
Plate 3.10, Fig. 5, ARE 458 (× 140), late Brigantian, Beadnell, Northumberland. Description: test lenticular, strongly elongate along diameter, laterally compressed. Sides almost subparallel but slightly swollen. Rounded periphery. Small spherical proloculus. Initially involute, coiling oscillates by up to 40° per whorl. Final 2–3 whorls evolute and more or less aligned in same plane. 6–7 whorls. Diameter 450–550 μm, width 170–220 μm, w/d = 0.34–0.41. Dark microgranular wall developed throughout. Lighter, fibrous, radiating layer very thin in first whorls, becoming much thicker in later revolutions. Tubular chamber relatively small in first three whorls, becoming higher and wider in final three whorls.
Range: Brigantian.

Nodasperodiscus gregorii (Dain)
Plate 3.10, Fig. 6, ARE (× 140), early Brigantian, Simonstone Limestone, Whitfield Gill, Askrigg. = *Archaediscus gregrorii gregorii* Dain, 1953. Description: discoidal with rounded periphery. 5–6 whorls. Coiling oscillating. Final 3–4 whorls evolute.
Range: Brigantian–Upper Namurian.

Asteroarchaediscus occlusus (Hallett)
Plate 3.10, Fig. 7, HR 3697 (× 140), late Brigantian, Scar Limestone, Trout Beck, Cumbria. = *Rugosarchaediscus occlusus* Hallett, 1970.
Range: Brigantian.

Pseudoammodiscus aff. *buskensis* (Brazhnikova)
Plate 3.10, Fig. 8, ARE 1143, early Asbian, Potts Beck Limestone, Little Asby Scar, Cumbria; Fig. 17, HR 3698, late Brigantian, Scar Limestone, Trout Beck, Cumbria. = *Ammodiscus buskensis* Brazhnikova, 1956. Description: $4\frac{3}{4}$ whorls from small proloculus. Tubular chamber grows rapidly in height from 10–15 μm initially to c. 70 μm in final whorl. Wall dark, 10–15 μm. Diameter 340 μm.

Howchinia bradyana (Howchin)
Plate 3.10, Fig. 9, (see also Plate 3.9, Fig. 13) (in B.G.S. Edinburgh), PS 1813, late Brigantian, Buccleuch Limestone, Archerbeck Borehole, depth 946 ft; Fig. 10, PS 1873, as Fig. 9 but depth 943 ft. = *Patellina bradyana* Howchin, 1888.
Range: Late Asbian–Brigantian.

Loeblichia paraammonoides (Brazhnikova)
Plate 3.10, Fig. 11, 12, R. Conil Coll. (× 140). = *Nanicella paraammonoides* Brazhnikova, 1956. Description: shell regularly coiled, flat and discoidal: strongly compressed along axis of rotation. Wide, shallow umbilici. Coiling planispiral or almost so. Numerous quadrate chambers, long, straight septa, approx. 20 in final whorl. Initially, whorls very narrow, becoming wider and higher in the later whorls. Spiral and septal sutures very indistinct initially but deepening in last whorls. 4–7 whorls. w/d = 0.13–0.28. Supplementary deposits in the form of chomata, often weak. Aperture narrow, slit-like basal.
Range: Brigantian.

Nodasperodiscus sp.
Plate 3.10, Fig. 13, HR 3697 (× 105), late Brigantian, Scar Limestone, Trout Beck, Cumbria. Description: test rounded, slightly laterally compressed. Central small proloculus. Final whorls with poorly formed nodosities. Marked increase in size of tubular chamber with growth. 5–6 whorls.
Range: Late Asbian–earliest Namurian.

Asteroarchaediscus? sp.
Plate 3.10, Fig. 14, HR 3697 (× 140), late Brigantian, Scar Limestone, Trout Beck, Cumbria. Description: test subspherical, well rounded. Central large proloculus.
Range: Late Asbian–earliest Namurian.

Valvulinella conciliata (Ganelina)
Plate 3.10, Fig. 15 (see also Plate 3.5, Fig. 16), ARE 527, late Brigantian, Cockleshell Limestone, Whitfield Gill, Askrigg. = *Tetrataxis conciliatus* Ganelina, 1956. Description: test conical, apex rounded, flanks convex. Umbilicus straight, occupying a quarter of basal diameter. Apical angle approx. 89°. Six whorls.
Range: Brigantian.

Textrataxis pressula pressula (Malakhova, 1956)
Plate 3.10, Fig. 16, ARE 479, late Brigantian, Upper Bath House Wood Limestone, River North Tyne, Barrasford, Northumberland. Description: test in form of broad, low cone, flattened from the apex. Umbilicus large and quite deep. Wall dark, microgranular to granular with a thin fibrous layer. Coiling regular with whorls gradually increasing in size. 5–7 whorls. Diameter 800–960 μm. Height 315–500 μm. h/d 0.37–0.46. Apical angle 110°–115°.
Range: Brigantian.

PLATE 3.11—Unless otherwise stated, the figured specimens are in the B.G.S. collections at Keyworth and are × 75.

Koskinotextularia cribriformis (Eickhoff, 1968)
Plate 3.11, Fig. 1, ARE 378, late Asbian, Trefor Rocks, Llangollen. Description: proloculus spherical or oval, often slightly displaced laterally. Chambers 9–13 in number, biserially arranged. Septa in the first 6–8 chambers are less hooked and overlap the central plane to a greater extent than the septa in more adult chambers. Septa in latter

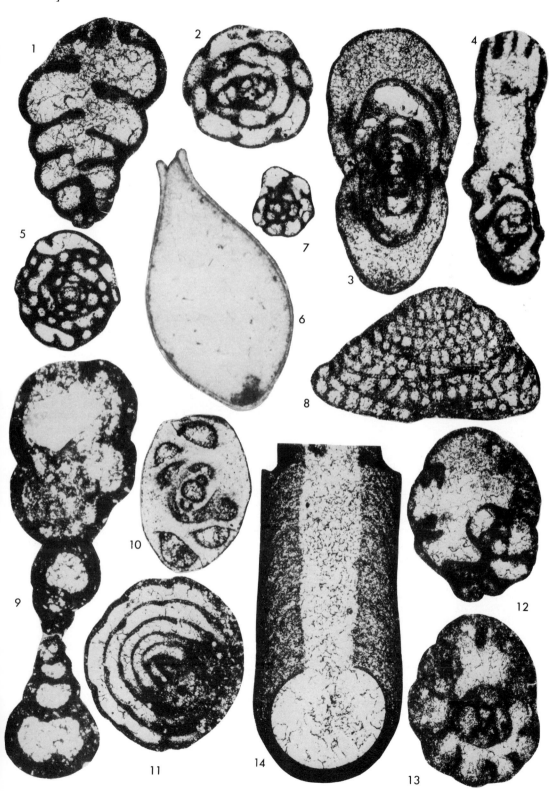

Plate 3.11

chambers tend to be more rounded and slightly globular at the ends. Aperture slit-like, cribrate. Length 0.81–1.04 mm; width 0.58–0.63 mm.
Remarks: differs from *K. obliqua* (Conil and Lys) by its regular expansion in width, the more feeble arrangement of sutures and thinner wall.
Range: Asbian–Brigantian.

Septabrunsiina mackeei (Skipp, 1966)
Plate 3.11, Fig. 2, ARE 837, Brigantian, Ravensholme Limestone, Red Syke, Pendle, Lancashire. Description: test small/medium, discoidal, largely evolute with broad shallow umbilicus. Proloculus small 10–15 μm. Coiling initially irregular, becomes planispiral in final 2–3 whorls. $3\frac{1}{2}$–5 whorls altogether. Chambers low, elongate and slightly lobate. $7\frac{1}{2}$–8 chambers in final whorl. Septa present throughout, less distinct initially, strong anterior orientation.
Range: Asbian–Brigantian.

Eostaffella sp.
Plate 3.11, Fig. 3, early Asbian, Potts Beck Limestone, Little Asby Scar, Cumbria. Description: test very large, elongate. Periphery well rounded. Well defined umbilici. Wall microgranular with tectum. Coiling planispiral. Spire expands gradually in height and rapidly in width throughout coiling. 5–6 whorls from central circular proloculus. Supplementary deposits in form of poorly developed pseudochomata on inner whorls.

Lituotubella glomospiroides (Rauser-Chernoussova, 1948)
Plate 3.11, Fig. 4, BLG 9996 (× 38), late Asbian, Beckermonds Scar Borehole, depth 82 m. Description: test large tubular, evolute, glomospirally coiled in early stages, uncoiled in later stages. Septal sutures not well developed. Glomospiral portion laterally compressed, 4–5 chambers/whorl. 5–6 pseudochambers in uniserial section. Aperture broad, coarsely cribrate in last two chambers. Flat apertural shield. Length 1.9 mm.
Range: Asbian.

Urbanella miranda matura (Vdovenko, 1972)
Plate 3.11, Fig. 5, AL 1367, Holkerian, South Bay, Tenby, 82 m below top of section.
Range: Holkerian.

Saccamminopsis fusulinaformis (McCoy)
Plate 3.11, Fig. 6, HR 3737 (× 37), Brigantian, Jew Limestone, Moulds Meaburn Quarry, near Appleby, Cumbria. = *Nodosaria fusulinaformis* McCoy, 1849; = *Saccammina cateri* Brady.
Range: Brigantian.

Endothyra obsoleta (Rauser Chernoussova, 1948)
Plate 3.11, Fig. 7, ARE 973, late Asbian, Danny Bridge Limestone, R. Clough, near Sedbergh, Yorkshire. Description: test involute. with a small number of convex chambers. 2–3 whorls, 6–7 chambers in final whorl. Septa long, broadly curved in the direction of coiling. Wall thin, dark, finely granular 10–15 μm thick. Supplementary deposits developed in the form of prominent nodosities on basal protuberances. Diameter 0.25–0.37 mm.
Remarks: differs from *E. similis* Rauser-Chernoussova and Reitlinger by having fewer, more convex chambers.
Range: Asbian–Brigantian.

Valvulinella latissima (Conil and Lys, 1964)
Plate 3.11, Fig. 8, ARE 1442, early Asbian, Strandhall Shore, Isle of Man. Description: test irregularly conical with base broader than height. Apical angle approx. 100°.
Remarks: differs from *V. lata* Grozdilova and Lebedeva by the larger apical angle of the cone.
Range: Asbian–Brigantian.

Forschiella prisca (Mikhailov, 1939)
Plate 3.11, Fig. 9, ARE 730, Asbian, Urswich Limestone, below Woodbine Shale, Stainton Quarry, Dalton-in-Furness, Cumbria. Description: gradual increase in height and width of whorls resulting in two large, deep umbilical depressions. Proloculus small. Axial portion of test often very thin. 4–5 whorls. Final whorl large and robust. Wall thickness increasing regularly up to 75–1500 μm in the last revolution. Aperture cribrate and complex. Diameter 900–1500 μm. Width 300–400 μm approx.
Range: Arundian–Lower Namurian.

Archaediscus itinerarius (Schlykova, 1951)
Plate 3.11, Fig. 10, BLG 9852 ($\times$ 140), early Brigantian, Beckermonds Scar Borehole, depth 4 m. Description: test involute, subspherical and laterally compressed. Periphery well rounded. 4–5 whorls. w/d = 0.61 to 0.78. Proloculus fairly large. Coiling sigmoidal. Tubular chamber with concave base.
Range: Brigantian.

Glomospiranella sp.
Plate 3.11, Fig. 11, ARE 933, Holkerian, Garsdale Limestone, R. Clough, near Sedbergh, Yorkshire. Description: Coiling irregular initially. Later whorls becoming generally orientated in the same plane. 6–7 whorls. Wall calcareous, microgranular to granular with some larger grains. Pseudosepta well defined in final whorls but quite indistinct or absent initially. Probably 10–13 pseudochambers in last revolution.

?*Janischewskina minuscularia* (Ganelina, 1956)
Plate 3.11, Fig. 12, ARE 335, Early Asbian, Tandinas Quarry, Anglesey. Description: test coiled, laterally compressed, involute. Umbilical depressions large, periphery rounded. $1\frac{1}{2}$–2 whorls. Chambers increase progressively in height from a spherical proloculus.
Remarks: differs from *J. calceus* Ganelina in the nature of the wall, smaller dimensions, less numerous and symmetrical nature of the whorls.

Cribranopsis sp.
Plate 3.11, Fig. 13, Early Asbian, Tandinas Quarry, Anglesey.

Earlandia, sp.
Plate 3.11, Fig. 14, ARE 1194, Early Asbian, Potts Beck Limestone, Groups Hollow, near Little Asby Scar, Cumbria. Description: test very large. Proloculus large spherical, slightly compressed. Tubular chamber cylindrical, with parallel sides. Width of proloculus and tubular chamber about the same. Wall very thick with dark microgranular outer layer grading into a less dense granular layer. Orientation of grains in the lighter layer tends to follow the rounded shape of the proloculus throughout the shell. Inner wall of proloculus dark and microgranular. Maximum length observed on incomplete specimens 1300 μm. Maximum external diameter of proloculus 575 μm. Maximum internal diameter of proloculus 450 μm.
Remarks: this large *Earlandia* is restricted to the Early Asbian of the British Isles.

PLATE 3.12—Unless otherwise stated, the figured specimens are × 75. Register numbers preceded by the letters HM are of specimens in the Hunterian Museum, Glasgow; the remainder are in the B.G.S. collections at Keyworth.

Endothyra excellens (Zeller)
Plate 3.12, Fig. 1, HM P433/1, Arnsbergian, Plean Limestone, Craigburn, Uddington, Lanarkshire. = *Plectogyra excellens* Zeller, 1953. Description: secondary deposits continuous on floor of chambers, well developed hook in last chamber. Proloculus small.
Range: late Asbian–early Namurian.

Bradyina cribrostomata (Rauser-Chernoussova and Reitlinger, 1937)
Plate 3.12, Fig. 2, Arnsbergian, Pike Hill Limestone, Throckley Borehole, Northumberland, depth 285 m. Description: slightly compressed laterally, chambers inflated, sutures depressed. 2–3 whorls, 6–7 chambers in last whorl; diameter 1.7–3.0 mm.
Remarks: differs from *B. potanini* Beninkov in greater number of chambers and more compressed test.
Range: Namurian.

Biseriella parva (Chernysheva)
Plate 3.12, Fig. 3, HM P421/1 Pendleian, Index Limestone, Kennox Water, Ayrshire (see also Plate 3.9, Fig. 4).
Range: late Asbian–early Namurian.

Endothyranopsis crassa (Brady)
Plate 3.12, Fig. 4, HM P 47531, Arnsbergian, Castlecary Limestone, Westerwood, Lanarkshire (see also Plate 3.7, Fig. 1).
Range: Holkerian–early Namurian.

Endostaffella sp.
Plate 3.12, Fig. 5, Wo 1844, Arnsbergian, Corbridge Limestone, Ouston Borehole, Northumberland.

Climacammina postprisca (Brazhnikov and Vinnichenko)
Plate 3.12, Fig. 6, Arnsbergian, Pike Hill Limestone, Throckley Borehole, Northumberland, depth 287 m. Description: biserial portion small compared with elongate uniserial, subparallel part. Septa short.
Range: Namurian.

Asteroarchaediscus sp.
Plate 3.12, Fig. 7, HR 3559 (× 140). Pendleian, Great Limestone, Brunton Bank Quarry, Northumberland.

Earlandia pulchra (Cummings, 1955)
Plate 3.12, Fig. 8, HM P538/1, Arnsbergian, Orchard Limestone, R. Nethan, Auchlochen House, Ayrshire. Description: tubular chamber with faint depressed sutures, no trace of septation internally.
Range: Early Namurian.

Cribrospira sp.
Plate 3.12, Fig. 9, HR 3577, Pendleian, Belsay Dene Limestone, Aydon near Corbridge, Northumberland.

Endothyra cf. *pandorae* (Zeller)
Plate 3.12, Fig. 10, HR 3558, same horizon and locality as Fig. 7. = *Plectogyra pandorae* Zeller, 1953. Description: test with broadly rounded periphery, umbilicate on one side only. Secondary deposits on floor of chambers with low nodes in last one or two chambers.
Remarks: differs from *E. excellens* in less extensive secondary deposits, thicker walls and less swollen chambers.
Range: Brigantian–early Namurian.

Warnantella subquadrata (Potievskaya and Vakarchuk)
Plate 3.12, Fig. 11, HM P 529/2; Fig. 12, HM P 529/4; Arnsbergian, Castlecary Limestone, Westerwood, Lanarkshire. = *Glomospira subquadrata* Potievskaya and Vakarchuk.
Remarks: characterised by its irregular tortuous coiling. Differs from *W. tenuiramosa* in its larger size.
Range; Brigantian–early Namurian.

Neoarchaediscus incertus (Grozdilova and Lebedeva)
Plate 3.12, Figs 13–20, GN 1184 (× 140), Yeadonian, *G. cancellatum* Band, Morton No. 1 Borehole, Lincolnshire. = *Archaedisacus incertus* Grozdilova and Lebedeva, 1954. Description: 4–6 whorls, diameter of test 0.2–0.31 mm, subparallel sides.
Range: late Asbian–late Namurian.

Asteroarchaediscus gregorii (Dain)
Plate 3.12, Fig. 14, GN 1184 (× 140), Yeadonian, *G. cancellatum* Band, Morton No. 1 Borehole, Lincolnshire (see also Plate 3.10, Fig. 6).
Range: Brigantian–late Namurian.

Archaediscus moelleri (Rauser-Chernoussova, 1948)
Plate 3.12, Fig. 15, HR 3605, Arnsbergian, Styford Limestone, Styford, Northumberland. Description: coiling irregular, tubular chamber initially high, grows slowly with coiling. W/d ratio 0.7.
Remarks: differs from *A. convexus* Grozdilova and Lebedeva, 1954 in having high initial chamber from the origin, and in having fewer whorls and more swollen test.
Range: Brigantian–early Namurian.

Pl. 3.12] **Carboniferous** 83

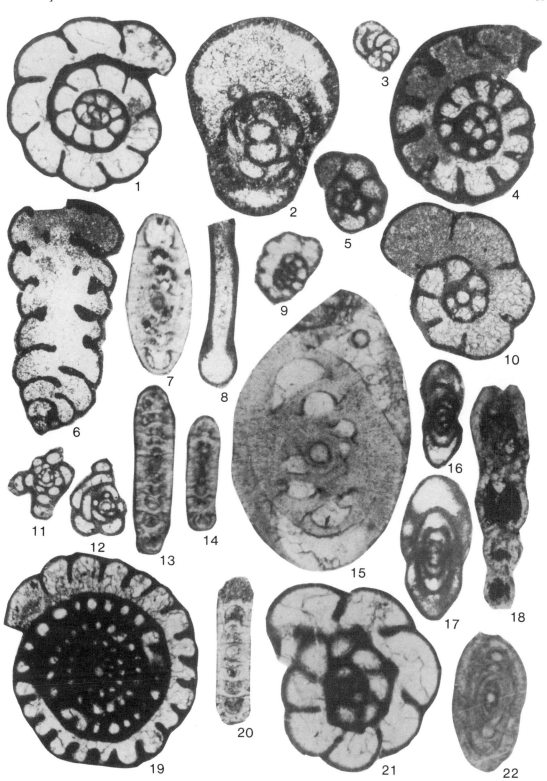

Plate 3.12

Eostaffella sp.
Plate 3.12, Figs 16, 17, 19. Arnsbergian, Fig. 16, HR 3600, Styford Limestone, Styford, Northumberland. Fig. 17, HR 3565, Corbridge Limestone, Corbridge, Northumberland. Fig. 19, BLG 6816, Newton Limestone, Hexham bypass Borehole No. 138a, Northumberland.

Lugtonia elongata (Cummings, 1955)
Plate 3.12, Fig. 18, HM P486, Arnsbergian, shale over Orchard Limestone, Poniel Water, Ayrshire. Description: about five pyriform linear chambers gradually increasing in size, separated by deep sutures.
Remarks: more elongate and cylindrical than *L. concinna* and with more chambers.
Range: Early Namurian.

Endothyra sp.
Plate 3.12, Fig. 21, ARE 633, Pendleian, Great Limestone, Greenleighton Quarry, Northumberland.

Eosigmoilina robertsoni (Brady)
Plate 3.12, Fig. 22, HM P515, Arnsbergian, Castlecary Limestone, Westerwood, Lanarkshire. = *Trochammina robertsoni* Brady, 1879.
Range: early Namurian (E_{2a} Zone).

REFERENCES

Armstrong, A. K. and Mamet, R. L. 1977. Carboniferous microfacies, microfossils and corals. Lisburne Group, Arctic Alaska. *U.S. Geol. Surv. Prof. Paper*, No. 849.

Brady, H. B. 1873a. On *Archaediscus karreri*, a new type of Carboniferous Foraminifera. *Ann. Mag. Nat. Hist.*, **12**(4), 286–290.

Brady, H. B. 1873b. In Notes on certain genera and species. *Mem. Geol. Surv. Scotland*, sheet 23, 93–107.

Brady, H. B. 1876. A monograph of Carboniferous and Permian Foraminifera (the genus *Fusulina* excepted). *Monogr. Palaeontogr. Soc.* 1–166, 12 pls.

Brazhinkova, N. E. 1962. *Quasiendothyra* and related forms in the Lower Carboniferous of the Donetz Basin and other areas of the Ukraine. *Akad. Sci. Uk., Trud. Inst. Geol., Strat. Paleont.*, **44**, 1–48.

Browne, R. G. and Pohl, E. R. 1973. Stratigraphy and genera of calcareous Foraminifera of the Frailey's facies (Mississippian) of central Kentucky. *Bull. Amer. Paleont.*, **64**, 173–239.

Bykova, E. V. and Polenova, E. I. 1955. Foraminifera and Radiolaria from the Volga–Ural region of the central Devonian basin, and their stratigraphical importance. *Trud. VNIGRI* n.s. **87**, 1–141.

Calver, M. A. 1969. Westphalian of Britain. *C. R. 6me Cong. Int. Strat. Géol. Carb.* (Sheffield 1967), **1**, 233–254.

Chernysheva, N. E. 1940. On the stratigraphy of the Lower Carboniferous Foraminifera in the Makarovski district of the south Urals. *Bull. Soc. Nat. Mosc.* n.s., **48**, 113–135.

Conil, R. 1976. Contribution a l'étude des forminifères du Dinantien en Irlande. *Ann. Soc. Géol. Belg.*, **99** 129–141.

Conil, R. and George, T. N. 1973. The age of the Caninia Oolite in Gower. *C. R. 7me Cong. Int. Strat. Géol. Carb.* (Krefeld 1971), **2**, 323–332.

Conil, R., Groessens, E. and Pirlet, H. 1976. Nouvelle charte stratigraphique du Dinantien type de la Belgique. *Ann. Soc. géol Nord.*, **96**, 363–371.

Conil, R. and Lees, A. 1974. Les transgressions viséennes dans l'Ouest de l'Irelande. *Ann. Soc. géol. Belg.*, **97**, 463–484.

Conil, R., Longerstaey, P. J. and Ramsbottom, W. H. C. 1980. Matériaux pour l'étude micropaléontologique du Dinantien de Grande-Bretagne. *Mém. Inst. géol. Univ. Louvain*, **30**, 1–186, 30pls (dated 1979).

Conil, R. and Lys, M.1964. Matériaux pour l'étude micropaléontologique du Dinantien de la Belgique et de la France (Avenois), Algues et Foraminifères. *Mém. Inst. géol. Univ. Louvain*, **23**, 1–279, 42 pls.

Conil, R. and Lys, M. 1970. Données nouvelles sur les Foraminifères du Tournaisien inférieur et des couches de passage du Fammenien au Tournaisien dans l'Avenois. *Cong. Coll. Univ. Liège*, **55**, 241–265.

Conil, R. and Lys, M. 1977. Les foraminifères du Viséan moyen V2a aux environs de Dinant. *Ann. Soc. géol. Belg.*, **99**, 109–142.

Conil, R. and Lys, M. 1978. Les transgressions dinantiennes et leur influence sur la dispersion et l'évolution des foraminifères. *Mém. Inst. géol. Univ. Louvain*, t., **29**, 9–55 (dated 1977).

Conil, R. and Pirlet, H. 1978. L'évolution des Archaediscidae viséens. *Bull. Soc. Belge. Géol.*, **82**, 241–300 (dated 1974).

Conil, R. 1981. Note sur quelques foraminifères du Strunien et du Dinantien d'Europe occidentale. *Ann. Soc. géol. Belg.*, **103**, 43–53.

Cummings, R. H. 1955a. *Nodosinella* Brady 1976 and associated Upper Paleozoic genera. *Micropaleontology*, **1**, 221–238.

Cummings, R. H. 1955b. New genera of Foraminifera from the British Lower Carboniferous. *Washington Acad. Sci. J.*, **45**, 1–8.

Cummings, R. H. 1956. Revision of the Upper Paleozoic textulariid foraminifera. *Micropaleontology*, **2**, 201–242.

Cummings, R. H. 1957. A problematic new microfossil from the Scottish Lower Carboniferous. *Micropaleontology*, **3** 407–409.

Cummings, R. H. 1961. The foraminiferal zones of the Archerbeck Borehole. *Bull. geol. Survey Gt. Br.*, **18**, 107–128.

Cushman, J. 1927. An outline of the reclassification of

the Foraminifera. *Contr. Cushman Lab. Foram. Res.*, **3**.

Dain, L. G. 1953. Tournayellidae. In Fossil Foraminifera of the U.S.S.R. *Trud. VNIGRI*, n.s., **74**, 7–54.

Davies, A. G. 1945. Micro-organisms in the Carboniferous of the Alport Boring. *Proc. Yorks. geol. Soc.*, **25**, 312–318.

Davies, A. G. 1951. *Howchinia bradyana* (Howchin) and its distribution in the Lower Carboniferous of England. *Proc. Geol. Assoc.*, **62**, 248–253.

Ehrenberg, C. 1854. *Microgeologie: Das Wirken des unsichtbaren kleinen Lebens auf der Erde.* Leipzig.

Eikhoff, G. 1968. Neue Textularien (Foraminifera) aus dem Waldecker Unterkarbon. *Pal. Zeitsch.*, **42**, 162–178.

Fewtrell, M. D. and Smith, D. G. 1978. Stratigraphic significance of calcareous microfossils from the Lower Carboniferous rocks of the Skipton area, Yorkshire. *Geol. Mag.*, **115**, 255–271.

Fewtrell, M. D. and Smith, D. G. 1981. The recognition and division of the Tournaisian Series in Britain. Discussion. *J. geol. Soc. Lond.*, **138**, (1).

Galloway, J. J. and Harlton, B. H. 1930. *Endothyranella*, a genus of Carboniferous Foraminifera. *J. Paleont.* **4**, 2–10.

Ganelina, R. A, 1956. Foraminifera of the Viséan sediments of the north-west region of the Moscow Syncline. *Trud. VNIGRI* n.s., **98**, 61–159.

Ganelina, R. A. 1966. Tournaisian and Lower Viséan Foraminifera of the Kama-Kinel basin. *Trud. VNIGRI* **250**, 64–151.

George, T. N., Johnson, G. A. L., Mitchell, M., Prentice, J. E., Ramsbottom, W. H. C., Sevastopulo, G. and Wilson, R. B. 1976. A correlation of Dinantian rocks in the British Isles. *Geol. Soc. Lond. Spec. Rep.*, **7**, 1–87.

Hallett, D. 1970; Foraminifera and algae from the Yoredale "Series" (Viséan-Namurian) of Northern England. *C. R. 6me Cong. int. Strat. Géol. Carb.* (Sheffield 1967), **3**, 873–900.

Holliday, D. W., Neves, R. and Owens, B. 1979. Stratigraphy and palynology of early Dinantian (Carboniferous) strata in shallow boreholes near Ravenstonedale, Cumbria. *Proc. Yorks. geol. Soc.*, **42**, 343–356.

Howchin, W. 1888. Additions to the knowledge of the Carboniferous Foraminifera. *J. Roy. Micr. Soc.* pt 2, 533–545.

Lipina, O. A. 1955. Foraminifera of the Tournaisian stage and uppermost Devonian of the Volga–Ural region and western slope of the Central Urals, *Akad. Sci. U.S.S.R., Trav. Inst. Sci. geol.*, **163**, 1–96.

Lipina, O. A. 1963a. On the evolution of the Tournaisian Foraminifera. *Vop. Mikropaleont.*, **7**, 13–21.

Lipina, O. A. 1963b. In Results of the second colloquium on the systematics of the endothyroid Foraminifera organised by the Commission on Micropaleontological Coordination. *Vop. Mikropaleont.*, **7**, 223–227.

Lipina, O. A. 1965. Taxonomy of the Tournayellidae. *Akad. Sci. U.S.S.R. Trud. Geol. Inst.*, **130**, 1–115.

Lys, M. 1976. Valorisation par microfaunas du Bashkirian inférieur (Namurian B) sous-zone R2 dans le Bassin Houiller du Nord de la France (groupes de Douai et Valenciennes). *Ann. Soc. géol. Nord*, **96**, 379–385.

Malakhova, N. P. 1963. A new foraminiferal genus from the Lower Viséan of the Urals. *Paleont. Zh.*, **3**, 111–112.

Malakhova, N. P. 1973. Tournaisian and Lower Viséan formations of the eastern slopes of the southern Urals. *Trav. Inst. Geol. Geochem. Akad. Sci. Central Urals SSR*, **82**, 5–14.

Mamet, B. 1970a. Precisions sur l'age Viséan du calcaire d'Ardengost (Pyr. Centrales). *C. R. Som. Soc. géol. Fr.*, **4**, 127–128.

Mamet, B. 1970b. Carbonate microfacies of the Windsor Group (Carboniferous), Nova Scotia and New Brunswick. *Geol. Surv. Canada, Paper*, **70–21**, 1–64.

Mamet, B. 1970c. Sur une microfaune tournaisienne du Massif Centrale. *C. R. Som. Soc. géol. Fr.*, **4**, 110.

Mamet, B. 1974a. Taxonomic note on Carboniferous Endothyraceae. *J. Foram. Res.*, **4**, 200–204.

Mamet, B. 1974b. Une zonation par foraminifères du Carbonifere inférieur de la Téthys occidentale. *C. R. 7me Cong. int. Strat. Géol. Carb.* (Krefeld 1971), **3**, 391–408.

Mamet, B. 1975. *Viseidiscus*, un nouveau genre de Planoarchaediscinae (Archaediscinae, Foraminifères). *Bull. Soc. géol. France, Suppl.*, **17**, 48–49.

Mamet, B. 1976. An atlas of microfacies in Carboniferous carbonates of the Canadian Cordillera. *Bull. geol. Surv. Can.*, **225**, 1–131. 95 pls.

Mamet, B., Choubert, G. and Hottinger, G. 1966. Notes sur la Carbonifère du Jebel Ouarkziz. Étude du passage du Viséan au Namurian d'après les foraminifères. *Notes Serv. géol. Maroc*, **27**, 7–21.

Marchant, T. R. 1974. Preliminary note on the micropalaeontology of the Dinantian, Dublin Basin, Ireland. *Ann. Soc. géol. Belg.* **97**, 447–461.

Mikhailov, A. V. 1935. Foraminifera from the Oka series of the Borovitch District, Leningrad Region. *Geol. Razv. Leningr. Izvestia*, **2–3**, 33–36.

Mikhailov, A. V. 1939a. On the characteristics of the genera of Lower Carboniferous Foraminifera. *Leningr. geol. Admin. Sbornik*, **3**, 47–62.

Mikhailov, A. V. 1939b. On Palaeozoic Ammodiscidae. *Leningr. geol. Admin. Sbornik*, **3**, 63–69.

Miklucho-Maklay, A. D. 1956. On the systematics of Palaeozoic Foraminifera. *Vest. Univ. Leningr. (Geol. Geog.)*, **1**, 57–66.

Möller, V., Von. 1878. Die Spiral-Gewundenen Foraminifera des Russischen Kohlenkalks. *Mém. Acad. Imp. Sci. St. Petersbourg*, (7)**25**, 1–147.

Möller, V., Von. 1879. Die Foraminiferen des Russischen Kohlenkalks. *Mém. Acad. Imp. Sci. St. Petersbourg*, (7)**27**, 1–131.

Phillips, J. 1846. On the remains of microscopic animals in the rocks of Yorkshire. *Proc. Yorks geol. poly. Soc.*, **2**, 277–279.

Pirlet, H. and Conil, R. 1978. See Conil and Pirlet 1978.

Plummer, H. J. 1930. Calcareous Foraminifera in the

Brownswood Shale near Bridgeport, Texas. *Bull. Univ. Texas*, **3019**, 5–21.

Ramsbottom, W. H. C., Calver, M. A., Eagar, R. M. C., Hodson, F., Holliday, D. W. Stubblefield, C. J. and Wilson, R. B. 1978. A correlation of Silesian rocks in the British Isles. *Geol. Soc. London Spec. Rep.*, **11**, 1–81.

Rauser-Chernoussova, D. M., Beljaev, G. M. and Reitlinger, E. A. 1936. Upper Palaeozoic Foraminifera of the Pechora district. *Trud. geol. Inst.*, **28**, 1–127.

Rauser-Chernoussova, D. M. *et al.* 1948. [A symposium of papers on stratigraphy of the Lower Carboniferous based on Foraminifera]. *Trud. Inst. geol.*, **62**, ser. geol., **19**, 1–243.

Reitlinger, E. A., 1958. The problem of the systemization and phylogeny of the superfamily Endothyridae. *Vop. Mikropaleont.*, **2**, 55–73.

Reitlinger, E. A., 1959. Foraminifera of the border stage of the Devonian and Carboniferous in the western part of central Kazakhstan. *Dokl. Acad. Sci. U.S.S.R.* **127**, 659–662.

Rozovskaya, C. E. 1961. On the systematics of the Endothyridae and Ozawainellidae. *Pal. Zh.*, **3**, 19–21.

Schlykova, T. I. 1969. A new genus of Foraminifera from the Lower Carboniferous. *Vop. Mikropaleont.*, **12**, 47–50.

Schubert, R. J. 1907. Beitrage zur einer naturlicheren Systematik der Foraminiferen. *Neues Jahrb. Min. Geol. Pal.* B-Band, **25**, 233–260.

Schubert, R. J. 1921. Palaeontologische Daten zur Stammengeschichte der Protozoen. *Pal. Zeitschr.*, **3**, 129–188.

Sheridan, D. J. R. 1972. Upper Old Red Sandstone and Lower Carboniferous of the Slieve Beagh Syncline and its setting in the north-west Carboniferous basin, Ireland. *Geol. Surv. Ireland Spec. Rep.*, **2**, 1–129.

Skipp, B. 1969. Foraminifera. In History of the Redwall Limestone of norther Arizona. *Mem. geol. Soc. Amer.*, **114**, 175–255.

Sollas, W. S. 1921, On *Saccamina carteri* Brady and the minute structure of the foraminiferal shell. *Quart. J. geol. Soc. London*, **77**, 193–211.

Suleimanov, I. S. 1945. Some new species of small foraminifers from the Tournaisian of the Ishimbayevo oil-bearing region. *Dokl. Acad. Sci. U.S.S.R.*, **48**, 124–127.

Thompson, M. L. 1942. New genera of Pennsylvanian fusulinids. *Amer. J. Sci.*, **240**, 403–420.

Vachard, D. 1978. Etude stratigraphique et micropaléontologique (algues et foraminifères) du Viséen de la Montagne Noire (Hérault, France). *Mem. Inst. géol. Louvain*, **29**, 111–195 (dated 1977).

Vdovenko, M. V. 1954. Some new species of Foraminifera in the Lower Viséan deposits of the Donentz Basin. *Notes Sci. Univ. Schevchenko Kieve.*, **13**, 64–76.

Vdovenko, M. V. 1967. Some representative Endothyridae, Tournayellidae and Lituolidae of the Lower Viséan of the Grand Donbass. *Akad. Sci. Ukr. S.S.R. Inst. geol. Sci.*, 18–27.

Vdovenko, M. V. 1970. New information on the systematics of the family Forschiidae. *Geol. J.*, **30**, 66–78.

Vdovenko, M. V. 1971. New genera of Foraminifera from Viséan strata. *Dokl. Akad. Sci. Ukr. S.S.R.*, **B, 33** 877–879.

Wedekind, P. R. 1937. Einführung in die Grundlagen der historischen Geologie. Band II, *Mikrobiostratigraphie die Korallen und Foraminiferenzeit*. 136pp. Stuttgart.

4

PERMIAN

J. Pattison

4.1 INTRODUCTION

The only marine Permian rocks in the British Isles are deposits of the late Permian Zechstein Sea, which covered most of north-central Europe including northeast England, and the contemporaneous Bakevellia Sea of northwest England and Northern Ireland (Fig 4.1). The fauna in the latter was less varied than that of the Zechstein Sea but similar enough for the term 'British Permian foraminifera' to be almost synonymous with 'British Zechstein foraminifera'. Any historical account of their study necessarily includes references to German and Polish work on Zechstein faunas.

The first references to Zechstein foraminifera were included in the separate, general Permian palaeontological works of H. B. Geinitz, R. Howse and W. King, all published in 1848, but all three assigned the foraminifera they described to the annelid genus *Serpula*. However, King's monograph on British Permian fossils (1850) contained descriptions by T. R. Jones of several foraminiferal species, including the first nodosariids to be recognised in Zechstein rocks. Reuss (1854), Richter (1855, 1861) and Schmid (1867) erected several more Zechstein nodosariid species, beginning a proliferation of names which was later criticised by Brady (1876). Schmid *op. cit.* also recorded more tubular forms, as did Jones, Parker and Kirkby (1869), and the confusion between the porcellanous and siliceous tests which started then has continued to the present-day. Brady's monograph included the first comprehensive review of Zechstein foraminifera, and the only one relating to the British faunas. In it he proposed the species *Nodosinella digitata*, the wall-structure of which, and its taxonomic significance, is another problem yet to be resolved.

The only published works on Zechstein foraminifera in the eighty years after 1876 came from Germany. Spandel (1898) produced a brief general summary which included several Zechstein 'firsts': records of *Ammobaculites, Lingulina* and '*Frondicularia*'; recognition of a distinction between the porcellanous and finely siliceous, tubular forms; and the discovery that the Zechstein '*Textularia*' of earlier authors were uniserial nodosariids (now referred to *Geinitzina*).

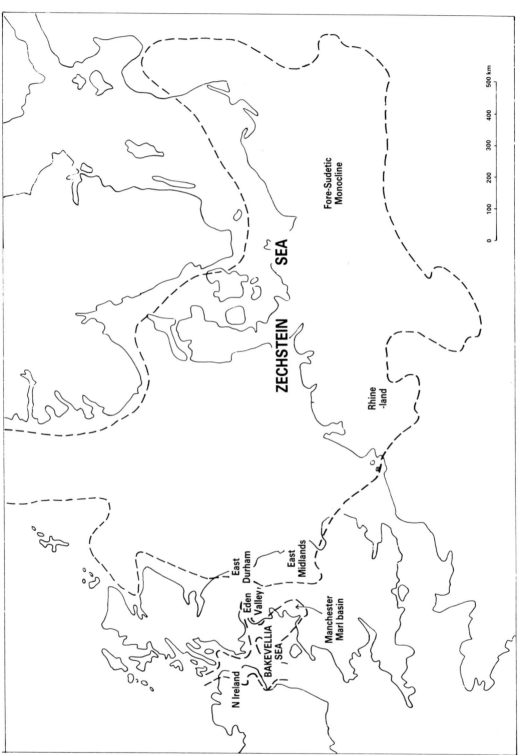

Fig. 4.1 — Map of north-central Europe showing the approximate maximum extent of the Zechstein and Bakevellia seas and source areas of 'Zechstein' foraminifera mentioned in the text.

Further studies of German Zechstein foraminiferal faunas were made by Paalzow (1935) and Brand (1937).

A revival of interest in Zechstein faunas in general and the microfossils in particular followed the drilling of deep boreholes in the north European plain from the 1950s onwards. Scherp (1962) described 119 foraminiferal species and subspecies, 66 of which were new, from the Zechstein rocks of one borehole in the German Rhineland. Among several Polish works on the subject, the most notable are those by Peryt and Peryt (1977), in which the emphasis was on palaeoenvironments, and Woszczynska (1968) who recorded the vertical distribution of the Zechstein foraminifera in some boreholes in northern Poland. Stratigraphical distribution was also the prime concern of the work by Mikluhu-Maclay and Uharskaja (in Suveizdis, 1975) on the Zechstein faunas in the Baltic area of the U.S.S.R. Both the last-named and the recent Polish studies have been based mostly on random thin-sections in contrast to the earlier British and German work on solid specimens.

The only publications on British Zechstein foraminifera in the last few decades have been limited studies of particular genera or small faunas. They include the work of Cummings (1955, 1956) on Upper Palaeozoic forms and wall-structures, which touched on some Zechstein species, Anderson (1964) on *Aschemonella*, and Pattison (1969) on the foraminifera of the Manchester Marl.

4.2 LOCATION OF IMPORTANT COLLECTIONS

(1) The British Museum (Natural History) has

    (a) the W. K. Parker Collection, including some of the specimens figured by Jones, Parker and Kirkby (1869);

    (b) the H. B. Brady Collection, including type material of *Nodosinella digitata* and *Textularia jonesi*;

    (c) other material which may include specimens figured by Jones (1850) in King's monograph.

(2) The British Geological Survey at Keyworth, Nottingham has Zechstein foraminifera from many localities in the British Isles within the general collections plus:

    (a) some material presented by both T. R. Jones and W. King, and

    (b) specimens figured by Anderson (1964) and Pattison (1969, 1970).

(3) Sunderland Museum has good collections of foraminifera from local Zechstein rocks including the Concretionary Limestone (Z_2) and classic Ford Formation reef localities (Z_1), as well as Marl Slate foraminifera (Z_1) figured by Bell *et al.* (1979).

4.3 STRATIGRAPHIC DIVISIONS

No formally-defined chronostratigraphic subdivisions are in use for the Zechstein, but for several decades it has been recognised (see e.g. Richter-Bernberg, 1955) that Zechstein rocks are the deposits of four or more sedimentary cycles identified as Z_1, Z_2, etc. The most complete Zechstein succession in the British Isles is in Yorkshire and the lithostratigraphic groups proposed for it by Smith *et al.* (1974) (from base to top: Don, Aislaby, Teesside, Staintondale and Eskdale) are the respective products of cycles Z_1 to Z_5. The successions in the areas of eastern England from which most of the known British Zechstein foraminifera have been collected are shown in Fig 4.2 together with the equivalent standard German sequence. Most of the English lithostratigraphical units have been re-named recently (Smith *et al.* 1986) and the new names are shown, with the superseded terms in brackets. All of the

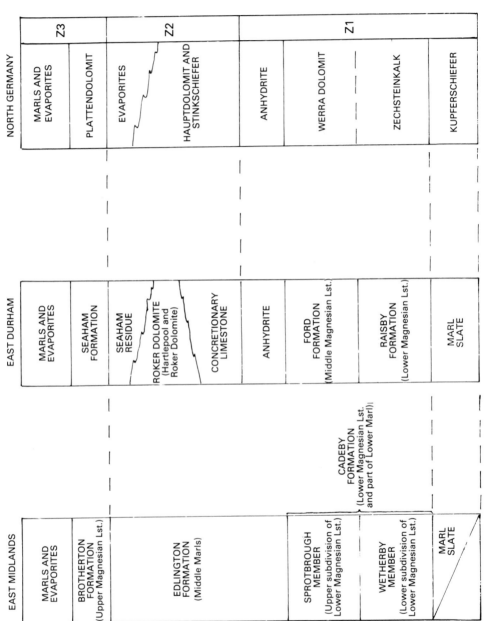

Fig. 4.2 — Permian successions in eastern England correlated with that in north Germany.

Permian formations in northwest England and Northern Ireland to have yielded marine fossils are correlated with Z_1, with the exception of the Belah Dolomite in the Eden Valley which has been tentatively dated as Z_3.

Correlation of Zechstein successions with Permian chronostratigraphic units established elsewhere in the world is hampered by the restricted nature of Zechstein faunas in general and the absence, in particular, of ammonoids and fusulinids. Recent work however, most notably on the conodonts and palynomorphs, suggests that the cycles Z_1 to Z_3 are equivalent to the Abadehian Stage of the Upper Permian (Kozur, 1981).

4.4 DEPOSITIONAL HISTORY, PALAEOECOLOGY AND FAUNAL ASSOCIATIONS

The Zechstein and Bakevellia seas occupied shallow epicontinental basins in dry low latitudes. The consequent high evaporation levels produced a depositional regime dominated by cyclic, evaporitic carbonate–sulphate–chloride sequences. They also resulted in environments generally inimical to marine life. The most favourable conditions were shortly after the initial marine transgression in early Z_1 time, with lesser re-invigorations during the carbonate phases of the Z_2 and Z_3 cycles. The foraminifera, although more tolerant of the variable salinites than many groups, were thus restricted in distribution both in time, to the early carbonate phases of each cycle, and geographically: they were concentrated in marine shelf environments and largely absent from both basin floor and hypersaline marginal lagoons.

The patchiness of Zechstein foraminiferal distribution has been exacerbated by diagenesis. It is quite likely that the dissolution and recrystallisation which accompanied the widespread dolomitisation destroyed all the calcitic organic remains in some beds, and in others they have left only moulds of the original structures (see Plate 4.1, Figs 4, 5 and 7).

Random sampling of Zechstein rocks for foraminifera is, therefore, largely unrewarding and both palaeoecological and biostratigraphical studies suffer as a result. For the former it is difficult to obtain an appreciation of foraminiferal distribution patterns at particular horizons, and virtually impossible in rocks of post-Z_1 age. As to Zechstein biostratigraphy, despite fairly extensive sampling by the writer and others, with existing knowledge it is impossible to produce meaningful foraminiferal species range-charts and that is why they are missing from this section.

The effects of hypersalinity are probably most manifest west of the Pennines where the known foraminiferal faunas, except in the Manchester Marl, are restricted to *Agathammina* and other porcellanous tubular genera.

The Zechstein 1 rocks east of the Pennines have yielded a number of distinct foraminiferal faunas. Foraminifera were absent from the typically euxinic conditions of the Marl Slate sea-bottom but conditions ameliorated enough, periodically and locally, during the deposition of that formation for the introduction of a shelly benthos including foraminifera (Bell, *et al.*, 1979). Strata of what Smith (1970) called the 'standard marine facies' within the Z_1 carbonate rocks contain sparse but relatively varied assemblages, including nodosariids and porcellanous tubular forms (especially *Agathammina*) associated with brachiopods and bryozoa. Reef dolomites, both in the Cadeby Formation patch reefs of Yorkshire and the Ford Limestone reef barrier of east Durham, have abundant *Orthovertella? gordiformis* and *Calcitornella* spp.

The largest and most varied foraminiferal faunas extracted from the British Zechstein have come from the grey marls and argillaceous limestones of the Cadeby Formation in north Nottinghamshire. They include abundant *Agathammina, Ammobaculites,*

Ammodiscus, Hyperammina and a large number of nodosariid forms. In that area there appears to be some faunal variation in response to changes in substrate and depth, as indicated in the following systematic section. It is noteworthy that the north Nottinghamshire fauna is one of several rich and varied Z_1 foraminiferal faunas to have been recorded from more or less argillaceous beds deposited in what were apparently comparable marginal positions in broad bights on the south side of the Zechstein (and Bakevellia) seas. The others are from the Manchester Marl (Pattison, 1969), the Rhineland (Scherp, 1962) and the Fore-Sudetic monocline in southwest Poland (Peryt and Peryt, 1977). It prompts the conjecture that in these areas the restrictive effects of hypersalinity, general in the Zechstein and Bakevellia seas, were offset by the influx of estuarine water.

4.5 STRATIGRAPHIC DISTRIBUTION

Because range charts would be misleading, the distribution of each species is indicated, under the respective descriptions, by the following symbols referring to recordings of the species, or a synonym, from particular formations. An asterisk indicates that the species is known by the writer to be common and/or widespread.

A   Mutterflöz of Poland (Peryt, 1976) — both sub-Kupferschiefer

B   'Upper Rotliegend' of north Germany (Plumhoff, 1966) — both sub-Kupferschiefer

C   Manchester Marl

D   Other Z_1 age deposits of the Bakevellia Sea

E   Cadeby Formation of N. Nottinghamshire in the E. Midlands

F   Raisby Formation of Durham

G   Other Cadeby Formation in eastern England

H   Ford Formation of Durham

I Z_1 rocks of Germany — above Kupferschiefer

J Z_1 rocks of Poland — above Kupferschiefer

K Z_1 rocks in Baltic area of U.S.S.R. (*in* Suveizdis, 1975)

L   Concretionary Limestone

M Z_2 rocks of Germany

N Z_2 rocks of Poland

O Z_2 rocks in Baltic area of U.S.S.R. (*in* Suveizdis, 1975)

P Z_3 rocks of Poland (Woszczynska, 1968)

SUBORDER TEXTULARIINA

Ammobaculites eiseli Paalzow
Plate 4.1, Fig. 6 (MPK 5331) ($\times$ 45); Fig. 18. (thin-sect. PL300) ($\times$ 50). = *Ammobaculites eiseli* (Spandel) Paalzow, 1935. *Haplophragmium eiseli* Spandel, 1898 was a *nomen nudum*. (syn. *A. procera* Sherp, 1962 but not *A. eiseli* (Spandel) Scherp). Description: early coiled part is 1.5–2 times the width of the first of the succeeding chambers and is eccentrically positioned in relation to rectilinear part; chambers in latter are broader than long; sutures distinct; wall, coarse-grained, bound by siliceous cement.
Remarks: rectilinear part has narrower chambers than in *A. eiseli* (Spandel) Scherp, 1962. Distribution: Z_1 (B, ?C, E*, H, I*, K), Z_2 (?O). Deep-water and relatively stenohaline.

Permian

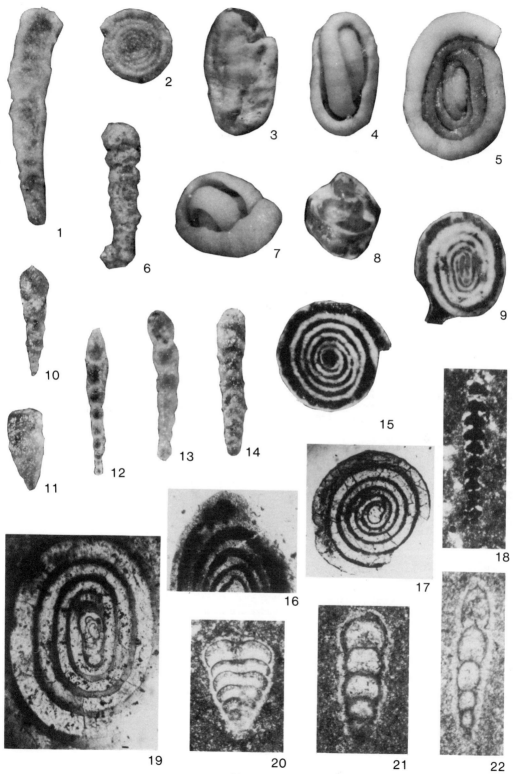

Plate 4.1

Ammodiscus roessleri (Schmid)
Plate 4.1, Fig. 2 (MPK 5329) ($\times$ 45); Fig. 16 (thin-sect. MPK 5336) ($\times$ 130). = *Serpula roessleri* Schmid, 1867. (syns ?*Trochammina incerta* (d'Orbigny) [part] Brady, 1876; *T. bradyna* s.l. Spandel, 1898). Description: discoidal, round to elliptical; proloculus spherical; tubular second chamber wound in evolute plane spiral with up to six whorls, increasing slowly in thickness; aperture is open end of second chamber; finely siliceous.
Remarks: it is a thinner disc and finer grained than *A. robustus* Vangerow, 1962 and lacks that species' involute final whorl. Distribution: Z_1 (A, B, C, E*, G, I*, J, K), Z_2 (?L, M). Associated with low-energy environments.

Hyperammina acuta Scherp, 1962
Plate 4.1, Fig. 1 (MPK 5328) ($\times$ 40). (syns *H. acuta conica* Scherp and *H. crescens* Scherp). Description: an elongate, straight or gently arcuate, flattened cone with a fine-grained, siliceous wall; width increases uniformly from the roughly pointed proximal end; aperture formed by the entire internal diameter of distal end.
Remarks: differs from *H. recta* Scherp, 1962 in its narrow proximal end. Distribution: Z_1 (?C, E*, G, I*, ?J, ?K), Z_2 (?M). Most common in argillaceous rocks.

SUBORDER MILIOLINA

Agathammina milioloides (Paalzow) *non* (Jones, Parker and Kirkby)
Plate 4.1, Fig. 5 (int. mould, MPK 5301) ($\times$ 25); Fig. 9 (MPK 5332) ($\times$ 35); Fig. 19 (thin-sect. MPK 5338, $\times$ 50). = *Glomospira milioloides* (Jones, Parker and Kirkby) [part] Paalzow, 1935. Description: as *A. pusilla* (see below), but early milioline whorls are followed by planispiral coiling. Distribution: Z_1 (D, E*, F, G, H, I, K). Facies as *A. pusilla* but probably more stenohaline.

Agathammina pusilla (Geinitz)
Plate 4.1, Fig. 3 (MPK 5330) ($\times$ 35); Fig. 4 (int. mould, MPK 5300) ($\times$ 25) = *Serpula pusilla* Geinitz, 1848 [part]. (syn. *Trochammina milioloides* Jones, Parker and Kirkby, 1869). Description: elongate, fusiform test consisting of tubular chamber with a hemispherical cross-section wound entirely in a milioline manner; with or without a globular proloculus; calcareous, porcellanous test. Distribution: Z_1 (A, D, E*, F*, G*, H*, I*, J*, K*). Shallow water and euryhaline.

Cyclogyra kinkelini (Spandel)
Plate 4.1, Fig. 15 (MPK 5335) ($\times$ 40); Fig. 17 (thin-sect. MPK 5337) ($\times$ 40). = *Cornuspira kinkelini* Spandel, 1898. (syn. *Trochammina incerta* d'Orbigny. Jones, Parker and Kirkby 1869). Description: calcareous, porcellanous test consisting of a spherical proloculus and a tubular second chamber with a hemispherical cross-section, which increases slowly and uniformly in diameter, wound in an evolute plane spiral with two to seven whorls. Distribution: Z_1 (B, ?D, E*, F, G, H, I, ?J, K). Characteristic of low-energy environments.

Orthovertella? gordiformis (Spandel)
Plate 4.1, Figs 7 (int. mould, MPK 5302) and 8 (MPK 5303) ($\times$ 40). = *Ammodiscus gordiformis* Spandel, 1898. (syns *Trochammina gordialis* Jones, Parker and Kirkby, 1869; *Hemigordius* sp. Pattison, 1969). Description: compact and globular, comprising a calcareous, porcellanous tube with a hemispherical cross-section wound tightly in a streptospiral; tube-width increases uniformly towards aperture but there may be shallow constrictions; suture not incised.
Remarks: more globular and tightly-coiled than *O.? mutabilis* (Scherp, 1962). Distribution: Z_1 (B, C, D, E*, F, G, H*, I, ?J, K). Common in high-energy environments.

SUBORDER LAGENINA

Dentalina permiana Jones 1850
Plate 4.1, Figs 12 (MPK 5305) ($\times$ 45) and 13 (MPK 5306) ($\times$ 55); Fig. 22 (thin-sect. PL291) ($\times$ 50). (syns *D. communis* d'Orbigny [part] Brady, 1876; *D. fallax* Franke and *D. farcimen* Soldani. Paalzow, 1935; *D.* sp. Pattison, 1969). Description: 7–9 barrel to bead-shaped chambers in a slightly-curved arc; sutures distinct and incised; non-radiate aperture in line with test axis; wall is calcareous, hyaline and simple.
Remarks: smaller and has more distinct sutures than *D. lineamargaritarum* Scherp, 1962. Distribution: Z_1 (?B, C, E*, F, G, H, I*, J, K), Z_2 (L). Widespread but especially common in low-energy and ?deeper-water environments.

Geinitzina acuta (Spandel)
Plate 4.1, Fig. 11 (MPK 5333), ($\times$ 60), Fig. 20 (thin-sect. MPK 5339) ($\times$ 50). = *Geinitzella acuta* Spandel, 1898. (syns *Textularia cuneiformis* Jones. and *T. triticum* Jones. Richter, 1855; *T. jonesi* [part] Brady, 1876; *G. postcarbonica* Spandel. Scherp, 1962 and Woszczynska, 1968). Description: uniserial, rectilinear, flattened and centrally-constricted with 8–10 chambers; spherical proloculus; most of other chambers are hour glass-shaped in transverse section, but earliest and latest may lack the central constriction; from front, at least the early part of the test is wedge-shaped but latest part may have parallel sides; aperture is centrally-placed slit on top of last chamber; calcareous, hyaline, double-layered wall.

Remarks: *G. kirkbyi* (Richter, 1861) is narrower and has more incised sutures and central constriction. Distribution: Z_1 (?A, ?B, ?C, E, F, G, H, I, J, K), Z_2 (?L), Z_3 (P). Usually found with other nodosariids but more tolerant of high-energy environments.

Nodosaria geinitzi Reuss, 1854
Plate 4.1, Fig. 10 (MPK 5304) ($\times$ 45); Fig. 21 (thin-sect. PL291) ($\times$ 50). (syns *N. radicula* (Linné) [? part] Brady, 1876; *N. cushmani* Paalzow, 1935; *N. scherpi* Mikluho-Maclay (in Suveizdis, 1975)). Description: globular proloculus; width and length of intermediate chambers about equal; final chamber elongate and may have pouting aperture; sutures distinct and moderately incised; calcareous, hyaline wall which is simple or, more rarely, double-layered. Remarks: longer chambers and more incised sutures than *N. permiana* (Spandel, 1898). Distribution: Z_1 (?A, B, E*, F, H, I*, J, ?K), Z_2 (L). Associated with low-energy and ? deeper-water environments.

Ichtyolaria? cavernula (Paalzow)
Plate 4.1, Fig. 14 (MPK 5334) ($\times$ 50). = *Spandelina cavernula* Paalzow, 1935. (syn. *Monogenerina* n. sp. *a Scherp*, 1962). Description: uniserial, rectilinear, flattened back and front; spherical proloculus; each of other chambers is sharply arched and clasps the preceding one; distinct sutures; aperture round to oval; calcareous, hyaline wall. Distribution: Z_1 (?A, B, E, I, J). One of the many rare members of the nodosariid assemblages in low-energy environments.

All specimens are from the collections of the British Geological Survey, Keyworth, Nottingham although those shown in Figs 18, 21 and 22 are currently missing.

Acknowledgements

The author would like to thank Prof. J. W. Murray and Mr J. Jones of the Geology Department of Exeter University for help with the plate. This paper is published by permission of the Director of the British Geological Survey (NERC).

REFERENCES

Anderson, F. W. 1964. *Aschemonella longicaudata* sp. nov. from the Permian of Derbyshire, England. *Geol. Mag.* **101**, 44–47.

Bell, J., Holden, J., Pettigrew, T. H. and Sedman, K. W. 1979. The Marl Slate and Basal Permian Breccia at Middridge, Co. Durham. *Proc. Yorks geol. Soc.*, **42**, 439–460.

Brady, H. B. 1876. Carboniferous and Permian foraminifera. *Palaeontogr. Soc. Monogr.* 1–166, 12 pls.

Brand, E. 1937. Über Foraminiferen im Zechstein der Wetterau. *Senckenbergiana*, **19**, 375–380.

Cummings, R. H. 1955. *Nodosinella* Brady, 1976, and associated Upper Palaeozoic genera. *Micropalaeontology*, **1**, 221–238.

Cummings, R. H. 1956. Revision of the Upper Palaeozoic textulariid foraminifera. *Micropaleontology*, **2**, 201–242.

Geinitz, H. B. 1848. *Die Versteinerungen des deutschen Zechsteingebirges*. Arnoldische. Dresden and Leipzig.

Howse, R. 1848. A catalogue of the fossils of the Permian system of the counties of Northumberland and Durham. *Trans. Tyneside Nat. Field Club*, **1**, 219–264.

Jones, T. R., Parker, W. K., and Kirkby, J. W. 1869. On the nomenclature of the foraminifera. XIII. The Permian *Trochammina pusilla* and its allies. *Ann. Mag. nat. Hist.*, ser. 4, **4**, 386–392.

King, W. 1848. *A catalogue of the organic remains of the Permian rocks of Northumberland and Durham.* Newcastle upon Tyne, 1–15.

King, W. 1850. A monograph of the Permian fossils of England. *Palaeontogr. Soc. Monogr*, 1–258, 29 pls.

Kozur, H. 1981. The correlation of the uppermost Carboniferous and Permian of middle and western Europe with the marine standard scale. In Depowski, S., Peryt, T. M. and Piatkowski, T. S. (eds) *Proceedings of International Symposium on Central European Permian* pp. 426–450. Wydawnicta Geologiszne, Warszawa.

Paalzow, R. 1935. Die Foraminifera im Zechstein des östlichen Thüringen. *Jb. preuss. geol. Landsanst.*, **56**, 26–45.

Pattison, J. 1969. Some Permian foraminifera from north-western England. *Geol. Mag.*, **106**, 197–205.

Pattison, J. 1970. A review of the marine fossils from the Upper Permian rocks of Northern Ireland and north-west England. *Bull. geol. Surv. Gt Br.*, **32**, 123–165.

Peryt, T. M. 1976. Ingresja morza Turynskiego (Gorny Perm) na obszarze monokliny Przedsudeckiej. *Rocznik Pol. Towarzystwa geol.*, **46**, 455–465.

Peryt, T. M. and Peryt, D. 1977. Otwornice cechsztynskie monokliny Przedsudeckiej i ich paleoekologia. *Rocznik Pol. Towarzystwa geol.*, **47**, 301–326.

Plumhoff, F. 1966. Marines Ober-Rotliegendes (Perm) im Zentrum des nordwest-deutschen Rotliegend-Beckens. Neue Beweise und Folgerungen. *Erdöl. Kohle Erdgas Petrochem.*, **10**, 713–720.

Reuss, A. E. 1854. Ueber Entomostracen und Foraminiferen im Zechstein der Wetterau. *Jber. Wetterauer Ges.* 59–77.

Richter, R. 1855. Aus dem thüringischen Zechstein. *Z. dt. geol. Ges.*, **7**, 523–533.

Richter, R. 1861. In Geinitz, H. B. *Dyas*. Wilhelm Engelmann. Leipzig.

Richter-Bernberg, G. 1955. Stratigraphische Gliederung des deutschen Zechsteins. *Z. dt. geol. Ges.* **105**, 843–854.

Scherp, H. 1962. Foraminiferen aus dem Unteren und Mittleren Zechstein Nordwestdeutschlands. *Fortschr. Geol. Rheinld West.* **6**, 265–330.

Schmid, E. E. 1867. Über die kleineren organischen Formen des Zechsteinkalks von Selters in der Wetterau. *Neues Jb. Geol. Petrefacten-Kunde.* 576–588.

Smith, D. B. 1970. The palaeogeography of the British Zechstein. In *Third Symposium on Salt*, Northern Ohio Geological Society, Cleveland. 20–23.

Smith, D. B., Brunstrom, R. G. W., Manning, P. I., Simpson, S. and Shotton, F. W. 1974. A correlation of Permian rocks in the British Isles. *Spec. Rep. geol. Soc. London* **5**.

Smith, D. B., Harwood, G. M., Pattison, J. and Pettigrew, T. H. 1986. A revised nomenclature for Upper Permian strata in eastern England. *in* Harwood, G. M. and Smith, D. B. (eds). The English Zechstein and related topics, *Geol. Soc. London Spec. Pub.* No. 22, 9–17.

Spandel, E. 1898. *Die Foraminiferen des deutschen Zechsteines*. Nürnberg, 1–15.

Suveizdis, P. (ed.) 1975. Permian deposits of Baltic area (stratigraphy and fauna) *Liet. geol. Moksly. Tyrimo Inst. Tr.* **29**, 1–305, [in Russian with English summary].

Vangerow, E. F. 1962. Über *Ammodiscus* aus dem Zechstein. *Paläont. Z.* **36**, 125–133.

Woszczynska, S. 1968. Wstepne wyniki badan mikrofauny osadow cechsztynu. *Kwart. geol.* **12**, 92–103.

5

Triassic

P. Copestake

5.1 INTRODUCTION AND HISTORY OF INVESTIGATION

British Triassic foraminifera are poorly known compared with those of the underlying Permian and overlying Jurassic. This is largely due to the continental red-bed facies which represents most of the British Triassic and therefore precludes the presence of foraminifera on palaeoenvironmental grounds. The uppermost part of the Triassic, in contrast, is developed in a marginal marine to marine facies represented by the uppermost Mercia Mudstone Group and Penarth Group (equivalent to the 'Rhaetic' of previous British nomenclature) of late Rhaetian age (Fig. 5.3). It is only in this part of the Triassic that foraminifera have thus far been recovered in Britain, and thus the discussions in this chapter are largely confined to this interval.

Published accounts of British Triassic foraminifera are few and these are from the uppermost Triassic Penarth Group. The most significant paper is that of Chapman (1894) who described a lower Penarth Group (lowermost Westbury Formation) fauna from Wedmore (Somerset) which is now known to be essentially typical of the Westbury Formation of southwest Britain. Since Chapman's time, Triassic foraminifera in Britain have been neglected, with only brief notes recording occurrences in the Rhaetian, including Barnard (1950) from Dorset, Ivimey-Cook (1965, 1974) from Banbury (Oxfordshire) and South Wales and Banner et al. (1971), also from South Wales. These later records are of limited foraminiferal assemblages in the upper Penarth Group and all report the presence of species which range upwards into the Lower Jurassic, such as early members of the *Lingulina tenera* plexus and *Eoguttulina liassica*. Of these papers, only Banner et al. (1971) actually figured any specimens from the Rhaetian (uppermost Langport Member and basal Lias).

Hallam and El Shaarawy (1982) produced a species list of foraminifera from southwest Britain in a paper discussing palaeoenvironmental comparisons between the Alpine and British Triassic sequences. The foraminifera are only very briefly discussed and their statements, firstly that the Westbury Formation is totally barren of foraminifera, and secondly that the richest Rhaetian assemblages occur in the Cotham Member (at Lavernock, South Wales) are at odds with the author's data from these units (see below). Furthermore, they omitted to study the Langport Member (which may explain their statment regarding the Cotham Member). Nevertheless, the British foraminiferal species list of Hallam and El Shaarawy compared with that from the Alpine Rhaetian clearly showed the dissimilarity of the two assemblages, the latter being rich in calcareous foraminifera. This is partly attributed by these authors to reduced salinity in the British Rhaetian compared to the Alpine region. Climatic factors must also have been important, however, with the British Rhaetian probably representing a cooler, clastic-dominated environment in contrast to the carbonate-dominated facies of the Alpine Rhaetian.

Although marine Rhaetian sediments occur also in the English Midlands, Yorkshire, Western Scotland and Northern Ireland, it is only from southwest Britain that published accounts or unpublished data are available. No published data exist concerning Rhaetian foraminifera from the offshore Mesozoic basins of the northwest European continental shelf, although potential must exist for the recovery of foraminifera in areas such as the Celtic Sea, Fastnet and southwest Approaches Basins where Penarth Group sequences similar to those in southwest Britain are known to occur. Such possibilities are fewer offshore east of Britain, where the equivalents of the onshore marine uppermost Triassic are either absent or developed in non-marine facies (see below for fuller discussion).

Onshore eastern Scotland (Dunrobin Pier Section), as in the offshore North Sea, the Late Triassic is non-marine and devoid of foraminifera. The western Scottish (Hebridean) Rhaetian sections are sandier than more typical sections further south, but have nevertheless been assigned to the Westbury and Lilstock Formations by Warrington et al. (1980). No foraminifera have been described from the Scottish Rhaetian, though foraminifera must potentially be present in western Scotland.

In the northern hemisphere Late Triassic foraminiferal assemblages appear to fall into three provinces, a Tethyan realm (incorporating southern Europe, the Mediterranean region and the Middle East), a more northerly Boreal realm (including northwest Europe, e.g. Britain and West Germany) and a northern Boreal, or Arctic realm (including Arctic Canada and Alaska).

Tethyan Triassic foraminifera have been described in many papers, including Oberhauser (1960, Ladinian and Carnian of Austria and Iran), Kristan-Tollmann (1964, Norian–Rhaetian of Austria), Styk (1979, Scythian–Rhaetian of the Polish Carpathians) and Zaninetti (1976, a summary of non-nodosariid assemblages from the Tethyan realm). These described Tethyan faunas are of warmer water, probable normal marine origin and bear no resemblance to the British or other north European Triassic foraminiferal associations.

Foraminifera from the Boreal region, including Britain, West Germany, France and Sweden have been generally neglected. The most significant account from this area, other than the British papers already mentioned, is that of Will (1969). This work describes an assemblage similar to that of the Westbury Formation and Cotham Member of southwest Britain, although it is less diverse. Common to the assemblages of the two areas are the dominantly agglutinating associations, including *Glomospira subparvula* and *Trochammina*, with the calcareous form *Eoguttulina*. In Britain this assemblage is characteristic of the lower Penarth Group (Westbury Formation and Cotham Member). However, the upper Penarth Group (Langport Member) contains a markedly richer fauna of predominantly calcareous foraminifera, the equivalent of which was not described by Will (1969), owing to the development of non-marine facies at the equivalent level in West Germany. Other smaller papers on Rhaetian foraminifera from West Germany are by Wicher (1951), Bartenstein (1962) and Ziegler (1964), again describing the lower Penarth Group type assemblage of agglutinating foraminifera. Rare foraminifera have been reported in the Rhaetian of Sweden. Brotzen (1950) reports an assemblage comparable to that of Chapman (1894) and that described herein from the British Westbury Formation and Cotham Member.

Foraminiferal records from the Triassic deposits north of Britain are limited, but differ markedly from those of northwest Europe and appear to belong to a northern or Arctic faunal province.

Souaya (1976) has described the Triassic foraminiferal sequence in oil exploration wells from Arctic Canada (Sverdrup Basin), characterised by a microfauna dominated by long-ranging agglutinating foraminifera (*Ammodiscus*, *Reophax* and abundant *Ammobaculites*). This fauna differs from the Late Triassic assemblages from Arctic North America (Alaska) described by Tappan (1955) in the abundance of agglutinating foraminifera. Tappan's assemblages contrast in being dominated by nodosariid foraminifera, though both areas do contain *Lingulina alaskensis* Tappan.

Until recently Triassic foraminifera were unknown south of the Tethyan carbonate area. Well preserved and diverse assemblages have now been described from the Australia–New Zealand region, however. Strong (1984) described a rich fauna of probable Norian age from New Zealand, which, although largely endemic in nature, contains species previously known from the Late Triassic of central and southern Europe (including *Grillina grilli* Kristan-Tollmann, *Frondicularia rhaetica* Kristan-Tollmann and *F. brizoides* Gerke). Several co-occurring species Strong (1984) refers to Lower Jurassic marker taxa such as '*Astacolus dorbignyi*' and '*Palmula deslongchampsi*' though it is evident from the figures and descriptions provided that the New Zealand material is incorrectly assigned to these two species. In addition, Strong's (1984) figures of '*Lingulina tenera*' show circular, frondicularian apertures and should not be referred to this species. From the Northwest Shelf, Western Australia, Apthorpe and Heath (1981) and Heath and Apthorpe (1986) respectively recorded Late Triassic (Norian–Rhaetian) and Middle Triassic (Anisian) assemblages. The latter assemblage is rich and well preserved and compares with that described by Goel *et al.*, (1981) from Spiti in the Himalayas. Early forms of *Ophthalmidium* are recorded and *Lingulina borealis* Tappan is dominant, a species previously described from the Late Triassic of Alaska (Tappan, 1955). The Northwest Shelf faunas also contain the earliest known forms similar to *Lingulina tenera*, plus forms close to *Frondicularia terquemi*, species which first appear later in northwest Europe, around the Triassic/Jurassic boundary.

The data in this chapter are from previously unpublished research by the author, integrated with the limited amount of available published information on British Triassic foraminifera (e.g. Chapman 1894; Ivimey-Cook 1974; Banner *et al.*, 1971).

5.2 LOCATION OF IMPORTANT COLLECTIONS

Owing to the general lack of previous study of Triassic foraminifera in Britain, substantial collections have yet to be made. The whereabouts of Chapman's (1894) collection is unknown.

(1) British Geological Survey, Keyworth: Collections of Ivimey-Cook from Rhaetian of South Wales (unpublished thesis) and Warwickshire (1965).

(2) BP Exploration, Glasgow:
Collection of present author from Rhaetian of South Wales, Somerset and Dorset.

5.3 STRATIGRAPHIC DIVISIONS

The Triassic of Europe is developed in two very different facies, a north European 'Germanic' sequence, largely of continental origin (red beds), and a predominatly marine, carbonate-dominated Tethyan or Alpine facies, developed in central and southern Europe.

The lack of macrofossils in the Germanic facies has led to a subdivision based mainly on lithostratigraphy, while the Tethyan Triassic is subdivided with more emphasis on biostratigraphy. It is upon the Tethyan area that the standard Triassic stages Scythian, Anisian, Ladinian, Carnian, Norian and Rhaetian have been founded. In the Germanic area, including Britain, a broad three-fold Triassic subdivision into 'Bunter' 'Muschelkalk' and 'Keuper' was used for many years. These terms have now been abandoned in Britain and a new nomenclature proposed which attempts to correlate with the standard Triassic stages (Warrington et al., 1980). These correlations are based almost entirely on palynology (numerous papers summarised in Warrington et al., 1980) and there are still many outstanding problems of correlation of the British Triassic with the standard stages.

A summary of British Triassic subdivisions and their correlation is shown in Fig. 5.1. The locations of these sections are shown in Fig. 5.2.

The Triassic is currently viewed as spanning a period of 45 million years (Forster and Warrington, 1985). The rocks of the uppermost British Triassic were formerly known as 'Rhaetic' (e.g. Richardson, 1905, 1906 and 1911), though use of this term in Britain is now discontinued as proposed by George et al., (1969), Pearson (1970) and Warrington et al., (1980). This is because of the confusion brought about by the similarity of 'Rhaetic' to the stage name Rhaetian (the former being used both in lithostratigraphic and chronostratigraphic senses), plus the fact that palynological evidence, in particular, indicates that the British 'Rhaetic' succession represents only the later part of Rhaetian time.

As a consequence the British Triassic Working Group (Warrington et al., 1980), advocated a set of new lithostratigraphic terms (previously used informally by Kent, 1970) for the section formerly known as 'Rhaetic'. The Penarth Group was established to embrace the former 'Rhaetic', including all rocks between the Mercia Mudstone Group and the basal Lias 'Paper Shale' of Richardson (1911). The group comprises a lower Westbury Formation (equivalent to the Westbury Beds of Richardson, 1911) and upper Lilstock Formation, the latter comprising a Cotham Member (equating to Richardson's, 1911, Cotham Beds) and Langport Member (equivalent to the 'White Lias', Langport Beds and Watchet Beds of Richardson, 1911). Fig 5.3 compares current uppermost Triassic nomenclature with that used previously.

Fig. 5.1 — Development and correlation of Triassic rocks in Britain. Data sources: Absolute ages from Forster and Warrington (1985); (1), (2), (3), (4), (9) Warrington et al., (1980); (5) Harrison (1971), Warrington (1971), Ivimey-Cook (1974); (6) Green and Melville (1956), Warrington et al., (1980); (7) Gaunt et al., (1980); (8) Warrington et al., (1980), Batten et al., (1986); (10) Deegan and Scull (1977), Fisher (1984), with some modifications herein; (11) Rhys (1974), Fisher (1984); (12) Bennet et al., (1985).
a = Pre-Planorbis Beds; b = Dunrobin Pier Conglomerate; W = Westbury Formation; L = Lilstock Formation.

Sec. 5.3] Stratigraphic Divisions

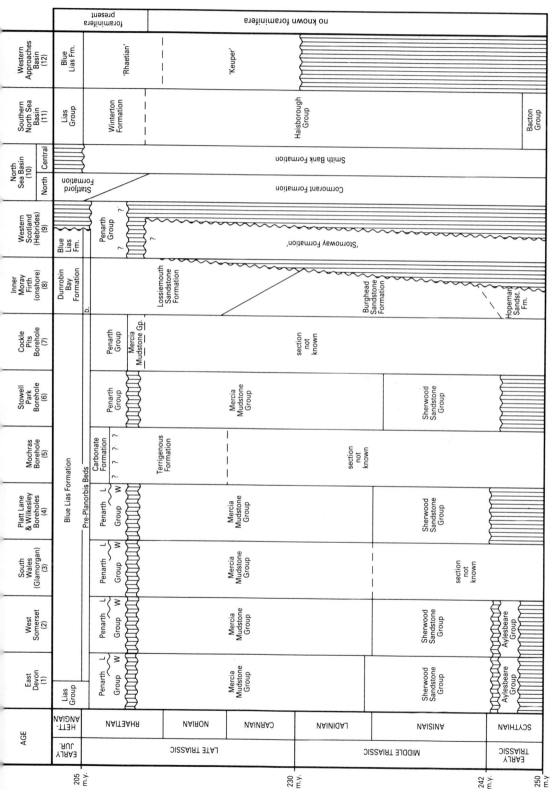

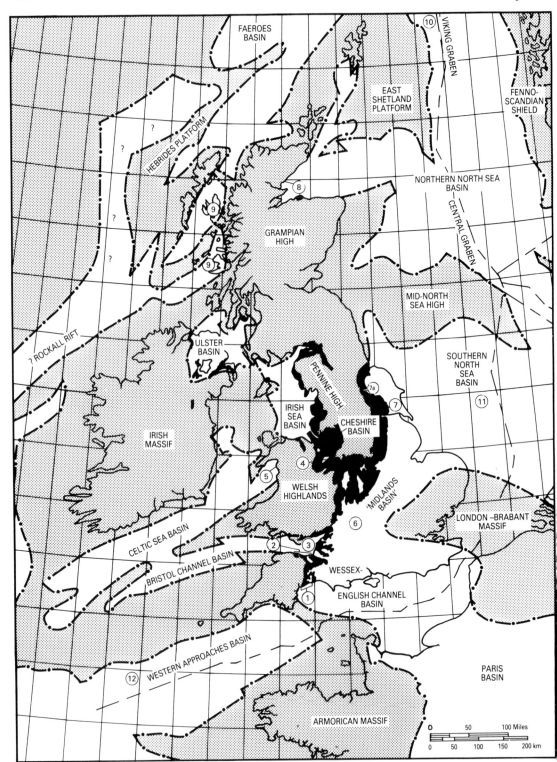

5.3.1 The Triassic/Jurassic boundary

Placement of the Triassic/Jurassic boundary in British Penarth Group-lowermost Lias sequences is difficult owing to the absence of Triassic ammonites in Britain. Following the recommendations of Warrington et al., (1980) and Cope et al., (1980), most British stratigraphers would now accept the boundary position at the first appearance of the Jurassic ammonite *Psiloceras*. This is a flood abundance horizon in Britain, being particularly well displayed in west Somerset (Palmer's, 1972, bed 21) and is consequently assumed to be a synchronous event. This recommendation superseded the previous boundary placement of George et al., (1969) at the lithostratigraphic horizon of the basal Lias Paper Shale (of Richardson, 1911). A further boundary position has been recently argued by Poole (1979) at the base of the Langport Member (White Lias), believing this to mark a significant non-marine/marine and faunal change (but see discussion of palaeoenvironments below).

Whilst accepting that a biostratigraphic criterion should be the basis for a system boundary, the base of *Psiloceras* unfortunately does not coincide with any significant micropalaeontological change. This makes the identification of this boundary difficult when utilising micropalaeontological criteria in the absence of ammonite information, as in subsurface well studies.

The first clearly defined incoming of Jurassic foraminifera (including *Reinholdella? planiconvexa* and abundant *Eoguttulina liassica*) occurs at the base of the Langport Member. This incursion is seen throughout southwest Britain, from the Devon coast to South Wales. Above this, a more pronounced Jurassic influx (including the first appearances of *L. tenera tenera* and *L. tenera collenoti*) occurs in the uppermost Langport Member. These latter taxa subsequently become abundant in the Pre-Planorbis Beds. These events in the Langport Member together comprise significant incomings of Jurassic foraminifera below the first ammonites, which is the suggested boundary of Cope et al., (1980). In fact no observable change in the foraminiferal faunas occurs around the appearance of *Psiloceras*. The lower of the two foraminiferal influxes does fall close to the Triassic/Jurassic boundary as suggested by Poole (1979), that is at the base of the Langport Member. This is also close to the boundary as favoured by Orbell (1973) on palynological evidence, within the Cotham Member. The upper of the two foraminiferal influxes (appearance of *L. tenera tenera* and *L. tenera collenoti*), on the other hand, occurs shortly beneath the basal Lias Paper Shale, which is the boundary position proposed by George et al., (1969).

Thus, there is biostratigraphic evidence in favour of a boundary placement significantly lower than the first appearance of Jurassic ammonites either at the base of the Langport Member, or near the top of the Langport Member. This includes the proposed type area for the boundary, the Somerset coast around Watchet (George et al., 1969).

Fig. 5.2 — Distribution of Triassic rocks in the British area and locations of successions displayed on Fig. 5.1. See text for details of sections studied for foraminifera.
(1) East Devon coast (Culverhole Point, Pinhay Bay, Tolcis Quarry); (2) West Somerset coast (Watchet, St Audrie's Slip); (3) South Wales (Lavernock Point, Glamorgan); (4) Platt Lane and Wilkesley Boreholes (Shropshire and Cheshire); (5) Mochras Borehole, Gwynedd, Wales; (6) Stowell Park Borehole, Gloucestershire; (7) Cockle Pits Borehole, North Humberside; (7a) Dibdale stream section, Northallerton, Yorkshire: (8) Inner Moray Firth Basin; (9) Western Scotland (Hebrides); (10) Northern North Sea Basin (Viking Graben, Central Graben); (11) Southern North Sea Basin; (12) Western Approaches and Celtic Sea Basins.
Triassic outcrop in black. Areas of limited or non deposition of Triassic rocks stippled. Mesozoic basins without ornament. Grid system equates to offshore exploration quadrants.

AGE (Warrington et al. 1980)		CURRENT LITHOSTRATIGRAPHIC TERMINOLOGY (Warrington et al. 1980; Whittaker and Green 1983)					PREVIOUS LITHOSTRATIGRAPHIC TERMINOLOGY (Richardson, 1911)			
EARLY JURASSIC	HETTANGIAN	planorbis Zone	BLUE LIAS	PRE-PLANORBIS BEDS	'MARLY BEDS'		LIAS	'OSTREA' BEDS / PAPER SHALE	WATCHET BEDS	
		LIAS			LANGPORT MEMBER	'WHITE LIAS LIMESTONE'			SUN BED	WHITE LIAS
									LANGPORT BEDS	
LATE TRIASSIC	RHAETIAN	PENARTH GROUP		LILSTOCK FORMATION	COTHAM MEMBER		COTHAM BEDS			
				WESTBURY FORMATION			WESTBURY BEDS			
		MERCIA MUDSTONE	BLUE ANCHOR FORMATION				SULLY BEDS			

(RHAETIC; UPPER / LOWER)

In proposing the appearance of *Psiloceras*, as a new definition of the boundary, Cope et al., (1980) assumed the sudden influx of this ammonite throughout Britain to be a synchronous event. Within Dorset, however, between the coast and the Winterborne Kingston Borehole, there is evidence that this event is time-transgressive (Ivimey-Cook, 1982). A similar diachroneity is also apparent in South Wales (Waters and Lawrence, 1987). This suggests that an evaluation of the total biota rather than only the ammonites should be undertaken when considering placement of the Triassic/Jurassic boundary in Britain. Pending such discussions, however, for the purposes of this publication, the boundary is retained at the base of *Psiloceras*. As a consequence, several more typically Jurassic taxa will be described and illustrated herein as they also characterise the uppermost Triassic in Britain so defined.

5.4 DEPOSITIONAL HISTORY; PALAEOECOLOGY; FAUNAL ASSOCIATIONS

The present state of knowledge suggests that foraminifera first appear in the British Triassic near the top of the Mercia Mudstone Group (Blue Anchor Formation). At this level rare specimens of *Glomospira subparvula*, *Eoguttulina liassica* and *Dentalina pseudocommunis* occur in west Somerset (St. Audrie's Slip).

The presence of foraminifera at this level supports Mayall's (1981) contention that marine conditions were first established in the British Triassic in the uppermost Mercia Mudstone Group (based on lithofacies and macrofaunal evidence). There is no strong evidence in Britain of marine conditions below this level. This contrasts with other parts of the world where marine Lower and Middle Triassic is well developed, and the Triassic sea level maximum is in the Norian (Haq et al. 1987).

The base of the Westbury Formation is generally considered to mark the main Rhaetian transgression in Britain and the rest of northwest Europe. The Westbury Formation in southwest Britain is characterised by a low diversity, exclusively agglutinating foraminiferal assemblage dominated by *Glomospirella* sp. 1 with subordinate *Ammodiscus auriculus*, *Glomospira subparvula* and rarer *Trochammina squamosa*, *Reophax helvetica* and *Bathysiphon* spp. The dominance of agglutinating foraminifera indicates a marine, dysaerobic palaeoenvironment for the black shales of the formation. Alternation of the dark shales with thin, ripple-bedded sandstones and the presence of disarticulated bivalve shells on limestone bedding planes are indicative of higher energy periods alternating with periods of quiet, but not necessarily deep, water sedimentation (Whittaker and Green, 1983). The absence of ammonites from the formation has been thought to be due to reduced salinity (Hallam and El Shaarawy, 1982) though there is no direct evidence of this on the basis of the foraminifera.

Comparability of the British Westbury Formation foraminiferal assemblages with those of equivalent beds in West Germany

Fig. 5.3 — Comparison of currently used lithostratigraphy (Warrington et al., 1980, Whittaker and Green, 1983) in the uppermost Triassic of Britain with that in previous usage (Richardson, 1911, based on Somerset). 'Marly beds' is an informal term introduced by Whittaker and Green (1983) to equate to Richardson's (1911) Watchet Beds. Placement of part of Richardson's Watchet Beds in the Pre-Planorbis Beds was suggested by Whittaker (1978) (1) equates to base Jurassic suggested by George et al., (1969). (2) equates to base Jurassic suggested by Poole (1979). (3) equates to base Jurassic suggested by Orbell (1973). Note that two significant changes occur in the foraminiferal faunas, one at position (2) and one just beneath (1) (see p. 103 for discussion).

('Contortaschichten'), Sweden and possibly Holland indicates that conditions of deposition of this unit became uniform over a wide area as a result of the initial Rhaetian transgression. The widespread nature of this event is suggestive of eustatic sea level rise rather than tectonic control. It is possible that this represents the mid-Rhaetian sea level rise proposed by Haq et al. (1987) (cycle 4.1), with the black shale Westbury Formation facies representing the maximum rate of sea level rise in the mid-part of the cycle.

The junction of the Cotham Member and Westbury Formation is sharp and irregular, indicating a marked change to higher energy conditions. Erosion at the contact, with overlying intraformational deformation and slumping may indicate a tectonic control on the environmental change. Polygonal desiccation cracks and ripple marks indicate shallow water, high-energy deposition with emergence accompanying this environmental change. The dominant sediment is grey calcareous mudstone, with subordinate limestones, siltstones and sandstones.

Although a significant environmental break heralded Cotham Member deposition, on the evidence of the foraminiferal assemblages, which essentially represent a continuation of those of the Westbury Formation, similar conditions of deposition to the latter formation appear to have been re-established above the basal beds of the Cotham Member. At Watchet, the assemblages are characterised by common to abundant *Glomospira subparvula* and *Glomospirella* sp. 1 with *Ammobaculites* sp. 1. At St Audrie's Slip, *Bathysiphon* is abundant. In contrast, the Cotham Member at Lavernock contains common *Eoguttulina liassica*. As the latter species is characteristic of the Langport Member in southwest Britain, the so-called Cotham Member sediments of Lavernock may prove, on foraminiferal evidence, to be Langport Member sediments.

The overlying Langport Member is developed as grey limestones with interbedded mudstones in Somerset and South Wales. Its Devon–Dorset correlative, the White Lias, is a massive unit of porcellaneous white limestones with minor shales and mudstones. These dominantly calcareous sequences are considered to be of warm, shallow water, lagoonal origin (Whittaker and Green, 1983). Throughout southwest Britain this facies is dominated by the abundance of a single, calcareous species, *Eoguttulina liassica*, associated with rare nodosariid foraminifera. This microfauna, showing high abundance of a single species, is characteristic of environmental stress and is consistent with the greater than normal marine salinity (i.e. hypersaliity) proposed for the White Lias by Hallam and El Shaarawy (1982). It is also clear from the foraminifera that the Langport Member was deposited with increased sea bed oxygenation relative to the Cotham Member and underlying Westbury Formation which are both dominated by agglutinating foraminifera. Further emergence and desiccation is indicated in later, Langport Member (White Lias) times by the bored, sun-cracked upper limestone of the unit (the 'Sun Bed' of Somerset). The White Lias appears to represent shallowing and could be interpreted as regressive carbonate build-up formed at a time of sea-level high stand. It may correspond to the top of cycle 4.1 of Haq et al. (1987).

The upper part of the Langport Member is represented by silty mudstones (the 'marly beds' of Whittaker and Green, 1983) in southwest Britain. These are equivalent to Richardson's (1911) Watchet Beds and are considered to be of deeper water origin than the preceding Langport Member lagoonal limestone sequence (Whittaker and Green, 1983). The foraminiferal faunas confirm this view, particularly in the uppermost part where the appearance of *Lingulina tenera tenera* occurs in Somerset and South Wales, accompanied by the appearance of *L. tenera collenoti*. Shortly above, at the base of the Blue Lias (sensu Whittaker, 1978, revised downwards from Richardson, 1911) occurs a marked paper shale unit. This, which incorporates Richardon's (1911) "Paper Shale" in its upper part, is generally considered to represent a transgressive feature at the base of the British Lias sequence (e.g. Whittaker and Green, 1983, p. 44). On foraminiferal evidence, this 'basal Jurassic' transgression is considered to have been initiated slightly earlier, in the underlying 'marly beds'. The Paper Shale unit may represent an increase in rate of the sea level rise, which caused temporary sea bed oxygen reduction and lack of benthos (including foraminifera). Above the Paper Shale, however, the foraminiferal faunas return, and become progressively more diverse into the Jurassic as the transgression proceeded. The basal Lias in Britain does not contain ammonites and has been referred to as the Pre-Planorbis Beds. The foraminiferal assemblages of this unit are abundant and dominated by *E. liassica*, *L. tenera tenera* and *L. tenera collenoti*. It is notable that there is no change in the foraminiferal associations accompanying the appearance of the earliest Jurassic ammonite *Psiloceras planorbis*. The latter appears as one of a succession of faunal events within a gradually deepening transgressive sequence initiated in the 'marly beds'.

While this deepening event in the uppermost Penarth Group to basal Jurassic is a major event in Europe, and probably of eustatic origin, it does not appear as such on the eustatic curve of Haq *et al*. (1987). On the latter, the Triassic/Jurassic boundary region is depicted as a period of sea level fall following a late Rhaetian high stand. The geological evidence from Britain clearly opposes this interpretation.

5.5 GEOGRAPHIC DISTRIBUTION

No stratigraphically based accounts have been published on British Triassic foraminifera. Although Rhaetian uppermost Mercia Mudstone Group and Penarth Group sediments are widespread in Britain, many sections remain unstudied micropalaeontologically. The data presented herein are based on work by the author on the following sections:

(1) Culverhole Point, Devon (Cotham Member, Westbury Formation)
(2) Pinhay Bay, Devon ('White Lias' (Langport Member))
(3) Tolcis Quarry, Axminster, Devon ('White Lias', Pre-Planorbis Beds)
(4) Watchet (Doniford Bay, east of Watchet Harbour), Somerset (Westbury Formation, Lilstock Formation, Pre-Planorbis Beds)
(5) St Audrie's Slip, Somerset (Westbury Formation, Lilstock Formation, Pre-Planorbis Beds)
(6) Lavernock Point, Glamorgan, South Wales (Westbury Formation, Lilstock Formation, Pre-Planorbis Beds)
(7) Platt Lane Borehole (BGS), Shropshire (Pre-Planorbis Beds)
(8) Wilkesley Borehole (BGS), Cheshire (Pre-Planorbis Beds).
(9) Cockle Pits Borehole (BGS), Humberside (Pre-Planorbis Beds)
(10) Dibdale Stream Section, Northallerton, Yorkshire (Westbury Formation, Lilstock Formation, Pre-Planorbis Beds).

The age ranges of these sections are summarised in Fig. 5.4. Information on the foraminifera in the Platt Lane, Wilkesley and Cockle Pits boreholes was obtained by study of prepared microfaunal slides in the collection of the British Geological Survey (BGS) (Keyworth).

This represents a spread of data points between the Devon coast area and Yorkshire; however, data from complete Penarth Group sequences are only available from southwest Britain (Somerset and Glamorgan coasts, sections 4–6). The other sections represent only parts of the Penarth Group – basal Lias interval (see Fig. 5.4). For instance, for the BGS Platt Lane, Wilkesley and Cockle Pits boreholes, although penetrating complete Penarth Group sequences, BGS

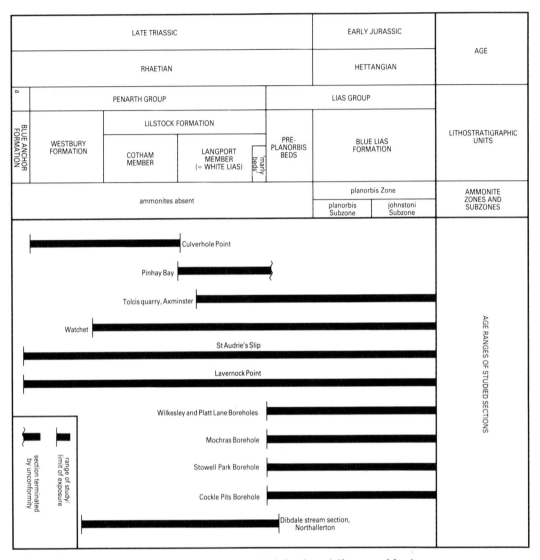

Fig. 5.4 — Age ranges of British Triassic sections studied for foraminifera. a = Mercia Mudstone Group.

faunal slides were only available for study from restricted parts of the latest Triassic section.

The Dibdale stream section, Northallerton, shows a succession from Westbury Formation through the the basal Lias (Fox-Strangways et al., 1886), but it unfortunately proved to be microfaunally barren.

Rhaetian sections are available between inland Somerset and Lincolnshire, including many BGS boreholes, but no studies have yet been undertaken. The same is true of Penarth Group sequences in western Scotland (e.g. Mull) and Northern Ireland, although sections in the Larne (County Antrim) area are reported to be barren of foraminifera (I. Boomer, pers. comm.).

Offshore west of Britain, there are no published accounts of late Triassic foraminifera. Probable Rhaetian sections are present in the Fastnet Basin (offshore southwest Ireland), but are reported to be barren of foraminifera (Ainsworth and Horton, 1986; Ainsworth et al., in press). As the latter authors do report typical Penarth Group ostracod assemblages, the apparent lack of foraminifera is unusual.

In the Western Approaches Basin the Penarth Group is also present. Work by the author on unreleased exploration wells indicates the presence of probable Lilstock Formation to Lias sequences, but with only rare foraminifera in the Penarth Group. Abundances of *Eoguttulina liassica* above the White Lias are probably indicative of the basal Lias Pre-Planorbis Beds (e.g. well 72/10-1a, Bennet et al., 1985).

In the southern North Sea, typical Penarth Group sediments referable to the Westbury Formation and Cotham Member are present (Kent, 1967), later to be included in the Winterton Formation of Rhys (1974).

In eastern Scotland onshore Inner Moray Firth (Sutherland), a non-marine sequence of Lossiemouth Sandstone Formation overlain by Dunrobin Pier Conglomerate is exposed at Dunrobin Castle Pier (Warrington et al., 1980) which represents the edge of the largely offshore Inner Moray Firth Basin; the conglomerate is dated palynologically as Rhaetian, equivalent in age to the Pre-Planorbis Beds of England and Wales (Batten et al., 1986).

This non-marine sequence is also developed offshore in the Inner Moray Firth Basin, for example in the Beatrice Field (Block 11/30). No foraminifera occur in these sections, although non-marine ostracods such as *Darwinula* are present.

In most of the central and northern North Sea (Central Graben, Outer Moray Firth and south Viking Graben) late Triassic foraminifera are absent due to unfavourable facies or lack of section. In the north Viking Graben, sediments representing the Rhaetian–Hettangian (dated palynologically) are developed as the non-marine Statfjord Formation of Deegan and Scull (1977), and are thus devoid of foraminifera.

In summary, it can be seen that although potential foraminifera-bearing sections referable to the Penarth Group are widespread in Britain onshore and offshore west of Britain, the majority of sections are as yet unstudied for foraminifera. Offshore east of Britain, equivalent sections in the North Sea are either absent or developed in non-marine facies.

5.6 STRATIGRAPHIC DISTRIBUTION

The nature of the vertical succession of foraminifera in the British Rhaetian is clearly related to the changing sequence of palaeoenvironments outlined previously. Nevertheless, these microfaunal changes appear to be correlatable over considerable distances and are, therefore, utilisable biostratigraphically (Fig. 5.5). Furthermore, as each lithostratigraphic unit of the Penarth Group to basal Lias can be characterised by distinctive foraminiferal assemblages, these associations may be used in a purely lithostratigraphic sense to aid the definition of rock units. The latter fact is of particular application in subsurface boreholes and wells where samples are often too small to allow recovery of good macrofossil assemblages.

The range charts presented in this chapter (Figs 5.6, 5.7) are based on all the sections listed on p. 107, but with emphasis on sections from the Somerset coast (Watchet and St Audrie's Slip) and Glamorgan (Lavernock Point).

A clear vertical sequence of foraminifera is observable in southwest Britain. The uppermost Mercia Mudstone Group contains only rare foraminifera (e.g. *Glomospira subparvula*). The Westbury Formation and Cotham Member are dominated by agglutinating foraminifera (*Glomospirella* sp. 1, *Glomospira subparvula*, *Bathysiphon* spp.). The overlying Langport Member (and equivalent White Lias), in contrast, contains abundant *Eoguttulina liassica*, with rare, typically Jurassic, species such as *Reinholdella? planiconvexa*.

A stratigraphically significant change occurring at the top of the Langport member is the appearance of *Lingulina tenera collenoti* (defining the uppermost Triassic to basal Jurassic JF1 foraminifera zone see Chapter 6), which becomes abundant in the overlying basal Lias Pre-Planorbis Beds, together with abundant *L. tenera tenera*. As disscussed above, this change occurs beneath the currently accepted Triassic/Jurassic boundary defined on the incoming of *Psiloceras*. This sequence occurs in Somerset and Glamorgan and is believed to be of correlative value at least in this region. The first appearance of *L. tenera* occurs shortly below Richardson's (1911) Paper Shale (revised by Whittaker, 1978 in Somerset) and suggests that the latter is a correlative marker bed. This shows that the base of the Lias Group, as defined by this shale, is a chronostratigraphic horizon.

The usefulness of this foraminiferal sequence for correlation beyond southwest Britain is difficult to assess owing to the general lack of published accounts of foraminifera from other British sequences or from adjacent areas such as mainland northwest Europe.

No data are available from France, but a number of papers from West Germany (Bartenstein, 1962; Ziegler, 1964; Will, 1969), Sweden (Brotzen, 1950) and Holland (Ten Dam, 1945) record foraminifera from the black shale facies 'Contortaschichten' which are directly comparable to those from the British Westbury Formation and Cotham Member. The distinctive agglutinating fauna typified by *Glomospira subparvula* and *Ammodiscus auriculus*, which is common to these areas, suggests that the Westbury Formation and Cotham Member may correlate with the 'Middle Rhaetian' of West Germany and the 'Rhat' of Sweden.

Correlation of the upper Penarth Group (Lilstock Formation) with West Germany and Sweden is not possible, owing to the development of non-marine facies in the latter two areas. The progressive sequence of marine foraminifera seen in the British Lilstock Formation to basal Lias has not been recorded outside the UK, but this is as likely to be due to lack of study as lack of suitable facies.

The relict Palaeozoic taxa recorded by Chapman (1894) from the lower Westbury Formation of Somerset have not been found in subsequent studies of British Rhaetian foraminifera. Chapman reported the presence of Carboniferous genera, such as *Stacheia* and *Nodosinella*, and also allocated specimens of *Haplophragmoides* and *Ammodiscus* to Carboniferous species. The present author's studies have not recovered representatives of *Stacheia* or *Nodosinella* although it is notable that Wicher and Bartenstein (1962) also mention the presence of *Stacheia* in the Triassic of Germany.

Chapman also reported the presence of species of the Carboniferous–Permian genus *Agathammina* (although he allocated the species to *Ammodiscus*), including *A. pusilla* (Geinitz, 1848) and *A. milioloides* (Jones,

---

Fig. 5.5 — Studied Penarth Group sections in Britain and their correlation using foraminiferal events.

Fig. 5.6 (p. 112) — Stratigraphic distribution of foraminifera in the Upper Triassic–lowest Jurassic of Britain. a = JF2 Zone; b = 'marly beds' of Whittaker and Green (1983); c = Mercia Mudstone Group.

Fig. 5.7 (p. 113) — Range summary of British Upper Triassic–lowest Jurassic foraminifera arranged in order of last occurrence.
a = JF2 Zone; b = 'marly beds' of Whittaker and Green (1983); c = Mercia Mudstone Group.

Sec. 5.6] Stratigraphic Distribution

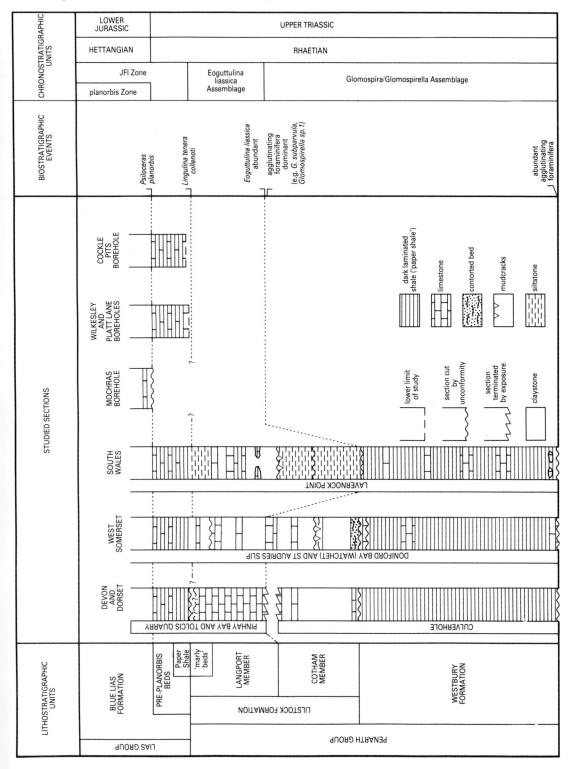

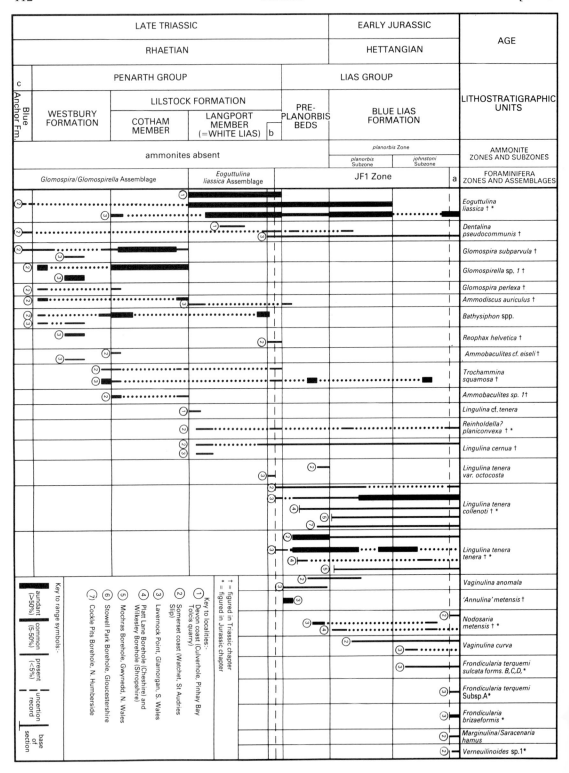

Sec. 5.6] Stratigraphic Distribution

AGE	LATE TRIASSIC			EARLY JURASSIC	
	RHAETIAN			HETTANGIAN	
LITHOSTRATIGRAPHIC UNITS	BLUE ANCHOR FM (c)	PENARTH GROUP		LIAS GROUP	
		WESTBURY FORMATION	LILSTOCK FORMATION	PRE-PLANORBIS BEDS	BLUE LIAS FORMATION
			COTHAM MEMBER / LANGPORT MEMBER (=WHITE LIAS) (b)		
AMMONITE ZONES & SUBZONES	ammonites absent			planorbis Zone (planorbis Subzone / johnstoni Subzone)	
FORAMINIFERA ZONES AND ASSEMBLAGES	Glomospira/Glomospirella Assemblage		Eoguttulina liassica Assemblage	JF1 Zone	(a)

Stratigraphic ranges:

- *Frondicularia terquemi* ssp. A
- *Frondicularia brizaeformis*
- *Marginulina/Saracenaria hamus*
- *Verneuilinoides* sp.1
- *Eoguttulina liassica*
- *Reinholdella? planiconvexa*
- *Lingulina cernua*
- *Dentalina pseudocommunis*
- *Lingulina tenera collenoti*
- *Lingulina tenera tenera*
- *Vaginulina curva*
- *Nodosaria metensis*
- *Trochammina squamosa*
- *Frondicularia terquemi sulcata* forms B,C,D.
- *Lingulina tenera* var. *octocosta*
- '*Annulina*' *metensis*
- *Ammodiscus auriculus*
- *Reophax helvetica*
- *Bathysiphon* spp.
- *Glomospirella* sp.1
- *Glomospira subparvula*
- *Ammobaculites* sp.1
- *Glomospira perplexa*
- *Ammobaculites* cf. *eiseli*

Parker and Kirkby, 1869). Chapman's specimens are considered herein to be referable to *Glomospira subparvula*, described by Bartenstein (1962) from the Rhaetian of West Germany. The latter is very similar to *Agathammina* in its apparent quinqueloculine habit, but differs in being agglutinating. *G. subparvula* is abundant in the Westbury Formation, which is the provenance of Chapman's assemblage. Other alleged Carboniferous species of *Ammodiscus* recorded by Chapman are *A. anceps* (Brady, 1876), *A. centrifugus* (Brady, 1873) and *A. robertsoni* (Brady, 1876). Examination of Chapman's figures suggests that the former two species should be referred to *Ammodiscus auriculus* Chapman (an apparently Rhaetian-restricted species known from Britain, West Germany and Holland), whilst the specimens of *A. robertsoni* belong to *G. subparvula*.

Thus it now seems that several of the 'Palaeozoic' taxa recorded by Chapman (1894) are referable instead to characteristic Rhaetian agglutinating foraminifera. Notwithstanding this, the presence of the relict Palaeozoic genera *Stacheia* and *Nodosinella* in the Rhaetian remains an open but important issue, which further study, or the discovery of Chapman's collection, should clarify.

The only taxon of Palaeozoic affinity in the author's material is an *Ammobaculites* form which is here compared to *A. eiseli* as figured by Pattison (1981) from the British Upper Permian.

Foraminiferal assemblages and zones

The vertical sequence of foraminifera seen in the British uppermost Triassic to basal Jurassic allows the recognition of three biostratigraphic units. The lower two of these are proposed for the first time here and appear to be of correlative potential with Britain and mainland western Europe. Nevertheless, as they are clearly related to facies they are termed assemblages rather than zones. These are succeeded by the basal Jurassic foraminiferal zone, which is widespread in Europe.

Glomospira/Glomospirella Assemblage
Definition of base: First appearance of *Glomospira subparvula* and *Glomospirella* sp. 1, seen to base of studied sections.
Definition of top: Last occurrence of *G. subparvula* and *Glomospirella* sp. 1. Base of overlying assemblage.
Characteristics: Assemblages almost exclusively agglutinating, also containing common to abundant *Ammodiscus auriculus*, *Trochammina squamosa*, *Bathysiphon* spp. and *Ammobaculites* sp. 1.
Lithostratigraphic equivalents: Upper Blue Anchor Fm, Westbury Fm, and Cotham Member.
Distribution: South Wales, Somerset, West Germany, Sweden, Holland.

Eoguttulina liassica Assemblage
Definition of base: Top of underlying assemblage. First appearance of common/abundant *Eoguttulina liassica* and other rare, but consistent calcareous foraminifera such as *Reinholdella? planiconvexa*.
Definition of top: First appearance of *Lingulina tenera collenoti*.
Characteristics: *Eoguttulina liassica* dominant, agglutinating foraminifera rare, though rare elements of the previous assemblage are occasionally present such as *Ammodiscus auriculus*, *Trochammina squamosa*.
Lithostratigraphic equivalents: Most of Langport Member (= White Lias) of Lilstock Formation, excluding up-

Distribution:	permost 'marly beds'. At present known in South Wales, Somerset and Devon/Dorset.

JF1 Zone

Definition of base:	First appearance of *Lingulina tenera collenoti*
Definition of top:	First appearance of *Frondicularia terquemi* sp. A (See Chapter 6)
Author:	Copestake and Johnson herein, Chapter 6.2. Equivalent to *Lingulina tenera collenoti* Zone of Copestake and Johnson (1984).
Characteristics:	First appearance of *L. tenera tenera* occurs near the base in Britain. This subspecies, with *collenoti* and *Eoguttulina liassica* are the dominant elements of the zone. *Nodosaria metensis* appears within the zone.
Lithostratigraphic equivalents:	Uppermost 'marly beds' (Langport Member, Lilstock Formation), Pre-Planorbis Beds to lower part of Blue Lias Formation (Lias Group). (*planorbis* Zone, intra *johnsoni* Subzone).
Distribution:	Widespread throughout Britain and Western Europe at the base of the Lias.

be Rhaetian-restricted in northwest Europe. A second category includes taxa such as *Eoguttulina liassica*, *Lingulina tenera tenera* and *Lingulina tenera collenoti* which first appear in the Rhaetian, but range up into the Jurassic, from which age of rocks they are better known.

No representative material of the 'Carboniferous taxa' reported by Chapman (1894) (e.g. *Stacheia*, *Nodosinella*) is at present available, as the whereabouts of Chapman's collection is unknown and as these taxa have not been found in subsequent studies. The suborder classification utilised is that of Loeblich and Tappan (1984).

5.7 INDEX SPECIES

A range of taxa are illustrated on plates 5.1, 5.2 and 5.3. These include species such as *Glomospira subparvula*, *Glomospirella* sp. 1 and *Ammodiscus auriculus* which appear to

SUBORDER TEXTULARIINA

Ammodiscus auriculus Chapman
Plate 5.1, Fig. 1 (× 220); Fig. 2 (× 194); Westbury Fm, St Audrie's Slip. = *Ammodiscus auricula* Chapman, 1894. Common synonym: *A. parvulus* Ten Dam, 1945. Description: test thin-walled, fine-grained, ovate, white, commonly flattened, comprising about three planispiral whorls.
Remarks: the compressed form, ovoid shape and fine-grained, delicate test are distinctive features. No citations of Chapman's species are known since the original description, though *A. parvulus* is clearly synonymous. The nature of the initial coil is obscure in all previous figures, and in reflected light, but SEM studies suggest that the species is completely planispiral. Distribution: Westbury Formation and Cotham Member of Watchet and St Audrie's Slip, Langport Member and Pre-Planorbis Beds of Lavernock, Westbury Formation of Wedmore (Chapman, 1894), lower 'Rhaet' of Holland (Ten Dam, 1945) and 'Mittelrhat' of West Germany (Wicher, 1951; Bartenstein, 1962). The species appears to be restricted to the Rhaetian of northwest Europe.

'*Annulina*' *metensis* Terquem, 1862
Plate 5.1, Fig. 3 (× 100); Fig. 8 (× 100); Pre-Planorbis Beds, Lavernock. Common synonym: *Ammodiscus wicheri* Bartenstein, 1962. Description: test agglutinated, flattened, discoidal, corpuscular, with depressed central area and raised outer rim, no aperture or chambers discernible.
Remarks: This peculiar taxon, most commonly reported from the Lower Jurassic, appears to belong to an unnamed foraminiferal genus. The lack of an aperture suggests that it belongs to the subfamily Psammosphaerinae. Loeblich and Tappan's (1964) interpretation of the species as either a radiolarian or echinoderm is not accepted here. Distribution: common from a single horizon in the Pre-Planorbis Beds of Lavernock. Previous records from the Rhaetian of West Germany (Bartenstein, 1962; Will, 1969); Lower Jurassic of West Germany (Franke, 1936; Bartenstein and Brand, 1937; Frentzen, 1941), France (Terquem, 1862) and England (Dorset) (Copestake, 1987). Total range: Rhaetian–Lower Jurassic.

Ammobaculites cf. *eiseli* Paalzow, 1935
Plate 5.1, Fig. 4 (× 80), Cotham Member, Doniford Bay (Watchet) (cf. *Ammobaculites eiseli* Paalzow, 1935). Description and remarks: test similar to *A. eiseli* in being coarse-grained, with large divergent-sided rectilinear portion of broad chambers; differs in its early coil which is smaller than *A. eiseli* and not offset relative to rectilinear portion. Distribution: Cotham Member of Watchet. *A. eiseli* has been previously recorded from the Upper Permian of England (Pattison, 1981) and West Germany (Paalzow, 1935).

Ammobaculites sp. 1
Plate 5.1, Fig. 5 (× 160); Fig. 6 (× 140); Fig. 7 (× 125); Cotham Member, Doniford Bay (Watchet). Description: megalospheric tests with involute, planispiral coil only, comprising six broad chambers; sutures depressed, straight; microspheric tests with evolute initial coil and narrow uniserial arrangement of at least three chambers.
Remarks: a species not previously described. Most recovered specimens are planispiral which, but for their association with *Ammobaculites*, appear close to the Jurassic species *Haplophragmoides kingakensis* Tappan, 1955. The uncoiled portions may be absent through breakage. Distribution: top and base of the Cotham Member of Doniford Bay (Watchet).

Glomospira perplexa Franke, 1936
Plate 5.1, Fig. 9 (× 312); Fig. 13 (× 320); Cotham Member, Doniford Bay (Watchet). Description: test comprising an irregularly coiled tube of constant diameter; finely agglutinating. Distribution: Westbury Formation and Cotham Member of Watchet and St Audrie's Slip. Most records are from the Lower Jurassic (e.g. Franke, 1936; Brouwer, 1969), though also described from the Rhaetian of Austria by Kristan-Tollmann (1964).
Total range: Rhaetian–Pliensbachian.

Glomospira subparvula Bartenstein, 1962
Plate 5.1, Fig. 10 (× 190); Fig. 11 (× 140); Fig. 12 (apertural view) (× 320); Fig. 14 (× 170); Fig. 15 (× 224); Fig. 16 (apertural view) (× 420); Cotham Member, Doniford Bay (Watchet). Description: test elongate, spindle-shaped, finely agglutinated, coiling in three planes approximately 120° apart, as seen in end (apertural) view; coiling mode imparts miliolid habit to test.
Remarks: an important member of Westbury Formation and Cotham Member assemblages in southwest Britain. The miliolid habit is very similar to species of *Agathammina* from the Permian, but the taxon differs from the latter in being agglutinated rather than calcareous (porcellanous). Previous authors have referred specimens of *G. subparvula* to Permian *Agathammina* species (e.g. Chapman, 1894; Will, 1969). Distribution: Westbury Formation and Cotham Member in West Somerset and Glamorgan, Westbury Formation of Somerset (Chapman, 1894 as *Ammodiscus pusilla* and *A. milioloides*), Rhaetian (Westbury Formation equivalent) of W. Germany (Bartenstein, 1962; Will, 1969, as *Agathammina pusilla*) and probably Sweden (Brotzen, 1950, as *Agathammina*). The species is apparently Rhaetian-restricted.

Triassic

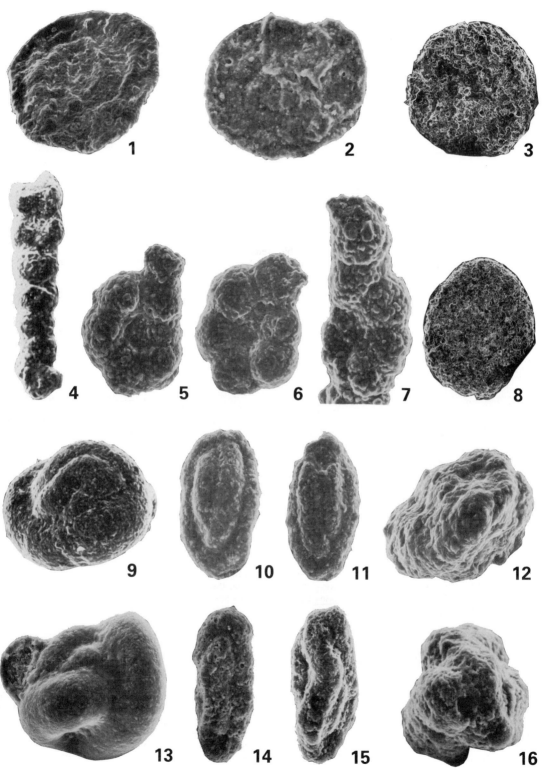

Plate 5.1

Glomospirella sp. 1
Plate 5.2, Fig. 1 (× 150); Fig. 2 (× 230); Fig. 3 (× 160); Fig. 4 (× 164); Fig. 2 Westbury Fm, St Audrie's Slip; Figs 1, 3, 4, Cotham Member, Doniford Bay (Watchet). Description: test ovate, finely agglutinated, comprising early streptospiral (glomospiral) coiled portion, followed by planispiral coil of one to three whorls.
Remarks: the species is an important constituent of Westbury Formation and Cotham Member assemblages in southwest Britain and occurs in association with *G. subparvula*. *Glomospirella* sp. 1 is similar to and easily confused with *Ammodiscus auriculus*, but the latter is distinct in being completely planispiral and finer grained. Distribution: recorded from the Westbury Formation and Cotham Member in West Somerset and Glamorgan and apparently restricted to the Rhaetian.

Reophax helvetica (Haeusler)
Plate 5.2, Fig. 7 (apertural view) (× 234); Fig. 8 (× 75); uppermost Langport Member, Doniford Bay (Watchet). = *Dentalina helvetica* Haeusler, 1881. Description: test agglutinated, with up to five chambers arranged uniserially, overlapping, early chambers appearing broader than high, later chambers equidimensional; sutures straight, slightly depressed. Distribution: in the Triassic, it has been observed in the Westbury Formation of Lavernock and the uppermost Langport Member ('marly beds') of Watchet. Previous records are from throughout the Jurassic (e.g. Bartenstein and Brand, 1937 as *R. multilocularis*); Lloyd, 1959; Norling, 1972).

Trochammina squamosa Ziegler, 1964
Plate 5.2, Fig. 5 (dorsal view) (× 115); Fig. 6 (dorsal view) (× 115); Fig. 11 (ventral view) (× 100); uppermost Langport Member ('marly beds'), Doniford Bay (Watchet). Common synonyms: *Trochammina squamata non* Jones and Parker, 1860; *T. quinquelocularis* Dain, 1972. Description: test coarsely agglutinated, comprising three whorls, with four to six large, broad chambers in final whorl; sutures depressed, recurved.
Remarks: this species is often referred to the Recent species *T. squamata*, though Ziegler (1964) erected the name *T. squamosa* for this Triassic–Jurassic homoemorph. Distribution: in the Penarth Group, it has been observed between the uppermost Westbury Formation and uppermost Langport Member ('marly beds') of Watchet and between the uppermost Westbury Formation and basal Blue Lias of Lavernock. The species ranges throughout the overlying Jurassic though most previous records are from the Upper Jurassic of Europe (Lloyd, 1959, as *T. squamata*) and U.S.S.R. (Dain, 1972, as *T. quinquelocularis*). The species is common in the Toarcian (Dunlin Group) and Upper Oxfordian (Heather Formation, Humber Group) in the northern North Sea.

SUBORDER LAGENINA

Dentalina pseudocommunis Franke, 1936
Plate 5.2, Fig. 9 (× 50), Pre-Planorbis Beds, Lavernock. Description: test uniserial, narrow, elongate, arcuate or nearly straight; initial chambers broad, later chambers high; sutures markedly oblique, initially flush, but depressed between adult chambers; aperture radiate.
Remarks: the species was originally erected by Franke (1936) for Lower Jurassic specimens similar to *D. communis* (d'Orbigny) (Upper Cretaceous–Recent). Distribution: infrequent in the uppermost Blue Anchor Formation, Langport Member and Pre-Planorbis Beds of Watchet, St Audrie's Slip and Lavernock, and White Lias of Dorset. Consistently common and widespread in the Lower Jurassic of northwest Europe (e.g. Barnard, 1950; Franke, 1936).

Eoguttulina liassica (Strickland)
Plate 5.2, Fig. 10 (× 47); Fig. 12 (× 50); Fig. 13 (apertural view) (× 272); Fig. 14 (× 52); Langport Member, St Audrie's Slip. = *Polymorphina liassica* Strickland, 1846. Common synonym: *Polymorphina metensis* Terquem, 1864. Description: test smooth, highly variable, elongate, comprising embracing chambers, spirally arranged, 90° apart, around longitudinal axis; variation includes forms with adult chambers becoming less enveloped, elongate, disposed at 180° (biserial) (typical of the genus *Polymorphina* d'Orbigny); sutures flush to depressed; aperture terminal, central, radiate.
Remarks: a wide range of variation is seen within the *Eoguttulina* populations of the Penarth Group, such variation being also seen in Jurassic assemblages. Owing to the integradational nature of the variation all the variants are included within a single species. The form which shows elongate, biserial adult chambers has been separated into *Polymorphina metensis* (e.g. Tappan, 1951), but is here considered to fall within the normal range of variation of *E. liassica*. Distribution: first occurs commonly at the base of the Langport Member, and continues as a prominent member of Langport Member to basal Blue Lias assemblages in Somerset, Pinhay Bay and Lavernock. Continues in sporadic abundance through the Lower Jurassic. The presence of a striated *Eoguttulina* from the upper Langport Member of Lavernock (Banner *et al.*, 1971) is not confirmed here. The common occurrence of *E. liassica* in the 'Cotham Member' of Lavernock (*sensu* Whittaker, 1978) appears anomalous; on the evidence of the foraminifera, these sediments (beds 4, C7–C10 of Richardson, 1911) are correlated with the Langport Member in Somerset. Reported also from the Rhaetian of West Germany (Will, 1969).

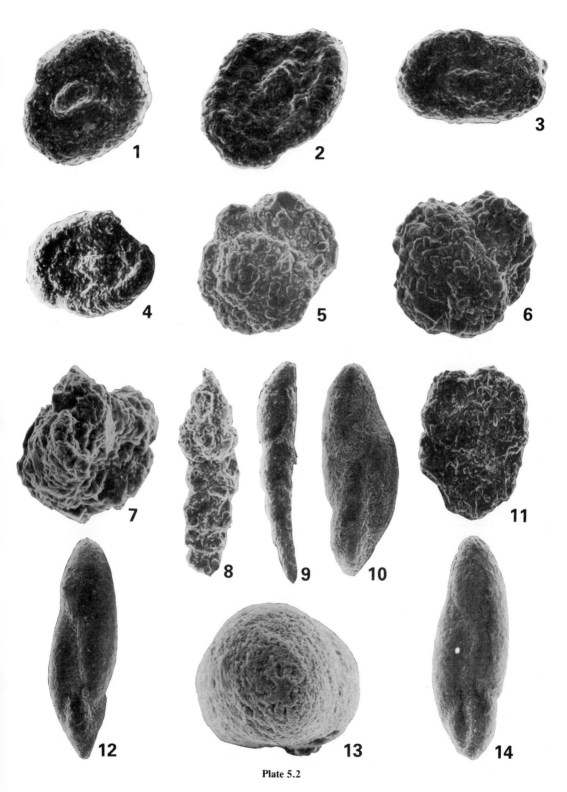

Plate 5.2

Lingulina cernua (Berthelin)
Plate 5.3, Fig. 1 (× 117); Fig. 2 (× 105); Fig. 3 (apertural view) (× 300); Fig. 4 (× 78); Langport Member, St Audrie's Slip. = *Frondicularia cernua* Berthelin, 1879. Description: test smooth, gently curved, moderately compressed to circular in section; chambers of variable height and width; sutures slightly arched to straight, depressed.
Remarks: the Rhaetian specimens are more compressed than more typical forms from the Lower Jurassic, but in this respect are similar to forms figured by Barnard (1956) (as *L.* cf. *cernua*). Distribution: first appears at the base of the Langport Member on the Somerset coast and at Lavernock, ranging through the Lower Jurassic in northwest Europe. Not previously recorded from the pre-Jurassic.

Lingulina tenera tenera Bornemann
Plate 5.3, Fig. 6 (× 75), Pre-Planorbis Beds, Doniford Bay (Watchet); Fig. 7 (× 105), Pre-Planorbis Beds, St Audrie's Slip. = *Lingulina tenera* Bornemann, 1854. Common synonym: *Geinitzinita tenera* (Bornemann). Description: test elongate, compressed, ornamented with marginal keel and two strong longitudinal costae on each side of test; aperture terminal, slit-shaped.
Remarks: the presence of two strong costae on either side of the test without intervening striations distinguishes this subspecies from other members of the *L. tenera* plexus. Intermediate forms are common, linking this subspecies with *L. tenera collenoti* and var. *octocosta*, particularly in the Pre-Planorbis Beds and Planorbis Zone of the Lower Jurassic. Distribution: in the author's data the *Lingulina tenera* plexus first appears in the uppermost Langport Member ('marly beds'), and at this horizon includes subspecies *tenera*, *collenoti* and var. *octocosta*. *L. tenera tenera* becomes abundant within the Pre-Planorbis Beds and continues to be common throughout the overlying Hettangian–Pliensbachian, becoming extinct in the Early Toarcian. In the Devon White Lias, a form similar to *tenera* occurs, but has distinct ornament (discontinuous ribbing). It is referred to *Lingulina* cf. *tenera*. Ivimey-Cook's (pers. comm.) reported record of *L. tenera* from the lower Langport Member of Lavernock is not confirmed by the author's data from this locality. Banner *et al.* (1971) recorded the subspecies (as both *Geinitzinita* spp. and *G. tenera*) from the upper Langport Member and Pre-Planorbis Beds of Lavernock. As all forms occurring in the Rhaetian range into the Jurassic, the statement of Banner *et al.* that species of the genus may help to distinguish between Rhaetian and Jurassic sediments is not accepted. Forms very close to *L. tenera tenera* have been recorded from sediments as old as Middle Triassic in Australia (Heath and Apthorpe, 1986) but these differ from European latest Triassic and Early Jurassic specimens in being more compressed, with more widely spaced costae and oblique secondary striae and in possessing apertural lips. Similar flattened forms with widely spaced ribs comparable to the Australian material have been described by Oberhauser (1960) from the Middle and Upper Triassic of Austria. Strong (1984) referred specimens from the Upper Triassic (Norian) of New Zealand to *L. tenera* but his figures show circular, radiate apertures and high-arched chambers indicating allocation to the genus *Frondicularia*. These Australasian records are important, however, in that they may represent evolutionary forerunners of the *L. tenera* plexus which appears abruptly in the latest Triassic of Britain. There are no other pre-Jurassic records of *L. tenera*.

Lingulina tenera collenoti (Terquem)
Plate 5.3, Fig. 5 (× 83); Fig. 8 (apertural view) (× 200); Fig. 9 (side view of Fig. 8) (× 105); Fig. 10 (× 78); Fig. 11, (× 75); Fig. 12 (× 108); Fig. 15 (× 70); Figs 8, 9 from Pre-Planorbis Beds, Lavernock; Figs 5, 10–12, 15 from Pre-Planorbis Beds, Doniford Bay (Watchet). = *Marginulina collenoti* Terquem, 1866a. Common synonyms: *Lingulina striata* Blake, 1876; *L. tenera* form A Barnard, 1956. Description: see Chapter 6.
Remarks: intermediates between true *collenoti* (many striae of approximately equal strength, rounded test margin) and *tenera* (two strong ribs on either side of test, marginal keel) are common in the British uppermost Rhaetian–basal Jurassic. These intermediates show two slightly stronger ribs plus a weakly developed keel (e.g. Plate 5.3, Fig. 11), but are included in *collenoti* on account of their overall multi-striated ornament and extremely elongate test form. The taxon is quite variable as the forms shown on Plate 5.3 illustrate. Distribution: appears abruptly in the uppermost Langport Beds ('marly beds') in southwest Britain (see also Banner *et al.*'s 1971 record and figures from offshore South Wales, as *Frondinodosaria* cf. *striata*, *Geinitzinita* cf. *lingula*) ranging to the *angulata* Zone, *complanata* Subzone of the Hettangian. Marker taxon for latest Rhaetian–basal Jurassic foraminifera zone. Recorded from latest Rhaetian in West Somerset, Lavernock and the Platt Lane Borehole. The only known pre-Jurassic record outside Britain is that from the latest Rhaetian of the Paris Basin (Bizon and Oertli, 1961).

Lingulina tenera Bornemann var. *octocosta* Brand
= *Frondicularia tenera octocosta* Brand (*in* Bartenstein and Brand, 1937). Common synonym: *Lingulina tenera* form G Barnard, 1956. Description: a variant of *L. tenera* with four (two pairs) of strong ribs on either side of the test; marginal keel; aperture terminal, slit-shaped. Distribution: this form appears within the general *L. tenera* plexus influx seen in the upper Langport Beds to Pre-Planorbis Beds of Somerset and Lavernock. The variant also occurs within the Lower Jurassic.

Nodosaria metensis Terquem, 1864
Plate 5.3, Fig. 16 (× 71), Pre-Planorbis Beds, Lavernock. Common synonym: *Nodosaria dispar* Franke, 1936. Description and remarks: see Chapter 6. Distribution: the species is very characteristic of the Lower Jurassic, in Britain and northwest Europe, but first appears within the mid-part of the Pre-Planorbis Beds at Lavernock, of latest Triassic age. Also recorded from the Rhaetian of Austria (Kristan-Tollmann, 1964).
Total range: Rhaetian–Bajocian.

Triassic

Plate 5.3

SUBORDER ROBERTININA

Reinholdella? planiconvexa (Fuchs)
Plate 5.3, Fig. 13 (ventral view) (× 122); Fig. 14 (edge view) (× 170); Langport Member, Doniford Bay (Watchet). = *Oberhauserella planiconvexa* Fuchs, 1970. Common synonym: *Discorbis advena* Bartenstein and Brand, 1937 *non* Cushman. Description and remarks: see Chapter 6.2. Distribution: the species first appears at the base of the Langport Member at Watchet, and continues in low numbers through the *planorbis* Zone before becoming abundant at the base of the *liasicus* Zone. The appearance of the species represents the first incoming of rare, typically Jurassic species which occurs at the base of the Langport Member in southwest Britain. *R? planiconvexa* has not been recorded below the Lower Jurassic outside Britain.

ACKNOWLEDGEMENTS

The author is grateful to BP Exploration for support of field collecting in Somerset and Devon, for provision of facilities and permission to publish. Thanks are also due to Drs P Bown and A. Lord (University College, London) for access to samples from the lower parts of the St. Audrie's Slip and Lavernock sections. The British Geological Survey allowed microfaunal preparations to be examined from the Platt Lane, Wilkesley, Stowell Park and Cockle Pits Boreholes. Dr H. C. Ivimey-Cook (British Geological Survey) was very helpful to the author in the preparation of the Triassic chapter in the first edition of this atlas. Liz Gordon, Fergie Norris and Mary Lou McDermott typed the manuscript and Alex Forsyth and Dave McCluskey draughted the figures.

REFERENCES

Ainsworth, N. R. and Horton, N. F. 1986. Mesozoic micropalaeontology of exploration well Elf 55/30-1 from the Fastnet Basin, offshore southwest Ireland. *J. micropalaeontol.* **5**, 19–29.

Ainsworth, N. R., O'Neill, M. and Rutherford, M. M. in press. Jurassic and latest Triassic biostratigraphy of the North Celtic Sea and Fastnet Basins. In Batten, D. J. and Keen, M. C. (eds) *Micropalaeontology, palynology and petroleum exploration, on- and offshore Europe.*

Apthorpe, M. C. and Heath, R. S. 1981. Late Triassic and Early to Middle Jurassic foraminifera from the North West Shelf, Australia. In *Fifth Australian Geological Convention: Geological Society of Australia Abstracts*, No 3, 66.

Banner, F. T., Brooks, M. and Williams, E. 1971. The geology of the Approaches to Barry, Glamorgan. *Proc. Geol. Ass.*, **82**, 231–247.

Barnard, T. 1950. Foraminifera from the Lower Lias of the Dorset coast. *Q. Jl. Geol. Soc. London*, **105**, 347–391.

Barnard, T. 1956. Some Lingulinae from the Lias of England. *Micropaleontology*, **2**, 271–282.

Bartenstein, H. 1962. Neue Foraminiferen aus Unterkreide und Oberkeuper NW Deutschlands und der Schweiz. *Senck. Leth.*, **43**, 135–149.

Bartenstein, H. and Brand, E. 1937. Mikropaläontologische Untersuchungen zur Stratigraphie des nordwest Deutschen Lias und Doggers. *Senckenb. Naturf. Ges., Abh.*, **439**, 1–224.

Batten, D. J., Trewin, N. H. and Tudhope, A. W. 1986. The Triassic–Jurassic junction at Golspie, inner Moray Firth Basin. *Scott. J. Geol.* **22**, 85–98.

Bennet, G., Copestake, P. and Hooker, N. P. 1985. Stratigraphy of the Britoil 72/10-1A well, Western Approaches. *Proc. Geol. Ass.*, **96**, 255–261.

Bizon, G. and Oertli, H. 1961. Contributions à l'étude micropaléontologique du Lias du Bassin de Paris. Septième Partie – Conclusions. In *Colloque sur le lias Français. Mem. Bur. Rech. Géol. Min.* **5**, 107–119

Brady, H. B. 1876. Carboniferous and Permian foraminifera. *Palaeontogr. Soc. Monogr.*, 1–66.

Brotzen, F. 1950. The geological results from the deep-borings at Hollviken. Part 2. Lower Cretaceous and Triassic. *Sver. Geol. Unders. Ser. C*, **43**, 1–48.

Brouwer, J. 1969. Foraminiferal assemblages from the Lias of NW Europe. *Ver. K. Ned. Akad, Wet., Afd. Natuurk; 1 Reeks, Deel.*, **25** 1–48.

Chapman, F. 1894. On Rhaetic Foraminifera from Wedmore, in Somerset. *Ann. Mag. Nat. Hist.*, **16**, 307–329.

Cope, J. C. W., Getty, T. A., Howarth, M. K., Morton, N. and Torrens, H. S. 1980. A correlation of Jurassic rocks in the British Isles. Part 1, Introduction and Lower Jurassic. *Spec. Rep. Geol. Soc. Lond.* No. 14, 1–72.

Copestake, P. 1987 in Dorset Jurassic. *Mesozoic and Cenozoic Stratigraphical Micropalaeontology of the Dorset Coast and Isle of Wight, Southern England. Field Guide for the XXth European Micropalaeontological Colloquium*, in Lord, A. R. and Bown, P. R. (eds) *British Micropalaeontological Society Guide Book*, pp. 1–78.

Copestake, P. and Johnson, B. 1984. Lower Jurassic (Hettangian–Toarcian) foraminifera from the Mochras Borehole, North Wales (UK) and their application to a worldwide biozonation. *Benthos 1983; 2nd International Symposium on Benthic Foraminifera*, Pau, April 1983, 183–184.

Dain, L. G. 1972. Foraminifera from Upper Jurassic deposits of Western Siberia. *Transactions of the All-Union Petroleum – Scientific – Geologic – Prospecting Institute (VNIGRI)*, **317**, Leningrad (in Russian).

Deegan, C. E. and Scull, B. J. (compilers) 1977. A proposed standard lithostratigraphic nomenclature for the Central and Northern North Sea *Rep. Inst. Geol. Sci. No 77/25; Bull. Norw. Petrol. Direct. No 1*, London HMSO, 1–36.

Fisher, M. J. 1984. Triassic. In Glennie, K. W. (ed.) *Introduction to the petroleum geology of the North Sea*. pp. 85–102, Blackwell Scientific, Oxford.

Forster, S. C. and Warrington, G. 1985. Geochronology of the Carboniferous, Permian and Triassic. In Snelling, N. J. (ed.) *The chronology of the geological record, Geological Society of London Memoir No. 10*, 99–113.

Fox-Strangways, C., Cameron, A. C. G. and Barrow, G. 1886. The geology of the country around Northallerton and Thirsk. *Mem. Geol. Surv. G. B.* No. 96, 1–75.

Franke, A. 1936. Die Foraminiferen des deutschen Lias. *Abh. Preuss. Geol. Landesanst., NF*, **169**, 1–140.

Frentzen, K. 1941. Die Foraminiferenfaunen des Lias, Doggers und unteren Malms des Umgebung von Blumberg (Oberes Wutach-Gebeit). *Beitr Naturkd. Forsch. Oberrhingeb.*, **6**, 125–402

Gaunt, G. D., Ivimey-Cook, H. C., Penn, I. E. and Cox, B. M. 1980. Mesozoic rocks proved by IGS boreholes in the Humber and Acklam areas. *Rep. Inst. Geol. Sci.* No. 79/13, 1–34.

George, T. N. *et al.* 1969. Recommendations on stratigraphical usage. *Proc. geol. Soc.* No. *1656*, 139–166.

Goel, R. K., Zaninetti, L. and Srivastava, S. S. 1981. Les foraminifères de l'Anisien (Trias Moyen) de la Localité de Guling, Vallée de Spiti (Himalaya, Inde Septentrionale). *Archives des Sciences (Genève)*, **34**, 227–234.

Green, G. W. and Melville, R. V., 1956. The stratigraphy of the Stowell Park Borehole (1949–51). *Bull. geol. Surv. Gt. Br.* **11** 1–66.

Hallam, A. and El Shaarawy, Z. 1982. Salinity reduction of the end-Triassic sea from the Alpine region into northwestern Europe. *Lethaia*, **15**, 169–178.

Haq, B. U., Hardenbol, J. and Vail, P. R. 1987. Chronology of fluctuating sea levels since the Triassic. *Science*, **235**, 1156–1166.

Harrison, R. K. 1971. The petrology of the Upper Triassic rocks in the Llanbedr (Mochras Farm) Borehole. In Woodland, A. W. (ed.) The Llanbedr (Mochras Farm) Borehole. *Rep. Inst. Geol. Sci.* No. 71/18, 37–72.

Heath, R. S. and Apthorpe, M. C. 1986. Middle and Early (?) Triassic foraminifera from the Northwest Shelf, Western Australia. *J. Foramin. Res.* **16**, 313–333.

Ivimey-Cook, H. C. In Edmonds, E. A., Poole, E. G. and Wilson, V. 1965. Geology of the country around Banbury and Edge Hill (sheet 201). *Mem. Geol. Surv. G.B.*

Ivimey-Cook, H. C. 1974. The Permian and Triassic deposits of Wales. In Owen, T. R. (ed.), *The Upper Palaeozoic and Post-Palaeozoic rocks of Wales*, 295–321. University of Wales Press.

Ivimey-Cook, H. C. 1982. Biostratigraphy of the Lower Jurassic and Upper Triassic (Rhaetian) rocks of the Winterborne Kingston Borehole, Dorset. In Rhys, G. H., Lott, G. K. and Calver, M. A. (eds) The Winterborne Kingston Borehole, Dorset, England. *Rep. Inst. Geol. Sci.* No, 81/3, 97–106.

Kent, P. E., 1967. Outline geology of the southern North Sea Basin. *Proc. Yorks. geol. Soc.* **36**, 1–22.

Kent, P. E., 1970. Problems of the Rhaetic in the East Midlands. *Mercian Geol.* **3**, 361–373.

Kristan-Tollmann, E. 1964. Die Foraminiferen aus den Rhatischen Zlambachmergeln der Fischerweise bei Aussee in Salzkammergut. *Jb. Geol. B.A.*, **10**, 1–189.

Lloyd, A. J. 1959. Arenaceous foraminifera from the type Kimmeridgian (Upper Jurassic). *Palaeontology*, **1**, 298–320.

Loeblich, A. R. and Tappan, H. 1964. Sarcodina. Chiefly 'Thecamoebians' and Foraminiferida. *Treatise on Invertebrate Paleontology, Part C, Protista 2, 1 and 2*, pp. 1–900.

Loeblich, A. R. and Tappan, H. 1984. Suprageneric classification of the Foraminiferida (Protozoa). *Micropaleontology*, **30**, 1–70.

Mayall, M. J., 1981. The Late Triassic Blue Anchor Formation and the initial Rhaetian marine transgression in southwest Britain. *Geol. Mag.* **118**, 377–384.

Norling, E. 1972. Jurassic stratigraphy and foraminifera of Western Scania, Southern Sweden. *Sver. geol. Unders. Afh. Ser. Ca.*, **47**, 1–120.

Oberhauser, R. 1960. Foraminiferen und Mikrofossilien 'incertae sedis' der Ladinischen und Karnischen Stufe der Trias aus den Ostalpen und aus Persien. *Jb. Geol. B.A.*, **5**, 5–46.

Orbell, G. 1973. Palynology of the British Rhaeto-Liassic. *Bull. Geol. Surv. G.B.*, **44**, 1–44.

Paalzow, R. 1935. Die Foraminiferen im Zechstein des ostlichen Thuringen. *Jb. Preuss. geol. Landesamt.*, **56**, 26–45,

Pattison, J. 1981. Permian. In Jenkins, D. G. and Murray, J. W. (eds) *Stratigraphical Atlas of Fossil Foraminifera*, (First Edn) 70–77, Ellis Horwood Chichester.

Palmer, C. P. 1972. The Lower Lias (Lower Jurassic) between Watchet and Lilstock in North Somerset (United Kingdom) *Newsl. Stratigr.*, **2**, 1–30.

Pearson, D. A. B. 1970. Problems of Rhaetian stratigraphy with special reference to the lower boundary of the stage. *Q. Jl geol. Soc. Lond.*, **126**, 125–150.

Poole, E. G. 1979. The Triassic–Jurassic boundary in Great Britain. *Geol. Mag.*, **116**, 303–311.

Rhys, G. H. (Compiler) 1974. A proposed standard lithostratigraphic nomenclature for the southern North Sea and an outline structural nomenclature for the whole of the (UK) North Sea. A report of the joint Oil Industry – Institute of Geological Sciences Committee on North Sea Nomenclature *Rep. Inst. Geol. Sci.* No. 74/8, 1–14.

Richardson, L. 1905. The Rhaetic and contiguous deposits of Glamorganshire. *Q. Jl geol. Soc. Lond.*, **61**, 385–424.
Richardson, L. 1906. On the Rhaetic and contiguous deposits of Devon and Dorset. *Proc. Geol. Ass.*, **19**, 401–409.
Richardson, L. 1911. The Rhaetic and contiguous Deposts of West, Mid and part of East Somerset. *Q. Jl geol. Lond.*, **67**, 1–72.
Souaya, F. J. 1976. Foraminifera of Sun – Gulf Global Linckens Island Well P-46, Arctic Archipelago, Canada. *Micropaleontology*, **22**, 249–306.
Strong, C. P. 1984. Triassic foraminifera from Southland Syncline, New Zealand. *New Zealand Geol. Surv. Palaeont. Bull.* **52**, 1–40.
Styk, O, 1979. Foraminiferida. In Karczewski, L. K., et al. *Atlas Skamienialosci Przewodnich 1 Charakterystycznych. 2a. Mezozoic, Trias. Budowa Geologiczna Polski.*, **3**, 16–27.
Tappan. H. 1951. Foraminifera from the Arctic Slope of Alaska: General introduction and Part I—Triassic foraminifera. *Prof. Pap. U.S. Geol. Surv.*, **236-A**, 1–20.
Tappan. H. 1955. Foraminifera from the Arctic slope of Alaska. Part II. Jurassic Foraminifera. *Prof. Pap. U.S. Geol. Surv.*, **236-B**, 21–90.
Ten Dam, A. 1945. Een nieuwe soort int het geslacht *Ammodiscus* Reuss in het Rhaet bij Winterswijk. *Geol. Mijnbouw*, **7**, 3.
Terquem, O. 1862. Récherches sur les Foraminifères de L'Étage Moyen et de L'Étage inférieur du Lias, Mémoire 2. *Mem. Acad. Imper. Metz*, **42**. (ser 2, 9), 415–466.
Warrington, G. 1971. Palynology of the New Red Sandstone Sequence of the South Devon Coast. *Proc. Ussher Soc.* **2**, 307–314.
Warrington, G., Audley-Charles, M. G., Elliott, R. E., Evans, W. B., Ivimey-Cook, H. C., Kent, P., Robinson, P. L., Shotton, F. W. and Taylor, F. M., 1980. A correlation of Triassic rocks in the British Isles. *Spec. Rep. Geol. Soc. London*, No. 13.
Waters, R. A., Lawrence, D. J. D. 1987. Geology of the South Wales coalfied, Part III. The country around Cardiff. 3rd Edition. *Mem. Geol. Surv.*
Whittaker, A. 1978. The lithostratigraphical correlation of the uppermost Rhaetic and lowermost Liassic strata of the W. Somerset and Glamorgan areas. *Geol. Mag.*, **115**, 63–67.
Whittaker, A. and Green, G. W. 1983. Geology of the country around Weston-Super-Mare. *Mem. Geol. Surv. G.B.*, 1–147.
Wicher, C. A. 1951. Zur mikropaläontologischen Gliederung des nichtmarinen Rhät. *Erdol. u. Kohle*, **4**, 755–760.
Wicher, C. A. and Bartenstein, H. 1962. Ausgewahlte Beispiele aus dem nord-deutschen Keuper. *Arbeitskreis Deutscher Mikropaläontologen, 1962. Leitfossilien der Mikropaläontologie.* Gebruder Borntraeger, Berlin - Nikolassee.
Will, H-J, 1969. Untersuchungen zur Stratigraphie und Genese des Oberkeupers in Nordwest Deutschland. *Beih. geol. Jb.*, **54**, 1–240.
Zaninetti, L. 1976. Les foraminifères du Trias. Essai de synthèse et correlation entre les domaines mesogeens européen et asiatique. *Riv. Ital. Paleont.*, **82**, 1–258.
Ziegler, J. 1964. Beschreibung einer Foraminiferenfauna aus dem Rhät vom Grossen Hassberg (Nordbayern). Bemerkungen zur Stratigraphie und Palaogeographie des Rhäts in Franken. *Geol. Bavar.*, **53**, 36–62.

6

Jurassic

6.1 INTRODUCTION

P. Copestake

The Jurassic System takes its name from the rocks of this age in the Jura Mountains of France and Switzerland. In northwest Europe, including Britain, the passage from the Triassic into the Jurassic is associated with a major change from continental to fully marine facies, via the transitional restricted marine uppermost Triassic (see Chapter 5). Marine environments subsequently dominated in the British area during the Early Jurassic, in which mainly argillaceous sediments (Lias Group), rich in benthonic foraminifera, were deposited over a wide area from western Scotland to Dorset, plus offshore areas. Non-marine facies were restricted to the Hettangian–Sinemurian of the Inner Moray Firth, northern North Sea and parts of the Fastnet–Celtic Sea Basins.

The Middle Jurassic (particularly the Aalenian–Bajocian) is more regressive in character than the Lower Jurassic and is characterised by carbonate facies in southern Britain (south of the Humber Estuary) (e.g. Inferior Oolite, Lincolnshire Limestone Formation), and clastic dominated, coastal/marginal marine facies in Yorkshire, western Scotland and northern North Sea (e.g. the Brent Group). Foraminifera are best developed in the carbonate facies but are markedly less diverse than in the marine Lower Jurassic. The Bathonian and Callovian, however, mark a return to marine shelf argillaceous deposition, and microfaunas again became abundant and diverse. These conditions, with rich foraminiferal assemblages, persisted through the Oxfordian and Kimmeridgian, punctuated by a regressive phase in the Middle Oxfordian when carbonate facies (e.g. Corallian) became widespread. The latest Jurassic (Portlandian) witnessed a regional shallowing in southern England (e.g. Portland Beds, Purbeck Beds) and Celtic Sea, with poor foraminiferal faunas recorded.

Throughout the British Jurassic a succession of foraminiferal assemblage changes is evident. Although the controls on these are not fully understood, certain major appearance and extinction events, such as those in the mid Toarcian and late Bajocian, are probably related to sea level changes. Such events are compared herein with published schemes of Jurassic sea level change (e.g. Hallam, 1981; Vail et al., 1977; Haq et al, 1987) and it appears in general that significant new appearances are related to transgressions and major extinctions to times of sea level drop. In addition, certain foraminiferal groups, such as the Robertinina appear to be particularly associated with transgressive episodes. Although evidence in the British area for a succession of relative sea level changes is compelling, it is not yet possible to ascertain the major controlling factor, whether it be tectonic, eustatic, or an interplay of both.

Most studies to date on northwest European Jurassic foraminifera have concentrated upon basinal marine clays, for ease of microfossil extraction. However, recent British work, for example on the Aalenian–Bajocian oolites of the Cotswolds (Morris, 1982) and on Lower Jurassic marginal marine sands and ironstones (R. L. A. Young unpublished), shows that significant foraminiferal associations can be found in 'unfavourable' rock types such as sandstones, ironstones and oolitic limestones. These studies are adding significantly to our knowledge of Jurassic foraminiferal palaeoecology, and are reviewed in the palaeoenvironmental sections of the chapter.

A refined biostratigraphy can be obtained in the Jurassic using foraminifera, with the degree of resolution increasing as documentation proceeds, in some instances matching that provided by ammonites. Foraminiferal zonations have been previously published for the northwest European Lower Jurassic (Copestake and Johnson, 1984) and the British Bathonian (Cifelli, 1959; Coleman, 1981 in the first edition), and in the present volume a new zonation for the British Lower Jurassic is proposed, which is known to be workable offshore as well as onshore. In addition the previously published Bathonian 'faunules' are compared in more detail with mainland Europe, while new assemblage zones are proposed for the British onshore and offshore Callovian.

It is now clear that foraminifera are of great value in the subdivision and correlation of Jurassic sequences over wide areas within and beyond Britain, particularly in the fully marine facies. This fact is of major economic importance as many petroleum reservoirs, plus the major source rocks, in the offshore and onshore oil provinces of Britain are of Jurassic age and are correlated using foraminifera and other microfossil groups. Ammonites and other macrofossils are generally not recovered or preserved in drilling samples.

In general, significantly more documentation of British offshore Jurassic foraminiferal assemblages is provided in each section of the present work compared to the first edition, reflecting the value of foraminifera to subsurface hydrocarbon exploration. In sections 6.2 and 6.3 many additional data are also included from onshore sections, as a result of recent and ongoing research. In section 6.2 a substantial number of additional taxa are described to provide more detail on longer ranging, but characteristic, forms. In section 6.3, new information is included on British Aalenian–Lower Bajocian foraminifera, about which little was known at the time of the first edition. Revisions in section 6.4 concern the provision of previously un-

Fig. 6.1.1 — Distribution of Jurassic rocks (outcrop and subsurface) in relation to tectonic elements in the British area (compiled from Ziegler, 1982; Stoneley, 1982; Sellwood et al., 1986; Bennet et al., 1985 and Masson and Miles, 1986). Basins studied for foraminifera: (1a), Viking Graben; (1b), Central Graben; (2), Moray Firth Basin; (3), Southern North Sea Basin; (4), Cleveland Basin; (5), 'Midlands Basin'–East Midlands Shelf; (6), Mendip High–Radstock Shelf; (7), South Wales–North Somerset (eastern Bristol Channel Basin); (8), Wessex–English Channel Basin; (9), Fastnet–Celtic Sea–Cardigan Basins; (10), Ulster Basin; (11), Mull–Morvern; (12), South Minch Basin (Skye–Raasay).

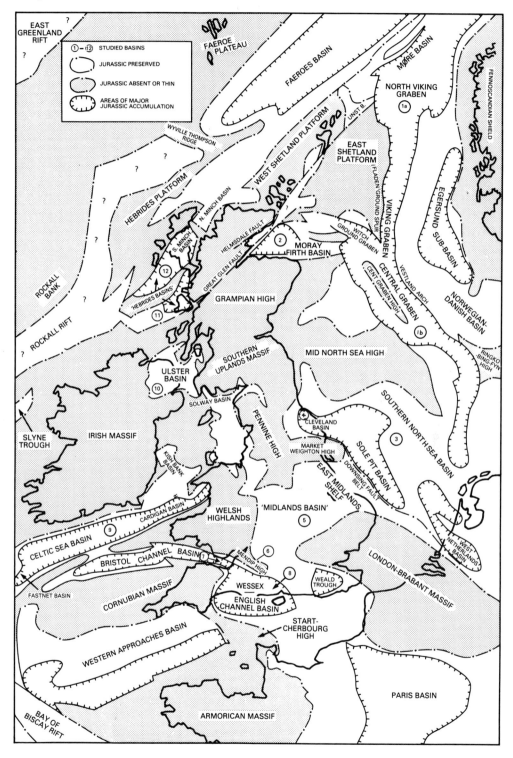

published information from onshore, including data by A. J. Lloyd on the foraminifera of the Dorset Kimmeridgian.

The foraminiferal geographic and stratigraphic distributions discussed in the Jurassic chapter are related to 12 major areas of deposition in the British area, some of which extend from onshore to offshore (see Fig. 6.1.1.). Terminology of the major Jurassic structural elements (basins and highs) is taken primarily from Ziegler (1982) (N.B. the informal term 'Midlands Basin' is used for the major depositional area between the Mendips and Market Weighton Highs). On the range charts, separate range bars are provided for each of these basins, each range bar thus representing a composite summary range for each area. The range data were compiled in this way to draw comparisons between basins and to provide a synthesis of information from numerous sections. It would not have been possible to provide individual range charts for each section studied. Notwithstanding this, it was considered useful to display the Callovian foraminiferal succession of the northern North Sea Heather Formation on a separate chart.

The absolute time scale used in the chapter is that of Kent and Gradstein (1985).

To achieve a consistency of format, all the figures were draughted by BP Exploration Drawing Office (particular thanks are due to Ross Munro).

6.2 THE HETTANGIAN TO TOARCIAN (LOWER JURASSIC)

P. Copestake, B. Johnson

6.2.1 Introduction and history of investigation

Foraminifera were first described from the British Lower Jurassic during the second half of the nineteenth century in several small, but important, papers. The earliest, that of Strickland (1846), named two of the most common Lower Jurassic species, namely *Spirillina infima* and *Eoguttulina liassica*, from the Hettangian (*planorbis* Zone) of Gloucestershire. In a subsequent paper Jones and Parker (1860) figured and discussed a small assemblage from a blue clay at Chellaston, near Derby. They believed the clay to be Triassic in age, though the microfauna is undoubtedly Early Jurassic, probably Toarcian in character. Most important of these early works was that of Blake (in Tate and Blake, 1876) on the Yorkshire Lias, since a considerable number of species are figured and indications are given of their stratigraphical distribution. Most of the forms described are from the Hettangian–Pliensbachian, probably because the Yorkshire Toarcian is generally impoverished in foraminifera. Much research on Lower Jurassic foraminifera was being conducted around this time by Brady, though he published only two brief papers (Brady, 1864, 1866). The second of these was based upon Late Pliensbachian and Toarcian material in Charles Moore's collection from southwest England. Brady indicated that he had prepared a monograph on British Lower Jurassic foraminifera which had been accepted by the Palaeontographical Society (Macfadyen, 1941), but unfortunately it was never published. This nineteenth century interest was completed in the 1890s by two brief studies of Crick and Sherborn (1891, 1892) upon material from the Pliensbachian and Toarcian of Northamptonshire. Interest subsequently waned for fifty years or so, and little was published other than lists of species from Hock Cliff, Gloucestershire (Richardson, 1908; Henderson, 1934) and localities in Lincolnshire (Trueman, 1918; Bartenstein and Brand, 1937).

The value of many of these early works is considerably diminished by their poor figures, brief descriptions and the use of Tertiary or Recent species names; recourse to the authors' original collections is necessary before their figures can be accurately identified. It is unfortunate that Blake's collection and the major part of Brady's have not yet been found.

The most comprehensive British works are those of Macfadyen (1941) and Barnard (1950a), on the Dorset coast Hettangian to Lower Pliensbachian, and of Barnard (1950b) on the Toarcian of Byfield (Northamptonshire). Macfadyen solved several taxonomic problems, whilst Barnard first applied the plexus concept to Jurassic foraminifera, an approach which he continued to pursue in several subsequent works (Barnard, 1956, 1957, 1960). These authors were among the first to grasp the problem of intraspecific variation in Jurassic foraminifera, and this aspect was further explored by Adams (1957) in the case of Toarcian microfaunas from Lincolnshire.

Documentation of British microfaunas has been continued more recently by Banner et al. (1971), and Coleman, in Warrington and Owens (1977), all based on offshore investigations, and by Horton and Coleman (1978) on Toarcian borehole material from Leicestershire. The latter author also published a brief description of Lower Jurassic foraminifera from the Winterborne Kingston Borehole, Dorset (Coleman, 1982). Several unpublished theses also contain important results, including two theses on the Mochras Borehole (Johnson, 1975; Copestake, 1978). A considerable amount of information from these latter two works is included in the present report. The work of the latter authors has led to a synthesis with published results from outside Britain, and the preliminary proposals of a zonation scheme of international application (Copestake and Johnson, 1984; Copestake, 1985). The most recent published work from onshore Britain includes a discussion of the foraminifera around the Triassic/Jurassic boundary (Copestake, 1985) and a description of foraminifera from the Dorset coast Jurassic for the 20th European Micropalaeontological Colloquium (Copestake, 1987).

Foraminiferal assemblages have been used extensively in the correlation and subdivision of Lower Jurassic sequences penetrated in offshore oil exploration wells in the North Sea and southwest of Britain since the early 1970s, but only recently has any work been published on the faunas. Nagy et al., (1984) and Nagy (1985a, 1985b) discussed the palaeoenviromental aspects of Pliensbachian–Toarcian foraminifera from the Statfjord Field area of the northern North Sea.

Details of the foraminifera from exploration wells in the Fastnet Basin (Fig. 6.1.1.) are beginning to be published (Ainsworth and Horton, 1986; Ainsworth et al., 1987, in press). The first of these papers described the biostratigraphy of the foraminifera of the Lower Jurassic from Elf operated well 55/30-1. The ranges indicated in these wells are very similar to those from the Mochras Borehole and indicate the uniformity of foraminiferal biostratigraphy in the Fastnet–Celtic Sea Basins.

Lower Jurassic foraminifera have been studied extensively in continental Europe since the last century, significant early workers being d'Orbigny (1849), Bornemann (1854) and Terquem (1858–1874). The copious works of Terquem are responsible for much of the taxonomic confusion regarding Jurassic foraminifera due to the poor quality of his figures and descriptions. The refiguring and redescription of a number of Terquem's types (Bizon, 1960; Ruget, 1976) has highlighted the inaccuracy of the type figures. Despite the latter works, many of Terquem's types still require redescription.

Two classic works from West Germany were published in the 1930s (Franke, 1936; Bartenstein and Brand, 1937) which are still valuable references today. Bartenstein and Brand (1937) were the first to show evolutionary changes within Jurassic foraminifera, which was the basis of their zonation, the first, based upon foraminifera, to be erected for the Jurassic. This paper was important in showing the stratigraphic value of Lower Jurassic foraminifera; the impetus for this development in micropalaeontology was, as it continues to be today, created by the need to subdivide and correlate subsurface sequences during exploration for oil. Work from West Germany has continued to recent times (e.g. Usbeck, 1952; Klingler, 1962; Welzel, 1968; Karampelas, 1978; Riegraf, 1985).

French workers have continued to produce useful papers, including a collection of regional reports on Lower Jurassic foraminifera (Colloque sur le Lias Français, 1961). Ruget has been the most active French worker in recent times (Ruget and Sigal, 1967, 1970; Ruget, 1967, 1976, 1985).

Lower Jurassic foraminifera comparable to those in Britain have been described from many parts of northern, central and western Europe, including Denmark (Nørvang, 1957; Bang, 1968, 1971, 1972), Sweden (Norling, 1972), Austria (Fuchs, 1970), Sicily (Barbieri, 1964), Portugal (Ruget and Sigal, 1970; Exton, 1979), Spain (Mira and Martinez-Gallego, 1981).

The previously mentioned papers all describe similar faunas of smaller calcareous benthonic foraminifera, dominated by nodosariids (group A of Gordon, 1970). In northern Europe and Arctic regions this fauna is replaced and dominated by agglutinating foraminifera, as described from Svalbard (Løfaldi and Nagy, 1980) and Arctic Canada (Souaya, 1976; Wall, 1983). A similar fauna has also been described from Alaska by Tappan (1955). Assemblages comparable to those from these northern areas are seen in the northern North Sea (U.K. sector and Norwegian sector, for latter see Nagy, 1985a, b) and in the Yorkshire coast Toarcian.

Lower Jurassic foraminifera are known from the North American continent in relatively few papers. In addition to those already mentioned from Arctic areas, papers by Gradstein (1976, offshore eastern Canada) and Cameron and Tipper (1985, Queen Charlotte Islands, western Canada) have described Lower Jurassic foraminifera closely comparable with western Europe.

In the Mediterranean area (southern Europe, northern Africa) and Middle East a Tethyan realm of carbonate deposition was dominated by large, complex-walled agglutinating foraminifera (group B of Gordon, 1970). South of this warm-water belt, however, assemblages comparable to western Europe have been described, for example, from Algeria (Maupin and Vila, 1976; Maupin, 1977) and offshore Morocco (Riegraf et al., 1984). Recent research shows that Lower Jurassic foraminifera of western European type (group A of Gordon, 1970) are widespread in the southern hemisphere, and have to date been reported from Papua New Guinea (Haig, 1979), Western Australia (Quilty, 1981; Apthorpe and Heath, 1981), and Argentina (Ballent, 1987a, b; A. Kielbowicz, pers. comm. 1984, 1985). This demonstrates the wide distribution of Lower Jurassic foraminiferal species of the type long familiar in Europe and highlights their great potential value to Lower Jurassic biostratigraphy and correlation worldwide.

6.2.2 Location of important collections

(1) British Museum (Natural History), London.
Specimens described by: Jones and Parker (1860), Brady (undescribed Lower Lias material from Chellaston, Hock Cliff and Stockton), Macfadyen (1941), Wood and Barnard (1946), Barnard (1950a,b, 1952), Adams (1962), Copestake (1986).

(2) Department of Geology, University College of Wales, Aberystwyth.
Assemblage slides and unpublished material of Jenkins (1958), Johnson (1975), Copestake (1978).

(3) British Geological Survey, Keyworth, near Nottingham.
Specimens described by Coleman *in* Warrington and Owens (1977), Horton and Coleman (1978), Coleman (1982).

Specimens described by Barnard (1956, 1957, 1960) from the Stowell Park Borehole.

Undescribed material from the Stowell Park, Wilkesley, Platt Lane, Burton Row, Hill Lane, Trunch and Cockle Pits British Geological Survey Boreholes.

(4) Geology Museum, Queen Square, Bath.
Specimens described by Brady (1866), plus additional undescribed material, all in the Charles Moore palaeontological collection.

(5) BP, Glasgow.
Unpublished material from numerous British localities (P. Copestake collection), plus many offshore oil exploration

wells in the northern North Sea, southern North Sea, Inner Moray Firth and Western Approaches.

(6) Department of Geology, University College of Wales, Swansea.
Material described by Banner *et al.* (1971).

(7) Department of Geology, Trinity College, Dublin.
Material described by Ainsworth and Horton (1986) and Ainsworth *et al.* (1987, in press) plus undescribed material from numerous oil exploration wells from the Celtic Sea and Fastnet Basins.

(8) Department of Geology, University College, London.
Unpublished material of F. M. Lowry from Ilminster.

Extensive collections of Lower Jurassic foraminifera from onshore and offshore oil exploration wells are also housed in the commercial micropalaeontological laboratories of oil companies (British Petroleum, Shell) and consultancy companies (e.g. Gearhart Geo Consultants, Paleoservices, Robertson Research), which have formed the basis of numerous confidential industry reports.

6.2.3 Stratigraphic divisions

The Lower Jurassic comprises the Hettangian, Sinemurian, Pliensbachian and Toarcian stages and is considered to span a period of 64 million years (Kent and Gradstein, 1985). The definitions of these stages are based on the boundaries of standard ammonite zones, following the scheme of ammonite zones and subzones proposed by Dean *et al.* (1961), with subsequent modifications (Howarth, in Cope *et al.*, 1980a; Donovan, in Whittaker and Green, 1983).

The position of the Triassic/Jurassic boundary has recently been revised upwards (Cope *et al.*, 1980a) to fall at the appearance of the first Jurassic ammonite in Britain, *Psiloceras planorbis*, defining a new base for the *planorbis* Subzone of the *planorbis* Zone. Previously, the base of this ammonite zone (and of the Jurassic) had been taken beneath *Psiloceras*, at the base of a unit without ammonites known as the Pre-Planorbis Beds (Donovan, 1956; George *et al.*, 1969).

Cope *et al.* (1980a) argued that the base of the Jurassic should be placed on the appearance of *Psiloceras*, assuming that this was a synchronous event throughout Britain. Notwithstanding the fact that foraminiferal associations of Jurassic aspect occur below *Psiloceras* (see Chapter 5), the revised base Jurassic boundary of Cope *et al.* (1980a) is followed here.

The Lower Jurassic is represented lithostratigraphically by predominantly argillaceous rocks referred to the Lias (Smith, 1799 MS) or Lias Group (Powell, 1984). Now that the base of the Jurassic has been moved upwards, the basal part of the Lias Group, the Pre-Planorbis Beds, is classified as belonging to the uppermost Triassic. The top of the Lias Group equates to the Lower/Middle Jurassic boundary.

Traditionally the Lias Group has been subdivided into the Lower, Middle and Upper Lias (Phillips, 1829), terms which have been persistently used in a chronostratigraphic sense, equating with the Hettangian–Lower Pliensbachian, the Upper Pliensbachian and the Toarcian respectively. As the terms 'Lower', 'Middle' and 'Upper' are no longer acceptable as formal lithostratigraphical subdivisions (Hedberg, 1976; Holland *et al.*, 1978), the use of Lower, Middle and Upper Lias should be discontinued. Although Powell (1984) has accordingly introduced a new lithostratigraphical nomenclature for the Lias Group in the Yorkshire Basin, for the rest of Britain, old informal names are still in use and an updated scheme is required.

Whilst ammonites have become the standard means of subdivision of the Jurassic, they are of little use in subsurface wells where, owing to the small size of rock samples generated by standard drill bits, and the rarity of core samples, microfossils are more important for biostratigraphy and correlation. Various microfossil groups are useful in subsurface correlation in the British Jurassic, particularly planktonic groups such as dinoflagellate cysts and radiolaria, together with calcareous nannofossils, foraminifera and ostracoda. In the Lower Jurassic, foraminifera and ostracoda enable the most refined subdivisions, owing to the poorly developed dinoflagellate populations (relative to higher parts of the Jurassic) and scarcity of radiolaria (from the UK area).

The apparent consistency of ranges of many foraminifera in the Lower Jurassic over wide areas has allowed the proposal of a preliminary zonation scheme for northwest Europe (Copestake and Johnson, 1984; Copestake, 1985) which is to be published in the near future.

A summary of British Lower Jurassic stages, ammonite zones and lithostratigraphy, onshore and offshore, is shown on Figs 6.2.2 and 6.2.3. Locations are shown in Fig. 6.2.1.

6.2.4 Depositional history; palaeoecology; faunal associations

6.2.4.1 Foraminiferal palaeoecology

Palaeoecological interpretations of Jurassic foraminiferal assemblages are difficult to make owing mainly to lack of direct modern-day analogues. This is due to drastic changes which took place in the nature of benthonic shelf populations during the Cretaceous and Tertiary. This has resulted in the replacement of nodosariids, which were dominant in the Jurassic, by other families of Rotaliina, nodosariids in the modern day occupying a deeper water setting than in the Jurassic.

In the Jurassic, three broad faunal realms can be differentiated on the basis of foraminifera. Around the former Tethyan ocean (equivalent to the modern day Mediterranean area, Middle East and southern Europe) warm water carbonate platform regimes existed in Jurassic times, occupied by larger, complex agglutinating foraminifera (group B1 of Gordon, 1970). North of this area, a Boreal realm was characterised by smaller benthonic foraminifera (group A of Gordon, 1970). Normal marine shelf assemblages in this region were dominated by nodosariids, with lesser amounts of Miliolina. Robertinidae, Spirillinina, Involutinina, Buliminacea, Textulariina, Polymorphinidae, and Cassidulinacea. Nodosariids seem to have preferred normal marine shelf conditions, particularly in areas of argillaceous sedimentation (Wall, 1960; Johnson, 1976; Nagy, 1985a, b), but were rare in shallow marine oolitic sediments and limestones (Bielecka and Pozaryski, 1954). An exception to the latter is the abundance of nodosariids within a Pliensbachian oolite horizon in the Amundsen Formation (North Sea).

Within nodosariid-dominated populations, major variations in species and generic abundance are seen in taxa which are long ranging, suggesting controlling palaeoenvironmental factors. In a statistical analysis of European Lias foraminiferal assemblages, Brouwer (1969) suggested that deeper shelf environments were dominated by *Lenticulina* (including *L. muensteri* and *L. varians*), intermediate depths by the *Marginulina prima* plexus and shallower shelf depths by the *Lingulina tenera* plexus. However, whilst Brouwer's (1969) analysis clearly indicated that distinctive assemblages exist within the overall nodosariid-dominated shelf areas, he has little direct evidence to support his environmental interpretations.

Occasionally the subordinate groups become dominant over nodosariids, and intervals mainly represented by *Spirillina* and *Ophthalmidium* can occasionally occur. Gordon (1970) has interpreted such floods as indicative of shallowing.

Polymorphinids (mainly represented by *Eoguttulina liassica*) occasionally occur in flood abundance. It is known that this species thrived in Rhaetian hypersaline carbonates (such as the White Lias) and it may therefore also indicate environments which deviated

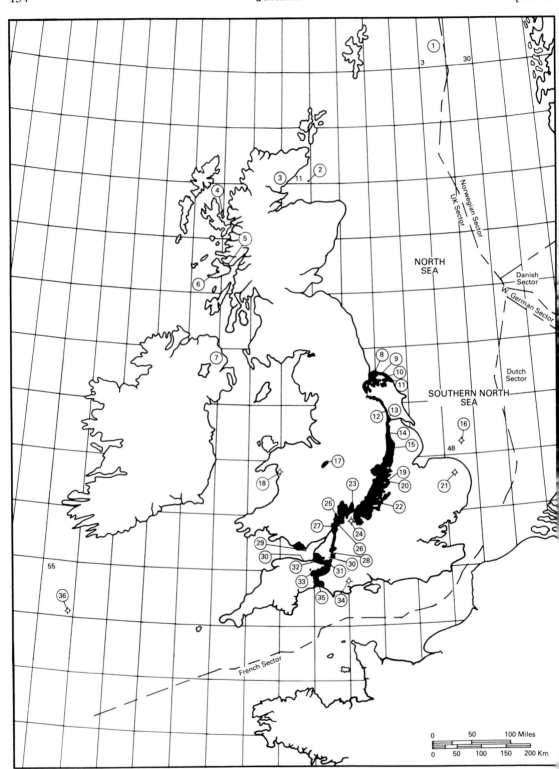

from normal marine salinity in the Early Jurassic. Brouwer (1969) also considered the species to be a possible indicator of lagoonal or shallow marine environments, based on analogy with modern day *Guttulina* distributions.

Robertinidae were represented by the genus *Reinholdella* in the Early Jurassic, and, although flood occurrences are frequently seen, they are difficult to interpret palaeoenvironmentally. Overall facies evidence suggests that flood levels may be associated with transgressions, for instance the abundance of *Reinholdella? planiconvexa* at the base of the *liasicus* Zone and of *R. macfadyeni, R. dreheri* and *R. pachyderma* in the *tenuicostatum–bifrons* Zones. This concurs with Brouwer's (1969) interpretation of the genus as a deep-water indicator in the Lower Jurassic.

There are, however, suggestions that individual species within the genus had different environmental preferences. Thus abundant *R.? planiconvexa* is generally not associated with diverse nodosariid populations, whereas *R. macfadyeni* and *R. pachyderma* occur with rich diverse nodosariids, for instance in the *tenuicostatum* Zone. It is likely that lowered oxygen levels may have induced the *R.? planiconvexa* floods.

Foraminiferal associations represented predominantly or exclusively by small, simple agglutinating species were also developed in the Early Jurassic, equivalent to Gordon's (1970) third type of shelf assemblage. This type has been consistently associated with regions of low oxygen levels and lowered pH, leading to unavailability of calcium carbonate, in both fossil and recent environments.

Such conditions are characteristic of two different environments, marginal marine areas of reduced salinity (marshes, estuaries, deltas) and deep marine areas (below the carbonate compensation depth). The crucial linking factor here is unavailability of calcium carbonate, which thus prevents colonisation by carbonate test-secreting foraminiferal groups. In the British Lower Jurassic there are two regions where this association predominates, in the Toarcian of the Yorkshire coast and the northern North Sea. The main taxa in these assemblages are small species of

Fig. 6.2.1 — Lower Jurassic localities in the British Isles and adjacent offshore areas. Lower Jurassic outcrop in black. Offshore grid system equates to petroleum exploration quadrants.
(1) North Viking Graben ('Brent Province'); (2) Beatrice Field, block 11/30; (3) Dunrobin Castle coast section; (4) Hallaig, Isle of Raasay; (5) Loch Aline; (6) Gribun, Isle of Mull; (7) Whitepark Bay; (8) Redcar; (9) Staithes; (10) Robin Hood's Bay; (11) Ravenscar; (12) Hotham Crossroads; (13) Cockle Pits Borehole; (14) Crosby Warren; (15) Bracebridge & Waddington; (16) Well 48/22-1; (17) Wilkesley & Platt Lane Boreholes; (18) Mochras Borehole; (19) Tilton; (20) Empingham; (21) Trunch Borehole; (22) Byfield; (23) Blockley; (24) Stowell Park Borehole; (25) Robin's Wood Hill; (26) Stonehouse; (27) Hock Cliff; (27) Radstock; (28) Cloford and Holwell; (29) Lavernock Point; (30) Watchet and St. Audrie's Slip; (31) Hornblotton Mill; (32) Hill Lane and Burton Row Boreholes; (33) Ilminster; (34) Winterborne Kingston Borehole; (35) Pinhay Bay - Watton Cliff; (36) Well 55/30-1 (Irish Sector).

Fig. 6.2.2 — Development and correlation of Hettangian–Sinemurian rocks in Britain. Data sources: Absolute ages from Kent and Gradstein (1985). (1) Deegan and Scull (1977), ages modified herein; (2) Neves and Selley (1975), Lee (1925), Hurst (1985); (3) Rhys (1974); (4) Powell (1984), Knox (1984), Howard (1985); (5) Cope *et al.* (1980a); (6) Cope *et al.* (1980a), Donovan and Kellaway (1984), Tutcher and Trueman (1925); (7) Trueman (1922), Palmer (1972a) Cope *et al.* (1980a), Whittaker and Green (1983); (8) Lang (1914, 1924 and 1932), Lang *et al.* (1923), Lang and Spath (1926), Palmer (1972b), Cope *et al.* (1980a); (9) Millson (1987); (10) Cope *et al.* (1980a); (11) Oates (1978), Lee and Bailey (1925), McLennan (1954); (12) Oates (1978), Hallam (1959).

Fig. 6.2.3 — Development and correlation of Pliensbachian–Toarcian rocks in Britain. Data sources: as for Fig. 6.2.2 except (2) Andrews and Brown (1987).

Jurassic

STAGE AND SUB-STAGES	AMMONITE ZONES AND SUBZONES		NORTHERN NORTH SEA BASIN		SOUTHERN NORTH SEA BASIN	CLEVELAND BASIN	EAST MIDLANDS SHELF — "MIDLANDS BASIN"
			NORTH VIKING GRABEN ①	MORAY FIRTH BASIN ②	③	④	⑤
198my UPPER SINEMURIAN	raricostatum	aplanatum		Ladys Walk Shale Member		Pyritous Shales	SANDROCK
		macdonnelli					
		raricostatoides				Siliceous Shales	LOWER LIAS CLAYS
		densinodulum		?			
	oxynotum	oxynotum					
		simpsoni					
	obtusum	denotatus					FRODINGHAM IRONSTONE
		stellare					
		obtusum					
LOWER SINEMURIAN	turneri	birchi	STATFJORD FORMATION — Nansen Member	DUNROBIN BAY FORMATION — Dunrobin Castle Member	LIAS GROUP	REDCAR MUDSTONE FORMATION — Calcareous Shales	
		brooki					
	semicostatum	sauzeanum					
		scipionianum					
		lyra					
	bucklandi	bucklandi					
		rotiforme					BLUE LIAS FORMATION
204my		conybeari					
HETTANGIAN	angulata	complanata					
		extranodosa					
	liasicus	laqueus	Eriksson Member	Dunrobin Pier Conglomerate Member			ANGULATA CLAYS & SALTFORD SHALES
		portlocki					
	planorbis	johnstoni					BLUE LIAS FORMATION
208my		planorbis					

Sec. 6.2] The Hettangian to Toarcian (Lower Jurassic) 137

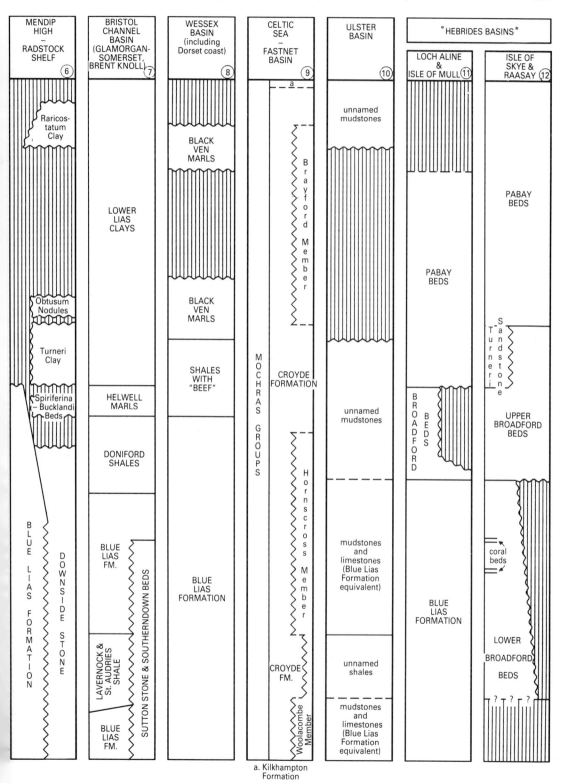

a. Kilkhampton Formation

SUB-STAGES	AMMONITE ZONES AND SUBZONES		NORTHERN NORTH SEA BASIN		SOUTHERN NORTH SEA BASIN	CLEVELAND BASIN		EAST MIDLANDS SHELF — "MIDLANDS BASIN"
			NORTH VIKING GRABEN ①	MORAY FIRTH BASIN ②	③	④		⑤
187my UPPER TOARCIAN	levesquei	aalensis	DRAKE FM.		LIAS GROUP	BLEA WYKE SST. FM.	Yellow Sandstone Member	CEPHALOPOD BEDS
		moorei						
		levesquei					Grey Sandstone Member	
		dispansum						
	thouarsense	fallaciosum					Fox Cliff Siltstone Member	COTTESWOLD SANDS
		striatulum				WHITBY MUDSTONE FORMATION	Peak Mudstone Member	
	variabilis							
LOWER TOARCIAN	bifrons	crassum	DUNLIN GROUP	'Cook Sand'			Alum Shale Member	clays
		fibulatum						
		commune						CEPHAL-OPOD BEDS
	falciferum	falciferum		COOK FM	ORRIN FORMATION		Jet Rock Member	FISH BEDS
		exaratum						
	tenuicostatum	semicelatum					Grey Shale Member	LEPTAENA BEDS
		tenuicostatum						
		clevelandicum						
193my		paltum						MARLSTONE ROCK BED
UPPER PLIENSBACHIAN	spinatum	hawskerense	BURTON FM.		MARLSTONE ROCK BED	CLEVELAND IRONSTONE FORMATION		
		apyrenum						
	margaritatus	gibbosus						Claystones, siltstones, sandstones
		subnodosus						
		stokesi						
LOWER PLIENSBACHIAN	davoei	figulinum	AMUNDSEN FM.		LIAS GROUP	STAITHES SANDSTONE FORMATION		clays with nodules
		capricornus						
		maculatum				REDCAR MUDSTONE FM.		
	ibex	luridum					Ironstone Shales	
		valdani						PECTEN IRONSTONE (top block)
		masseanum	? ?					
	jamesoni	jamesoni						PECT. IRONST. (bottom)
		brevispina						
		polymorphus					Pyritous Shales	
198my		taylori		DUN. BAY FM. / Ladys Walk Shale Member				

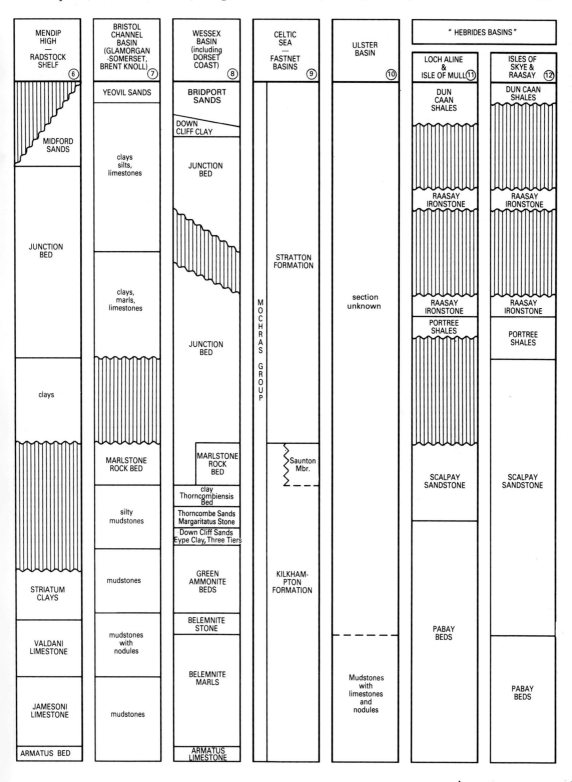

Trochammina (*T. squamosa*, *T. canningensis*), *Ammobaculites* (*A. alaskensis*, *A. vetustus*), *Lagenammina* (*L. jurassica*) and *Verneuilinoides* sp. 1. This is the *Trochammina–Verneuilinoides* assemblage of Nagy (1985a, b), described from the Toarcian Drake Formation in the Statfjord Field (Norwegian sector of the North Sea), interpreted as representative of a prodelta muddy environment. The observed reduction in frequency of *Verneuilinoides* and relatively increased numbers of *Lagenammina* and *Ammobaculites* towards the top of the formation probably reflects increased deltaic influence prior to the outbuilding of the Brent Group delta in the Middle Jurassic (Nagy *et al* 1984). The diversity is much higher than would be expected in a brackish environment, and a fully marine setting is envisaged; an important factor is the association with a high content of terrestrially derived organic material, causing conditions of reduced oxygen and pH.

From the Pliensbachian of the Statfjord area (northern North Sea), Nagy (1985a, b) interpreted an assemblage with up to 45% *Ammodiscus* (nodosariids comprising the rest of the association) as being indicative of increased deltaic influence, based mainly on the analogy with abundant *Ammodiscus* in lagoonal environments in Svalbard (Lower Jurassic) and Yorkshire (Middle Jurassic). The associated abundance of nodosariids in the North Sea case suggests, however, that this particular species of *Ammodiscus* (*A. siliceus*) preferred normal marine conditions.

Involutinids are occasionally common in the Lower Jurassic (mainly *Involutina liassica*, with *Semiinvoluta* sp.), recorded from limestones (Brodie, 1853) as well as shales; such horizons may be indicative of regressive carbonate buildups or reefs (e.g. Wicher, 1952) as in the Tethyan Triassic.

6.2.4.2 Depositional history and faunal associations

Through the Early Jurassic an overall progressive increase in diversity of the foraminiferal associations is seen in Britain and mainland Europe, related to a continuation of the marine transgression which began in the latest Triassic (Rhaetian). Superimposed upon this overall trend of sea level rise, a series of alternating rises and falls in sea level have been proposed (Hallam, 1981; Haq *et al.*, 1987). From the point of view of the foraminiferal microfauna, transgressions appear to relate, in general, to new appearances of taxa and regressions (sea level drop) to species extinctions. Both types of change can also create unconformities, either by sediment starvation in basinal locations during sea level rise or by erosion due to reduction in base level during sea level fall.

Sedimentary facies changes would also be controlled by such events, as well as by variations in local tectonics and rates of subsidence. Some interrelationships are now becoming apparent linking microfaunal variations and evolution with larger scale depositional events in the British Early Jurassic.

The general trend of increasing salinity seen in the latest Triassic, as a result of the late Rhaetian transgression, continued into the earliest Jurassic, allowing the progressive development of more diverse foraminiferal faunas. The Paper Shale unit seen throughout Britain at the base of the Lias Group immediately precedes a marked increase in foraminiferal diversity noted in the Pre-Planorbis Beds. These events occur shortly beneath the appearance of *Psiloceras* and appear to be the result of a major transgressive pulse close to the base of the Jurassic.

The appearances of widespread bituminous paper shale horizons mark the beginnings of several transgressive sequences in the Jurassic, and have been interpreted as reflecting sea-bed anoxicity during sea level rise (Hallam and Bradshaw, 1979); such paper shales typically contain only rare foraminifera. Thus, by early Hettangian times a broad shallow marine shelf sea was established over most of Britain and northwest Europe, colonised by a low (but gradually increasing) diversity foraminiferal fauna suggesting normal oxygen levels. The typical

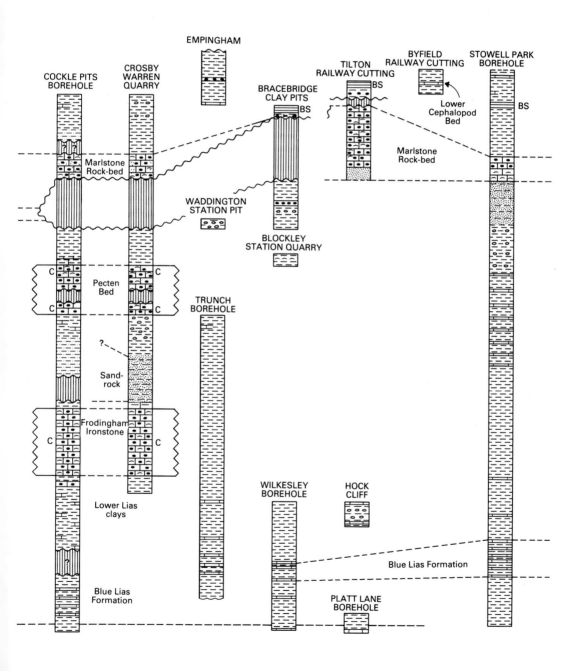

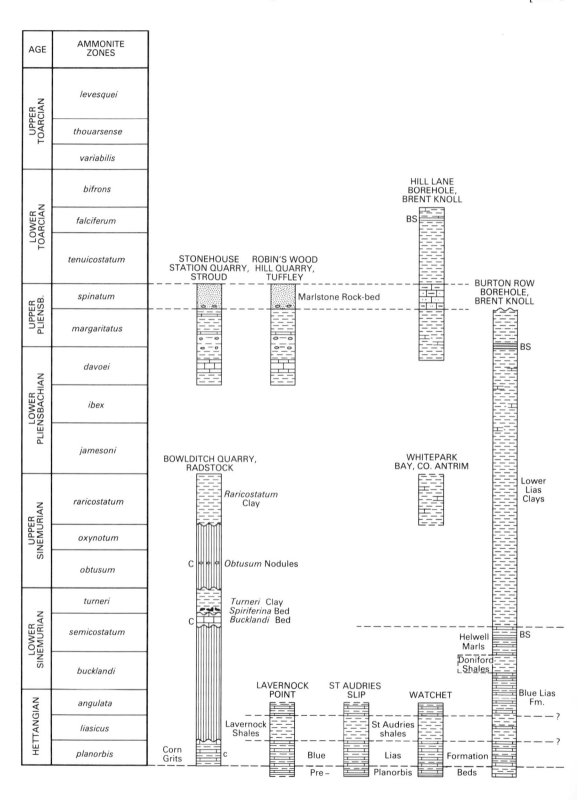

Sec. 6.2] The Hettangian to Toarcian (Lower Jurassic)

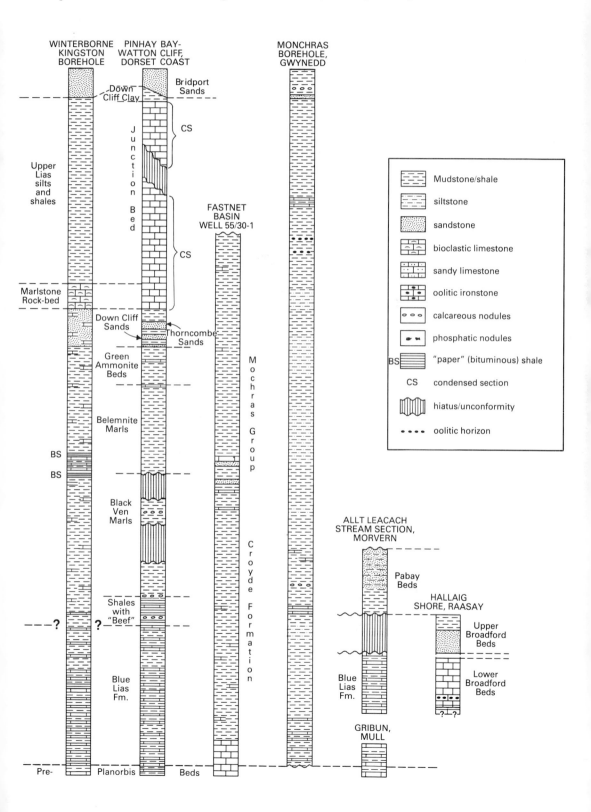

changes revealed in outcrop and onshore borehole data (Hallam, 1961, 1978, 1981, 1984; Donovan et al., 1979; Morton, 1987) and the second being to study the relationships between subsurface sequences deduced from analysis of seismic data (Vail et al., 1977, 1984; Vail and Todd, 1981; Haq et al., 1987).

Changes in the foraminiferal assemblages seem to be related to sea level changes, in that well-documented transgressions mark the arrival of new species and subspecies (evolutionary appearances and/or migration) whereas regressions and times of lowered sea level appear to equate to extinctions. This relationship has been discussed by Hallam (1961, 1987), with reference to a range of fossil groups, including foraminifera, but we now have many more data to hand regarding the distribution of Lower Jurassic foraminifera and reported sea level changes. The new data set presented herein on the stratigraphic distribution charts show that several major assemblage turnover events are recognisable.

Tectonic changes also exercised important control on lithological and thickness variations in the Lower Jurassic onshore and offshore Britain, with positive, high areas (often fault-bounded fault blocks or horsts) ('swells' sensu Hallam, 1958) separating basinal (or graben) areas of major subsidence. Thus, thick, continuous basinal sequences (as in the Celtic Sea and Southern North Sea Basins) are separated by condensed and attenuated sequences on highs (as in the Dorset, Radstock–Mendip and Market Weighton areas).

Certain of these highs appear to have been emergent in the Hettangian–earliest Sinemurian, including probable islands of Carboniferous Limestone in the Mendip–Glamorgan area, where nearshore calcareous sandstones were deposited (Downside Stone in the Mendips, Southerndown Beds and Sutton Stone in Glamorgan).

For the Lias Group of southern Britain, it has been proposed that periods of greater transgression coincided with deposition of more argillaceous units, with regressive periods being characterised by the development of more calcareous units (Donovan et al., 1979). The Blue Lias Formation is interrupted in the *liasicus* Zone by a widespread shale unit in Somerset (St. Audrie's Shales, Palmer, 1972a), South Wales (Lavernock Shales, Trueman, 1920; Waters and Lawrence, 1987), Avon (Saltford Shales, Donovan, 1956), the east Midlands, Humberside, Cleveland (Angulata Clays, Getty, in Cope et al., 1980a) and Northern Ireland (unnamed argillaceous units, Getty, in Cope et al., 1980a) which has been interpreted as indicating a time of raised sea level (Donovan et al., 1979). This lithofacies change (at the base of the *liasicus* Zone) coincides with an influx of new species, including important zonal indices such as *Planularia inaequistriata* and *Frondicularia terquemi* subsp. A (Figs 6.2.6, 6.2.7), plus a widespread abundance of *Reinholdella* (mainly *R.? planiconvexa*). The latter may indicate reduced oxygen levels associated with this transgression, although bituminous shales are not recorded at this time. Shortly afterwards, miliolids (particularly *Ophthalmidium liasicum* and '*Spiroloculina*' sp. A) became abundant, this being the first appearance of miliolids in the British Jurassic. Abundances of miliolids in the Jurassic have been considered as typical of shallow water conditions (Gordon, 1970) though these particular forms in the *liasicus* Zone appear to be associated with increased water depth. The foraminiferal evidence agrees with the interpretations of Donovan et al. (1979) and Hallam (1981) of a transgressive event in the earliest *liasicus* Zone. Haq et al. (1987) place a peak of Hettangian sea level slightly younger, within the *liasicus* Zone.

The seismic stratigraphers recognise a major sea level drop around the end of the Hettangian (basal Sinemurian sequence boundary of Vail and Todd (1981) revised to intra-*angulata* Zone by Haq et al., 1987), at which time one would expect prograding regressive deposits (carbonate buildups or deltas) followed by a widespread major unconformity. Such large scale features are not seen in the British region, even though this interpretation is apparently derived from the British onshore sections (Haq et al., 1987).

There is some evidence for a regression at this time in Britain, but on a more modest scale. There is a suggestion of an erosional break marked by a conglomerate in South Wales (in the nearshore facies) (Hallam, 1961), an unconformity on the Radstock Shelf (though this could be of any age between *angulata* Zone and late *bucklandi* Zone) plus an increase in limestone in the *angulata* Zone of southern Britain signifying a return to Blue Lias facies. The interpretation of an end-Hettangian regression in the northern North Sea (Vail and Todd, 1981) is based on the recognition of prograding clinoform seismic reflectors in the upper Statfjord Formation, however this event cannot be dated with any accuracy owing to impoverished microfossil yields in the sediments in question (unpublished data), as was also admitted by Vail and Todd (1981). The evidence for a similar break in the Inner Moray Firth, also cited by Vail and Todd (1981), is equally tentative, as the non-marine Dunrobin Bay Formation is extremely difficult to date owing to a lack of fossils.

If fossil extinctions are to be expected at times of regression, it is notable that several index foraminifera become extinct at or near the top of the *angulata* Zone (*Lingulina tenera collenoti*, *L. tenera substriata* (at most localities), *F. terquemi* subsp. A). *Dentalina langi* both appeared and became extinct within the late *angulata* Zone.

The *bucklandi* Zone and lower *semicostatum* Zone represent the upper part of the Blue Lias Formation. This interval is characterised in Britain by an increased content of limestone and it is in this facies that *Involutina liassica* becomes common. This genus is a well known indicator of shallow water carbonate accumulations in the Tethyan Triassic and its abundance at this time in Britain may be similarly controlled. This species has been observed to be concentrated in limestones of the *bucklandi* Zone (see Brodie, 1853) which concurs with the view that this period of carbonate concentration in Britain is indicative of shallowing (Donovan et al., 1979). In western Scotland, the deposition of coral beds in the *bucklandi* Zone may be indicative of this shallowing period. On the eustatic sea level curve of Haq et al. (1987), the *bucklandi* Zone was a time of low, but increasing, sea level.

An important facies change occurs in western Europe within the *semicostatum* Zone (*scipionianum–lyra* Subzones), where Blue Lias deposition is terminated by a return to argillaceous sedimentation, including organic-rich shales (e.g. the Shales with "Beef" of Dorset). A significant deepening event is indicated by this change (Hallam, 1981) which also saw increased onlap onto the London Platform (Donovan et al., 1979) and transgression of offshore argillaceous facies onto the former islands in Glamorgan. In western Scotland, this event is marked in Skye by an abrupt change form sandy deposition to shale (Hallam, 1961). In eastern England, the deposition of the Frodingham Ironstone, a condensed, iron-rich calcareous unit, was initiated in the late *semicostatum* Zone; the lack of clastic sediment in this unit probably is due in part to suppression of

Fig. 6.2.6 — Stratigraphic distribution of foraminifera in the British Hettangian–Sinemurian (part 1). Numbered ranges refer to U.K. areas as in Fig. 6.1.1: (1) Viking Graben; (2) Moray Firth Basin; (3) Southern North Sea Basin; (4) Cleveland Basin; (5) 'Midlands Basin'–East Midlands Shelf; (6) Mendip High–Radstock Shelf; (7) South Wales–North Somerset (eastern Bristol Channel Basin); (8) Wessex Basin; (9) Fastnet–Celtic Sea–Cardigan Basins; (10) Ulster Basin; (11) Mull–Morvern; (12) South Minch Basin (Skye-Raasay). Unnumbered ranges are summary ranges of longer ranging taxa.

148 **Jurassic** [Ch. 6]

The Hettangian to Toarcian (Lower Jurassic)

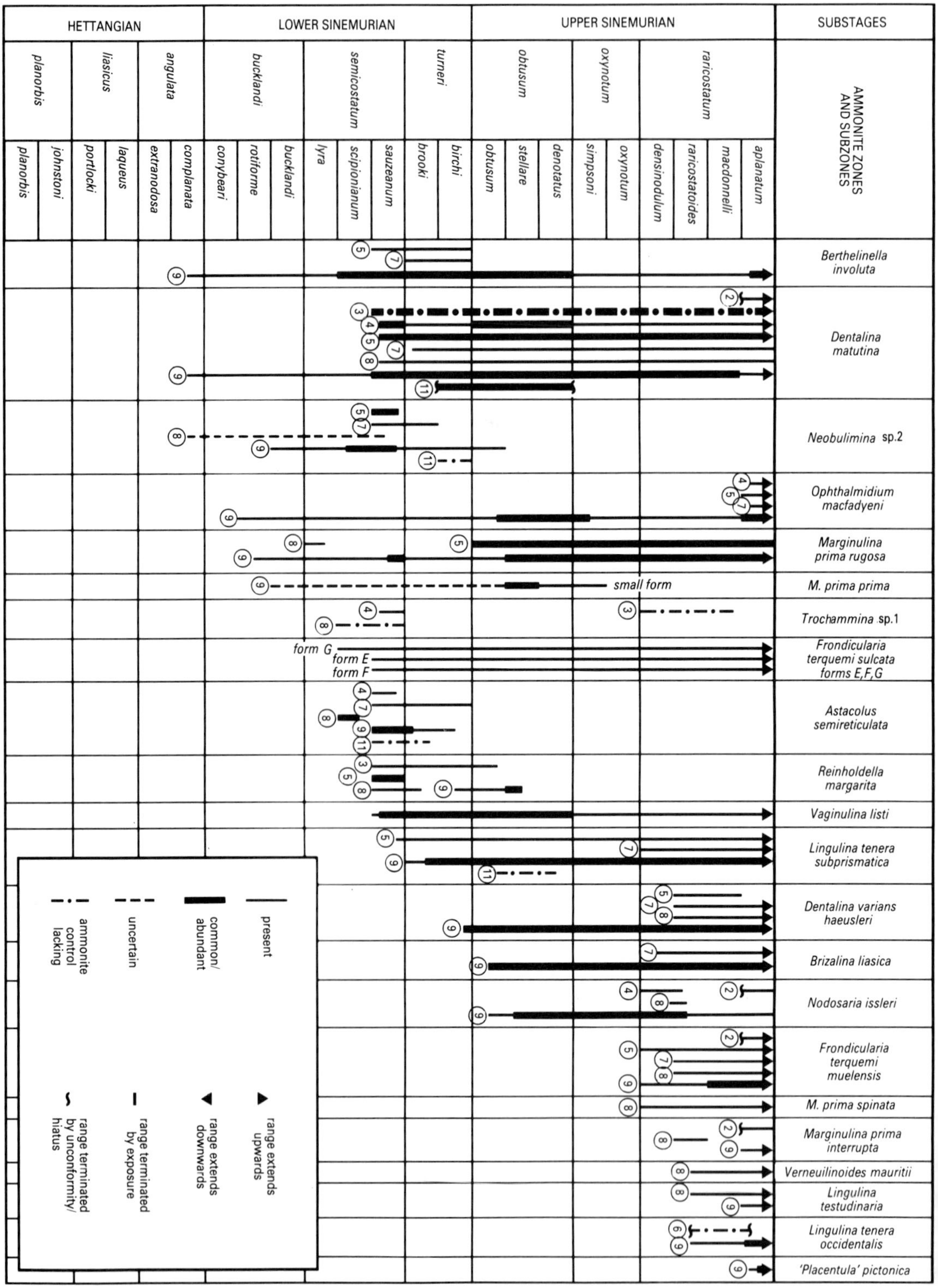

150 Jurassic [Ch. 6]

SUBSTAGES	HETTANGIAN			LOWER SINEMURIAN				UPPER SINEMURIAN					
AMMONITE ZONES AND SUBZONES	planorbis	liasicus	angulata	bucklandi	semicostatum	turneri	obtusum	oxynotum	raricostatum				
	planorbis	johnstoni	portlocki	laqueus	extranodosa	complanata	conybeari	rotiforme	bucklandi	lyra	scipionianum	sauzeanum	brooki	birchi	obtusum	stellare	denotatus	simpsoni	oxynotum	densinodulum	raricostatoides	macdonnelli	aplanatum

Taxa (right column):
- Berthelinella involuta
- Dentalina matutina
- Neobulimina sp.2
- Ophthalmidium macfadyeni
- Marginulina prima rugosa
- M. prima prima
- Trochammina sp.1
- Frondicularia terquemi sulcata forms E,F,G
- Astacolus semireticulata
- Reinholdella margarita
- Vaginulina listi
- Lingulina tenera subprismatica
- Dentalina varians haeusleri
- Brizalina liasica
- Nodosaria issleri
- Frondicularia terquemi muelensis
- M. prima spinata
- Marginulina prima interrupta
- Verneuilinoides mauritii
- Lingulina testudinaria
- Lingulina tenera occidentalis
- 'Placentula' pictonica

Legend:
- ▬ present
- ▬▬ common/abundant
- - - - uncertain
- -·-·- ammonite control lacking
- ▶ range extends upwards
- ◀ range extends downwards
- | range terminated by exposure
- ∫ range terminated by unconformity/hiatus

terrestrial clastic input by rising sea level.

This sea level rise is regarded as one of the most significant of the whole Jurassic (Hallam, 1981). In the northern North Sea (Viking Graben), marine deposition took place for the first time in the Jurassic, where the marine topmost calcareous unit (Nansen Member) of the Statfjord Formation onlaps non-marine sediments beneath. Although referred to by Vail and Todd (1981) as 'basal Sinemurian', this transgression can in fact be dated as late *semicostatum* Zone or younger on the basis of its foraminiferal content (authors' data). Haq *et al*. (1987) show a eustatic peak at the base of the *semicostatum* Zone, but it is not clear whether this equates to the transgression discussed here.

This major intra-*semicostatum* Zone transgression coincides also with a major incursion of foraminiferal taxa, including several first appearances of stratigraphically significant species, such as *Astacolus semireticulata*, *Reinholdella margarita* and *F. terquemi sulcata* forms F and G. In addition, there are numerical increases in longer ranging species such as *Neobulimina* sp. 2, *Dentalina matutina*, *Vaginulina listi* and *Lingulina tenera subprismatica*.

This transgressive phase appears to have extended through the succeeding *turneri* Zone, at which time the Turneri Clay onlapped the Radstock Shelf. The typical late *semicostatum* Zone foraminiferal assemblages carry over into the *turneri* Zone and the interval appears to be palaeoenvironmentally consistent, at least in England. The geology is more complex in western Scotland where an intra-*turneri* Zone hiatus locally omits the whole of this zone (as in Morvern), or separates upper *turneri* Zone shales above from a lower *turneri* Zone sandstone below (as in Skye). The evidence from Scotland would appear to support the proposed minor sea level drop of Haq *et al*. (1987) at which time an unconformity may be expected.

The evidence in the succeeding *obtusum* and *oxynotum* Zones is similarly equivocal. This was a time of postulated shallowing in Europe according to Hallam (1978, 1981); thus in Dorset and across southern England, including the Radstock Shelf, there is a widespread break omitting the *oxynotum* Zone (see Cope *et al*., 1980a). This hiatus also occurs in the subsurface around the northwest margins of the London Platform (Donovan *et al*., 1979). In this latter area a clear angular unconformity can be seen between the *obtusum* Zone and *raricostatum* Zone, suggesting either a sea level drop or tectonic change. This evidence conflicts with the eustatic peak postulated by Haq *et al*. (1987) in the earliest *oxynotum* Zone.

The foraminiferal evidence shows a significant turnover event within the *obtusum* Zone, in the early part of which several important species became extinct (*Neobulimina* sp. 2, *Astacolus semireticulata*, *Reinholdella margarita*, *Planularia inaequistriata*). This is suggestive of regression, and is quickly succeeded within the zone by new appearances such as *Brizalina liasica*, *Nodosaria issleri* and *Lenticulina muensteri* (in abundance), which may indicate renewed transgression.

The *raricostatum* Zone has been considered to represent the beginning of a major latest Sinemurian transgression in Europe and beyond (Hallam 1961, 1978). This is clearly seen around the margins of the London Platform (Donovan *et al*., 1979) and elsewhere in Britain (Cope *et al*., 1980a) where strong onlap occurs at this time and *raricostatum* Zone sediments overlie pre-*oxynotum* Zone section. In eastern Scotland, at Dunrobin Castle, fully marine *raricostatum* Zone clays with abundant foraminifera occur above the non-marine Dunrobin Bay Formation, a situation also seen offshore in

Fig. 6.2.7 — Stratigraphic distribution of foraminifera in the British Hettangian–Sinemurian (part 2). Range numbering as in Fig. 6.2.6.

the Inner Moray Firth (e.g. Beatrice Field). In the northern North Sea (Viking Graben) the basal Dunlin Group is consistently of Early Pliensbachian age.

In contrast, Haq et al. (1987) propose a major sea level drop in the *raricostatum* Zone, which is supported by the developement of a break at this time in some areas, such as the Witney Borehole (Oxfordshire), Rugby (Warwickshire), plus North and South Humberside (Cope et al., 1980a). The development of a sandy facies (Sandrock) between Nottinghamshire and South Humberside may be further local evidence of regression at this time.

Within the foraminiferal fauna, important new evolutionary appearances are seen in the *raricostatum* Zone of species which range into the Pliensbachian such as *Lingulina tenera occidentalis*, *Marginulina prima interrupta* and *M. prima spinata*. This supports the view of Hallam (1981) of a latest Sinemurian transgression, extending into the Early Pliensbachian. On the other hand, at the Sinemurian/Pliensbachian boundary, several extinctions occur, plus there is evidence in some areas of shallowing indicated by local abundances of nubeculariids (particularly *Ophthalmidium* and *Nubecularia*, as in the Mochras Borehole).

This pronounced faunal turnover occurred across the Sinemurian/Pliensbachian boundary throughout northwest Europe, and diverse foraminiferal faunas are generally recorded in the Lower Pliensbachian as in eastern England (e.g. Lincolnshire), southern North Sea Basin and southern England (Dorset). In other areas of Britain, such as the Celtic Sea Basin (Mochras Borehole) and Fastnet Basin (Ainsworth and Horton, 1986), less diverse foraminiferal faunas are recorded, particularly in the *jamesoni* and *ibex* Zones.

In the *ibex* Zone (*luridum* Subzone) and *davoei* Zone in eastern England (e.g. Crosby Warren, Robin Hood's Bay) agglutinating foraminifera suddenly become dominant (including *Haplophragmoides* sp. 1, *H. lincolnensis*). This may suggest reduced oxygenation due to increased input of terrestrial organic matter in this area. This possible regressive event equates with a proposed sea level drop (Haq et al., 1987) beginning in the *luridum* Subzone.

Transgressive events have been proposed by Hallam (1981) and Haq et al. (1987) at the beginning of the Late Pliensbachian (*margaritatus* Zone), although it is difficult to reconcile such a sea level rise with the frequent and widespread development of sandy sediments seen in Britain at this time, such as in Dorset (Three Tiers, Downcliff Sands, Thorncombe Sands), Yorkshire (Staithes Formation) and western Scotland (Scalpay Sandstone). Locally in Yorkshire and in Skye and Raasay, sand deposition commenced as early as *luridum* Subzone or *davoei* Zone; this may indicate some merging with the *luridum* Subzone sea level drop discussed above.

The British Upper Pliensbachian is characterized by the development of oolitic ironstones, which was initiated in Yorkshire in the *margaritatus* Zone (Cleveland Ironstone Formation), and continued into the *spinatum* Zone. The *spinatum* Zone was also a time of ironstone development in the Midlands and southern England, where the Marlstone Rock Bed was deposited. The latter ranges in age into the *tenuicostatum* Zone. The origin of the ironstones has been much debated (e.g. Hallam and Bradshaw, 1979), although there is considerable uncertainty regarding their environmental implications. An oolitic horizon is also developed in the northern North Sea (Viking Graben) in the *margaritatus* Zone (top of the Amundsen Formation of the Dunlin Group).

The overlying *spinatum* Zone, represented by the Burton and lower Cook Formations, is unusual for the British area in being developed in claystones; this is also the case in the Celtic Sea Basin (e.g. the Mochras Borehole) where the zone is developed in an expanded argillaceous facies.

Foraminifera are abundant in the Upper

Pliensbachian but differ from those of the Lower Pliensbachian in that species disappear (e.g. *Vaginulinopsis denticulatacarinata*) or decline across the boundary (e.g. *M. prima interrupta*). Other taxa increase in abundance in the *margaritatus* Zone, such as *Brizalina liasica* (especially in the southern North Sea) and *Lenticulina muensteri acutiangulata* (southern and northern North Sea, Midlands and southern England).

These abundance variations may be suggestive of environmental change, but can nevertheless be used for correlation over wide areas. The Mochras Borehole Upper Pliensbachian sequence, with its dominance of argillaceous sediment, without suggestion of sand or ironstone development, suggests different environmental conditions; this is confirmed by the relative abundance in this section of *Ophthalmidium* and *Spirillina*, genera possibly indicative of shallow water (Gordon, 1970).

In eastern England (Yorkshire, North Humberside, South Humberside) an erosional unconformity is developed at the base of the *spinatum* Zone, probably controlled by tectonic movements on the Market Weighton High (Howard, 1985). An unconformity is also seen in the northern North Sea (Viking Graben) within the Burton Formation, and can be correlated with that in eastern England on the basis of age diagnostic foraminifera. This may be equivalent to the '189' unconformity of Vail and Todd (1981) recognised in the Viking Graben on seismic evidence, and may have been caused either by tectonic activity or eustatic sea level drop. In both regions the break is onlapped by the sediments of the *spinatum* Zone, which may reflect higher sea level at this time (possibly causally linked to formation of the Marlstone Rock Bed ironstone). There is no evidence of the intra-*spinatum* Zone sea level fall of Haq et al. (1987).

The base *spinatum* Zone unconformity coincides with, and may be linked genetically to, several important extinctions which occur synchronously over a wide area, including *Dentalina matutina* and *Haplophragmoides lincolnensis*, whilst also terminating the abundance of *L. muensteri acutiangulata*. *Spinatum* Zone foraminifera are not well known from Britain owing to the unfavourable condensed ironstone facies. Faunas in the Marlstone Rock Bed comprise mainly *Lenticulina muensteri*, a form associated with nearshore/shallow marine sands and oolites in the Jurassic (R. L. A. Young, pers. comm.). The shales between the thin ironstones in Yorkshire (Cleveland Ironstone Fm.) are characterised by a predominantly agglutinating foraminiferal association, probably indicative of reduced sea bed oxygenation. This association is also developed in the northern North Sea *spinatum* Zone.

In the only other known British argillaceous development of the zone, the Mochras Borehole, calcareous foraminifera are predominant, with nodosariids in the lower part and *Spirillina* spp. and *Ophthalmidium* spp. (*northamptonensis* and *macfadyeni*) becoming common in the upper part. The latter two genera may be indicative of local shallowing.

At the base of the Lower Toarcian occurs a local bituminous shale development in the Cleveland Basin, suggesting restricted circulation possibly linked to continued sea level rise. The Toarcian is well developed in argillaceous facies only in Yorkshire and the Mochras Borehole. Elsewhere in Britain the stage is represented by highly condensed limestones (e.g. Junction Bed of Dorset–Somerset, Cephalopod Bed and Cephalopod Limestone from Avon to Northamptonshire), ironstone (Raasay Ironstone of Skye–Raasay) or sandstones (Cotteswold Sands of Avon, Gloucestershire). Most of the Upper Toarcian is absent between Gloucestershire and Yorkshire.

Lowermost Toarcian (*tenuicostatum* Zone) foraminiferal assemblages are rich and constant in character from the southern North Sea to the Celtic Sea, probably due to good geographic communication at a time of maximum transgression. In content, the foraminiferal assemblages are most similar to those of the Upper Pliensbachian and many species range across the boundary. A regional variation is seen within the *Reinholdella* populations, which increase in abundance in the *tenuicostatum* Zone; in eastern and central England and the southern North Sea, *Reinholdella macfadyeni* attained flood abundance from the base of the zone, whilst in the Celtic Sea Basin (Mochras Borehole), *R. dreheri* and *R. pachyderma* were common. Of significance are the rare appearances in the *tenuicostatum* Zone of species more characteristic of the later Toarcian and early Bajocian such as *Citharina colliezi, Palmula deslongchampsi* and *Vaginulina/ Citharina clathrata*.

A fundamental turnover in the foraminiferal associations occurs across the *tenuicostatum/falciferum* Zone boundary, at which horizon many taxa, both long and short ranging, became extinct (see Figs 6.2.8, 6.2.9). The *Frondicularia terquemi*, *Lingulina tenera* and *Marginulina prima* plexus groups underwent major declines and the extinction of subspecies such as *F. terquemi* subsp. B, *L. tenera occidentalis* and *M. prima interrupta* occurred.

This event signalled the most marked microfaunal turnover of the Early Jurassic and coincides with a major black bituminous or paper shale development across Britain and mainland Europe at the base of the *falciferum* Zone (*exaratum* Subzone) (e.g. the Jet Rock Member in Yorkshire, Posidonienscheifer in West Germany). In the northern North Sea, an equivalent facies occurs within the Cook Formation.

This regional black shale development is considered to reflect anoxic sea bed conditions related to a major sea level rise (Hallam and Bradshaw, 1979; Hallam, 1981) locally associated with onlap of basement highs (Horton *et al.*, 1980). Haq *et al.* (1987) place this sea level peak one ammonite zone too high, at the beginning of the *bifrons* Zone. Unlike other transgressive peaks in the Early Jurassic, this *falciferum* Zone event caused more foraminiferal extinctions than new appearances; this is probably related to the widespread stagnation. Notwithstanding this, a number of calcareous species survived the anoxic event and became abundant above, including *Reinholdella macfadyeni* and *Lenticulina muensteri*.

In the *falciferum* Subzone, normal marine conditions became re-established in Britain and from this time onwards new species appearances are seen, with the overall balance of species becoming more Middle Jurassic in affinity. In this subzone *Citharina colliezi* and *V./C. clathrata* become common and the earliest *Lenticulina quenstedti* appears. This faunal change is seen at the same level across a wide area of Britain (Midlands, Wessex Basin, Celtic Sea Basin) and appears to constitute a faunal replenishment following the *exaratum* Subzone anoxic event. These calcareous microfaunas are best represented in the East Midlands, Ilminster and Mochras Borehole sections.

To the north of the Market Weighton High, agglutinating foraminifera became dominant in the Early Toarcian and continued as such through the Late Toarcian. This situation is also seen in the northern North Sea (Cook and Drake Formations), and is probably related in these areas to oxygen-deficient sea bed conditions.

The *bifrons* Zone is interpreted by Hallam (1981) as a time during which sea level continued to rise following the early *falciferum* Zone transgressive event. Although this author (Hallam, 1978, 1981) has tentatively suggested a shallowing event at the end of *bifrons* Zone times (i.e. at the Early/Late Toarcian boundary), the regional geology is more complex, with condensed limestones (e.g. Dorset), sands (Avon–Gloucestershire) and ironstone (western Scotland) all being deposited at this time. A correlative sand development is also seen at the top of the Cook Formation in the northern North Sea; it is of latest Early Toarcian age.

Within the Late Toarcian, three cycles of eustatic sea level rise and fall are proposed by Haq et al. (1987), while Hallam (1981) indicates a deepening phase through the *variabilis* and *thouarsense* Zones, followed by a latest Toarcian shallowing (within the *levesquei* Zone). Both these works agree in their interpretation of a latest Toarcian shallowing, presumably represented by sand influxes in Dorset (Bridport Sands) and Yorkshire (Blea Wyke Sandstone Formation). Lack of regional expression of such 'eustatic' changes is evident, however. Thus, in the Mochras Borehole, marine conditions persisted throughout the Late Torcian, allowing the development of rich foraminiferal assemblages represented mainly by the *Lenticulina muensteri* plexus, *Spirillina* spp., *Reinholdella dreheri*, *R. macfadyeni* and *Palmula deslongchampsi*, among the calcareous foraminifera. Agglutinating foraminifera (*Lagenammina jurassica* Barnard, *Ammobaculites* spp.) are also occasionally common. A comparable calcareous assemblage occurs in the Down Cliff Clay of Dorset (*levesquei* Zone).

There is no evidence of shallowing in the northern North Sea where agglutinating foraminiferal assemblages (dominated by *Trochammina* spp. and *Verneuilinoides* sp. 1, comparable to those of Yorkshire, occur in the Drake Formation (uppermost Dunlin Group).

6.2.5 Stratigraphic distribution
6.2.5.1 Sections studied
A considerable amount of information is available on the distribution of Lower Jurassic foraminifera from many sections throughout Britain, including outcrops, onshore boreholes and offshore oil exploration and production wells. To include separate range charts for each studied section is not feasible owing to the large number of sections involved. Therefore on the range charts (Figs 6.2.6–6.2.9) and stratigraphic columns, data have been combined from groups of localities which fall within discrete basins or shelf areas. The range charts include separate range bars for 12 basinal regions which represent major areas of Jurassic accumulation in Britain (see Fig. 6.1.1); these basins extend from present day onshore to offshore areas. The lithostratigraphic columns represent the same basinal areas, but also include sections from shelf or high areas where Jurassic deposition was attenuated. Most of these data are from the senior author's research, but all previously published work is also included in this stratigraphic compilation. Basinal regions not included in the range charts include those where the Lower Jurassic is poorly developed (e.g. Central Graben, Faeroes Basin) or from which wells were not released at the time of writing (e.g. Western Approaches Basin).

The localities from which data are available are listed below and lithological columns for each studied section are shown in Figs 6.2.4 and 6.2.5. The numbers given here are those used in Figs 6.2.2, 6.2.3 and 6.2.6 to 6.2.9. Localities with previously published foraminiferal data are indicated by author.

Northern North Sea Basin (i.e. north of Mid North Sea High) (1a, 2).
(1a) North Viking Graben ('Brent Province', including Statfjord Field, Norway block 33/9, see Nagy, 1985a, b).
(2) Moray Firth Basin, offshore (Beatrice Field, Block 11/30), and onshore (Dunrobin Castle coast section).

Southern North Sea Basin (3)
Well 48/22-1

Fig. 6.2.8 — Stratigraphic distribution of foraminifera in the British Pliensbachian–Toarcian (part 1). Range numbering as in Fig. 6.2.6.

Fig. 6.2.9 — Stratigraphic distribution of foraminifera in the British Pliensbachian–Toarcian (part 2). Range numbering as in Fig. 6.2.6.

156 **Jurassic** [Ch. 6]

SUB-STAGES	AMMONITE ZONES AND SUBZONES		Lingulina tenera tenera	Lingulina tenera tenuistriata	Lingulina tenera pupa	Eoguttulina liassica	Reinholdella? planiconvexa	Nodosaria metensis	Frondicularia brizaeformis	Verneuilinoides sp.1	Frondicularia terquemi sulcata forms C, E, F, G	Frondicularia terquemi bisostata	F. terquemi terquemi	Lenticulina muensteri muensteri	Lenticulina varians varians	Ophthalmidium liasicum	Ophthalmidium northamptonensis
UPPER TOARCIAN	levesquei	aalensis						⑨									⑨
		moorei															
		levesquei															
		dispansum				⑨											
	thouarsense	fallaciosum															
		striatulum															
	variabilis															⑨	
LOWER TOARCIAN	bifrons	crassum								①							
		fibulatum															
		commune															
	falciferum	falciferum												large form			
		exaratum															
	tenuicostatum	semicelatum															⑤
		tenuicostatum			④												
		clevelandicum							④		form C		large form				
		paltum			①				⑤		Ophthalmidium northamptonensis						
UPPER PLIENSBACHIAN	spinatum	hawskerense									form E					⑤	
		apyrenum			③				⑨		form G						
	margaritatus	gibbosus							①								
		subnodosus			⑤				⑤		form F						
		stokesi			⑦												
LOWER PLIENSBACHIAN	davoei	figulinum								①							
		capricornus															
		maculatum															
	ibex	luridum			⑧								small form				
		valdani															
		masseanum															
	jamesoni	jamesoni															
		brevispina															
		polymorphus					⑨										
		taylori															

The Hettangian to Toarcian (Lower Jurassic)

Jurassic

SUB-STAGES	AMMONITE ZONES AND SUBZONES		Frondicularia terquemi muelensis	Verneuilinoides mauritii	Lingulina testudinaria	L. tenera occidentalis	'Placentula' pictonica	Vaginulinopsis denticulatacarinata	Haplophragmoides lincolnensis	Haplophragmoides sp.1	Tristix liasina	Lingulina tenera subsp. A
UPPER TOARCIAN	levesquei	aalensis										
		moorei										
		levesquei										
		dispansum										
	thouarsense	fallaciosum	⑨									
		striatulum										
	variabilis		?									
LOWER TOARCIAN	bifrons	crassum										④
		fibulatum										
		commune										
	falciferum	falciferum										⑤
		exaratum	⑤			⑨						
	tennicostatum	semicelatum										
		tenuicostatum	④									
		clevelandicum										
		paltum								④		⑧
UPPER PLIENSBACHIAN	spinatum	hawskerense										
		apyrenum				⑨		① ④ ⑤	① ⑤		①	⑤
	margaritatus	gibbosus										
		subnodosus					⑤?	③			⑨	
		stokesi					① ③					
LOWER PLIENSBACHIAN	davoei	figulinum		⑧		⑤					⑧	
		capricornus										
		maculatum		⑨								
	ibex	luridum										
		valdani						④				
		masseanum										
	jamesoni	jamesoni										
		brevispina		⑧								
		polymorphus	②		⑨		②					
		taylori										

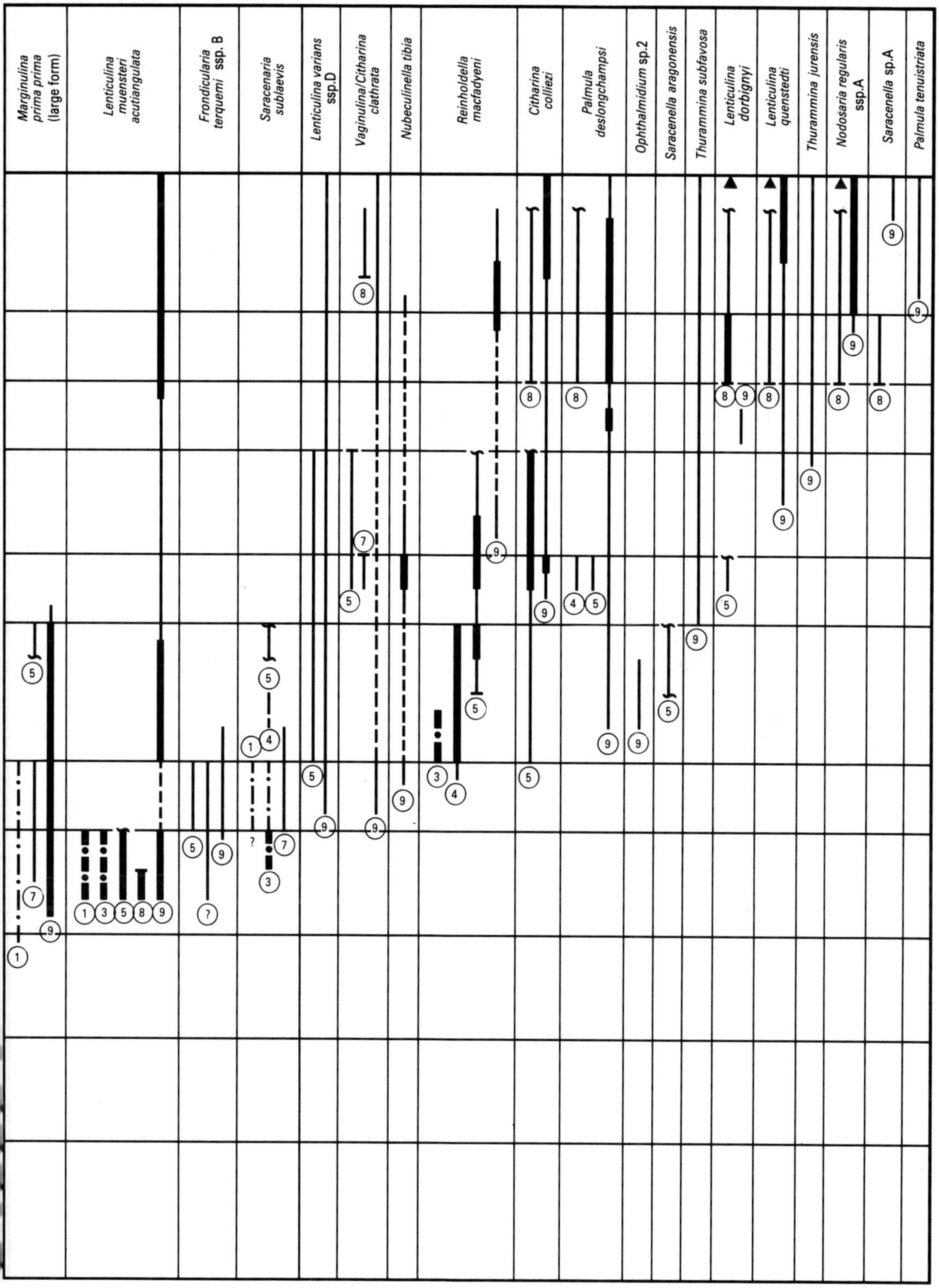

Marginulina prima prima (large form)
Lenticulina muensteri acutiangulata
Frondicularia terquemi ssp. B
Saracenaria sublaevis
Lenticulina varians ssp.D
Vaginulina/Citharina clathrata
Nubeculinella tibia
Reinholdella macfadyeni
Citharina colliezi
Palmula deslongchampsi
Ophthalmidium sp.2
Saracenella aragonensis
Thurammina subfavosa
Lenticulina dorbignyi
Lenticulina quenstedti
Thurammina jurensis
Nodosaria regularis ssp.A
Saracenella sp.A
Palmula tenuistriata

Cleveland Basin (4)
  Redcar coast section (Cleveland)
  Robin Hood's Bay to Hawsker Bottoms coast section (Yorkshire).
  Staithes coast section (Yorkshire).
  Ravenscar coast section (Yorkshire).
  Cockle Pits Borehole (North Humberside).
  Hotham Crossroads Pits (North Humberside).

East Midlands Shelf–'Midlands Basin' (5)
  Trunch Borehole (Norfolk)
  Crosby Warren, Scunthorpe (Lincolnshire)
  Bracebridge Heath Clay Pit, Lincoln (Lincolnshire)
  Waddington Station Pit, Lincoln (Lincolnshire) (Copestake, 1986)
  Empingham (Leicestershire) (Horton and Coleman, 1978)
  Tilton Railway Cutting (Leicestershire)
  Wilkesley Borehole (Cheshire)
  Platt Lane Borehole (Shropshire)
  Byfield Railway Cutting (Northamptonshire) (Barnard, 1950b)
  Stowell Park Borehole (Gloucestershire)
  Hock Cliff (Gloucestershire)
  Robin's Wood Hill, Gloucester (Gloucestershire)
  Stonehouse Station Quarry (Gloucestershire)
  Blockley Station Quarry (Gloucestershire) (Copestake, 1986)

Mendip High–Radstock Shelf (6)
  Bowldish Quarry, Radstock (Avon)
  Cloford and Holwell Quarries, Shepton Mallet (Somerset) (Copestake, 1982)

South Wales–North Somerset (eastern Bristol Channel Basin) (7)
  Lavernock Point (Glamorgan)
  St Audrie's Slip (Somerset)
  Watchet (Somerset)
  Burton Row and Hill Lane Boreholes, Brent Knoll (Somerset)
  Hornblotton Mill, banks of River Brue (Somerset)

Wessex Basin (8)
  Ilminster Bypass, Seavington St. Michael, Ilminster (Somerset) (F. M. Lowry pers. comm.)
  Winterborne Kingston Borehole, Dorset (Coleman, 1982)
  Pinhay Bay to Watton Cliff coast section (Devon–Dorset) (Barnard, 1950a; Jenkins, 1958; Macfadyen, 1941; Copestake, 1987)

Cardigan Basin–Celtic Sea Basin–Fastnet Basin (9)
  Mochras Borehole (Gwynedd) (Johnson, 1975, 1976; Copestake, 1978)
  Fastnet Basin (Ainsworth and Horton, 1986; Ainsworth et al., 1987, in press)

Ulster Basin (10)
  Whitepark Bay (Co. Antrim) (McGugan, 1965) (plus unpublished data of authors)

Hebrides Basins (including South Minch Basin) (12)
  Allt Leacach stream section, Loch Aline, Morvern (unpublished data of authors and J. F. Gregory, pers. comm.)
  Gribun, Isle of Mull
  Hallaig coast section, Isle of Raasay

6.2.5.2 *Stratigraphic distribution*

Foraminifera are of great value in subdividing and correlating the Lower Jurassic. Although the foraminifera were exclusively benthonic during the Early Jurassic (apparent planktonic species are not recorded until the Middle Jurassic), certain species show very consistent stratigraphic ranges over wide geographic areas. Ranges are frequently constant not only within Britain but also in other European countries. One problem, however, is that of variable species definition between workers, owing partly to the considerable degree of morphological variation.

Nevertheless, as documentation of foraminiferal occurrences from European and other countries increases, the international range consistency of Lower Jurassic foraminifera (in both the northern and southern hemispheres) is becoming increasingly evident.

The stratigraphic application of Jurassic foraminifera was first seriously attempted by Bartenstein and Brand (1937), resulting from the need to correlate subsurface sequences in oil exploration onshore West Germany. With the absence of macrofauna such as ammonites due to destruction by the drill bit, it was realised that foraminifera (and other microfossils) survived intact owing to their small size, and indeed occurred in great abundance. As with other fossil groups, appearances ('bases'), extinctions ('tops') and abundance changes were used to date and correlate sequences and thereby interpret subsurface stratigraphy.

Bartenstein and Brand (1937) erected the first zonal scheme based on Jurassic foraminifera, although the zones were generally long ranging. This provided the necessary stimulus, and subsequent works laid greater emphasis on the stratigraphic value of Lower Jurassic foraminifera, including studies in England (Barnard, 1948, 1950a, b, 1956, 1957, 1960), Denmark (Nørvang, 1957; Bang, 1968, 1971, 1972), France (Colloque sur le Lias Français, 1961; Ruget, 1985) and Sweden (Norling, 1972). In addition to Bartenstein and Brand (1937), foraminiferal zonation schemes have been proposed by Barnard (1948, two broad zones for the Hettangian–Sinemurian of England), Bang (1971, Hettangian–Pliensbachian of Denmark, based on boreholes, but not dated by ammonites), Norling (1972, borehole and limited outcrops in Southern Sweden, little ammonite control, Hettangian–Toarcian) and Horton and Coleman (1978, local assemblage zones for the Lower Toarcian of eastern England). More recently, Exton and Gradstein (1984) have also noted the wide geographic consistency in stratigraphic ranges in their proposal of a foraminiferal zonation of the Lower Jurassic (Sinemurian–Toarcian) of offshore Eastern Canada and Portugal. Finally, the present authors, as a result of a synthesis of data from Great Britain with all previously published data, have outlined a preliminary zonation scheme for the whole Lower Jurassic (Copestake and Johnson, 1984; Copestake, 1985).

Since the latter works, many more (largely unpublished) data are now available to the authors as a result of ongoing research regarding foraminiferal distribution in Britain. This has formed the basis for the present work and demonstrates sufficiently consistent ranges (against well documented ammonite zones) over a wide area to allow the proposal herein of a zonation scheme for the British region. This is more detailed than our previous scheme (Copestake and Johnson, 1984) which is of wider geographic scope but of less precision due to the poor range documentation of species against ammonite zones in many published works. The scheme proposed here is known from experience to be workable both onshore and offshore. The zones and subzones, equivalence to ammonite zones and defining criteria (first appearances and last occurrences) are shown in Fig. 6.2.10. Details of the known geographic distribution of the index taxa are shown in Figs 6.2.6 to 6.2.9 and are described in the species descriptions.

Using this zonation it is not possible, with the present definition of the Triassic/Jurassic boundary, to precisely define the base of the Jurassic (Hettangian) on foraminiferal criteria. The base of JF1 zone occurs in the uppermost Penarth Group, slightly beneath the base of the Lias Group (see Chapter 5 for a fuller discussion of the foraminiferal changes around the boundary).

The bases of the Lower Sinemurian, Upper Sinemurian, Lower Pliensbachian, Upper Pliensbachian and Lower Toarcian are all recognisable on foraminiferal criteria (including 'tops'). The base of the Upper Toarcian is not precisely definable in most areas, the JF15/JF16 boundary being placed one ammonite zone above the base (*thouarsense* Zone). However, in the northern North Sea, the

Jurassic

STAGES/ SUBSTAGES	AMMONITE ZONES	AMMONITE SUBZONES	FORAMINIFERA ZONES	FIRST APPEARANCES ('bases')	LAST OCCURRENCES ('tops')	OTHER EVENTS
UPPER TOARCIAN	levesquei	aalensis				
		moorei				
		levesquei				
		dispansum	JF 16			← P. tenuistriata
	thouarsense	fallaciosum				
		striatulum		*← N. regularis ssp. A		common
	variabilis					?← Verneuilinoides sp.1 (Northern North Sea)
LOWER TOARCIAN	bifrons	crassum	JF 15			
		fibulatum		*← L. dorbignyi		
		commune	JF 14		consistent L. tenera plexus ←	
	falciferum	falciferum		*← consistent/common C. colliezi, V./C. clathrata P. deslongchampsi	M. prima prima *← M. prima spinata S. sublaevis M. prima interrupta, F. terquemi plexus, L. tenera occidentalis	← L. quenstedti ← R. macfadyeni •
		exaratum	JF 13			
	tenuicostatum	semicelatum				
		tenuicostatum	JF 12 b			
		clevelandicum				
		paltum		*← R. macfadyeni •	F. terquemi ssp.B	← F. brizaeformis
UPPER PLIENSBACHIAN	spinatum	hawskerense	a			
		apyrenum		← S. sublaevis	*← D. matutina	← H. lincolnensis
	margaritatus	gibbosus				L. muensteri acutiangulata ←
		subnodosus	JF 11	*← F. terquemi ssp.B		consistent ← D. varians haeusleri
		stokesi	JF 10		*← V. denticulata-carinata	L. tenera tenuistriata common
LOWER PLIENSBACHIAN	davoei	figulinum				
		capricornus	b			← L. tenera subprismatica
		maculatum				
	ibex	luridum		← H. lincolnensis		
		valdani	JF 9			
		masseanum				
	jamesoni	jamesoni	a			
		brevispina				
		polymorphus				
		taylori		*← V. denticulatacarinata	*← Nissleri	← common D. matutina V. listi, A. speciosus
UPPER SINEMURIAN	raricostatum	aplanatum	JF 8			
		macdonnelli				
		raricostatoides		F. terquemi muelensis, M. prima spinata, *M. prima interrupta, L. tenera occidentalis		consistent ← D. varians haeusleri
		densinodulum				
	oxynotum	oxynotum				
		simpsoni	JF 7			
	obtusum	denotatus			*← R. margarita	← Neobulimina sp.2
		stellare			*← A. semirecticulata	
		obtusum	JF 6	← N. issleri		
LOWER SINEMURIAN	turneri	birchi	JF 5			consistent/• L. tenera subprismatica ←
		brooki				
		sauzeanum		*← R. margarita D. matutina •		← V. listi •
	semicostatum	scipionianum		*← A. semireticulata		
		lyra	b		*← I. liassica •	← Neobulimina sp.2 •
	bucklandi	bucklandi	JF 4			
		rotiforme	a			
		conybeari			*← D. langi, L. tenera substriata, F. terquemi ssp.A	M. prima incisa,
HETTANGIAN	angulata	complanata	JF 3	*← D. langi, L. tenera substriata, consistent P. inaequistriata	consistent L. tenera collenoti	M. prima insignis, I. liassica (•)
		extranodosa				
	liasicus	laqueus	JF 2	*← P. inaequistriata	common L. tenera collenoti	← R? planiconvexa •
		portlocki		*← F. terquemi ssp.A		
	planorbis	johnstoni				
		planorbis	JF 1	*← L. tenera collenoti		← L. tenera plexus •
RHAETIAN						

base of common *Verneuilinoides* sp. 1 appears to equate to the Lower/Upper Toarcian boundary.

With the data at present available, the top of the Lower Jurassic (Toarcian/Aalenian boundary) in Britain is not definable on the basis of consistently occurring foraminifera species. The top of the JF16 zone, defined on the extinctions of *Lenticulina dorbignyi* and *Nodosaria regularis* subsp. A, occurs in the *discites* Zone of the Lower Bajocian. Lack of expression of the Lower/Middle Jurassic boundary in the foraminiferal assemblages is typical of the Boreal province, not just Great Britain.

6.2.6 Index species

The species included here are those considered to be the most important for zonation and correlation purposes in the British and north European Lower Jurassic, based on our current state of knowledge at the time of writing. Recently published work indicates that certain of these species occur with similar ranges elsewhere in the Northern and also the Southern Hemisphere (e.g. Canada, Argentina, Australia).

Seventy-six species and subspecies are described and figured. The list incorporates not only forms with restricted stratigraphic ranges, but also taxa which are longer ranging, but characteristic (and often common) members of Lower Jurassic foraminiferal assemblages, such as *Lingulina tenera*, *Frondicularia brizaeformis* and *Nodosaria metensis*. Fifteen taxa described in the first edition of this book are not included in this edition. This is to allow additional taxa of greater stratigraphic value to be introduced together with representatives of the background fauna not treated in the first edition.

Non-referenced species records cited in the text are from the senior author's research, while those from the Mochras Borehole are from the unpublished research of both authors. The suprageneric classification followed is that of Loeblich and Tappan (1984). The new genera proposed by Loeblich and Tappan (1986) for dentaline and palmate nodosariids based on surface ornament and wall structure are not utilised herein.

Most taxa are illustrated by scanning electron micrographs. However, several species are also photographed in reflected light to aid their identification by light microscope. Additional taxa are shown in transmitted light views to demonstrate internal chamber arrangement; this is particularly important for specific and generic identification of the Miliolina. Note that the transmitted light views are not thin sections but were obtained by mounting the specimens in Canada balsam followed by photography on a petrographic microscope.

Fig. 6.2.10 — Foraminiferal biozonation of the British Lower Jurassic. Defining events indicated by asterisks (∗). ● = species abundance.

SUBORDER TEXTULARIINA

Ammodiscus siliceus (Terquem)
Plate 6.2.1, Fig.1, reflected light (× 40), Upper Sinemurian, Dorset coast. = *Involutina silicea* Terquem, 1862. Common synonym: *Ammodiscus asper* (Terquem, 1863). Description: test comprising agglutinated tube of four to eight whorls, arranged in circular to ovate planispire; flattened or concave in vertical section; composed of fine to coarse sand grains set in siliceous cement.
Remarks: many authors separate *A. siliceus* and *A. asper*, with the latter allegedly having fewer whorls and coarser texture (Terquem, 1863) and being ovate rather than circular. Lower Jurassic *Ammodiscus* populations contain both end members plus intermediates and are considered to represent a single variable species. Distribution: ranges throughout the Lower Jurassic and into the Aalenian. A characteristic member of Lower Jurassic assemblages, frequently becoming common to abundant, as in the Upper Sinemurian of Robin Hood's Bay and *birchi* Subzone of Dorset (Copestake, 1987). Its oldest record appears to be from the Triassic (Kristan-Tollmann, 1964, as *A. incertus*), but in Western Europe it first occurs in the Hettangian.

Haplophragmoides sp. 1
Plate 6.2.1., Fig.2, reflected light, (× 70); Fig.3 (× 70), Upper Pliensbachian, Bracebridge, Lincoln. Description: test fine grained, white, involute, four to five chambers in final whorl, compressed.
Remarks: the species is most similar in chamber number and chamber height/width ratio to *H. kingakensis* Tappan, 1955, but differs from the latter in being finer grained and compressed rather than inflated. Distribution: *luridum* Subzone to top *margaritatus* Zone at Crosby Warren, Robin Hood's Bay and in northern North Sea, ranging into the *spinatum* Zone at Staithes.

Haplophragmoides lincolnensis Copestake, 1986
Plate 6.2.1., Fig.4, side view (× 90), Fig.9, edge view (× 47), Lower Pliensbachian, Waddington Station Pit, Lincolnshire. Common synonyms: *Trochamminoides* sp. Pietrzenuk, 1965; *Haplophragmoides emaciatus* Bartenstein and Brand, 1937 *non Haplophragmium emaciatum* Brady, 1884. Description: test flattened, bioconcave, subtriangular in outline, 10–12 tubular chambers, three per whorl; sutures depressed.
Remarks: this is an unusual species on account of its three tubular chambers per whorl: it is a valuable stratigraphic marker in being morphologically distinctive, short ranging and widespread. Distribution: intra-Pliensbachian restricted in the UK area (range upper *ibex* Zone–top *margaritatus* Zone); its extinction ('top') is an important marker event in the North Sea (Copestake, 1986). The species is most common in eastern England (Lincolnshire, Yorkshire coast) and occurs also at Blockley and in Dorset (Winterborne Kingston Borehole, Coleman 1982, as *H. emaciatus*). The species appears to be absent from western Britain (e.g. Celtic Sea Basin). Recorded also from northwest Germany (Bartenstein and Brand, 1937), East Germany (Pietrzenuk, 1961, as *Trochamminoides* sp.) and France (Brouwer, 1969, as *Trochamminoides* sp.), but these records are not linked to ammonite zones.

Thurammina jurensis (Franke)
Plate 6.2.1, Fig.5 (× 46), Upper Toarcian, Mochras Borehole, North Wales. = *Thurammina jurensis* Franke, 1936. Description: small, spherical, with large, blunt, radiating spikes. Distribution: Toarcian/Aalenian from northwest Germany/eastern Holland, France, south Germany/Austria/Swiss Jura (Brouwer, 1969). Recorded from *bifrons* Zone of south Germany (Riegraf, 1985), Toarcian of West Germany (Franke, 1936; Karampelas, 1978), Toarcian of France (Payard, 1947) and uppermost Toarcian (*levesquei* Zone) of Portugal (Exton, 1979) and offshore Eastern Canada (Exton and Gradstein, 1984). Ranges into the Aalenian in West Germany (Bartenstein and Brand, 1937; Issler, 1908, as *Storthosphaera albida*) and Switzerland (Wernli and Septfontaine, 1971). The only British record is from the *bifrons–levesquei* Zones of the Mochras Borehole. Not known from the British Middle Jurassic.
Total range: upper Lower Toarcian (*bifrons* Zone)–Aalenian (*opalinum* Zone).

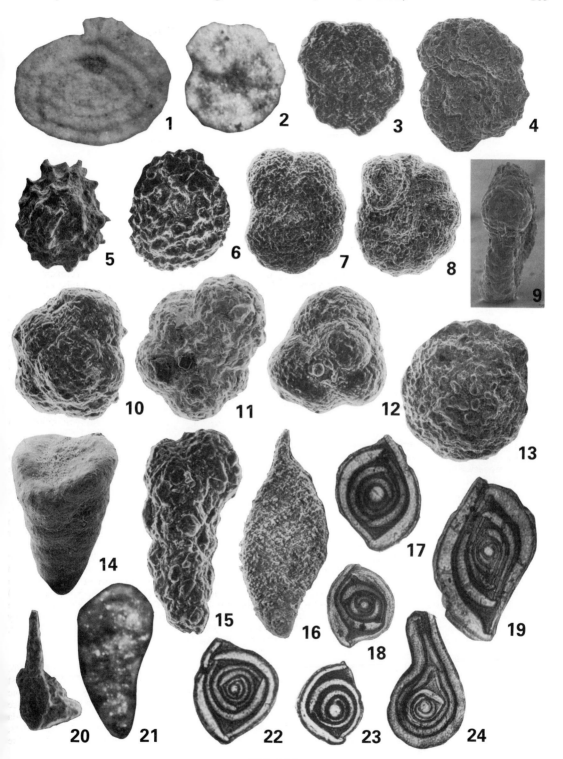

Plate 6.2.1

Thurammina subfavosa Franke, 1936
Plate 6.2.1, Fig.6 (× 85), Upper Toarcian, Mochras Borehole, North Wales. Description: small, with honeycomb surface ornamentation, produced by short spikes connected by narrow ridges. Distribution: recorded from Toarcian of Mochras Borehole and West Germany (Franke, 1936), with no known Aalenian records.

Trochammina sp. 1
Plate 6.2.1, Fig.7, dorsal view (× 94); Fig.8, ventral view (× 94), Lower Sinemurian, Dorset coast. Common synonym: *Trochammina nana* form a Brand, in Bartenstein and Brand, 1937. Description: chambers narrow, elongated, arranged in low trochospire, up to ten in last whorl; sutures swept back; margin rounded, lobate; narrow, deep umbilical hollow.
Remarks: this is a well known species from the Sinemurian of Europe, but has been usually referred to the Recent species *Trochammina nana* Brady. Wishing to separate these Jurassic forms from the Recent forms, Tappan (1955) erected *T. gryci*. However, Tappan's type material from the Upper Jurassic of Alaska clearly differs from the European Sinemurian form which has narrow, recurved chambers and lower spire with planar dorsal side. True *T. gryci*, comparable with Tappan's types, is frequent in the Upper Oxfordian–Lower Kimmeridgian of the northern North Sea. A new name is thus required for the European Sinemurian form, which is provisionally held in open nomenclature. Distribution: Sinemurian restricted in Britain, recorded from the southern North Sea (48/22-1), Dorset coast and Robin Hood's Bay, being restricted in the latter two localities to the upper *semicostatum* Zone. Known also from the Sinemurian of West Germany (Bartenstein and Brand, 1937), Sweden (Norling, 1972) and Denmark (Bang, 1968, 1972). It is uncertain whether Ainsworth *et al*.'s (1987) record of '*T. gryci*' from the Lower Sinemurian of the Fastnet and North Celtic Sea Basins is equivalent to *Trochammina* sp.1.

Trochammina canningensis Tappan, 1955
Plate 6.2.1., Fig.10, dorsal view (× 130); Fig.11, ventral view (× 130); Fig.12, dorsal view of three chambered form (× 120); Fig.11, Lower Sinemurian, Mochras Borehole, N. Wales; Figs 10, 12, Bracebridge, Lincoln.
Common synonym: *T. globigeriniformis non* Parker and Jones. Description: test high- to low-spired, chambers globular, producing globigeriniform test.
Remarks: Tappan (1955) erected this species for Jurassic homoeomorphs of the Tertiary–Recent species *T. globigeriniformis*. Low-spired forms with five chambers in the final whorl and an open umbilicus (see Figs 10,11) are homoeomorphs of *Globigerina ciperoenis* Bolli from the Oligocene–Miocene. Distribution: Hettangian–Kimmeridgian of northwest Europe and Alaska. Frequent in the Upper Toarcian of the northern North Sea.

Verneuilinoides sp. 1 Wall, 1983.
Plate 6.2.1., Fig.15, (× 206), well Britoil 211/18a-A43, North Sea; Fig.21 (× 160). Common synonym: *V. mauritii* Franke, 1936, Bartenstein and Brand, 1937, Riegraf, 1985 *non* Terquem, 1866a. Description: test small, narrow, agglutinating, medium grained, tapered, comprising three rows of globular chambers; sutures depressed.
Remarks: this species has consistently been erroneously referred to *V. mauritii* by previous authors (e.g. Franke, 1936; Bartenstein and Brand, 1937; Usbeck, 1952; Riegraf, 1985; Nagy, 1985a, b), but differs from this species in having depressed sutures and a more narrow, tapering, more coarsely grained test. Distribution: Hettangian–Toarcian of Britain and West Germany (see above) and Aalenian of Arctic Canada (Wall, 1983). In the northern North Sea the species is common in the Upper Toarcian Drake Member (Nagy, 1985a, b), and its extinction ('top') marks the top of the Toarcian below the Aalenian–Bathonian Brent Group throughout the North Viking Graben.

Verneuilinoides mauritii (Terquem)
Plate 6.2.1, Fig.14, side view (× 110), Lower Pliensbachian, Dorset coast. = *Verneuilina mauritii* Terquem, 1866a. Synonym: *Verneuilina georgiae* Terquem, 1866b. Description: test smooth, finely arenaceous, usually high trochospiral, circular in cross-section and sometimes nail-shaped in side view; sutures flush; chambers indistinct.
Remarks: test more conical and ventrally broader than *V. favus* (Bartenstein, in Bartenstein and Brand, 1937). Terquem's (1866a) types were redescribed by Bizon (1960). Distribution: restricted to the top Upper Sinemurian (*raricostatum* Zone) to Lower Pliensbachian in Britain (Mochras Borehole, Fastnet Basin (Ainsworth and Horton, 1986; Ainsworth *et al.*, 1987) and Dorset coast (Jenkins, 1958)). Occurs also in the Upper Pliensbachian in the Paris Basin (Bizon and Oertli, 1961) and Portugal (Exton, 1979). Recently figured from immediately south of the Tethyan area (High Atlas of Morocco) by Exton and Gradstein (1984). Reported as young as Late Bathonian in Switzerland (Wernli, 1971).

SUBORDER INVOLUTININA

Involutina liassica (Jones)
Plate 6.2.1, Fig.13 (× 40), Lower Sinemurian, well Burmah 48/22-1, southern North Sea. = *Nummulites liassicus* Jones in Brodie, 1853. Synonym: *I. turgida* Kristan-Tollmann, 1957. Description: test large, planispiral, lenticular, final whorl only visible, earlier whorls obscured by numerous irregular pillars filling umbilicus; surface pitted, rugose or smooth. Distribution: this is the only known *Involutina* species in the northwest European Jurassic. Common in the Rhaetian of Tethyan realm (see Zaninetti, 1976), but not appearing in northwest Europe until the *angulata* Zone. In British areas north of the Mendips, the species is most abundant in the *angulata–bucklandi* Zones (e.g. Wilkesley, Redcar, Hock Cliff, southern North Sea), but in the Wessex Basin (e.g. Barnard, 1950a; Macfadyen, 1941; Copestake, 1987) it is commonest in the *davoei* Zone. Rare in Britain above the Lower Pliensbachian, but ranging as high as *tenuicostatum* Zone (e.g. Mochras Borehole). Widespread in Britain, including the Fastnet Basin (Ainsworth and Horton, 1986), but not yet recorded from Scotland. Ranging from Hettangian to basal Toarcian in Europe (Brouwer, 1969), with records from France (Bizon, 1961), Denmark (Bang, 1972), Sweden (Norling, 1972), West Germany (Bartenstein and Brand, 1937) and Austria (Fuchs, 1970). Diagnostic of the uppermost Sinemurian–Pliensbachian of offshore Eastern Canada (Exton and Gradstein, 1984).

SUBORDER MILIOLINA

Nubeculinella tibia (Jones and Parker)
Plate 6.2.1, Fig.10 (× 70) Lower Toarcian, Mochras Borehole, North Wales. = *Nubecularia lucifuga* Defrance var. *tibia* Jones and Parker, 1860 *partim*. Description: irregularly coiled initial portion with approximately six chambers; uncoiled portion has subtubular chambers which taper distally.
Remarks: coiled portion rarely found. Distribution: questionable Lower Lias records exist (Adams, 1962), but the British (Mochras Borehole and Lincolnshire (Adams, 1962)) and West German acmes are in the Toarcian. Occurs also in the Aalenian of South Germany (unpublished data).

Ophthalmidium liasicum (Kübler and Zwingli)
Plate 6.2.1, Fig.22, microspheric specimen (× 202), transmitted light; Fig.23, megalospheric specimen (× 202); Hettangian, Mochras Borehole, North Wales. = *Oculina liasica* Kübler and Zwingli, 1866. Description: test small, thin, asymmetrical, most adult chambers of 3/4 whorl, imparting to test a generally circular outline, only final chamber of 1/2 whorl; broad flange area between whorls; chambers narrow, only slightly wider proximally than distally.
Remarks: the dominance of chambers of 3/4 whorl distinguishes the species from other species with spiroloculine, bisymmetrical tests. The species is the type species of the genus. Distribution: Hettangian (*liasicus* Zone)–Toarcian (*variabailis* Zone) of Switzerland, West Germany (Kübler and Zwingli, 1866), Burton Row and Mochras Boreholes. In the latter section it is abundant at the bases of the *liasicus* and *tenuicostatum* Zones. Recorded also from *tenuicostatum* Zone of Crosby Warren and Kirton-in-Lindsey (R. L. A. Young, pers. comm.). Reported to be common in the Lower Toarcian (*falciferum–bifrons* Zones) of Portugal (Exton and Gradstein, 1984).

Ophthalmidium macfadyeni Wood and Barnard, 1946.
Plate 6.2.1, Fig.18 (× 128), asymmetrically coiled, transmitted light; Fig.19 (× 125), symmetrically coiled, transmitted light; Lower Pliensbachian, Mochras Borehole, North Wales. Common synonym: *O. carinatum* Macfadyen, 1941 *non* Kübler and Zwingli. Description: test bilaterally symmetrical, eight to 12 chambers in adult, mostly of 1/2 whorl in length, markedly wider at proximal ends, distally produced as apertural necks.
Remarks: Pliensbachian–Toarcian specimens are clearly distinguished from *O. northamptonensis* by greater size and symmetry (see Wood and Barnard, 1946). Sinemurian specimens are smaller, and show similar variation trends towards asymmetry as *O. northamptonensis*. Distribution: in Britain ranging from Lower Sinemurian (*bucklandi* Zone) to Toarcian (*falciferum* Zone) and recorded from Mochras Borehole, Dorset coast (Jenkins, 1958; Macfadyen, 1941), Burton Row Borehole, Hill Lane Borehole, Cockle Pits Borehole, Trunch Borehole. First appears in the latter five sections in the *rariocostatum* Zone.

Ophthalmidium northamptonensis Wood and Barnard, 1946.
Plate 6.2.1, Fig.17 (× 192), symmetrically coiled, transmitted light, Hettangian, Mochras Borehole, North Wales.
?Synonym: *O. minutum* Fuchs, 1970. Description: test small, irregular, discoidal, six to ten chambers in megalospheric forms, ten or 11 in microspheric forms; chambers of variable length, not narrowing markedly distally; apertural neck typically short.
Remarks: there is considerable variation in chamber length, the degree to which this is developed separating specimens into symmetrical, intermediate and asymmetrical types. These types occur at random, not in a progressive stratigraphic series from asymmetrical to symmetrical as suggested by Wood and Barnard (1946); symmetrical forms are generally infrequent. Hettangian to Lower Sinemurian specimens are smaller and more delicate than those from younger levels (e.g. Toarcian type material), and are close to *O. minutum* Fuchs from the 'Lower Lias' of Austria. Distribution: Hettangian (*liasicus* Zone)–Upper Toarcian (*levesquei* Zone), with acmes at bases of *jamesoni*, *tenuicostatum* and *variabilis* Zones (Mochras Borehole); Lower Toarcian of Byfield (Wood and Barnard, 1946) and *exaratum* Subzone of Tilton (R. L. A. Young, pers. comm,); 'Lower Lias' of Austria (Fuchs, 1970) (if *O. minutum* is synonymous).

Ophthalmidium sp. 2 (Ruget and Sigal)
Plate 6.2.1, Fig. 16 (× 99), Lower Toarcian, Zambujal, Portugal = *Spirophthalmidium* sp. 2 Ruget and Sigal 1970. Description: test with agglutinated external coating; up to nine chambers (megalospheric generation), arranged symmetrically; sutures depressed; periphery rounded.
Remarks: distinguished from other species of *Ophthalmidium* by its thick test with an agglutinated outer layer. *Spiroloculina aspera* is superficially similar but possessses a rugose surface ornament (Terquem and Berthelin, 1875). Distribution: recorded from the Pliensbachian to basal Toarcian (*tenuicostatum* Zone) of Portugal (Ruget and Sigal, 1970; Exton, 1979, as *Ophthalmidium* sp. 1), the basal Toarcian (*tenuicostatum* Zone) of the Mochras Borehole, the *davoei* Zone of the Hill Lane Borehole and basal Toarcian of the Fastnet and North Celtic Sea Basins (Ainsworth *et al.*, 1987).

Praeophthalmidium orbiculare (Burbach)
Plate 6.2.1., Fig.24 (× 157.5), transmitted light, Lower Sinemurian, Mochras Borehole, North Wales. = *Ophthalmidium orbiculare* Burbach, 1866. Common synonym: *O. ovale* Burbach, 1866. Description: test outline flask-shaped, comprising up to five chambers; second chamber of one to four whorls, followed by up to three chambers of one whorl; each chamber produced at distal (apertural) end.
Remarks: this is the only known species of the genus, and is distinguished from superficially similar species of *Ophthalmidium* by having a minimum chamber length of one whorl, rather than 1/2 whorl; this imparts a characteristic circular test outline to the species. Structural variations produce two possible test outlines, originally separated by Burbach (1866) as *O. orbiculare* and *O. ovale*, but included by Knauff (1966) and the present authors in a single variable species. Distribution: appears in the Hettangian (*liasicus* Zone) and ranges throughout the Lower Jurassic, with records from Britain (Mochras Borehole), West Germany (Knauff, 1966; Karampelas, 1978), East Germany (Pietrzenuk, 1961), Italy, France and South Germany (Riegraf, 1985, as *O. liasicum*). The specimens illustrated from the Middle Triassic of Australia (Heath and Apthorpe, 1986) appear closer to *O. liasicum*. Reported from the Pliensbachian of offshore Morocco (Riegraf *et al.*, 1984, as *O. liasicum*).

SUBORDER SPIRILLININA

'*Placentula*' *pictonica* (Berthelin)
Plate 6.2.2, Fig.1, dorsal view (× 277.5), Lower Pliensbachian, Mochras Borehole, North Wales; Fig.2, edge view (× 290), Upper Pliensbachian, same locality; Fig.3, ventral view (× 202.5), Upper Sinemurian, same locality. = *Placentula pictonica* Berthelin, 1879. Description: test conical, patelline, with four to six chambers per whorl; small, apical depression occurs on the dorsal side; numerous curved growth lines radiate from the umbilicus on the ventral side.

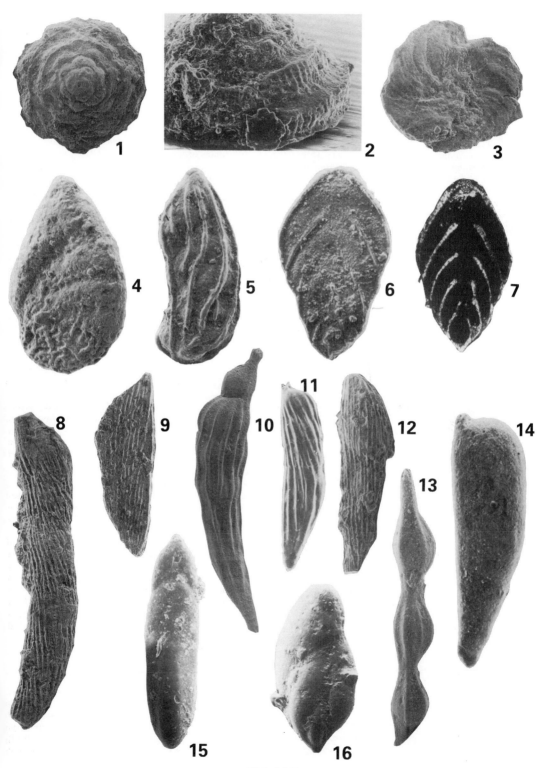

Plate 6.2.2

Remarks: species represents a new genus, soon to be fully described. Distribution: in Britain, yet observed only in the Mochras Borehole (uppermost Sinemurian–top Pliensbachian) and Winterton, South Humberside (*davoei* Zone) (R. L. A. Young and M. D. Brasier, pers comm.). Previous records from the *margaritatus* Zone of Paris Basin (Berthelin, 1879) and Portugal (J. Exton, pers. comm.).

SUBORDER LAGENINA

Astacolus semireticulata (Fuchs)
Plate 6.2.2, Fig.4 (× 139.5), Lower Sinemurian, Mochras Borehole, North Wales. = *Lenticulina* (*Lenticulina*) *semireticulata* Fuchs, 1970. Synonyms: *Lenticulina* sp. 26 Bang, 1968; *Astacolus semireticulata* Norling, 1972. Description: test compressed; early chambers with irregular reticulate ornament, later chambers smooth with depressed sutures; aperture radiate, with apertural chamberlet.
Remarks: *Lenticulina dorbignyi* differs in its less compressed test and more regular reticulate ornament which completely covers the test surface. Distribution: in Britain the species is known from Raasay, Morvern, Robin Hood's Bay, Radstock, Burton Row Borehole, Hornblotton Mill, Mochras Borehole, Dorset coast (Copestake, 1987) and Fastnet Basin (Ainsworth and Horton, 1986; Ainsworth *et al.*, 1987), restricted to the upper *semicostatum* Zone–upper *turneri* Zone interval. In Sweden (Norling, 1972), Denmark (Bang, 1968, 1972) and Austria (Fuchs, 1970) it occurs between the upper Hettangian and Lower Sinemurian.

Astacolus speciosus (Terquem)
Plate 6.2.2, Fig.5 (× 90), Lower Sinemurian, Mochras Borehole, North Wales. = *Cristellaria speciosa* Terquem, 1858. Common synonyms: *Marginulina quadricostata* Terquem, 1863, *Marginulina radiata* Terquem, 1863, *Lenticulina* (*Astacolus*) *neoradiata* Neuweiler (in Bach *et al.*, 1959). Description: test compressed, auriculate, with two to seven oblique, intermittent ribs on each side, disappearing near apertural face; marginal ribs usually parallel to test outline; periphery keeled, aperture marginal.
Remarks: this is one of the several variable species characteristic of the Lower Jurassic, and includes a range of forms which have oblique ribbing and uncoiling as diagnostic features. Examination of fossil assemblages shows that several of Terquem's 'species' are merely morphotypes within the normal variation range. Terquem's (1863) erroneous description of '*M*'. *radiata* as having a circular cross-section (see refiguring of type material by Bizon, 1960) led to the erection of 'new' species *L*. (*A*.) *homoradiata* Seibold and Seibold (1953) and *L*. (*A*.) *neoradiata* Neuweiler (in Bach *et al.*, 1959) on the basis of an ovate cross-section, which '*M*'. *radiata* in fact has. Narrow, elongate forms with ovate cross-sections are referrable to *Vaginulinopsis* whereas more compressed forms with rectangular or ovate cross-sections can be placed in *Astacolus*. Distribution: in Britain, first appears in the Hettangian (*angulata* Zone) and thereafter occurs throughout the Lower Jurassic. Particularly abundant from the upper *semicostatum* Zone to *obtusum* Zone. Also recorded from the Middle Jurassic (see Section 6.3) and Upper Jurassic (Seibold and Seibold, 1953).

Berthelinella involuta (Terquem)
Plate 6.2.2, Fig.6 (× 144); Fig.7, transmitted light, (× 120); Lower Pliensbachian, Mochras Borehole, North Wales. = *Frondicularia involuta* Terquem, 1866. Common synonyms: *Frondicularia varians* Terquem, 1866, *Frondicularia paradoxa* Berthelin, 1879. Description: test smooth, flattened oval in outline, rectangular in section with truncated margin; either early biserial arrangement is developed or chambers are uniserial, frondicularian throughout; sutures raised between adult chambers; aperture terminal, central, slit-shaped with lip.
Remarks: biserial early chambers may or may not be developed (see Plate 6.2.2, Fig.7) as noted by Berthelin (1879) and may be up to three in number; completely uniserial forms, which are part of the normal range of variation, equate to *F. varians* Terquem. Terquem's (1866) original figure indicated a completely frondicularian chamber arrangement, though Bizon's (1960) refiguring of his type proved the presence of initial biserial chambers, identical with *F. paradoxa* Berthelin. The species' distinctive features have led to the erection of a monotypic genus *Berthelinella* (Loeblich and Tappan, 1957). Distribution: total range Hettangian–Bathonian, particularly characteristic of the Lower Jurassic (most common between *semicostatum* and *ibex* Zones). The species has been used for zonal purposes in the Lower Pliensbachian of West Germany (Bartenstein and Brand, 1937) and Denmark (Bang, 1971), but in Britain it is commonest in the Lower Sinemurian (*semicostatum* Zone). In Portugal and Eastern Canada it does not range above the Lower Pliensbachian (Exton and Gradstein, 1984). Recorded from the Bathonian of the Bath area by Cifelli (1959) (as *Plectofrondicularia paradoxa*). Appears south of the Tethyan area in offshore Morocco (Sinemurian–Pliensbachian, Riegraf *et al.*, 1984).

Citharina colliezi (Terquem)
Plate 6.2.2, Fig.9 (× 35), Upper Toarcian, Mochras Borehole, North Wales. = *Marginulina colliezi* Terquem, 1866a. Synonym: *Marginulina flabelloides* Terquem, 1868. Description: large, compressed, multichambered (10 to 16 chambers); many longitudinal striae which are interrupted over the sutures; thick dorsal margin. Distribution: British records from Bristol Channel (Lloyd *et al.*, 1973), probable age Toarcian to Early Bajocian; possibly from undifferentiated 'Middle and Upper Lias' of Somerset (?= *Planularia reticulata*, Brady, 1866) and from the Toarcian of the Mochras Borehole, Byfield (Barnard, 1950b), Empingham (Horton and Coleman, 1978), north Lincolnshire (Adams, 1955, unpublished thesis), Midlands/Yorkshire (Brouwer, 1969) and Ilminster (F. M. Lowry, pers. comm.).

In Britain the species becomes common over a wide area in the *faciferum* Subzone. European records: Toarcian to Oxfordian of Sweden (Norling, 1972), Holland (Brouwer, 1969), Portugal (Exton, 1979), West Germany (Bartenstein and Brand, 1937), France (Bizon, 1961; Ruget, 1985).

Dentalina langi Barnard, 1950
Plate 6.2.2 Fig.8 (× 46), Lower Sinemurian, Redcar, Cleveland. Description: test large, robust, arcuate, ornamented with numerous oblique, fine ribs; final two or three chambers large, inflated.
Remarks: this species is readily identifiable in being the only ribbed *Dentalina* of the Hettangian. Of later evolving species, *D. matutina* has much coarser costae while *D. tenuistriata* has numerous finer striations and a more slender test. Distribution: a good marker for the uppermost Hettangian (*angulata* Zone), to which it is restricted in England, Scotland and West Germany (Usbeck, 1952). In France, Ruget (1985) also records its extinction in the *angulata* Zone, but reports its appearance in the *liasicus* Zone. In Britain the species is restricted to the *angulata* Zone (*complanata* Subzone), with records from Dorset (Barnard, 1950a), Burton Row Borehole, Wilkesley Borehole, Redcar and Loch Aline.

Dentalina matutina (d'Orbigny)
Plate 6.2.2, Fig.10 (× 36); Fig.11 (× 24), Upper Pliensbachian, Bracebridge, Lincoln. Synonym: *D. primaeva* d'Orbigny, 1849. Description: robust, elongate, ornamented with coarse, oblique costae; aperture produced, radiate.
Remarks: distinguished from *D. langi* and *D. tenuistriata* by its coarser, fewer ribs/costae. Intermediates exist with the smooth species *D. terquemi*. Distribution: first appears commonly in the late *semicostatum* Zone over a wide area, ranging as high as the *margaritatus* Zone in Britain (including the North Sea) and up to the top *spinatum* Zone in mainland Europe (e.g. France, Ruget, 1985; West Germany, Karampelas, 1978), though most common in Upper Sinemurian. Widespread throughout northwest Europe (Brouwer, 1969). Also reported from offshore Morocco (Riegraf et al., 1984).

Dentalina tenuistriata Terquem, 1866.
Plate 6.2.2, Fig.12 (× 100), Lower Pliensbachian, Mochras Borehole, North Wales. Description: test narrow, elongate, comprising four to eight chambers ornamented with numerous (18–20) longitudinal striations paralleling test margin; sutures constricted.
Remarks: tests are fragile, and the types (refigured by Bizon, 1960 and Ruget, 1976) are fragmentary. Nevertheless, the combination of ornament and test form is diagnostic. Complete specimens have been figured by Franke (1936) and Pietrzenuk (1961). It is readily separable from *D. matutina* by its much finer ornament and depressed sutures. Distribution: most consistent and common in the Pliensbachian throughout Europe, though ranging from Hettangian to Middle Jurassic.

Dentalina terquemi d'Orbigny, 1849.
Plate 6.2.2, Fig.14 (× 87), Upper Pliensbachian, Bracebridge, Lincoln. Synonym: *Vaginulina hausmanni* Bornemann, 1854. Description: test large, robust, smooth, cylindrical, arcuate, comprising ten to 12 chambers, sutures horizontal, flush, proloculus apiculate.
Remarks: d'Orbigny's (1849) types were not figured until 1936 (Macfadyen, 1936). The species has been well described by Barnard (1950a) and well figured by Ruget and Sigal (1967). *Dentalina torta* Terquem is a similar smooth species, but has oblique, depressed sutures. *D. matutina* is closely related to *terquemi* but differs in possessing coarse ribs. Distribution: British range Hettangian to top Pliensbachian, with acmes in the *semicostatum* Zone, *obtusum* Zone and Upper Pliensbachian (including the northern North Sea). The latter two acmes are also documented in mainland Europe (Ruget and Sigal, 1967; Brouwer, 1969). Recorded also from the Middle Triassic of Australia (Heath and Apthorpe, 1986).

Dentalina varians haeusleri (Schick)
Plate 6.2.2, Fig.13 (× 75), Lower Pliensbachian, Dorset coast. = *Nodosaria* (*Dentalina*) *haüsleri* Schick, 1903. Synonym: *D. varians* Terquem, 1866a *partim*. Description: ornamented with longitudinal ribs paralleling the test margin; sutures deeply constricted; aperture marginal, produced, radiate.
Remarks: distinguished from end chambers of *D. matutina* in having fewer ribs which are parallel, rather than oblique, to the test margin. Distribution: although long-ranging (Lower Sinemurian–Lower Toarcian), the subspecies is most often recorded in the Upper Sinemurian in northwest Europe (Brouwer, 1969). In Britain it is commonest in the *raricostatum* Zone (at which level it first appears at several localities) but has been recorded throughout the Upper Sinemurian and Lower Pliensbachian, its youngest British records being from the Lower Toarcian (*bifrons* Zone) of Empingham (Horton and Coleman 1978) and Ilminster (F. M. Lowry, pers. comm.). This *raricostatum* Zone acme probably extends at least to South Germany, where the taxon is common in the uppermost Sinemurian (Karampelas, 1978).

Eoguttulina liassica (Strickland)
Plate 6.2.2, Fig.16 (× 135), Lower Sinemurian, Mochras Borehole, North Wales. = *Polymorphina liassica* Strickland, 1846. Synonyms: *Polymorphina metensis*, *P. bilocularis*, *P. polygona*, *P. simplex* Terquem, 1864, *Polymorphina oolithica* Terquem, 1874. Description: test smooth, elongate, chambers usually strongly embracing, variable in shape, sutures oblique to vertical, depressed or flush; aperture terminal, radiate.
Remarks: Early Jurassic *Eoguttulina* populations are very variable particularly in chamber shape, size and habit (embracing or elongate, resulting in globular or compressed tests respectively). All forms intergrade within assemblages and have the same stratigraphic range, and are thus not worthy of the specific separation for instance by

Terquem (1864, 1874), followed by Riegraf (1985). The angle between the chambers is difficult to ascertain, but appears to range from 90° (suggesting the genus *Eoguttulina*) to 180° (suggesting *Polymorphina*). Distribution: known range Rhaetian–Kimmeridgian, particularly abundant and widespread in the Boreal Lower Jurassic.

Frondicularia brizaeformis Bornemann, 1854
Plate 6.2.2, Fig.15 (× 48), Lower Sinemurian, Dorset coast. Common synonyms: *Frondicularia intumescens* Bornemann, 1854, *F. major* Bornemann, 1854, ?*Frondicularia xiphoidea* Kristan-Tollmann, 1964. Description: test large, smooth, robust; central part of test thickened, globular proloculus followed initially by rapidly flaring chambers, test thereafter parallel-sided; margin usually rounded, sutures mostly flush, aperture produced, radiate.
Remarks: tests are quite variable (see Barnard, 1957), leading to the erection of many 'species' by previous authors, of no stratigraphic value. These variants were grouped by Barnard (1957) as a single variable species, with the name *brizaeformis* having priority. *Frondicularia terquemi terqemi* differs in being more compressed, sometimes sulcate, with a smaller proloculus and sharper margin. *F. xiphoidea* Kristan-Tollmann, 1964, described from the Tethyan Rhaetian is similar, but has a circular rather than radiate aperture. *Lingulina borealis* Tappan, 1951, also from the Triassic, differs in having a slit-shaped aperture (see Heath and Apthorpe, 1986). Distribution: a characteristic species of Boreal Lower Jurassic assemblages in Europe. Range from Hettangian (*liasicus* Zone) to Lower Toarcian, being rare above the Upper Pliensbachian. Acme of occurrence Lower Sinemurian.

The *Frondicularia terquemi* plexus
Flattened, sulcate, multi-ribbed frondicularias are diagnostic of the Lower Jurassic. The high degree of variability in rib number and character caused the erection of many new 'species' by early workers such as Bornemann and Terquem. It is, however, evident that these morphotypes are closely related. Barnard (1957) included them all within a single plexus of evolving forms, showing progressive evolution from strongly ribbed to smooth tests through the Lower Jurassic of England. Barnard used letters (A to K) for the various forms and recognised no distinct subspecies. Nørvang (1957) noted the same trend in rib reduction in the Danish Lias, but distinguished five subspecies with different ranges. A study of many British sequences indicates that Barnard's (1957) scheme is more accurate, though it was based on a single section, the Stowell Park Borehole. Although all of Barnard's forms are currently recognised, additional ones are known, plus some of the ranges differ from those indicated by Barnard. Those of Barnard's forms which have consistent stratigraphic differentiation are regarded here as subspecies. The various coarsely ribbed forms (B, C, D, E, F and G) are included in *F. terquemi sulcata*, though it is believed that some may prove to warrant subspecific separation. Several of Barnard's forms (A, B, C, D) were grouped into a 'new' species *F. densicostata* by Hohenegger (1981), allegedly restricted to the Hettangian, and separated from Sinemurian *F. 'sulcata'*. As discussed herein, certain of these forms (B,C,D) range into the Sinemurian, whereas form A is regarded as a distinct Hettangian restricted subspecies. Forms at different stages of rib reduction are grouped as *F. terquemi bicostata* (two ribs) and *F. terquemi terquemi* (smooth) while three further subspecies are recognised with specific ranges (subsp. A, subsp. B, *muelensis*). The plexus has been referred to various names, e.g. *F. sulcata* plexus (Barnard, 1957), *F.* gr. *sulcata* (Ruget, 1967), *Spandelina bicostata* group (Nørvang, 1957) and *F. bicostata* (Brouwer, 1969), though the name *terquemi* has priority.

Frondicularia terquemi bicostata d'Orbigny
Plate 6.2.3, Fig.1 (× 80), reflected light, Lower Pliensbachian, Dorset coast; Fig.2, small form (× 207), Lower Sinemurian, Mochras Borehole. = *Frondicularia bicostata* d'Orbigny, 1849. Synonym: *F. sulcata* form H Barnard, 1957. Description: test with two distinct central ribs, occasionally bordered by two or more weaker lateral striations; test size variable.
Remarks: d'Orbigny's (1849) holotype was not figured until 1936 (Macfadyen, 1936). The subspecies first evolved in the Hettangian (*liasicus* Zone) as a small form, by a process of rib reduction from the four- to five-ribbed *F. terquemi* subsp. A, before becoming extinct in the *ibex* Zone. A larger form developed by a similar process in the *semicostatum* Zone from *F. terquemi sulcata*, becoming extinct in the *spinatum* Zone. Distribution: British range from Hettangian to top Pliensbachian. Widespread in Europe; also recorded from Alaska (as *F. lustrata* Tappan, 1955) and Argentina (Ballent, 1987b). Most previous records appear to be of the large form.

Plate 6.2.3

Frondicularia terquemi muelensis Ruget and Sigal
Plate 6.2.3, Fig.7 (× 98), Lower Toarcian, Bracebridge, Lincoln. = *Frondicularia muelensis* Ruget and Sigal, 1970, *F. sulcata* form K Barnard, 1957. Description: test compressed, non-sulcate, ornamented with numerous fine, parallel striae.
Remarks: the small size and compressed, non-sulcate test form distinguish the subspecies from similar multi-ribbed forms of *sulcata*. Distribution: widespread in Europe during the Pliensbachian, with records from West Germany (= *F. bicostata sulcata*, Dreyer, 1967, Plate 9, Fig. 3), France (= *F. bicostata*, Brouwer, 1969, Plate 5, Figs 5, 6), Portugal (Ruget and Sigal, 1970; Exton, 1979) and South Germany (authors' observation). First appears in Britain in the *raricostatum* Zone; in the Cocklepits and Mochras Boreholes and at Bracebridge the upper limit of its range is top *tenuicostatum* Zone, (though questionably occurring at Mochras as high as *thouarsense* Zone). Additional British records are from the Trunch and Stowell Park Boreholes, the Dorset coast (A. Jenkins coll.), Burton Row Borehole, and Moray Firth coast (Dunrobin Castle).

Frondicularia terquemi subsp. A Barnard
Plate 6.2.3, Fig.3 (× 162) Hettangian, Mochras Borehole, North Wales. = *F. sulcata* form A Barnard, 1957. Description: test small, ornamented with four to five weak longitudinal ribs; six to eight chambers.
Remarks: distinguished from most other members of the *F. terquemi* plexus by its small size, and from subsp. *bicostata* by rib number. A small form of subsp. *bicostata* evolved from subsp. A (intermediates between the two occur in the *liasicus* Zone). The taxon is recognised as a distinct chronosubspecies owing to its short stratigraphic range. Distribution: Hettangian restricted in Britain (*johnstoni–complanata* Subzones), observed in the Stowell Park (Barnard, 1957), Mochras, Wilkesley and Burton Row Boreholes, Glamorgan and West Somerset. Recorded by the authors from the top *angulata* Zone of South Germany: also known from northwest Germany (as *F. sulcata*, Bartenstein and Brand, 1937, Plate 1A, Fig. 12) and Austria(?) (as *F. bicostata*, Brouwer, 1969, Plate V, Fig. 10).

Frondicularia terquemi subsp. B
Plate 6.2.3, Fig.4 (× 57); Fig.5 (× 60); Lower Toarcian, Mochras Borehole, North Wales. Description: robust subspecies with four to six longitudinal, coarse, high costae confined to central region of test and bordered by lateral smooth areas; margin keeled.
Remarks: differs from other members of the *F. terquemi* plexus in its robustness and degeneration of ribs laterally away from the central region. Distribution: known in Britain from Tuffley (Gloucester), and in the Mochras and Hill Lane Boreholes, with total range from *margaritatus* Zone (*subnodosus* Subzone) to *tenuicostatum* Zone. Recently recorded also from the Upper Pliensbachian of the Fastnet and North Celtic Sea Basins (Ainsworth *et al.*, 1987). On the continent, also recovered from West Germany (= *F. baueri*, Franke, 1936, Plate 7, Fig.10; Wicher, 1938, Plate 25, Fig.6), Austria (?) (= *F. bicostata*, Brouwer, 1969, Plate 5, Figs 13 – 15) and France (= *F. bicostata*, Espitalié and Sigal, 1960, Plate 3, Fig.9; *F.* sp. 3 Seronié-Vivien *et al.*, 1961) in the *margaritatus* and *spinatum* Zones, and from the *tenuicostatum* Zone of South Germany (authors' observation).

Frondicularia terquemi sulcata Bornemann.
Plate 6.2.3, Fig.6, form B (× 106), Hettangian, Mochras Borehole; Fig.8, form C (× 100), Lower Sinemurian,

Mochras Borehole; Fig.9, form D, Lower Sinemurian, Mochras Borehole; Fig.10, form E (× 80), reflected light, Lower Pliensbachian, Dorset coast; Fig.12, form E (× 18); Fig.13, form F (× 90), Upper Sinemurian, Mochras Borehole; Fig.11, form F/G intermediate (× 80), Lower Pliensbachian, Dorset coast; Fig.14, form G (× 87), Upper Sinemurian, Mochras Borehole; Fig.15, form G (× 122), Upper Sinemurian, Mochras Borehole. = *Frondicularia sulcata* Bornemann, 1854. Common synonyms: *F. dubia* Bornemann, 1854; *F. pulchra* Terquem, 1858; *F. sulcata* forms B, C, D, E, F, G Barnard, 1957; *F. quadricosta* Terquem, 1958; *F. baueri* Burbach, 1886. Description: test robust, with between four and 12 coarse, longitudinal ribs which are either parallel, divergent or convergent; median sulcus occasionally present.

Remarks and Distribution: grouped in this subspecies are the large, many ribbed forms of the *F. terquemi* plexus. The subspecies so conceived is highly variable in terms of rib number, rib orientation and test width, and many 'species' names are available for the various forms. Various authors have attempted to make sense of the variation and evolution, but without consensus (Barnard, 1957: Nørvang, 1957; Norling, 1966; Ruget, 1967, 1985). The morphological forms which appear to be the most useful stratigraphically are those of Barnard (1957), although there are some range inconsistencies between localities. The first appearances of these forms seem to be the most consistent stratigraphic features. The salient features and consistent (not total) British ranges of these forms are as follows:

form B; six ribs; *liasicus* Zone–*semicostatum* Zone;
form C; large test, ten ribs; *angulata* Zone–*spinatum* Zone;
form D; up to eight convergent ribs; *liasicus* Zone–*bucklandi* Zone;
form E; seven to nine convergent ribs; upper *semicostatum* Zone–*margaritatus* Zone;
form F; six to seven main ribs; upper *semicostatum* Zone–*raricostatum* Zone;
form G; four main ribs; upper *semicostatum* Zone–*raricostatum* Zone.

A full assessment of the applicability of these ranges outside Britain is not possible as other workers (e.g. Ruget, 1967, 1985) have not recognised Barnard's forms. Once their geographic and stratigraphic distribution is more fully documented, some of these forms may prove to warrant subspecific taxonomic status.

Frondicularia terquemi terquemi d'Orbigny
Plate 6.2.3, Fig.16, small form (× 225), Upper Sinemurian, Mochras Borehole, North Wales. = *Frondicularia terquemi* d'Orbigny, 1849 Synonym: *Frondicularia sulcata* Bornemann form J Barnard, 1957. Description: test smooth or very faintly striated, occasionally with slight median longitudinal sulcus, sutures flush, periphery rounded or keeled; test size variable.
Remarks: d'Orbigny's (1849) holotype was first figured by Macfadyen (1936). This subspecies evolved initially in the Hettangian by a process of rib reduction from a small form of subspecies *bicostata* and again in the Late Sinemurian (*oxynotum* Zone) from a large form of *bicostata*. The subspecies' almost smooth test is distinctive. Distribution: Hettangian–Toarcian in Britain and N.W. Europe, most frequent in the Pliensbachian (e.g. Barnard, 1957; Exton, 1979). Most, if not all, published records are of the large form.

Lenticulina dorbignyi (Roemer)
Plate 6.2.4, Fig.1 (× 113), Upper Toarcian, Mochras Borehole, North Wales. = *Peneroplis d'Orbignii* Roemer, 1839. Description: loosely coiled, six to eight chambers in final whorl: reticulate ornament covering whole test.
Remarks: both astacoline (Franke, 1936) and flabelline forms (Brouwer, 1969; Ruget, 1985; plus Mochras data) are reported. Ornament regular to irregular. Distribution: British records include the Mochras Borehole, Bristol Channel (Lloyd et al., 1973, probable age Toarcian to Early Bajocian) possibly from the undifferentiated 'Middle and Upper Lias' of Somerset (?= *Cristellaria costata*, Brady, 1866), the *bifrons–thouarsense* Zones of Ilminster (F. M. Lowry, pers. comm.) and the *levesquei* Zone of the Dorset coast. Ranges as high as the *discites* Zone (Lower Bajocian) in Britain. Apart from an isolated Upper Pliensbachian record from Northwest Germany/East Holland (Brouwer, 1969), European occurrences from West Germany (Franke, 1936; Riegraf, 1985), Holland (Brouwer, 1969), France (Ruget, 1985) and Switzerland (Wernli and Septfontaine, 1971) fall within a Toarcian to Lower Bajocian range. The species first appears in Britain and the Paris Basin (Bizon and Oertli, 1961) in the *bifrons* Zone, though Ruget (1985) indicated its occurrence in France as old as *falciferum* Zone. Recently reported from the Upper Toarcian of British Columbia, Canada (Cameron and Tipper, 1985) and Northwest Australia (M. Apthorpe, pers. comm). In Portugal and Eastern Canada the species appears later, in the *thouarsense* Zone, ranging to the *concavum* Zone (Aalenian) (Exton and Gradstein, 1984). A questionable specimen was figured by Maupin and Vila (1976) from the Toarcian of Algeria. Total range: Lower Toarcian (? *falciferum* Zone, Ruget, 1985, consistent from *bifrons* Zone) to Lower Bajocian (*discites* Zone), but disappearing in West Germany, France and Switzerland at the top of the Aalenian.

Lenticulina quenstedti (Guembel)
Plate 6.2.4, Fig.5 (× 178), Upper Toarcian, Dorset coast. = *Cristellaria quenstedti* Guembel, 1862. Description: see Section 6.3, 6.4. Remarks and Distribution: the species first appears in Britain (in considerable numbers) in the Lower Toarcian and ranges to the Oxfordian. These Toarcian records have been questioned by Exton and Gradstein (1984). However, the British Toarcian specimens are typical of the species, and the characteristic raised sutures which merge with a circular umbilical rib are well developed. Known British Lower Jurassic occurrences are from the *falciferum* Zone of Tilton (Leicestershire), the *falciferum* Zone of the Mochras Borehole, the *falciferum–thouarsense* Zones of Ilminster (F. M. Lowry, pers. comm.) and the *thouarsense* Zone of Dorset (Watton Cliff). In Central Europe the species first appears in the Middle Jurassic (Bajocian) (Wernli and Septfontaine, 1971) in which stage it also becomes more widespread and common in Britain.

Lenticulina varians (Bornemann) plexus.
Plate 6.2.4, Fig.3 (× 150), Hettangian, Mochras Borehole, North Wales. = *Cristellaria varians* Bornemann, 1854. Common synonyms: *Cristellaria major* Bornemann, 1854, *Cristellaria varians suturaliscostata* Franke, 1936. Description: test compressed, comprising up to 15 high, broad chambers, with strong tendency to uncoiling, sutures flush or raised, margin rounded or keeled, aperture radiate.
Remarks: a variable species, which shows a strong trend towards uncoiling, uncoiled forms occurring throughout the Lower Jurassic. A wide variety of forms have been described by Barnard (1950a, b), most of which belong to this plexus; only one form, form D of Barnard (1950b) appears to have stratigraphic separation, and this is considered herein as a subspecies within the plexus. Forms with raised sutures equate to Franke's (1936) *suturaliscostata*. Extensive synonymies embracing the various forms of the plexus were given by Macfadyen (1941) and Nørvang (1957). It differs from the *L. muensteri* plexus in being more loosely coiled, more compressed and in lacking umbilical plugs or bosses. Distribution: Rhaetian–Oxfordian, first appearing in Britain in the *liasicus* Zone. Common throughout the Lower Jurassic, ranging into the Middle and Upper Jurassic.

Lenticulina varians subsp. D Barnard
Plate 6.2.4, Fig.4 (× 77), Upper Toarcian, Mochras Borehole, North Wales. =*Lenticulina varians* Bornemann form D Barnard, 1950b. Description: test palmate, comprising small, loosely coiled lenticuline portion followed by flabelline portion of one to five chambers; sutures raised.
Remarks: differs from *Eoflabellina chicheryi* Payard, 1947 in having raised sutures and no keel. Distribution: Toarcian of Mochras Borehole, Byfield (Barnard, 1950b), Ilminster (F. M. Lowry, pers. comm.), France (?= *Falsopalmula chicheryi*, Magné et al., 1961; Champeau, 1961; *F.* cf. *centro-gyrata*, Magné et al., 1961) and possibly Western Canada (Cameron and Tipper, 1985, as *Falsopalmula varians*). May range into the Lower Aalenian (Brouwer, 1969).

Lenticulina muensteri acutiangulata (Terquem)
Plate 6.2.4, Fig.6 (× 56), Upper Pliensbachian, Mochras Borehole, North Wales. = *Robulina acutiangulata* Terquem, 1864. Description: test smooth with sharply keeled periphery and protruding umbilical boss; sutures flush; seven to eight chambers in final whorl.
Remarks: the combination of keel, umbilical boss and lack of raised sutures distinguishes this from other members of the *L. muensteri* plexus. Distribution: large typical forms appear in abundance at the base of the *subnodosus* Subzone (*margaritatus* Zone) throughout Britain, as in Dorset (Copestake, 1987), eastern England (Bracebridge), northern North Sea and southern North Sea; this horizon is of important correlative value over a wide area of northwest Europe, being also recorded in south Germany (Karampelas, 1978). This acme ranges to the top of the *margaritatus* Zone.
Total range: Lower Pliensbachian–Oxfordian, throughout Europe.

Plate 6.2.4

Lenticulina muensteri muensteri (Roemer)
Plate 6.2.4, Fig.2 (× 120), Upper Sinemurian, Mochras Borehole, North Wales. = *Robulina muensteri* Roemer, 1839. Common synonyms: *Cristellaria matutina* d'Orbigny, 1849, *C. gottingensis* Bornemann, 1854, *Marginulina sigma* Terquem, 1866, *Lenticulina* forms D, F, G, I Barnard, 1960. Description: test large, robust, smooth, margin angular, sutures flush, merging into smooth umbilical area; seven to ten chambers in final whorl in planispiral forms; uncoiled forms with well developed, parallel-sided portion of three to six chambers.
Remarks: the *Lenticulina muensteri* plexus is a polymorphic species group common in the European Jurassic, containing at least four closely related subspecies (*muensteri, acutiangulata, polygonata, subalata*). Included within subspecies *muensteri* are smooth forms without keeled margins, raised sutures or thickened umbilical bosses. An uncoiled variant with a uniserial portion of variable length appears to be part of the normal variation of subspecies *muensteri* populations and occurs at most horizons, possibly as a palaeoecological adaption. It has been unjustifiably separated from *muensteri* by previous authors, usually as either *L. matutina* or *Marginulina sigma*. Distribution: the *L. muensteri* plexus first becomes common from within the *obtusum* Zone in Britain, which is a datum of widespread correlative value (e.g. Dorset coast, Mochras Borehole); thereafter it comprises a common constituent of Upper Sinemurian and Pliensbachian assemblages, becoming predominant in the Toarcian. Widespread throughout north-west Europe and beyond (e.g. Argentina, Ballent 1987b).

The *Lingulina tenera* Bornemann plexus
Members of this species plexus are probably the most diagnostic foraminiferal constituent of the Boreal Lower Jurassic (particularly the Hettangian–Pliensbachian), in terms of abundance and consistency of occurrence. At first sight, a bewildering range of variation is apparent; this was addressed by the studies of Barnard (1956) from the Lias of England and Nørvang (1957) from the Lias of Denmark. Barnard recognised a series of forms (A to J) whereas Nørvang referred successive morphotypes to a number of related chronosubspecies, each showing morphological intergradation. The scheme followed here recognises most of Nørvang's subspecies, but updates the known ranges; discussion of the actual evolutionary development of the group is beyond the scope of the present work. As with the *F. terquemi* plexus, forms and subspecies additional to those of Barnard and Nørvang can be recognised, primarily because Barnard's study was based on only one section, the Stowell Park Borehole, and Nørvang seems to have recovered significantly fewer forms in Denmark.

The approach adopted herein is to recognise stratigraphically significant morphotypes within the plexus as subspecies (chronosubspecies); ten subspecies are identified, all of which intergrade with other subspecies. These include eight of the subspecies recognised by Nørvang, with two additional subspecies, *occidentalis* and subsp. A. These subspecies have consistent stratigraphic ranges over a wide area of northwest Europe. Nørvang's subspecies *carinata* is not considered to have regional stratigraphic significance. Most of Barnard's 'forms' can be identified with these named subspecies.

Lingulina tenera collenoti (Terquem)
Plate 6.2.4, Fig.7 (× 75), Hettangian, Galboly, Northern Ireland. = *Marginulina collenoti* Terquem, 1866a. Synonyms: *Lingulina striata* Blake, 1876, *L. tenera* form A Barnard, 1957. Description: test elongated, ornamented with fine, often discontinuous striae; periphery usually rounded, without keel.
Remarks: the largest and most elongated form in the *L. tenera* plexus. Intermediates occur with subsp. *tenuistriata*. See also comments in Chapter 5. Distribution: index for the basal Lias throughout Britain, e.g. South Wales, Yorkshire (Tate and Blake, 1876) and Stowell Park (Barnard, 1957) Wilkesley, Cockle Pits, Burton Row and Mochras Boreholes, Watchet, Mull (Scotland) and Northern Ireland. Total British range from uppermost Triassic (Penarth Group) to *angulata* Zone, most common in the Pre-Planorbis Beds and *planorbis* Zone. Also occurs in the *planorbis* Zone of Denmark (Nørvang, 1957), France (Terquem, 1866a; Bizon, 1960; Ruget, 1985) and East Germany (Pietrzenuk, 1961).

Lingulina tenera occidentalis (Berthelin)
Plate 6.2.4, Fig.8 (× 142), Lower Pliensbachian, Mochras Borehole, North Wales = *Frondicularia occidentalis* Berthelin, 1879. Description: test with arched, raised sutural ribs between two widely spaced longitudinal ribs; seven to nine chambers.
Remarks: the sutural ribs are distinctive. The taxon is linked by intermediates with *L. tenera tenera* in the Mochras Borehole sequence. Distribution: in Britain only yet recorded from Mochras Borehole and possibly Radstock (Avon) (*raricostatoides–macdonnelli* Subzones), ranging from the *raricostatum* to *tenuicostatum* Zones. Also present between the *jamesoni* and *margaritatus* Zones in the Paris Basin (Berthelin, 1879; Bizon and Oertli, 1961), between the *spinatum* and *tenuicostatum* Zones in south Germany (Riegraf, 1985) and from the Toarcian (probably Lower) of Russia (Mamontova, 1957).

Lingulina tenera pupa (Terquem)
Plate 6.2.4, Fig.13 (× 75), Upper Pliensbachian, Bracebridge, Lincoln. = *Marginulina pupa* Terquem, 1866. Synonyms: *Lingulina tenera* forms I, H Barnard, 1956, *Geinitzina tenera* subsp. *praepupa* Nørvang, 1957. Description: test inflated, pupiform with typically constricted final chamber(s); on each side of test ornamented with two main longitudinal ribs between which are numerous slightly weaker ribs; most ribs continuous; no marginal keel.
Remarks: typical forms, as Terquem's types, possess two main ribs plus weaker interstitial ribs. In the Lower Pliensbachian there is a tendency for all ribs to become of equal strength; this is *pupa* sensu Nørvang (1957). Forms

with weaker interstitial ribs were separated by Nørvang into a new subspecies *praepupa*. However, the latter forms are comparable to Terquem's types and must be regarded as typical *pupa*. In Britain these forms also have the same stratigraphic range as *pupa* sensu Nørvang, and thus are not considered herein as a separate subspecies. Intermediate forms provide an evolutionary link with subsp. *tenuistriata*. Distribution: British range Hettangian–Upper Toarcian (*levesquei* Zone), acme Lower Pliensbachian (particularly *jamesoni* Zone). The earliest occurrence is a figured form from the *angulata* Zone of Dorset (Barnard, 1950a). Widespread throughout northwest Europe. Similar forms from the Lower Cretaceous of Germany (Bartenstein and Brand, 1937; Michael, 1967) and the North Sea are probably unrelated homoeomorphs.

Lingulina tenera subprismatica (Franke)
Plate 6.2.4, Fig.11 (× 133), Lower Sinemurian, Mochras Borehole, North Wales. = *Nodosaria subprismatica* Franke, 1936. Synonym: *Frondicularia tenera prismatica* Brand, 1937. Description: test narrow, elongate, parallel-sided, with hexagonal cross-section, basal spine and oval aperture.
Remarks: test form (tending to become *Nodosaria*-like), aperture shape and basal spine separate the subspecies from other members of *L. tenera* plexus. Distribution: British range from *turneri* Zone to Lower Pliensbachian. This appears to be a consistent range throughout Europe, with records from Denmark (Nørvang, 1957), Sweden (Norling, 1972), West Germany (Franke, 1936), East Germany (Pietrzenuk, 1961), Portugal (Ruget and Sigal, 1970; Exton, 1979), Poland (Kopik, 1960), Italy and France (Brouwer, 1969). Recorded also in the Lower Pliensbachian of Western Canada (Cameron and Tipper, 1985) and offshore Morocco (as *L. acuformis*, Riegraf *et al.*, 1984). Post-Lower Pliensbachian records are known from the Upper Pliensbachian of Portugal (Exton, 1979) and France (*margaritatus* Zone) (Ruget, 1985).

Lingulina tenera substriata (Nørvang)
Plate 6.2.4, Fig.12 (× 128), Hettangian, Mochras Borehole, North Wales. = *Geinitzinita tenera substriata* Nørvang, 1957. Description: test keeled, indistinctly sulcate; ornamented with irregular, often discontinuous, non-parallel ribs, of which two are usually prominent.
Remarks: distinguished by its keel, sulcus and less elongated form from *L. tenera collenoti* and by its irregular ribbing from *L. tenera tenuistriata* (Nørvang). Distribution: observed in Britain in the southern North Sea (48/22-1), Wilkesley Borehole, Burton Row Borehole, Dorset (Copestake, 1987), Mochras Borehole and Loch Aline. Restricted to the *complanata* Subzone except at Mochras where there are questionable specimens in the *bucklandi* and *planorbis* Zones. The subspecies is reported from the Hettangian of the Fastnet Basin by Ainsworth and Horton (1986), although their figured specimen clearly belongs to the *Frondicularia terquemi* plexus. These authors (Ainsworth *et al.*, 1987) have subsequently recorded the taxon from the lowermost Sinemurian of the same region. In Denmark (Nørvang, 1957; Bang, 1971) and Sweden (Norling, 1972) it ranges throughout the Hettangian (*planorbis* to top *angulata* Zones).

Lingulina tenera tenera Bornemann
Plate 6.2.4, Fig.9 (× 130), Lower Sinemurian, Mochras Borehole, North Wales. = *Lingulina tenera* Bornemann, 1854. Synonyms: *Geinitzina tenera* subsp. *carinata* Nørvang, 1957, *Lingulina tenera* Bornemann forms, B, D, E, G, J Barnard, 1957. Description: test with median sulcus between two longitudinal ribs of variable strength and height; periphery keeled; megalopheric tests pupiform, microspheric tests tapering; sutures flush.
Remarks: the two strong ribs with keel and absence of interstitial ribs or striae are diagnostic. Forms with high ribs and a flange-like keel were separated by Nørvang (1957) as a new subspecies *carinata*, restricted to the Pliensbachian in Denmark and Sweden (Norling, 1972). In Britain, such forms occur as old as Early Sinemurian, and are not of regional stratigraphic significance. Intermediates between *tenera* and other members of the plexus (*tenuistriata*, *occidentalis* and *subprismatica*) are common. See Chapter 5 for comments on Triassic records of the subspecies. Distribution: main range Rhaetian–Toarcian, with a distinct acme in Britain in the upper *anguluta* Zone (Mochras Borehole, Redcar, Loch Aline). A homoeomorph occurs in the Valanginian–Barremian (e.g. West Germany, Michael, 1967). Widespread in northwest Europe, with records also from Argentina (Ballent, 1987a, b), Australia (Quilty, 1981), offshore Morocco (Riegraf *et al.*, 1984) and Algeria (Maupin, 1977).

Lingulina tenera tenuistriata (Nørvang)
Plate 6.2.4, Fig.10 (× 130), Lower Sinemurian, Mochras Borehole, North Wales. = *Geinitzina tenera* subsp. *tenuistriata* Nørvang, 1957. Synonyms: *Lingulina tenera* form F Barnard, 1956, *Geinitzina tenera* subsp. *pupoides* Nørvang, 1957, *Lingulina tenera concosta* Kristan-Tollmann, 1964. Description: test with median sulcus and two strong ribs on each side of test, plus numerous finer, often discontinuous interstitial striations.
Remarks: the presence of two stronger ribs with numerous striations is distinctive; the interstitial striations are weaker and less continuous than the ornament of subsp. *pupa*. High degree of variation is typical of this subspecies, including convergent ribbing and the development of four main ribs, together with intermediate forms with other subspecies. A wide range of forms have been illustrated by Nørvang (1957) and Ruget (1985). Nørvang's (1957) subspecies *pupoides* is practically inseparable from *tenuistriata*, and Kristan-Tollmann's (1964) *concosta* from the Rhaetian is also identical. This is the most abundant member of the plexus in the Celtic Sea–Cardigan Basins. Distribution: British range from Rhaetian to Upper Toarcian, most abundant in the *bucklandi–semicostatum* Zones. Infrequent above the Sinemurian in Britain, as also in mainland Europe (e.g. Denmark, Nørvang, 1957; Sweden, Norling, 1972; France, Ruget, 1985). Recorded also from offshore Morocco (Riegraf *et al.*, 1984).

Lingulina tenera subsp. A
Plate 6.2.4, Fig.14 (× 120), Upper Pliensbachian, Bracebridge, Lincoln. Description: test broad, flattened, with slight median sulcus bordered by two longitudinal raised areas, which may be striated; margin rounded, smooth.
Remarks: the almost totally smooth test is distinctive. The form appears suddenly at the base of the Upper Pliensbachian, but is clearly related to *Lingulina tenera* by virtue of its sulcus and vestigial remnants of two longitudinal ribs. Distribution: main range Upper Pliensbachian, observed in the Cleveland Basin (Staithes, Hawsker Bottoms), Lincolnshire (Bracebridge) and offshore in the northern North Sea (Viking Graben) and southern North Sea (e.g. 48/22-1), occurring most consistently in the *margaritatus* Zone. An isolated specimen was recovered from the *bifrons* Zone at Ravenscar.

Lingulina testudinaria Franke, 1936
Plate 6.2.4, Fig.15 (× 116), Upper Sinemurian, Dorset coast. Description: test sharply keeled; ornamented with two longitudinal, central ribs (bordering a median sulcus) and transverse sutural ribs; cross-section elliptical/subrhomboidal; aperture slit-shaped and sometimes centrally constricted.
Remarks: ribbing pattern, combined with test shape and keeled margin is unique. Distribution: in Britain the species is known to occur in the Mochras Borehole and in Dorset (W. A. Macfadyen Coll.) with a total range of *raricostatum* Zone (topmost *raricostatoides* Subzone) to *davoei* Zone. The species occurs throughout the Lower Pliensbachian in West Germany (Franke, 1936; Bartenstein and Brand, 1937), but in Austria it is reported from around the Lower/Upper Sinemurian boundary (Fuchs, 1970). In France the reported range is Upper Pliensbachian–Upper Toarcian (*variabilis* Zone) (Ruget, 1985), though Ruget's figured specimens possess more ribs than typical forms.

Marginulina prima incisa Franke
Plate 6.2.5, Fig.2 (× 110), Hettangian, Mochras Borehole, North Wales. = *Marginulina incisa* Franke. 1936, plate 8, Fig. 11 only. Description: test elongate, narrow with six to eight oblate to spherical chambers, final chamber globular; early sutures flush, later ones depressed; ornamented with up to nine continuous, longitudinal ribs connecting with eccentric produced, radiate aperture.
Remarks: the morphotype, characterised by its globular final chamber, lack of thickened face and ribs which reach the aperture, has a distinct range over a wide area, and is thus recognised as a separate subspecies. It has been included in synonymy with *M. prima prima* by previous authors (e.g. Nørvang, 1957; Brouwer, 1969). Distribution: consistent in Britain between the upper *angulata* to mid-*semicostatum* Zones, with an acme in the *angulata–bucklandi* Zones. Franke's (1936) original record was from the Pliensbachian of West Germany. Not separated by other authors.

Marginulina prima insignis (Franke)
Plate 6.2.5, Fig.1 (× 90), Lower Sinemurian, Mochras Borehole, North Wales. = *Dentalina insignis* Franke, 1936. Description: test narrow, elongate, curved dorsal margin, oval in section, sutures indistinct; ribs oblique, fusing on thickened apertural face to form series of umbrella-like arches.
Remarks: the combination of umbrella-like fusion of ribs on the apertural face, dorsal curvature and oblique ribs is diagnostic. The thickened apertural face and horizontal sutures separate the taxon from *Dentalina matutina*. Distribution: in Britain, consistent between the *angulata* and *bucklandi* Zone, with a widespread acme in the latter zone (Loch Aline, Mochras Borehole, Stowell Park Borehole). Recorded from the Upper Pliensbachian in Denmark (Nørvang, 1957).

Marginulina prima interrupta (Terquem)
Plate 6.2.5, Fig.3 (× 100), Lower Toarcian, Bracebridge, Lincoln. = *Marginulina interrupta* Terquem, 1866a. Description: ten to 12 ribs, interrupted across depressed sutures; chambers arranged in a short, straight series, parallel-sided.
Remarks: interrupted ribs and nodosarian chamber arrangement are distinctive. The test is broader and less elongate than in other subspecies of *M. prima*. Distribution: known British range is from *raricostatum* Zone to *tenuicostatum* Zone, with records from Dorset (Copestake, 1987), Mochras Borehole, Crosby Warren, Bracebridge, Cockle Pits Borehole and Dunrobin Castle. A common member of *tenuicostatum* Zone assemblages in eastern England (e.g. Cockle Pits Borehole, Scunthorpe and Lincoln). Widespread in northern hemisphere, occurring in West Germany (Bartenstein and Brand, 1937), Sweden (Norling, 1972), Denmark (Nørvang, 1957), Portugal (Ruget and Sigal, 1970), France (Bizon, 1961; Ruget, 1985), Poland (Kopik, 1960), Russia (Gerke, 1961) and Alaska (Tappan, 1955), with an identical stratigraphic range.

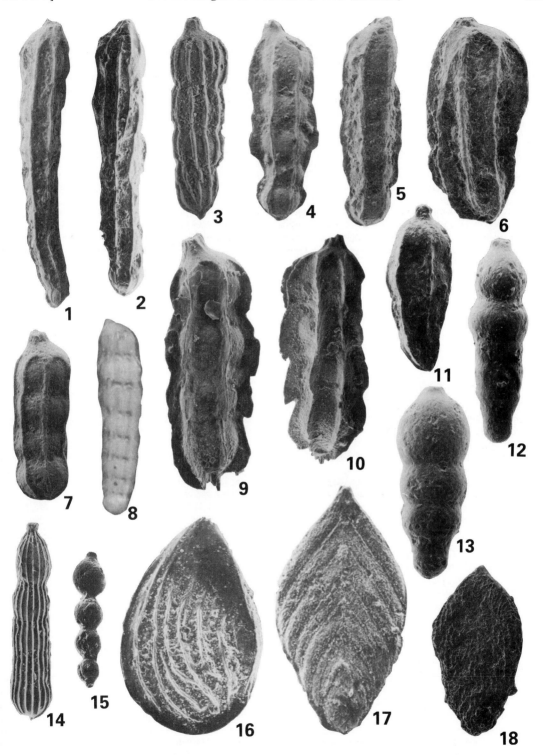

Plate 6.2.5

Marginulina prima praerugosa Nørvang, 1957
Plate 6.2.5, Fig.5 (× 105), Upper Sinemurian, Mochras Borehole, North Wales. Synonym?: *Dentalina burgundiae* Blake, 1876 *non* Terquem, 1864. Description: test elongate, straight, parallel-sided, of up to six chambers; five to six ribs; no thickening on apertural face; aperture marginal.
Remarks: this subspecies is separated from subspecies *prima* and *rugosa* by absence of a thickened apertural face and its marginal aperture, and from subspecies *insignis* and *incisa* by its shorter, straighter test. It appears to be the evolutionary forerunner of subspecies *rugosa* (Nørvang, 1957). Distribution: in Britain it occurs between the uppermost *liasicus* Zone to top *tenuicostatum* Zone, being most consistent and widespread in the Sinemurian. In Denmark (Nørvang, 1957) and Sweden (Norling, 1972) it does not range above the Sinemurian, whereas in West Germany it ranges from lower Hettangian to uppermost Sinemurian (Karampelas, 1978).

Marginulina prima prima d'Orbigny, 1849
Plate 6.2.5, Fig.7 (× 100), Upper Sinemurian, Mochras Borehole, North Wales, Fig.8 (× 80), reflected light, Lower Pliensbachian, Dorset coast. Common synonym: *Marginulina burgundiae* Terquem, 1864. Description: test with six to ten longitudinal ribs fusing together as umbrella-like arches on apertural face, previous arches visible between earlier chambers; sutures depressed; aperture eccentric.
Remarks: *M. burgundiae* appears to represent the microscopic generation, typified by an early coil. Variations in test size, degree of lateral compression and general morphology have induced previous authors (e.g. Terquem, 1858; Franke, 1936; Pietrzenuk, 1961) to recognise a number of varieties of little stratigraphical importance. The arches on the apertural face are wider and better developed than in subspecies *rugosa*, and are consequently visible at the sutures between earlier chambers. Distribution: in Britain and mainland Europe typical large forms first appear and occur commonly in the Late Pliensbachian, becoming extinct in the Early Toarcian (*falciferum* Zone). British records are from the Mochras Borehole, Burton Row Borehole, Bracebridge and northern North Sea. A smaller form occurs in the Sinemurian of the Mochras Borehole. Recorded also from the Upper Pliensbachian of Western Canada (Cameron and Tipper, 1985), offshore Morocco (Riegraf *et al.*, 1984) and Argentina (Ballent, 1987b). Common between the uppermost Sinemurian and basal Toarcian in West Germany (Karampelas, 1978).

Marginulina prima rugosa Bornemann
Plate 6.2.5, Fig.4 (× 124), Lower Pliensbachian; Fig.6 (× 105), Upper Sinemurian, both from Mochras Borehole, North Wales. = *Marginulina rugosa* Bornemann, 1854. Description: test divergent-sided, with up to 12 sharp ribs, fusing on slightly thickened apertural face; aperture marginal; sutures indistinct, nearly flush.
Remarks: this subspecies appears to be the antecedent of *M. prima prima* and is separated from the latter by its less thickened apertural face and more marginal aperture. Bornemann's (1854) holotype was refigured by Rabitz (1963) and the taxon well described by Nørvang (1957). Distribution: first appears in the Lower Sinemurian in Britain (e.g. Dorset coast, Mochras Borehole), but is most frequent in the Upper Sinemurian, where it first appears in Denmark (Nørvang, 1957). Rare in the Pliensbachian, becoming extinct in the earliest Toarcian. In West Germany it is common from the basal Sinemurian to mid-Upper Pliensbachian (Karampelas, 1978). Figured from the Upper Pliensbachian of southern Spain (Mira and Martinez-Gallego, 1981, as *M. prima*).

Marginulina prima spinata Terquem
Plate 6.2.5, Fig.9 (× 151); Fig.10 (× 170); Upper Sinemurian, Dorset coast. = *M. spinata* Terquem, 1858. Description: six or seven ribs, notched and typically projected as spines over sutures, but not interrupted.
Remarks: notched and spinose ribs are diagnostic. Also usually has fewer ribs and chambers than *M. prima interrupta*. Distribution: main British range from *raricostatum* Zone to *tenuicostatum* Zone, but recorded as young as *falciferum* Zone at Empingham (Horton and Coleman, 1978). Recorded also from southern North Sea, Risby Warren (Adams, 1957), Tilton, Cockle Pits Borehole, Trunch Borehole, Mochras Borehole, Dorset (Macfadyen, 1941; Copestake, 1987). Continental range mainly Sinemurian to top Lower Pliensbachian, occurring in Sweden (Norling, 1971), West Germany (Bartenstein and Brand, 1937), East Germany (Pietrzenuk, 1961) and France (Ruget, 1985). In southern Germany (Usbeck, 1952) and Poland (Gazdzicki, 1975) the subspecies is said to occur in the Hettangian/Lower Sinemurian. Recorded also from the Pliensbachian of offshore Morocco (Riegraf *et al.*, 1984).

Nodosaria issleri Franke, 1936
Plate 6.2.5, Fig.11 (× 110), Upper Sinemurian, Mochras Borehole, North Wales. Synonym: *N. aequalis*, Issler 1908. Description: six to eight sharp ribs which pass uninterrupted over flush sutures, disappearing at middle of final chamber; test occasionally slightly curved; aperture produced on short, stout neck.
Remarks: combination of flush sutures, semi-smooth final chamber and produced aperture distinguish the species from the related *N. radiata* (Terquem), *N. mitis* (Terquem and Berthelin) and *Pseudonodosaria multicostata* (Bornemann). Distribution: ranges throughout Upper Sinemurian in Mochras Borehole and also present in this substage in the Trunch Borehole, Dorset coast, Dunrobin Castle and Fastnet Basin sections (Ainsworth and Horton, 1986). Restricted to this substage also in West Germany (Bartenstein and Brand, 1937; Karampelas, 1978), East Germany (Pietrzenuk, 1961), Denmark (Nørvang, 1957) and France (Ruget, 1985).

Nodosaria metensis Terquem, 1863
Plate 6.2.5, Fig.14 (× 36), Upper Pliensbachian, Bracebridge, Lincoln. Synonyms: *Nodosaria dispar* Franke, 1936, *N. metensis* var. *robusta* Barnard, 1950a, *Nodosaria reineckei* Hagenmeyer in Bach *et al.*, 1959. Description: test large, robust, up to six chambers, ornamented with eight to 23 fine to coarse, continuous ribs; proloculus large, spherical,

succeeding chambers typically narrower, final chamber elongated, equalling proloculus in diameter; sutures depressed, aperture radiate.
Remarks: rib number and strength are highly variable, ranging from fine and numerous to coarse and few. The species is a characteristic member of Lower Jurassic assemblages. Distribution: total range Rhaetian–Bajocian, most consistent in the Hettangian–Pliensbachian, with a particular abundance in the Sinemurian. Widespread in the Boreal Lower Jurassic of Europe (e.g. Brouwer, 1969).

Nodosaria regularis Terquem subsp. A
Plate 6.2.5, Fig.15 (× 17.5), Upper Toarcian, Mochras Borehole, North Wales. Synonyms: *N. regularis*, Franke, 1936, Magné *et al.*, 1961, Wernli and Septfontaine, 1971, Exton, 1979. Description: large subspecies with regularly enlarging, subspherical chambers connected by short, constricted necks.
Remarks: stratigraphically distinct and three to four times larger than *N. regularis regularis* Terquem. Distribution: British records from uppermost Toarcian (*thouarsense–levesquei* Zones) of Mochras Borehole and Dorset and possibly from the undifferentiated 'Middle and Upper Lias' of Somerset (as *N. radicula*, Brady, 1866). Ranges as high as Lower Bajocian (*discites* Zone) in Britain (Lyme Bay Borehole, see section 6.3). European Range: uppermost Toarcian to Lower Bajocian of France (Magné *et al.*, 1961), southern France/Switzerland (Wernli and Septfontaine, 1971), West Germany (Bartenstein and Brand, 1937; Frentzen, 1941), East Germany (Stoermer and Weinholz, 1965) and *levesquei* Zone of Portugal (Exton, 1979). Exton and Gradstein (1984) report the species from the uppermost Toarcian (*thouarsense–levesquei* Zones) of Grand Banks, Eastern Canada. The subspecies is a nominate taxon for the uppermost Toarcian–Aalenian foraminiferal subzone (Copestake and Johnson, 1984; Copestake, 1985).

Palmula deslongchampsi (Terquem)
Plate 6.2.5, Fig. 17 (× 75), Upper Toarcian, Ilminster, Somerset. = *Flabellina deslongchampsi* Terquem, 1864. Synonym: *Flabellina jurensis* Franke, 1936. Description: small lenticuline portion followed by rapidly flaring, embracing flabelline portion; distinct, depressed sutures. Distribution: British records span an age range of basal Toarcian to Upper Bajocian (*garantiana* Zone), occurrences being from the Toarcian of Whitby, Risby Warren (Adams, 1955), 'Midlands/Yorkshire' (Brouwer, 1969), Mochras Borehole, Bristol Channel (Lloyd *et al.*, 1973), Ilminster (F. M. Lowry, pers. comm.), Dorset coast and Aalenian–Upper Bajocian of offshore Dorset (Lyme Bay Borehole) (see section 6.3). Consistently appears in Western Europe at the base of the Toarcian (Copestake and Johnson, 1984; Copestake, 1985), though Norling (1972) documents its presence in the Pliensbachian of 'SW Europe'.

Palmula tenuistriata (Franke)
Plate 6.2.5, Fig. 18 (× 70), Upper Toarcian, Mochras Borehole, North Wales. = *Flabellina tenuistriata* Franke, 1936. Description: initial small coiled portion with two to three comma-shaped chambers followed by asymmetrical, flabelline portion with four to five chambers; raised sutures; subreticulate ornament. Distribution: known in Britain only from the *levesquei* Zone of the Mochras Borehole. Excellent marker in Europe for Upper Toarcian and Aalenian (Copestake and Johnson, 1984; Copestake, 1985). Recorded in Britain, West Germany (Bartenstein and Brand, 1937), France (Wernli, 1971; Ruget, 1985), Switzerland (Wernli and Septfontaine, 1971) and Portugal (Exton, 1979). No post-Toarcian records are known from Britain.

Planularia inaequistriata (Terquem)
Plate 6.2.5, Fig. 16 (× 81), Lower Sinemurian, Dorset coast. = *Marginulina inaequistriata* Terquem, 1863. Synonym: *P. choiseulensis* Ruget and Sigal, 1967. Description: test robust, ornamented with irregular, often bifurcating ribs; periphery keeled; sutures flush. Distribution: widespread and frequent in Britain and mainland Europe (Denmark, Sweden, West Germany, Holland, Austria and France) and also occurs in Alaska (= *Vaginulina curva*, Tappan, 1955, Plate 22, Figs 12, 16–19). First appears at the base of the *liasicus* Zone in Britain, but reported from the *planorbis* Zone in France (Bizon and Oertli, 1961; Ruget, 1985). Typical forms disappear in the *obtusum* Zone (*stellare* Subzone) in Britain (data herein, plus Ainsworth *et al.*, 1987) and parts of France (Bizon and Oertli, 1961), though Ruget (1985) extends the range to the top of the Sinemurian (but also figures a form from the Lower Pliensbachian (Plate 32, Fig. 12)). Nørvang (1957) and Norling (1972) also carry the range of the species towards the top of the Sinemurian, though ammonite control is not consistent in their studies. Typical forms with relatively few, strong ribs pass into forms with more numerous, fainter ribs (e.g. *Planularia ornata* (Terquem)) in the Sinemurian. It is possible that workers have included this morphotype, which ranges into the Pliensbachian, within their definition of *inaequistriata*.

Pseudonodosaria vulgata (Bornemann)
Plate 6.2.5, Fig. 12 (× 135), Fig. 13 (× 152), Lower Pliensbachian, Mochras Borehole, North Wales. = *Glandulina vulgata* Bornemann, 1854. Synonyms: *Glandulina tenuis* Bornemann, 1854, *Orthocerina pupoides* Bornemann, 1854, *Glandulina irregularis* Franke, 1936, *Pseudonodosaria vulgata* vars. *pupiforme, humulis, regularis* Barnard, 1950b. Description: test small, generally parallel-sided, smooth (occasionally faintly striated), comprising three to eight overlapping chambers, of variable width; early sutures flush, depressed in adult; aperture radiate, slightly produced.
Remarks: The high degree of variation, particularly in chamber width and development of flush sutures allows the recognition of many morphotypes. These were studied by Barnard (1950b) who included several 'species' of Bornemann (1854) and Franke (1936) within a single variable species *vulgata*, none of the morphotypes having stratigraphic significance. Norling (1972) also included several Middle and Upper Jurassic 'species' in synonymy with

vulgata, and extended the range into the Upper Cretaceous. Distribution: typically common throughout the Lower Jurassic of Europe, first appearing in Britain in the *liasicus* Zone, with an acme in the Upper Sinemurian.

Saracenaria sublaevis (Franke)
Plate 6.2.6, Fig. 1 (× 63), Lower Toarcian, Bracebridge, Lincoln. = *Cristellaria (Saracenaria) sublaevis* Franke, 1936. Description: test triangular in cross-section, and comprises a coil of three to five chambers followed by an uncoiled portion of two to six chambers; margins subangular; ventral surface slightly convex. Distribution: despite its reported Hettangian to Lower Toarcian range, its acme during the Upper Pliensbachian has been used zonally, e.g. in West Germany (Bartenstein and Brand, 1937) and Sweden (Norling, 1972). Known in Britain from northern North Sea, southern North Sea, Mochras Borehole, Bracebridge and Tilton between the *spinatum* and topmost *tenuicostatum* Zones. Recorded from the Lower Toarcian in the Fastnet Basin (Ainsworth *et al.*, 1987). Its range in the southern North Sea (well 48/22-1) may extend down into the *margaritatus* Zone as the species occurs in association with abundant *L. muensteri acutiangulata*. Also recorded from Denmark (Nørvang, 1957), Holland (Brouwer, 1969), France (Bizon and Oertli, 1961; Ruget, 1985), East Germany (Pietrzenuk, 1961), Poland (Kopik, 1960), Spain (Brouwer, 1969), Sicily (Barbieri, 1964), Portugal (Exton, 1979) and Algeria (Maupin, 1977). The species is recorded from the Lower Pliensbachian of Spain (Brouwer, 1969) and Portugal (Exton, 1979), but appears in most areas in the Upper Pliensbachian, with an acme in the *spinatum–tenuicostatum* Zones, becoming extinct at the top of the latter zone.

Saracenella aragonensis (Ruget)
Plate 6.2.6, Fig. 2 (× 82), Fig. 3, apertural view (× 158), Lower Toarcian, Bracebridge, Lincoln. = *Lenticulina (Saracenaria Saracenella) aragonensis* Ruget, 1982. Description: test smooth, elongate, triangular in section, test sides concave between swollen, rounded test margins; 10 to 12 chambers, initial three to four in curvilinear series, subsequent chambers rectilinear, uniserial; final chamber occasionally inflated, spherical.
Remarks: the species' elongate form and concave sides with swollen margins are diagnostic. Distribution: an excellent marker species, originally described from the *semicelatum* Subzone (*tenuicostatum* Zone) of Spain (Ruget, 1982); subsequently recorded from the same subzone in south Germany (Riegraf, 1985) and eastern England (Bracebridge); also occurs in the *tenuicostatum* Subzone at Crosby Warren. Brouwer's (1969) record (as *Saracenaria trigona*) from the Paris Basin is given only as 'Lower–Middle Toarcian', not in terms of ammonite zones.

Saracenella sp. A
Plate 6.2.6, Fig. 9 (× 55), Upper Toarcian, Mochras Borehole, North Wales. Description: (microspheric) test smooth, elongate, sutures flush; seven to ten 'pagoda-like' chambers, aperture produced; (megalospheric) smaller, more robust with fewer chambers and a shorter, more curved initial portion. Distribution: occurs in the Upper Toarcian of Mochras Borehole (*moorei–aalensis* Subzones) and Dorset coast (topmost Junction Bed, *thouarsense* Zone).

Tristix liasina (Berthelin)
Plate 6.2.6, Fig. 13 (× 172), Lower Pliensbachian, Mochras Borehole, North Wales = *Rhabdogonium liasinum* Berthelin, 1879. Description: test uniserial, triangular in cross-section with rounded or angular margins; sutures depressed or flush; test sometimes coarsely perforated; aperture produced, radiate. Remarks: this is the only known Liassic *Tristix* species. It is common in the Swedish Pliensbachian (Norling, 1972), but the Swedish material differs from that so far recorded from Britain in having mainly angular or keeled, but not rounded, margins, flush rather than depressed sutures and fewer chambers. Distribution: British range is from *ibex* Zone (*luridum* Subzone) to *falciferum* Zone (*exaratum* Subzone), with occurrences in the Mochras Borehole, Dorset (Macfadyen, 1941) and Empingham (Horton and Coleman, 1978). Swedish range is from *jamesoni* Zone to Toarcian (Norling, 1972). Restricted to the Lower Pliensbachian in Poland (Kopik, 1960), the *margaritatus* Zone in the Paris Basin (Berthelin, 1879) and the *davoei* to early *margaritatus* Zone interval in Portugal (Exton, 1979).

Vaginulina/Citharina clathrata (Terquem)
Plate 6.2.6, Fig. 4 (× 70), Upper Toarcian, Mochras Borehole, North Wales; Fig. 5 (× 70), Upper Toarcian, Dorset coast. = *Marginulina longuemari* var. *clathrata* Terquem, 1864. Synonyms: *Marginulina proxima* Terquem, 1868, *Vaginulina infraopalina* Brand, 1949. Description: test robust, keeled, triangular to curved-triangular; three to six continuous costae paralleling dorsal margin.
Distribution: a characteristic Toarcian/Aalenian species (= *V. proxima* Barnard, 1948, *C. infraopalina*, Brand and Fahrion, 1962). Norling (1972) (= *C. clathrata* and *C. infraopalina*) established a zone with *C. clathrata* in the Swedish Toarcian/Aalenian. Although ranging from uppermost Pliensbachian to Upper Jurassic, its acme throughout Europe is in the Toarcian/Aalenian. British records from the Mochras Borehole, possibly from the undifferentiated 'Middle and Upper Lias' of Somerset (? = *V. striata* Brady, 1866), Ilminster (F. M. Lowry, pers. comm.), Byfield (Barnard, 1950b), Empingham (Horton and Coleman, 1978), Midlands/Yorkshire (Brouwer, 1969), Tilton, Stowell Park Borehole, Hill Lane Borehole and Dorset coast. First appears (commonly) in several areas in the *falciferum* Subzone above the Early Toarcian anoxic event.

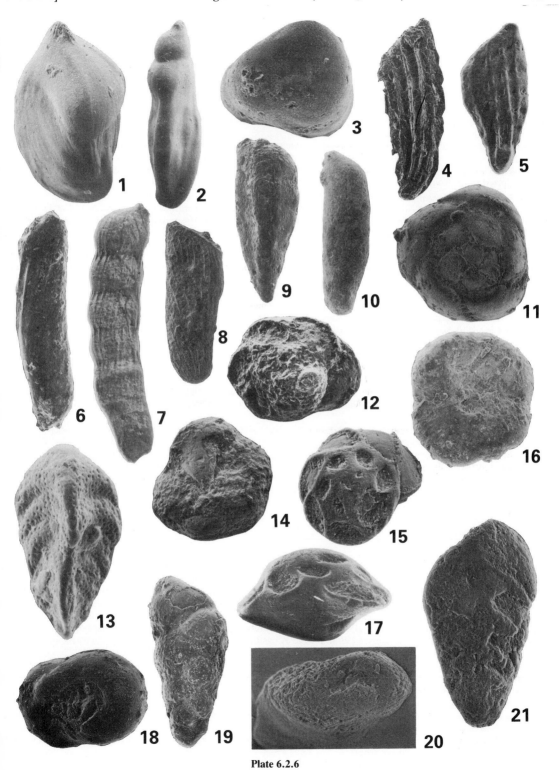

Plate 6.2.6

Vaginulina listi (Bornemann)
Plate 6.2.6, Fig. 10 (× 44), Lower Sinemurian, Mochras Borehole, North Wales. = *Cristellaria listi* Bornemann, 1854. Synonym: *Marginulina hybrida* Terquem, 1866a, ?*M. lumbricalis* Terquem, 1866b. Description: test smooth, oval in section, parallel-sided, with curved initial portion followed by elongate rectilinear portion; up to 12 broad chambers, increasing markedly in height in adult; sutures flush.
Remarks: distinguished from uncoiled variants of *L. muensteri* by the lack of an initial coil. Terquem's *M. lumbricalis* (actually a *Vaginulina*) may be an elongate variant of the species. Distribution: first appears in Britain in the upper *semicostatum* Zone, ranging to the top of the Toarcian, particularly common in the Upper Sinemurian. Total reported range from Hettangian to Lower Bajocian; widespread in the Lower Jurassic of Europe.

Vaginulinopsis denticulatacarinata (Franke)
Plate 6.2.6, Fig. 6 (× 57), Lower Pliensbachian, well 48/22-1, southern North Sea; Fig. 7 (× 44), Lower Pliensbachian, Crosby Warren, Scunthorpe, South Humberside. = *Cristellaria (Astacolus) denticulata-carinata* Franke, 1936. Synonym: *Marginulina bergquisti* Tappan, 1955. Description: test with small indistinct initial coiled portion and long parallel-sided uniserial portion; vertical and dorsal margins rounded, lower margin denticulate; aperture radiate; test predominantly smooth, with occasional oblique ribs.
Remarks: the well developed uncoiled portion and denticulate lower margin are diagnostic. Oblique ribs have only been seen in British specimens. Distribution: excellent marker for the Lower Pliensbachian in Europe, though there are two possible records from the Upper Pliensbachian, in Sweden (Norling, 1972) and the Fastnet and Celtic Sea Basins (Ainsworth *et al.*, 1987, as '*Astacolus* cf. *A. denticula-carinata*'). British range throughout Lower Pliensbachian, with occurrences from Gloucestershire (Blockley Station quarry), East Midlands (Bracebridge, Crosby Warren), Cleveland–Southern North Sea Basin (Robin Hood's Bay and 48/22-1), East Anglia (Trunch Borehole), Northern North Sea and Inner Moray Firth (onshore and offshore). Most common and widespread in Britain in the *jamesoni* Zone.

Vaginulinopsis exarata (Terquem)
Plate 6.2.6, Fig. 8 (× 57), Lower Sinemurian, Mochras Borehole, North Wales. = *Marginulina exarata* Terquem, 1866b. Synonym: *V. subporrecta* Bizon, 1960. Description: test elongate, ornamented with numerous, fine, oblique striations; long, rectilinear, parallel-sided uncoiled portion with flush early sutures becoming depressed between adult chambers.
Remarks: finer ornamentation and more elongate test than *Astacolus speciosus*. Differs from *Dentalina langi* in having an initial coil. Distribution: common in the British *semicostatum* Zone (e.g. Mochras and Stowell Park Boreholes) and thought to occur between the Hettangian and Upper Pliensbachian in Europe. A marker for the Lower Sinemurian in Sweden (Norling, 1972), northern Germany (Klingler, 1962) and France (Bizon and Oertli, 1961), but occurs in the Upper Sinemurian of Denmark (Nørvang, 1957) and the Pliensbachian of southern Germany (Klingler, 1962).

SUBORDER ROBERTININA

Reinholdella macfadyeni (Ten Dam)
Plate 6.2.6, Fig. 15, dorsal view (× 105); Fig. 17, edge view (× 102), Lower Toarcian, Bracebridge, Lincoln. = *Asterigerina macfadyeni* Ten Dam, 1947. Description: convex dorsal surface with up to 3.5 whorls; raised, merging spiral and septal sutures; ventral surface smooth, planar/slightly convex with umbilical plug; limbate marginal keel.
Distribution: earliest British record is from the Upper Pliensbachian of Yorkshire, but the species becomes common and widespread at the base of the Toarcian, e.g. Yorkshire, Crosby Warren, Lincoln, Tilton, Empingham (Horton and Coleman, 1978), Byfield (Barnard, 1950b), Mochras Borehole, a boring at Pattishall (Northamptonshire) (type description, Ten Dam, 1947) and Dorset (Winterborne Kingston Borehole, Coleman, 1982). Particularly abundant in the *tenuicostatum–bifrons* Zones of the Cleveland–Southern North Sea Basin and Midlands. It is notably absent from the Toarcian at Ilminster (F. M. Lowry, pers. comm.). In the Celtic Sea Basin (Mochras Borehole) it is more abundant in the Upper Toarcian. European range between Upper Pliensbachian and Lower Aalenian from southern Germany/Austria/Switzerland and France (Brouwer, 1969).

Reinholdella margarita (Terquem)
Plate 6.2.6, Fig. 12, dorsal view (× 100), Fig. 14, ventral view (× 76), Upper Sinemurian, Mochras Borehole, North Wales. = *Rotalina margarita* Terquem, 1866b. Synonym: *Epistomina liassica* Barnard, 1950a. Description: test high-spired, with broad elevated spiral and septal sutures and chamber margins; open, crater-like umbilicus on ventral side; seven or eight chambers in last whorl.
Remarks: *R. macfadyeni* differs in its thinner, less elevated sutures and closed umbilicus. Distribution: British range

from upper *semicostatum* Zone to lower *obtusum* Zone, with observed occurrences in the Mochras, Stowell Park and Trunch Boreholes, Dorset coast (= *Epistomina liassica* Barnard, 1950a), southern North Sea (e.g. 48/22-1) and Fastnet Basin (Ainsworth and Horton, 1986). Its absence from coeval sediments in western Scotland is notable. Also recorded from France (upper *semicostatum* Zone to *obtusum* Zone) (Bizon, 1961; Ruget and Sigal, 1967) and Sweden (Lower Sinemurian, Norling, 1972). Its first appearance in the upper *semicostatum* Zone is within a significant foraminiferal influx related to an apparent transgression.

Reinholdella? planiconvexa (Fuchs)
Plate 6.2.6, Fig. 11, dorsal view (pyritised cast) (× 158), Lower Sinemurian; Fig. 16, ventral view (×145), Upper Sinemurian, Mochras Borehole, North Wales. = *Oberhauserella planiconvexa* Fuchs, 1970. Description: test smooth, discoidal to ovate; plano-convex in vertical section; marked umbilical hollow and usually clearly visible supplementary aperture on ventral side; two to 2.5 whorls with five to six chambers in final whorl.
Remarks: differs from *R. dreheri* (Bartenstein) in exhibiting an open umbilicus and greater variability in shape. The two apertures which are characteristic of *Reinholdella* are visible at high magnification. However, thin section studies do not indicate the presence of a toothplate, hence the questionable generic allocation. Distribution: first appears in Britain in the Rhaetian (see Chapter 5) and ranges to the basal *jamesoni* Zone, with a widespread acme at the base of the *liasicus* Zone (Celtic Sea, Somerset coast, Wilkesley Borehole), associated with a transgressive event. Occurs consistently up to the *semicostatum* Zone. To date not recorded from Scottish sections. In mainland Europe, the species is abundant in the Hettangian–Lower Sinemurian (Brouwer, 1969, as *Conorboides* sp.), but restricted to the Hettangian in West Germany (Bartenstein and Brand, 1937; Klingler, 1962, both as *Discorbis advena*), Austria (Fuchs, 1970) and France (Brouwer, 1969).

SUBORDER ROTALIINA

Brizalina liasica (Terquem)
Plate 6.2.6, Fig. 21 (× 160); Fig. 20, apertural view (× 190), Lower Toarcian, Bracebridge, Lincoln. = *Textilaria liasica* Terquem, 1858. Synonyms: *Bolivina rhumbleri* Franke, 1936, *Bolivina rhumbleri amalthea* Brand, 1937. Description: test smooth, compressed, sometimes with thickened median area; early sutures flush, later ones depressed; aperture an elongate slit; internal toothplate.
Remarks: this is the only known Liassic species of *Brizalina*. The species is probably aragonitic in composition (Witthuhn, 1968), which necessitates either generic emendation or erection of a new genus. Distribution: consistent British range from *jamesoni* Zone to top *tenuicostatum* Zone, with an acme in the Upper Pliensbachian, as in the southern North Sea. Widespread onshore England and offshore in the southern North Sea and Celtic Sea Basins. Not known from northern North Sea or Scotland. Only rarely reported above the *tenuicostatum* Zone, in the *commune* Subzone of Yorkshire and questionably in the *dispansum* Subzone of the Mochras Borehole. Widespread and common in the Pliensbachian of Western Europe (e.g. France, Bizon and Oertli, 1961, Ruget, 1985; Denmark, Nørvang, 1957, Bang, 1968; Sweden, Norling, 1972; West Germany, Bartenstein and Brand, 1937), but reported to range into the Aalenian in Sweden (Norling, 1972) and Switzerland (Wernli, 1971). The Upper Pliensbachian acme of the species in the offshore British area (and offshore Holland) matches the coeval abundance in Portugal and offshore Eastern Canada reported by Exton and Gradstein (1984). Recently described from the Pliensbachian of offshore Morocco (Riegraf et al., 1984).

Neobulimina sp. 2 Bang, 1968
Plate 6.2.6, Fig. 18, apertural view (× 215), Fig. 19 (× 125), Lower Sinemurian, Mochras Borehole, North Wales. = *Gaudryina gradata* form a Brand, 1937. Description: initial triserial portion comprising less than one-third test length followed by biserial chamber arrangement; surface hispid or smooth; test sides divergent throughout; periphery rounded; aperture wide, 'u'-shaped; internal toothplate. Remarks: the only known Liassic species of the genus, but probable aragonitic composition may require the erection of a new genus. Distinguished from *Brizalina liasica* by its triserial initial portion, hispid ornament and thicker, non-compressed test with rounded margins.
Distribution: the species is common in the *semicostatum* Zone at which level it is most widespread, apparently associated with the intra-*semicostatum* Zone transgression in Britain. Known occurrences demonstrating this acme are from the Stowell Park Borehole, Burton Row Borehole, Mochras Borehole, Fastnet Basin (Ainsworth and Horton, 1986) and Dorset (Winterborne Kingston Borehole, Coleman, 1982). Though recorded only from this latter zone in Dorset (Copestake, 1987), Norling (1972) describes its apparent presence in the Dorset *angulata* Zone. The earliest positive record in Britain is from the *bucklandi* Zone of the Mochras Borehole, where it ranges to the *obtusum* Zone. Occurs also in the Allt Leacach section (western Scotland) (?*turneri* Zone). Ranging through, but restricted to the Lower Sinemurian in mainland Europe, i.e. Denmark (Bang, 1968, 1971 and 1972), Sweden (Norling, 1972) and West Germany (Bartenstein and Brand, 1937, as *Gaudryina gradata* form a).

Acknowledgements

The authors are grateful to BP Exploration and Gearhart Geo Consultants for provision of facilities and permission to publish. The British Geological Survey allowed microfaunal preparations to be examined from the Cockle Pits, Trunch, Stowell Park, Brent Knoll, Hill Lane, Platt Lane and Wilkesley Boreholes and also released unpublished information on the Trunch Borehole.

Thanks are due to Dr. R. L. A. Young for provision of unpublished data from several localities in eastern England, to Mr J. F. Gregory for allowing reference to additional information from Loch Aline (Allt Leacach stream section), to Dr. F. M. Lowry for allowing access to information and material (including the provision of photographs of selected species) from the Toarcian of Ilminster and to Mr J. Exton for provision of material from Portugal. Fergie Norris and Liz Gordon typed the manuscript and Ross Munro and Alex Forsyth draughted the figures. Special thanks are due to Paula for her unending tolerance and assistance with manuscript compilation.

6.3 THE AALENIAN TO CALLOVIAN (MIDDLE JURASSIC)

P. H. Morris, B. E. Coleman

6.3.1 Introduction and history of investigation

British Middle Jurassic foraminifera are poorly documented with the Aalenian to Bajocian interval being largely ignored by workers until recently. Morris (1980, 1982) described in detail foraminifera from the Lower Inferior Oolite (early Aalenian, *opalinum – murchisonae* Zones) of the Cotswolds and related distribution closely to carbonate lithofacies. Description of late Aalenian–Bajocian species is based almost entirely on British Geological Survey borehole material from condensed sequences of Inferior Oolite penetrated offshore and onshore Dorset (Lyme Bay, Winterborne Kingston and Horsecombe Vale Boreholes; Coleman, in Penn *et al.*, 1979, Penn, 1982). Bajocian foraminifera have also been described from the Lincolnshire Limestone sequence (*discites – laeviuscula* Zones; Packer, 1986) and marginal marine beds of the Deltaic Series (*discites – laeviuscula* Zones; Nagy *et al.*, 1981).

Descriptions of Bathonian foraminifera appeared in the latter part of the last century (e.g. Jones, 1884), however, Cifelli (1959) published the first detailed account of faunas from the Upper Fuller's Earth Clay and Frome Clay (*morrisi – aspidoides* Zones) of Bath and the Dorset coast.

Bathonian assemblages were further described from more complete sequences penetrated by the Winterborne Kingston and Horsecombe Vale Boreholes (Coleman, in Penn, 1982, Penn *et al.*, 1979). More recently, Copestake (1987) listed Late Bathonian assemblages from several Dorset localities and made some stratigraphic comments.

Oxford Clay foraminifera of Callovian age have been described from several brickpits of the Midlands area (Barnard *et al.*, 1981; Medd, 1983) and also from cores of the Milton Keynes and Worlaby Boreholes (Coleman in Horton *et al.*, 1974, Medd in Richardson, 1979). Cordey (1962) documented late Callovian (*lamberti* Zone) assemblages from the Staffin Bay Shales, and Skye and Gordon (1967) described Callovian species from the Brora Argillaceous Series, Sutherland.

There are no previously published descriptions of Middle Jurassic foraminifera from any part of the North Sea area, despite the abundant occurrence of Middle Bathonian – Callovian assemblages through the Tarbert and Heather Formations (see section 6.3.4.5.). Similarly there are no publications from the northern and western continental margins although Ainsworth *et al.* (1987) published an outline of foraminiferal distribution from Aalenian–Bathonian sequences of the North Celtic Sea and Fastnet basins, offshore Ireland.

Outside Britain there is extensive literature on Boreal Middle Jurassic foraminifera. Work on the French 'Oolithique' was initiated by Terquem (1868–1886) in describing Aalenian–Bathonian species from the Moselle district. More recent papers include Nouet (1958—Bathonian of the Boulonnais), Prestat

(1967—Bathonian – Callovian of the Paris Basin) and Wernli and Septfontaine (1971—Aalenian – Callovian of the Jura Mountains and Swiss Prealpes). Details of foraminifera from the Dogger Group, onshore North Germany are included in papers by Bartenstein and Brand (1937), Ziegler (1959), Munk (1978) and Ohmert (in press).

High-latitude Middle Jurassic assemblages which have bioprovincial links with the North Sea Basin are described in papers by Løfaldi and Nagy (1980, 1983, Callovian of Spitsbergen) and Dain (1972, Callovian of Western Siberia). In addition there are several relevant papers from North America and Western Canada. These include Tappan (1955), Wall (1960) and Brooke and Braun (1972, 1981). Gradstein (1976) and Ascoli (1976) described foraminifera of Middle Jurassic sequences penetrated by exploration wells including Western Atlantic sites.

The location, chronostratigraphy and lithostratigraphy of British Middle Jurassic sections analysed for foraminifera, are presented in Figs 6.3.1 and 6.3.2.

6.3.2 Location of important collections
(1) British Museum (Natural History).
  Cifelli 1959; on permanent loan from the Harvard Museum of Comparative Zoology.
  Cordey 1962, including unpublished material of Gordon, 1967 (Oxford Clay).
(2) British Geological Survey, Keyworth.
  Coleman's borehole material of Penn *et al.*, 1979; Penn, 1982; Horton *et al.*, 1974; and Richardson, 1979.
(3) University College, London.
  Barnard *et al.*, 1981 (Oxford Clay).
(4) University College of Wales, Aberystwyth.
  Morris 1980, 1982 (Lower Inferior Oolite).

6.3.3 Stratigraphic divisions
Stratigraphic division of the British Middle Jurassic succession is summarised in Fig. 6.3.3 For details of stage nomenclature, lithostratigraphy and ammonite zonation the reader is referred to Cope *et al.* (1980b).

It should be noted that in this edition the Aalenian and Bajocian are treated as separate stages, which together with the Bathonian and Callovian comprise the British Middle Jurassic. According to Kent and Gradstein (1985) the Middle Jurassic subsystem ranges from 187 my (base *opalinum* Zone) to 163 my BP (top *lamberti* Zone), spanning 24 million years.

An important feature of Aalenian–Bathonian stratigraphy is the predominance of shallow and marginal marine sequences (e.g. Inferior Oolite Group, Great Estuarine Group, Brent Group), which display considerable lateral thickness and lithological variation. Whilst established ammonite zonal schemes have been applied to these deposits (see Torrens, 1969; Parsons, 1974, 1976; Cope *et al.*, 1980b) the paucity of zonal species has hindered correlation within several major rock units. Because of this limitation, there has been a long-standing requirement to use other fossil groups for biostratigraphy (e.g. brachiopod faunas of the Inferior Oolite—Buckmann, 1895, 1897, 1901) despite the attendant problems of facies control on distribution. The use of microfossil groups including dinocysts and ostracods has, however, largely been ignored until quite

Fig. 6.3.1 — Middle Jurassic outcrop and localities: 1, Dorset Coast; 2, Lyme Bay Borehole; 3, Winterborne Kingston Borehole; 4, Seabarn Farm Borehole; 5, Frome Borehole; 6, mid-Cotswolds; 7, Naggridge No. 1 Borehole; 8, Horsecombe Vale 15 Borehole; 9, Atworth; 10, Chickerell; 11, Calvert; 12, Woodham; 13, Bletchley; 14, Stewartby; 15 Peterborough; 16, Hutton's Ambo; 17, Milton Keynes Boreholes; 18, South Humberside; 19, Worlaby Borehole; 20, Staffin Bay; 21, Brora.

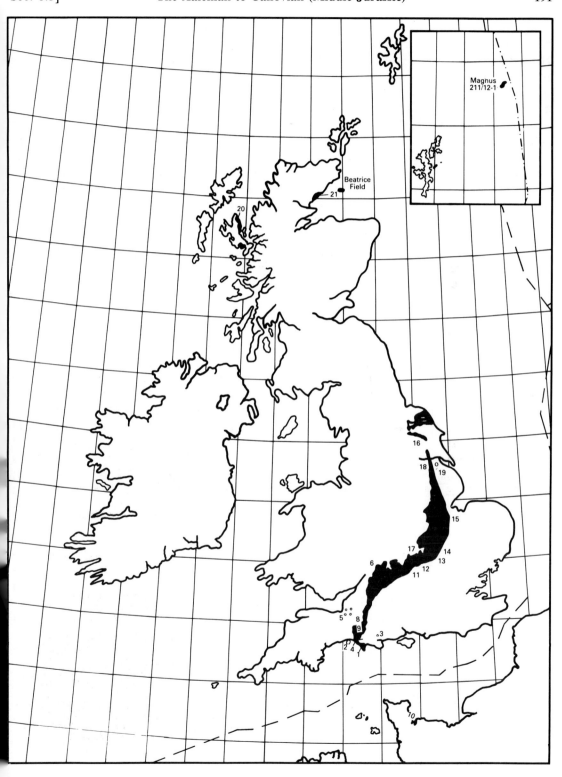

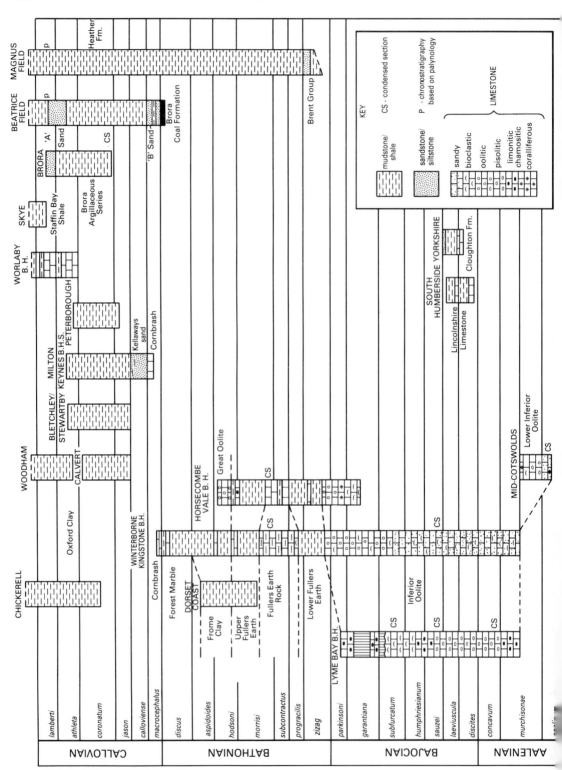

recently (see Woollam and Riding, 1983; Bate and Robinson, 1978). The utilisation of foraminifera in Aalenian–Bajocian stratigraphy has been almost totally neglected, with outcrop studies limited to basal Inferior Oolite limestones, the Lincolnshire Limestone and Deltaic Series (Morris, 1982; Packer, 1986; Nagy et al., 1981). Work on Bathonian outcrop and borehole material in Southern England (Cifelli, 1959; Coleman, in Penn et al., 1979; Penn, 1982) has established a localised foraminiferal zonation (referred to as 'faunules', section 6.3.4.3) which can be related to regional events. This work highlights both the biostratigraphic potential of foraminiferal studies and the need for further work, especially on the Aalenian–Bajocian interval.

6.3.4 Depositional history, palaeoecology and faunal associations

6.3.4.1 Aalenian

The widespread development of strongly regressive facies is a characteristic feature of Aalenian deposition in both onshore and offshore basins. Relatively continuous marine deposition is represented in the Lower Inferior Oolite of the Weald Basin, which formed the main depocentre at this time (Sellwood et al., 1986). Facies within the Aalenian of the Weald Basin and adjoining areas (including the Cotswolds) comprise shallow water carbonates (mainly grainstones), deposited in foreshoal, shoal and lagoonal environments (Mudge, 1978; Sellwood et al., 1986). Foraminifera associated with these deposits are known only in the Cotswolds and Dorset area (Coleman, in Penn et al., 1980; Penn, 1982; Morris, 1982) (see Fig. 6.3.2). Assemblages from basal Inferior Oolite limestones (*opalinum* Zone) compare closely with faunas from the Late Toarcian, with little evidence of a major faunal turnover at the Early – Middle Jurassic boundary (Morris, 1982). Low-diversity assemblages dominated by nodosariids continue into Aalenian beds with the following species occurring abundantly: *Lenticulina ex gr. muensteri*, *L. varians*, *L. dorbignyi*, *Citharina clathrata*, *C. colliezi*, *Pseudonodosaria ex gr. vulgata* and *Eoguttulina liassica*. Basal beds (Scissum Beds) also yield high numbers of the miliolid *Ophthalmidium strumosum* which may have its inception near the base of the Aalenian. In higher beds (*murchisonae* Zone) the large agglutinant species, *Ammobaculites coprolithiformis* and *Tetrataxis* sp. Coleman, also appear, associated with a general increase in diversity (Morris, 1982). The incoming of large numbers of small rotaline foraminifera, including *Reinholdella dreheri*, *Conicospirillina* cf. *trochoides*, *Spirillina infima* and *Paalzowella feifeli*, coincides with the development of oncolitic facies, indicating that these faunas were phytal dwelling (Morris, 1982).

In the Dorset area, faunas from condensed Inferior Oolite limestone sequences of the *murchisonae* – *concavum* Zones suggest a continued increase in diversity through the British Aalenian with the appearance of *Lenticulina dictyodes* and *L. exgaleata*, together with the reappearance of *Lenticulina quenstedti* at the base of the *concavum* Zone (Coleman, in Penn et al., 1980; Penn 1982). Foraminiferal assemblages described from the calcareous Dogger (alpha, beta units) of

Fig. 6.3.2 — Chronostratigraphy of British Middle Jurassic sections analysed for foraminifera: Lyme Bay Borehole (Coleman in Penn et al., 1980); Dorset Coast (Ciffeli, 1959; Copestake, 1987); Winterborne Kingston Borehole (Coleman in Penn, 1982); Horsecombe Vale Boreholes (Coleman in Penn et al., 1979); mid-Cotswolds (Morris, 1982); Humberside (Packer, 1986); Yorkshire (Nagy et al., 1981) Chickerell, Calvert, Woodham, Bletchley, Stewartby (Barnard et al., 1981); Milton Keynes Borehole (Coleman in Horton et al., 1974); Peterborough (Barnard et al., 1981); Worlaby Borehole (Medd in Richardson, 1979); Skye (Cordey, 1962); Brora (Gordon, 1967); Beatrice Field (Copestake unpub.) Magnus Field (this chapter)

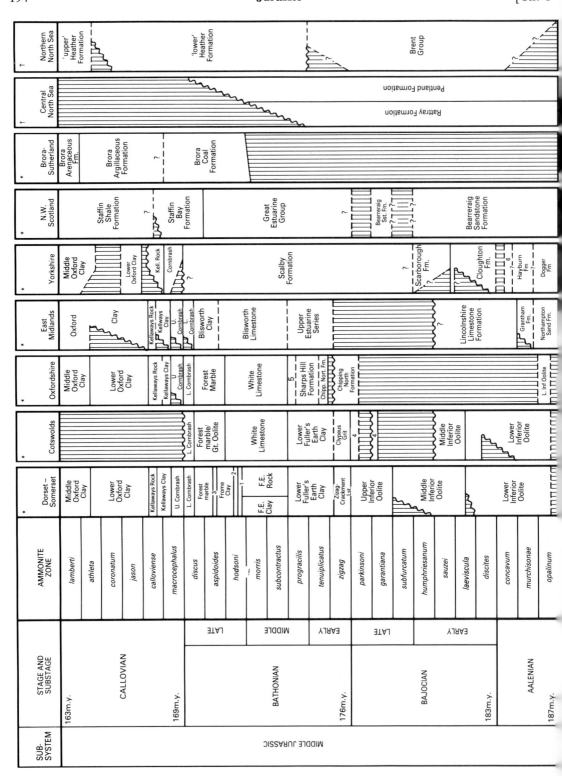

France and Switzerland (Wernli and Septfontaine, 1971) compare with those of the British Aalenian in displaying a predominance of species ranging from the Early Jurassic including *Lenticulina ex gr. muensteri*, *L. subalata*, *L. dorbignyi*, *Citharina 'colliezi'* and *C. clathrata*. A major turnover is evident in these rocks, however, at the top *concavum* Zone – base of the *discites* Zone with *Lenticulina dorbignyi*, *Nodosaria regularis* ssp. A and *N. pulchra* becoming locally extinct. The reappearance of *Lenticulina quenstedti* is also recorded at this level.

The Aalenian and Bajocian of the Central and Northern North Sea is composed of non-marine, fluvial–deltaic sediments (e.g. Brent Group) comparable to the Deltaic Series of Yorkshire (Hancock and Fisher, In Illing and Hobson 1981). These deposits are intercalated with volcanics of the Rattray Formation in the Outer Moray Firth Basin and South Viking Graben. Further south, in the Southern North Sea, near-shore, marine sediments of Aalenian-Bajocian age are represented by the lower West Sole Group. Associated foraminiferal assemblages, are typically nodosariid dominated with prominent *Lenticulina ex gr. muensteri*, *Citharina clathrata* and *Eoguttulina liassica*.

In the North Celtic Sea and Fastnet Basin, foraminifera from Aalenian well sections include *Lenticulina gottingensis*, *L. polygonata*, *Nodosaria regularis* ssp. A Copestake and Johnson, and *Reinholdella macfadyeni* (Ainsworth *et al.*, 1987).

There are no descriptions of Aalenian–Bajocian foraminifera for the ammonitiferous Bearreraig Sandstone Formation, west Scotland, although foraminifera here have been obtained from the Skye section (Copestake pers. comm.)

The development of deep marine and/or restricted conditions through the *opalinum–concavum* Zones is indicated in the German Dogger (Braunjura, alpha, beta units), where the aforementioned calcareous assemblages are replaced by agglutinating taxa of the genera *Reophax*, *Ammobaculites* and *Trochammina* (Ziegler, 1959; Munk, 1978). The presence of a number of species including *Trochammina sablei*, *T. topagorukensis*, *T. canningensis* and *Haplophragmoides kingakensis* attests to Aalenian bioprovincial links with North America and Canada (see Tappan, 1955; Brooke and Braun, 1972).

6.3.4.2 Bajocian

The pattern of regressive sedimentation that had become established during the Aalenian continued into the Bajocian, with marine deposition occurring predominantly in inner-shelf, tidal-dominated settings. In the Weald and Wessex–English Channel basins, however a relatively thick (c. 180 m) sequence of carbonates was deposited, comprising the Inferior Oolite (Sellwood *et al.*, 1986). Descriptions of Inferior Oolite foraminifera are limited to borehole material obtained from condensed subcrop sections of the Dorset area, sited near the Wessex–English Channel basin margin (Fig. 6.3.2, Coleman, in Penn *et al.*, 1979; Penn, 1982). Faunas from the *discites–parkinsoni* Zones of Dorset include a high proportion of long-ranging nodosariids, including *Lenticulina ex gr. muensteri*, *L. quenstedti*, *Citharina clathrata*, *C. colliezi* and *Nodosaria* spp. Large numbers of small rotaline foraminifera are also present including *Reinholdella dreheri*, *Paalzowella fiefeli* and *Spirillina infima*, which appear to range through the Inferior Oolite Group (Fig. 6.3.4). Evidence of a faunal turnover is seen in the *discites–laeviuscula* Zones, corresponding approximately to the Lower/Middle Inferior Oolite boundary. At this stratigraphic level *Nodosaria regularis* ssp. A, *N. pulchra*,

Fig. 6.3.3 — Middle Jurassic stages, ammonite zones and British lithostratigraphic nomenclature (*modified from Cope *et al.*, 1980b; † Deegan and Scull, 1977). 1, Upper Fuller's Earth Clay; 2, Wattonensis Beds; 3, Boueti Beds; 4, Upper Inferior Oolite (*Trigonia–Clypeus* Grits); 5, Hampen Marly Formation; 6, Eller Beck Formation.

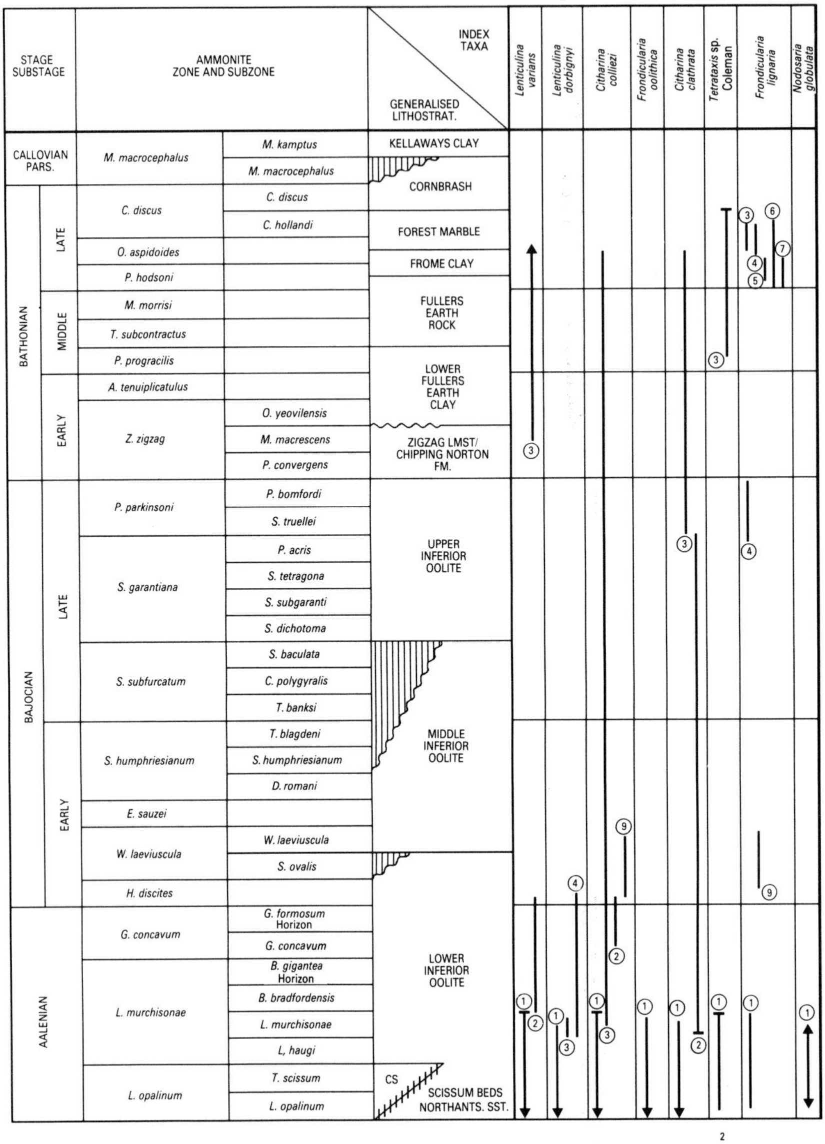

Fig. 6.3.4 — Range chart of selected Aalenian, Bajocian and Bathonian foraminifera

Frondicularia lignaria, *Falsopalmula deslongchampsi* and *Conicospirillina* cf. *trochoides* have their extinctions.

Within the Upper Inferior Oolite (*parkinsoni* Zone) the appearance of new taxa is notable including *Lenticulina tricarinella*, *Saracenaria oxfordiana* and *Frondicularia obliqua* (Fig. 6.3.4). Radiation, particularly in the Nodosariacea is also documented in the Late Bajocian of France and Germany (Wernli and Septfontaine, 1971; Ziegler, 1959; Munk, 1978) and may be related to regional controls on deposition.

Bajocian foraminifera are also described from shallow marine carbonates of the Lincolnshire Limestone Formation, South Humberside (Packer, 1986). Microfaunal assemblages from the *discites – laeviuscula* Zones (Fig. 6.3.2), compare with Inferior Oolite assemblages, indicating a similar palaeoecology. Foraminiferal assemblages from the Cloughton Formation (Yons Nab Beds) of the Yorkshire Coast (Nagy *et al.*, 1981) show the increased influence of pro-deltaic sedimentation within the *discites – laeviuscula* interval. Diverse nodosariid and spirillinid assemblages ('*Citharina*' assemblages) obtained from the basal beds compare with carbonate-associated faunas of the Inferior Oolite and Lincolnshire Limestone Formation (Nagy *et al.*, 1981) The replacement of these assemblages by textulariids (*Ammodiscus*

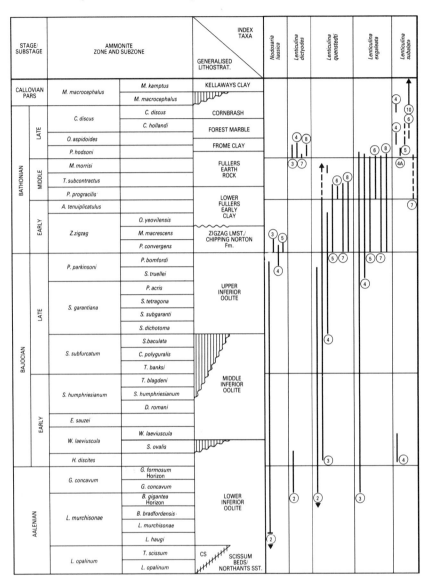

Fig. 6.3.4 — Range chart of selected Aalenian, Bajocian and Bathonian foraminifera (continued)

yonsnabensis, Nagy, Lofaldi and Bomstad, 1981, *Trochammina* sp., *Lagenammina* sp. and *Ammobaculites coprolithiformis*) in the succeeding beds reflect brackish water conditions linked to deltaic progradation (Nagy *et al*, 1981).

The faunal associations of the British Bajocian are broadly similar to those described from the French and German Dogger (gamma and delta units; Wernli and Septfontaine, 1971; Bartenstein and Brand, 1937; Ziegler, 1959; Munk, 1978; Ohmert, in press). Comparison of these faunas, however, reveal regional biofacies variation within the Bajocian and significant differences in the timing of local 'inceptions' and 'extinctions'

The Aalenian to Callovian (Middle Jurassic)

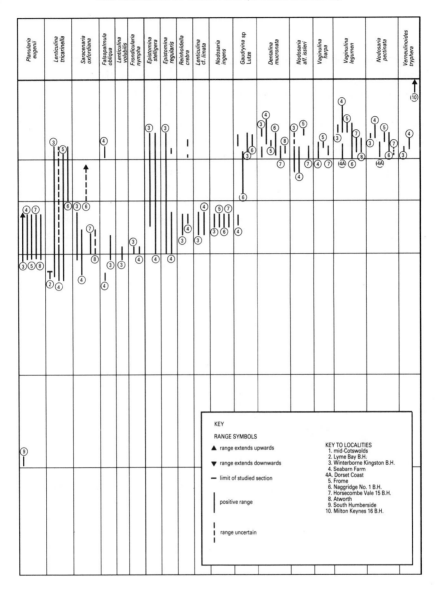

of benthonic taxa. In the French Jura, carbonate-associated taxa of the Aalenian (alpha and beta units) persist through the Bajocian with a preponderance of nodosariid taxa (*Lenticulina* spp., *Citharina* spp.) in interbedded marls (Wernli and Septfontaine, 1971). Changes in microfaunal composition indicate the presence of a hiatus at topmost Aalenian (*concavum* Zone) corresponding to the beta–gamma unit junction.

At this level *Lenticulina quenstedti* makes its first appearance, whilst *Lenticulina dorbignyi*, *Nodosaria pulchra*, *N. regularis* ssp. A and other taxa appear to become extinct (Wernli and Septfontaine, 1971). Within the Late Bajocian (*subfurcatum – garantiana* Zones) several new taxa made their appearance including *Lenticulina tricarinella*, *Saracenaria cornucopiae*, *Textularia agglutinans* and *Flabellammina althoffi*

(Wernli and Septfontaine, 1971). Again, these faunal changes can be correlated with a widespread lithostratigraphic break dividing the delta 1 and delta 2 units.

In northern Germany textulariid assemblages of deepwater/restricted aspect persisted into the Early Bajocian (*discites* Zone) with faunas dominated by agglutinated genera including *Trochammina*, *Haplophragmoides*, *Ammodiscus* and *Ammobaculites* (Munk, 1978; Ziegler, 1959). Calcareous assemblages became established within the *laeviuscula – sauzei* Zones; however, *Lenticulina quenstedti* does not appear within these assemblages until the *humphriesianum* Zone (Ziegler, 1959). In southern Germany, the Lower Bajocian (*discites – laeviuscula* Zones) is represented by a condensed shallower marine sequence, with *Lenticulina* spp. and *Citharina* spp. predominant (Ohmert, in press). In these shallow marine assemblages *Lenticulina quenstedti* appears at the base of the *laeviuscula* Zone.

Facies control on Bajocian foraminiferal distribution is also evident between the French and German Dogger with the 'late' appearance of *Lenticulina polymorpha* (= *Lenticulina tricarinella*) and other nodosariids in the *parkinsoni* Zone of North Germany (Ziegler, 1959; Munk, 1978). Further, *Lenticulina dorbignyi* which appears to become extinct at the *concavum* level (Aalenian) in Britain and France ranges into the *subfurcatum* Zone in North Germany (Ziegler, 1959).

6.3.4.3 Bathonian

During the Bathonian, sedimentation in northern England continued to be influenced by fluvial – deltaic systems which built southwards from the mid-North Sea High (Anderton *et al.*, 1979; Leeder, 1983). The switchover to marine deposition occurred in the East and Central Midlands, where marginal marine, lagoonal/shoal facies are represented in the Upper Estuarine Series and Blisworth Limestone/Clay Formations. No foraminiferal assemblages have been described from these lithostratigraphic units although marine ostracod faunas are known from the Upper Estuarine Series (Bate, 1967).

Fully marine Bathonian sequences of the Great Oolite Group were deposited to the south of the London Platform in the Weald and Wessex – English Channel basins. Higher subsidence rates in the Wessex – English Channel basins resulted in thick basinal mudstone accumulations (Fuller's Earth – Frome Clay) in the Dorset area (Sellwood *et al.*, 1985, 1986, 1987). Detailed work on Bathonian outcrop in the Bath and Dorset areas (Cifelli, 1959) provided the basis for originally defining several correlatable 'faunules'. This work was later amended and applied to BGS boreholes in the Dorset area (Coleman, in Penn *et al.*, 1979; Penn, 1982). An outline of this revised scheme is presented below:

Faunule A (*parkinsoni—zigzag* Zones)—characterised by *Epistomina stelligera*, *E. regularis*, *Planularia eugenii*, *Nodosaria ingens*, *Frondicularia nympha* and *Lenticulina* cf. *volubilis*. This assemblage displays Late Bajocian affinities, with associated taxa including *Lenticulina tricarinella*, *L. exgaleata*, *L. quenstedi*, *Saracenaria oxfordiana* and *Vaginulina clathrata*. Regionally Faunule A compares closely with shallow shelf assemblages described from the *humphriesianum – zigzag* Zones of the French and North German Dogger (delta – epsilon 1 units; Wernli and Septfontaine, 1971; Ziegler, 1959; Munk, 1978).

Faunule A associates with fore-shoal – back-shoal facies of the Upper Inferior Oolite – basal Fuller's Earth.

Faunule B_1 (*tenuiplicatus – progracilis* Zones)—low-diversity assemblages composed of long-ranging taxa (e.g. *Lenticulina quenstedti*, *L. exgaleata*), with the development of horizons rich in agglutinating species. Faunule B_1 associates with fore-shoal, basin mudstones of the Lower Fuller Earth and reflects the onset of deep-water sedimentation. Comparable trends are seen in the *tenuiplicatus – progracilis* Zones of the Dogger (lower epsilon unit) of North Germany (Munk, 1978) and can be related to a eustatic sea-level rise proposed for this interval (Haq et al., 1987).

Faunules $B_2 – C_3$ (subcontractus – basal aspidoides Zones)—represents the reappearance of a diverse nodosariid fauna, including long-ranging taxa of faunules A and B_1 (e.g. *Lenticulina tricarinella*, *L. dictyodes* and *L. exgaleata*). Species appearing for the first time comprise *Nodosaria* aff. *issleri*, *Vaginulina legumen* and *Ophthalmidium carinatum*. Local subdivision of this faunule has been undertaken (Coleman, in Penn et al., 1979), with differentiation into fore-shoal, shoal assemblages (above) associated with the White Limestone and Fuller's Earth Rock, and agglutinating-dominated assemblages which occur with deeper basinal facies of the Upper Fuller's Earth – Lower Frome Clay. In the latter assemblages, *Verneuilinoides tryphera* and *Gaudryina* sp. 2 Lutze are characteristic. A number of nodosariid taxa including *Lenticulina dictyodes*, *L. tricarinella* and *L. exgaleata* have their extinctions within the basal *aspidoides* Zone. Faunules $B_2 – C_3$ reflect a regressive phase in deposition which resulted in the progradation of oncolitic limestones, into the basin centre (Sellwood et al., 1987). Assemblages broadly compare in composition and diversity with those described from the Dogger, epsilon 2 (*aspidoides* Zone) of North Germany (Ziegler, 1959; Munk, 1978).

Faunule C_4 (aspidoides Zone)—characterised by the increased dominance of inner shelf species, including *Trocholina nodulosa*, *Epistomina stelligera*, *E. regularis*, *Reinholdella crebra*, *Lenticulina subalata* and *Tetrataxis* sp. Coleman. This faunule is associated with the emplacement of mobile oolite shoals (Great Oolite Formation) resulting from shoaling during the *aspidoides* Zone (see Sellwood et al., 1985, 1987).

Faunule D (discus Zone)—characterised by newly appearing nodosariid species: *Nodosaria pectinata* and *N.* aff. *issleri*, with prominent *Massilina dorsetensis* and *Cornuspira liasina*. Faunule D associates with mixed fore-shoal, shoal and back-shoal facies of the Forest Marble and Lower Cornbrash, reflecting the culmination of regression through the *hodsoni – discus* Zones (Sellwood et al., 1985, 1987). A regional sea-level fall through the corresponding interval proposed by Hallam (1981) and Haq et al. (1987), may be one explanation for these facies changes within the Wessex – English Channel and Weald basins.

North of the mid-North Sea High, in the Moray Firth Basin and Central Graben, Bathonian sedimentation continued to be dominated by fluvial – deltaic systems until the onset of marine transgression in the early Callovian (see section 6.3.4.5). In the Northern North Sea, however, marine deposition became established within the Early – Middle Bathonian, as represented by the Tarbert and (lower) Heather Formations. The foraminiferal assemblages associated with these deposits continue into the Callovian and are described in section 6.3.4.5 with reference to well 211/12-1 (Magnus Field).

Whilst marine and marginal marine Bathonian is known in other Jurassic basins, including the South Minch and Fastnet basins (Hudson, 1983; Ainsworth et al., 1987), little is known of the foraminiferal associations. In the latter basin however, shallow shelf assemblages are described in association with fore-shoal – shoal carbonates (Ainsworth et al., 1987). Species listed for the Late Bathonian include *Epistomina parastelligera*, *E. regularis*, *Lenticulina exgaleata*, *L. quenstedti* and *L. tricarinella*, comparable to faunules $B_1 – C_4$ of southern England (Coleman, in Penn et al., 1979; Penn, 1982). Diverse epistominid assemblages including *Epistomina stelligera*, with *Frondicularia nympha* and *Planularia beierana* occur in the preceding Early Bathonian interval and compare with inner-shelf associations of southern England (Faunule A) and the continent (Dogger epsilon 1; Ziegler, 1959; Munk, 1978).

6.3.4.4 Callovian

The Callovian of British onshore and offshore basins is generally characterised by a sequence of marine mudstones and bituminous shales deposited in a partially restricted, inner to outer-shelf environment. The transgressive nature of Callovian deposits and their regional uniformity is attributed to a major eustatic rise in sea-level which was initiated during the Late Bathonian (Hallam, 1975, 1978; Haq et al., 1987). Well and seismic data from the Central and Northern North Sea provides evidence of rifting throughout the Callovian, with localised erosion and onlap over fault-bounded structural highs (Bowen, 1975; Badley et al., 1988).

Foraminifera associated with Callovian facies can be clearly differentiated into agglutinating and calcareous-dominated assemblages, reflecting basin-wide changes in O_2/H_2S levels comparable to the Kimmeridge Clay Formation (see Tyson et al., 1979). Where benthonic conditions favoured agglutinant taxa, e.g. Viking Graben, the composition of assemblages suggests strong bioprovincial links with arctic Eurasia, and indicate the re-establishment of a northern sea-way connection at this time, in accordance with macrofossil evidence (Hallam, 1978; see section 6.3.4.5).

In the Central and East Midlands the distribution of Callovian foraminifera through the Lower and Middle Oxford Clay has been documented in both outcrop and borehole studies (Barnard et al., 1981; Medd, 1983; Coleman, in Horton et al., 1974; Medd, in Richardson, 1979). From these sources three assemblage zones can be recognised:

Assemblage Zone 1 (*macrocephalus – calloviense* Zones)—composed predominantly of agglutinant taxa including *Ammobaculites agglutinans*, *Verneuilinoides tryphera*, *Gaudryina* sp. and *Trochammina globigeriniformis*. Calcareous accessories include *Eoguttulina liassica*, *Lenticulina ectypa* and *Lenticulina varians*.

Assemblage Zone 1 microfaunas replace Late Bathonian taxa at the Cornbrash/Kellaways Beds junction reflecting the onset of deeper marine or partially restricted sedimentation.

Assemblage Zone 2 (*calloviense – coronatum* Zones)—characterised by abundant epistominids (chiefly *Epistomina parastelligera*) and *Epistomina nuda*. Accessory taxa include *Lenticulina subalata* and *Lenticulina ectypa*.

The appearance of Assemblage Zone 2 faunas coincides with the Kellaways Beds/Lower Oxford Clay junction (upper *calloviense* Zone). The predominance of calcareous species suggest the development of a more oxygenated benthic environment.

Assemblage Zone 3 (*coronatum – lamberti* Zones)—characterised by a highly diverse calcareous fauna dominated by nodosariid species including *Planularia eugenii*, *Lenticulina ex gr. muensteri* and *L. ectypa*. A strong miliolid element is also characteristic of this zone with *Ophthalmidium dyeri*, *O. strumosum* and *Nubeculinella bigoti sensu* Cordey common in the upper part of the zone.

The base of this assemblage zone approximates lithostratigraphically to the Lower/Middle Oxford clay boundary (*coronatum – athleta* interval, Fig. 6.3.3) with assemblages of this type continuing through the Middle Oxford Clay into the Oxfordian.

The presence of diverse, calcareous assemblages suggests that well oxygenated benthonic conditions prevailed during deposition of Middle Oxford Clay claystones.

Scottish Callovian foraminifera are documented from the Staffin Shale, Skye (*lamberti* Zone—Cordey, 1962) and the

Brora Argillaceous Formation (*coronatum* Zone—Gordon, 1967). In the former study nodosariid-dominated assemblages range through the *lamberti* Zone into the Oxfordian, and are broadly comparable with faunas of Assemblage Zone 3 (above). Similarly, in the Brora section nodosariids predominate, with species common to both sections including *Lenticulina ex gr. muensteri*, *Lenticulina ectypa*, *Frondicularia franconica* and *Epistomina parastelligera*.

Comparison of British (onshore) and north German Callovian assemblages (Dogger epsilon 3 and zeta units: Ziegler, 1959; Munk, 1978) reveal similarities in microfaunal trends consistent with increased circulation through the *jason* – *athleta* interval. The appearance of abundant '*Globigerina*' *helvetojurassica* and diverse radiolaria in the German Callovian is a significant difference, however, reflecting deeper marine conditions and/or stronger bioprovincial links with Tethys (see Munk, 1978; Riegraf, 1986).

6.3.4.5 North Sea, Middle Bathonian – Callovian Foraminifera

In common with onshore sections, the Middle Bathonian – Callovian of the North Sea Basin (where preserved) is characterised by a transgressive sequence of inner to outer shelf sediments represented by the Tarbert and lower Heather Formations, which overlie non-marine Brent Group sediments. Foraminifera, together with other microfossil groups (radiolaria, ostracods, dinocysts), are frequently abundant and are used extensively in the oil industry for zonation purposes. An outline of foraminiferal distribution is presented here, with reference to the Beatrice Field area, Quadrant 11 and 12, (Copestake unpublished) and the Magnus Field, well 211/12-1 (Fig. 6.3.6).

In the Inner Moray Firth, Beatrice Field area, the stratigraphy is similar to the onshore section (Brora) with the Brora Coal Formation of Bathonian age, barren of foraminifera. Above this occur marginal to shallow marine sands, the B and A Sands (part of Beatrice Field reservoir), devoid of foraminifera, with an interbedded marine shale age equivalent to the Heather Formation. The mid-shale contains a low diversity, calcareous foraminiferal assemblage with low numbers of *Epistomina parastelligera*, *Lenticulina ex gr. muensteri*, *Frondicularia franconica* and *Marginulina batrakiensis*, consistent with faunas described from the onshore, Brora Argillaceous Formation (*coronatum* Zone – Gordon, 1967), and indicates deposition in a similar but probably more marginal marine environment. The occurrence of *Lenticulina exgaleata*, however, in these assemblages suggests a broader range than that evident from southern England borehole data (section 6.3.4.3). Above the A Sand occurs a marine shale of late Callovian age, representing the basal part of the main Heather Formation section in the Inner Moray Firth. This upper Callovian shale contains a restricted assemblage of agglutinating foraminifera associated with *Lenticulina ex gr. muensteri*.

Foraminiferal assemblages from the Bathonian – Callovian of the Magnus Field area differ from the Moray Firth Basin and onshore in containing a high proportion of agglutinating taxa throughout, reflecting deeper water or partially restricted shelf conditions (Fig. 6.3.6). In the lower Heather Formation the following assemblages can be differentiated:

Assemblage 1 (Middle/Late Bathonian – early Callovian)—composed predominantly of agglutinating taxa with the following species stratigraphically restricted: *Verneuilinoides tryphera*, *Verneuilinoides* sp. 2, *Haplophragmoides* cf. *dolininae*, *Haplophragmium pokrovkaensis* and *Ammodiscus* cf. *cheradospira*. Occuring abundantly, but not restricted to this assemblage are *Haplophragmoides infracalloviensis*, *Haplophragmoides*

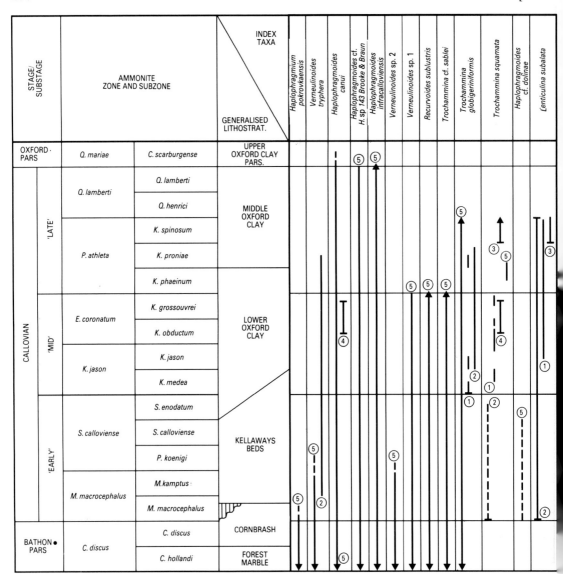

Fig. 6.3.5 — Range chart of selected Callovian foraminifera

sp. cf. *H.* sp. 143 Brooke and Braun, *Haplophragmoides canui*, *Recurvoides sublustris*, *Ammobaculites* cf. *agglutinans*, *Verneuilinoides* sp. 1 and *Trochammina* spp. (notably *T. canningensis* and *T. globigeriniformis*). Calcareous taxa which occur as minor accessories include *Epistomina* cf. *tenuicostata*, *Lenticulina* cf. *inflathiformis* and *Marginulinopsis phragmites*.

Assemblage 2 (mid-Callovian)—characterised by the top downhole occurrences of *Verneuilinoides* sp. 1, *Haplophragmoides infracalloviensis*, *Haplophragmoides* sp. cf. *H.* sp. 143 Brooke and Braun and *Recurvoides sublustris*, with the absence of Assemblage 1 taxa.

Assemblage 3 (late Callovian – Middle

Sec. 6.3] The Aalenian to Callovian (Middle Jurassic)

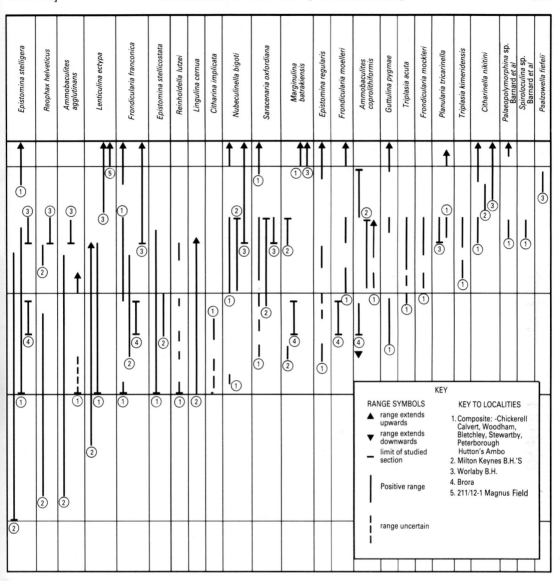

Oxfordian)—a number of agglutinating taxa including *Haplophragmoides canui*, *H.* cf. *topogorukensis*, *Ammobaculites deceptorius*, *A.* cf. *agglutinans* range from latest Callovian into the Middle Oxfordian (upper Heather Formation). Associated are common calcareous taxa (mainly nodosariids), including *Lenticulina ex gr. muensteri*, *L. ectypa*, *Frondicularia nitikini* and *Eoguttulina liassica*.

The above Heather Formation assemblages show taxonomic affinities with Early, Middle and Late Jurassic faunas described from high-latitude localities including Svalbard, western Siberia and western Canada (Løfaldi and Nagy, 1980, 1983; Dain, 1972; Tappan, 1955; Wall, 1960; Brooke and Braun, 1972; 1981). Faunal composition closely compares with assemblages described from the Late Bathonian – Early Oxfordian

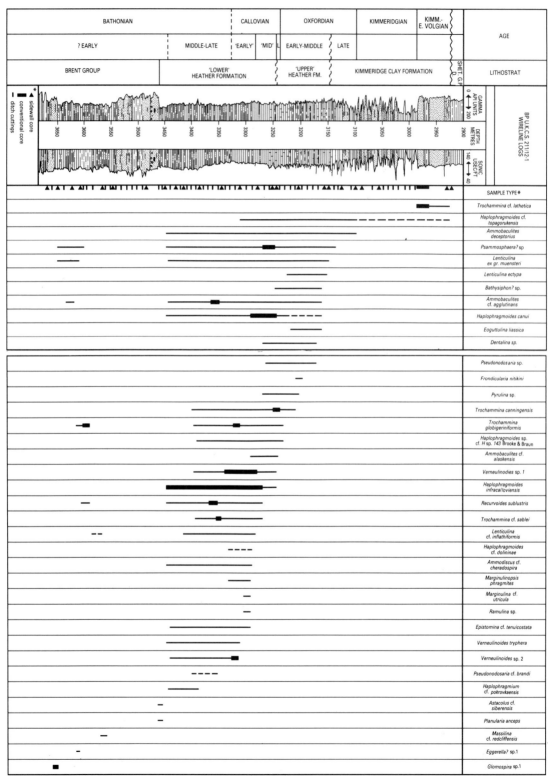

of East Svalbard, which also display a similar upward trend of increasing diversity with the incoming of nodosariids (Løfaldi and Nagy, 1980). Of note is the high abundance of *Haplophragmoides* including *H. canui* and *H.* aff. *excavatus* (possibly conspecific with *H. infracalloviensis*), together with *Recurvoides* and *Ammobaculites* through the Late Bathonian – Callovian, comparable to Assemblages 1 and 2 (above).

Western Canadian, early-mid Callovian assemblages from the Rierdon Formation, whilst dominated by nodosariids (reflecting shallow shelf conditions), have many taxa in common with Assemblages 1 and 2. These include *Ammodiscus cheradospira*, *Verneuilinoides tryphera* and *Marginulinopsis phragmites*, which may have regional zonal significance (see Wall, 1960; Brooke and Braun, 1972).

Further comparison can be made between North Sea Middle Jurassic taxa and species comprising agglutinating dominated assemblages from the Oxfordian – Volgian of southern Svalbard, western Siberia and western Canada (Løfaldi and Nagy, 1980; Dain, 1972; Brooke and Braun, 1981). In accordance with regional data, many North Sea taxa, including *Haplophragmoides infracalloviensis*, *H. canui*, *Recurvoides sublustris* and *Ammobaculites* cf. *agglutinans*, have long total ranges and reappear at higher levels in the North Sea Basin (e.g. Volgian and Upper Ryazanian of the Kimmeridge Clay Formation).

6.3.5 Stratigraphic distribution

The chronostratigraphic ranges of selected taxa from British onshore and offshore Middle Jurassic sequences are presented in Figs. 6.3.4. and 6.3.5. It should be emphasised that, because little is known of distribution through the Aalenian – Bajocian interval, the ranges shown may be conservative in many instances. Whilst the majority of Middle Jurassic species are long ranging and display complex lithofacies associations (which often give the impression of local zonal significance), major lithostratigraphic units have distinct faunal associations as previously discussed (section 6.3.4). Further, certain elements do not range across lithostratigraphic boundaries although similar biofacies are juxtaposed (e.g. *Lenticulina dorbignyi* tops near to the Lower – Middle Inferior Oolite boundary). Whilst regional comparisons suggest that the timing of such 'events' are variable regionally (section 6.3.4) local extinctions and inceptions are likely to have chronostratigraphic application within individual basins in many instances.

6.3.6 Index species

Middle Jurassic species included in this edition fall mainly into two categories: (1) species displaying a stratigraphically restricted range within the British succession, or that have reliably defined inceptions and extinctions of correlative value (e.g. *Lenticulina dorbignyi*, *Nodosaria regularis* ssp. A., *Verneuilinoides tryphera*); and (2) species which constitute an important component of assemblages at one or more horizons whilst displaying overall a long total range (e.g. *Lenticulina ex gr. muensteri*, *Spirillina infima*, *Haplophragmoides infracalloviensis*); identification of these species is considered important to the characterisation of assemblages, essential to Middle Jurassic biozonation.

Note: The first edition species entries and plates have been incorporated into this edition with minor amendment. New species, from either the Lower Inferior Oolite or the North Sea, Heather Formation, have been placed before and after the original entries, resulting in a 'non-sequential' taxonomic listing.

Fig. 6.3.6 — Distribution of Middle–Late Jurassic foraminifera in B.P. U.K.C.S. well 211/12-1 (Magnus Field)

SUBORDER TEXTULARIINA

Haplophragmoides cf. *cushmani* Loeblich and Tappan, 1946.
Plate 6.3.1, Fig. 1, PM/BJ 12 (× 130). Lower Inferior Oolite, Cotswolds. Description: test involute, globose to compressed, with up to eight chambers exposed in final whorl. Aperture consists of an interiomarginal slit in juveniles becoming loop-shaped in adults. Test wall coarsely arenaceous.
Remarks: the Cotswolds material differs from the original types and Liassic material of Copestake (unpublished) in being involute and in attaining a greater size. Distribution: *H. cushmani* was originally described from the Early Cretaceous of Texas, ranging from Pliensbachian to Callovian in North West Europe (Brouwer, 1969). Recorded from the *murchisonae* Zone (Aalenian) of the Cotswolds (Morris, 1982).

Reophax sterkii Haeusler, 1890
Plate 6.3.1, Fig. 2, PM/BJ 1 (× 80). Lower Inferior Oolite, Cotswolds. Description: test large, composed of up to four chambers, which are initially low becoming higher, with a pyriform final chamber. Aperture simple, on a short apertural neck. Test wall coarse, often incorporating *Ophthalmidium strumosum* in the Cotswolds material. Distribution: a long ranging Jurassic species recorded from the *murchisonae* Zone of the Cotswolds (Morris, 1982).

Tetrataxis sp. Coleman 1978
Plate 6.3.1, Fig. 3, PM/BJ 96 (× 130); Fig. 4, PM/BJ 97 (× 130); Lower Inferior Oolite, Cotswolds. See Plate 6.3.6, Fig. 7.

SUBORDER MILIOLINA

Ophthalmidium strumosum (Gümbel)
Plate 6.3.1, Fig. 5, PM/BJ 14 (× 250); Fig. 6, PM/BJ 14 (× 220); Lower Inferior Oolite, Cotswolds = *Guttulina strumosa* Gumbel, 1862. See Plate 6.3.6, Fig. 14. Note: this species occurs in high numbers through the Scissum Beds (*opalinum–murchisonae* Zones) and may have its inception at the base of the Middle Jurassic.

The Aalenian to Callovian (Middle Jurassic)

Plate 6.3.1

SUBORDER ROTALIINA

Astacolus speciosus (Terquem)
Plate 6.3.2, Fig. 1, PM/BJ 55 (× 90) Fig. 2, PM/BJ 55 (× 65) Lower Inferior Oolite, Cotswolds = *Cristellaria speciosa* Terquem, 1858. Description: test compressed biconvex, citharine, but with an initial coil of up to three chambers in microscopic generation. Successive chambers broaden obliquely to overlap along inner margin. Sutures are oblique, becoming progressively depressed. Test ornamented with numerous longitudinal ribs tending to converge towards the dorsal angle. Aperture radiate, produced at dorsal margin.
Remarks: *A. speciosus* resembles *Marginulina radiata* Terquem, but differs in being compressed. Distribution: originally described from the Hettangian, *A. speciosus* occurs within the *murchisonae* Zone of the Cotswolds (Neuweiler 1959; Morris 1982).

Citharina colliezi (Terquem)
Plate 6.3.2, Fig. 3, PM/BJ 53 (× 65); Fig. 4, PM/BJ 53 (× 80) Lower Inferior Oolite, Cotswolds = *Marginulina flabelloides* Terquem, 1868. Description: test compressed, parallel sided, with globular proloculus followed by an incurving portion with successive chambers flaring to produce a subtriangular test. Test ornamented with numerous discontinuous ribs which curve towards dorsal apex. Aperture radiate, produced at dorsal margin. Size of prolocus and breadth of test vary between microspheric and megalospheric generations. Distribution: a long ranging Jurassic species that occurs commonly through the Aalenian and Bajocian (Morris 1982; Coleman, in Penn *et al.*, 1980, Penn, 1982; Packer, 1986).

Citharina clathrata (Terquem)
Plate 6.3.2, Fig. 5, PM/BJ 51 (× 33); Fig. 6, PM/BJ 52 (× 65); Lower Inferior Oolite, Cotswolds = *Marginulina longuemeri* var. *clathrata* Terquem. Description: test variable, large, robust, compressed, subtriangular in outline. Chambers low, broadening with little increase in height. Test ornamented with coarse longitudinal ribs, numbering 6–19 in juveniles, increasing by intercalation and bifurcation to produce in excess of 20 in adults.
Remarks: marked variation is evident between generations with microspheric forms smaller and narrower resembling *Vaginulina*. This spectrum of variation has led to a complex synonymy, with *Vaginulina proxima* (Terquem) and *Citharina lepida* (Schwager) likely to be conspecific. Distribution: a long ranging Jurassic species that occurs abundantly in Aalenian and Bajocian regressive facies (Morris, 1982; Coleman, in Penn *et al.*, 1980, Penn, 1982).

Conicospirillina cf. *trochoides* (Berthelin)
Plate 6.3.2, Fig. 7, PM/BJ 86 (× 140); Fig. 8, PM/BJ 86a (× 140); Fig. 9, PM/BJ 87 (× 150); Lower Inferior Oolite, Cotswolds = *Spirillina trochoides* Berthelin, 1879. Description: test comprised of up to six whorls arranged in a low trochospire. Dorsal side of test evolute with depressed spiral suture. Ventral surface infilled with calcite plug. Surface of whorls coarsely perforate with pores arranged concentrically. Aperture comprising an interiomarginal slit extending back from terminal opening (evident only in well preserved specimens).
Remarks: Liassic forms of *C. trochoides* differ from Aalenian specimens in showing involution of the ventral side as a result of expansion of the final whorl (Copestake, unpublished). Distribution: *C.* cf. *trochoides* appears restricted to the Aalenian and Early Bajocian, occuring abundantly in oncolitic carbonate facies (Morris, 1982; Packer, 1986).

The Aalenian to Callovian (Middle Jurassic)

Plate 6.3.2

Frondicularia irregularis Terquem, 1870
Plate 6.3.3, Fig. 1 PM/BJ 34 (× 70); Lower Inferior Oolite, Cotswolds. Description: test highly variable, compressed with sides initially divergent with later chambers often decreasing in width to give a lanceolate appearance. Sutures straight or gently arched, constricted or flush.
Remarks: similar to *Frondicularia franconica* Gümbel 1862 (Plate 6.3.8, Fig. 3) but differs in lacking an apertural neck and chevron-shaped chambers. Distribution: originally described from the Late Bajocian of France, this species occurs in English Aalenian (Morris, 1982).

Frondicularia lignaria Terquem, 1866
Plate 6.3.3, Fig. 2, PM/BJ 36 (× 70); Fig. 3, PM/BJ 36 (× 110); Lower Inferior Oolite, Cotswolds. See Plate 6.3.8, Fig. 5.

Frondicularia oolithica Terquem, 1870
Plate 6.3.3, Fig. 4, PM/BJ 40 (× 190); Fig. 5 PM/BJ 40 (× 70); Fig. 6, PM/BJ 39 (× 35); Lower Inferior Oolite, Cotswolds. Description: test smooth, compressed with proloculus followed by a series of chevron-shaped chambers. Width variable depending on generation. Aperture consists of a narrow concentric slit.
Remarks: similar to *F. franconica* Gumbel but differs in possessing a flush slit-shaped aperture. Distribution: ranges from the Bajocian to Oxfordian on the Continent (Bartenstein and Brand 1937) but recorded only from the Aalenian in Britain (Morris, 1982).

Lagena sp. A Copestake, 1978
Plate 6.3.3, Fig. 7, PM/BJ 32 (×140), Lower Inferior Oolite, Cotswolds. Description: test spherical to pear-shaped, tapering distally to a thin aperture neck which is terminated by a thickened collar. Apex of test ornamented with 8–12 deep sulci. The lower portion of the test is smooth. Aperture a minute circular opening, often obscured. Distribution: originally described from the Lower Lias of Mochras (Copestake, 1978) occurs rarely into Aalenian, *murchisonae* Zone (Morris, 1982).

Lenticulina dorbignyi (Roemer)
Plate 6.3.3, Fig. 8, PM/BJ 57 (× 140); Fig. 9, PM/BJ 57 (× 190); Lower Inferior Oolite, Cotswolds = *Peneroplis d'orbignyi* Roemer, 1839. Description: test variable, compressed, consisting of an initial lenticuline portion, with later chambers uncoiling and broadening. Sutures oblique, raised, interconnected by a coarse reticulate ornamentation in initial portion which gives way to prominent longitudinal ribbing in adult chambers. Aperture radiate.
Remarks: size of initial coil and width of test vary between generations with megalopsheric forms appearing marginuline. Ontogenetic variation typical for the genus (see *Lenticulina ex gr. muensteri* (Roemer)). Distribution: stratigraphically important within the Middle Jurassic being restricted to the Aalenian and Early Bajocian, *opalinum–discites* Zones; (Morris, 1982; Coleman, in Penn *et al.*, 1980; Penn, 1982).

Plate 6.3.3

Lenticulina ex gr. muensteri (Roemer)
Plate 6.3.4, Fig. 1, PM/BJ 68 (× 60); Fig. 2, PM/BJ 64 (× 66); Fig. 3, PM/BJ 67 (× 66); Fig. 4, PM/BJ 75 (× 66); Lower Inferior Oolite, Cotswolds = *Robulina muensteri* Roemer, 1839. Common synonyms include *Lenticulina subalata* (Reuss), *Lenticulina acutiangulata* (Terquem) and *Lenticulina polygonata* (Franke). Description: test extremely variable, robust, biconvex, coiled throughout or uncoiling at any stage of ontogeny. Chambers, triangular 4–13 in final whorl, depending on stage of uncoiling. Sutures curved, depressed or raised with the latter connected to umbilical boss.
Remarks: geometric variation within the group can be explained in terms of uncoiling within ontogenetic series, each with a fixed coil diameter (Morris, 1982). Ornamental variation of sutural ribs is also evident (Gordon, 1966), which combine to produce a highly complex morphogroup. Distribution: a widespread Jurassic species which generally predominates in regressive facies, notably of the Aalenian, Bajocian and Bathonian.

Lenticulina varians (Bornemann)
Plate 6.3.4, Fig. 5, PM/BJ 61 (× 130); Fig. 6, PM/BJ 60 (× 66); Lower Inferior Oolite, Cotswolds = *Cristellaria varians* Bornemann, 1854. Description: test highly variable, compressed, with an initial coiled portion tending to uncoil with occasional development of flabelline chambers in adults. Pronounced keel and raised sutural ribs well developed in adults. Aperture radiate, marginal. Distribution: a long ranging Jurassic species which occurs commonly in regressive facies of the Aalenian and Bathonian (Morris, 1982; Coleman, in Penn *et al.*, 1980; Penn, 1982).

Lingulina esseyana Deeke, 1886
Plate 6.3.4, Fig. 7, PM/BJ 43 (× 66); Fig. 8., PM/BJ 43 (× 155); Lower Inferior Oolite, Cotswolds. Description: test compressed, with a globular proioculus followed by a nodosarine portion, with later chambers becoming compressed. Sutures depressed, often with sutural ribs. Surface of test hispid. Aperture a small flush oval opening. Distribution: common in the Early Jurassic and ranging into the Aalenian, *murchisonae* Zone (Copestake, 1978; Morris, 1982).

Lingulina longiscata longiscata (Terquem)
Plate 6.3.4, Fig. 9, PM/BJ 44 (× 130); Fig. 12 (× 140); Lower Inferior Oolite, Cotswolds = *Frondicularia longiscata* Terquem, 1870. Description: test compressed, with globular proloculus followed by a series of 5–9 low chevron-shaped chambers, which increase gradually in width. Test ornamented with numerous striae, tending to converge towards apex. Aperture a narrow elongate slit. Distribution: a long ranging Jurassic species that occurs sporadically through the Aalenian–Bathonian (Morris 1982; Coleman, in Penn *et al.*, 1980, Penn 1982; Copestake, 1978).

Lingulina cf. *longiscata alpha* Adams, 1955
Plate 6.3.4, Fig. 10, PM/BJ 45 (× 130); Fig. 11, PM/BJ 45 (× 165); Lower Inferior Oolite, Cotswolds. Description: test compressed with a broadening series of chambers which display considerable overlap. Final chambers may reduce in width to give a lanceolate appearance. Test ornamented with numerous striae converging towards apex. Aperture radiate.
Remarks: differs from *L. longiscata alpha* Adams, 1955 in displaying a radiate aperture. Distribution: the present form is recorded from the Aalenian, *murchisonae* Zone (Morris, 1982).

The Aalenian to Callovian (Middle Jurassic)

Plate 6.3.4

Marginulina cf. *scapa* Lalicker, 1950
Plate 6.3.5, Fig. 1, PM/BJ 76 (× 65); Lower Inferior Oolite, Cotswolds. Description: test variable compressed, straight or gently curved composed of a series of obliquely arranged chambers. Sutures depressed in adult chambers. Aperture radiate, well produced at dorsal margin.
Remarks: rate of increase in chamber width highly variable. Distribution: recorded from the Aalenian (Morris, 1982) this species may range into the Late Jurassic.

Nodosaria globulata Barnard, 1950
Plate 6.3.5, Fig. 2, PM/BJ 19 (× 140); Lower Inferior Oolite, Cotswolds. Description: test large, consisting of a rectilinear series of up to seven drum-like chambers each displaying considerable overlap. As a result of this the final chamber appears high and globular. Test ornamented with 13 or 14 short discontinuous ribs that radiate from below the apertural face leaving a smooth crown. Aperture circular, flush or slightly produced. Distribution: a long ranging Jurassic species occurring commonly through the Lower Inferior Oolite, *murchisonae* Zone (Morris, 1982).

Nodosaria hortensis Terquem, 1866
Plate 6.3.5, Fig. 3, PM/BJ 20 (× 140); Lower Inferior Oolite, Cotswolds. Description: test variable, rectilinear or curved, consisting of up to ten bead-shaped chambers. Insertion of chamber with greater or lesser width common, giving an irregular appearance. Test ornamented with 11–14 discontinuous ribs. Aperture radiate, often with short neck. Distribution: a long ranging Jurassic species that occurs commonly through the *murchisonae* Zone of the Lower Inferior Oolite (Morris, 1982).

Nodosaria metensis (Terquem)
Plate 6.3.5, Fig. 4, PM/BJ 24 (× 140); Lower Inferior Oolite, Cotswolds = *Dentalina metensis* Terquem, 1858. Description: test consists of arcuate series of 4–10 drum-like chambers that increase gradually in size. Ornamentation is highly distinctive comprising of 14–17 ribs that coalesce on the apertural face to produce smooth crown (visible on final chamber). Aperture a small circular opening flush with chamber. Distribution: a long ranging Jurassic species that occurs rarely in the Lower Inferior Oolite, *murchisonae* Zone (Morris, 1982).

Nodosaria mutabilis Terquem, 1870
Plate 6.3.5, Fig. 5, PM/BJ 25 (× 70). Lower Inferior Oolite, Cotswolds. Description: test consists of a rectilinear series of 2–5 globular chambers of uniform size, although emplacement of reduced chamber common. Test ornamented with 10–14 strong ribs. Sutures deeply constricted. Aperture a small flush, elongate opening. Distribution: long ranging, occurring commonly in the *murchisonae* Zone of the Lower Inferior Oolite (Morris, 1982).

Nodosaria obscura Reuss var. *liassica* Barnard, 1950b
Plate 6.3.5, Fig. 6, PM/BJ 29 (× 70). Lower Inferior Oolite, Cotswolds. Description: test comprised of an initial conical portion followed by a series of uniform, parallel-sided chambers. Final chamber commonly irregular in adults. Test ornamented with 8–10 continuous ribs which tend to overlap those of the previous chambers. Aperture a circular opening with prominent lip. Distribution: originally described from the Early Toarcian (Barnard, 1950b), this species occurs commonly through the *murchisonae* Zone of the Lower inferior Oolite (Morris 1982).

Paalzowella fiefeli (Paalzow)
Plate 6.3.5, Fig. 7, PM/BJ 92 (× 130); Fig. 8, PM/BJ 93 (× 130); Lower Inferior Oolite, Cotswolds = *Trocholina fiefeli* Paalzow, 1932. See Plate 6.3.6, Fig. 6.

Reinholdella dreheri (Bartenstein)
Plate 6.3.5, Fig. 9, PM/BJ 90 (× 170); Fig. 10, PM/BJ 89 (× 130); Lower Inferior Oolite, Cotswolds = *Discorbis dreheri* Bartenstein, 1937. Description: test smooth, plano-convex, chambers arranged in a low trochospire. Dorsal side evolute, with 4–5 chambers in final whorl. Ventral side involute, with clear umbilical plug. Sutures oblique and depressed on dorsal surface, radial on ventral side. Final chamber usually broken off, but in complete individuals primary aperture is a low interiomarginal umbilical arch, situated adjacent to a loop-shaped secondary aperture.
Remarks: broken specimens of *R. media* (Kapterenko-Chernousova) figured by Norling (1972) may be conspecific. Distribution: Aalenian to Early Bathonian, *zigzag* Zone (Morris, 1982; Coleman in Penn *et al.*, 1980; Penn, 1982).

Spirillina infima (Strickland); emend. Barnard, 1952
Plate 6.3.5, Fig. 11, PM/BJ 82a (× 190). Lower Inferior Oolite, Cotswolds = *Orbis infimus* Strickland, 1846. Description: test evolute, planispiral or slightly trochoid with a small globular proloculus succeeded by a spirally wound second chamber of up to nine whorls. Test surface covered with pseudo-pores (visible in well preserved specimens). Aperture simple, at open end of tube. Distribution: a long ranging Jurassic species that is particularly abundant in oncolitic limestones and marls of the *murchisonae* Zone, Cotswolds (Morris, 1982).

Plate 6.3.5

SUBORDER TEXTULARIINA

Ammobaculites agglutinans (d'Orbigny)
Plate 6.3.6., Fig. 1 MPK 2206 (× 100), Upper Fuller's Earth, Bath = *Spirolina agglutinans* d'Orbigny, 1846. Description: test small, finely agglutinated, early portion planispirally coiled, later rectilinear, circular in cross-section; coil generally of four chambers with depressed umbilicus followed by well developed linear portion of 5–6 globular chambers which remain fairly constant in breadth, not exceeding the diameter of the coil, sutures distinct, depressed; aperture, simple, terminal. Distribution: common Jurassic species but does not become an important component of the fauna until Late Bathonian *retrocostatum* Zone, (Cifelli, 1959; Coleman, in Penn *et al.*, 1979). See also *Ammobaculites* cf. *agglutinans*—North Sea Middle Jurassic foraminifera.

Ammobaculites coprolithiformis (Schwager)
Plate 6.3.6, Fig. 2, MPK 22198 (× 60), Lower Fuller's Earth, Bath = *Haplophragmium coprolithiformis* Schwager, 1867. Description: test generally robust, coarsely agglutinated, early portion planispirally coiled, later rectilinear, circular in cross section; coil composed of 4–5 chambers followed by linear portion of 2–4, rarely five chambers; sutures usually indistinct, each chamber enveloping the previous one; aperture simple, terminal, produced on short neck. Distribution: appears in the Lower Inferior Oolite, *murchisonae* Zone, ranging throughout the Middle Jurassic (Morris, 1982; Coleman, in Penn, 1982; Gordon, 1970; Coleman, in Horton *et al.*, 1974).

Ammobaculites fontinensis (Terquem)
Plate 6.3.6, Fig. 3 MPK 2392 (× 75), Lower Fuller's Earth, Dorset = *Haplophragmium fontinense* Terquem, 1870. Description: test variable in size, finely agglutinated, compressed, planispiral in early part; coil evolute, composed of numerous globular chambers gradually increasing in size and arranged in 2–3 whorls; linear portion composed of 1-5 low disc-shaped chambers with depressed sutures; aperture simple, terminal. Distribution: recorded from the Aalenian, occurring sporadically through the Bajocian, and Bathonian. (Morris, 1982; Coleman in Penn *et al.*, 1979; Penn 1982; Cifelli, 1959).

Gaudryina? sp. Lutze, 1960
Plate 6.3.6, Fig. 4, MPK 2375 (× 166), Frome Clay, Dorset. Description: test small, finely agglutinated; initial chambers globular, arranged in small trochospire with four chambers in the first whorl; final 4–5 chambers biserially arranged; increasing gradually in size, somewhat compressed; sutures depressed, particularly in later part of test to give slightly lobulate outline; aperture simple at base of final chamber. Distribution: sporadic Bathonian–Callovian species, although common in the Frome Clay, *aspidoides* Zone (Coleman, in Horton *et al.*, 1974, Penn *et al.*, 1979).

Miliammina jurassica (Haeusler)
Plate 6.3.6, Fig. 5, SAB 515, F1 (× 100), Lower Oxford Clay, Milton Keynes = *Trochammina jurassica* Haeusler, 1882. Description: test small, oval, finely agglutinated; composed of 5–9 elongated chambers arranged in quinqueloculine fashion; each chamber forms a half whorl and is slightly swollen at its initial end, tapering distally; sutures indistinct except in final chambers; aperture simple at end of final chamber. Distribution: described originally from the Middle Oxfordian of the Swiss Jura, this species occurs in the English mid-Callovian, *jason–coronatum* Zones (Coleman, in Horton *et al.*, 1974).

Plate 6.3.6

SUBORDER ROTALIINA

Paalzowella fiefeli (Paalzow)
Plate 6.3.6, Fig. 6, MPK 2317 (× 25), Middle Inferior Oolite, Lyme Bay = *Trocholina fiefeli* Paalzow, 1932. Description: small calcareous, conical test with chambers arranged in 4–7 whorls, visible on the spiral side; spiral suture generally elevated to give step-like appearance; ventral side involute with umbilical depression and fine radial markings extending from umbilicus to margin; three chambers in final whorl. Distribution: ranges throughout the Middle Jurassic occurring commonly in the Lower Inferior Oolite, *murchisonae* Zone, and sporadically through the Bajocian–Bathonian (Morris, 1982; Coleman, in Penn *et al.*, 1980, Penn, 1982).

SUBORDER TEXTULARIINA

Tetrataxis sp. Coleman, 1978
Plates 6.3.6, Fig. 7, MPK 2368 (× 250); Forest Marble, Dorset. Synonym: *Trochammina haeusleri* (Galloway) Cifelli, 1959. Description: test calcareous, granular, conical chambers arranged in trochoid spiral with four to a whorl in early portion, later three; sutures flush, indistinct; aperture a slit base of last chamber in umbilical region. Distribution: occurs abundantly in the Aalenian; *murchisonae* Zone, ranging into the Late Bathonian (Morris, 1982; Coleman, in Penn *et al.*, 1980, Penn, 1982; Cifelli, 1959).

Triplasia althoffi (Bartenstein)
Plate 6.3.6, Fig. 8, MPK 2380 (× 85), Frome Clay, Dorset; Fig. 9, SAB 471 F1 (× 65); Middle Oxford Clay, Milton Keynes, = *Flabellammina althoffi* Bartenstein, 1937. Common synonyms: *Triplasia variabilis* (Brady) Bartenstein and Brand, 1937; *Triplasia bartensteini* Loeblich and Tappan, 1952. Description: test large, agglutinated, highly variable in outline; planispiral coil of 3–5 chambers followed by linear portion of 2–9 compressed, triangular or, rarely, quadrangular chambers with excavated sides; sutures straight initially, later strongly convex, slightly depressed; aperture oval, terminal, produced on a short neck.
Distribution: Late Bajocian to Callovian, *athleta* Zone (Coleman, in Penn *et al.*, 1980, Penn, 1982; Cifelli, 1959; Gordon, 1970; Coleman, in Horton *et al.*, 1974).

Trochammina canningensis Tappan, 1955
Plate 6.3.6, Fig. 11, MPK 2386 (× 190), Lower Fuller's Earth, Dorset. See Plate 6.3.12, Figs 1,2.

Verneuilinoides tryphera Loeblich and Tappan, 1950
Plate 6.3.6, Fig. 17, MPK 2261 (× 150), Frome Clay, Faulkland. Description: test small, finely agglutinated, triserial, flaring distally; chambers subglobular, increasing very gradually in size as added; sutures distinct, depressed; aperture crescent-shaped at base of final chamber. Distribution: Late Bathonian to Callovian (Coleman, in Penn *et al.* 1980, Penn, 1982, Horton *et al.*, 1974). Restricted to the Middle Bathonian–early Callovian, Heather Formation, North Sea Basin (see Plate 6.3.12, Figs 6, 7).

SUBORDER MILIOLINA

Cornuspira liasina Terquem, 1866
Plate 6.3.6, Fig. 10, MPK 2198 (× 150), Forest Marble, Dorset. *Cornuspira orbicula* Terquem and Berthelin, 1875. Description: test porcellaneous, imperforate, planispiral; large spherical proloculus followed by regular, undivided tube of 5–8 whorls, increasing very gradually in width; aperture simple, open end of tube. Distribution: occurs rarely in the Aalenian: *murchisonae* Zone, becoming common within the Bajocian–Callovian (Morris 1982; Coleman, in Penn *et al.*, 1979, 1980, Penn, 1982, Horton *et al.*, 1974).

Massilina dorsetensis Cifelli, 1959
Plate 6.3.6, Fig. 12, MPK 2264 (× 130), Forest Marble, Bath. Description: test suboval, slightly biconvex; composed of eight chambers, each one half-whorl in length increasing in size as added, the first six arranged in quinqueloculine series, the last two opposite; sutures flush, producing a smooth surface on both sides of test; aperture simple, without tooth. Distribution: Late Bathonian, Forest Marble, *discus* Zone (Coleman, in Penn *et al.*, Penn, 1982; Cifelli, 1959).

Ophthalmidium carinatum Kubler and Zwingli, 1870
Plate 6.3.6, Fig. 13, MPK 2759 (× 100) Wattonensis Beds, Lyme Bay. Description: test compressed, planispiral, biconvex in cross-section; only later chambers are clearly visible each one $\frac{1}{2}$ whorl in length; initial chamber arrangement appears to be typical for the genus (see *Ophthalmidium strumosum*), periphery may be keeled although this is a variable feature; aperture simple, produced on short neck. Distribution: Bajocian ? to Bathonian (Coleman, in Penn *et al.*, 1979, 1980).

Ophthalmidium strumosum (Gümbel)
Plate 6.3.6, Fig. 14, SAB 461 F2 (× 200), Middle Oxford Clay, Milton Keynes = *Guttulina strumosa* Gümbel, 1862. Description: test smooth, highly compressed, planispiral, generally an elongate oval in outline; periphery rounded 2–5 chambers visible in reflected light, each $\frac{1}{2}$ whorl in length and swollen at its proximal end; initial chamber arrangement typical of proloculus followed by tube-like second chamber approximately $\frac{1}{2}$ whorl in length and third chamber one complete turn in length; occasionally a chamber appears perpendicular to main planispire; aperture simple, at end of long tapering neck. Distribution: Aalenian to late Callovian (Morris, 1982; Coleman, in Horton *et al.*, 1974, see Plate 6.3.1, Figs 5, 6).

Ophthalmidium sp. A. Coleman, 1974
Plate 6.3.6, Fig. 15, SAB 463 F1 (× 150), Middle Oxford Clay, Milton Keynes. Description: test compressed, elliptical, with all chambers visible and separated by depressed sutures; proloculus gives rise to tube-like second chamber which forms approximately $\frac{1}{2}$ whorl; third chambers form $\frac{3}{4}$ whorl about the prolocus; this and subsequent chambers are widest at their proximal ends, tapering distally, each forming $\frac{1}{2}$ whorl; there are generally 6–8 chambers present; aperture simple, at end of long tapering neck. Distribution: late Callovian, *athleta–lamberti* Zones (Coleman, in Horton *et al.*, 1974; Medd, in Richardson, 1979).

Spirophthalmidium concentricum (Terquem and Berthelin)
Plate 6.3.6, Fig. 16, MPK 2395 (× 150), Lower Fuller's Earth, Dorset = *Spiroloculina concentricum* Terquem and Berthelin, 1875. Description: test compressed, planispire; outline variable, generally elongate oval; initial chamber arrangement not clearly visible but appears to be small planispire; later chambers $\frac{1}{2}$ whorl in length with two final chambers distinctly larger than preceeding ones and separated from them by depressed sutures; aperture simple at end of long neck. Distribution: Late Bajocian to Late Bathonian (Coleman in Penn *et al.*, 1979, 1980, Penn, 1982; Cifelli, 1959).

SUBORDER ROTALIINA

Citharina flabellata (Gümbel)
Plate 6.3.7, Fig. 1, MPK 2760 (× 70), Callovian, Lyme Bay = *Marginulina flabellata* Gümbel, 1862. Description: test compressed, uniserial, composed of 5–10 chambers gradually increasing in size to give triangular outline; sutures generally depressed; ornamented by variable numbers of narrow, continuous ribs which sometimes bifurcate; aperture simple, terminal, produced on a prominent neck surrounded by a notched collar.
Remarks: differs from *Citharina colliezi* (Terquem) in possessing continuous rather than discontinuous ribs. Distribution: Callovian: *jason* to *athleta* Zones (Gordon, 1970; Coleman, in Horton *et al.*, 1974; Medd, in Richardson, 1979).

Citharinella nikitini (Uhlig)
Plate 6.3.7, Fig. 2, SAB 443 F1 (× 40), Middle Oxford Clay, Milton Keynes = *Frondicularia nikitini* Uhlig, 1883. Description: test large, highly compressed, quadrate; proloculus followed by asymmetrical or citharine-type chamber; subsequent chambers symmetrical, chevron-shaped, uniserial, 7–12 in number; central depression gives slightly bilobed cross section; margin rounded, slightly lobulate, sutures distinct, depressed; surface ornamented by numerous, discontinuous arcuate costae; aperture simple, terminal, situated on short neck surrounded by notched collar.
Remarks: differs from *Palmula anceps* (Terquem) in having only one citharinid-type chamber and a larger number of chevron chambers, giving a more symmetrical outline. The extent of the ornament is highly variable, some specimens being almost smooth. Distribution: rare in the Aalenian: *murchisonae* Zone (Morris, 1982), common in the late Callovian: *athleta* and *lamberti* Zones (Coleman, in Horton *et al.*, 1974; Medd, in Richardson, 1979).

Citharinella moelleri (Uhlig)
Plate 6.3.7, Fig. 3, MPK 2761 (× 50), Middle Oxford Clay, Milton Keynes = *Frondicularia moelleri* Uhlig, 1883. Description: test large, highly compressed, quadrate; second chamber citharine-type, remaining chambers chevron shaped, uniserially arranged to give a symmetrical outline; sutures indistinct; surface ornamented by variable numbers of strong, continuous costae; aperture simple, produced on short neck surrounded by notched collar. Distribution: late Callovian, *athleta* Zone (Coleman, in Horton *et al.*, 1974).

Citharinella sp. A Coleman, 1974
Plate 6.3.7, Fig. 4, SAB 443 F2 (× 50), Middle Oxford Clay, Milton Keynes. Description: *Citharinella* with later chambers strongly overlapping earlier ones to give broad, almost semicircular outline to test; sutures limbate, depressed in later part of test; surface ornamented by variable number of discontinous costae, generally absent from final one or two chambers; aperture simple, produced on short neck, surrounded by notched collar. Distribution: late Callovian, *athleta* Zone (Coleman, in Horton *et al.*, 1974; Medd, in Richardson, 1979).

Dentalina pseudocommunis Franke, 1936
Plate 6.3.7, Fig. 5, MPK 2242 (× 76) = *Dentalina communis* d'Orbigny: Coleman, 1978; Lower Fuller's Earth, Bath. Description: test large, uniserial, chambers increasing gradually in size giving slender, gently flaring, outline; suture oblique, generally indistinct at first, later depressed. Aperture radiate, terminal. Distribution: common Middle Jurassic species.

Dentalina filiformis (d'Orbigny)
Plate 6.3.7, Fig. 6, MPK 2762 (× 100), Middle Oxford Clay, Milton Keynes = *Nodosaria filiformis* d'Orbigny, 1826. Description: test slender, fragile, uniserial, composed of elongate chambers constricted at sutures; sutures horizontal but aperture, when visible, is marginal rather than central; the species is therefore referred to the genus *Dentalina*. Distribution: Callovian: *athleta* Zone (Coleman in Horton *et al.*, 1974).

Dentalina intorta Terquem, 1870
Plate 6.3.7, Fig. 7, MPK 2214 (× 100), Forest Marble, Bath. Description: test uniserial, composed of 4–9 chambers, all visible; sutures oblique, distinct, generally depressed; chambers fairly constant in breadth but increase gradually in height, later chambers strongly overlapping previous ones, aperture radiate, terminal. Distribution: Bajocian–Bathonian (Coleman, in Penn *et al.*, 1979, 1980, Penn, 1982; Cifelli, 1959).

Dentalina mucronata Neugeboren, 1856
Plate 6.3.7, Fig. 8, MPK 2215 (× 100), Forest Marble, Bath. Description: variable species of *Dentalina* with chambers broader than high and strongly oblique, depressed sutures; initial part of test usually slightly curved; aperture radiate, terminal. Remarks: Sometimes difficult to distinguish from *Vaginulina legumen* (Linné). See Cifelli (1959) for full variation. Distribution: Upper Fuller's Earth Clay–Forest Marble, *hodsoni–discus* Zones (Coleman in Penn *et al.*, 1979, 1980, Penn, 1982; Cifelli, 1959).

Epistomina cf. *nuda* Terquem, 1883
Plate 6.3.7, Fig. 9, Dorsal view, SAB 519 F1, (× 150); Fig. 10, central view, SAB 519 F2 (× 150); Lower Oxford Clay, Milton Keynes. Description: test small, trochoid, smooth biconvex; peripheral margin rounded, slightly lobate, chambers poorly defined on spiral side with six chambers in final whorl visible on umbilical side; sutures occasionally depressed, otherwise flush; secondary apertures narrow, crescent-shaped, visible close to and parallel with margin; primary aperture at base of final chamber.
Remarks: this species is distinct from that attributed to *B. nuda* by Pazdro (1969), the latter being a stratigraphically old form of *E. stelligera*. Distribution: mid-Callovian: *jason–coronatum* Zones (Coleman, in Horton *et al.*, 1974).

Plate 6.3.7

Epistomina regularis Terquem, 1883
Plate 6.3.7, Fig. 11, MPK 2763 (× 130), Fig. 12, MPK 2896 (× 130); Lower Fuller's Earth, Stowell. Synonym: *Epistomina mosquensis*, Uhlig. Bartenstein and Brand, 1937 pars. Description: test trochoid, biconvex; peripheral margin acute, keeled; all chambers visible on the spiral side with generally 6–7 in final whorl; sutures raised, frequently ornamented with bosses and pits; on umbilical side only chambers of last whorl visible, sutures radial, usually masked by variable reticulate ornament; secondary apertures broad, crescent-shaped, parallel with margin; primary aperture low arch at base of final chamber. Distribution: Bathonian–Callovian (Coleman in Penn *et al.*, 1979, 1980, Penn, 1982; Cordey, 1962).

Epistomina stellicostata Bielecka and Pozaryski, 1954
Plate 6.3.7, Fig. 13, SAB 473 F5 (× 100), Middle Oxford Clay, Milton Keynes. Description: test trochoid, biconvex, peripheral margin acute, keeled; on spiral side all chambers are visible with 7–8 chambers in final whorl; sutures raised, the costae in the central part being very broad, the chambers being reduced to round depressions; costae often bare small pits. On umbilical side the sutures are radial with pitted costae forming an irregular network covering a central umbilical disc; secondary apertures crescent-shaped, parallel with margin; primary aperture indistinct at base of final chamber. Distribution: Callovian, *jason–athleta* Zones (Barnard *et al.*, 1981; Coleman, in Horton *et al.*, 1979).

Epistomina stelligera (Reuss)
Plate 6.3.7, Fig. 14, MPK 2377 (× 160), Frome Clay, Dorset; Fig. 15, SAB 467 F2 (× 100), Middle Oxford Clay, Milton Keynes = *Rotalia stelligera* Reuss, 1854. Common synonym: *Brotzenia parastelligera* Hofker, 1954. Description: test trochoid, bioconvex; peripheral margin acute, occasionally keeled; on spiral side early chambers are not visible; sutures limbate or flush, 5–8 chambers in final whorl; on umbilical side chambers of last whorl are marked by small triangular areas between broad sutural ribs which converge to form an umbilical boss; secondary apertures elongate, parallel with test margin; primary aperture at base of final chamber. Distribution: Bathonian–Callovian (Coleman, in Penn *et al.*, 1980, Penn, 1982, Horton *et al* 1979; Medd, in Richardson, 1979; Gordon, 1967).

Palmula deslongchampsi Terquem, 1863
Plate 6.3.8, Fig. 1, MPK 2285 (× 75), Lower Inferior Oolite, Lyme Bay. Description: test generally large, highly compressed, quadrate; initial planispiral portion enveloped by later equivalent, chevron-shaped chambers; sutures distinct, often marked by low ribs, aperture radiate, terminal, central. Distribution: Aalenian–Early Bajocian (Coleman, in Penn *et al.*, 1980, Penn, 1982).

Palmula obliqua (Terquem, 1863)
Plate 6.3.8, Fig. 2, MPK 2764 (× 100), Frome Clay, Frome. Description: test highly compressed, early chambers arranged in loose planispiral followed by variable number of chevron-shaped chambers which do not extend to the coiled portion; aperture radiate, terminal, central. Remarks: treated here as distinct species since it tends to be characteristic of particular stratigraphical horizons. However, it is probably a morphological variant of *Planularia beierana* Gümbel. Distribution: occurs most commonly at the top of Bajocian–basal Bathonian, *parkinsoni–zigzag* Zones, ranging into the Late Bathonian, *hodsoni* Zone (Coleman, in Penn *et al.*, 1979, Penn, 1982).

Frondicularia franconica Gümbel, 1862
Plate 6.3.8, Fig. 3, SAB 452 F1 (× 110), Middle Oxford Clay, Milton Keynes. Description: test compressed, uniserial, smooth oval or bilobate in cross-section; margins rounded, unkeeled; chambers broad exhibiting variable growth rate to give irregular outline; sutures depressed or flush, usually chevron-shaped but may be transverse in later chambers; aperture radiate, terminal, central, elevated on a short neck. Distribution: Callovian: *jason–lamberti* Zones (Barnard *et al.*, 1981; Coleman, in Horton *et al.*, 1974; Medd, in Richardson, 1979; Gordon, 1967).

Frondicularia nympha Kopik, 1969
Plate 6.3.8, Fig. 4, MPK 2408 (× 115), Lower Fuller's Earth, Dorset. Description: test compressed, uniserial with variable outline; margins occasionally rounded but generally keeled; composed of two to eight chevron-shaped chambers with thickened, raised sutures; aperture radiate, terminal, central, elevated on short neck. Distribution: Late Bajocian–Early Bathonian (Coleman, in Penn *et al.*, 1979, Penn, 1982).

Frondicularia lignaria Terquem, 1866
Plate 6.3.8, Fig. 5, MPK 2217 (× 90). Upper Fuller's Earth, Bath. = *Frondicularia oolithica* Terquem, Coleman, 1978. Description: test compressed, uniserial, elongate to quadrate; spherical proloculus followd by six to 11 chevron-shaped chambers, sutures distinct, depressed, often marked by low ribs; aperture oval, terminal, central. Distribution: Aalenian–Bathonian (Morris, 1982; Coleman, in Penn *et al.*, 1979, 1980, Penn, 1982; see Plate 6.3.3, Figs 2, 3).

Guttulina pera Lalicker, 1950
Plate 6.3.8, Fig. 6, SAB 394, F1 (× 150), Lower Oxford Clay, Milton Keynes. Description: test small, globular in outline, early chambers strongly embraced by later ones added in planes 144° apart; sutures distinct, depressed; aperture radiate, terminal. Distribution: Callovian; *jason* Zone (Coleman, in Horton *et al.*, 1974).

Lenticulina dictyodes (Deecke)
Plate 6.3.8, Fig. 7, MPK 2381 (× 114), Upper Fuller's Earth, Dorset = *Cristellaria dictyodes* Deecke, 1884. Description: test somewhat compressed, elongate, with initial 4–6 chambers arranged in small coil, later 2–4 chambers uncoiling; sutures may be slightly depressed but generally concealed by distinctive, fine, mesh-like ornament which covers entire surface of test; aperture radiate, terminal. Distribution: Aalenian–Bajocian: (Coleman, in Penn *et al.*, 1979, 1980, Penn, 1982).

Lenticulina exgaleata Dieni, 1985
Plate 6.3.8, Fig. 8, MPK 2402 (× 70), Lower Fuller's Earth, Dorset; Fig. 9, MPK 2897 (× 130), Lower Fuller's Earth, Stowell, Somerset = *Cristellaria galeata* Terquem, 1870 (preoccupied). Description: chambers arranged in slightly involute, biconvex planispiral, most close coiled with occasionally uncoiled in later portion; sutures deeply depressed, generally with strong ribs along the margins extending to the keel; development of both ribs and keel variable.
Remarks: a tendency towards uncoiling and a triangular, *Saracenaria*-like cross section, is prevalent in the Bathonian examples. Distribution: Late Aalenian–Late Bathonian, becoming most common in the Lower Fuller's Earth Clay and Fuller's Earth Rock (Coleman in Penn *et al.*, 1979, 1980, Penn, 1982).

Lenticulina cf. *limata* Schwager, 1867
Plate 6.3.8, Fig. 10 MPK 2404; (× 78), Lower Fuller's Earth, Dorset. Description: test compressed, smooth; initial planispiral coil of 3–5 chambers followed by slightly arcuate, uncoiled portion; sutures indistinct, flush; later chambers broader than high, fairly uniform in size, resulting in a parallel sided outline; final chamber occasionally smaller than preceeding one; aperture radiate, terminal peripheral. Distribution: Early Bathonian, *zigzag* Zone (Coleman, in Penn, 1982).

Lenticulina major (Bornemann)
Plate 6.3.8, Fig. 11 SAB 437, F2 (× 90), Middle Oxford Clay, Milton Keynes = *Cristellaria major* Bornemann, 1854. Description: test generally robust, somewhat compressed; initial coil of 3–5 chambers followed by rectilinear portion of up to seven chambers, each one broader than high, increasing very little in size to give a parallel sided outline; sutures marked by strongly limbate ribs. Distribution: Callovian: *jason–athleta* Zones (Coleman, in Horton *et al.*, 1974).

Lenticulina quenstedti (Gümbel)
Plate 6.3.8, Fig. 12, MPK 2247 (× 130), Lower Fuller's Earth, Bath = *Cristellaria quenstedti* Gümbel, 1862. Common synonym: *Cristellaria polonica* Wisniowski, 1890. Description: test biconvex, involute planispiral; chambers increasing gradually in size; sutures marked by distinct sharp ribs which generally converge to form a circular umbilical rib; keel well developed especially in older chambers; aperture radiate, terminal, peripheral.
Remarks: some Lower Bathonian specimens are difficult to distinguish from *L. exgaleata* and *L. tricarinella*. Distribution: late Aalenian, *concavum* Zone to late Callovian, *lamberti* Zone. Occurs commonly through the Late Bajocian–Early Bathonian, *parkinsoni–tenuiplicatus* Zones (Coleman, in Penn *et al.*, 1979, 1980, Penn, 1982, Horton *et al.*, 1974; Medd, in Richardson, 1979).

Lenticulina subalata (Reuss)
Plate 6.3.8, Fig. 13, MPK 2246 (× 67), Twinhoe Beds, Bath = *Cristellaria subalata* Reuss, 1854. Description: test robust, biconvex, involute, planispiral; periphery acute, sometimes angular, often keeled; sutures marked by rounded ribs which converge to form an umbilical boss; aperture radiate, terminal, peripheral. Distribution: common Jurassic species, occurring frequently in the Late Bathonian, *hodsoni–aspidoides* Zones (Coleman, in Penn *et al.*, 1979, Penn, 1982; Cifelli, 1959).

Lenticulina tricarinella (Reuss)
Plate 6.3.8, Figs.14, MPK 2208 (× 100), Upper Fuller's Earth, Bath; Fig. 15, MPK 2226 (× 100), Lower Fuller's Earth, Bath = *Cristellaria tricarinella* Reuss, 1863. Common synonym: *Cristellaria polymorpha* Terquem, 1870 Description: test compressed; chambers arranged in loose planispire with parallel sides; periphery keeled, ornamented with sharp sutural ribs and lateral keeling; degree of coiling variable, occasionally the lateral keels are replaced by non-persistent oblique costae along margins of test; aperture radiate, terminal peripheral. Distribution: Late Bajocian to Late Bathonian (Coleman, in Penn *et al.*, 1979, 1980, Penn, 1982).

Lenticulina volubilis Dain 1958
Plate 6.3.8, Fig. 16, MPK 2410 (× 112), Lower Fuller's Earth, Dorset. Description: test compressed, involute planispiral with tendency for later chambers to uncoil; distinct keel; sutures slightly depressed, bordered by costae-bearing pits or depressions and short secondary costae which branch off upwards; aperture radiate, terminal peripheral, situated on short neck. Distribution: Late Bajocian–Early Bathonian (Coleman, in Penn, 1982).

Lingulina longiscata (Terquem)
Plate 6.3.9, Fig. 1, MPK 2220 (× 60), Upper Fuller's Earth, Bath = *Frondicularia longiscata* Terquem, 1870. Common synonym: *Frondicularia nodosaria* Terquem, 1870. Description: test highly compressed, uniserial, lanceolate, oval proloculus followed by up to ten chambers which increase gradually in size; sutures usually strongly arched, depressed; margins rounded, lobulate; surface ornamented by large numbers of fine striae which do not continue from one chamber to the next; aperture elongate slit, terminal, central, produced on short neck Distribution: a common Early-Middle Jurassic species occurring in Aalenian, Bajocian and Bathonian strata (Morris 1982; Coleman, in Penn *et al.*, 1980, Penn, 1982; see Plate 6.3.4, Figs 9–11).

Plate 6.3.9

Lenticulina ectypa (Loeblich and Tappan)
Plate 6.3.9, Fig. 2, SAB 376 F1 (× 150), Lower Oxford Clay, Milton Keynes = *Astacolus ectypus* Loeblich and Tappan, 1950. Description: test compressed, involute, planispiral, with later chambers uncoiling but generally maintaining contact with coil on ventral margin; sutures deeply depressed, exaggerated by inflation of chambers just anterior to sutures and presence of sharp sutural ribs; aperture radiate, terminal, peripheral, produced on a short neck. Distribution: Callovian: *calloviense–lamberti* Zones, ranging into the Oxfordian (Barnard *et al.*, 1981; Coleman, in Horton *et al.*, 1974; Medd, in Richardson, 1979; Morris, this chapter).

Nodosaria hortensis Terquem, 1866
Plate 6.3.9, Fig. 3, MPK 2212 (× 144), Upper Fuller's Earth, Bath. Description: test uniserial, highly variable, ornamented by 8–14 continuous costae which vary in continuity; aperture radiate, terminal, central.
Remarks: similar to *N. fontinensis* Terquem which occurs commonly in the Bajocian and Lower Bathonian; differs in more robust nature of test, stronger costae which are generally fewer in number and chambers which are less spherical. Distribution: Aalenian to Late Bathonian, occurring commonly in the Late Bajocian–Early Bathonian (Morris, 1982; Penn, 1982).

Nodosaria ingens (Terquem)
Plate 6.3.9, Fig. 4, MPK 2253 (× 80), Forest Marble, Frome = *Dentalina ingens* Terquem, 1870. Common synonym: *Nodosaria guttifera* Bartenstein and Brand non d'Orbigny, 1937. Description: test uniserial composed of spherical proloculus followed by 2–3 inflated chambers deeply constricted at the sutures; aperture radiate, terminal, central. Distribution: appears restricted to Early Bathonian, *zigzag* Zone (Coleman, in Penn *et al.*, 1979, Penn, 1982).

Nodosaria aff. *issleri* Franke *sensu* Cifelli, 1959
Plate 6.3.9, Fig. 5 MPK 2239 (× 60) Upper Fuller's Earth, Bath. Description: test robust, uniserial, tapered, slightly arcuate, composed of 4–6 chambers, increasing in breadth and height as added; sutures depressed, transverse; test ornamented with 8–12 ribs which extend from base of proloculus to aperture. Aperture situated on a short neck. Distribution: Late Bathonian, *morrisi–discus* Zones (Coleman in Penn *et al.*, 1979, Penn, 1982; Cifelli, 1959).

Nodosaria opalina Bartenstein, 1937
Plate 6.3.9, Fig. 6, MPK 2213 (× 142), Forest Marble, Bath. Description: test generally small, uniserial, tapered; sutures depressed or flush; ornamented by large number of fine ribs which are sometimes difficult to distinguish; aperture radiate, terminal, central. Distribution: common Middle Jurassic species.

Nodosaria pectinata (Terquem)
Plate 6.3.9, Fig. 7, MPK 2254 (× 100), Forest Marble, Frome = *Dentalina pectinata* Terquem, 1870. Description: test uniserial, nearly always arched, composed of slightly elongate chambers constricted at sutures and ornamented by large number of costae; aperture not seen as specimens are invariably broken. Distribution: Late Bathonian, *hodsoni–discus* Zones with a distinctive occurrence in the Boueti Bed at the base of the Forest Marble (Coleman, in Penn *et al.*, 1979, Penn, 1982).

Planularia eugenii (Terquem)
Plate 6.3.9, Fig. 8 SAB 477 F1 (× 85), Lower Oxford Clay, Milton Keynes; Fig. 9, MPK 2765 (× 114), Lower Fuller's Earth, Dorset = *Cristellaria eugenii* Terquem, 1864. Description: test compressed, variable in outline but generally elongate with parallel margins; initial 4–5 chambers arranged in loose planispire, later chambers uncoiling; entire surface ornamented by variable number of ribs which may or may not be continuous; aperture radiate, terminal, peripheral.
Remarks: some specimens resemble *P. beierana* (Gümbel) but are distinguished by presence of ornament. Distribution: characteristic of top Bajocian and early Bathonian. Reappears in the Callovian (Lower Oxford Clay) although this may prove to be a separate species (Coleman, in Penn *et al.*, 1979, Penn, 1982; Cifelli, 1959; Coleman, in Horton *et al.*, 1974).

Planularia beierana (Gümbel)
Plate 6.3.9, Fig. 10, MPK 2249 (× 150), Lower Fuller's Earth, Bath = *Marginulina beierana* Gümbel, 1862. Description: test highly compressed; initial chambers arranged in loose planispire; later uncoiled chambers may or may not retain contact with the spiral portion; sutures generally flush but may be depressed or marked by lower ribs; aperture radiate, terminal, peripheral.
Remarks: *Fasopalmula obliqua* (Terquem) is probably a flabelline variant of this species. Distribution: common Middle Jurassic species.

Pseudolamarckina rjasanensis (Uhlig)
Plate 6.3.9, Fig. 11, SAB 473 F3 (× 150); Fig. 12, SAB 473 F4 (× 150); Middle Oxford Clay, Milton Keynes = *Pulvinulina rjasanensis* Uhlig, 1883. Description: test trochoid, plano-convex or concavo-convex; chambers semi-circular, arranged in 2–3 whorls with five chambers in final whorl, coiling generally dextral, initial chambers not usually visible, sutures on spiral side marked by low ribs; margin rounded–acute; on umbilical side sutures are depressed and form characteristic branchings near umbilicus; aperture small, loop-shaped, umbilical. Distribution: late Callovian, *athleta* Zone (Coleman, in Horton *et al.*, 1974).

Reinholdella crebra Pazdro, 1969
Plate 6.3.9, Fig. 13, MPK 2387 (× 110), Lower Fuller's Earth, Dorset; Fig. 14, MPK 2766 (× 150), Lower Fuller's Earth, Lyme Bay. Description: test trochoid, plano-convex, smooth; 12–16 semicircular chambers arranged in 2–3 whorls; coiling generally sinistral; margin acute; sutures on umbilical side radial, slightly depressed; umbilicus closed; aperture is deep, loop-shaped indentation on last septal suture, extending almost to middle of final chamber; previous apertures closed becoming less visible in earlier chambers, all perpendicular to suture and almost perpendicular to one another. Distribution: Bathonian, *zigzag–aspidoides* Zones (Coleman, in Penn *et al.*, 1979, Penn, 1982).

Saracenaria oxfordiana Tappan, 1958
Plate 6.3.9, Fig. 15, MPK 2767 (× 120), Lower Oxford Clay, Milton Keynes. Common synonym: *Saracenaria triquetra* Gümbel, 1862. Description: initial 4–7 chambers arranged in small planispiral, later 4–5 chambers uncoiling, triangular in cross-section, periphery keeled; margins of apertural face acute, elevated; sutures distinct, depressed; aperture radiate, terminal, peripheral.
Remarks: differs from *S. cornucopiae* (Schwager) in larger size of coil, broader apertural face and small numbers of chambers in linear portion. Distribution: Late Bajocian to Early Bathonian; reappears in the Callovian, ?*jason* Zone and ranges into the Oxfordian (Coleman, in Penn *et al.*, 1979, Penn, 1982, Horton *et al.*, 1974; Medd, in Richardson, 1979; Barnard *et al.*, 1981).

Tristix oolithica (Terquem)
Plate 6.3.9, Fig. 16, MPK 2453 (× 125); Fig. 17, MPK 2269 (× 90); Frome Clay, Lyme Bay = *Tritaxia oolithica* Terquem, 1856. Common synonym: *Tristix suprajurassica* (Paalzow) 1932. Description: test uniserial, composed of spherical proloculus followed by 5–8 triangular or rarely quadrangular chambers which gradually increase in size to give flared outline; final chamber sometimes smaller than preceeding one; sutures strongly depressed; margins of test rounded or keeled; aperture radiate, terminal central, on short neck. Distribution: Aalenian to mid Callovian (Morris, 1982; Coleman, in Horton *et al.*, 1974).

Citharina clathrata (Terquem) *eypensa* (Cifelli)
Plate 6.3.9, Fig. 18 MPK 2209 (× 75), Upper Fuller's Earth, Bath. Description: test large, robust, uniserial, composed of 5–7 chambers increasing rapidly in breadth as added, to give triangular outline; sutures oblique, slightly depressed; ornamented by numerous strong costae which are generally continuous over the surface of the test; aperture radiate, terminal peripheral.
Remarks: as stated by Cifelli, this subspecies succeeds *C. clathrata* (Terquem) *sensu stricto* in the Bathonian. It is distinguished by a broader, more divergent test and more irregular development of the ribs (see also *Citharina clathrata*). Distribution: Bathonian (Coleman, in Penn *et al.*, 1979, Penn, 1982; Cifelli, 1959).

Vaginulina harpa (Roemer) *sensu* Cifelli, 1959
Plate 6.3.9, Fig. 19, MPK 2244 (× 40), Frome Clay, Bath. Description: test large compressed, uniserial, composed of 8–9 chambers which at first gradually increase in breadth but then remain fairly constant to give a parallel side outline; sutures oblique, slightly depressed, ornamented by large number of fine ribs which are continuous over several chambers; aperture radiate, terminal, peripheral. Distribution: Late Bathonian: *hodsoni–aspidoides* Zones (Coleman, in Penn *et al.*, 1979, Penn, 1982; Cifelli, 1959).

Vaginulina legumen (Linné)
Plate 6.3.9, Fig. 20, MPK 2238 (× 45), Upper Fuller's Earth, Bath = *Nautilus legumen* Linné, 1758. Description: test large, robust, uniserial, oval in cross-section, curved or straight; chambers broad, increasing very gradually in size giving smooth outline to test; sutures oblique, generally flush although may be depressed in later part of test; aperture radiate, terminal peripheral. Distribution: Late Bathonian: *hodsoni-discus* Zones (Coleman, in Penn *et al.*, 1979, Penn, 1982; Cifelli, 1959).

NORTH SEA MIDDLE JURASSIC FORAMINIFERA (TEXTULARIINA)

All specimens from well 211/12-1 (Magnus Field).

Ammobaculites cf. *agglutinans* (d'Orbigny)
Plate 6.3.10, Fig. 1 MAG. 1, (3350 m) (× 155); Fig. 2 MAG 2 (3350 m) (× 155); Fig. 3, MAG 3 (3350 m) (× 150), (pyritised infill) = *Spirolina agglutinans* d'Orbigny, 1846. Description: test small, coarsely agglutinated; initial portion planispirally coiled becoming rectilinear. Coiled portion evolute, with depressed umbilicus; chambers in intial whorl extremely elongate, showing through as a spiral tube. Later whorl composed of 6–7 globular chambers of almost constant breadth, with last 2–3 chambers departing from coil to produce a rectilinear portion.
Remarks: differs from *A. agglutinans* in having fewer uncoiled chambers and being coarser. Distribution: long ranging, occuring sporadically in the Upper Brent Group (Bathonian), becoming particularly common in the lower Heather (Middle Bathonian–mid-Callovian).

Haplophragmoides canui Cushman, 1930
Plate 6.3.10, Fig. 4, MAG 4, (3310 m) (× 77); Fig. 5, MAG 5, (3310 m) (× 70); Fig. 6, MAG 6 (3310 m) (× 110). Description: test medium size, finely agglutinated, involute with six to eight inflated chambers in final whorl which increase gradually in height. Sutures radiate, often thickened.
Remarks: compressed specimens appear similar to *H. infracalloviensis* Dain, but differ in possessing fewer chambers in the last whorl, and in having thickened sutures. Distribution: occurs abundantly in the lower Heather Formation (Middle Bathonian–mid-Callovian), ranging into the upper Heather Formation (Late Oxfordian).

Haplophragmoides cf. *H. (Cribrostomoides) dolininae* Bulynnikova, 1972
Plate 6.3.10, Fig. 7, MAG 7 (3290 m) (× 130); Fig. 8, MAG 8 (3290 m) (× 130), Fig. 9 MAG 9 (3290 m) (× 110) (pyrite infill). Description: test medium size, involute with 9–10 chambers in final whorl. Final whorl appears unstable, coiling out of coiling plane to produce a low trochospire. Test wall finely agglutinated. Aperture not visible.
Remarks: differs from original figures (Dain, 1972) in possessing 2–3 additional chambers. Due to test irregularity this species approaches *Recurvoides sublustris* Dain, but lacks a streptospiral initial portion. Distribution: occurs rarely in the upper part of the lower Heather Formation (early mid-Callovian).

Haplophragmoides infracalloviensis Dain, 1948.
Plate 6.3.10, Fig. 10, MAG 10 (3410 m) (× 77); Fig. 11, MAG 11 (3410 m) (× 77); Fig. 12, MAG 12 (3210 m) (× 110). Description: test large, compressed, involute with wide umbilicus which is moderately raised to form a spiral ridge, with adjoining septae raised. Chambers subtriangular, 10–11 chambers in final whorl. Wall fine to coarsely agglutinated.
Remarks: can be distinguished from crushed specimens of *H. canui* by its greater number of chambers per whorl and less prominant sutural thickening. Distribution: ranges through the Middle Bathonian–Callovian of the Heather Formation, occurring abundantly throughout the lower Heather Formation.

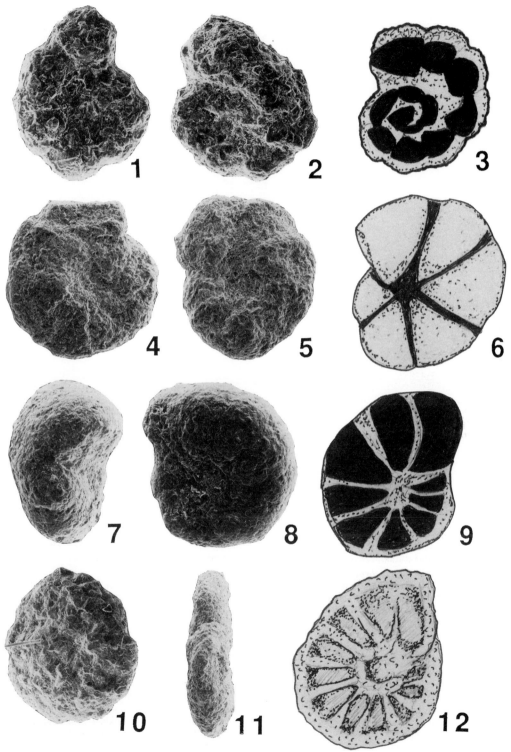

Plate 6.3.10

Haplophragmoides sp. cf. *Haplophragmoides* sp. 143 Brooke and Braun 1981.
Plate 6.3.11, Fig. 1 MAG 13 (3330 m) (× 105); Fig. 2, MAG 14 (3330 m) (× 90); Fig. 3 MAG 15, (3160 m) (× 120) (pyrite infill). Description: test large, compressed, involute, with wide umbilicus which is raised to form spiral ridge; adjoining septae straight, slightly raised. Chambers, subtriangular, low, slowly increasing in height, with 11–14 in final whorl. Test well fine to coarsely agglutinated.
Remarks: similar to *H. infracalloviensis* Dain, but differs in possessing more chambers per whorl with chamber height increasing more gradually, and is also more evolute, displaying often two or more whorls. Present material possesses more chambers and is more evolute than *Haplophragmoides* sp. 143 Brooke and Braun, 1981. Distribution: occurs commonly through the lower Heather Formation (Middle Bathonian–mid-Callovian).

Haplophragmium cf. *pokrovkaensis* Kosyreva 1972
Plate 6.3.11, Fig. 4, MAG 16 (3280 m) (× 120); Fig. 5, MAG 17 (3280 m) (× 110); Fig. 6, MAG 18 (3280 m) (× 165). Description: test medium–large, composed of initial streptospiral coiled portion of 3–4 globular chambers becoming rectilinear. Rectilinear portion composed of 3–7 chambers which initially increase slowly in height. Test wall fine to very coarse.
Remarks: inclusion in *Haplophragmium* follows Dain (1972). The present material compares closely with figured specimens of Dain (1972) differing only in the number of chambers in the rectilinear portion. Distribution: appears restricted to the lower part of the lower Heather Formation (Middle–Late Bathonian), in the Northern North Sea.

Recurvoides sublustris Dain, 1972
Plate 6.3.11, Fig. 7, MAG 19 (3350 m) (× 175); Fig. 8 MAG 20 (3350 m) (× 110); Fig. 9, MAG 21 (3330 m) (× 120). Description: test medium, irregular, with initial streptospiral portion which subsequently becomes trochospiral, occasionally planispiral (superficially resembling *Haplophragmoides*). Degree of involution variable. Chambers subtriangular with thickened septae, up to 15 present in final whorl. Periphery rounded to subangular. Test wall medium-fine grained. Distribution: appears in the Brent Group (Bathonian), becoming common throughout the lower Heather Formation (Middle Bathonian–mid-Callovian).

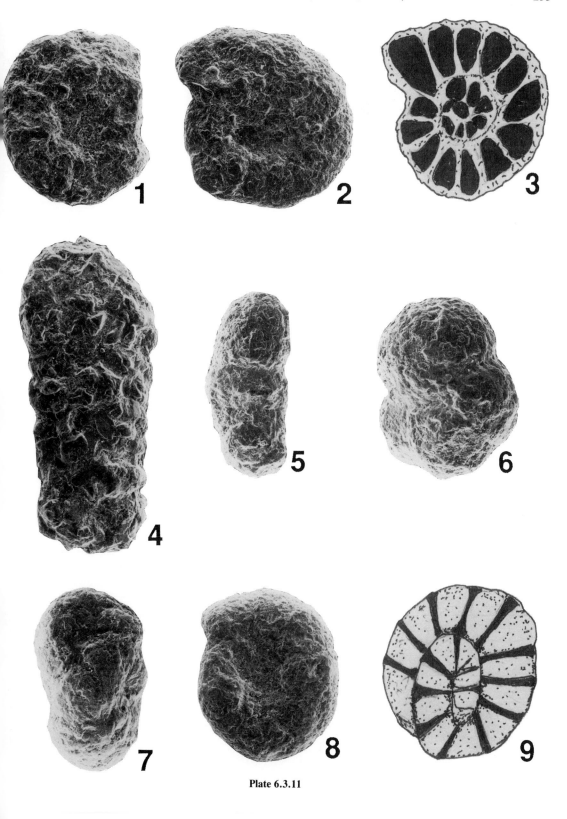

Plate 6.3.11

Trochammina canningensis Tappan, 1955
Plate 6.3.12, Fig. 1. MAG 22 (3311.5 m) (× 295); Fig. 2, MAG 23 (3311.5 m) (× 355). Description: test small, globular, highly variable, ranging from involute forms composed of up to five inflated chambers visible, or evolute forms, which appear assymetric with 1–2 early chambers visible.
Remarks: differs from *T. globigeriniformis* (Parker and Jones) in being more involute with spherical rather than elongate globular chambers; also in the great degree of test assymetry, which is especially pronounced in evolute specimens. Distribution occurs commonly in the lower Heather Formation (Middle Bathonian–mid-Callovian), ranging into the Oxfordian upper Heather Formation.

Trochammina globigeriniformis (Parker and Jones)
Plate 6.3.12, Fig. 3, MAG 24 (3311.5 m) (× 235). Description: test small, arranged in a low trochospire, with 4–5 globular chambers in final whorl.
Remarks: see Lloyd (1959) for full description. Distribution: long ranging, appearing in the Brent Group (Bathonian) occurring commonly in the lower Heather Formation (Middle Bathonian–mid-Callovian).

Trochammina squamata Jones and Parker, 1860
Plate 6.3.12, Fig. 4, MAG 25 (3311.5 m) (× 355); Fig. 5. MAG 26 (3311.5 m) (× 295). Description: test small, compressed, with chambers arranged in a flat trochospire. Chambers, subtriangular with depressed, straight sutures, 5–6 in final whorl.
Remarks: see Lloyd (1959) for description of similar material. Distribution: long ranging, occurring commonly in the lower Heather Formation (Middle Bathonian–mid-Callovian).

Verneuilinoides tryphera Loeblich and Tappan, 1950
Plate 6.3.12, Fig. 6, MAG 27 (3311.5 m) × 195; Fig. 7, MAG 28 (3311.5 m) (× 235). Description: test small, finely agglutinated, triserial, flaring distally; chambers subglobular, increasing gradually in size. Sutures distinct, depressed. Distribution: restricted to the Middle Bathonian–early Callovian of the lower Heather Formation.

Verneuilinoides sp. 1
Plate 6.3.12, Fig. 8, MAG 29 (3290 m) (× 145); Fig. 9 MAG 30 (3290 m) (× 105). Description: test medium, fine to medium grained, triserial, uniformly tapered, with subglobular chambers. Sutures slightly depressed, often indistinct. Aperture not visible.
Remarks: differs from *Verneuilinoides* sp. 2, in being smaller, and uniformly tapered. Distribution: common to abundant throughout the lower Heather Formation (Middle Bathonian–mid-Callovian).

Verneuilinoides sp. 2
Plates 6.3.12, Fig. 10, MAG 31, (3331.5 m) (× 145); Fig. 11, MAG 32, (3331.5 m) (× 120); Fig. 12, MAG 33 (3311.5 m) (× 120). Description: test medium size, sometimes large, medium grained, triserial, with test flaring throughout (accentuated with crushing) or becoming parallel sided. Chambers sub-globular, with depressed sutures. Aperture not visible.
Remarks: generally larger than *Verneuilinoides* sp. 1, with wider and bulkier upper test portion. Distribution restricted to the Middle Bathonian–early Callovian of the lower Heather Formation.

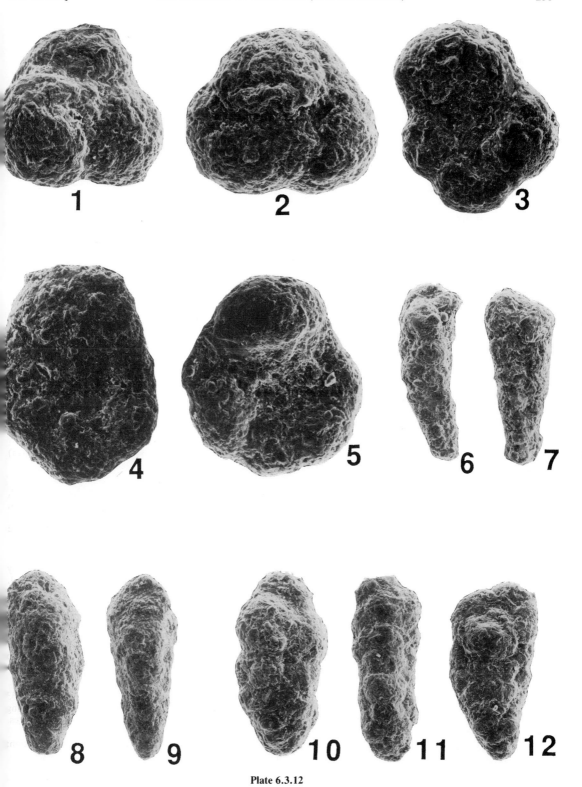

Plate 6.3.12

ACKNOWLEDGEMENTS

The authors would like to express their gratitude to B.P. Petroleum Development Limited for permission to publish biostratigraphic and lithological data from well 211/12-1, Magnus Field. Thanks are due also to B.P. Exploration for permission to publish general biostratigraphic and lithostratigraphic data from the Beatrice Field area, and for technical support with regard to drafting of diagrams. GeoStrat Ltd. have also provided financial support for photography and typing, which is gratefully acknowledged. The authors are indebted to Phil Copestake and Robin Dyer for constructive criticism of the text, and to Hilde Ann Quigley for typing the manuscript.

This paper is published with the permission of the directors of GeoStrat Ltd.

6.4 THE OXFORDIAN TO PORTLANDIAN

D. J. Shipp

6.4.1 Introduction

The pioneer work on the study of Jurassic foraminifera was carried out in the latter half of the nineteenth century mainly by continental workers who initially concentrated their attention on the Lias. Some of the more important papers in this category include those by Borneman (1854), Terquem (1855 onwards), Berthelin (1879) and Häusler (1881). English workers of this time also dealt mainly with the Lias and their papers include those of Strickland (1846), Jones and Parker (1860), Brady (1867) and Crick and Sherborn (1891).

Interest in Upper Jurassic foraminifera generally developed later than in Lias forms, although two of the earlier papers (those of Gümbel (1862) and Schwager (1865), are both important and deal with the Oxford Clay. Other contributions from this period include those of Häusler (1883), Deeke (1886) and Wisniowski (1890 and 1891).

In Britain, Whittaker (1866) and Crick (1887) published unillustrated lists of Oxford Clay foraminifera. A short paper on adherent foraminifera from the Oxford Clay was produced by Sherborn (1888). Early papers on the Kimmeridge Clay include those by Blake (1875), Chapman (1897) and Woodward (1895).

Although these early workers laid the foundations of Jurassic micropalaeontology they are to some extent responsible for the confusion that now exists in Jurassic nomenclature. This is partly due to the fact that they often erected a number of separate species names for what were probably variations of the same species and partly because their work was often poorly illustrated.

After the war, Barnard initiated further studies on Late Jurassic foraminifera with his papers on the Oxford Clay (1952 and 1953). Both are useful and well illustrated. More recent papers on the British Late Jurassic have included those on the Oxford Clay by Cordey (1962, 1963a) and Gordon (1967); the Corallian by Gordon (1965); the Ampthill Clay by Gordon (1962), and the Kimmeridgian by Lloyd (1959, 1962). These are generally stratigraphic studies. Palaeoecology and palaeogeography have mainly been neglected, except by Gordon (1970) and Barnard et al. (1981).

A very useful publication which includes details on the Jurassic of Dorset was produced as a field guide for the XXth European Micropalaeontological Colloquium Copestake 1987.

Two papers dealing with the rarely published topic of offshore sections of Late Jurassic microfossils were produced by Colin et al. (1981) and Ainsworth et al. (1987) on the Celtic Sea Basin.

Continental work has not neglected the Late Jurassic in recent times with useful papers being produced by Lutze (1960), and Munk (1980) from Germany, Bizon (1958), Guyader (1968), and Barnard and Shipp (1981) from France; Norling (1972) from Sweden; Bielecka and Pozaryski (1954), Bielecka (1960) from Poland and Løfaldi and Nagy (1980 and 1983) from Svalbard.

Moving further afield, a significant contribution on foraminifera from the Late Jurassic of Western Siberia was produced by Dain (1972) while a number of papers on Canadian Late Jurassic foraminifera: Ascoli (1976, 1981); Ascoli *et al*. (1984); Gradstein (1976 and 1978) and Wall (1983), are of relevance to British microfaunas when the effects of continental drift are considered.

Three useful papers of a different kind where the authors revise the works of 'classic' micropalaeontologists are Seibold and Seibold (1955), revising the work of Gümbel (1862), Seibold and Seibold (1956) and Bizon (1960), revising some of Terquem's type species.

6.4.2 Collections of Upper Jurassic foraminifera

(1) British Museum (Natural History)
   - Barnard: Oxford Clay (Barnard, 1953)
   - Cordey: Oxford Clay (Cordey, 1962, 1963b)
   - Gordon: Oxford Clay and Corallian (Gordon, 1965)
   - Lloyd: Kimmeridge Clay (Lloyd, 1959, 1962)
   - Shipp and Murray: Oxford Clay and Corallian (Shipp and Murray, 1981—1st edition, *Stratigraphical Atlas of foraminifera*).

(2) British Geological Survey
   - Medd: Oxford, Ampthill and Kimmeridge Clays (Medd, 1979).

A large collection of Upper Jurassic Foraminifera is also present in the Micropalaeontology Department of University College, London.

6.4.3 Stratigraphic Divisions

The Upper Jurassic consists of the Oxfordian, Kimmeridgian and Portlandian stages. The Oxfordian is subdivided into lower, middle and upper, while the Kimmeridgian is subdivided into lower and upper as is the Portlandian. In the boreal area (to the north of the Market Wheaton axis) the term Volgian is preferable to the Portlandian and upper Kimmeridgian. The relationships between these terms, the Upper Jurassic ammonite zones and lithostratigraphy can be seen in Figs 6.4.4 (onshore) and 6.4.5 (offshore). Figs 6.4.1 and 6.4.2 show the outcrop of the Late Jurassic and the onshore localities providing data on the foraminifera. Fig 6.4.3 shows the section details at these localities.

The Corallian is a term used in Britain for the mixed deposits of the middle and upper Oxfordian which lie between the predominantly argillaceous Oxford Clay of early Oxfordian age and the Kimmeridge Clay of the Kimmeridgian. The term Portlandian includes the Purbeck Beds up to the base of the mid-Purbeck Cinder Bed.

Ammonite zonal schemes for the Upper Jurassic of Britain have been developed by a succession of workers. Foremost are the publications of Arkell (1933–1956) which provide the basis of our present zones. Mention must also be made of Brinkman (1929) and Spath (1939) who have both contributed to our knowledge of Jurassic ammonites. More recently Callomon (1955, 1964, 1968b, c), Callomon and Cope, (1971) and Cope (1967, 1974a, b, 1978) together with Sykes (1975) and Sykes and Callomon (1979) have given much attention to refining the details of the Upper Jurassic zonation. In addition a standard reference for correlating the Upper Jurassic rocks of British Isles has been produced by Cope *et al*. (1980b).

6.4.4 Depositional history; palaeoecology; faunal associations

The Oxford Clay has a rich benthic fauna of epifaunal and infaunal elements and this demonstrates that the sea was oxygenated down to the bottom (Hudson and Palframan, 1969; Duff, 1975). The presence of a rich benthic

Sec. 6.4] The Oxfordian to Portlandian

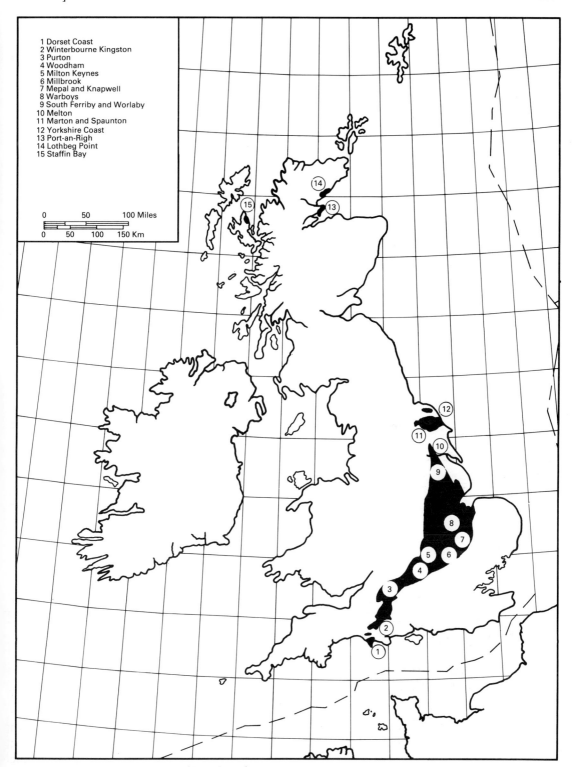

Fig. 6.4.1 — Localities and outcrop of Late Jurassic.

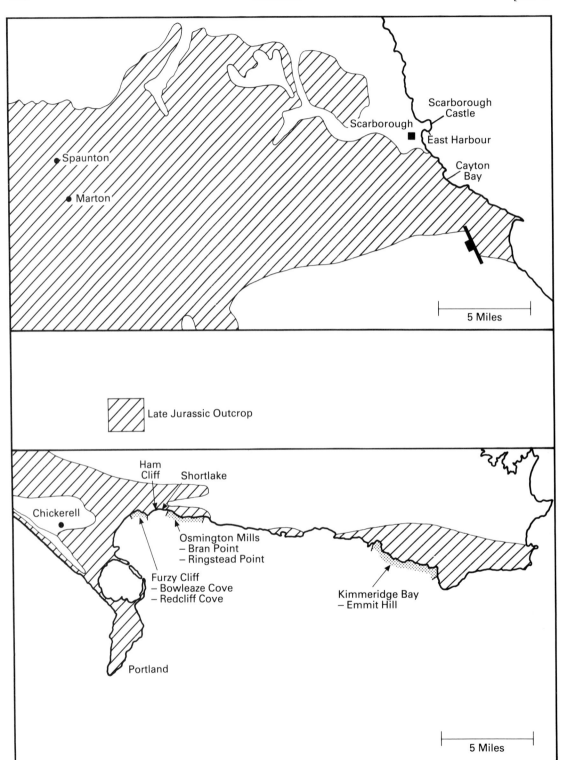

Fig. 6.4.2 — Localities in Dorset and Yorkshire.

foraminiferal fauna supports this interpretation. The early Oxfordian is thus a time of stable shelf conditions and a large and varied microfauna recovered from the upper Oxford Clay reflects this stability.

As stated by Fürsich (1976) the Corallian forms a shallow water deposit representing a variety of environments ranging from offshore shelf to intertidal between the deeper water Oxford and Kimmeridge Clays. Several authors have worked on the environment of the Dorset Corallian; Arkell, 1933, 1935; Wilson 1968a, b; Talbot, 1973, 1974; Brookfield, 1973 and Fürsich, 1976. Fürsich also worked on the Corallian of Yorkshire and Normandy and considered that Talbot's reconstruction was the most satisfactory. Shipp (1978) agreed with this conclusion and used Talbot's model as a basis for his investigation of the palaeoecology of the Corallian foraminifera.

In comparison with those of the Dorset coast, the Corallian microfaunas at Millbrook and Warboys contain a more varied nodosariid assemblage together with a greater number of forms, such as *Ophthalmidium*, *Quinqueloculina* and *Spirillina*, thought to indicate shallow water conditions. The greater variety of nodosariids suggests a more stable environment. *Epistomina*, thought to represent open marine shelf conditions, is absent from the Millbrook and Warboys sections although it occurs in parts of the Dorset section. Thus in the central area of England the depth of water appears to have been less than in the deeper part of the Dorset section, although still in the general offshore shelf category. Conditions were apparently also more restricted here with less open marine influence. The main connection with the open sea at this time is thought to have been in the south, with more normal marine conditions more likely in the Dorset area.

The Kimmeridgian marks a return to stable shelf conditions with the argillaceous deposits of the Kimmeridge Clay Formation as the principal lithotype. Offshore, sands, such as those of Piper and Claymore, are more common but the Kimmeridge Clay is still the dominant lithology. The varied microfaunas seen onshore in the Kimmeridgian (Lloyd, 1958) are less common offshore in the North Sea where the rich organic clays are believed to have been deposited under dysaerobic or anaerobic conditions. (Rawson and Riley, 1982). This resulted in reduced, predominantly agglutinating microfaunas, where oxygen levels are low and a complete absence of foraminifera in many sections. The underlying Heather Formation, however, contains microfaunas which are more comparable to those seen onshore in the Oxford Clay in that calcareous benthonic forms are more common, although agglutinating forms still usually predominate. The Oxfordian and Kimmeridgian represent the periods of maximum eustatic sea level rise according to Hallam (1969, 1978). In addition, regional subsidence in the Kimmeridgian is thought to have led to further deepening (Hallam and Sellwood, 1976). Bituminous shale developed at certain times (Morris, 1980).

Townson (1975) has suggested that the Portland Group was deposited under generally shallow conditions with periods when low oxygen levels prevailed on the sea bed. This may account for the low numbers of foraminifera present (Shipp, 1978; Copestake, pers. comm.). Ostracods recorded from this section occur in a greater number of samples than the foraminifera and appear to have been more tolerant of the rather harsh conditions that existed at this time. Much of the Portlandian is unrepresented onshore and the shallow deposits of the Portland Sand, Stone and Purbeck Beds are principally confined to southern England. Offshore in the North Sea the Kimmeridge Clay Formation persists into the Early Cretaceous.

Throughout the Late Jurassic, deposition in the Celtic Sea basin is generally represented by non-marine 'Purbeck-Wealden' sediments and foraminifera are usually unrecorded (Colin *et al.*, 1981). Ainsworth *et al.*

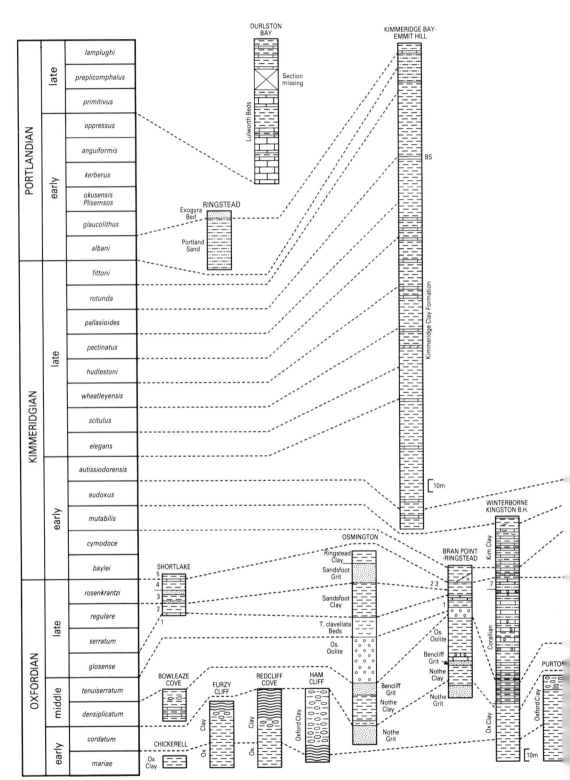

Sec. 6.4]   **The Oxfordian to Portlandian**   243

Fig. 6.4.3 — Chronostratigraphy of British Upper Jurassic sections analysed for foraminifera. Chickerell, Ham Cliff, Purton, Woodham, Millbrook, Cayton Bay (Shipp 1978, Barnard, Cordey and Shipp 1981); Furzy Cliff (Shipp 1978, Copestake 1987); Redcliff Cove (Barnard 1953, Shipp 1978, Barnard, Cordey and Shipp 1981, Copestake 1987); Shortlake, Ringstead (Shipp 1978); Bowleaze Cove (Shipp 1978, Copestake 1987); Durleston Bay (Robertson Research 1974); Osmington (Gordon 1965, Robertson Research 1974, Shipp 1978, Barnard, Cordey and Shipp 1981, Copestake 1987); Kimmeridge Bay – Emmit Hill, Lloyd 1958, 1959, 1962, Robertson Research 1974, Copestake 1987); Winterbourne Kingston (Medd 1982); Worlaby (Medd 1979); Mepal and Knapwell (Gordon 1962); South Ferriby, Melton, Marton, Spaunton, East Harbour, Scarborough Castle, Port-on-righ (Robertson Research 1974); Lothbeg Point (Robertson Research 1974, Gregory, in press); Staffin Bay (Cordey 1962, Shipp 1978).

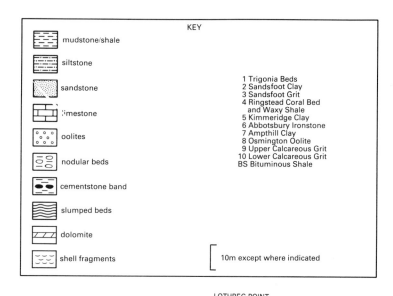
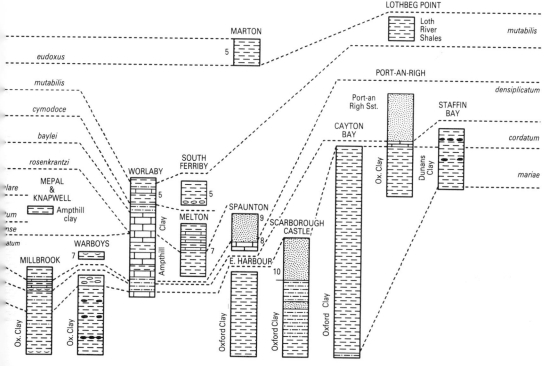

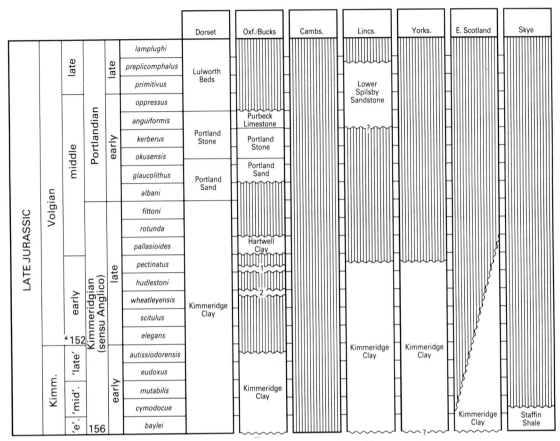

Fig. 6.4.4 — Late Jurassic Stages, ammonite zones and British lithostratigraphic nomenclature (modified from Cope et al., 1980b). (1) Shotover Grit Sands; (2) Wheatley Nodule Clays. * Million years B.P. (after Kent and Gradstein, 1985).

(1987), however do record rare foraminifera in the early Kimmeridgian.

The environmental significance of the various groups of foraminifera found in the Oxford Clay has been tentatively inferred by Barnard et al. (1981) as follows:

(a) Small agglutinated specimens (*Trochammina*, *Gaudryina*, small *Ammobaculites*, etc.). These probably represent shallow water with reduced salinity or less oxygenated bottom conditions, especially when they occur alone.
(b) Large agglutinated specimens (*Ammobaculites coprolithiformis* mainly, but also *Triplasia*). These appear to represent generally deeper water conditions.
(c) Species of *Ophthalmidium* are thought to represent a shallow water environment.
(d) *Epistomina* appears to reflect normal marine and hence possibly deeper water conditions.
(e) Adherent species are thought to reflect breaks in the succession which may be due to slow rate of deposition and/or to increased bottom currents.
(f) Nodosariacea, (excluding the Polymorphinidae) primarily indicate deeper water, but are not restricted to it, occurring frequently with the other groups.
(g) Polymorphinidae (mainly *Eoguttulina*). As yet the significance of this group is

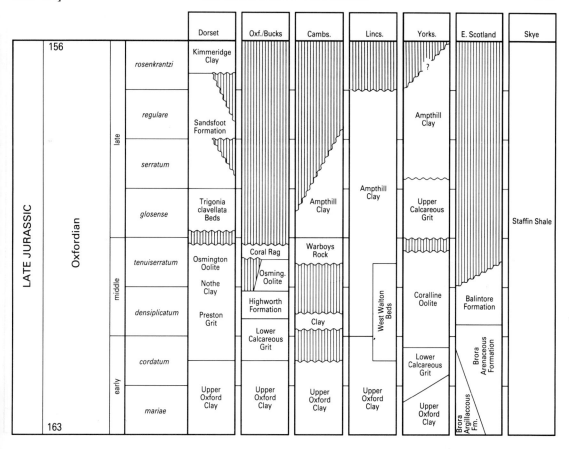

not clear, but as they often occur in large numbers, there may be some distinct factor, such as salinity or clarity of water, affecting their distribution.

It is considered that these groups are likely to have similar environmental significance in the other deposits of the Late Jurassic. The Oxfordian is characterised by three types of assemblage that correspond to those described by Gordon (1967) from the Callovian of Brora:

(a) Those dominated by Textulariina, notably *Lagenammina difflugiformis, Reophax horridus, Ammobaculites suprajurassicus, Reophax sterkii* and *Ammobaculites coprolithiformis*. Such assemblages are characterised by their low diversity. It is possible that calcareous forms have been removed by solution during diagenesis.

(b) Those dominated by calcareous benthonics especially nodosariaceans (60%) with some Textulariina (40%). The dominant species is *Lenticulina muensteri*. Other common forms are *Citharina flabellata, Dentalina guembeli, Lenticulina quenstedti, Marginulina scapha, Ammobaculites corprolithiformis, Eoguttulina liassica* and *Epistomina (Brotzenia) parastelligera*. Diversity is moderate.

(c) An assemblage with an even mixture of calcareous benthonics and Textulariina characterised particularly by the dominance of *Epistomina (Brotzenia) parastelligera*.

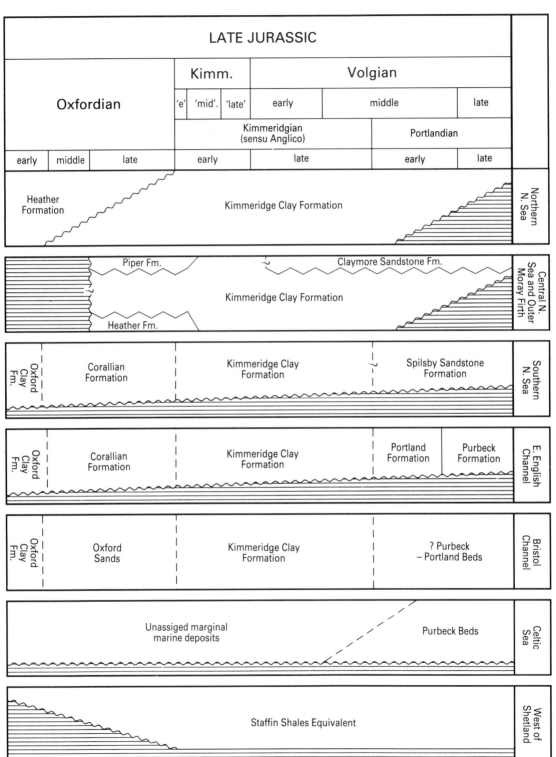

Fig. 6.4.5 — Late Jurassic offshore lithostratigraphy (based mainly on Cope *et al.*, 1980a).

The Oxford Clay of Skye (Cordey, 1962) extends through the *mariae* into the *cordatum* Zone. Lagenina, especially nodosariaceans, are dominant. *Lenticulina muensteri* is common to abundant in every sample. Other species common in one or two samples include *Planularia beierana, Epistomina (Brotzenia) mosquensis, Trocholina granulata* and *Conorboides pygmaea*. These assemblages are of type (b) above.

The Upper Oxford Clay of Warboys Pit, Huntingdonshire, spans the *mariae* and *cordatum* Zones. The assemblages are dominated by *Lenticulina quenstedti* and *L. muensteri* together with *Bullopora rostrata* (Barnard, 1952). These assemblages again are of type (b) above.

In addition, Barnard *et al.* (1981) obtained a large and varied fauna from the English Upper Oxford Clay from the Oxfordian. Preservation of the foraminifera is generally good and 54 species were recorded. The fauna is dominated by the superfamily Nodosariacea, which constitutes over 75% of the species and a much greater percentage of the total population. The genus *Lenticulina* occurs in virtually every sample. Agglutinated foraminifera form the second most important element of the fauna with the species *Ammobaculites coprolithiformis* being especially prominent. Agglutinated forms often dominate individual samples in which calcareous forms are few or absent. Species belonging to the Epistomininae and Ophthalmidiidae, although less significant in the total fauna, often swamp individual samples with genera such as *Ophthalmidium* and *Epistomina* frequently occurring in large numbers.

The Corallian of the Dorset coast east of Weymouth has been examined by Gordon (1965). Two species dominate the assemblages at most levels: *Ammobaculites coprolithiformis* and *Lenticulina muensteri*. Gordon commented on the similarity of the assemblages throughout this sequence. This seems surprising considering the varied environments of deposition inferred by Talbot (1973). However, examination of the details of Gordon's sampling shows that he restricted himself to clays and sands. Thus, not all the depositional environments were sampled for foraminifera. The Corallian assemblages are mainly of type (b).

Two levels within the Ampthill Clay, thought to be within the *cautisnigrae* and *decipiens* Zones, yielded assemblages which were considered to be similar to those of the Dorset Corallian (Gordon, 1962).

Shipp (1978) also studied the Corallian and recorded a microfauna, which although it does not contain as many species as seen in the Oxford Clay, is nevertheless numerous and diverse. Forty-eight species were identified of which eight are agglutinating, while of the remaining 40 species the vast majority (26) are representatives of the Nodosariidae.

The fauna is dominated by four species: *Lenticulina muensteri, Ammobaculites coprolithiformis, Citharina serratocostata* and *Spirillina tenuissima*. *Lenticulina muensteri* constitutes almost 50% of the specimens recovered from the Dorset Coast and Warboys section and a quarter of the specimens from Millbrook. *Ammobaculites coprolithiformis* forms 23% of the fauna from Dorset and 14% of the fauna from Millbrook although it is much less numerous at Warboys.

Citharina serratocostata and *Spirillina tenuissima* are most common at Millbrook and Warboys but occur in much lower numbers on the Dorset Coast. Ophthalmidiidae, Polymorphinidae and Epistomininae are generally less common than in the Oxford Clay.

Lloyd (1958, 1959, 1962) has described a similarly varied microfauna from the Kimmeridge Clay of Dorset. Nodosariacea again form the dominant group with Textulariinae, Epistomininae and Polymorphinidae also forming significant elements. J. Exton (pers. comm.) has supplied the following information on this section: In the upper part of the Kimmeridge Clay (*scitulus* to *hudlestoni* Zones) the assemblages are dominated by small, flattened agglutinated forms of *Ammobaculites*, *Trochammina*, *Textularia* and *Haplophragmoides*. Nodosariids (mostly *Lenticulina muensteri* and *Marginulina* sp.) and other calcareous groups are rare. In the interval from the *pectinatus* to the *rotunda* Zone, nodosariids are almost as abundant as agglutinated genera. The abundance and size of the faunas also show change. From the *scitulus* to lower *pectinatus* Zone the agglutinating faunas are rich in numbers of individuals but low in species diversity (bituminous shales are abundant at this level). From the upper *pectinatus* Zone to the *fittoni* Zone there is an increase in the size of the individuals associated with an increase in the proportion of quartz silt and calcareous content of the sediment. There may have been increased oxygen availability in the environment.

The Kimmeridge Clay of Humberside has been investigated in the Worlaby G borehole. Only the *baylei* and *cymodoce* Zones are adequately represented. The assemblages are similar to those of the underlying Ampthill Clay, being dominated by *Trochammina squamata*, *Ammobaculites* sp., *Epistomina porcellanea* (given as *Brotzenia*) and *Lenticulina muensteri* (Medd, 1979).

Details of Portlandian foraminifera are as yet unpublished apart from information provided by Copestake in the Southern England Field Guide (1987). He records *Lenticulina muensteri*, *Citharina serratocostata*, *Marginulina costata* and *Lenticulina varians* from the *albani* Zone and *Citharina serratocostata*, *C. flabelloides*, *Lenticulina muensteri* and *Saracenaria oxfordiana* from the *okusensis* Zone of the Isle of Portland, Portland Sand.

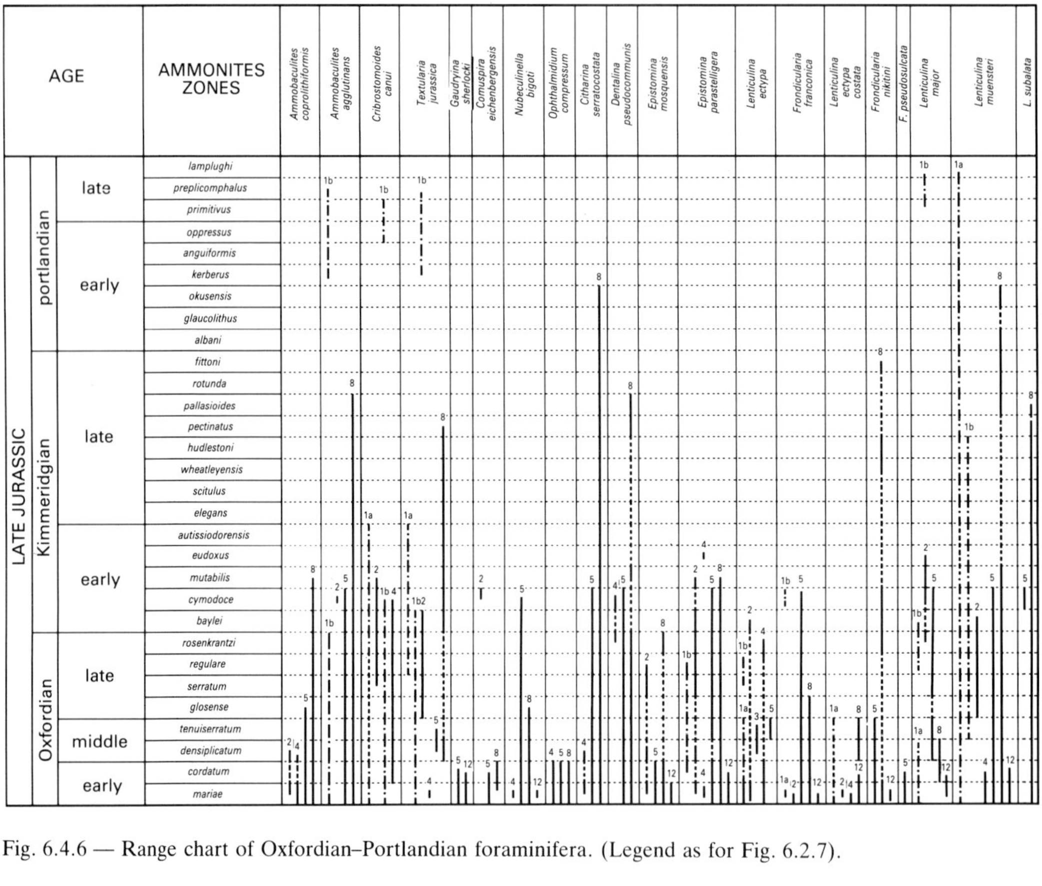

Fig. 6.4.6 — Range chart of Oxfordian–Portlandian foraminifera. (Legend as for Fig. 6.2.7).

Shipp (1978) also examined samples from the *albani* Zone of the Portland Sand of Ringstead Bay. The foraminifera seen are very sparse with only a few specimens of *Lenticulina muensteri* having been recovered from the samples immediately below the Exogyra Bed.

In addition P. Copestake has supplied data on foraminifera recovered from the Portland Sand and Portland Stone of the Vale of Wardour, Wiltshire. *Lenticulina muensteri* is again the most common species. The remainder of the microfauna consists of other species of nodosariids, mainly of *Lenticulina*, a single specimen of *Trocholina* and, of more relevance, a specimen of *Globigerina*: the only record of a planktonic species from the Late Jurassic of Britain. A number of species have also been recorded from the Portlandian offshore where microfaunas are more varied (Fig. 6.4.6).

Most of the Upper Jurassic foraminiferal faunas described from Britain are from muddy sediments. Many of the assemblages are dominated by *Lenticulina* and *Ammobaculites*. Other relatively common genera are *Epistomina, Citharina, Marginulina, Dentalina* and *Eoguttulina*. None of these genera is common in any modern shelf environment, but *Ammobaculites* is common in some brackish areas (see Murray (1973) for a summary). *Lenticulina, Marginulina* and *Dentalina* are all known from modern shelf, slope and deep-sea, cold-water, normal marine environments.

6.4.5 Stratigraphic distribution

In general foraminiferal assemblages are very

uniform through the Late Jurassic with no gross stratigraphic changes. Where assemblages are different they can be related to environmental and facies changes rather than any evolutionary cause. Consequently stratigraphic subdivision is usually based on the ranges of individual species as opposed to more general overall changes to the microfaunas. Some exceptions do occur, however. An example is the reappearance of foraminifera in the early Kimmeridgian below the 'hot' shale of the Kimmeridge Clay Formation in wells drilled in the central part of the North Sea. In general, however, such gross changes are geographically limited and, as in this case, still related to major changes in the benthonic environment.

The overall environment of the Late Jurassic is relatively stable and similar conditions persist into the Early Cretaceous with the result that the overall composition of Early Cretaceous foraminiferal microfaunas are very comparable to those of the Late Jurassic. Major changes do not occur until the Barremian and younger when planktonic foraminifera become increasingly prominent.

6.4.6 Geographic distribution

Regional variation can be seen in the Late Jurassic microfaunas of the British Isles with assemblages recovered from onshore locations being generally much more varied and richer in calcareous benthonic forms than microfaunas recovered from the relatively deeper depositional environments seen in offshore sections, particularly within the Graben areas. In these deeper locations agglutinating foraminifera generally predominate, probably because they are more tolerant of the lower oxygen conditions believed to have been present on the sea floor. As mentioned previously the Celtic Sea basin was essentially non-marine during the Late Jurassic with foraminifera being absent.

The closest Late Jurassic foraminiferal assemblages to Great Britain have been described from the Boulonnais (Barnard and Shipp, 1981), Normandy (Bizon, 1958) and the Seine Estuary (Guyader, 1968).

The assemblages recorded are essentially similar to those seen in Dorset. Comparable microfaunas have been described from Germany (Lutze, 1960; Munk, 1980), Poland (Bielecka and Pozaryski, 1954; Bielecka, 1960), South Sweden (Norling, 1972) and Western Siberia (Dain, 1972), although in the latter case pre-existing European nomenclature has generally not been followed. All these microfaunas are essentially examples of the mixed shelf assemblages described by Gordon (1970) containing simple agglutinating forms together with nodosariids and other calcareous benthonic types such as epistominids and ophthalminids. This type of assemblage appears to be confined to areas to the north and south of the Tethyan region as similar assemblages have been reported by Said and Barakat (1958) from Egypt and by Macfadyen (1935) from the Somali Peninsula. Representatives of these mixed shelf assemblages have also been recorded from the Grand Banks and Scotian Shelf of eastern Canada by Ascoli (1976, 1981), Ascoli et al. (1984) and Gradstein (1976, 1978). Epistominids are especially prominent at these locations and form the basis for stratigraphic subdivision. Loeblich and Tappan (1950) have described a comparable microfauna (although generally ignoring European nomenclature) from South Dakota in the U.S.A. suggesting that the mixed shelf assemblage continued in a belt from Western Siberia to North America to the north of the Tethyan ocean. As described by Gordon (1970) the microfaunas recorded from the Tethyan belt contain agglutinating forms with complex internal morphology and greater numbers of planktonic forms than seen in the mixed shelf assemblages.

On examining localities to the north of those already described, the proportion of simple agglutinating forms is seen to increase until they become the dominant constituent, as seen in the Arctic slope of Alaska (Tappan, 1955) and Svalbard (Løfaldi and Nagy, 1980, 1983).

6.4.7 Index species

The species selected include both rare short-ranging and common long-ranging forms chosen to give an overall impression of Late Jurassic assemblages. The classification is that of Loeblich and Tappan (1984). The ranges are based on Barnard (1952, 1953), Cordey (1962, 1963a), Barnard et al. (1981), Gordon (1962, 1967), Lloyd (1959, 1962) together with details from Copestake (1987) and previously unpublished data from Lloyd (1958, PhD thesis, U.C.L.) and Gregory (in press). In addition, data from two studies by the Robertson Group on the onshore and offshore Jurassic respectively have also been included. The latter includes details from nearly 50 Jurassic sections from the Moray Firth and Central Graben areas of the British sector and a further 15 wells from the Central Graben area of the Norwegian Sector. It must be stated, however, that recovery in many of these sections was limited. As a result of problems with confidentiality no details are available from the southern North Sea offshore area although a number of the illustrations will be seen to be of specimens from the 48 block of the British North Sea. No foraminifera have been recorded from the Celtic Sea basin from the Late Jurassic (Colin et al., 1981). The stratigraphic ranges are shown in Fig. 6.4.6 where the numbers correspond to the basins shown in Fig. 6.1.1. The quoted British range is that for the Late Jurassic only. The majority of species whose range includes the *mariae* Zone range down into the Middle Jurassic (see section 6.3).

SUBORDER TEXTULARIINA

Ammobaculites agglutinans (d'Orbigny)
Plate 6.4.1, Fig. 1 (× 75), late Kimmeridgian 48 block southern North Sea = *Spirolina agglutinans* d'Orbigny, 1846. Description: agglutinated test with early portion an evolute planispiral and later portion rectilinear. About five chambers in the last whorl of the coil and generally 3–5 chambers in the uncoiled portion. Test oval in cross-section and never greatly compressed. Grains closely set with little cement. Sutures depressed, often indistinct. Margin may be slightly lobulate. Aperture central, terminal, often difficult to see. Uncoiled portion often reduced; length 0.25–0.45 mm.
Remarks: forms assigned by Lloyd (1959) as *A.* cf. *hockleyensis*, Cushman and Applin are very similar. Lloyd recognised three variants of *A. agglutinans* and also considered that *A. infravolgensis* Myatliuk and *A. suprajurassicus* (Schwager) were probably synonymous to *A. agglutinans*. Distribution: British range *mariae–preplicomphalus*; Europe, early Kimmeridgian of Central Poland (Bielecka and Pozaryski, 1954), prominent in the Callovian–Kimmeridgian of Svalbard (Løfaldi and Nagy, 1980), as *A. suprajurassicum* (Schwager), early Kimmeridgian of northern France (Barnard and Shipp, 1981).

Ammobaculites coprolithiformis (Schwager)
Plate 6.4.1, Fig. 2 (× 56), early Kimmeridgian 48 block southern North Sea, Figs 3, 4 (× 19), Osmington Oolite Series, *tenuiserratum* Zone, Osmington, Dorset. = *Haplophragmium coprolithiforme* Schwager, 1867. Description: test commences with a planispiral whorl of 2–6 chambers, followed by an uncoiled uniserial portion of up to six chambers; these form approximately one-third and two-thirds the length of the test respectively; initial part compressed, later circular in cross-section; final chamber often inflated, conical or pyriform. Some specimens remain close-coiled; length 0.6–1.5 mm occasionally up to 3.0 mm.
Remarks: this form is common throughout the Late Jurassic, especially so in the Oxfordian sections on the Dorset Coast. Distribution: British range *mariae–mutabilis*; Europe, Oxfordian and Kimmeridgian of Poland (Bielecka and Pozaryski, 1954), Oxfordian of Germany (Lutze, 1960), as *Haplophragmium aequale*, also recorded from the Middle Jurassic of the French Jura under the same name (Wernli, 1971); North America, Kimmeridgian and Portlandian of the Scotian shelf (Ascoli, 1976).

Haplophragmoides canui Cushman
Plate 6.4.1, Figs 5, 6 (× 76), Nothe Clay, *densiplicatum* Zone, Bowleaze Cove, Dorset. = *Cribrostomoides canui* (Cushman), 1910. Description: planispiral, umbilicate, involute, sometimes evolute in the last half whorl. Nine or ten chambers in the final whorl; aperture circular or oval, centrally placed on apertural face; average diameter 0.55 mm.
Remarks: this species assigned to *Ammobaculites laevigatus* Lozo by Lloyd (1959). Distribution: British range *mariae–primitivus*; Europe, Oxfordian of Normandy (Guyader, 1968), Callovian–?Kimmeridgian of Svalbard (Løfaldi and Nagy, 1980), North America, Callovian–Portlandian of Scotian Shelf (Ascoli, 1976, 1981), late Volgian of Svedrup Basin (Wall, 1983), as *H.* cf. *canui*.

Textularia jurassica (Guembel)
Plate 6.4.1, Fig. 7 (× 190), Ampthill Clay, *tenuiserratum* Zone, Millbrook. = *Textilaria jurassica* Guembel, 1862. Description: test only slightly flaring, initial planispiral whorl of 4–5 chambers seen in some specimens; test usually consists of 18–25 biserially arranged chambers; aperture an interiomarginal arcuate slit; average length 0.4 mm.
Remarks: *Textularia pugiunculus* (Schwager) recorded by Gordon (1965) is thought to be a variant of this species. Distribution: British range *mariae–preplicomphalus*; Europe: Oxfordian of Southern Germany (Siebold and Siebold, 1960), Kimmeridgian of Sweden (Norling, 1972.)

Gaudryina sherlocki (Bettenstaedt)
Plate 6.4.1, Fig. 8 (× 110), Oxfordian 48 block southern North Sea. = *Gaudryinella sherlocki* Bettenstaedt, 1952. Description: initial trochoid portion of 3–5 chambers (often difficult to see), later reducing to three chambers per whorl and finally two; aperture varies from simple, central terminal to crescentric at base of final chamber; average length 0.35 mm.
Remarks: variation shows a tendency towards a uniserial arrangement but it is never fully developed and the final chamber is always in contact with the antepenultimate one. Distribution: British range *mariae–cordatum*. Europe: Oxfordian of Germany (Lutze, 1960) and northern France (Guyader, 1968) (both as *Gaudryina* sp. 2), Oxfordian of southern Germany (Seibold and Seibold, 1960) as *Gaudryinella uvigeriniformis*.

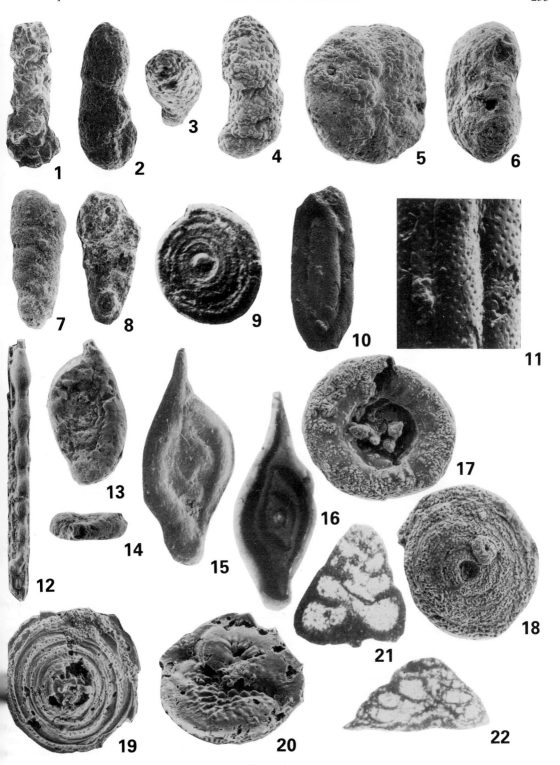

Plate 6.4.1

SUBORDER INVOLUTININA

Miliospirella lithuanica Grigelis, 1958
Plate 6.4.1, Fig. 10 (× 145), Oxford Clay, *mariae* Zone Warboys; Fig. 11 detail of same specimen (× 600). Description: test with proloculus followed by nonseptate enrolled second chamber. Plane of coiling changing regularly such that successive chambers are added approximately 120° apart giving triloculine appearance. Aperture simple at open end of tube, wall coarsely perforate; average length 0.3 mm, average width 0.125 mm.
Remarks: the presence of pores distinguishes this species from the otherwise similar *Agathammina antiqua* Said and Barakat (1958). The species was originally described from the late Callovian of Russia. It is illustrated in the Treatise (Loeblich and Tappan, 1964).
According to Cordey (1963b) the undescribed Brady collection in the British Museum (Natural History) contains several specimens mounted on a slide labelled '*Spirillina*' (p. 8220). These specimens were from the Bradford Clay, Bradford, and although it was not possible for Cordey to examine them internally he considered them to be externally identical to *M. lithuanica*. Distribution: British range *mariae–cordatum*.

SUBORDER MILIOLINA

Cornuspira eichenbergensis Kubler and Zwingli, 1870
Plate 6.4.1, Fig. 9 (× 125), Oxford Clay, *cordatum* Zone Ham Cliff. Description: porcellanous test consisting of proloculus followed by simple, undivided, planispirally coiled, evolute second chamber, aperture simple at open end of tube, diameter between 0.2 mm and 0.45 mm.
Remarks: only recorded in Britain in the Oxford Clay, being rare at Woodham and Warboys but common at Ham Cliff (Barnard *et al.*, 1981) and in the Kintradwell Boulder Bed of Scotland in the *cymodoce* zone (Gregory, in press). Distribution: British range *mariae–cymodoce*. Europe: common in Callovian of N. W. Germany (Lutze, 1960).

Nubeculinella bigoti Cushman, 1930
Plate 6.4.1, Fig. 12 (× 32). Nothe Clay, *densiplicatum* Zone, Bowleaze Cove, Dorset. Description: adherent test with initial coil of globose proloculus with a small tube of 2–7 chambers wound planispirally around it, followed by up to 12 flask-shaped hemispherical chambers; length of individual chamber 0.2–0.35 mm.
Remarks: usually attached to bivalve shells but also to echinoid spines, large foraminifera tests, etc. Often damaged with initial coil missing. Its presence may indicate a slow rate of sediment deposition. Common in Upper Oxford Clay, *mariae–cordatum* (Barnard *et al.*, 1981). Distribution: British range *mariae–cymodoce*; Europe: originally described from the *cordatum* Zone of Normandy (Cushman, 1930). It has also been recorded from this region by Guyader (1968). Adams (1962) gives its overall range as Lias–Kimmeridgian.

Ophthalmidium compressum Barnard, Cordey and Shipp, 1981
Plate 6.4.1, Fig. 15 (× 85), Fig. 16, transmitted light (× 110); Oxford Clay, *mariae* Zone Warboys. Description: compressed porcellanous test with variable outline, periphery acute and keeled, aperture has a slight lip, is simple,

circular, at the end of a long tapering neck. In the megalospheric generation, the second chamber, a narrow tube, extends perpendicularly from the spherical proloculus, curves over and then embraces it for a third of a whorl. The third chamber widens rapidly initially, gradually narrowing distally. This chamber completes one turn about the proloculus and second chamber, and then continues tangentially for about 40–50 μm. The fourth chamber of half a whorl arises abruptly from the third, rapidly reaches its maximum breadth and then tapers off. The remainder of the chambers are similar to the third. Rare microspheric forms are distinguished by the larger number of chambers (8–9) and the small proloculus (11–15 μm). Average length (excluding neck) 0.4 mm; average width 0.2 mm. Remarks: the only other species of *Ophthalmidium* with a keel similar to *O. compressum* in the Upper Jurassic is *O. carinatum marginatum* Wisniowski which was described by Bielecka (1960). According to Bielecka the chamber arrangement can only be seen in thin section suggesting that the chambers are more inflated than in *O. compressum*. The presence of a keel and the flattened nature of the test allows the ready separation of this species from other ophthalmids in the Upper Jurassic. The long neck is fairly fragile and is frequently absent. Distribution: only recorded from Britain *mariae–cordatum*.

Ophthalmidium strumosum (Guembel)
Plate 6.4.1, Figs 13, 14 (× 84), Nothe Clay, *densiplicatum* Zone, Bowleaze Cove, Dorset. = *Guttulina strumosum* Guembel, 1862. Description: chambers half-whorl in length tapering slightly from the base towards the oral end, sutures depressed; aperture terminal circular on short neck; length 0.3–0.67 mm
Remarks: the distinctive feature of this species is its inflated later chambers. Distribution: British range *cordatum–tenuiserratum*; Europe Oxfordian of Germany (Lutze, 1960), (Seibold and Seibold, 1955), Oxfordian of France, (Bastien and Sigal, 1962) as *Spirophthamidium strumosum*; North America recorded from Bathonian, Callovian and Oxfordian of the Scotian Shelf (Ascoli, 1976, 1981),

Paalzowella feifeli (Paalzow)
Plate 6.4.1, Figs 19 and 20 (× 160), Ampthill Clay, *tenuiserratum* Zone, Millbrook; Fig. 21 (× 175), Fig. 22 (× 215), vertical sections of high- and low-spired forms, Nothe Grit, *cordatum* Zone, Ham Cliff. *Trocholina feifeli* Paalzow, 1932. Description: test conical, of single tubular chamber, spirally enrolled, almost involute on umbilical side, evolute on spiral side; radial striations present on umbilical surface of some specimens; aperture at open end of tube, simple; diameter 0.2–0.3 mm.
Remarks: the height of the spire is quite variable. *Paalzowella* sp. A from the Middle Jurassic of France (Wernli, 1971) appears to be identical. Distribution: British range *cordatum–glosense*; Europe Oxfordian of Germany (Seibold and Seibold, 1960), as *P. feifeli feifeli*, (Lutze, 1960), early Kimmeridgian of Germany (Munk, 1980), as *P. feifeli feifeli*, Oxfordian of France (Bastien and Sigal, 1962), as *P.* aff. *feifeli*; North America, Oxfordian and Kimmeridgian of the Grand Banks (Gradstein, 1977, 1978), Oxfordian and Kimmeridgian of Scotian Shelf (Ascoli, 1976).

Trocholina nodulosa Seibold and Seibold, 1960
Plate 6.4.1, Figs 17, 18 (× 163), Elsworth Rock Series, *densiplicatum* Zone, Millbrook. Description: test plano-convex, consisting of proloculus and an undivided second chamber of 4–5 whorls, spiral side convex, umbilical surface concave and pustulate, apical angle approximately 120° average diameter 0.3 mm.
Remarks: probably closely related to *T. granulata*. Distribution: British range *densiplicatum–tenuiserratum*; Europe, Oxfordian of Germany (Lutze, 1960) and (Seibold and Seibold, 1960).

SUBORDER LAGENINA

Dentalina pseudocommunis Franke, 1936
Plate 6.4.2, Fig. 2 (× 90), Oxford Clay, *mariae* Zone, Warboys. Description: test uniserial, arcuate consisting of 6–10 chambers becoming progressively larger, proloculum often apiculate; sutures oblique, slightly depressed; aperture terminal radial; length 0.4–1.0 mm occasionally longer.
Remarks: *D. guembeli* Schwager appears to be very similar but the original specimens have more depressed sutures producing a lobate outline. Many later records of *D. guembeli* are thought to belong to *D. pseudocommunis* although with this rather variable foraminifer the latter may in fact be a junior synonym. This is a common form throughout the Upper Jurassic. Identical forms have also been recorded from the Lias (Barnard, 1950a). Distribution: British range *mariae–pallasioides*. Europe, Kimmeridgian of Poland (Bielecka and Pozaryski, 1954), as *D. communis*; early Kimmeridgian of Germany (Munk, 1980), as *D. communis*; Kimmeridgian of France (Barnard and Shipp, 1981).

Citharina serratocostata (Guembel)
Plate 6.4.2, Fig. 1 (× 50), late Kimmeridgian 48 block southern North Sea. = *Marginulina serratocostata* Guembel, 1862. Common synonym: *Citharina flabellata* (Guembel). Description: test compressed, triangular in outline, 7–16 chambers, sutures usually flush except at inner margin where they may be depressed; aperture marginal, terminal, radiate or simple; ornament of coarse, sometimes branching ribs, length 0.5–1.5 mm, width, 0.2–0.45 mm.
Remarks: *C. serratocostata* is thought to be synonymous with *C. flabellata* (Gordon, 1962 and Wernli, 1971). This is a highly variable species and continuous variation to forms identical to *Falsopalmula anceps* (Terquem) and approaching *Frondicularia nikitini* Uhlig in appearance has been described (Shipp, 1978). Distribution: British range *mariae–okusensis*. Europe, Oxfordian of northern France (Guyader, 1968), as *Vaginulina flabellata*.

Frondicularia franconica Guembel, 1862
Plate 6.4.2, Figs 3, 4 (× 47), Oxford Clay, *mariae* Zone, Millbrook. Common synonym: *Frondicularia irregularis*, Terquem. Description: test compressed, uniserial, composed of 4–12 chambers, each a broad chevron shape, sutures generally flush; aperture central, terminal, radiate, length 0.25–1.0 mm. Remarks: this form is very variable. Chambers are added somewhat irregularly so the periphery is sometimes stepped. Especially common in Oxford Clay. Distribution: British range *mariae–cymdoce*; Europe, Bathonian—southern Sweden (Norling, 1972), Callovian—Oxfordian of Germany (Lutze, 1960), early Kimmeridgian of northern France (Barnard and Shipp, 1981).

Frondicularia nikitini Uhlig, 1883
Plate 6.4.2, Fig. 5 (× 65), late Kimmeridgian 48 block southern North Sea. Description: test usually large, compressed, of five to ten chevron-shaped chambers, periphery smooth to lobate; sutures distinct, depressed; aperture circular terminal usually radiate; ornament varied, typically consists of approximately 15–20 discontinuous arcuate costae interrupted by the sutures, length 0.8–1.8 mm.
Remarks: variation in the ornament has been described by Cordey (1962). Chevron-shaped chambers sometimes do not appear until the fourth or fifth chamber after the proloculus giving forms identical to *Citharinella exarata* Loeblich and Tappan. See also remarks of *Citharina serratocostata*. Distribution: British range *mariae–fittoni*; Europe, Callovian and Oxfordian of Germany (Lutze, 1960), early Kimmeridgian of Germany (Munk, 1980).

Frondicularia pseudosulcata Barnard, 1952, emend. Barnard, 1957
Plate 6.4.2, Fig. 6 (× 41), Oxford Clay, *mariae* Zone, Warboys. Description: test compressed, rhomboid to elongate-rhomboid of 6–10 chevron-shaped chambers, sutures flush to slightly compressed, proloculus often apiculate; aperture central terminal, radiate, ornament of approximately ten costae more continuous in the central region of the test; length 0.9–1.2 mm.
Remarks: the costae are more continuous than in *Frondicularia nikitini* especially on the early part of the test where the apertures are less distinct. The Lias species *Frondicularia sulcata* (see Barnard, 1957) differs from *F. pseudosulcata* in having more strongly developed costae. Distribution: British range *mariae–cordatum*.

Lenticulina ectypa (Loeblich and Tappan)
Plate 6.4.2, Fig. 7 (× 130), Oxford Clay, *tenuiserratum* Zone, Millbrook. = *Astacolus ectypus* Loeblich and Tappan, 1950. Description: test planispiral, coiled to partly uncoiled, keeled 8–12 chambers in the final whorl. Sutures deeply depressed on proximal side of each suture line, often ribbed on distal side; aperture simple, circular, peripheral; length 0.25–0.7 mm.
Remarks: can be distinguished from *Lenticulina quenstedti* by its longitudinal section (Lutze, 1960 and Cordey, 1962) but also by the deep recess on the proximal side of each rib. Rare in Corallian (Shipp, 1978). Distribution: British range *mariae–baylei*; Europe, Callovian of Germany (Lutze, 1960). Originally described from the Callovian of the U.S.A. (Loeblich and Tappan, 1950); North America Oxfordian of Sverdrup Basin (Wall, 1983).

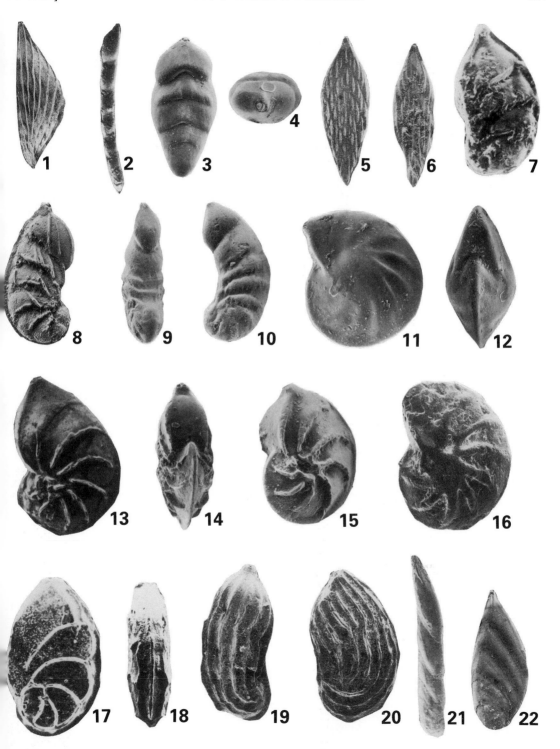

Plate 6.4.2

Lenticulina ectypa (Loeblich and Tappan) *costata* Cordey, 1962
Plate 6.4.2, Fig. 8 (× 95), Ampthill Clay, *tenuiserratum* Zone, Warboys. Distribution: as *L. ectypa* but ribs more prominent and an additional ornament of 2–10 costae arising from each of these ribs; keel more pronounced; length 0.25–0.9 mm. Distribution: only recorded from Britain range *mariae–tenuiserratum*.

Lenticulina major (Bornemann)
Plate 6.4.2, Figs 9, 10 (× 40), Oxford Clay, *mariae* Zone, Weymouth. = *Cristellaria major* Bornemann, 1854. Common synonym: *Lenticulina suprajurassica* Gordon. Description: test loosely coiled, planispiral, compressed with 4–8 chambers in the uniserial portion, sutures flush or marked by low ribs, aperture terminal, marginal, radiate; length 0.5–1.8 mm.
Remarks: the variation of this species includes forms described by Gordon (1962, 1965) under *Lenticulina suprajurassica*. This species is distinguished by its short evolute coil. Some specimens lack the depressions in the test noted by Gordon (1962) and have a smooth outline approaching *Vaginulina barnardi* in general appearance but lacking the faint striations discernible on this latter species. Distribution: British range *mariae–eudoxus* with an additional record offshore from the *preplicomphalus* Zone; Europe, Callovian and Oxfordian of Germany (Lutze, 1960). Originally described from the Lias.

Lenticulina muensteri (Roemer)
Plate 6.4.2, Figs 11, 12 (× 53), Oxford Clay, *mariae* Zone, Millbrook. = *Robulina muensteri* Roemer, 1839. Description: test biconvex, planispiral involute, rarely becoming evolute with uniserial portion, periphery subangular to carinate; sutures limbate, usually flush or slightly raised, sigmoid; umbilici with clear bosses showing earlier chambers underneath, flush or rarely raised; aperture marginal, radiate; diameter 0.3–1.0 mm, uncoiled forms length 0.8–1.1 mm.
Remarks: this is possibly the commonest species in the Late Jurassic and has been recorded from virtually every horizon and locality where nodosariids have been recorded (under various names). It is closely related to *Lenticulina subalata* (Reuss). Distribution: British range *mariae–lamplughi*, this is one of the few species to have been recorded from the Portland Sand (Shipp, 1978); Europe, Oxfordian of Germany (Lutze, 1960), (Seibold and Seibold, 1960), early Kimmeridgian of Germany (Munk, 1980), Kimmeridgian of France (Barnard and Shipp, 1981), Callovian and Oxfordian of southern Sweden (Norling, 1972).

Lenticulina quenstedti (Guembel)
Plate 6.4.2, Fig. 13 (× 70), Ampthill Clay, *tenuiserratum* Zone, Warboys; Figs 14, 15 (× 50), Oxford Clay, *mariae* Zone, Warboys. = *Cristellaria quenstedti* Guembel, 1862. Description: test planispiral, biconvex, involute with 8–10 chambers in the final whorl, sometimes a later uncoiled portion of 2–3 chambers; periphery keeled, less marked on uncoiled portion; strong ribs on sutures meet an umbilicus which is bounded by a rib connecting the others at right angles; diameter 0.25–0.85 mm, maximum length uncoiled 1.0 mm.
Remarks: the distinctive feature of this species is its ornament. It is closely related to *Lenticulina brueckmanni* (Mjatliuk). Distribution: British range *mariae–tenuiserratum*; Europe, Bathonian–Oxfordian of southern Sweden (Norling, 1972), Callovian and Oxfordian of Germany (Lutze, 1960), as cf., early Kimmeridgian of Germany (Munk, 1980); Oxfordian of France, (Bastien and Sigal, 1962), as *Cristellaria* plexus – *quenstedti*; North America, Oxfordian–Kimmeridgian of Grand Banks (Gradstein, 1977, 1978), Callovian, Oxfordian and Kimmeridgian of Scotian Shelf (Ascoli, 1976, 1981).

Lenticulina subalata (Reuss)
Plate 6.4.2, Fig. 16 (× 75), late Kimmeridgian 48 block southern North Sea. = *Cristellaria subalata* Reuss, 1854. Description: test planispiral, involute, 8–10 chambers in last whorl, occasionally uncoiled, margin sharp with broad keel, aperture terminal, marginal, radiate; raised boss and thick radial ribs.
Remarks: similar to *Lenticulina muensteri* (Roemer) to which it appears to be related. It can be distinguished, however, by the raised opaque boss and sutural ribs. Similar forms exist from the Lias–Early Cretaceous. Distribution: British range *mariae–okusensis*, Europe, Callovian and Oxfordian of Germany (Lutze, 1960), Kimmeridgian of Poland (Bielecka and Pozaryski, 1954), Kimmeridgian of northern France (Barnard and Shipp, 1981).

Lenticulina tricarinella (Reuss)
Plate 6.4.2, Fig. 17 (× 100), *densiplicatum* Zone, Millbrook, Fig. 18 (× 95), *mariae* Zone, Woodham, (peripheral view). = *Cristellaria tricarinella* Reuss, 1863. Description: test planispiral, evolute, flaring, laterally compressed with flattened sides; final whorl of 5–6 chambers; periphery with angular keel and angled shoulders on each side; aperture terminal, marginal, radiate; length 0.3–0.6 mm, rarely up to 0.72 mm.
Remarks: a distinctive but long ranging species being found in the Middle Jurassic (Cifelli, 1959); it was originally described from the Lower Cretaceous. Distribution: British range *mariae–preplacomphalus*; Europe, Callovian and Oxfordian of Germany (Lutze, 1960) and (Seibold and Seibold, 1960), early Kimmeridgian of Germany (Munk, 1980), Oxfordian of France (Bastien and Sigal, 1962) as *Cristellaria* plexus—*flexuosa*, Kimmeridgian of Poland (Bielecka and Pozaryski, 1954), Callovian and Oxfordian of southern Sweden (Norling, 1972); N. America, Oxfordian and Kimmeridgian of Grand Banks (Gradstein, 1977, 1978), Oxfordian and Kimmeridgian of Scotian Shelf (Ascoli, 1976, 1981).

Marginulina costata (Batsch, 1791)
Plate 6.4.2, Fig. 19 (× 95), late Kimmeridgian 48 block southern North Sea. = *Nautilus (orthoceras) costatus* Batch. Description: initial planispire of 3–4 chambers is followed by a curvilinear series of 3–5 chambers, sutures distinctly compressed resulting in a lobulate outline, sutures oblique on linear portion, test circular in cross-section, ornament of numerous ribs running parallel to the growth axis, about 15–20 ribs, number remaining fairly constant throughout the length of the test, aperture terminal peripheral, mounted on small boss. Length 0.45–0.55 mm. Width 0.15–0.2 mm.
Remarks: *Marginulina costata* (Batsch) differs from *Marginulina undulata* (Terquem) in having a rounded cross-section, being less wide and more lobulate. *Marginulina undulata* (Terquem) has an oval cross-section, entire periphery and a generally wider test in lateral view. Distribution: British range *rosenkrantzi–preplicomphalus*; Europe, Kimmeridgian of France (Barnard and Shipp, 1981), as *M. radiata*.

Marginulina undulata Terquem, 1858
Plate 6.4.2, Fig. 20 (× 120), late Kimmeridgian 48 block southern North Sea. Description: initial planispiral coil of about five chambers followed by a rectilinear or slightly curvilinear series of about three chambers, the latter number may be reduced; on the coil the sutures are flush but may become depressed distally; umbilical margin may be slightly lobulate, periphery usually not so; sutures are oblique on uncoiled portion; ornament of approximately 20 ribs which are parallel to the direction of growth; a somewhat larger rib may run around the periphery to form a keel; more ribs are present in the distal portion, the extra ones sometimes being added by branching; aperture terminal, peripheral, mounted on a small boss. Distribution: British range *eudoxus–fittoni*; Europe, early Kimmeridgian of France (Shipp, 1971).

Planularia angustissima (Wisniowski)
Plate 6.4.2, Fig. 21 (× 90), Oxford Clay, *cordatum* Zone, Warboys. = *Cristellaria angustissima* Wisniowski, 1891. Description: narrow smooth test, compressed, of approximately 12 chambers, 3–6 of which are in contact with the proloculus which is completely enclosed; sutures distinct, depressed; aperture simple, terminal borne on short neck; some specimens bear rare striae near aperture; length 0.38–0.7 mm.
Remarks: a distinctive species recognised by its narrow test. Distribution: British range *mariae–mutabilis*; Europe, Oxfordian of France (Bastien and Sigal, 1962), and (Guyader, 1968) as *Marginulina angustissima*.

Planularia beierana (Guembel)
Plate 6.4.2, Fig. 22 (× 54), Oxford Clay, *mariae* Zone, Millbrook. = *Marginulina beierana* Guembel, 1862. Description: test smooth, compressed, of 4–10 chambers, 3–6 of which are in contact with proloculus; in microspheric generation proloculus completely enclosed, sutures distinct, usually depressed, poorly developed keel present on early chambers of some specimens; aperture simple, circular, terminal on short neck; some specimens bear ornament of 1 or 2 fine striae on apertural margin of last few chambers; length 0.2–0.75 mm.
Remarks: a highly variable species, the shape depends on the varied growth rate. The variation was studied by Cordey, 1962. A common Late Jurassic species most prominent in the Oxford Clay. Distribution: British range *mariae–rotunda*; Europe, Kimmeridgian of Poland (Bielecka, 1960), Oxfordian of France (Bastien and Sigal, 1962), as *Cristellaria* plexus–*treptensis*, Oxfordian and Kimmeridgian of France (Guyader, 1968), Kimmeridgian of France (Barnard and Shipp, 1981), early Kimmeridgian of Germany (Munk, 1980); N. America, Oxfordian and early Kimmeridgian of Scotian Shelf (Ascoli, 1981).

Pseudonodosaria radiata (Barnard)
Plate 6.4.3, Figs 1, 2 (× 80). Oxford Clay, *mariae* Zone, Millbrook. = *Pseudoglandulina radiata* Barnard, 1952. Description: test a uniserial series of 5–10 chambers; sutures initially flush but may be constricted between later chambers; central radiate aperture; distinctive ornament of 20–24 longitudinal costae which are slightly twisted with respect to the growth axis; average length of 0.5 mm.
Remarks: a rare but stratigraphically restricted species, its distinctive feature is its spirally arranged costae. Distribution: only recorded from Britain, range *mariae–cordatum*.

Pseudonodosaria vulgata (Bornemann)
Plate 6.4.3, Fig. 3 (× 50), Ampthill Clay, *tenuiserratum* Zone, Millbrook. = *Glandulina vulgata* Bornemann, 1854. Common synonyms: *P. humilis* (Roemer), *P. oviformis* (Terquem), *P. tenuis* (Bornemann). Description: test uniserial of 4–10 chambers, chambers strongly overlapping, especially earlier ones; outline smooth initially becoming obate as later chambers may be inflated; aperture central, terminal, radiate, length 0.25–0.65 mm occasionally larger.
Remarks: this is a highly variable species common throughout the Jurassic and Cretaceous. It includes many forms previously described as separate species (Lutze, 1960) and (Norling, 1972). Distribution: it has also been widely recorded on the continent. British range *mariae–preplicomphalus*; Europe, Oxfordian of Germany (Seibold and Seibold, 1956), as *Pseudoglandulina vulgata*, Oxfordian and Kimmeridgian France (Guyader, 1968), Kimmeridgian France (Barnard and Shipp, 1981), Lias–Cretaceous (Norling, 1972).

Saracenaria oxfordiana Tappan, 1955
Plate 6.4.3, Fig. 4 (× 38), Oxford Clay, *mariae* Zone, Millbrook. Common synonym: *Saracenaria triquetra* (Guembel). Description: test planispiral with small initial coil followed by uncoiled portion of 2–6 chambers, keeled; sutures marked by low ribs; aperture radiate, terminal, sometimes on short neck; apertural face flattened; length 0.2–0.45 mm.
Remarks: can be distinguished from other *Saracenaria* by presence of ribs. Guembel first used the specific name *triquetra* for an Eocene form (Tappan, 1955). Distribution: British range *mariae–lamplughi*; Europe, early Oxfordian of S. Sweden (Norling, 1972), early Kimmeridgian of Germany (Munk, 1980), Oxfordian–?Kimmeridgian of Svalbard (Løfaldi and Nagy, 1980); North America Oxfordian and early Kimmeridgian of northern Alaska (Tappan, 1955).

Tristix triangularis Barnard, 1953
Plate 6.4.3, Figs 5, 6 (× 44), Oxford Clay, *mariae* Zone, Millbrook. Common synonyms: *T. acutangulus* (Reuss), *T. oolithica* (Terquem). Description: test a rectilinear series of 4–8 chambers, triangular in cross-section; sutures slightly depressed, arched towards the aperture; edges of the test often keeled; aperture central, terminal, simple or radiate; length 0.3–0.9 mm.
Remarks: according to Ellis and Messina the types of *T. acutangulus* (Reuss) and *T. oolithica* Terquem should be referred to genera other than *Tristix*. Gordon (1965) has described the variation of this species and included forms with rounded and carinate margins. Rare specimens are quadrate. Distribution: British range *mariae–fittoni*. Europe, Callovian and Oxfordian of Germany (Lutze, 1960), as *T. acutangulus*, Oxfordian and Kimmeridgian of France (Guyader, 1968), as *T. suprajurassica*, Kimmeridgian of France (Barnard and Shipp, 1981), Oxfordian of S. Sweden (Norling, 1972), as *T. oolithica*.

Vaginulina barnardi Gordon, 1965
Plate 6.4.3, Fig. 7 (× 30), Ampthill Clay, *tenuiserratum* Zone, Warboys. Description: test uniserial, straight or curved, slightly compressed; 9–12 chambers, rarely up to 16; sutures oblique flush; aperture terminal, marginal radiate; ornament of fine, longitudinal striations, length 0.3–1.0 mm occasionally longer.
Remarks: striations are always present but may be very faint. *Vaginulina pasqueti* Bizon is possibly synonymous. P. Copestake (pers. comm.) has recorded *V. pasqueti* from the Portlandian of the Vale of Wardour. Distribution: only recorded from Britain, range *cordatum–regulare*.

Vaginulina 'prima'
Plate 6.4.3, Fig. 8 (× 55), late Kimmeridgian 48 block southern North Sea. Lloyd in the unpublished section of his Ph.D. (1958) records several variants under the name *V. prima*. To avoid confusion with the very similar Liassic forms the species name has been used here in inverted commas. Description: uniserial test of 8–10 chambers, sutures distinct depressed, ornament of 6–12 strong ribs, aperture simple, terminal mounted on a short neck. Length 0.8–1.4 mm.
Remarks: this is a prominent species of *Vaginulina* in the Kimmeridgian of Dorset. Distribution: only recorded from Britain as such, *mutabilis–fittoni*.

SUBORDER ROBERTININA

Epistomina mosquensis Uhlig, 1883
Plate 6.4.3, Fig. 12 (× 60), late Kimmeridgian 48 block southern North Sea. Common synonym: *Brotzenia mosquensis* (Uhlig). Description: test biconvex, with umbilical side more convex than dorsal, trochospiral; 6–8 chambers in outer whorl; intercameral and spiral sutures thickened and raised, those on spiral side curved, those on umbilical side radial about an umbilical ring; a weakly developed peripheral double keel, between which are the occluded peripheral apertures; diameter 0.35–0.82 mm.
Remarks: close to *E. ornata* but the latter is more flattened with strongly developed peripheral keels. Distribution British range *mariae–rosenkrantzi*; Europe, Callovian and Oxfordian of Germany (Lutze, 1960), Oxfordian of Northern France (Guyader, 1968), Callovian and Oxfordian of southern Sweden (Norling, 1972); North America Oxfordian and Kimmeridgian of Grand Banks (Gradstein, 1976, 1978), Oxfordian and Kimmeridgian of Scotian Shelf (Ascoli, 1976, 1981).

Epistomina ornata (Roemer)
Plate 6.4.3, Figs 9, 10, 11 (× 95). Sandsfoot Clay, *serratum/regulare* Zone, Shortlake, Dorset. = *Planulina ornata* Roemer, 1841. Description: test flattened, biconvex with two peripheral keels, trochospiral; 6–7 chambers in outer whorl, intercameral and spiral sutures thickened, raised and tuberculate; on umbilical side umbo is filled with a lattice of raise tuberculate ribs; on both surfaces the exposed parts of the chamber walls are finely punctate and covered with low hemispherical tubercules; peripheral apertures between keels closed by a porous plate except in last chamber; septal foramen round and areal; diameter 0.3–0.52 mm.
Remarks: distinguished by its distinctive ornament. Distribution: British range *glosense–mutabilis*; Europe, late Kimmeridgian of Poland (Bielecka and Pozaryski, 1954), as *E. stellicostata* var. *granulosa*.

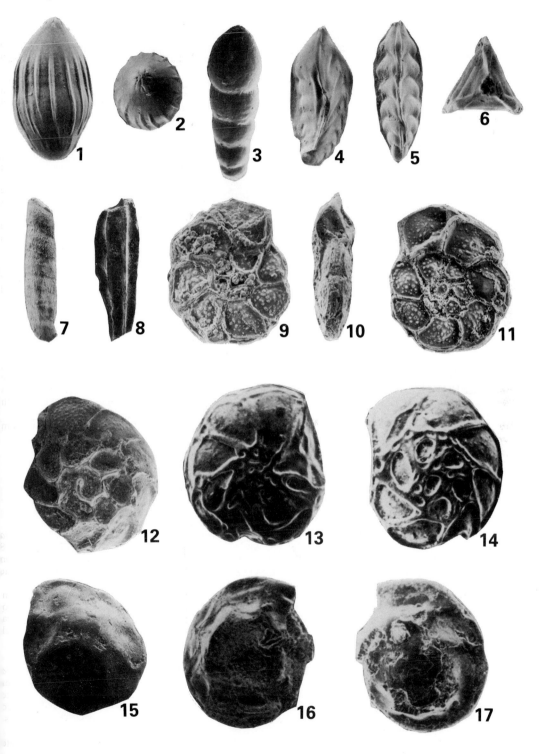

Plate 6.4.3

Epistomina parastelligera (Hofker)
Plate 6.4.3, Fig. 15 (× 120), early Kimmeridgian 48 block southern North Sea. = *Brotzenia parastelligera* Hofker, 1954. Description: test biconvex, trochospiral with 6–8 chambers in final whorl, generally smooth with intercameral sutures on spiral side, slightly raised with occluded peripheral secondary apertures on umbilical surface; diameter 0.25–0.5 mm.
Remarks: a common species distinguished by its general lack of ornament. *B. porcellanea* as used by Lloyd (1962) is considered to represent *E. parastelligera* of the present author. Lloyd's specimens of *B. parastelligera* are strongly ribbed. Distribution: British range *mariae–eudoxus*; Europe, Callovian and Oxfordian of Germany (Seibold and Seibold, 1960) and (Lutze, 1960), Callovian and Oxfordian of Poland (Bielecka, 1960), Callovian and Oxfordian of southern Sweden (Norling, 1972).

Epistomina reticulata (Reuss)
Plate 6.4.3, Figs 13, 14 (× 130), Sandsfoot Clay, *serratum/regulare* Zone, Shortlake. = *Rotalia reticulata* Reuss, 1863. Description: test biconvex, trochospiral with approximately eight chambers in final whorl, intercameral and spiral sutures raised to form reticulate pattern over central area, sutures curved backwards on dorsal surface, radial on umbilical surface, occluded secondary apertures lie between peripheral keels on umbilical side; average diameter 0.3 mm.
Remarks: similar to *E. parastelligera* but distinguished by raised sutures. Distribution: British range *tenuiserratum–regulare*.

Epistomina tenuicostata Bartenstein and Brand, 1951
Plate 6.4.3, Figs 16, 17 (× 130), Sandsfoot Clay, *serratum/regulare* Zone, Shortlake, Dorset. Common synonyms: *Brotzenia tenuicostata*, *Voorthuysenia tenuicostata*. Description: test biconvex, trochospiral with 7–8 chambers in final whorl; spiral side less convex than umbilical side; sutures limbate and curved backwards on spiral side, limbate and radial on umbilical side; apertures are present close to the periphery on the umbilical side; diameter 0.2–0.4 mm, rarely larger. Distribution: British range *cordatum–regulare*, originally described from the Lower Cretaceous of northwest Germany.

ACKNOWLEDGEMENTS

The author wishes to thank Dr A. J. Lloyd for permission to refer to the unpublished portions of his PhD thesis, although Dr Lloyd would like to point out that the data are now nearly 30 years old. Likewise the author thanks F. J. Gregory for access to his M.Sc. data and the Robertson Group for permission to use information from their two Jurassic studies.

REFERENCES FOR CHAPTER 6

Adams, C. G. 1955. Foraminifera from the Upper Lias of North Lincolnshire. *Unpublished PhD Thesis, University of Nottingham.*
Adams, C. G. 1957. A study of the morphology and variation of some Upper Lias foraminifera. *Micropaleontology*, **3**, 205–206.
Adams, C. G. 1962. Calcareous adherent foraminifera from the British Jurassic and Cretaceous and the French Eocene. *Palaeontology*, **5**, 149–170.
Ainsworth, N. R. and Horton, N. F. 1986. Mesozoic micropalaeontology of exploration well Elf 55/30-1 from the Fastnet Basin, offshore southwest Ireland. *J. micropalaeontol.*, **5**, 19–29.
Ainsworth, N. R., O'Neill, M., Rutherford, M. M. Clayton, G., Horton, N. F. and Penney, R. A. 1987. Biostratigraphy of the Lower Cretaceous, Jurassic and uppermost Triassic of the North Celtic Sea and Fastnet Basins. In Brooks, J. and Glennie, K. (eds) *Petroleum Geology of North West Europe*, 611–622, Graham & Trotman.
Ainsworth, N. R., O'Neill, M., and Rutherford, M. M. (in press). Jurassic and latest Triassic biostratigraphy of the North Celtic Sea and Fastnet Basins. In Batten, D. J. and Keen, M. C. (eds) *Micropaleontology, palynology and petroleum exploration, on and offshore Europe*.
Anderton, R., Bridges, P. H., Leeder, M. R. and Sellwood, B. W. 1979. *A dynamic stratigraphy of the British isles. A study in crustal evolution*. George Allen & Unwin, London, 1–301.
Andrews, I. J. and Brown, S. 1987. Stratigraphic evolution of the Jurassic, Moray Firth. In Brooks, J. and Glennie, K. (eds) *Petroleum geology of North West Europe*, 785–795, Graham & Trotman.
Apthorpe, M. C. and Heath, R. S. 1981. Late Triassic and Early to Middle Jurassic foraminifera from the North West Shelf, Australia. In *Fifth Australian Geological Convention: Geological Society of Australia Abstracts*, No. 3, 66.
Arbeitskreis Deutscher Mikropaläontologen, 1962. *Leitfossilien der Mikropaläontolgie*. Gebruder Borntraeger, Berlin-Nikolassee.
Arkell, W. J. 1933. *The Jurassic System in Great Britain*. Clarendon Press, Oxford, 681 pp.
Arkell, W. J. 1935. The Corallian Beds of Dorset. *Proc*

References

Dorset Nat. Hist. Archaeol. Soc., **57**, 59–93.
Arkell, W. J. 1947. The geology of the country around Weymouth, Swanage, Corfe and Lulworth. *Mem. geol. Surv. Gt. Br.*, 1–386.
Arkell, W. J. 1956. *Jurassic geology of the World*. Oliver & Boyd, Edinburgh and London.
Ascoli, P. 1976. Foraminiferal and ostracod biostratigraphy of the Mesozoic–Cenozoic, Scotian Shelf, Atlantic Canada. *First International Symposium on Benthonic Foraminifera of Continental Margins, Part B: Paleoecology and Biostratigraphy. Maritime Sediments Special Publication* 1, Pt. B, 653–671.
Ascoli, P. 1981. Foraminiferal–ostracod Late Jurassic biozonation of the Scotian Shelf. *Geological Survey of Canada, Ottawa*, Open File Report 753, 1–32.
Ascoli, P. 1984. Epistominid biostratigraphy across the Jurassic–Cretaceous boundary on the northwestern Atlantic Shelf. *Benthos '83, 2nd International Symposium Benthic Foraminifera*, 27–34.
Ascoli, P., Poag, C. W. and Remane, J. 1984. Microfossil zonation across the Jurassic–Cretaceous boundary on the Atlantic margin of North America. *Geological Association of Canada, Special Paper* 27, 31–48.
Bach, H., Hagenmeyer, P. and Neuweiler, F. 1959. Neubeschreibung und Revision einiger Foraminiferenarten und - unterarten aus dem Schwabischen Lias. *Geol. Jb*, **76**, 427–452.
Badley, M. E., Price, J. D., Rambelh Dahl, C. and Agdestein, T. 1988. The structural evolution of the northern Viking Graben and its bearing upon extensional modes of basin formation. *J. geol. Soc. London*, **145**, 455–472.
Bairstow, L. 1969. In Torrens, H. S. (ed) *International Field Symposium on the British Jurassic Excursion No 3: Guide for north-east Yorkshire*, 1–47, University of Keele.
Ballent, S. C. 1987a Bioestratigrafia y edad de las microfaunas del Jurasico inferior de Argentina. *IV Congreso Latinoamericano de paleontologia, Bolivia*, **1**, 331–342.
Ballent, S. C. 1987b. Foraminiferos y ostracodos del Jurasico inferior de Argentina. *Revista del Museo de la Plata (Nueva Serie)*, **9**, 43–130.
Bang,I. 1968. Biostratigraphical investigation of the pre-Quarternary in the Oresund Boreholes mainly on the basis of foraminifera. In Larsen, G. *et al* (1968) pp. 86–88.
Bang, I. 1971. Jura aflejringerne i Ronde Nr. 1 (2103-2164M). Biostratigrafi pa grundlag af foraminiferen. *Dan. Geol. Unders*. 3 Rapp, No.39, 74–80.
Bang, I. 1972. Jura-biostratigrafi i Novling Nr. 1 pa grundlag af foraminiferer. *Dan. Geol. Unders*. 3 Rapp, No. 40, 119–123.
Banner, F. T., Brooks, M. and Williams, E. 1971. The geology of the approaches to Barry, Glamorgan. *Proc. Geol. Assoc.*, **82**, 231–247.
Barbieri, F. 1964. Micropaleontologia del Lias e Dogger del pozzo Ragusa (Sicilia). *Rev. Ital. Pal.*, **70**, 709–830.
Barnard, T. 1948. The uses of foraminifera in Lower Jurassic stratigraphy. *Int. Geol. Congress 'Report of the 18th Session, Great Britain 1948'*, **15**, 34–41, London.

Barnard, T. 1950a. Foraminifera from the Lower Lias of the Dorset Coast. *Quart. J. geol. Soc. London*, **105**, 347–391.
Barnard, T. 1950b. Foraminifera from the Upper Lias of Byfield, Northamptonshire. *Quart. J. geol. Soc. London*, **106**, 1–36.
Barnard, T. 1952. Foraminifera from the Upper Oxford Clay (Jurassic) of Warboys, Huntingdonshire. *Proc. Geol. Ass.*, **63**, 336–350.
Barnard, T. 1953. Foraminifera from the Upper Oxford Clay (Jurassic) of Redcliff Point, near Weymouth, England. *Proc. Geol. Ass.*, **64**, 183–197.
Barnard, T. 1956. Some Lingulinae from the Lias of England. *Micropalaeontology*, **2**, 271–282.
Barnard, T. 1957. *Frondicularia* from the Lower Lias of England. *Micropalaeontology*, **3**, 171–181.
Barnard, T. 1960. Some species of *Lenticulina* and associated genera from the Lias of England. *Micropaleontology*, **6**, 41–55.
Barnard, T. and Shipp, D. J. 1981. Kimmeridgian foraminifera from the Boulonnais. *Rev. Micropal.*, **24**, 3–26.
Barnard, T., Cordey, W. G. and Shipp, D. J. 1981. Foraminifera from the Oxford Clay (Callovian–Oxfordian of England). *Rev. Esp. Micropaleontol.*, **13**, 383–462.
Bartenstein, H. and Brand, E. 1937. Mikropaläontologische Untersuchungen zur Stratigraphie des nordwestdeutschen Lias und Doggers. *Senckenb. Naturf. Ges., Abh.*, **439**,1–224.
Bastien, M. Th. and Sigal, J. 1962. Contribution à l'étude paléontologie de l'Oxfordien Supérieur de Trept (Isère). Part II. Foraminifères. *Trav. Lab Geol. Lyon N.S.*, No. 8., 83–123.
Bate, R. H. 1967. The Bathonian Upper Estuarine Series of Eastern England Pt. 1, Ostracoda. *Bull. British Musuem (Nat. Hist.) Geol., London* **14**, 21–66.
Bate, R. H. and Robinson, E. 1978. *A stratigraphic index of British Ostracoda*. Seel House Press, Liverpool, 1–538.
Bennet, G., Copestake, P. and Hooker, N. P. 1985. Stratigraphy of the Britoil 72/10-1A well, Western Approaches. *Proc. Geol. Ass.*, **96**, 255–261.
Berthelin, G. 1879. Foraminifères du Lias moyen de la Vendée. *Rev. Mag. Zool.*, **3**, 24–41.
Bielecka, W. 1960. Micropalaeontological stratigraphy of the Lower Malm in the vicinity of Chrzanowa (southern Poland). *Inst. Geol. Prace*, **33**, 1–155.
Bielecka, W. and Pozaryski, W. 1954. Micropalaeontological Stratigraphy of the Upper Malm in Central Poland. *Inst. Geol. Prace.* **12**, 143–206.
Bignot, G. and Guyader, J. 1966. Découverte de Foraminifères planctoniques dans l'Oxfordien du Havre (Seine–Maritime). *Rev. Micropal.*, **9**, 104–110.
Bizon, J. J. 1958. Foraminifères et Ostracodes de l'Oxfordien de Villers sur mer (Calvados). *Rev. Inst. Fr. Pet.*, **13**, 1–145.
Bizon, G. 1960. Revision de quelques espèces-types de foraminifères du Lias du Bassin Parisien de la collection Terquem. *Rev. Micropal.*, **3**, 3–18.
Bizon, G. 1961. Contributions à l'étude micropaléontologique du Lias du Bassin de Paris. Deuxième Partie.

Lorraine, Region de Nancy et Thionville. In *Colloque sur le Lias Français*, 433-436.
Bizon, G. and Oertli, H. 1961. Contributions à l'étude micropaléontologique du Lias du Bassin de Paris. Septième Partie – Conclusions. In *Colloque sur le Lias Français*, (1961) pp. 107-119.
Blake, J. F. 1875. On the Kimmeridge Clay of England. *Quart. J. geol. Soc. Lond.*, **31**, 196-237.
Blake, J. F. 1876. Class Rhizopoda. In Tate, R. and Blake, J. F. (1876) pp. 449-473.
Bornemann, J. G. 1854. *Uber die Liasformation in der Umgegend von Gottingen und ihre organichen Einschlusse*. A. W. Schade, Berlin.
Bowen, J. M. 1975. Brent Oil-field. In Woodland, A. W. (ed), *Petroleum and the continental shelf of North West Europe*., London: Applied Science Publishers, 353-362.
Brady, H. B. 1864. On *Involutina liassica* (*Nummulites liassicus*, Rupert Jones). *Geol. Mag.*, **1**, 193-196.
Brady, H. 1866. Foraminifera. In Moore, C. On the Middle and Upper Lias of the South West of England. *Proc. Somerset Arch. and Nat. Soc.*, **13**, 119-230.
Brady, H. B. 1867. Synopsis of the middle and upper Lias of Somerset. *Proc. Somerset Arch. and Nat. Soc.*, **13**.
Brand, E. and Fahrion, H, 1962. Dogger, N. W. Deutschlands. In *Leitfossilien der Mikropälaontologie*, pp. 123-158.
Brinkmann, R. 1929. Statistisch-biostratigraphische Untersuchungen an mittel-jurassichen Ammoniten uber Artbegriff und Stammesentwicklung. *Abhandl. Ges. Wiss. Göttingen, Math-phys., Kl., N.F.*, **13**, 1-249.
Brodie, P. B. 1853. Remarks on the Lias at Fretherne near Newham and Purton near Sharpness; with an account of some new Foraminifera discovered there. *Proc. Cotteswold Nat. Fld. Cl.*, **1**, 241-246.
Brooke, M. M., and Braun, W. K. 1972. Briostratigraphy and microfanuas of the Jurassic system of Saskatchewan. *Sask. Dep. Min. Res., Rep. 161*.
Brooke, M. M. and Braun, W. K. 1981. Jurassic microfaunas and biostratigraphy of northeastern British Columbia and adjacent Alberta. *Geol. Surv. of Canada Bull.* **283**.
Brookfield, M. 1973. Palaeogeography of the Upper Oxfordian and Lower Kimmeridgian (Jurassic) in Britain. *Palaeogeogr. Palaeoclimatol., Palaeoecol.*, **14**, 137-167.
Brouwer, J. 1969. Foraminiferal assemblages from the Lias of N.W. Europe. *Ver. K. Ned. Akad, Wet. Afd. Natuurk; 1 Reeks, Deel.*, **25**, 1-48.
Buckman, S. S. 1895. The Bajocian of the mid-Cotswolds. *Q. J. geol. Soc. London*, **51**, 388-462.
Buckman, S. S. 1897. Deposits of Bajocian age in the northern Cotswolds; the Cleeve Hill Plateau. *Q. J. geol. Soc. London*, **53**, 607-629.
Buckman, S. S. 1901. The Bajocian and contiguous deposits in the north Cotswolds; the main hill-mass. *Q. J. geol. Soc. London*, **57**, 126-155.
Burbach, O. 1866. Bietrage zur Kenntnis der Foraminiferen des mittleren Lias von grossen Seeberg bei Gotha. II – Die Milioliden. *Z. ges. Naturwiss, 59 ser. 4*, **5**, 493-502.

Callomon, J. H. 1955. The Ammonite Succession in the Lower Oxford Clay and Kellaways Beds at Kidlington, and the Zones of the Callovian Stage. *Phil. Trans. R. Soc., B.* **239**, 215-264.
Callomon, J. H. 1964. Notes on the Callovian and Oxfordian Stages. *Compte rendu et Mem. Coll. Jur. Luxembourg, 1962. Inst. Gr.-duc Sect. sci. nat. phys. math.*, 269-291.
Callomon, J. H. 1968a. In Hallam, A. The Lias. In Sylvester-Bradley, P. C. and Ford, T. (eds). *Geology of the East Midlands*, Leicester University Press, pp. 188-208.
Callomon, J. H. 1968b. The Kellaways Beds and Oxford Clay. In Sylvester-Bradley, P. C. and Ford, T. (eds) *Geology of the East Midlands*, Leicester University Press, pp. 264-280.
Callomon, J. H. 1968c. The Corallian Beds, the Ampthill Clay and the Kimmeridge Clay. In Sylvester-Bradley, P. C. and Ford, T. (eds) *Geology of the East Midlands*, Leicester University Press, pp. 291-299.
Callomon, J. H. and Cope J. C. W. 1971. The stratigraphy and ammonite succession of the Oxford and Kimmeridge Clays in the Warlingham Borehole. *Bull. geol. Surv. Gt. Br.*, **36**, 147-176.
Cameron, B. E. B. and Tipper, H. W. 1985. Jurassic stratigraphy of the Queen Charlotte Islands, British Columbia. *Geol. Survey of Canada Bulletin*, **365**, 1-49.
Champeau, H. 1961. Contributions a l'étude micropaléontologique du Lias du Bassin de Paris. Troisième Partie. Étude de la microfauna des niveaux marneux du Lias dans le sud-est du Bassin de Paris. In *Colloque sur le Lias Français*, 437-443.
Chapman, F. 1897. Notes on the Microzoa from the Jurassic beds at Hartwell. *Proc. Geol. Ass.*, **14**, 96-97.
Cifelli, R. 1959. Bathonian foraminifera of England. *Bull. Comp. Zool. Harvard*, **121**, 265-368.
Coleman, B. E. 1974. Foraminifera of the Oxford Clay and Kellaways Beds. Appendix 3. The geology of the new town of Milton Keynes. *Rep. Inst. Geol. Sci.*, No. 74/1b.
Coleman, B. E. 1982. Lower and Middle Jurassic foraminifera from the Winterborne Kingston Borehole, Dorset. In Rhys, G. H., Lott, G. K. and Calver, M. A. (eds) The Winterborne Kingston Borehole, Dorset, England. *Rep. Inst. Geol. Sci. No 81/3*, 82-88.
Colin, J. P., Lehmann, R. A. and Morgan, B. F. 1981. Cretaceous and Late Jurassic biostratigraphy of the North Celtic Sea Basin, offshore Southern Ireland. In Neale, J. W. and Brasier, M. D. (eds) *Microfossils from Recent and Fossil Shelf Seas*. Br. Micropaleontol. Soc., Ellis Horwood Ltd., 122-155.
Colloque sur le Lias Français 1961. *Mem. Bur. Rech. Géol. Min.*, **5**.
Cope, J. C. W. 1967. The palaeontology and stratigraphy of the lower part of the Upper Kimmeridge Clay of Dorset. *Bull. Brit. Mus. (Nat. Hist.) Geol.*, **15**, (1), 1-79.
Cope, J. C. W. 1974a. New information on the Kimmeridge Clay of Yorkshire. *Proc. Geol. Assoc. London*,

85, 211–21.
Cope, J. C. W. 1974b. Upper Kimmeridgian ammonite faunas of the Wash area and a subzonal scheme for the lower part of the Upper Kimmeridgian. *Bull. geol. Surv. G.B.* **47**, 29–37.
Cope, J. C. W. 1978. The ammonite faunas and stratigraphy of the upper part of the Upper Kimmeridge Clay of Dorset. *Palaeontology*, **21**, 469–533.
Cope, J. C. W. 1984. The Mesozoic history of Wales. *Proc. Geol. Assoc.*, **95**, 373–385.
Cope, J. C. W., Getty, T. A., Howarth, M. K., Morton, N. and Torrens, H. S.1980a. Correlation of Jurassic rocks in the British Isles. Part One: Introduction and Lower Jurassic. *Geol. Soc. Lond. Special Report* No 14, 1–73.
Cope, J. C. W., Duff, K. L., Parsons, C. F., Torrens, H. S., Wimbledon, W. A., Wright, J. K. 1980b. A correlation of Jurassic rocks in the British Isles. Part Two: Middle and Upper Jurassic. *Geol. Soc. Lond. Special Report* No. 15, 1–109.
Copestake, P. 1978. Foraminifera from the Lower and Middle Lias of the Mochras Borehole. Unpublished Ph.D. thesis, University College of Wales, Aberystwyth.
Copestake, P. 1982. Lower Sinemurian foraminifera and Ostracoda from two fissure deposits in the Eastern Mendips (Somerset, England). *J. micropaleontol.*, **1**, 149–153.
Copestake, P. 1985. Foraminiferal biostratigraphy in the Lower Jurassic. In Michelsen, O. and Zeiss, A. (eds) *Proc. International Symposium on Jurassic Stratigraphy, Erlangen, 1984, Geol. Surv. Denmark*, **1**, 192–206.
Copestake, P. 1986. *Haplophragmoides lincolnensis* sp. nov., a widespread foraminiferal index species in the Pliensbachian (early Jurassic) of Europe. *Rev. Española de Micropal.*, **17**, 403–411.
Copestake, P. 1987. In Lord, A. R. *et al.* Dorset Jurassic. Mesozoic and Cenozoic stratigraphical micropalaeontology of the Dorset coast and Isle of Wight, Southern England Field Guide for the XXth European Micropalaeontological Colloquium, Lord A. R. and Bown, P. R. (eds) *British Micropalaeontological Society Guide Book*, **1**, 1–78.
Copestake, P. and Johnson, B. 1984. Lower Jurassic (Hettangian–Toarcian) foraminifera from the Mochras Borehole, North Wales (U.K.) and their application to a worldwide biozonation. *Benthos '83; Second Int. Symp. Benthic Foraminifera Pau, April 1983*, 183–184.
Cordey, W. G.1962. Foraminifera from the Oxford Clay of Staffin Bay, Isle of Skye, Scotland. *Senckenbergiana leth.*, **43**, 375–409.
Cordey, W. G. 1963a. The genera *Brotzenia* Hofker 1954, and *Voorthuysenia* Hofker 1954 and Hofker's Classification of the *Epistomariidae*. *Palaeontology*, **6**, 653–657.
Cordey, W. G. 1963b. Oxford Clay foraminifera from England (Dorset–Northamptonshire) and Scotland. Unpublished PhD Thesis, University College, London.
Crick, W. D. 1887. Note on some foraminifera from the Oxford Clay at Keystone near Thrapston. *Northants. Nat. Hist. Soc.*, **4**, 233.
Crick, W. D. and Sherborn, C. D. 1891. On some Liassic foraminifera from Northamptonshire. *J. Northampt. Nat. Hist. Soc.* **6**, 1–15.
Crick, W. D. and Sherborn, C. D. 1892. The Leda-ovum beds of the Upper Lias, Northamptonhire. *J. Northampt. Nat. Hist. Soc.*, **7**, 67–72.
Cushman, J. A. 1930. Note sur quelques foraminifères Jurassique de Auberville (Calvados) *Bull. Soc. Linn. Normandie (8)*, **2** (1929), 134.
Dain, L. G. 1972. Foraminifera of Upper Jurassic deposits of Western Siberia. *Translations of the VNIGRI*, No. 317, 1–466.
Dean, W. T., Donovan, D. T. and Howarth, M. K. 1961. The Liassic ammonite zones and subzones of the north-west European Province. *Bull. Br. Mus. nat. Hist. (Geol.)*, **4**, 435–505.
Deeke, W. 1886. Les Foraminifères de l'Oxfordien des environs de Montbeliard. *Mem. Soc. emulation (3)*, **16**, 283–335.
Deegan, C. E. and Scull, B. J. (Compilers) 1977. A proposed standard lithostatigraphic nomenclature for the Central and Northern North Sea. *Rep. Inst. Geol. Sci. No. 77/25; Bull. Norw. Petrol. Direct.* No. 1, 1–36, HMSO London.
Donovan, D.. T. 1956. The zonal stratigraphy of the Blue Lias around Keynsham, Somerset. *Proc. Geol. Assoc. London*, **66**, 182–212.
Donovan, D. T. and Kellaway, G. A. 1984. Geology of the Bristol district: the Lower Jurassic rocks. *Mem. Br. Geol. Surv.*, 1–69.
Donovan, D. T., Horton, A. and Ivimey-Cook, H. C. 1979. The transgression of the Lower Lias over the northern flank of the London Platform. *J. geol. Soc. London*, **136**, 165–173.
D'Orbigny, A. 1849. *Prodrome de paléontologie stratigraphique universelle des animaux mollusques et rayonnes.* V. Masson, Paris 1–392.
Dreyer, E. 1967. Mikrofossilien der Rhät und Lias von S. W. Brandenburg. *Jb. Geol.*, **1**, 491–531.
Drexler, E. 1958. Foraminiferen und Ostracoden aus dem Lias alpha von Siebeldingen/Pfalz. *Geol. Jb.*, **75**, 475–554.
Duff, K. L. 1975. Palaeoecology of a Bituminous Shale – The Lower Oxford Clay of Central England. *Palaeontology*, **18**, 443–482.
Espitalié, J. and Sigal, J. 1960. Microfaunes du Domerien du Jura Meridional et du detroit de Rodez. *Rev. Micropal.*, **3**, 52–59.
Evans, D. J. 1973. The Stratigraphy of the Central Part of the Bristol Channel. Unpublished PhD Thesis, University College, Swansea.
Exton, J. 1979. Pliensbachian and Toarcian microfauna of Zambujal, Portugal: systematic palaeontology. *Carleton University Geological Paper 79–1*, i–viii,1–104.
Exton, J. and Gradstein, F. M. 1984. Early Jurassic stratigraphy and micropalaeontology of the Grand Banks and Portugal. In Westermann, G. E. G. (ed.) Jurassic–Cretaceous biochronology and palaeogeography of North America. *Geol. Association of Canada Special Paper* **27**, 13–30.
Franke, A. 1936. Die foraminiferen des deutschen Lias.

Abh. Preuss. Geol. Landesant; N.F. (169), 1–140.
Frentzen, K. 1941. Die foraminiferenfaunen des Lias, Doggers und unteren Malms des Umgeburg von Blumberg (Oberes Wutach–Gebiet). *Beitr. Natrukd. Forsch. Oberrhingeb.*, **6**, 125–402.
Fuchs, W. 1970. Eine Alpine, Tiefliassische foraminiferenfauna von Hernstein in Niederosterreich. *Verh. Geol. B.A.*, (1), 66–145.
Fürsich, F. T. 1976. The use of microinvertebrate associations in interpreting Corallian (Upper Jurassic) environments *Palaeogeogr., Palaeoclimatol., Palaeoecol.*, **20**, 235–256.
Gaunt, G. D., Ivimey-Cook, H. C., Penn, I. E. and Cox, B. M. 1980. Mesozoic rocks proved by IGS boreholes in the Humber and Acklam areas. *Rep. Inst. Geol, Sci. No. 79/13*, 1–34.
Gazdzicki, A. 1975. Lower Liassic ('Gresten Beds') microfacies and foraminifers from the Tatra Mts. *Acta Geol. Polonika*, **25**, 385–398.
George, T. N. et al. 1969. Recommendations on stratigraphical usage. *Proc. geol. Soc. London*, **1656**, 139–166.
Gerke, A. A. 1961. Foraminifera of the Triassic and Liassic deposits of the petroliferous region of north central Siberia (in Russian). *Trudy nauchno-issled. Inst. Geol. Arkt.*, **120**, 1–518.
Getty, T. A. 1972. Revision of the Jurassic family Echioceratidae. Unpublished PhD thesis, University of London.
Gordon, W. A.1962. Some Foraminifera from the Ampthill Clay, Upper Jurassic of Cambridgeshire. *Palaeontology*, **4**, 520–537.
Gordon, W. A. 1965. Foraminifera from the Corallian Beds, Upper Jurassic of Dorset, England. *J. Paleont.*, **39**, 838–863.
Gordon, W. A. 1966. Variation and its significance in Classification of some English Middle and Upper Jurassic Nodosariid Foraminifera. *Micropaleontology*, **12**, 325–333.
Gordon, W. A. 1967. Foraminifera from the Callovian (Middle Jurassic) of Brora, Scotland, *Micropaleontology*, **13**, 445–464.
Gordon, W. A. 1970. Biogeography of Jurassic foraminifera. *Bull geol. Soc. Am.*, **81**, 1689–1704.
Gradstein, F.M. 1976. Biostratigraphy and biogeography of Jurassic Grand Banks Foraminifera. First International Symposium on Benthonic Foraminifera of Continental Margins. Part B: Paleoecology and Biostratigraphy. *Maritime Sediments Spec. Pub.* 1, 557–583.
Gradstein, F. M. 1978. Jurassic Grand Banks Foraminifera: *Journal of Foraminiferal Research* **8**, 2, 97–109
Gradstein, F. M., Williams, G. L., Jenkins, W. A. M. and Ascoli, P. 1975. Mesozoic and Cenozic stratigraphy of the Atlantic Continental Margin, Eastern Canada. In C. J. Yorath, E. R. Parker and D. J. Glass (eds), *Canada's Continental Margin and Offshore Petroleum Exploration. Can. Soc. Petrol. Geol. Mem. No. 4*, 103–133.
Green, G. W. and Melville, R. V. 1956. The stratigraphy of the Stowell Park Borehole (1949–51). *Bull. geol. Surv. Gt. Br.*, **11**, 1–66.

Gregory, J. F. (in press) Lower Kimmeridgian foraminifera from the Helmsdale/Brora outlier, East Sutherland, Scotland. In Batten, D. and Keen, M. C. (eds). *Micropalaeontology, palynology and petroleum exploration, on- and offshore Europe*. Aberdeen University Symposium 1987.
Grigelis, A. A. 1958. *Globigerina oxfordiana* sp. — occurrence of Globigerines in the Upper Jurassic deposits of Lithuania. (Russian) *Nauchnye Doklady Vysshey Shkoly, Geol-Geogr. Nauki. Moscow, 1958*, No. 3, 110, 111.
Gümbel, C. W. 1862. Die Streitbergerer Schwammergel und ihre Foraminiferen-Einschlusse. *Jahres. Ver. Vaterl. Naturk. Wurttemberg, Jarg. 18*, 192–238.
Guyader, J. 1968. Le Jurassique Supérieur de la Baie de la Seine-Étude Stratigraphique et Micropaléontologique. *Thèse de Docteur, University of Paris*.
Haig, D. W. 1979. Early Jurassic foraminiferids from the Western Highlands of Papua New Guinea. *N. Jb. Geol. Palaont. Mh. No. 4*, 208–215.
Hallam, A. 1958. The concept of Jurassic axes of uplift. *Sci. Prog.*, **46**, 441–449.
Hallam, A. 1959. Stratigraphy of the Broadford Beds of Skye, Raasay and Applecross. *Proc. Yorks. Geol. Soc.*, **32**, 165–184.
Hallam, A. 1961. Cyclothems, transgressions and faunal change in the Lias of north west Europe. *Trans. Edinburgh Geol. Soc.*, **18**, 132–174.
Hallam, A. 1969. Tectonism and Eustasy in the Jurassic. *Earth Sci. Rev.*, **5**, 45–68.
Hallam, A. 1975. *Jurassic environments*. Cambridge University Press, 1–269.
Hallam, A. 1978. Eustatic cycles in the Jurassic. *Palaeogeog., Palaeoclimatol., Palaeoecol.*, **23**, 1–32.
Hallam, A. 1981. A revised sea-level curve for the early Jurassic. *J. geol. Soc. Lond.*, **138**, 735–743.
Hallam, A. 1984. Pre-Quaternary sea-level changes. *Ann. Rev. Earth Planet. Sci.*, **12**, 205–243.
Hallam, A. 1987. Radiations and extinctions in relation to environmental change in the marine Lower Jurassic of northwest Europe. *Paleobiology*, **13**, 152–168.
Hallam, A and Bradshaw, M. 1979. Bituminous shales and oolitic ironstones as indicators of transgressions and regressions. *J. geol. Soc. Lond.*, **136**, 157–164.
Hallam, A. and Sellwood, B. W. 1976. Middle Mesozoic sedimentation in relation to tectonics in the British area. *J. Geol.*, **84**, 301–321.
Haq, B. U., Hardenbol, J. and Vail, P. R. 1987. Chronology of fluctuating sea-levels since the Triassic. *Science*, **235**, 1156–1167.
Häusler, R. 1881. Untersuchungen uber die mikroskopischen Strukturverhaltniss der Aargauer Jurakalke mit besonderer berucksichtigung ihre Foraminiferen fauna. Diss. Universitat Zurich, 47 pp.
Häusler, R. 1883. Notes on some Upper Jurassic *Astrorhizidae* and *Lituolidae. Quart. J. geol. Soc. London*, **39**, 25–28.
Heath, R. S. and Apthorpe, M. C. 1986. Middle and Early(?) Triassic foraminifera from the Northwest Shelf, Western Australia. *J. Foramin. Res.*, **16**, 313–333.
Hedberg, H. D. (ed.) 1976. *International stratigraphic guide: a guide to stratigraphic classification, termin-*

ology and procedure by International Subcommission on Stratigraphic Classification of IUGS Commission on Stratigraphy. John Wiley & Sons, 1–200.

Henderson, I. J. 1934. The Lower Lias at Hock Cliff, Fretherne. *Proc. Bristol Nat. Soc.*, **4**, 549–564.

Hewitt, R. A. and Hurst, J. M. 1977. Size changes in Jurassic liparoceratid ammonites and their stratigraphical and ecological significance. *Lethaia*, **10**, 287–301.

Hofker, J. 1952. The Jurassic genus *Reinholdella* Brotzen (1948) (Foram.). *Paläont. Z.*, **26**, 15–29.

Hohenegger, J. 1981. *Icthyolaria densicostata* n. sp., a new species of Foraminifera characteristic of the Lower Lias. *Stuttgarter Beitr. Naturk* Ser. B, Nr. 74, 1–33.

Holland, C. G. *et al.* 1978. A guide to stratigraphical procedure. *Spec. Rep. geol. Soc. London* No. 10.

Horton, A. *et al.* 1974. The geology of the new town of Milton Keynes. *Inst. Geol. Sci. Report*, No. 74/16, 90–102.

Horton, A., Ivimey-Cook, H. C., Harrison, R. K. and Young, B. R. 1980. Phosphatic ooids in the Upper Lias (Lower Jurassic) of Central England. *J. geol. Soc. London*, **137**, 731–740.

Horton, A. and Coleman, B. E. 1978. The lithostratigraphy and micropalaeontology of the Upper Lias at Empingham, Rutland. *Bull. Geol. Surv. G.B.*, **62**, 1–12.

House, M. R. 1985. Are Jurassic sedimentary microrhythms due to orbital forcing? *Proc. Ussher Soc.*, **6**, 299–311.

Howard, A. S. 1985. Lithostratigraphy of the Staithes Sandstone and Cleveland Ironstone Formations (Lower Jurassic) of North-East Yorkshire. *Proc. Yorks. Geol. Soc.*, **45**, 261–275.

Howarth, M. K. 1955. Domerian of the Yorkshire coast. *Proc. Yorks. geol. Soc.*, **30**, 147–175.

Howarth, M. K. 1962. The Jet Rock Series and the Alum Shale Series of the Yorkshire coast. *Proc. Yorks. geol. Soc.*, **33**, 381–422.

Howarth, M. K. 1968. In: Hallam, A. The Lias. In Sylvester-Bradley, P. C. and Ford, T. (eds) *Geology of the East Midlands*. Leicester University Press, 188–208.

Howarth, M. K. 1978. The stratigraphy and ammonite fauna of the Upper Lias of Northamptonshire. *Bull. Br. Mus. Nat. Hist. (Geol.)*, **29**, 235–288.

Howarth, M. K. 1980. The Toarcian age of the upper part of the Marlstone Rock Bed of England. *Palaeontology*, **23**, 637–656.

Hudson, J. D. 1983. Mesozoic sedimentation and sedimentary rocks in the Inner Hebrides. *Proc. Roy. Soc. Edinb.*, **83B**, 47–63.

Hudson, J. D. and Palframan, D. F. B. 1969. The ecology and preservation of the Oxford Clay fauna at Woodham, Buckinghamshire. *Quart. J. geol. Soc. Lond.*, **124**, 387–418.

Hurst, A. 1985. The implications of clay mineralogy to palaeoclimate and provenance during the Jurassic in N. E. Scotland. *Scott. J. Geol.*, **21**, 143–160.

Illing, L. V. and Hobson, G. D. (eds.) 1981. Petroleum Geology of the Continental Shelf of North-West Europe. *Proceedings Second Conference on Petroleum Geology of the Continental Shelf of North-West Europe* Heydon & Son Ltd.

Issler, A. 1908. Bietrage zur Stratigraphie und Mikrofauna des Lias in Schwaben. *Palaeontographica*, **55**, 1–103.

Ivimey-Cook, H. C. 1975. The stratigraphy of the Rhaetic and Lower Jurassic in East Antrim. *Bull. geol. Surv. G.B.*, **50**, 51–69.

Ivimey-Cook, H. C. 1982. Biostratigraphy of the Lower Jurassic and Upper Triassic (Rhaetian) rocks of the Winterborne Kingston Borehole, Dorset. In Rhys, G. H., Lott, G. K. and Calver, M. A (eds) The Winterborne Kingston Borehole, Dorset, England. *Rep. Inst. Geol. Sci.* No.81/3, 97–106.

Jenkins, D. A. 1958. Liassic Foraminifera. Unpublished M.Sc. Thesis, University College of Wales, Aberystwyth.

Johnson, B. 1975. Upper Domerian and Toarcian foraminifera from the Llanbedr (Mochras Farm) Borehole, North Wales. Unpublished Ph.D. thesis, University College of Wales, Aberystwyth.

Johnson, B. 1976. Ecological ranges of selected Toarcian and Domerian (Jurassic) foraminiferal species from Wales. 1st International Symposium on Benthonic Foraminifera of Continental Margins Part B: Paleoecology and Biostratigraphy. *Maritime Sediments Spec. Pub.*, **1** 545–556.

Jones, T. R. 1853. On the Lias at Fretherne, near Newham and Purton, near Sharpness, with an account of some new foraminifera discovered there. *Ann. Mag. Nat. Hist. Ser. 2*, **12**, 272–277.

Jones, T. R. 1884. Notes on the Foraminifera and Ostracoda from the Deep Boring at Richmond. *Quart. J. geol. Soc. London*, **11**, 765–777.

Jones, T. R. and Parker, W. K. 1860. On some fossil foraminifera from Chellaston near Derby. *Quart. J. geol. Soc. London*, **16**, 452–456.

Kaptarenko-Chernousova, O. K., Golak, L. M., Zerneckij, B. F., Krajeva, E. J. and Lipnik, E. S. 1963. Atlas of Characteristic foraminifera from the Jurassic, Cretaceous and the Palaeogene of the Ukranian Platform. (In Russian). *Trudy Akad, Nauk USSR, Ser. strat. paleont.* **V.45**, 1–200.

Karampelas, G. A. 1978. Foraminifera eines vollstandingen Liasprofils aus der Lengenbrucker Senke (Nordbaden). *Jb. Geol. Landesant. Baden-Wurttemberg*, **20**, 43–66.

Kent, D. V. and Gradstein, F. M. 1985. A Cretaceous and Jurassic geochronology. *Geol. Soc. America Bulletin*, **96**, 1419–1427.

Klingler, W, 1962. Lias Deutschlands. In: *Leitfossilien der Mikropaläontologie*, 73–122.

Knauff, W. 1966. *Praeophthalmidium* n. gen. (Foram.). Eine entwicklungsgeschichtliche Untersuchung. *Neues Jb. Geol. Paläontol. Abhandlungen*, **125**, 96–102.

Knox, R. W. O'B. 1984. Lithostratigraphy and depositional history of the late Toarcian sequence at Ravenscar, Yorkshire. *Proc. Yorks. Geol. Soc.*, **45**, 99–108.

Kopik, J. 1960. Micropalaeontological characteristic of Lias and Lower Dogger in Poland (in Polish). *Kuratalnik Geol.*, **4**, 921–935.

Kristan-Tollmann, E. 1964. Die Foraminiferen aus den

Rhatischen Zlambachmergeln der Fischerwiese bei Aussee in Jalzkammergut. *Jahrb. Geol. B. A.*, **10**, 1–189.

Kübler, J. and Zwingli, H. 1866. Mikroskopische Bilder aus der Urwelt der Schweiz Heft II. *Neujarhsbl. Burgersbibliothek Winterthur*, i–ix, 1–28.

Lang, W. D. 1914. The geology of Charmouth cliffs, beach and foreshore. *Proc. Geol. Assoc. London*, **25**, 293–360.

Lang, W. D. 1924. The Blue Lias of the Devon and Dorset coasts. *Proc. Geol. Assoc. London*, **35**, 169–185.

Lang, W. D. 1932. The Lower Lias of Charmouth and the Vale of Marshwood. *Proc. Geol. Assoc. London*, **43**, 97–106.

Lang, W. D. and Spath, L. F. 1926. The Black Marl of Black Ven and Stonebarrow, in the Lias of the Dorset coast. *Q. J. geol. Soc. London*, **82**, 144–187.

Lang, W. D., Spath, L. F. and Richardson, W. A. 1923. Shales-with-'Beef', a sequence in the Lower Lias of the Dorset Coast. *Q. J. geol. Soc. London*, **79**, 47–99.

Larsen, G., Buch, A., Christensen, O. B. and Bang, I. 1968. Oresund. Helsingor-Halsingborgslinien. Geologisk Rapport. *Dan. Geol. Unders. Rapp.* 1–90.

Lee, G. W. in: Read, H. H., Ross, G. and Phemister, J. 1925. The geology of the country around Golspie, Sutherlandshire. *Mem. geol. Surv.*

Lee, G. W. and Bailey, E. B. 1925. The pre-Tertiary geology of Mull, Loch Aline and Oban. *Mem. geol. Surv. G.B.*, **140**.

Leeder, M. R. 1983. Lithospheric stretching and North Sea Jurassic clastic sourcelands. *Nature*, **305** (5934), 510–513.

Lloyd, A. J. 1958. Foraminifera from the type Kimmeridgian. Unpublished PhD Thesis, University College, London.

Lloyd, A. J. 1959. Arenaceous foraminiferal faunas from the Type Kimmeridgian. *Palaeontology*, **1**, 298–320.

Lloyd, A. J. 1962. Polymorphinid, Miliolid and Rotaliform Foraminifera from the Type Kimmeridgian. *Micropaleontology*, **8**, 369–383.

Lloyd, A. J., Savage, R. J. G., Stride, A. A. and Donovan, D. T. 1973. The geology of the Bristol Channel floor. *Phil. Trans. R. Soc.*, **A274**, 595–626.

Loeblich, A. R. and Tappan, H. 1950. North American Jurassic Foraminifera I. The Type Redwater Shale (Oxfordian) of South Dakota. *J. Paleont.*, **24**, 39–60.

Loeblich, A. R. and Tappan, H. 1957. Eleven new genera of Foraminifera. *Bull. U.S. natl. Hist. Mus.*, **215**, 223–232.

Loeblich, A. R. and Tappan, H. 1964. Sarcodina. Chiefly 'Thecameobians' and Foraminiferida, Treatise on Invertebrate Paleontology (ed. Moore, R. C.) Part C, Protista 2, **1 and 2**, *Geol. Soc. Am. University of Kansas Press*.

Loeblich, A. R. and Tappan, H. 1984. Suprageneric classification of the Foraminiferida (Protozoa). *Micropaleontology*, **30**, 1–70.

Loeblich, A. R. and Tappan, H. 1986. Some new and revised genera and families of hyaline calcareous Foraminiferida (Protozoa). *Trans. Am. Microsc. Soc.*, **105**, 239–265.

Løfaldi, M. and Nagy, J. 1980. Foraminiferal stratigraphy of Jurassic deposits on Kongsoya, Svalbard. *Norsk Polarinstitutt, Skrifter*, No. 172, 63–96.

Løfaldi, M. and Nagy, J. 1983. Agglutinating foraminifera in Jurassic and Cretaceous dark shales in Southern Spitsbergen. In Verdenius, J. A., Van Hinte, J. E. and Fortuin, A. R. (eds) *Proceedings of the First Workshop on Arenaceous Foraminifera*. University of Amsterdam/IKU publication No. 108, 91–107.

Lutze, G. F. 1960. Zur Stratigraphie une Paläontologie des Callovian und Oxfordian in Nordwest Deutschland. *Geol. Jahrb.*, **77**, 391–532.

Macfadyen, W. A. 1935. Jurassic Foraminifera. *The Mesozoic paleontology of British Somaliland*. Pt. II, 7–20.

Macfadyen, W. A. 1936. D'Orbigny's Lias Foraminifera. *Jl. R. microsc. Soc.*, **56**, 147–153.

Macfadyen, W. A. 1941. Foraminifera from the Green Ammonite Beds, Lower Lias, of Dorset. *Phil Trans. R. Soc.*, **B231**, 1–173.

McGugan, A. 1965. Liassic foraminifera from Whitepark Bay, Co. Antrim. *Irish Nat. J.*, **15**, 85–87.

McLennan, R. M. 1954. The Liassic sequence in Morvern. *Trans-geol. Soc. Glasgow*, **21**, 447–455.

Magné, J., Seronié-Vivien, R. M. and Malmoustier, J. 1961. Le Toarcian de Thouars (Deuxsèvres). In *Colloque sur le Lias Français*, 357–370.

Mamontova, E. V. 1957. Toarcian foraminifera from the Northern Slopes of the Caucasus Mountains (Kuban-Laba) (in Russian). *Leningrad State Univ., Geologiya, Uchenye Zapiski: [Sci. Notes] No. 225 (Ser. Geol. Sci., vypusk 9)* Leningrad.

Masson, D. G. and Miles, P. R. 1986. Development and hydrocarbon potential of Mesozoic sedimentary basins around margins of North Atlantic. *American Association of Petroleum Geologists Bulletin*, **70**, 721–729.

Maupin, C. 1977. Données micropaléontologiques nouvelles et precisions stratigraphiques sur le Lias du Kef Ben Chike Bou Rouhou et du Kef Toumiette Nord (Chaine Calcaire Kabyle-Nord du Constantinois-Algerie). *Rev. Micropal.*, **20**, 91–99.

Maupin, C. and Vila, J.-M. 1976. Microfossiles du Lias Supérieur au Djebel Youssef (Hautes plaines setifiennes, Algerie). *Rev. Micropal.*, **19**, 162–165.

Medd, A. W. 1979. In: Richardson, G. The Mesozoic Stratigraphy of two boreholes near Worlaby, Humberside. *Bull. geol. Surv. G.B.*, **58**, 1–24.

Medd, A. W. 1982. Upper Jurassic foraminifera of the Winterborne Kingston Borehole, Dorset. In: Rhys, G. H., Lott, G. K. and Calver, M. A. (eds) The Winterborne Kingston borehole, Dorset, England. *Rep. Inst. Geol. Sci.*, No 81/3, 51–52.

Medd, A. W. 1983. Foraminifera from the Lower Oxford Clay (Callovian stage), Cambridgeshire. *Rev. Esp. Micropaleontol.*, **15**, 221–240.

Michael, E. 1967. Die Mikrofauna des neu-deutschen Barrême; Teil 1, Die Foraminiferen des neudesutschen Barrême. *Palaeontographica, Suppl.*, **12**, 1–153.

Millson, J. A. 1987. The Jurassic evolution of the Celtic Sea Basins. In: Brooks, J. and Glennie, K. (eds) *Petroleum geology of North West Europe*, 599–610, Graham & Trotman.

Mira, F. and Martinez-Gallego, J. 1981. Foraminifera del Lias Margoso (Carixiense superior Domeriense inferior y medio) en el sector central de las Cordilleras Beticas. *Rev. Esp. Micropal.*, **13** (3), 313–342.

Morris, K. A. 1980. Comparison of major sequences of organic-rich mud deposition in the British Isles. *J. geol. Soc. Lond.*, **137**, 157–170.

Morris, P. H. 1980. A microfaunal and sedimentological analysis of some Lower Inferior Oolite sections of the Cotswolds. Unpublished PhD thesis, University of Wales.

Morris, P. H. 1982. Distribution and Palaeoecology of Middle Jurassic Foraminifera from the Lower Inferior Oolite of the Cotswolds. *Palaeogeog. Palaeoclimatol. Palaeoecol.*, **37**, 319–347.

Morter, A. A. and Gallois, R. W. 1979. Provisional report on the I.G.S. Borehole drilled at Trunch, Norfolk. *Internal unpublished report of the British Geological Survey*, 1–247.

Morton, N. 1987. Jurassic subsidence history in the Hebrides, N.W. Scotland. *Marine and Petroleum Geology*, **4**, 226–242.

Mudge, D. C.. 1978. Stratigraphy and sedimentation of the Lower Inferior Oolite of the Cotswolds. *J. geol. Soc. London*, **135**, 611–627.

Munk, C. 1978. Feinstratigraphische und mikropalaontologische untersuchungen an Foraminiferen-Faunen im Mittlersen und Oberen Dogger (Bajocian–Callovian) der Frankenalb. *Erlanger geol. Abh.*, **105**, 1–72.

Munk, C. 1980. Foraminifera from the Lower Kimmerdige (*Platynota* Beds) of the northern and middle Franconian Alb, Germany—fauna and palaeoecology *Facies*, **2**, 149–218.

Murray, J. W. 1973. *Distribution and ecology of living benthic foraminiferids*. London, Heinemann. 288 pp

Nagy, J. 1985a. Jurassic foraminiferal facies in the Statfjord area, Northern North Sea—I. *Jour. Petr. Geol.*, **8**, 273–295.

Nagy, J. 1985b. Jurassic foraminiferal facies in the Statfjord area, Northern North Sea—II. *Jour. Petr. Geol.*, **8**, 389–404.

Nagy, J., Dypvik, H. and Bjaerke, T. 1984. Sedimentological and palaeontological analyses of Jurassic North Sea deposits from deltaic environments. *Jour. Petr. Geol.*, **7**, 169–188.

Nagy, J. and Løfaldi, M. 1981. Agglutinating foraminifera in Jurassic dark shale facies in Svalbard. In Neale, J. W. and Brasier, M. D. (eds) *Microfossils from recent and fossil Shelf Seas*, Br. Micropalaeontol. Soc., Ellis Horwood Ltd., 114–121.

Nagy, J., Løfaldi, M. and Bomstad, K. 1981. Marginal marine microfaunas of the Jurassic (Bajocian) Yons Nab Beds of the Yorkshire coast. In Verdenius, J. G., Van Hinte, J. E. and Fortuin, A. R. (eds) *Proceedings of the First Workshop on Arenaceous Foraminifera*, University of Amsterdam/IKU publication, No. 108, 111–128.

Neaverson, E. 1921. The foraminifera of the Hartwell Clay and subadjacent beds. *Geol. Mag.*, **58**, 454–473.

Neves R. and Selley, R. C. 1975. A review of the Jurassic rocks of north-east Scotland. *Proc. Jurassic Northern North Sea Symposium Stavanger*, **5**, 1–29.

Norling, E. 1966. On the genus *Ichthyolaria* Wedekind, 1937. *Sver. Geol. Unders.*, *Ser. C (613)*, 1–24.

Norling, E. 1972. Jurassic stratigraphy and foraminifera of Western Scania, Southern Sweden. *Sver. geol. Unders. Afh. Ser. Ca (47)* 1–120.

Nørvang, A. 1957. The Foraminifera of the Lias series in Jutland, Denmark. *Mede. Dansk. geol. Foren.*, **13**, 1–135.

Nouet, G. 1958. Caractères stratigraphiques et micropaléontologiques du Bathonien de la Basse Normandie au Boulonnais. *Rev. Micropal.*, **1**, 17–21.

Oates, M. J. 1976. The Lower Lias of Western Scotland. Unpublished PhD thesis, University of London.

Oates, M. J. 1978. A revised stratigraphy for the Western Scottish Lower Lias. *Proc. Yorks. Geol. Soc.*, **42**, 143–156.

Ohmert, W. (in press). The *ovalis* zone (Lower Bajocian) in the type area, southwestern Germany. In: *Proceedings 2nd International Symposium on Jurassic Stratigraphy (Lisbon, 1987)*.

Paalzow, R. 1917. Bietrage zur Kenntnis der Foraminiferen fauna der Schwammergerl des Unteren Weissen Jura Suddeutschland. *Abh. Naturhist. Ges. Nurnberg.*, **19**, 203–248.

Packer, S. R. 1986. Foraminifera and ostracods from the Lincolnshire Limestone Formation (Bajocian), Kirton-in-Lindsey, South Humberside. Unpublished MSc Thesis, University of Hull.

Palmer, C. P. 1971. The stratigraphy of the Stonehouse and Tuffley claypits in Gloucestershire. *Proc. Bristol Nat. Soc.*, **32**, 58–68.

Palmer, C. P. 1972a. The Lower Lias (Lower Jurassic) between Watchet and Lilstock in North Somerset (United Kingdom). *Newsl. Stratig.*, **2**, 1–30.

Palmer, C. P. 1972b. A revision of the zonal classification of the Lower Lias of the Dorset coast of south west England. *Newsl. Stratig.*, **2**, 45–54.

Parsons, C. F. 1974. The Bajocian stratigraphy of eastern England. *Proc. Yorks. Geol. Soc.*, **40**, 115–116.

Parsons, C. F. 1976. A stratigraphical revision of the *humphriesianum–subfurcatum* Zone rocks (Bajocian stage, Middle Jurassic) of Southern England. *Newsl. Stratig.*, **5**, 114–142.

Payard, J.-M. 1947. Les Foraminifères du Lias superieur de Detroit Poitevin. *Thèses Faculté Sci. Univ. Paris*, 1–255.

Pazdrowa, O. 1969. Middle Jurassic Epistominidae (foraminifera) of Poland. *Stud. Geol. Pol.* **27**, 7–88.

Penn, I. E., Dingwall, R. G. and O'B. Knox, R. W. 1980 The Inferior Oolite (Bajocian) sequence from a borehole in Lyme Bay. *Rep. Inst. Geol. Sci.* No. 79/3.

Penn, I. E. 1982. Middle Jurassic stratigraphy and correlation of the Winterborne Kingston borehole, Dorset. In: Rhys, G. H., Lott, G. K., and Calver, M. A. (eds) The Winterborne Kingston Borehole, Dorset, England. *Rep. Inst. Geol. Sci. London*, 81/3, 53–76.

Penn, I. E., Merriman, R. J. and Wyatt, R. J. 1979. A proposed type section for the Fuller's Earth (Bathonian) based on Horsecombe Vale, No. 15 Borehole, near Bath, with details of contiguous strata, Part 1. In The Bathonian Strata of Bath–Frome area. *Rep. Inst. Geol. Sci.* No. 78/22.

Penn, I. E., Dingwall, R. G. and O'B. Knox, R. W. 1980 The Inferior Oolite (Bajocian) sequence from a borehole in Lyme Bay. *Rep. Inst. Geol. Sci.* No. 79/3.

Phillips, J. 1829. *Illustrations of the geology of Yorkshire, or a description of the strata and organic remains Part 1. The Yorkshire Coast.* York.

Pietrzenuk, E. 1961. Zur Mikrofauna einiger Liasvorkommen in der Deutschen Demokratischen Republik. *Freiberg. Forschungsh.*, **C 113**, Pal. 1–129.

Poole, E. G. and Whiteman, A. J. 1966. Geology of the country around Nantwich and Whitchurch. *Mem. geol. Surv. G.B.*, 1–154.

Powell, J. H. 1984. Lithostratigraphical nomenclature of the Lias Group in the Yorkshire Basin. *Proc. Yorks. Geol. Soc.*, **45**, 51–57.

Prestat, B. 1967. Etude micropaléontologique du passage Bathonian–Callovian dans le centre SW du Bassin de Paris. *Colloque Jurassique Luxembourg, 1967.*

Quilty, P. G. 1981. Early Jurassic foraminifera from the Exmouth Plateau, Western Australia. *Jour. Paleont.*, **55**, 985–995.

Rabitz, G. 1963. Foraminiferen des Göttinger Lias. *Paläontol. Z.*, **37**, 198–224.

Rawson, P. F. and Riley, L. A., 1982. Latest Jurassic–Early Cretaceous events and the 'Late Cimmerian unconformity' in the North Sea area. *Bull. Am. Assoc. Pet. Geol.* **66**, 12, 2628–2648.

Rhys, G. H. (Compiler) 1974. A proposed standard lithostratigraphic nomenclature for the southern North Sea and an outline structural nomenclature for the whole of the (UK) North Sea. A report of the joint Oil Industry—Institute of Geological Sciences Committee on North Sea Nomenclature. *Rep. Inst. Geol. Sci.* No. 74/8, 1–14.

Richardson, G. 1979. The Mesozoic stratigraphy of two boreholes near Worlaby, Humberside. *Bull. Geol. Surv. G. B.*, **58**, 1–24.

Richardson, L. 1905. The Rhaetic and Contiguous deposits of Glamorganshire. *Q. J. geol. Soc. Lond.*, **61**, 385–424.

Richardson, L. 1908. On the section of Lower Lias at Hock Cliff, Fretherne, Gloucestershire. *Proc. Cotteswold Nat. Fld. Cl.*, **16**, 135–142.

Riegraf, W. 1985. Mikrofauna, Biostratigraphie und Fazies im Unteren Toarcian Sudwestdeutschlands und Vergleiche mit benachbarten Gebieten. *Tubinger mikropaläont. Mitt.* Nr. 3, 1–232.

Riegraf, W. 1986. Callovian (Middle Jurassic) radiolaria and sponge spicules from southwest Germany. *Stuttgarter Beitr. Naturk. Ser. B. Nr. 123*, 1–31.

Riegraf, W., Luterbacher, H. and Leckie, R. M. 1984. Jurassic foraminifers from the Mazagan Plateau, Deep Sea Drilling Project Site 547, Leg 79, off Morocco. In: Hinz, K., Winterer, E. L. et al., *Initial Reports of the Deep Sea Drilling Project*, **79**, Washington (U.S. Govt. Printing Office).

Robertson Research International Limited, 1974. The *Jurassic of Northwest europe: onshore project*. Petroleum Industry Report.

Roberston Research International Limited, 1975. The *Jurassic of Northwest Europe: offshore project*. Petroleum Industry Report.

Ruget, C. 1967. Variations morphologiques chez quelques espèces de Frondiculaires costulées du Lias de Lorraine (Lotharingien). *Rev. Micropal.*, **10**, 22–36.

Ruget, C. 1976. Revision des Foraminifères de la Collection Terquem: Dentalina, Marginulina, Nodosaria. *Cahiers de Micropaléontologie*, **4**, 1–118.

Ruget, C. 1980. Evolution et biostratigraphie des Lagenides (foraminifères) dans le Lias de l'Europe Occidentale. *Bull. Geol. Soc. France Ser. 7*, **22**, (4), 623–626.

Ruget, C. 1982. Foraminifères du Lias moyen et superieur d'Obon (Chaines iberiques, province de Tereul, Espagne). *Geobios*, No. 15, 53–91.

Ruget, C. 1985. Les foraminifères (Nodosariides) du Lias de l'Europe Occidentale. *Docum. Lab. Géol. Lyon* No. 94, 1–273.

Ruget, C. H. and Sigal, J. 1967. Les Foraminifères du Sondage de Laneuveville-devant—Nancy (Lotharingien de la region type). *Est. Sc. de la Terre*, **12**, 33–70.

Ruget, C. H. and Sigal, J. 1970. Le Lias moyen de Sao Pedro de Muel II. Les Foraminifères. *Communic. dos Serv. Geolog. de Portugal*, **54**, 79–108.

Rutten, M. G. 1956. Depositional Environment of Oxford Clay at Woodham Clay Pit. *Geol. Mijnb. N.S.*, **18**, 344–347.

Said, R. and Barakat, M. G. 1958. Jurassic microfossils from Gebel Maghara, Sinai, Egypt. *Micropaleontology*, **4**, 231–272.

Schwager, C. 1865. Beitrage zur Kenntnis der mikroskopischen Fauna Jurassicher Schichten. *Jb. Ver. Vaterl. Naturkde. Wurtt. Jahrg.*, **21**, 82–151.

Seibold, E. and Seibold, I. 1953. Foraminiferenfauna und Kalkgehalt eines Profils im gebankten unteren Malm Schwabens. *Neues Jahrb. Geol. Paläont. Abh.*, **98**, 28–86.

Seibold, E. and Seibold, I. 1955. Revision der Foraminiferen-Bearbeitung G. W. Gümbel's (1862) aus den Streitberger Schwammumergeln (Oberfranken), Unterer Malm. *Neues Jahrb. Geol. Paläont. Abh.*, **101**, 91–134.

Seibold, E. and Seibold, I. 1956. Revision der Foraminiferen-Bearbeitung C. Schwager's (1865) aus den Impressaschichten (Unterer Malm) Suddeutschland. *Neues. Jahrb. Geol. Paläont., Abh.*, **103**, 91–154.

Seibold, E. and Seibold, I. 1960. Foraminiferen der Bank—und Schwamm-Fazies im unteren Malm Suddeutschland. *Neues Jahrb. Geol. Paläont. Abh.*, **109**, 309–438.

Sellwood, B. W., Scott, J., James, B., Evans, S. R. and Marshall, J. 1987. Regional significance of 'dedolomitisation' in Great Oolite reservoir facies of Southern England. In: Brooks, J. and Glennie, K. (eds) *Petrol. Geol. of North West Europe*, Graham & Trotman, 129–137.

Sellwood, B. W., Scott, J. and Lunn, G. 1986. Mesozoic basin evolution in Southern England. *Proc. Geol. Assoc.*, **97**, 259–290.

Sellwood, B. W., Scott, J., Mikkelsen, P. and Akroyd, P. 1985. Stratigraphy and sedimentology of the Great Oolite Group in the Humbly Grove Oilfield, Hampshire. *Marine and Petroleum Geology*, **2**, 44–55.

Seronié-Vivien, R. M., Magné, J. and Malmoustier, J. 1961. Le Lias des Bordures Septrionale et Orientale du Bassin d'Aquitaine. In *Colloque sur le Lias Français*, 757–791.

Sherborn, C. D. 1888. Notes on *Webbina irregularis* (d'Orbigny) from the Oxford Clay at Weymouth. *Proc. Bath. Nat. Hist. Antiq. Fld. Cl.*, **6**, 322–333.

Shipp, D. J. 1971. Foraminifera from the Kimmeridgian of the Boulonnais. Unpublished M.Sc. thesis, University College, London.

Shipp, D. J. 1978. Foraminifera from the Oxford Clay and Corallian of England and the Kimmeridgian of the Boulonnais, France. Unpublished Ph.D. thesis University College, London.

Souaya, F. J. 1976. Foraminifera of Sun-Global Linckens Island Well P-46, Arctic Archipelago, Canada. *Micropaleontology*, **22**, 249–306.

Spath, L. F. 1939. The ammonite Zones of the Upper Oxford Clay of Warboys, Huntingdonshire. *Bull. Geol. Surv. Gt. Br.*, **1**, 82–98.

Stoneley, R. 1982. The structural development of the Wessex Basin. *J. geol. Soc. London*, **139**, 543–554.

Strickland, H. E. 1846. On two species of microscopic shells found in the Lias. *Quart. J. geol. Soc. London*, **2**, 30–31.

Strong, C. P. 1984. Triassic foraminifera from Southland Syncline, New Zealand. *New Zealand Geological Survey Palaeontological Bulletin*, **52**, 1–37.

Sykes, R. M. 1975. The stratigraphy of the Callovian and Oxfordian stages (Middle and Upper Jurassic) in northern Scotland. *Scott. J. Geol.*, **11**, 51–78.

Sykes, R. M. and Callomon, J. H. 1979. The *Amoeboceras* Zonation of the Boreal Upper Oxfordian. *Palaeontology*, **22**, 893–903.

Talbot, M. R. 1973. Major sedimentary cycles in the Corallian Beds. *Palaeogeog., Palaeoclimatol., Palaeoecol.*, **14**, 293–317.

Talbot, M. R. 1974. Ironstones in the Upper Oxfordian of southern England. *Sedimentology*, **21**, 433–450.

Tappan, H. 1955. Foraminifera from the Arctic slope of Alaska. Part II. Jurassic Foraminifera. *Prof. Pap. U.S. Geol. Surv.*, **236-B**, 21–90.

Tate, R. and Blake, F. J. 1876. *The Yorkshire Lias*. London, J. Van Voorst.

Ten Dam, A. 1947. A new species of *Asterigerina* from the Upper Liassic of England. *J. Paleont.*, **21**, 396–397.

Terquem, O. 1855. Paléontologie de l'étage inférieur de la Formation Liassique de la province de Luxembourg, grandduché (Hollande) et de Hettange du départment de la Moselle. *Mem. Soc. géol. France (2)*, **5**, (1854), pt. 2, mem. 3.

Terquem, O. 1858. Première mémoire sur les Foraminifères du Lias du Department de la Moselle. *Mem. Acad. Imper. Metz.*, **39**, 563–654.

Terquem, O. 1862. Réchercnes sur les Foraminifères de L'Etage Moyen et de L'Etage inférieur du Lias, Memoire 2. *Mem. Acad. Imper. Metz.*, **42**, (ser. 2, 9), 415–466.

Terquem, O. 1863. Troisième mémoire sur les Foraminifères du Lias des départements de la Moselle, de la Côte d'Or, du Rhone de la Vienne et du Calvados. *Mem. Acad. Imper. Metz.*, **44** (2) (ser. 2, 11), 361–438.

Terquem, O. 1864. Quatrième mémoire sur les Foraminifères du Lias comprenant les polymorphines de Départements de la Cote d'Or et de l'Indre, 233–305, Metz.

Terquem, O. 1866a. Cinquième mémoire sur les Foraminifères du Lias des Départements de la Moselle, de la Côte d'Or et de l'Indre, 313–454, Metz.

Terquem, O. 1866b. Sixième mémoire sur les Foraminifères du Lias des Départements de l'Indre et de la Moselle, 459–532, Metz.

Terquem, O. 1868. Première mémoire sur les Foraminifères du Système Oolithique Étude, du Fullers-Earth de la Moselle. *Bull. Soc. Hist. Nat. Moselle Metz.*, **11**, 1–138.

Terquem, O. 1870a. Deuxième Mémoire sur les Foraminifères du Systeme Oolithique. Zone a *Ammonites parkinsoni* de la Moselle. *Mem. Imper. Acad. Metz.*, **50**, 403–456.

Terquem, O. 1870b. Troisième Mémoire sur les Foraminifères du Systeme Oolithique comprenant les genres *Frondicularia, Flabellina, Nodosaria*, etc. de la Zone *Ammonites parkinsoni* de Fontoy (Moselle) *Mem. Acad. Imper. Metz.*, **51**, 299–380.

Terquem, O. 1874. Quatrième mémoire sur les Foraminifères du systeme oolothique. *Mem. Acad. Imper. Metz.*, **55**, 279–333.

Terquem, O. 1883. Cinquième Mémoire sur les Foraminifères du Systeme Oolithique de la zone à *Ammonites parkinsoni* de Fontoy (Moselle). *Mem. Acad. Imper. Metz.*, **64**, 279–338.

Terquem, O. 1886. Les Foraminifères et les Ostracodes de Fuller's Earth des environs de Varsovie. *Mem. Soc. Géol. France, ser. 3*, **4**, 1–112.

Terquem, O. and Berthelin, G. 1875. Étude microscopique des marnes du Lias Moyen d'Essey-lès-Nancy, zone inférieure de l'assise a *Ammonites margaritatus*. *Mem. Soc. Géol. France, Ser. 2*, **10**, 1–126.

Torrens, H. S. 1969. The stratigraphical distribution of Bathonian ammonites in Central England. *Geol. Mag.*, **106**, 63–76.

Townson, W. G. 1975. Lithostratigraphy and Deposition of the type Portlandian. *J. geol. Soc. London*, **131**, 619–638.

Trueman, A. E. 1918. The Lias of South Lincolnshire. *Geol. Mag.*, **5**, 64–73, 101–111.

Trueman, A. E. 1922. The Liassic rocks of Glamorgan. *Proc. Geol. Assoc.*, **33**, 245–284.

Tutcher, J. W. and Trueman, A. E. 1925. The Liassic rocks of the Radstock district, Somerset. *Quart. J. geol. Soc. London*, **81**, 595–666.

Tyson, R. V., Wilson, R. L. and Downie, C. 1979. A stratified water column environmental model for the Kimmeridge Clay. *Nature*, **277**, 377–380.

Usbeck, I. 1952. Zur Kenntnis von Mikrofauna und Stratigraphie im unteren Lias alpha Schwabens. *Neues Jb. Geol. Paläont. Abh.*, **95**, 371–476.

Vail, P. R., Hardenbol, J. and Todd, R. G. 1984. Jurassic unconformities, chronostratigraphy and sea level changes from seismic stratigraphy and biostratigraphy. In *Interregional unconformities and hydrocarbon accumulation*. American Association of Petroleum Geologists Memoir 36, 129–144.

Vail, P. R. and Todd, R. G. 1981. Northern North Sea Jurassic unconformities, chronostratigraphy and sea-level changes from seismic stratigraphy. In Illing, L. V. and Hobson, G. D. (eds), *Petroleum geology of the Continental Shelf of North West Europe*, 216–235, Institute of Petroleum, London.

Vail, P. R., Mitchum, R. M., Todd, R. G., Widimier, J. M., Thompson, S., Sangree, J. B., Bubb, J. N. and Hatlelid, W. G. 1977. Seismic stratigraphy and global changes of sea-level. In *Seismic stratigraphy—applications to hydrocarbon exploration*. American Association of Petroleum Geologists Memoir, 26, 49–212.

Wall, J. H. 1960. Jurassic microfaunas from Saskatchewan. *Sask. Dep. Min. Resources Rep.* **53**, 1–229.

Wall, J. H. 1983. Jurassic and Cretaceous foraminiferal biostratigraphy in the eastern Sverdrup Basin, Canadian Arctic Archipelago. *Bull. Canadian Petr. Geol.*, **31**, 246–281.

Warrington, G. and Owens, B. (Compilers) 1977. Micropalaeontological biostratigraphy of offshore samples from south-west Britain. *Rep. Inst. Geol. Sci.* No. 77/7, 1–49.

Warrington, G. *et al.* 1980. A correlation of Jurassic rocks in the British Isles. *Geol. Soc. Lond., Special Report*, **13**, 78pp.

Waters, R. A. and Lawrence, D. J. D. 1987. Geology of the South Wales Coalfield, Part III, the country around Cardiff. *Mem. Br. Geol. Surv.*, Sheet 263, 1–114.

Weedon, G. P. 1986. Hemipelagic shelf sedimentation and climatic cycles: the basal Jurassic (Blue Lias) of South Britain. *Earth and Planet. Sci. Lett.*, **76**, 321–335.

Welzel, E. 1968. Foraminiferen und Fazies des frankischen Domeriums. *Erlanger geol. Abh.*, **69**, 1–77.

Wernli, R. 1971. Les Foraminifères du Dogger du Jura meridional (France). *Arch. Soc. Genève*, **24**, 261–368.

Wernli, I. and Septfontaine, M. 1971. Micropaléontologie comparée du Dogger du Jura meridional (France) et des Préalpes Medianes Plastiques romandes (Suisse). *Eclog. Geol. Helv.*, **64**, 437–458.

Whittaker, W. 1866. On some borings in Kent. A contribution to the deep-seated geology of the London Basin. *Quart. J. geol. Soc. London.*, **42**, 26–48.

Whittaker, A. and Green, G. W. 1983. Geology of the country around Weston-super-Mare. *Mem. Geol. Surv. G.B.*, 1–147.

Wicher, C. A. 1938. Mikrofaunen aus Jura und Kreide, insbesondere Nordwestdeutschlands. I. Teil: Lias Alpha bis epsilon. *Abh. Preuss. Geol. L.A.*, **193**, 1–16.

Wicher, C. A. 1952. *Involutina, Trocholina* und *Vidalina*—Fossilien des Riffbereichs. *Geol. Jb.*, **66**, 257–284.

Williamson, M. A. 1987. A quantitative foraminiferal biozonation of the Late Jurassic and Early Cretaceous of the East Newfoundland Basin. *Micropaleontology*, **33**, No. 1, 37–65.

Wilson, R. C. C. 1968a. Upper Oxfordian Palaeogeography of southern England. *Palaeogeog., Palaeoclimatol., Palaeoecol.*, **4**, 5–28.

Wilson, R. C. C. 1968b. Carbonate facies variation within the Osmington Oolite Series in southern England. *Palaeogeog., Palaeoclimatol., Palaeoecol.*, **4**, 89–123.

Wimbledon, W. A. and Cope, J. C. W. 1978. The ammonite faunas of the English Portlandian Beds and the zones of the Portlandian Stage. *J. geol. Soc. London.*, **135**, 183–190.

Wisniowski, T. 1890. Mikrofauna i ow ornatowych okolicy Krakowa. 1.—Otwornice gornego Kelloeayu w Grojco. *Pam. Akad. Um. Krakowie, Wydz. Mat.—Przyr.*, **17**, 181–242.

Wisniowski, T. 1891. Mikrofauna i ow ornatowych okolicy Krakowa. 2-Gabki gornego Kellawayu w Grojcu craz howe otwornice tych samych warstw. *Pam. Akad. Um. Krakowie, Wydz. Mat.-Przyr.*, **21**, 327–330.

Witthuhn, W. 1968. Schalensubstanz sub Schalenstruktur der gattung *Bolivina* Orb. (Foram.) aus dem Mittleren Lias Nordwestdeutschlands. *Beih. Ber. naturhist. Ges. Hannover.* **5**, 445–455.

Wood, A. and Barnard, T. 1946. Ophthalmidium. A study of nomenclature, variation and evolution in the foraminifera. *Quart. Jl. geol. Soc. London.*, **102**, 77–113.

Woodland, A. W. (ed.) 1971. The Llanbedr (Mochras Farm) Borehole. *Rep. Inst. geol. Sci. London* 71/18, 1–115.

Woodward, H. B. 1895. The Jurassic rocks of Britain (Yorkshire excepted) Vol. V. Middle and Upper Oolithic Rocks. *Mem. geol. Surv. G. B.*

Woollam, R. and Riding, J. B. 1983. Dinoflagellate Cyst Zonation of the English Jurassic. *I.G.S. Report*, No. 83/2, 1–42.

Zaninetti, L. 1976. Les foraminifères du Trias. *Riv. Ital. Paleont.* **82** (1), 1–258.

Ziegler, J. H. 1959 Mikropaläontologische Untersuchungen zur stratigraphie des Braunjura in Nordbayern. *Geol. Bavarica*, No. **40**, 9–128.

Ziegler, P. A. 1975. North Sea Basin history in the tectonic framework of Northwest Europe. In: Woodland, A. W. (ed.), *Petroleum and the Continental Shelf of Northwest Europe 1, Geology*. Appl. Sci. Publ., London, 131–149.

Ziegler, P. A. 1981. Evolution of sedimentary basins in Northwest Europe. In: Illing, L. V. and Hobson, G. D. (eds), *Petroleum Geology of the Continental Shelf of Northwest Europe. Proc. 2nd Conf., Inst. Pet.*, Heyden & Son Limited, London, 3–39.

Ziegler, P. A. 1982. Geological Atlas of Western and Central Europe. *Shell International Petroleum Maatschappij B. V.*, 1–130, Elsevier.

7

Cretaceous

M. B. Hart, H. W. Bailey, S. Crittenden,
B. N. Fletcher, R. J. Price, A. Swiecicki

7.1 INTRODUCTION

The Cretaceous System, proposed by d'Omalius d'Halloy in 1822, takes its name from Creta—the Latin for chalk. Chalk is the most distinctive rock type of the System, not only in Britain (Fig. 7.1), but over a considerable area of the globe. Upper Cretaceous sediments at one time covered the whole of northwest Europe, save for a few isolated areas of ancient massifs like the Baltic Shield. Today, sediments of Cretaceous age are limited (onshore) to a narrow band down eastern England, together with a wide tract of country across southern England. Offshore, Cretaceous strata are more extensive, covering, or known from, large areas of the North Sea Basin, the English Channel, the Celtic Sea and the South West Approaches Basin.

7.2 COLLECTIONS OF CRETACEOUS FORAMINIFERA

British Museum (Natural History), London
This museum contains the largest, and most important, collections of Cretaceous foraminifera. Many of the suites of specimens contain material from specific publications, although there is also a large amount of reference material from all over the British Isles. The most important collections are those of Barnard (1962, 1963), Barr (1962), Carter and Hart (1977), Chapman (1891–1898), Jukes-Browne (1898) and Williams-Mitchell (1948). Important samples of material, or selected foraminifera, have been donated to the collections by Adams, Blow, Brady, Curry, Davis, Elliot, Heron-Allen, Hodgkinson, James, Khan, Owen and Rowe.

British Geological Survey, Keyworth
The collections of the Geological Survey include material from shallow and deep boreholes, from both onshore and offshore. They have recently been moved from London and Leeds, both collections now being concentrated in the Keyworth site.

Department of Geology, Imperial College, London
The material from the Channel Tunnel Site Investigation (Albian–Turonian) is housed in this department under the supervision of Dr S. Radford.

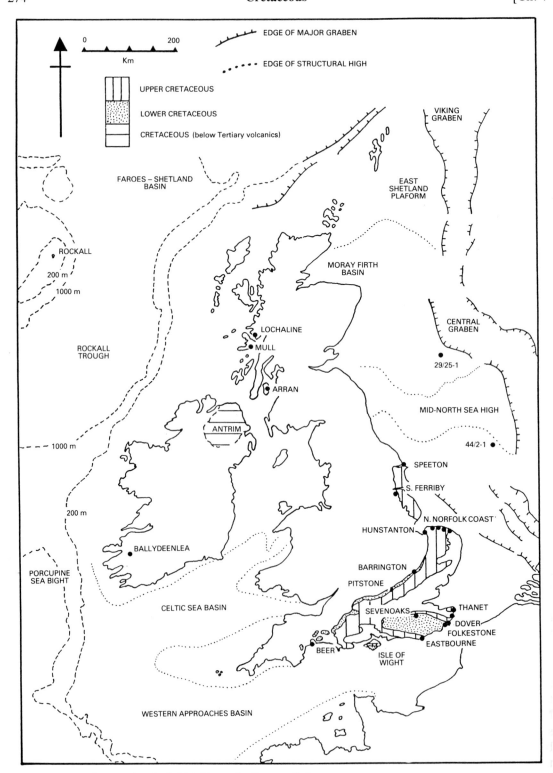

Fig. 7.1 — Locality map for the British Isles and adjacent off-shore areas.

There are other, smaller, collections at many universities and colleges throughout the British Isles, including University College London (Prof. F. T. Banner), Hull University, University College of Wales, Aberystwyth (Prof. J. Haynes), Southampton University (Prof. J. W. Murray) and Polytechnic South West, formerly Plymouth Polytechnic (Prof. M. B. Hart).

7.3 STRATIGRAPHIC DIVISIONS

The Cretaceous Period lasted approximately from 135–65 Ma. In recent years several attempts have been made to produce an acceptable time scale for the various stages within the Cretaceous (Casey, 1964; Bandy, 1967; Obradovich and Cobban, 1975; Kauffman, 1970; van Hinte, 1976; Harland et al., 1982; Birkelund et al., 1984; Hallam et al., 1985; Haq et al., 1987) but there has been little, if any, agreement. In Table 7.1 the stages used in this account are listed, together with some recent proposals for their duration.

7.4 HISTORY OF INVESTIGATIONS

British Cretaceous foraminifera were little used for stratigraphic, or other purposes, before 1948. All these early publications were monographic in nature, typified by the publications of Burrows et al. (1890) on the microfauna of the Red Chalk of Yorkshire, and Chapman (1891–1898) on the foraminifera of the Gault Clay.

Williams-Mitchell's (1948) account of the Chalk foraminifera from the Portsdown borehole was the first to suggest that a series of microfaunal zones could be established in the Upper Cretaceous. This stratigraphic approach was extended to the Upper Cretaceous of Northern Ireland by the work of McGugan (1957). There then followed a series of publications (Barnard, 1958, 1963a, b; Barnard and Banner, 1953) on selected taxa. The works of Burnaby (1962) and Jefferies (1962, 1963) showed how statistical methods could help in palaeoecological interpretation, but more than that they showed (particularly Jefferies (1962)) how a small part of the succession *should* be investigated in detail. From then on parts of the Cretaceous succession began to receive more detailed treatment, and this work is continuing.

The Speeton Clay of Yorkshire was fully investigated by Fletcher (1966, 'Lower Cretaceous foraminifera from the Speeton Clay, Yorkshire', unpublished thesis, University of Hull; 1973) and the foraminifera of the Red Chalk were described by Dilley (1969). The Lower Cretaceous of the North Sea Basin has recently been documented by Crittenden (1988, 'The lithostratigraphy and biostratigraphy (Foraminifera) of the Early Cretaceous of the southern North Sea Basin', unpublished thesis, Plymouth Polytechnic). In the south, the Gault Clay was investigated by Hart (1970, 'The distribution of the Foraminifera in the Albian and Cenomanian of SW England'; unpublished thesis, University of London; 1973), Carter and Hart (1977), Price (1975, 'Biostratigraphy of the Albian Foraminifera of north-west Europe'; unpublished thesis, University of London; 1976, 1977a, b) and Harris (1982, 'Albian microbiostratigraphy (foraminifera and ostracoda) of S.E. England and adjacent areas', unpublished thesis, Plymouth Polytechnic).

The Upper Cretaceous succession has been the subject of numerous major projects during the last twenty years (Barr, 1962, 1966a; Bailey, 1975, 1978, 'A foraminiferal biostratigraphy of the Lower Senonian of southern England;' unpublished thesis, Plymouth Polytechnic; Hart, 1970, unpublished thesis; Carter and Hart, 1977; Swiecicki, 1980. 'A foraminiferal biostratigraphy of the Campanian and Maastrichtian Chalks of the United Kingdom', unpublished thesis, Plymouth Polytechnic; Hart and Bailey, 1979; Bailey and Hart, 1979; Ball, 1985, 'A foraminiferal biostratigraphy of the Upper Cretaceous of the southern North Sea Basin (U.K. Sector)'; unpublished thesis, Plymouth

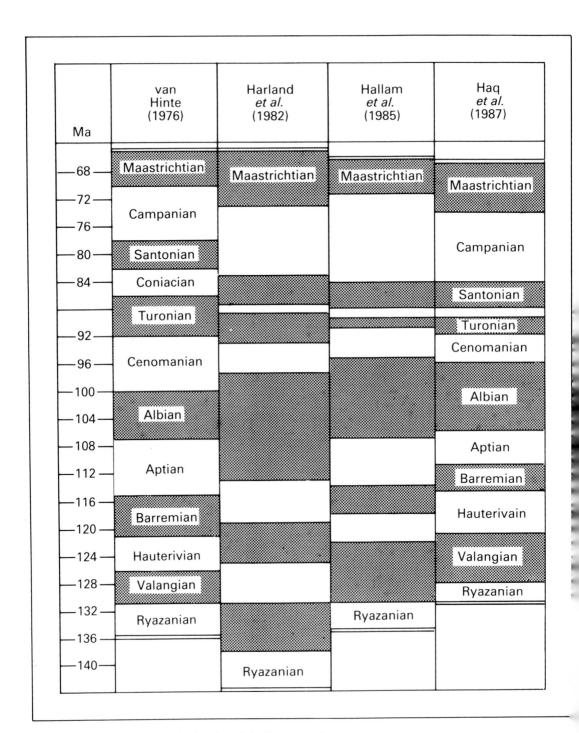

Table 7.1. Recent estimates for the duration of the Cretaceous Stages

Ryazanian—Based on the Ryazan Beds of the Moscow Platform (Sazonov, 1951), this stage was applied to the British succession by Casey (1973) and Rawson et al. (1978). In both accounts the problems of relating the Ryazanian to the Berriasian stage are discussed.

Valanginian—The type section in the Seyon Gorge, Valangin, Neuchatel, Switzerland, contains a succession of shallow-water, calcareous sediments. The fauna is rather limited, the foraminifera having little in common with the faunas recorded from Britain and the rest of northwest Europe.

Hauterivian—Renevier (1874) established this stage using a succession of shallow water carbonate sediments at Hauterive, Neuchatel, Switzerland. The microfauna, as above, is rather poor, and Kemper (1973a) has provided some important data concerning the correlation of this section with that in northwest Europe.

Barremian—Following work by Busnardo (1965) the type succession is based on the Angles (Basse-Alpes, France) road section. The limestones contain a good ammonite fauna but correlation of the microfauna again relies on data from northwest Germany (Neale, 1974; Fletcher, 1973).

Aptian—The succession of Apt (Basse-Alpes, France) has been fully described by Breistoffer (1947). The sub-stages proposed for that section are not used in Britain, Casey (1961) having proposed a simple grouping of zones into lower and upper sub-stages. The Aptian microfanua of the British Isles is relatively poorly known.

Albian—This stage, proposed by d'Orbigny (1842) for the Aptian–Cenomanian interval, takes its name from the Aube region of Eastern France. The foraminifera of the Aube have been exceptionally well described by Magniez–Jannin (1975) and this has allowed detailed correlation with the British Isles (Price, 1977a).

Cenomanian—Introduced by d'Orbigny for the marginal sands, gravels, marls and chalks of the Sarthe area of France, the boundaries of the stage have posed many problems (Rawson et al., 1978). The microfauna of the type area has been discussed by Marks (1967) and Carter and Hart (1977).

Turonian—The type area, from Saumur to Montrichard (along the Loire–Cher Valley), has been fully investigated in recent years (e.g. Lecointre, 1959). The foraminifera of the Cher Valley (just outside the designated type area) have been described by Butt (1966), Bellier (1971) and Carter and Hart (1977).

Coniacian—Introduced by Coquand (1857) and based on the succession near Cognac (Aquitaine); a detailed correlation with the British Isles is not yet possible (Rawson et al., 1978).

Santonian—Introduced by Coquand (1857) and based on the succession near Saintes (Aquitaine), the Santonian, like the Coniacian, has proved difficult to correlate precisely with the British succession by means of the macrofauna. The microfauna of the British Isles is quite diagnostic, with many zonally important species of planktonic foraminifera being recorded (Bailey, 1978, unpublished thesis), despite statements to the contrary in Rawson et al. (1978).

Campanian—The Campanian stage was introduced by Coquand in 1857, using sections described in 1856, notably those at Aubeterre-sur-Dronne in Charente. The deposition of the Campanian (and Maastrichtian) in SW France has recently been completely reviewed (Séronie-Vivien, 1972). The microfauna of the type sections is rather poor and correlation with other regions is difficult.

Maastrichtian—This stage was introduced by Dumont (1850) and based on sections near the town of Maastricht, Holland. The early definitions were rather unsatisfactory and recently (Voigt, 1956; Deroo, 1966; Schmid, 1959, 1967) the situation has been clarified. The type sections are, however, in a rather different facies to those normally encountered elsewhere in NW Europe and the British Isles.

Polytechnic; Hart and Ball, 1986; Leary, 1987, 'The Late Cenomanian anoxic event; implications for foraminiferal evolution', unpublished thesis, Plymouth Polytechnic; Hart and Swiecicki, 1987; Jarvis et al., 1988a,b) although views on the nature of the 'chalk palaeoenvironment' are as varied as ever.

Elsewhere in Europe, Cretaceous foraminifera have been investigated by many workers and some of their results are applicable to Britain. Of particular note are the publications of Robaszynski et al. (1980) in Belgium and northern France; Stenestad (1969) in Denmark; Berthelin (1880), Marie (1938, 1941), Goel (1965), and Magniez-Jannin (1975) in France; Bartenstein (1952, 1974, 1976a, b, 1977, 1978a, b, 1979), Bartenstein and Bettenstaedt (1962), Bartenstein et al., (1957), Bartenstein and Brand (1951), Bartenstein and Kaever (1973), Bettenstaedt (1952), and Aubert and Bartenstein (1976) in Germany; Fuchs (1967), and Ten Dam (1947, 1948a, b, 1950) in Holland; Gawor-Biedowa (1969, 1972) in Poland; Brotzen (1934a, b, 1936, 1942, 1945, 1948) in Sweden; Caron (1966) and Klaus (1960a, b, c) in Switzerland; Bartenstein and Kovatcheva (1982) in Bulgaria; and Neagu (1965, 1969) in Rumania. Recently many European micropalaeontologists have collaborated (under the auspices of IGCP No. 58 'Mid-Cretaceous Events') to produce a monograph of the mid-Cretaceous planktonic foraminifera (Robaszynski and Caron, 1979) together with a companion volume on the Late Cretaceous planktonic foraminifera (Robaszynski et al., 1984).

In America the Cretaceous foraminifera have been investigated by many specialists, including J. A. Cushman. Nearly every publication on Cretaceous foraminifera will include an extensive list of his references and none has been singled out for special mention here. Other workers on the Cretaceous include Loeblich and Tappan (1961, 1964), Eicher (1965, 1966, 1967, 1969) Eicher and Worstell (1970), Douglas (1969a, b) Sliter (1968), Pessagno (1967), Frizzell (1954), Plummer (1931), Olsson (1964), Petters (1977a, b), Ascoli (1976), Masters (1977), McNeil and Caldwell (1981), Nyong and Olsson (1984), Olsson and Nyong (1984) and Ascoli et al., (1984).

7.5 DEPOSITIONAL HISTORY

The Late Cimmerian movements at the end of the Jurassic (Ziegler, 1975) effectively terminated active sedimentation over large areas of Europe, and in particular, revived a series of positive areas between Britain and Poland (the London–Brabant, Rhenish, and Bohemian massifs). This formed a highly effective barrier between the 'Boreal' realm of the North Sea Basin, Danish–Polish Furrow and Russian Platform, and the 'Tethyan' area of Alpine Europe, and the evolving North Atlantic Ocean. This interconnected chain of massifs controlled surface water movements and faunal distributions well into the Late Cretaceous, even though they were probably inundated by the Early Turonian.

The British Isles, sitting astride the London-Brabant Massif, was effectively part of two depositional basins, and these were only occasionally interconnected in the pre-Aptian interval. During and after the Aptian the connection was almost continuous. Apart from a short-lived marine phase near the Jurassic–Cretaceous boundary (Casey, 1973) the southern basin remained essentially non-marine until the early Aptian (see Ziegler, 'Palaeogeographic atlas of north-west Europe'). There followed a marine or near-marine phase during the Aptian–Albian interval, during which time the Atherfield Clay, Lower Greensand, Gault Clay and Upper Greensand were deposited over southern England and the English Channel Basin. This interval also marks the first appearance of the foraminifera in the Cretaceous of southern England, although the fauna is very restricted in parts of the Lower Greensand. In the North Sea Basin (which effectively includes parts of Yorkshire, Humberside and Lincolnshire) the pre-Albian interval is characterised by the Speeton Clay Formation of the Cromer Knoll Group (Rhys, 1975). This marine clay succession is fossiliferous throughout and shows a marked similarity to faunas from north Germany (Lott, Fletcher and Wilkinson, 1986). The uppermost Lower Cretaceous in the North Sea Basin is characterised by the Red Chalk Formation. This thin, red-coloured, limestone succession contains a rich foraminiferal fauna that contains large numbers of small, simple, planktonic foraminifera. The same planktonic fauna can also be recognised in the uppermost part of the Gault Clay of southern England. The Upper Albian, therefore, marks the point in the succession when the two basins of deposition became faunally united. From the Cenomanian onwards the British fauna is, in most respects, uniform, although a few regional trends can be identified. As early as the Late Cenomanian the fauna becomes recognisable as typical of the whole of northern Europe, with many species being quite widespread in occurrence. At several levels (notably in the Upper Cenomanian, Turonian, Santonian, and Maastrichtian (especially in the North Sea Basin)), planktonic foraminifera of international biostratigraphic importance are found, despite statements to the contrary (Rawson et al., 1978, p.25). The apparently uniform chalk succession of the Cenomanian–Maastrichtian interval means that the evolution of such benthonic genera as *Gavelinella*, *Bolivinoides*, and *Stensioeina* is of great value in inter-regional correlation (Bailey and Hart, 1979; Hart and Swiecicki, 1987).

Since all the British Cretaceous sediments were being deposited in an extensive shelf sea, changes in sea level were clearly of fundamental importance. The eustatic changes now known from the Cretaceous (Hancock, 1976; Hancock and Kauffman, 1979; Cooper, 1977; Haq et al. 1987) have had a profound effect on the distribution, and evolution, of both the benthonic and planktonic foraminifera (Hart and Bailey, 1979; Swiecicki, 1980, unpublished thesis; Hart, 1980a, b). In the last few years it has also been suggested that the presence of an expanded oxygen minimum zone and dysaerobic bottom waters may have controlled the evolution and distribution of the fauna (Schlanger and Jenkyns, 1976; Hart and Bigg, 1981; Schlanger et al., 1987; Hart, 1985; Hart and Ball, 1986; Leary, 1987, unpublished thesis; Jarvis et al., 1988b).

7.6 STRATIGRAPHIC DISTRIBUTION OF CRETACEOUS FORAMINIFERA

Many sections of Cretaceous strata are available for investigation in the British Isles, and those chosen for inclusion in this account are probably some of the best exposed, and at present yield the best, most typical, biostratigraphic data. This has inevitably meant that successions in marginal areas such as Devon, Northern Ireland, northwest Scotland, and to some extent Central England, have been omitted from the discussion. In Devon and Dorset the faunas of the Albian to Coniacian interval have been thoroughly investigated by Hart (1970, unpublished thesis; 1973, 1975), Hart and Weaver (1977), Hart et al. (1979), Bailey (1975, 1978, unpublished thesis), and Leary (1987, unpublished thesis). In Northern Ireland the Cretaceous succession is poorly exposed and much of the Chalk is very hard, having been affected by Tertiary volcanism. The work of McGugan (1957) is the most complete account of the foraminifera of that area, although Barr (1966b) has described a Cretaceous fauna from a cave infill in Eire. The Scottish sections have been considered occasionally over the years, but unpublished work by one of us (MBH) on material from Mull and Lochaline has failed to provide data worthy of inclusion in this account. This area is currently being totally reinvestigated by Ms Sharon Braley (Polytechnic South West).

The successions considered in this account, therefore, are those at Speeton, Folkestone, Dover, southeast Kent, Scratchells Bay/Alum Bay (Isle of Wight), Ather-

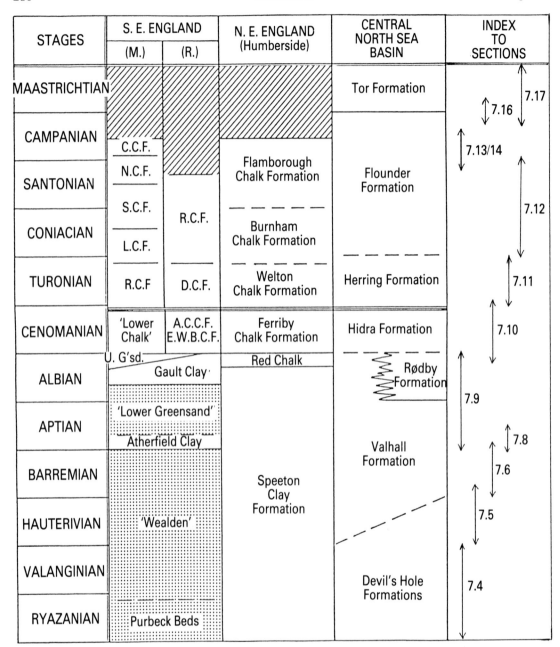

Fig. 7.2 — Stages of the Cretaceous and lithostratigraphic nomenclature for the British Isles. The right-hand column gives the distribution of the various range charts used in the text. There have been two recent attempts at providing a lithostratigraphy of the chalk group (M — Mortimore, 1986; R — Robinson, 1986) and as there has been no time to assess their relative merits both are included in this figure. The abbreviations are: (M):CCF — Culver Chalk Formation, NCF — Newhaven Chalk Formation, SCF — Seaford Chalk Formation, LCF — Lewes Chalk Formation, RCF — Ranscombe Chalk Formation, (R):RCF — Ramsgate Chalk Formation, DCF — Dover Chalk Formation, ACCF — Abbots Cliff Chalk Formation, EWBCF — East Wear Bay Chalk Formation. The double line that runs across all the columns in the latest Cenomanian is the *A. plenus* Marls/Black Band. In the Isle of Wight the Upper Greensand (UG'sd.) appears as a facies in the latest Albian. The stippled ornament indicates a predominantly non-marine facies while the ruled ornament indicates a lack of strata of that age in the area indicated.

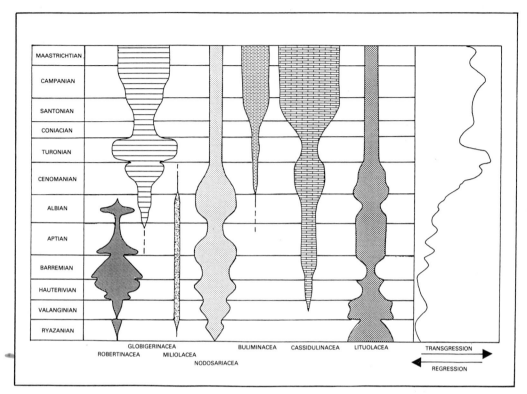

Fig. 7.3 — Schematic distribution of the major superfamilies of Foraminifera throughout the Cretaceous.

field Point (Isle of Wight), the north Norfolk coast, and the southern North Sea Basin (well 44/2-1). These are shown in Fig. 7.2, together with the appropriate lithostratigraphic terminology.

Throughout the Cretaceous succession the fauna changed in response to the environmental changes brought about by marked changes in sea level (see Fig. 7.3). These fluctuations drastically changed the nature of the fauna from a Lituolacea/Nodosariacea/Robertinacea-dominated fauna in the Early Cretaceous to a Lituolacea/Cassidulinacea/Buliminacea/Globigerinacea fauna in the Late Cretaceous. This is summarised in Fig. 7.3 which gives an indication of the relative dominance of all the superfamilies throughout the succession. The use made of these various superfamilies is also interesting. The Nodosariacea are important for biostratigraphic work in the Early Cretaceous as there is little else available, while in the Late Cretaceous (though still present in reasonable numbers) they are little used, as other groups appear more attractive. The Lituolacea are mainly used for zonation in the mid-Cretaceous where the *Arenobulimina, Plectina*, and *Flourensina* lineages are well-developed. In the Early Cretaceous and Late Cretaceous the Lituolacea are rarely used, even though some genera and species appear to be quite distinctive at certain levels in the succession (Adams *et al.*, 1973). In the marginal facies of the mid-Cretaceous, *Orbitolina* has been recorded widely (Hart *et al.* 1979) and recent work by Schroeder *et al.* (1986) has clarified the stratigraphic position of this genus in the UK succession.

The Globigerinacea are perhaps the most useful for international correlation (as distinct from inter-regional) and many key species are now being recorded from the British succession for the first time. Magniez-Jannin (pers. comm. 1981) has found specimens of

Rotalipora appenninica (Renz) and *Planomalina buxtorfi* (Gandolfi) in the uppermost Albian of Folkestone. Other distinctive species such as *Rotalipora reicheli* (Mornod), *R. greenhornensis* (Morrow), *Praeglobotruncana helvetica* (Bolli), *Marginotruncana coronata* (Bolli), *M. renzi* (Gandolfi), *M. sigali* (Reichel), *Dicarinella concavata* (Brotzen), *Rosita fornicata* (Plummer), *R. contusa* (Cushman), and *Globotruncanella havanensis* (Voorwijk) are also now being recorded, and while their stratigraphic distribution must be viewed after consideration of the palaeogeography and palaeocirculation patterns, many do appear to have real value.

In the mid-Late Cretaceous, apart from the Turonian (Hart and Bailey, 1979) the fauna is dominated by the benthonic foraminifera, and an accurate interregional correlation (Bailey and Hart, 1979) can be effected using members of the *Gavelinella*, *Stensioeina*, and *Bolivinoides* lineages. These have been well documented by Koch (1977), Bailey (1978, unpublished thesis) and Swiecicki (1980, unpublished thesis). The prolonged period of chalk sedimentation seems to have allowed their gradual evolution in the northwest European area, and a very refined stratigraphy is now possible.

The palaeoenvironments represented by these faunas are all very similar. Throughout the Cretaceous, marine sedimentation was typical of a shallow-water to deeper-water shelf area. In no case can one recognise the passage from marine to non-marine deposition by means of foraminifera. Most workers have therefore concentrated on estimating or guessing the depth of water over the shelf at any particular time. Various estimates range from 9.0 m to 600 m in the mid-Late Cretaceous (Burnaby, 1962; Hart and Bailey, 1979; Kennedy, 1970; Hancock, 1976; Hancock and Kauffman, 1979). Recently the problem of 'anoxic events' in the Cretaceous succession has been considered by Schlanger and Jenkyns (1976) and Hart and Bigg (1981), especially with reference to the Black Band in Yorkshire. More recently there has been considerable interest in the world wide, Late Cenomanian, oxygen minimum zone expansion and the recognition of the typical $\delta^{13}C$ signal preserved in many oceanic and shelf successions (Hilbrecht and Hoefs, 1986; Hart and Ball, 1986; Schlanger *et al.*, 1987; Jarvis *et al.*, 1988a,b; Hart and Leary, 1989). The effects this event has wrought on the foraminiferal fauna are quite significant, with detailed models being provided by Leary (1987, unpublished thesis) and Jarvis *et al.* (1988a,b).

7.7 BIOSTRATIGRAPHY

The range charts (Figs 7.4–7.6, 7.8–7.14, and 7.16–7.17) present the stratigraphic distribution of foraminifera as seen in several key sections in southern and eastern England, including one composite section from the southern North Sea Basin (with the approval of Shell UK Exploration and Production Ltd and ESSO Exploration and Production UK Ltd). A summary chart (Fig. 7.18) is also included for the full Cretaceous succession.

Within the Lower Cretaceous succession planktonic foraminifera are extremely rare, and, even when present, are of little value for region or international correlation. In the post-Albian succession, however, a biostratigraphy can be created using planktonic taxa, parts of which are of direct value in international correlation.

Throughout the whole Cretaceous succession, however, the benthonic foraminifera can be used for zonation (Carter and Hart, 1977) and correlation (Barteinstein, 1977). In the Early Cretaceous the almost separated basins of northern and southern England require separate consideration and a single zonation for detailed correlation may probably never emerge. This is certainly not the case for the post-Albian interval where a zonation, developed in the chalk facies of south-east England, can be applied to many areas of northwest Europe (France, Germany, Holland, Belgium, Denmark and

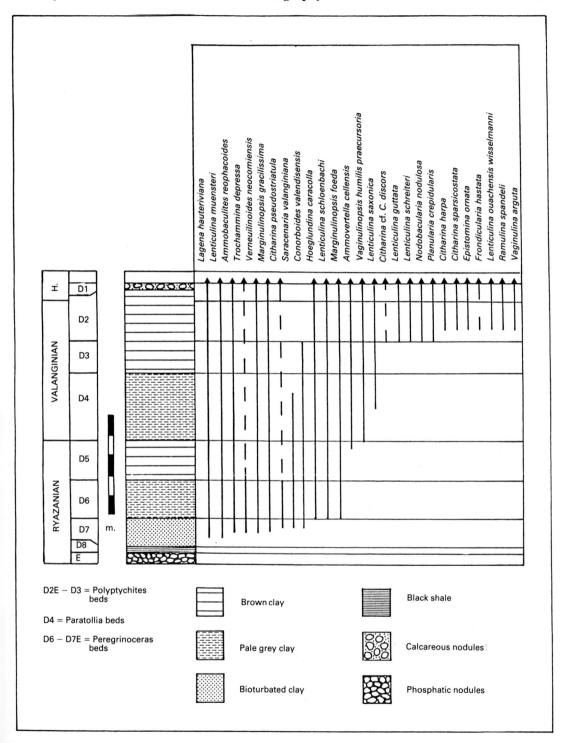

Fig. 7.4 — The Speeton Clay Formation (Ryazanian and Valanginian) of Speeton, Yorkshire.

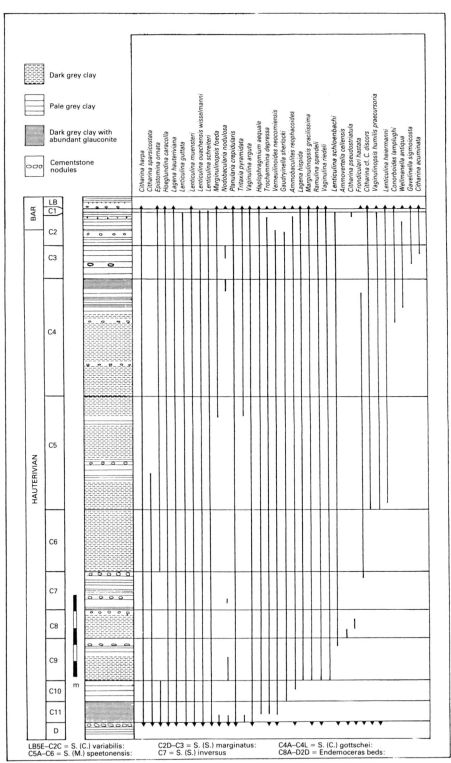

Fig. 7.5 — The Speeton Clay Formation (Hauterivian and earliest Barremian) of Speeton, Yorkshire.

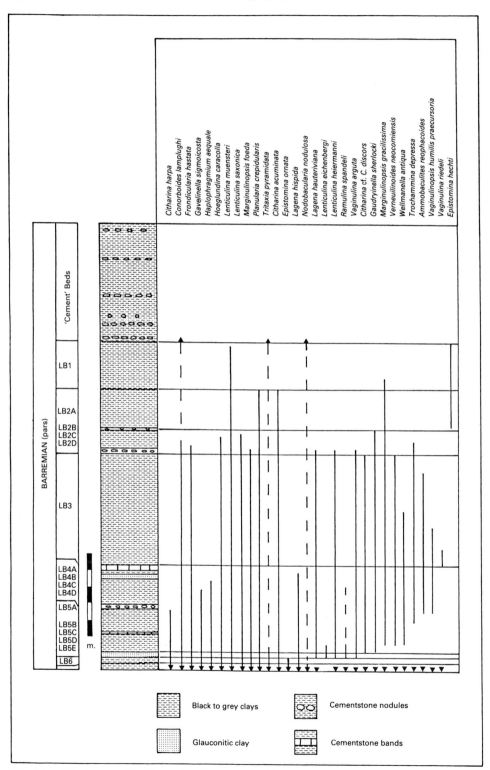

Fig. 7.6 — The Speeton Clay Formation (Barremian) of Speeton, Yorkshire.

286   **Cretaceous**   [Ch. 7

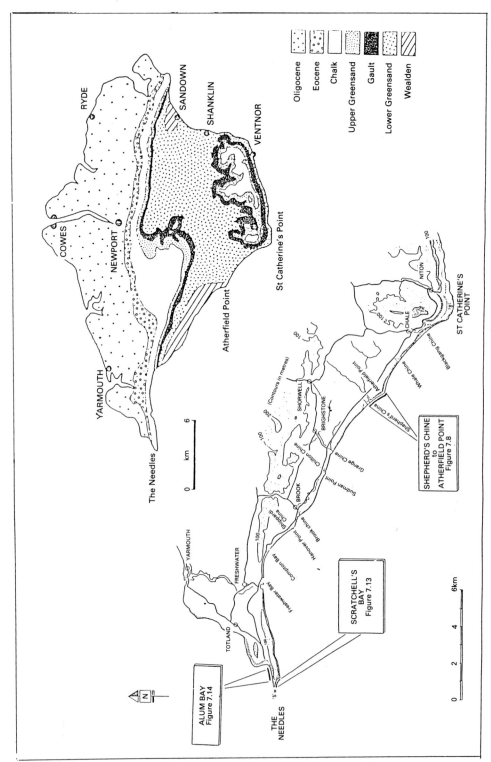

Fig. 7.7 — Geological map of the Isle of Wight and location of key sections.

Sec. 7.7] Biostratigraphy 287

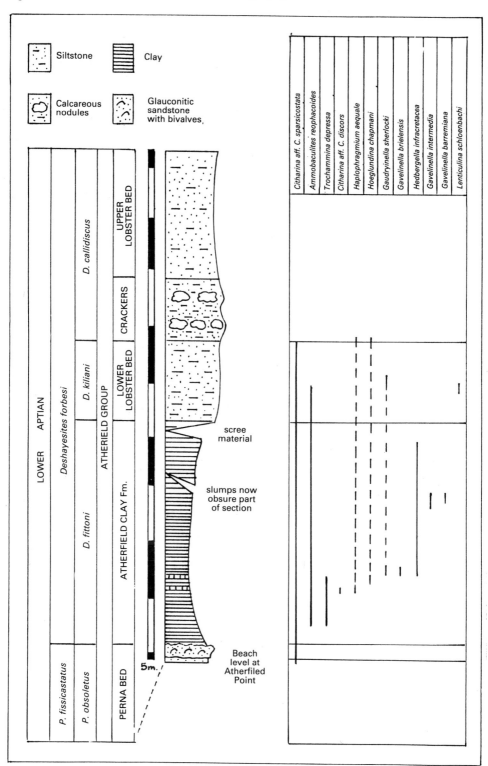

Fig. 7.8 — The Atherfield Clay Formation (Early Aptian) of Atherfield Point, Isle of Wight.

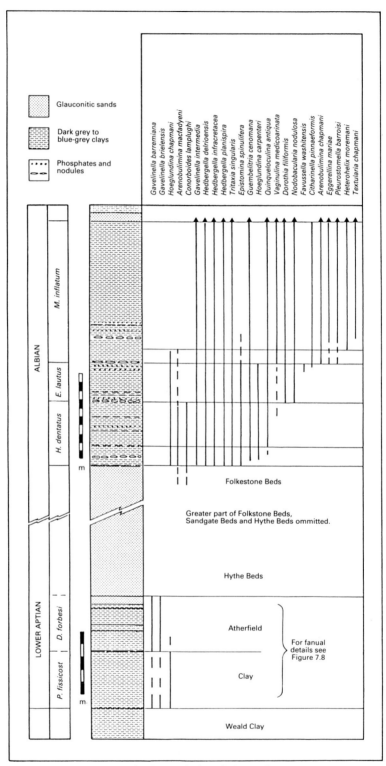

Fig. 7.9 — The Atherfield Clay Formation (Early Aptian) and Gault Clay Formation (Middle and Late Albian) of Sevenoaks and Folkestone, Kent.

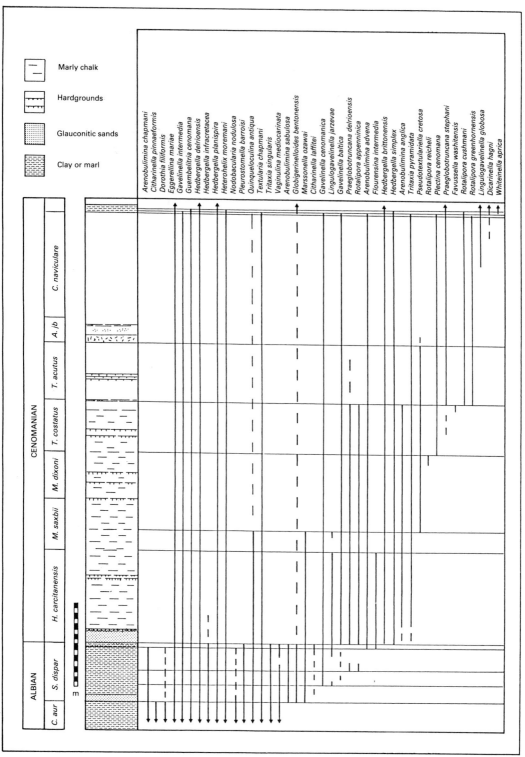

Fig. 7.10 — The uppermost Gault Clay Formation (Late Albian) and Lower Chalk (Cenomanian) of Folkestone and Dover, Kent.

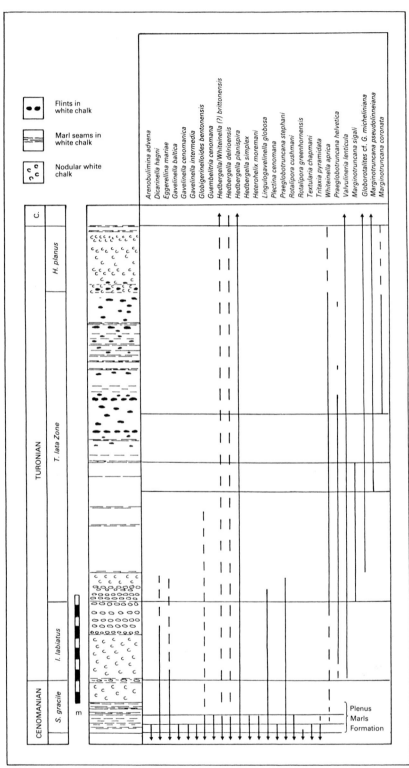

Fig. 7.11 — The Middle Chalk (Turonian) of Dover, Kent.

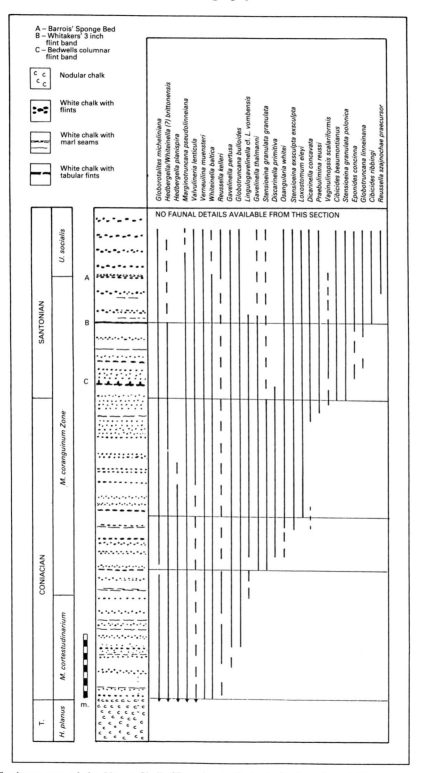

Fig. 7.12 — The lower part of the Upper Chalk (Turonian to Santonian) of south east Kent.

292 Cretaceous [Ch. 7

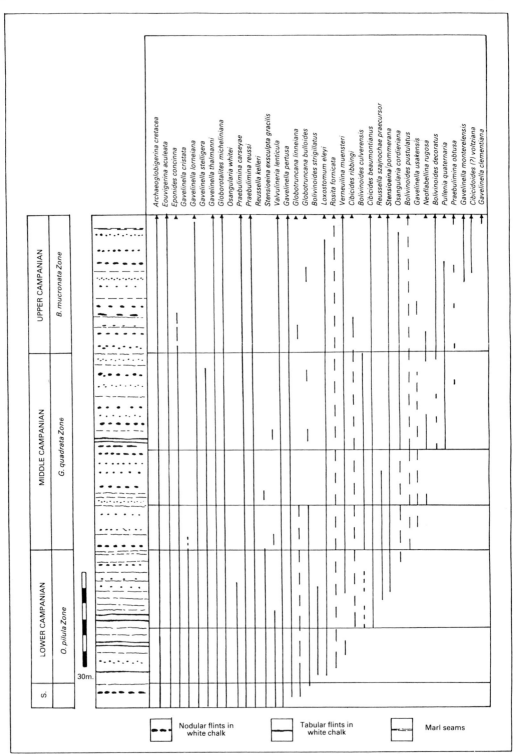

Fig. 7.13 — The middle part of the Upper Chalk (Santonian to late Campanian) of Scratchell's Bay, Isle of Wight. N.B. this section is accessible only by boat, and then only in favourable weather conditions.

Sec. 7.7] Biostratigraphy 293

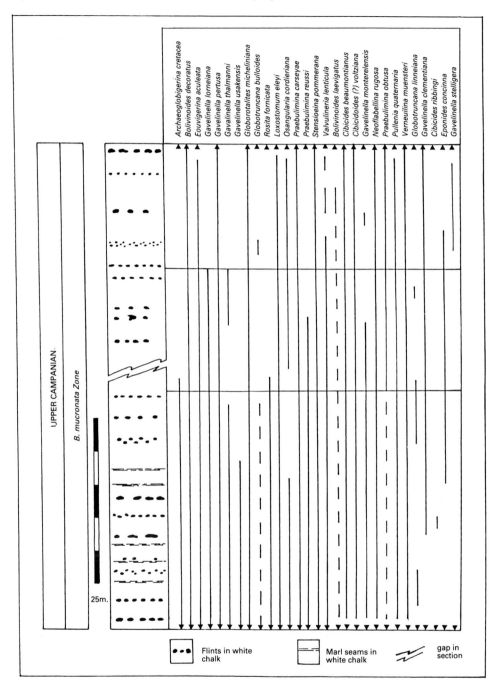

Fig. 7.14 — The middle part of the Upper Chalk (Late Campanian) of Alum Bay, Isle of Wight. N.B. While this section extends southwards through the Needles into Scratchell's Bay (Fig. 7.13) on no account should this be attempted on foot.

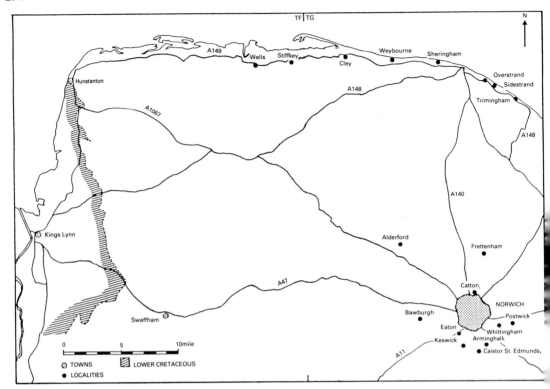

Fig. 7.15 — Upper Cretaceous localities in Norfolk, some of which are used to compile Fig. 7.16.

Sweden). The Late Cretaceous (post-Albian) planktonic (UKP.1–UKP.17) and benthonic (UKB.1–UKB.22) are now formally defined and illustrated in Figs 7.19–7.25. Of necessity, some of the taxa mentioned in these zonations are not figured/described in this index, for reasons given previously. All are, however, very well known from the literature.

7.7.1 Planktonic zonation
At Folkestone the base of the chalk succession is marked by an erosion surface cut into the micaceous silts of latest Albian age. Recently Magniez-Jannin (1981) has recorded the presence of *Planomalina buxtorfi* (Gandolfi) and *Rotalipora appeninica* (Renz) immediately below this major lithological change. As the fauna is very restricted it is impossible to satisfy the criteria for the recognition of a *P. buxtorfi* Zone. Accordingly the first appearance of *R. appeninica* is taken as the beginning of the Late Cretaceous planktonic foraminiferal zonation in the U.K.

UKP.1 *Rotalipora appeninica* (Renz) Interval Zone
Age: Early Cenomanian (see Fig. 7.19)
Definition: taken from the first appearance of *R. appeninica* to the first appearance of *Rotalipora reicheli* Mornod.
R. appeninica is relatively rare in strata of earliest Cenomanian age in the U.K. The first appearance is coincident with a major influx of *Globigerinelloides bentonensis* (Morrow), which appears to occur at the same level all over southeast England. The same influx is used as a substitute Albian-cenomanian boundary over much of the North Sea Basin, where it appears to be—within the limits of resolution—synchronous. The *R. appeninica* Zone contains a poorly

preserved, low-diversity, planktonic fauna that is dominated by tiny hedbergellids. Counts based on the 250–500 μm size fraction (see Figure 7.19) indicate that throughout this zone there is a very impoverished planktonic fauna, with a maximum level of 12%.

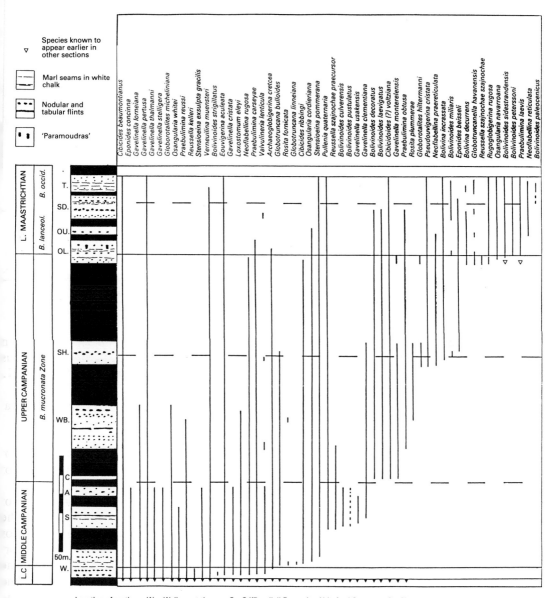

Fig. 7.16 — The upper part of the Upper Chalk (Early Campanian to Early Maastrichtian) of the North Norfolk coast. This is a composite section based on several localities; for a detailed description of the section see Swiecicki (Thesis, 1980).

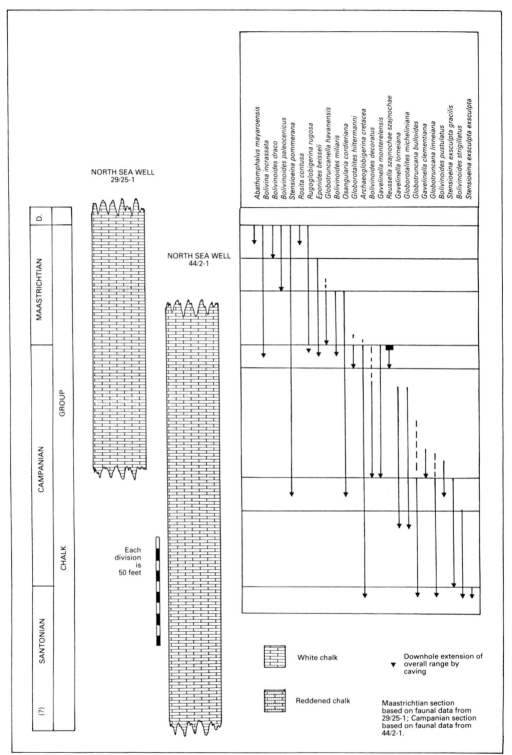

Fig. 7.17 — The upper part of the Chalk Group in the southern North Sea Basin (Maastrichtian to Campanian).

UKP.2 Rotalipora reicheli Mornod Taxon Range Zone
Age: early Middle Cenomanian (see Fig. 7.19)
Definition: the total range of *R. reicheli*.
R. reicheli is a distinctive species which appears abruptly, in moderate numbers, just below the base of the Middle Cenomanian (based on the ammonite zonation of Kennedy, 1969). After this initial influx the species is very rare and, at the present time, has only been found from localities south of the River Thames. Within the upper levels of this zone there is a distinctive occurrence of *Favusella washitensis* (Carsey). The top of the zone is, however, best marked by the appearance of *Rotalipora cushmani* (Morrow) in considerable numbers.

UKP.3 Rotalipora cushmani (Morrow) Taxon Range Zone
Age: Middle and Late Cenomanian (see Fig. 7.19)
Definition: the total range of *R. cushmani*.
R. cushmani appears immediately above a distinctive flood of *F. washitensis*. This level is also associated with major changes in the benthonic population (the mid-Cenomanian non-sequence of Carter and Hart, 1977). Throughout the zone *R. cushmani* can be found in association with *Praeglobotruncana stephani* (Gandolfi) and *Rotalipora greenhornensis* (Morrow). Within the *A. plenus* Marls, immediately below the level of maximum $\delta^{13}C$ isotope excursion (Jarvis *et al.*, 1988b; Schlanger *et al.*, 1987; Hilbrecht and Hoefs, 1986) *R. cushmani* becomes extinct a few centimetres above the extinction of *R. greenhornensis* (see Hart, 1985). *R. cushmani* does not occur throughout the North Sea Basin and is only rarely found in Lincolnshire and Humberside. In situations where the first appearance downhole of *R. cushmani* can be identified it may substitute for the Cenomanian–Turonian boundary, although as can be seen in Figure 7.19 *R. cushmani* becomes extinct a little way below that level.

UKP. 4 Whiteinella archaeocretacea Pessagno Interval Zone
Age: latest Cenomanian and very earliest Turonian (see Figs 7.19 and 7.20)
Definition: from the extinction of *R. cushmani* to the first appearance of diagnostic *Praeglobotruncana helvetica* (Bolli).
With the extinction of the *Rotalipora* fauna within the *A. plenus* Marls the planktonic fauna becomes dominated by forms transitional between *Hedbergella* and *Whiteinella* (including *W. aprica* (Loeblich and Tappan), *W. archaeocretacea* Pessagno, and *Hedbergella brittonensis* Loeblich and Tappan). Associated with this non-keeled fauna (for full discussion see Leary, 1987 (unpublished thesis) and Jarvis *et al.*, 1988b) are rare *Praeglobotruncana stephani*, *Dicarinella hagni* (Scheibnerova), *D. algeriana* (Caron) and very rare, early marginotruncanids. Also present, especially in the hard nodular chalks above the *A. plenus* Marls, are early forms of *P. helvetica* that are normally ascribed to *P. praehelvetica* (Trujillo). The boundary between these morphotypes and the 'true' *helvetica* is very subjective as the typical tethyan form of the species (large form, seven chambers in final whorl, strong single keel, etc.) is never seen in the U.K. Appearing at about this level are early forms of *Dicarinella imbricata* (Mornod).

UKP.5 Praeglobotruncana helvetica (Bolli) Interval Zone
Age: early Turonian (see Fig. 7.20)
Definition: from the first appearance of *P. helvetica* to the first appearance of *Marginotruncana sigali* (Reichel).
As indicated above there are problems related to the recognition of the first true *P. helvetica*. The range of the taxon is further complicated by its general rarity, especially north of the River Thames. Indeed, it is most abundant in the successions of southeast Devon (Hart and Weaver, 1977). While it would be useful if the full range of the species could be used as a taxon range zone, this is not

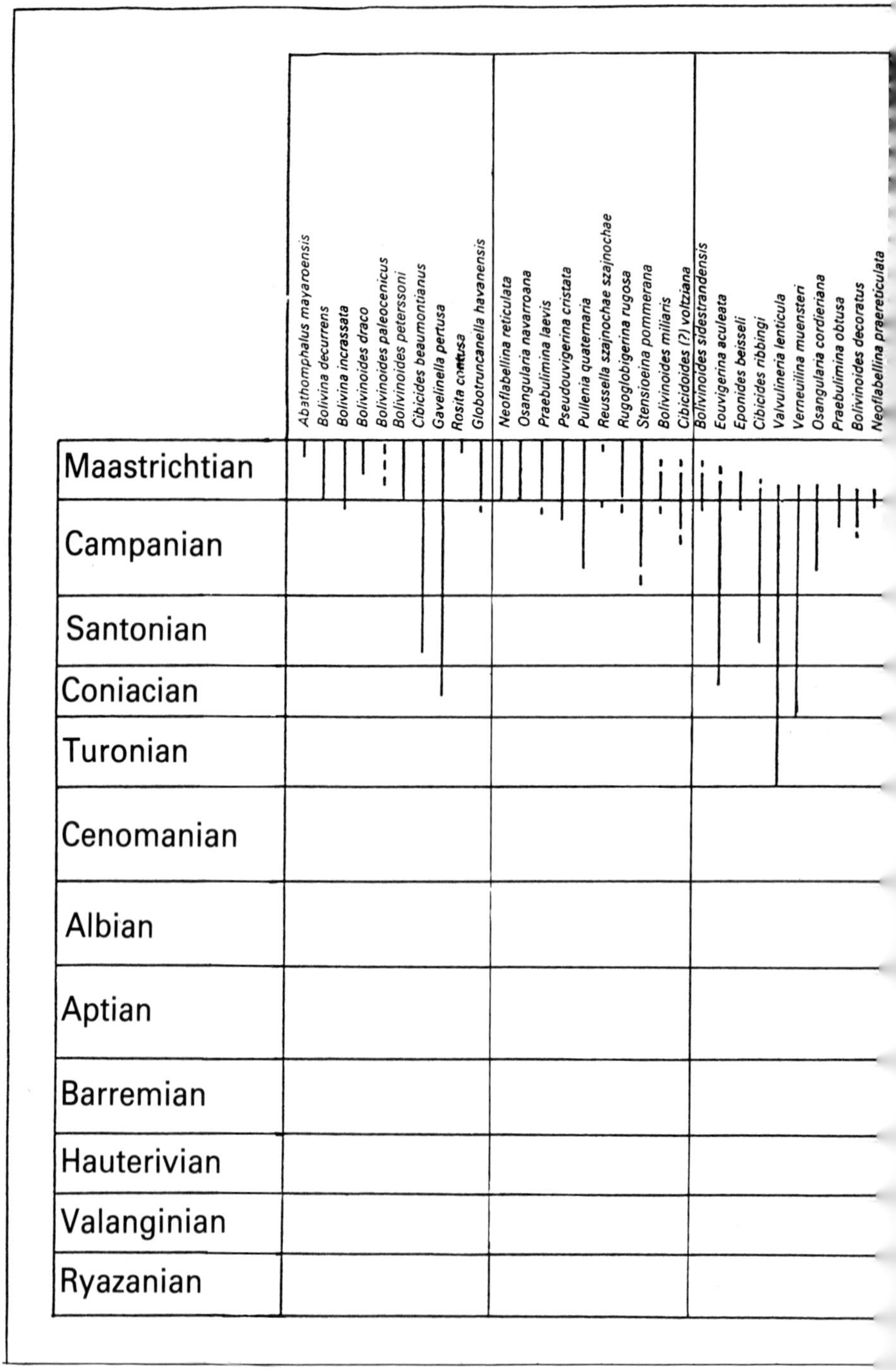

Fig. 7.18 — Summary of the distribution of Cretaceous Foraminifera.

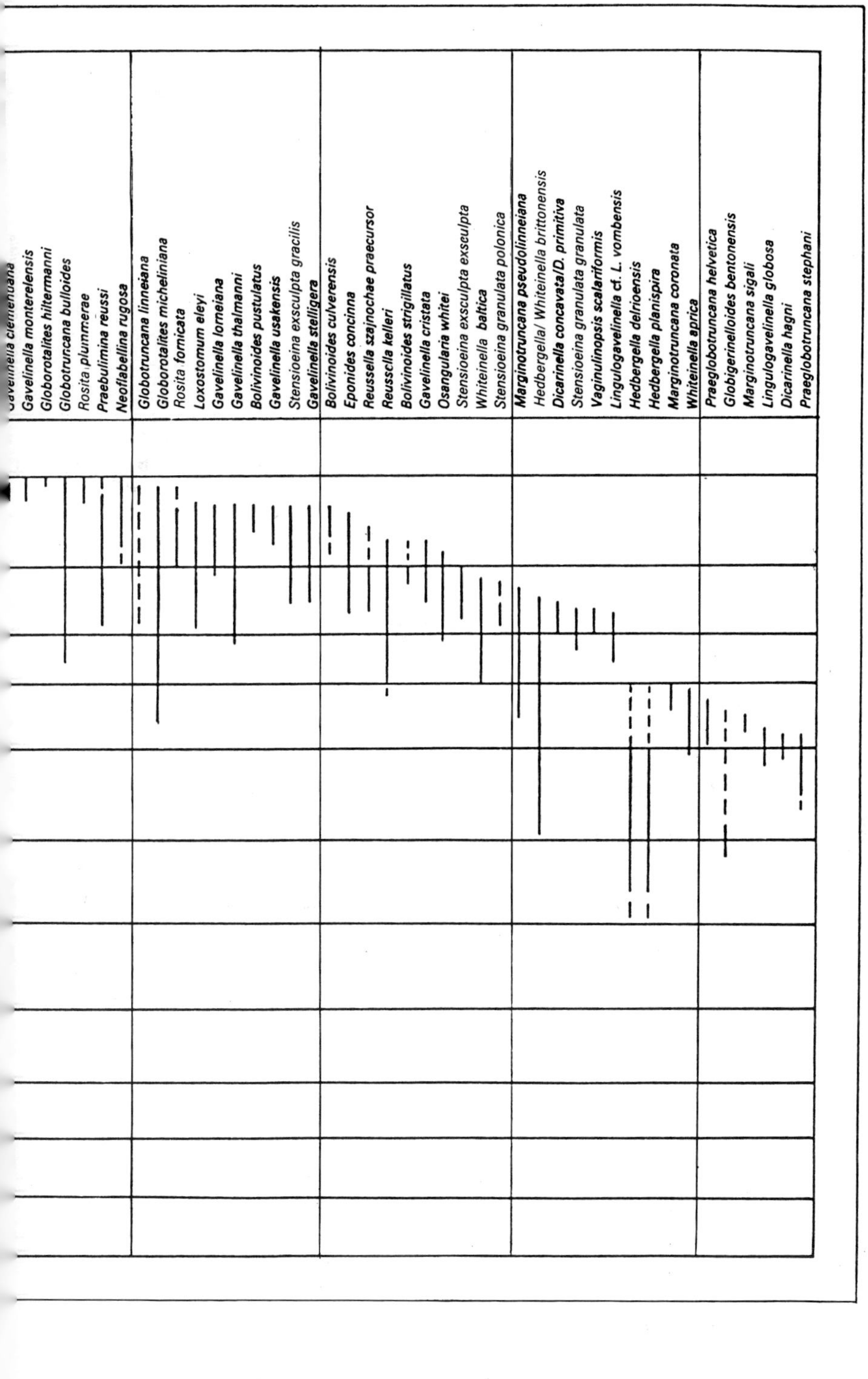

Biostratigraphy chart (species range chart, Sec. 7.7, p. 299)

Species (left to right across the chart):

- Gavelinella clementiana
- Gavelinella monterelensis
- Globorotalites hiltermanni
- Globotruncana bulloides
- Rosita plummerae
- Praebulimina reussi
- Neoflabellina rugosa
- Globotruncana linneiana
- Globorotalites micheliniana
- Rosita fornicata
- Loxostomum eleyi
- Gavelinella lorneiana
- Gavelinella thalmanni
- Bolivinoides pustulatus
- Gavelinella usakensis
- Stensioeina exsculpta gracilis
- Gavelinella stelligera
- Bolivinoides culverensis
- Eponides concinna
- Reussella szajnochae praecursor
- Reussella kelleri
- Bolivinoides strigillatus
- Gavelinella cristata
- Osangularia whitei
- Stensioeina exsculpta exsculpta
- Whiteinella baltica
- Stensioeina granulata polonica
- Marginotruncana pseudolinneiana
- Hedbergella / Whiteinella brittonensis
- Dicarinella concavata/D. primitiva
- Stensioeina granulata granulata
- Vaginulinopsis scalariformis
- Lingulogavelinella cf. L. vombensis
- Hedbergella delrioensis
- Hedbergella planispira
- Marginotruncana coronata
- Whiteinella aprica
- Praeglobotruncana helvetica
- Globigerinelloides bentonensis
- Marginotruncana sigali
- Lingulogavelinella globosa
- Dicarinella hagni
- Praeglobotruncana stephani

	Eggerellina mariae	Guembelitria cenomana	Hedbergella simplex	Heterohelix moremani	Rotalipora cushmani	Arenobulimina advena	Gavelinella baltica	Gavelinella cenomanica	Gavelinella intermedia	Plectina cenomana	Textularia chapmani	Tritaxia pyramidata	Quinqueloculina antiqua	Rotalipora greenhornensis	Pseudotextulariella cretosa	Praeglobotruncana delrioensis	Arenobulimina anglica	Favusella washitensis	Rotalipora appeninica	Rotalipora reicheli	Lingulogavelinella jarzevae	Marssonella ozawai	Flourensina intermedia	Arenobulimina chapmani	Arenobulimina sabulosa	Citharinella laffitei	Dorothia filiformis	Hedbergella infracretacea
Maastrichtian																												
Campanian																												
Santonian																												
Coniacian																												
Turonian																												
Cenomanian	│	│	│ │	│	│ │ │	│ │	│	│ │	│ │	│ │	│ │	│ │	│	│ :	│ .	│ .	│ .	│ .	│ │ │	│								
Albian	│			│		│				│							.	.									│
Aptian																												│
Barremian											│		│															│
Hauterivian																												
Valanginian																												
Ryazanian																												

Biostratigraphy

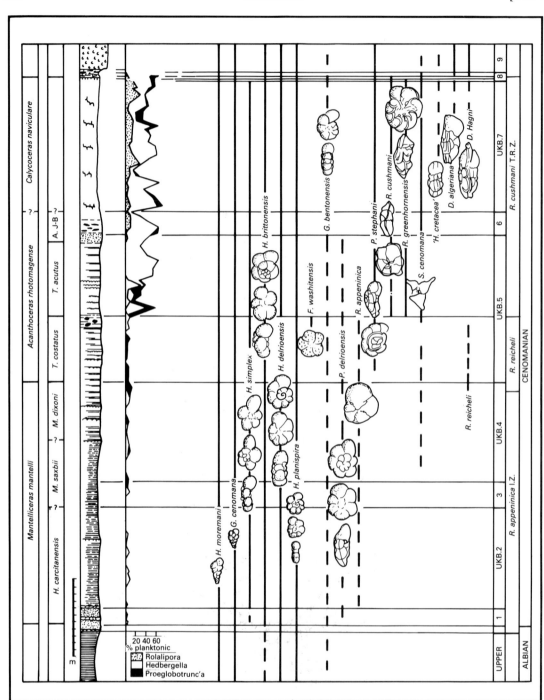

Fig. 7.19 — Planktonic zonation of the Lower Chalk (Cenomanian). The section is that between Folkestone and Dover, Kent.

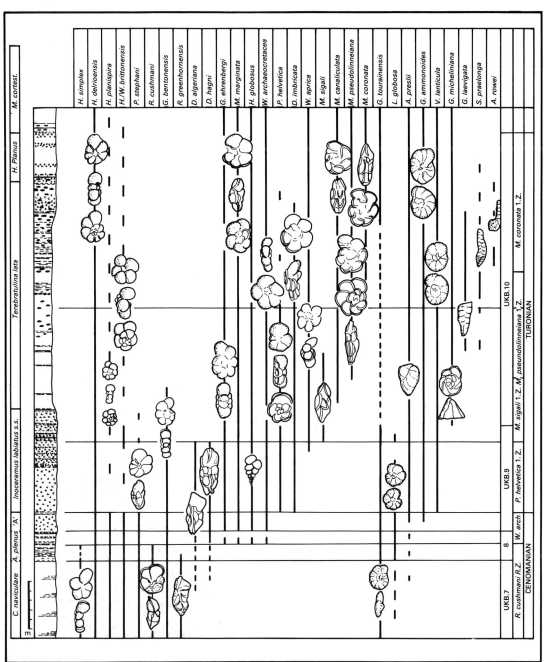

Fig. 7.20 — Planktonic and benthonic zonation of the Middle Chalk (Turonian). The section is that around Dover, Kent.

possible because of its scarcity. It must also be stressed that use of the *P. helvetica* Interval Zone *does not* indicate synchronism with the *P. helvetica* Zone used in Tethyan successions. Towards the top of its range in the U.K., the widely accepted shallowing of the seaway (Hancock, 1976; Hancock and Kauffman, 1979) was probably a limiting factor.

UKP.6 *Marginotruncana sigali* (Reichel) Interval Zone
Age: mid-Turonian (see Fig. 7.20)
Definition: first appearance of *M. sigali* to the first appearance of *M. pseudolinneiana* Pessagno.
Within the mid-Turonian there is a very distinctive occurrence of single-keeled planktonic taxa, although—as before—this zone cannot be traced into northern England. As *M. pseudolinneiana* is very common in chalks of Late Turonian age it is easier to use that appearance rather than the extinction of *M. sigali*.

UKP.7 *Marginotruncana pseudolinneiana* Pessagno Interval Zone
Age: mid- to Late Turonian (see Fig. 7.20)
Definition: first appearance of *M. pseudolinneiana* to the first appearance of *M. coronata* (Bolli).
The first appearance of *M. pseudolinneiana* marks a very distinctive horizon in the U.K. chalk succession. Above this level species of *Marginotruncana* and *Globotruncana* are generally present in reasonable numbers although, for the most part, the U.K. succession does not yield zonally diagnostic taxa.

UKP.8 *Marginotruncana coronata* (Bolli) Interval Zone
Age: Late Turonian (see Fig. 7.20)
Definition: first appearance of *M. coronata* to the first appearance of *Whiteinella baltica* (Douglas and Rankin).
The Late Turonian is characterised by a *M. coronata/M. pseudolinneiana* assemblage although this fauna is much reduced by the onset of the hard nodular chalks which typify the Turonian–Coniacian boundary interval. The relationship between *W. baltica* and *W. alpina* Porthault was fully discussed by Robaszynski and Caron (1979) but in Tethyan areas the range of these whiteinellids with four chambers in the final whorl is drawn from Late Cenomanian to Santonian.

In the U.K., typical *W. baltica* (identical to the holotype of Douglas and Rankin, 1969) are only found in strata of Coniacian and Santonian age.

UKP.9 *Whiteinella baltica* Douglas and Rankin Interval Zone
Age: Early Coniacian (see Fig. 7.21)
Definition: first appearance of *W. baltica* to the first appearance of *Dicarinella primitiva* (Dalbiez).
While this interval is characterised by the dominance of the planktonic fauna by *W. baltica* there are some interesting occurrences of other taxa in this zone. None of these can be used for stratigraphic correlation as their occurrence simply record the northward migration of Tethyan taxa. Notable are the records of *Marginotruncana sinuosa* and *M. renzi*, although the appearance of *Globotruncana bulloides* Vogler is a little more diagnostic as it continues as an important element of the fauna up into the Campanian.

UKP.10 *Dicarinella primitiva* (Dalbiez) Interval Zone
Age: Late Coniacian to earliest Santonian (see Fig. 7.21)
Definition: first appearance of *D. primitiva* to first appearance of *Globotruncana linneiana* (d'Orbigny).
It is tempting to omit this zone and rely on an extended *W. baltica* Zone, ranging through to the first appearance of *G. linneiana*. North of East Anglia that approach is necessary as *D. primitiva* and *D. concavta* (Brotzen) are not present. The evolutionary relationship between *primitiva*, *concavata*, and *asymmetrica*, which is so stratigraphically useful in Tethyan regions, is not seen in the

U.K., and, as shown by Hart and Bailey (1979) and Hart and Ball (1986) this is probably the result of a water depth restriction.

UKP. 11 Globotruncana linneiana (d'Orbigny) Interval Zone
Age: Santonian (see Fig. 7.21)
Definition: first appearance of *G. linneiana* to the first appearance (in considerable numbers) of *Rugoglobigerina pilula* Belford. While *G. linneiana* is quite commonly encountered, the planktonic fauna is dominated by *W. baltica* and *Archaeoglobigerina bosquensis* Pessagno.

UKP.12 Rugoglobigerina pilula Belford Range Zone
Age: Early Campanian (see Fig. 7.22)
Definition: from the first appearance of considerable numbers of *R. pilula* up to the final extinction of the taxon.
Appearing with *R. pilula* are *Rosita fornicata* (Plummer) and *Archaeoglobigerina cretacea* (d'Orbigny), thereby making this a distinctive faunal boundary. Robaszynski *et al.* (1984) have erected the genus *Rosita* for the *G. fornicata, R. plummerae* Gandolfi, *G. contusa* (Cushman) plexus, and while the name *Rosita* is used in the charts the species concerned appear in the species listings where *Globotruncana* would be found. This, it is thought, will aid their easy location.

UKP.13 Globotruncana rugosa (Plummer) Interval Zone
Age: Middle Campanian (see Fig. 7.22)
Definition: from the first appearance of *G. rugosa* to the first appearance of *R. plummerae* Gandolfi.
Also appearing within this zone are *Hedbergella holmdelensis* Olsson and *Globigerinelloides multispina* (Lalicker), both of which are characteristically very small, although neither should be mis-identified. Rare specimens of *Globotruncana arca* (Cushman) are to be found in the upper half of the zone.

UKP. 14 Rosita plummerae Gandolfi Range Zone
Age: Late Campanian (see Fig. 7.22)
Definition: the range of *G. plummerae*.
The zonal index appears to have an almost identical range to that of *Globotruncana austinensis* Gandolfi, although this is not the case elsewhere in Europe, North Africa and the Atlantic Ocean.

UKP. 15 Globotruncanella havenensis (Voorwijk) Interval Zone
Age: Early Maastrichtian (see Fig. 7.22)
Definition: from the first appearance of large numbers of *G. havanensis* to the first appearance of the *Rosita patelliformis* Gandolfi/*Rosita contusa* (Cushman) plexus.
The zonal marker is a quite distinctive species with a widespread distribution in Europe. This interval is also characterised by the appearance of abundant rugoglobigerinids (*R. rugosa* (Plummer) and *R. milamensis* Smith and Pessagno).

UKP. 16 Rosita patelliformis Gandolfi Interval Zone
Age: mid-Maastrichtian (see Fig. 7.22)
Definition: from the first appearance of the *R. patelliformis/R. contusa* plexus to the first appearance of *Abathomphalus mayaroensis* (Bolli).
The gradation from the *patelliformis* morphotype to the fully developed *contusa* morphotype is probably a function of water depth (Hart and Ball, 1986) with the large, typically high-spired, *contusa* only being recorded in the central parts of the North Sea Basin.

UKP. 17 Abathomphalus mayaroensis (Bolli) Interval Zone
Age: latest Maastrichtian (see Fig. 7.22)
Definition: from the first appearance of *A.mayaroensis* to the first appearance of typically Danian taxa.

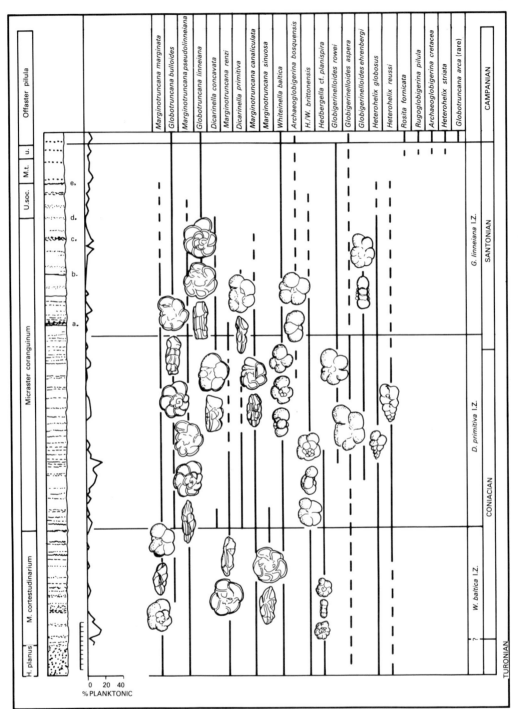

Fig. 7.21 — Planktonic zonation of the lower part of the Upper Chalk (Coniacian and Santonian). The section is that of southeast Kent. a. Bedwells Columnar Flint Band; b. Whitaker's 3" Flint; c. Barrois Sponge Bed; d. Peake's Sponge Bed; e. *Echinocorys elevata* Band.

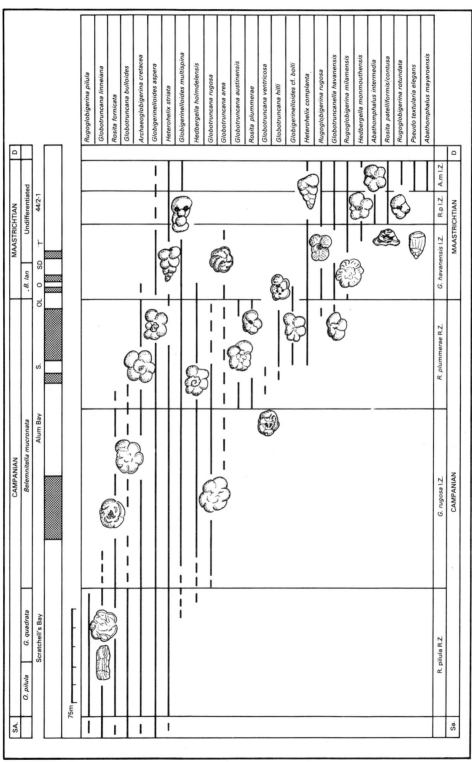

Fig. 7.22 — Planktonic zonation of the upper part of the Upper Chalk (Campanian and Maastrichtian). The section is that based on Scratchell's Bay, Alum Bay and the North Norfolk coast.

This zone is very characteristic of the latest Maastrichtian in Europe, North Africa and the low-latitude oceanic realm. The distinctive 'bow-tie' morphology and ornamentation of the species ensures that little confusion exists over the first appearance of the zonal marker, although the co-occurrence of many other typical latest Maastrichtian taxa ensures a positive identification of this interval.

7.7.2 Benthonic zonation

As indicated previously the Late Cretaceous chalk facies of the U.K. and northwest Europe (Bailey and Hart, 1979; Hart and Swiecicki, 1987) can readily be zoned using benthonic taxa, many of which have quite widespead distributions.

UKB.1 Plectina mariae (Franke)/ *Bulbophragmium aequale folkestoniensis* (Chapman) Concurrent Range Zone.
Age: earliest Cenomanian (see Fig. 7.23).
Definition: the concurrent distribution of the two zonal markers.
This zone (Zone 7 of Carter and Hart, 1977) contains the first, purely Cenomanian, benthonic fauna that is dominated by *Arenobulimina advena* (Cushman), *Flourensina intermedia* Ten Dam, *Marssonella ozawai* Cushman, and a wide spectrum of other diagnostic taxa. This zone coincides with the 'Glauconitic Marl' and the abundance of the large Lituolacea is facies-controlled. Away from southeast England this zone is grouped within UKB.2.

UKB.2 Flourensina intermedia Ten Dam/*Arenobulimina anglica* Cushman Concurrent Range Zone.
Age: Early Cenomanian (see Fig. 7.23)
Definition: F. intermedia in association with abundant *A. anglica*.
The characteristic agglutinated fauna of UKB.1 continues but *B. aequale folkestoniensis* is replaced by the less coarsely agglutinated *B. aequale aequale* (Reuss).

UKB.3 Marssonella ozawai Cushman/ *Pseudotextulariella cretosa* (Cushman) Concurrent Range Zone.
Age: Early Cenomanian (see Fig. 7.23)
Definition: M. ozawai in association with small *P. cretosa*.
F. intermedia is absent, but a wide range of taxa continue up from below. The *P. cretosa* are typically smaller than those found in the overlying zone.

UKB. 4 Pseudotextulariella cretosa (Cushman) Interval Zone
Age: Early Cenomanian (see Fig. 7.23)
Definition: an interval of strata characterised by *P. cretosa* following the extinction of *M. ozawai* and preceeding the appearance of *Plectina cenomana* Carter and Hart.
Large *P. cretosa* characterize this interval, together with robust specimens of *A. advena*, *Tritaxia pyramidata* Reuss and *Gavelinella* spp.

UKB. 5 Plectina cenomana Carter and Hart Interval Zone.
Age: Middle Cenomanian (see Fig. 7.23)
Definition: from the first appearance of *P. cenomana* to the appearance *in flood abundance* of the species *Flourensina mariae* Carter and Hart.
This zone, which occupies almost the whole of the Middle Cenomanian, is quite distinctive, even though there is a dramatic change in the numbers of benthonic foraminifera at the level of the mid-Cenomanian nonsequence (Carter and Hart, 1977).

UKB.6 Flourensina mariae Carter and Hart Acme Zone
Age: latest Middle Cenomanian (see Fig. 7.23)
Definition: a very thin zone characterised by the appearance of large and abundant specimens of the very distinctive *F. mariae*. Subsequent to the recognition of *F. mariae* by

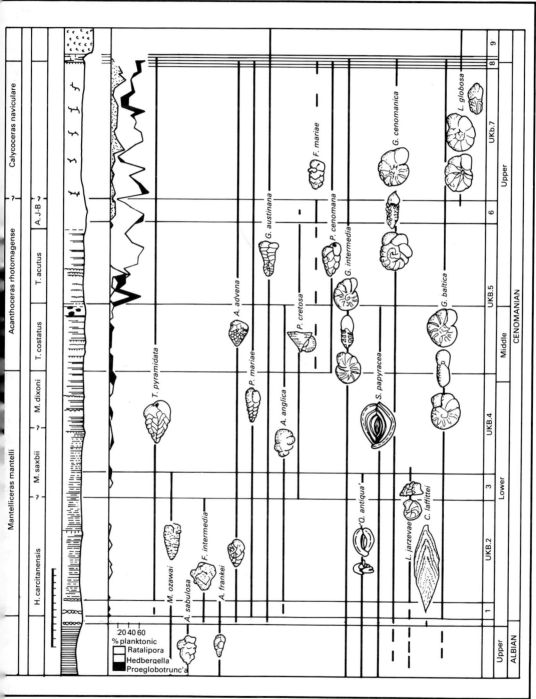

Fig. 7.23 — Benthonic zonation of the Lower Chalk (Cenomanian). The section is that between Folkestone and Dover, Kent.

Carter and Hart (1977), Barnard and Banner (1980) have placed this species in the new genus, *Crenaverneuilina*.

While this is clearly a facies-controlled occurrence of large agglutinated benthonic foraminifera, it is remarkably widespread and laterally consistent. *P. cretosa* dies out just above its base and there is evidence of a return to shallower water conditions at this level. The fauna is sometimes abraded and in some sections this zone is terminated above by a weakly developed non-sequence.

UKB. 7 Lingulogavelinella globosa (Brotzen) Interval Zone
Age: Late Cenomanian (see Fig. 7.23)
Definition: from the termination of the *F. mariae* Acme Zone up to the level of the sub-*plenus* erosion surface and the appearance of a distinctive, high diversity, benthonic fauna.
L. globosa appears just above the extinction of *P. cretosa*, within the *F. mariae* Acme Zone. Although this zone is characterised by a diverse benthonic fauna the total assemblage is dominated by planktonic foraminifera.

UKB. 8 Arenobulimina preslii (Reuss) Interval Zone
Age: latest Cenomanian (see Fig. 7.23)
Definition: this zone embraces the characteristic interval of the *A. plenus* Marls, with its distinctive fauna; see Carter and Hart (1977), Jefferies (1962, 1963) and Leary (1987, unpublished thesis).
The *A. plenus* Marls (the Plenus Marl Formation of Jarvis *et al.*, 1988b) can be divided lithologically into eight distinctive beds of chalk and marl. Each of these has a distinctive fauna, the characteristics of which can be attributed to the development of dysaerobic bottom water at the maximum development of the Late Cenomanian anoxic event (see Jarvis *et al.*, 1988b, for a full account and review).

UKB. 9 Gavelinella tourainensis Butt Interval Zone
Age: very latest Cenomanian and earliest Turonian (see Fig. 7.20)
Definition: from the top of the *A. plenus* Marls, with its distinctive assemblage to the first appearance of *Globorotalites micheliniana* (d'Orbigny).
Gavelinella tourainensis appears in the latest Cenomanian and ranges through the *A. plenus* Marls and into the Turonian. Just above the top of Zone UKB. 9 it disappears, only to reappear in the latest Turonian and Coniacian. Recent work by Leary (1987, unpublished thesis) and Koutsoukos (pers. comm.) has shown that this morphotype of *Gavelinella* (with a variable-sized umbilical boss) seems to occur in large numbers at times of oxygen depletion; its occurrence is almost typical of late Cretaceous 'anoxic events'. The remaining benthonic fauna of this zone is sparse with only *A. preslii*, *Gavelinella ammonoides* (Reuss) and other long-ranging taxa.

UKB. 10 Globorotalites micheliniana (d'Orbigny) Interval Zone
Age: Turonian (see Fig. 7.20)
Definition: from the first appearance of *G. micheliniana* to the first appearance of *Reussella kelleri* Vasilenko.
The Turonian fauna of the U.K. is dominated by planktonic taxa and the benthonic fauna is a low-diversity assemblage of long-ranging taxa. At the base of this zone small specimens of *Globorotalites* appear which have been variously attributed to *G. hangensis* Vassilenko, *G. minuta* Goel, *G. subconica* (Morrow), *G. multisepta* (Brotzen) and *G. cushmani* Goel.

Within Zone UKB. 10 there is a distinctive occurrence of a large, complex, lituolid (*Coskinohragma* sp.) that has been described from all over southern England and northern

France (Hart, 1982; Robaszynski *et al.*, 1980).

UKB. 11 *Reussella kelleri* Vasilenko Interval Zone
Age: Early Coniacian (see Fig. 7.24)
Definition: from the first appearance of *R. kelleri* to the first appearance of *Stensioeina granulata granulata* (Olbertz).
The chalks of latest Turonian age are very hard and nodular and this creates severe problems of sample preparation at this level. Precise ranges are therefore difficult to ascertain but the zonal taxon would appear to coincide with the stage boundary. Also appearing at this level (or close to it) are *Verneuillina muensteri* Reuss and *Gavelinella pertusa* (Marsson). The appearance of *Stensioeina* spp., *Gavelinella* spp., *Osangularia* spp., etc., marks the onset of the typical latest Cretaceous benthonic fauna, which then persists to the K/T boundary.

From this level onwards the zonation is dominated by the progressive appearance of species associated with the genera *Stensioeina*, *Gavelinella* and *Bolivinoides*. As all these are well-known evolutionary lineages it is possible to erect a tight zonation with wide applicability in Europe and North Africa.

UKB. 12 *Stensioeina granulata granulata* (Olbertz) Interval Zone
Age: mid-Coniacian (see Fig. 7.24)
Definition: from the first appearance of *S. granulata granulata* to the first appearance of *S. exsculpta exsculpta* (Reuss).
At the base of this zone *Gavelinella thalmanni* (Brotzen) also appears, followed closely by *Osangularia* sp. and abundant *Lingulogavelinella* sp. cf. *L. vombensis* (Brotzen).

UKB. 13 *Stensioeina exsculpta exsculpta* (Reuss) Interval Zone.
Age: latest Coniacian to earliest Santonian (see Fig. 7.24)
Definition: from the first appearance of *S. exsculpta exsculpta* to the first appearance of *Sensioeina granulata polonica* Witwicka.
Immediately above the base of the zone *Loxostomum eleyi* (Cushman) appears, closely followed by *Vaginulinopsis scalariformis* Porthault. The former taxon can be used in many parts of northwest Europe and North America (see Hart, 1987) as an alternative when *S. exsculpta exsculpta* is either rare, poorly preserved or absent.

UKB. 14 *Stensioeina granulata polonica* Witwicka Interval Zone.
Age: Santonian (see Fig. 7.24)
Definition: from the first appearance of *S. granulata polonica* to the first appearance of *Bolivinoides strigillatus* (Chapman).
Appearing with *S. granulata polonica* is *Cibicides beaumontianus* (d'Orbigny), closely followed by *Praebulimina carseyae* (Plummer), *Reussella szajnochae praecursor* De Klasz and Knipscheer, *Gavelinella stelligera* (Marie) and *Gavelinella cristata* (Geol). All these first appearances are quite diagnostic and several alternative zonations could be used with equal success (see Bailey *et al.*, 1983, 1984).

UKB. 15 *Bolivinoides strigillatus* (Chapman) Taxon Range Zone
Age: latest Santonian to earliest Campanian (see Figs 7.24 and 7.25)
Definition: the range of *B. strigillatus* (Chapman).
Just before the first appearance of *B. strigillatus*, *S. granulata perfecta* Koch appears, while immediately above the base of the zone *S. exsculpta gracilis* Brotzen appears. At, or about, the Santonian/Campanian boundary *Bolivinoides culverensis* Barr and *Pullenia* sp. appear and could be used for finer stratigraphic resolution.

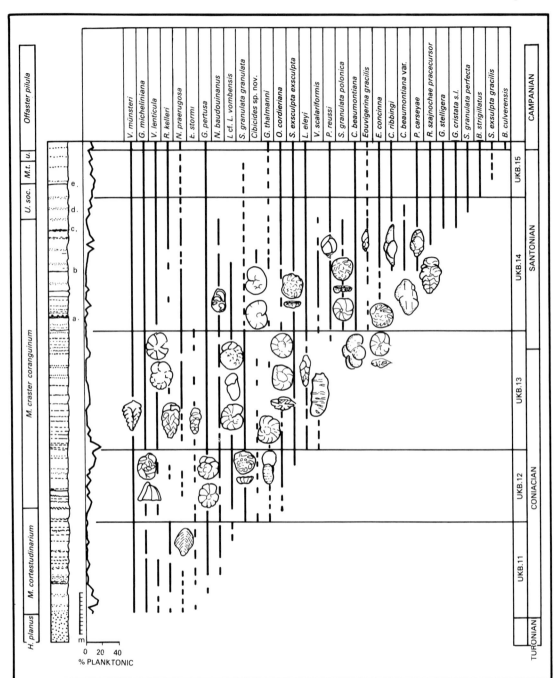

Fig. 7.24 — Benthonic zonation of the lower part of the Upper Chalk (Coniacian and Santonian). The section is that of southeast Kent. a. Bedwells Columnar Flint Band; b. Whitaker's 3″ Flint; c. Barrois Sponge Bed; d. Peake's Sponge Bed; e. *Echinocorys elevata* Band.

Fig. 7.25 — Benthonic zonation of the upper part of the Upper Chalk (Campanian and Maastrichitian). The section is that based on Scratchell's Bay, Alum Bay and the North Norfolk coast.

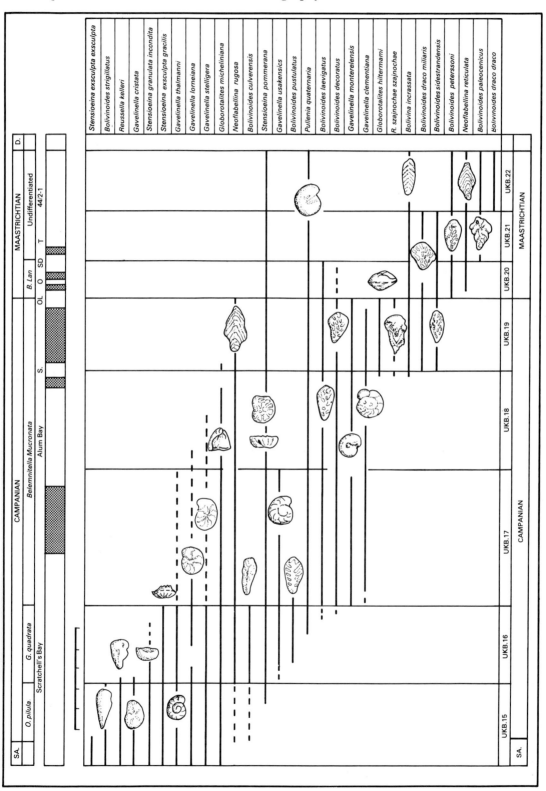

UKB. 16 Bolivinoides culverensis Barr Interval Zone.
Age: Early Campanian (see Fig. 7.25)
Definition: from the extinction of *B. strigillatus* to the extinction of *B. culverensis*.
The appearance of *Gavelinella usakensis* (Vasilenko), followed by *Bolivinoides pustulatus* Reiss and *Pullenia quaternaria* (Reuss) make this a distinctive zone, even when *B. culverensis* is rare.

UKB. 17 Gavelinella monterelensis (Marie)/ *Gavelinella usakensis* (Vasilenko) Concurrent Range Zone.
Age: mid-Campanian (see Fig. 7.25)
Definition: the co-occurrence of the two zonal markers.
G. monterelensis is a very distinctive mid- to Late Campanian species, occurring throughout most of its range with *Bolivinoides decoratus* (Jones).

UKB. 18 Bolivinoides decoratus (Jones) Interval Zone.
Age: mid to Late Campanian (see Fig. 7.25)
Definition: from the extinction of *G. usakensis* to first appearance of *Bolivinoides sidestrandensis* Barr and *B. miliaris* Hiltermann and Koch.
This very characteristic zone is also marked by the abundance of *Gavelinella clementiana* (d'Orbigny) and the first appearance, at the very top of the zone, of *Globorotalites hiltermanni* Kaever, *Reussella szajnochae szajnochae* (Grzybowski) and *Bolivina incrassata* Reuss.

UKB. 19 Bolivinoides miliaris Hiltermann and Koch Interval Zone
Age: latest Campanian (see Fig. 7.25)
Definition: from the first appearance of *B. miliaris* to the first appearance of *B. peterssoni* Brotzen.
The zonal indicator has the same overall range as *B. sidestrandensis* and either species can be used to identify this interval.

UKB. 20 Bolivinoides peterssoni Brotzen Interval Zone
Age: Early Maastrichtian (see Fig. 7.25)
Definition: from the first appearance of *B. petersonni* to the first appearance of *B. paleocenicus* (Brotzen).
The characteristic species *Neoflabellina reticulata* (Reuss) appears just above the base of this zone, although the transition from *N. praereticulata* Hiltermann can make positive identification difficult.

UKB. 21 Bolivinoides paleocenicus (Brotzen) Interval Zone.
Age: Middle Maastrichtian (see Fig. 7.25)
Definition: from the first appearance of *B. paleocenicus* to the first appearance of *B. draco* (Marsson).

UKB. 22 Bolivinoides draco (Marsson) Taxon Range Zone.
Age: Late Maastrichtian (see Fig. 7.25)
Definition: the range of *B. draco*.

7.8 INDEX SPECIES

The classification used in this account is that proposed by Loeblich and Tappan (1964, 1974), but also includes genera incorporated into the literature since that time. In the following section the species are grouped by suborder (Textulariina, Miliolina and Rotaliina), and then discussed in alphabetical order.

The species selected for inclusion are, in most cases, the most stratigraphically useful, and those which would probably be encountered in any work on the Cretaceous succession. However, in an attempt to provide a balanced distribution throughout the succession, some stratigraphically useful (e.g. *Marssonella kummi* Zedler, *Orbitolina* sp., *Plectina mariae* (Franke), *Dicarinella canaliculata* (Reuss), *D. imbricata* (Mornod), *Globotrun-*

cana arca (Cushman), *Lingulina denticulco-carinata* (Chapman), *Marginotruncana renzi* (Gandolfi), *Racemiguembelina fructicosa* (Egger)) species have had to be omitted.

SUBORDER TEXTULARIINA

Ammobaculites reophacoides Bartenstein, 1952
Plate 7.1, Figs 1, 2 (× 50), Speeton Clay, Speeton, Yorkshire, Barremian (LB3). Description: test free, small, initial coil followed by 2–4 uniserial chambers; uniserial chambers broader than high, elliptical cross-section, sutures radial and indistinct in coiled portion, distinct, horizontal, depressed in uniserial part; aperture terminal rounded.
Remarks: commonly crushed, thereby giving a variable outline. When crushed it is very similar to *A. subcretacea* Cushman and Alexander.
Range: Ryazanian to earliest Albian.

Ammovertella cellensis Bartenstein and Brand, 1951
Plate 7.1, Figs 3, 4 (× 20), Speeton Clay, Speeton, Yorkshire, Valanginian (D2E). Description: test free or attached, long, tubular with constrictions and bends; test may or may not branch in an irregular manner; large amount of cement giving a smooth test; aperture at the end of tube, or branching tubes.
Remarks: this distinctive species shows great variation in external shape.
Range: mid- to Late Valanginian, rare in Hauterivian; branching forms commoner in the Late Valanginian.

Arenobulimina advena (Cushman)
Plate 7.1, Fig. 5 (× 60), Lower Chalk, Eastbourne, Sussex, Cenomanian. = *Hagenowella advena* Cushman, 1936. Description: test free, trochospiral, with last three chambers occupying over half of test; chambers slightly inflated, sutures distinct, depressed; interior of test divided by complex partitions; aperture an interiomarginal loop, set in a hollow in face of last chamber.
Remarks: the complexity of the internal partitions varies throughout the Cenomanian, being relatively simple in the Early Cenomanian and becoming progressively more complex in the Late Cenomanian.
Range: very uppermost Albian to Cenomanian.

Arenobulimina anglica Cushman, 1936
Plate 7.1, Fig. 6 (× 70), Lower Chalk, Eastbourne, Sussex, Middle Cenomanian. Description: test free, trochospiral; last whorl occupies over half of test; chambers rounded, slightly inflated, sutures distinct, depressed; last chamber rounded, almost terminal; agglutinated material fine grained, giving a saccharoidal appearance to the test; aperture an interiomarginal loop, set in hollow in last chamber.
Remarks: the sugary appearance is characteristic. In the late Middle Cenomanian forms with slightly crenulated chamber margins have been recorded.
Range: Early to Middle Cenomanian.

Arenobulimina chapmani Cushman, 1936
Plate 7.1, Fig. 7 (× 70), Bed XIII, Gault Clay, Folkestone, Kent, Upper Albian. Description: test free, trochospiral, proximal end pointed, widening rapidly distally, rounded in cross section; adult test consists of five whorls, four chambers in each except the last which has five; sutures depressed; aperture a loop shaped in a trough like depression, running from top of test to edge of last chamber.
Remarks: characterised by its rapidly tapering test and large last whorl. It is easily distinguished from its predecessor, *A. macfadyeni* Cushman, being larger, more inflated, and more coarsely agglutinated.
Range: Late Albian.

Arenobulimina macfadyeni Cushman, 1936
Plate 7.1, Figs 8, 9 (× 45), Bed IV, Gault Clay, Folkestone, Kent, Lower Albian. Description: test free, small, trochospiral, proximal end sharply pointed, gradually widening distally; longitudinal outline triangular, rounded in cross-section; test surface finely agglutinated; adult test comprises 4–5 whorls with three chambers in each; sutures depressed; aperture a loop-shaped opening, surrounded by a raised lip.
Remarks: the smallest and most finely agglutinated of all the Albian arenobuliminids.
Range: Early and Middle Albian, with occasional, atypical and/or reworked specimens in the earliest Late Albian (Carter and Hart, 1977; Price, 1977a, b).

Arenobulimina sabulosa (Chapman)
Plate 7.1, Fig. 10 (× 55), Bed XIII, Gault Clay, Folkestone, Kent, Upper Albian. = *Bulimina preslii* Reuss var. *sabulosa* Chapman, 1892. Description: test free, stout, trochospiral, subtriangulate in longitudinal outline, subquadrate in cross-section; adult test comprises 4–5 whorls, quadriserial, the last chamber covering the complete upper surface of the test; sutures depressed, but often obscured by the very coarse agglutinated material; aperture loop-shaped.
Remarks: the sub-quadrate cross section and coarsely agglutinated test are distinctive.
Range: Latest Albian.

Dorothia filiformis (Berthelin)
Plate 7.1, Figs 11, 12 (× 115), Bed XI, Gault Clay, Folkestone, Kent, Upper Albian. = *Gaudryina filiformis* Berthelin, 1880. Description: test free, narrow and elongate, proximal portion trochospiral with four chambers per whorl; distal portion biserial, thin and elongate, with almost sub-parallel sides; aperture an interiomarginal slit developed on the last chamber, rarely visible.
Remarks: easily recognised by its small size and almost parallel sides.
Range: Early to Late Albian.

Pl. 7.1] **Cretaceous** 317

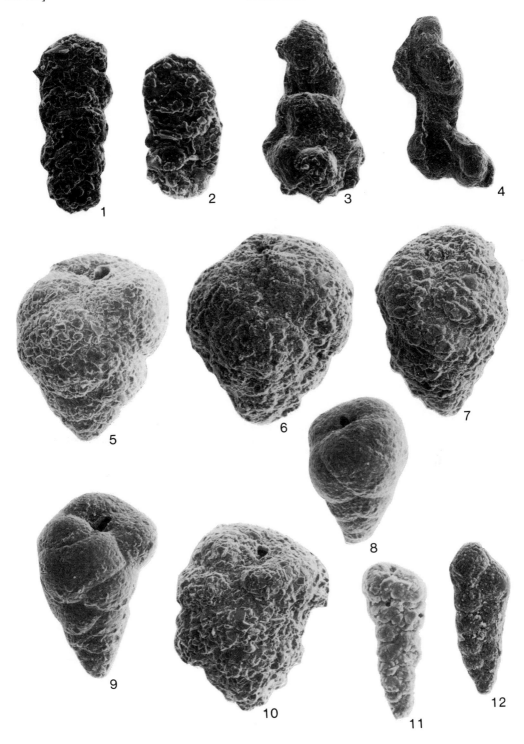

Plate 7.1

Eggerellina mariae Ten Dam, 1950
Plate 7.2, Figs 1, 2 (× 70), Cambridge Greensand, Barrington, Cambridgeshire, Lower Cenomanian. Description: test free, varying from short pyramidal to long and narrow, triserial, conical to ovoid in longitudinal outline; chambers very often very inflated and embracing; wall very finely agglutinated; aperture narrow, hook-shaped, interiomarginal, and extending about half way up chamber face.
Remarks: very variable in outline within the Cenomanian and Turonian, as noted by Carter and Hart (1977).
Range: Late Albian to Early Turonian.

Flourensina intermedia Ten Dam, 1950
Plate 7.2, Fig. 3 (× 60), Cambridge Greensand, Barrington, Cambridgeshire, Lower Cenomanian. Description: test free, trochospiral, coarsely agglutinated; chamber margins irregular, with depressed sutures; distinctly triserial, giving markedly triangular cross-section; aperture loop-shaped on face of final chamber.
Remarks: can be confused with the quadriserial *Arenobulimina sabulosa* (Chapman), although there should be no problem if viewed from the proximal end.
Range: Latest Albian to Early Cenomanian.

Gaudryinella sherlocki Bettenstaedt, 1952
Plate 7.2, Figs 4, 5 (× 65), Speeton Clay, Speeton, Yorkshire, Barremian (LB4B). Description: test free, elongate, subcylindrical; earlier portion of test triserial, followed by biserial that may tend to become uniserial; chambers subglobular, giving lobate periphery; sutures distinct, depressed in biserial part of test; aperture terminal, central, flush, circular or ovate.
Remarks: there is little or no variation, any recorded normally being due to compression during preservation.
Range: Barremian to Early Albian, although it is most abundant in the Early to mid-Barremian.

Haplophragmium aequale (Roemer)
Plate 7.2, Figs 6, 7 (× 60), Speeton Clay, Speeton, Yorkshire, Hauterivian (C8). = *Spirolina aequalis* Roemer, 1841. Description: test free, elongate, slightly compressed, early stage close-coiled, planispiral, later portion uniserial, 3–5 chambers; chambers distinct, inflated, sutures distinct, depressed, straight; aperture rounded, terminal.
Remarks: there is considerable variation in test outline, particularly in smaller specimens in which the uncoiled portion has not developed fully.
Range: Hauterivian to earliest Barremian.

Marssonella ozawai Cushman, 1936
Plate 7.2, Figs 8, 9 (× 60), Cambridge Greensand, Barrington, Cambridgeshire, Lower Cenomanian. Description: test free, elongate trochospiral, conical; proximal portion triserial, later biserial; biserial portion of test with sub-parallel sides, rounded in cross-section; sutures flush to slightly depressed; aperture a small, low opening at the inner margin of the last chamber.
Remarks: Carter and Hart (1977) indicated that this species can be differentiated from the closely related *M. trochus* (d'Orbigny) by its coarser agglutination and almost sub-parallel sides. Rare transitional forms between the two species are found in the Late Albian.
Range: Latest Albian to Early Cenomanian.

Plectina cenomana Carter and Hart, 1977
Plate 7.2, Fig. 10 (× 60), Lower Chalk, Eastbourne, Sussex, Middle Cenomanian. Description: test free, trochospiral; 2–4 whorls, with up to five chambers per whorl, reducing to three chambers per whorl in later growth stages; overall appearance triserial, rapidly tapering; sutures distinct, depressed; aperture rounded, occasionally oval, positioned in a slight depression in final chamber.
Remarks: the relationship between this species, *Plectina mariae* (Franke) and *Arenobulimina frankei* Cushman is rather unclear.
Range: Middle to Late Cenomanian.

Pseudotextulariella cretosa (Cushman)
Plate 7.2, Figs 11, 12, (× 50), Lower Chalk, Eastbourne, Sussex, Middle Cenomanian. = *Textulariella cretosa* Cushman, 1932. Description: test free, very large; subconical, with an earlier triserial and later biserial growth form; chambers internally complex with horizontal and vertical marginal partitions; aperture interiomarginal.
Remarks: a very distinctive species, but its distribution is probably facies controlled.
Range: in southeast England the range is Early to Middle Cenomanian, although rare, small, individuals have been found in the Red Chalk (Late Albian) of Yorkshire and Humberside.

Textularia chapmani Lalicker, 1935
Plate 7.3, Fig. 1 (× 50), Lower Chalk, Eastbourne, Sussex, Lower Cenomanian. Description: test free, small, biserial, with 8–10 chambers; test widening rapidly, with broad distal portion, triangulate in longitudinal outline, sub-compressed; sutures indistinct, slightly depressed; aperture a low arch at base of last chamber.
Remarks: the rapidly widening growth form is distinctive.
Range: Late Albian to Late Cenomanian.

Cretaceous

Plate 7.2

Tritaxia pyramidata Reuss, 1862
Plate 7.3, Figs 2, 3 (× 30), Lower Chalk, Folkestone, Kent, Lower Cenomanian (Fig. 2) and Upper Cenomanian (Fig. 3). Description: test free, triserial, triangular in cross-section; sides generally concave but may be straight; up to nine triserially arranged chambers overlapping ¼–½ their widths; sutures indistinct in early stages becoming more depressed later; aperture circular, may become terminal.
Remarks: while this is a distinctive species in the upper part of its range it can be confused with *Tritaxia singularis* Magniez-Jannin in the Early Cretaceous. The latter species differs from *T. pyramidata* mainly in having more excavated sides, and a more coarsely agglutinated test. *T. pyramidata* shows a wide range of variation in both shape and size. In the Late Cenomanian some individuals tend to become uniserial.
Range: typically Late Albian and Cenomanian, but has been variously recorded from the Hauterivian onwards.

Tritaxia singularis Magniez-Jannin, 1975
Plate 7.3, Fig. 4 (× 50), Bed XIII, Gault Clay, Folkestone, Kent, Upper Albian. Description: test free, triserial, pyramidal, triangular in cross-section; sides characteristically excavated, giving sharp edges to the test; sutures distinct, depressed; aperture circular, normally terminal.
Remarks: can be confused with early forms of *T. pyramidata*.
Range: Albian.

Trochammina depressa Lozo, 1944
Plate 7.3, Figs 5, 6 (× 75), Speeton Clay, Speeton, Yorkshire, Hauterivian (C4K). Description: test free, trochoid, compressed, lobulate periphery; chambers distinct, 5–6 in final whorl; chambers very slightly inflated, sutures distinct, depressed; aperture a small opening at base of last chamber.
Remarks: any variation is usually caused by degree of compression and asymmetrical deformation during preservation. *T. murgeanui* Neagu from the Barremian of Rumania may be synonymous.
Range: Ryazanian to Early Barremian.

Verneuilina muensteri Reuss, 1854
Plate 7.3, Figs 7, 8 (× 40), Upper Chalk, Redbournbury, Hertfordshire, Santonian. Description: test elongate, triangular in cross-section, increasing uniformly and rapidly in size; sides flat to slightly concave; chambers distinct; sutures flush to slightly raised, strongly curved; aperture loop-shaped opening towards the centre of the inner margin of the apertural face.
Remarks: specimens referred to this species are also close to forms described as *V. limbata* Cushman. This trivial name is already occupied (Terquem, 1883), but some authors have followed Cushman, and used *V. limbata* for this species.
Range: Coniacian to Early Maastrichtian.

Verneuilinoides neocomiensis (Mjatliuk)
Plate 7.3, Fig. 9 (× 40), Speeton Clay, Speeton, Yorkshire, Valanginian (D2E). = *Verneuilina neocomiensis* Mjatliuk, 1939. Description: test free, narrow, elongate, triangular in cross-section; chambers arranged triserially in 7–8 rows; chambers distinct, giving a lobulate outline in uncrushed specimens; sutures distinct, depressed; aperture a slit at base of last chamber.
Remarks: there is considerable variation in the degree of taper of the test, although many specimens are crushed. It is very similar to *V. subfiliformis* Bartenstein, from which it differs in being shorter and broader.
Range: Ryazanian to Barremian.

SUBORDER MILIOLIINA

Nodobacularia nodulosa (Chapman)
Plate 7.3, Fig. 14 (× 46), Bed XIII, Gault Clay, Folkestone, Kent, Upper Albian. = *Nubecularia nodulosa* Chapman, 1891. Description: test free, porcellanous, always fragmentary, generally 1–3 chambers, joined by narrow tube; chambers vary from nodulose to pyriform; aperture simple, terminal, circular.
Remarks: Chapman's specimens (BMNH No. P.4597) from the Albian are smaller and more delicate than the more robust forms normally encountered in the earliest Cretaceous. There is a great deal of varation.
Range: Ryazanian to Albian.

Quinqueloculina antiqua (Franke)
Plate 7.3, Figs 12, 13 (× 40), Bed XIII, Gault Clay, Folkestone, Kent, Upper Albian. = *Miliolina (Quinqueloculina) antiqua* Franke, 1928. Description: test free, in typical 'quinqueloculine' coil; wall finely porcellanous; aperture on short neck, with tooth not normally visible.
Remarks: Price (1977a) has retained both *Q. antiqua* and *Q. ferussaci* (d'Orbigny) in his zonation of the Albian, although Carter and Hart (1977) used only the former name for all Albian quinqueloculines. The Polish form described by Gawor-Biedowa (1972) as *Q. kozlowski* may also be synonymous.
Range: Middle Albian to mid-Cenomanian; rare atypical forms have been recorded in the Late Cenomanian.

Pl. 7.3] **Cretaceous** 321

Plate 7.3

Wellmanella antiqua (Reuss)
Plate 7.3, Figs 10, 11 (× 45), Speeton Clay, Speeton, Yorkshire, Barremian (LB5B). = *Hauerina antiqua* Reuss, 1863. Description: test free, broadly ovate, periphery rounded; chambers inflated, strongly curved, often indistinct, generally triloculine, but may have four chambers per whorl; sutures generally indistinct; aperture slit-like, at the base of the last chamber.
Remarks: there is considerable variation in the shape and arrangement of the chambers, and in the size and form of the aperture.
Range: Late Hauterivian to mid-Barremian.

SUBORDER ROTALIINA (of Loeblich and Tappan, 1964)

Abathomphalus mayaroensis (Bolli)
Plate 7.4, Figs 1–3 (× 100), Shell/Esso 44/2–1, North Sea, Maastrichtian. = *Globotruncana mayaroensis* Bolli, 1951. Description: test free, low trochospiral coil, spiral side weakly convex, umbilical surface typically concave; outline sub-circular, lobate; periphery truncated, with widely spaced, double keel; chambers distinct, 2½–3 whorls, 5–6 chambers in final whorl; sutures straight, radial, depressed on umbilical side, but crescentric, raised, slightly beaded on spiral side; umbilicus shallow, covered by a tegilla. Primary aperture interiomarginal, extraumbilical.
Remarks: distinguished by its shallow umbilicus, and apertural characteristics.
Range: Latest Maastrichtian.

Archaeoglobigerina cretacea (d'Orbigny)
Plate 7.4, Figs 4, 5 (× 100), Upper Chalk, Alderford, Norfolk, upper Middle Campanian. = *Globigerina cretacea* d'Orbigny, 1841. Description: test free, low trochospiral coil of 2½–3 whorls, outline subcircular, lobate; periphery weakly truncated, initially rounded; chambers distinct, subglobular, with weakly developed, faint keels bordering a raised imperforate band; sutures radial, depressed; umbilicus broad, deep, covered by a tegilla with accessory apertures; primary aperture umbilical.
Remarks: *Globigerina cretacea* d'Orbigny is referred to the genus *Archaeoglobigerina* on the basis of its non-truncate, globular chambers, and near-radial, depressed sutures.
Range: Latest Santonian and Campanian.

Bolivina decurrens (Ehrenberg)
Plate 7.4, Figs 6, 7 (× 100), Upper Chalk, Overstrand, Norfolk, Lower Maastrichtian. = *Grammostomum decurrens* Ehrenberg, 1854. Description: test free, elongate, slender 2–3 times as long as wide, initially bluntly pointed, occasionally spinose; test compressed, margins sub-acute, spinose; chambers biserial throughout, initially indistinct, but becoming inflated; aperture elongate, elliptical slit, extending from basal suture of final chamber to occupy a sub-terminal position.
Remarks: easily distinguished by its spinose periphery and elongated, slender, form.
Range: Maastrichtian.

Bolivina incrassata Reuss, 1851
Plate 7.4, Figs 8, 9 (× 45), Upper Chalk, Trimingham (Fig. 8) and Overstrand (Fig. 9), Norfolk, Lower Maastrichtian. Description: test free, elongate, varying in length from a long slender form to one that is very stout and robust, test compressed, with sub-rounded margins, periphery entire, occasionally becoming slightly lobate; chambers numerous, sutures distinct, slightly depressed, steeply inclined; aperture elongate, wide, ovate, highly inclined subterminal opening.
Remarks: highly variable and many subspecies have been recognised (Koch, 1977).
Range: Latest Campanian and Maastrichtian.

Bolivinoides culverensis Barr, 1967
Plate 7.4, Figs 10, 11 (× 100), Upper Chalk, Overstrand, Norfolk, Lower Maastrichtian. Description: test free, elongate, moderately flaring; proloculus sub-globular, followed by 7–8 pairs of biserial chambers; sutures indistinct, slightly curved; test surface possessing raised, broad, elongate lobes, usually three per chamber; initial portion of test possesses only faintly raised nodules.
Remarks: represents transitional forms between the ancestral *B. strigillatus* (Chapman) and the descendant *B. decoratus* (Jones).
Range: Transitional forms between *B. strigillatus* and *B. culverensis* occur sporadically in the latest Early Campanian, and the species occurs abundantly throughout the Middle Campanian.

Plate 7.4

Bolivinoides decoratus (Jones)
Plate 7.4, Fig. 12 (× 75), Upper Chalk, Caister St. Edmunds, Norfolk, Upper Campanian. = *Bolivina decorata* Jones, 1875. Description: test free, elongate, rhomboid outline; periphery sub-rounded; cross-section compressed, elliptical; globular proloculus, followed by 7–9 pairs of biserial chambers, sutures oblique, indistinct, obscured by strongly developed lobes; aperture narrow, loop-shaped.
Remarks: the erection of a lectotype (Barr,1966a) has clarified the taxonomic position of this widely recorded species.
Range: Late Campanian and Early Maastrichtian.

Bolivinoides draco (Marsson)
Plate 7.4, Fig. 13 (× 100), Shell/Esso 44/2-1, North Sea, Maastrichtian. = *Bolivina draco* Marsson, 1878. Description: test free, rhomboidal, compressed; margins sub-acute to acute, often carinate; initial end bluntly rounded, followed by 6–7 pairs of biserially arranged chambers; test surface covered by strongly developed, longitudinally elongated lobes, four on each chamber, which coalesce to form longitudinal ribs; aperture wide, loop-shaped, bordered by thin lip, and possessing an internal tooth-plate.
Remarks: separated from *B. miliaris* Hiltermann and Koch by its more regular rhomboid test, and the fusion of the lobes into ridges.
Range: Latest Early Maastrichtian and the Late Maastrichtian.

Bolivinoides laevigatus Marie, 1941
Plate 7.5, Figs 1, 2 (× 110), Upper Chalk, Eaton, Norfolk, Upper Campanian. Description: test free, elongate; cross-section compressed, periphery sub-acute; globular proloculus followed by 7–9 pairs of biserial chambers; chambers distinct, sutures slightly depressed; test surface possessing weakly developed circular to elongate nodes, generally 2–3 per chamber; aperture narrow, loop-shaped.
Remarks: a distinctive, weakly ornamented, species.
Range: Late Campanian and Early Maastrichtian.

Bolivinoides miliaris Hiltermann and Koch, 1950
Plate 7.5, Fig. 3 (× 100), Upper Chalk, Trimingham, Norfolk, Lower Maastrichtian. Description: test free, kite-shaped, compressed, margins acute; proloculus globular, followed by 7–8 pairs of slightly inflated, biserial chambers; sutures indistinct, except at the periphery; test surface initially pustulose, later possessing elongate, narrow, lobes, generally three per chamber; aperture loop-shaped, occasionally bordered by an indistinct lip.
Remarks: *B. miliaris* is morphologically intermediate between *B. decoratus* (Jones) and *B. draco* (Marsson).
Range: Latest Campanian and Early Maastrichtian.

Bolivinoides paleocenicus (Brotzen)
Plate 7.5, Fig. 4 (× 200), Upper Chalk, Sidestrand, Norfolk, Lower Maastrichtian. = *Bolivina paleocenica* Brotzen, 1948. Description: test free, kite-shaped, compressed, periphery sub-acute; 6–7 pairs of uninflated, biserial chambers; sutures near periphery are distinct, depressed, but in centre of test are obscured by raised network of intersecting narrow ridges; aperture narrow, loop-shaped.
Remarks: a highly distinctive species, recognised by the compressed nature of the test and the surface sculpture.
Range: moderately common in the late Early Maastrichtian, and rare in the Late Maastrichtian.

Bolivinoides peterssoni Brotzen, 1945
Plate 7.5, Fig. 5 (× 100), Upper Chalk, Overstrand, Norfolk, Lower Maastrichtian. Description: test free, elongate, compressed, periphery slightly lobate, cross-section elliptical, periphery sub-acute; proloculus globular, followed by 7–8 pairs of slightly inflated, biserial chambers; sutures distinct, depressed; median part of test surface possessing well-developed, elongate lobes which extend across the chamber faces perpendicular to the sutures; aperture narrow, loop-shaped, basal opening.
Remarks: this distinctive species may be distinguished from its probable ancestor (*B. laevigatus* Marie) by its broader, markedly compressed form and surface sculpture, which is even more restricted to the median region.
Range: Maastrichtian.

Bolivinoides pustulatus Reiss, 1954
Plate 7.5, Fig 6 (× 150), Upper Chalk, Alderford, Norfolk, upper Middle Campanian. Description: test free, cross-section elliptical, periphery sub-acute; globular proloculus followed by 6–8 pairs of slightly inflated, biserial chambers; sutures indistinct, depressed, obscured by numerous weakly raised, somewhat elongate, pustules that often merge with one another; aperture narrow, loop-shaped, slit.
Remarks: may be conspecific with *B. texana* Cushman.
Range: Late Middle and early Late Campanian.

Plate 7.5

Bolivinoides sidestrandensis Barr, 1966
Plate 7.5, Fig. 7 (× 100), Upper Chalk, Overstrand, Norfolk, Lower Maastrichtian. Description: test free, elongate, periphery rounded to sub-acute; globular proloculus followed by 7–8 pairs of inflated biserial chambers; sutures oblique, depressed, generally obscured by the well-defined, narrow, elongate, lobes that extend across the chambers; aperture narrow, loop-shaped, bordered by a faint lip.
Remarks: *B. delicatulus regularis* (sensu Koch, 1977) is a common synonym of this species. For full taxonomic discussion of this species see Swiecicki (1980, unpublished thesis).
Range: Late Campanian and Early Maastrichtian.

Bolivinoides strigillatus (Chapman)
Plate 7.5, Fig. 8 (× 100), Upper Chalk, Scratchells Bay, Isle of Wight, Lower Campanian. = *Bolivina strigillata* Chapman, 1892. Description: test free, elongate, cross-section sub-circular, only slightly compressed; subglobular proloculus followed by 6–8 pairs of slightly inflated, biserial chambers; sutures indistinct, obscured by the 2–3 raised lobes, that extend the length of each chamber; aperture narrow, loop-shaped, with an indistinct lip.
Remarks: *B. strigillatus*, originally described from the Upper Santonian phosphatic chalk of Taplow, England, is the earliest known species of the genus.
Range: Latest Santonian and Early Campanian.

Cibicides beaumontianus (d'Orbigny)
Plate 7.5, Figs 9, 10 (× 85), Upper Chalk, Culver Cliff, Isle of Wight, Upper Santonian. = *Truncatulina beaumontiana* d'Orbigny, 1840. Description: test attached, plano-convex, shape variable, margins distinctly rounded; chambers distinct, sub-globular, 5–6 in final whorl; sutures distinct, depressed, straight; wall smooth, to very slightly rugose on the ventral side; aperture an equatorial slit to semi-circular hole, extending back along the dorsal side, following the spiral suture.
Remarks: the attached habit leads to considerable morphological variation, especially on the dorsal side, along which the attachment is made.
Range: Santonian to Maastrichtian.

Cibicides ribbingi Brotzen, 1936
Plate 7.6, Figs 1–3 (× 85), Upper Chalk, Scratchells Bay, Isle of Wight, Middle Campanian. Description: test attached, plano-convex, ventral side only slightly inflated; peripheral outline extremely variable, especially in later chambers, which show a tendency to become irregular; margin accutely angled; chambers distinct, sutures flush on dorsal side, slightly depressed on the ventral side, straight, radiate; aperture a slit on the inner margin of the final chamber, extending round towards the umbilicus.
Remarks: there is a possibility that this species is an ecophenotype of *C. beaumontianus* (d'Orbigny), as suggested by Barr (1966b), although this was not supported by Bailey (1978, unpublished thesis) nor Swiecicki (1980, unpublished thesis). Range: Santonian to Early Maastrichtian.

Cibicidoides (?) voltziana (d'Orbigny)
Plate 7.6, Figs 4–6 (× 80), Upper Chalk, Alum Bay (Figs 4, 6) and Claister St. Edmunds (Fig. 5), Upper Campanian. = *Rotalina voltziana* d'Orbigny, 1840. Description: test free, trochospiral, 3–4 whorls, plano-convex, involute, with moderate to large calcite boss; umbilical side flat, may have small, central calcite boss; periphery sub-acute; chambers indistinct, dorsally inflated in final whorl; sutures distinct, slightly depresssed; aperture an interiomarginal, broad, low, arch, extending onto umbilical surface.
Remarks: a distinctive species, although its generic position is still under discussion.
Range: Late Campanian and Early Maastrichtian.

Citharina acuminata (Reuss)
Plate 7.6, Figs 7, 8 (× 30), Speeton Clay, Speeton, Yorkshire, Barremian (LB2 – Fig. 7 and LB4D – Fig. 8). = *Vaginulina acuminata* Reuss, 1863. Description: test free, compressed, elongate, slender, delicate, dorsal edge nearly straight; sutures strongly oblique, flush with surface, frequently obscured by ornament, which comprises 6–9 closely spaced, longitudinal, fine ribs; aperture radiate, on a sub-cylindrical neck.
Remarks: the only variation concerns the placing of the fine ribs on the surface of the test.
Range: Late Hauterivian to Late Aptian but most abundant in the Barremian.

Citharina cf. *C. discors* (Koch)
Plate 7.6, Figs 9, 10 (× 18), Speeton Clay, Speeton, Yorkshire, Lower Barremian (LB3) (Fig. 9) and Hauterivian (C4E) (Fig. 10). = *Vaginulina discors* Koch, 1851. Description: test free, large, compressed, test outline sub-triangular; chambers low, broad, gently curving, extending to base proximally; 12–15 chambers following an elliptical proloculus that may have a small spine; sutures distinct oblique, often obscured by the strong, longitudinal costae; aperture radiate, terminal, on a short neck.
Remarks: differs from *C. discors* var. *gracilis* Marie in that the proloculus is not so elongated and drawn out. *C. d'orbigny* Marie differs in the greater number of costae and their strong divergence.
Range: Valanginian to Early Barremian, but most common in the Late Hauterivian.

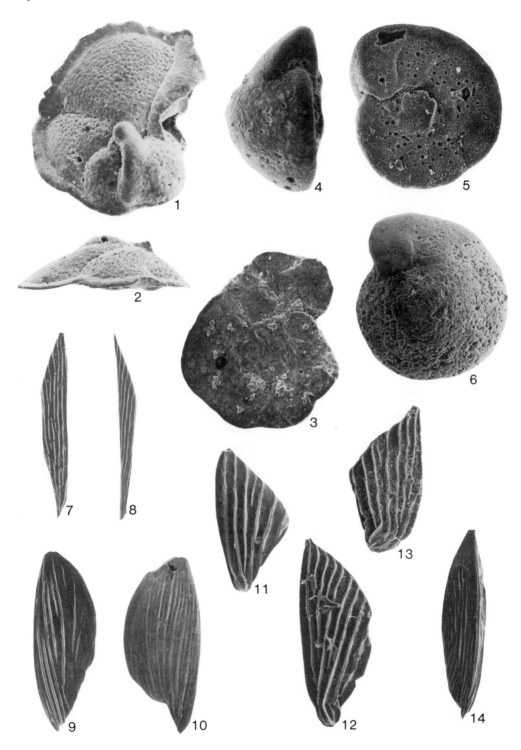

Plate 7.6

Citharina harpa (Roemer)
Plate 7.6, Fig. 11 (× 15), Speeton Clay, Speeton, Yorkshire, Hauterivian (C4E). = *Vaginulina harpa* Roemer, 1841. Description: test free, robust, sub-triangular in outline, compressed, dorsal margin has prominent thin keel; chambers numerous, parallel sided, curved; proloculus elliptical, sometimes with a short spine; sutures limbate, oblique, flush with surface; ornament comprises 10–16 strong, thin ribs, parallel to the dorsal margin, with occasional short costae between the main ribs; aperture terminal, radiate, on a short neck.
Remarks: there is some variation in the overall shape and strength of the ornamentation.
Range: Late Valanginian to the Hauterivian/Barremian boundary.

Citharina pseudostriatula Bartenstein and Brand, 1951
Plate 7.6, Figs 12, 13 (× 40), Speeton Clay, Speeton, Yorkshire, Hauterivian (C2A). Description: test free, small, triangular in outline, compressed; globular proloculus followed by 6–12 curving chambers, sutures inclined, limbate, flush; ornament comprises longitudinal costae, which bifurcate, as many as 18 being seen on the final chamber; aperture radiate, marginal, on a small tubular extension.
Remarks: the overall outline of the test is very variable. This species is very similar to *C. rusocostata* Bartenstein and Brand, but differs from it in having bifurcating ribs.
Range: Late Ryazanian and Early Valanginian but with some rare occurrences in the Hauterivian.

Citharina sparsicostata (Reuss)
Plate 7.6, Fig. 14 (× 16), Speeton Clay, Speeton, Yorkshire, Hauterivian (D2C). = *Vaginulina sparsicostata* Reuss, 1863. Description: test free, large, compressed, triangular in outline, dorsal margin typically with strong keel extending from proloculus to aperture; oval proloculus followed by up to 16 chambers; sutures limbate, flush, sometimes obscured by ornament of short costae; aperture radiate, protruding at dorsal angle.
Remarks: there is a great variation in size, shape and ornamentation. In many specimens there may be a trend to longer, more continuous, costae. *C. cristellaroides* (Reuss) differs in having slightly inflated chambers, and much weaker, more regularly spaced ornament which tends to be confined to the chambers.
Range: Early Hauterivian.

Citharinella laffittei Marie, 1938
Plate 7.7, Fig. 1 (× 25), uppermost Gault Clay, Arlesey, Bedfordshire, Upper Albian. Description: test free, lanceolate; early chambers indistinct, rapidly becoming chevron-shaped, symmetrical; 10–12 chambers, with the last one extending from a third to one half the length of the test; chambers ornamented by regular numerous, discontinuous, short, narrow striations; aperture terminal, radiate.
Remarks: very similar to both *C. karreri* (Berthelin) and *C. lemoinei* Marie. However it is much narrower than *C. karreri* and lacks the more continuous ribs at the proximal end. *C. lemoinei* is a much smoother species, lacking the prominent ribbing across the sutures.
Range: Late Albian, but rare except in the very uppermost levels.

Citharinella pinnaeformis (Chapman)
Plate 7.7, Fig. 2 (× 25), Bed XI, Gault Clay, Folkestone, Kent, Upper Albian. = *Frondicularia pinnaeformis* Chapman, 1893. Description: test free, lanceolate; early chambers indistinct, becoming chevron-shaped, symmetrical; 5–16 chambers with the last extending over half the length of the test; test surface smooth except for two prominent, sharply-edged, sub-parallel, longitudinal ridges which extend the entire length of the test; aperture terminal, radiate.
Remarks: the two prominent ribs are most distinctive and this species can usually be identified even from fragmentary material.
Range: Latest Middle and early Late Albian.

Conorboides lamplughi (Sherlock)
Plate 7.7, Figs 3–5 (× 90), Bed III, Gault Clay, Folkestone, Kent, Middle Albian. = *Pulvinuline lamplughi* Sherlock, 1914. Description: test free, biconvex to plano-convex, trochoid, 2–3 whorls, with 5–6 chambers in the final whorl; periphery acute, slightly lobulate; sutures on dorsal side distinct, slightly raised, curved, meeting the periphery tangentially; thickened sutures of early chambers fuse to form an umbonal boss; ventral sutures depressed, radial; aperture interiomarginal.
Remark: *Lamarckina hemiglobosa* Ten Dam differs in having a low, very broadly rounded, dorsal surface. In *C. lamplughi* there is some variation in the height of the spire, and some forms have a more acute apical angle.
Range: Late Hauterivian to Middle Albian.

Conorboides valendisensis Bartenstein and Brand, 1951.
Plate 7.7, Figs 6–8 (× 85), Speeton Clay, Speeton, Yorkshire, Valanginian (D4). Description: test free, small, trochoid, 2–3 whorls plano-convex or concavo-convex, periphery acute: initial chambers indistinct on dorsal side; sutures on dorsal side narrow, gently curved, flush or slightly raised; aperture a low, interiomarginal, umbilical arch with short flap.
Remarks: there is little variation and no difficulty in identification. Most specimens are poorly preserved, the majority being infilled with pyrite. Range: Ryazanian and Valanginian.

Cretaceous

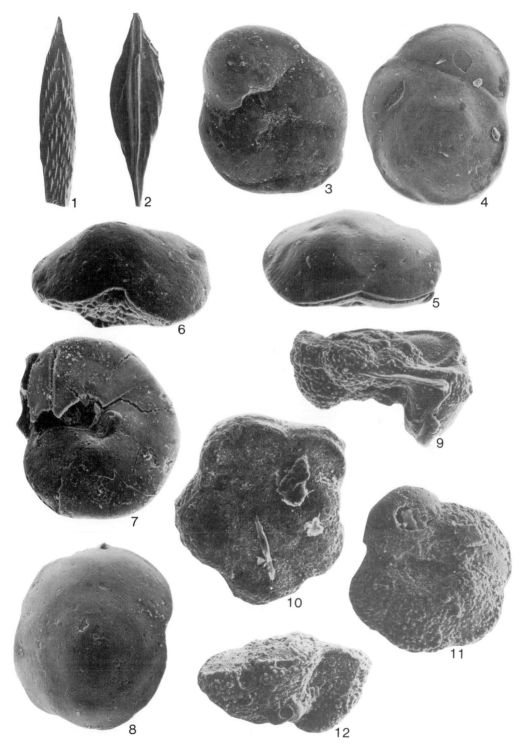

Plate 7.7

Dicarinella concavata (Brotzen)
Plate 7.7, Figs 9, 10 (× 60), Upper Chalk, Kelveden, Essex, Santonian. = *Rotalia concavata* Brotzen, 1934. Description: test free, low trochospire, dorsal side flat to slightly concave, ventral side strongly convex; angular periphery with two distinct, closely set keels; chambers distinct, rapidly increasing in size in final whorl, of 5–6 chambers; sutures distinct, slightly elevated and curved on the spiral side, radial and deeply constricted on the umbilical side; aperture an interiomarginal low arch, umbilical to slightly extra-umbilical.
Remarks: this distinctive species occurs only sporadically in the U.K., but nevertheless is of great stratigraphic value.
Range: Latest Coniacian and Early Santonian.

Dicarinella hagni (Scheibnerova)
Plate 7.7, Figs 11, 12 (× 80), Middle Chalk, Shillingstone, Dorset, Lower Turonian. = *Praeglobotruncana hagni* Scheibnerova, 1962. Description: test free, trochospiral, symmetrical biconvex to plano-convex; 2–2½ whorls, 5–8 chambers in the final whorl; chambers on umbilical side triangular, inflated to globular, sutures, radial depressed; chambers on spiral side petaloid, sutures raised, curved forwards; peripheral margin with two closely-spaced, parallel, keels; aperture extraumbilical–umbilical.
Remarks: this species has had a confused taxonomic history, fully discussed in Robaszynski and Caron (1979). It appears to be restricted to more northerly latitudes.
Range: Latest Cenomanian and Early Turonian.

Dicarinella primitiva (Dalbiez)
Plate 7.8, Figs 1, 2 (× 100), Upper Chalk, Borehole 25, Thames Barrier Site, Santonian. = *Globotruncana ventricosa primitiva* Dalbiez, 1955. Description: test free, trochospiral, biconvex to slightly umbilico-convex; 2–2¼ whorls, 5–6 chambers in final whorl; chambers on umbilical side inflated, sutures radial, depressed; chambers on spiral side petaloid, sutures oblique, slightly raised; equatorial periphery lobate, margin marked by two very close keels; primary aperture extraumbilical–umbilical.
Remarks: very close to *D. concavata* (Brotzen), which is more plano-convex, and has chambers that are more inflated ventrally.
Range: Coniacian to Early Santonian.

Eouvigerina aculeata (Ehrenberg)
Plate 7.8, Figs 3, 4 (× 220), Upper Chalk, Redbournbury, Hertfordshire, Santonian. = *Loxostomum aculeata* Ehrenberg, 1854. Description: test free, small, biserial, margins rounded; chambers distinct, elongated, slightly pyriform, final chamber centrally positioned; sutures slightly depressed, oblique; terminal aperture bordered by a lip, usually at the end of a short neck.
Remarks: the stratigraphic value of this species has been largely devalued by its unclear taxonomic status. Specimens have been variously referred to *E. americana* Cushman, *E. cretacea* (Heron-Allen and Earland), and *E. serrata* (Chapman).
Range: Coniacian to Early Maastrichtian.

Epistomina hechti Bartenstein and Bolli, 1957
Plate 7.8, Figs 5–7 (× 70), Speeton Clay, Speeton, Yorkshire, Barremian (LB2). Description: test free, trochoid, biconvex, composed of 2½–3 whorls, the last having 9–10 chambers; periphery acute; sutures straight, radiating from central boss; dorsal structures curved; apertures of two types, a small oval hole on the terminal face, and a long, slit-like, latero-marginal depression parallel to the periphery on the ventral surface.
Remarks: very close to *Hoeglundina chapmani* Ten Dam, but differs from it in not having raised sutures and depressed chambers in the early whorls.
Range: Early and mid-Barremian.

Epistomina ornata (Roemer)
Plate 7.8, Figs 8–10 (× 40), Speeton Clay, Speeton, Yorkshire, Hauterivian (C4). = *Planulina ornata* Roemer, 1841. Description: test free, lenticular, biconvex, trochoid; 1½–2 whorls, 10 chambers in final whorl; chambers on dorsal side curved, depressed, chambers on ventral side triangular in shape, depressed; dorsal sutures tuberculate, raised, ventral sutures gently curved, radiating from central umbilical pit; areal aperture rarely seen, latero-marginal apertures seen on ventral surface, near periphery.
Remarks: similar in general appearance to *E. spinulifera* (Reuss) but lacks the lobulate outline and the peripheral spines. The surface ornament is highly variable as seen in Pl. 7.8, Figs 8–10; when well-developed it can completely envelope the specimen.
Range: occurs throughout the Hauterivian but is most abundant in the Late Hauterivian.

Pl. 7.8] Cretaceous 331

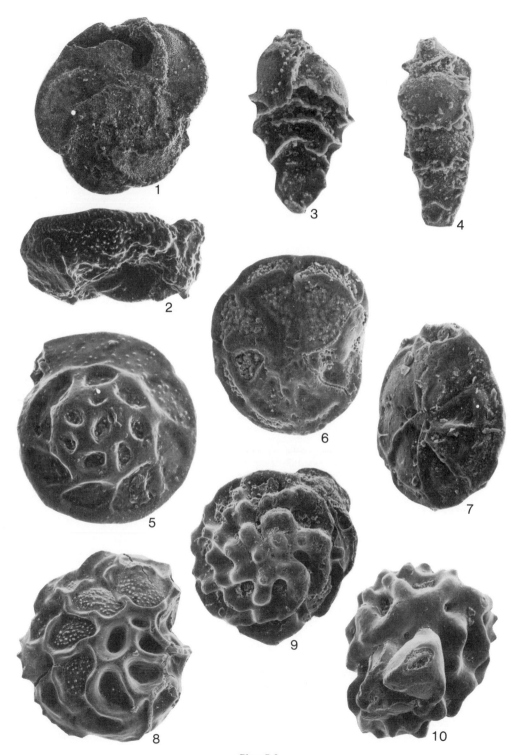

Plate 7.8

Epistomina spinulifera (Reuss, 1862)
Plate 7.9, Figs 1–3 (× 45), Bed IV, Gault Clay, Folkestone, Kent, Middle Albian. – *Rotalia spinulifera* Reuss, 1862. Description: test free, unequally biconvex, lenticular, trochospiral; 2–2½ whorls, 7–10 chambers in the final whorl; periphery with sharp, wide, crenulate, spinose keel; all sutural ribs high and sharp, strongly arcuate; chamber surfaces depressed; aperture often damaged but thin, fairly wide, tooth-plate seen running nearly perpendicular to vertical axis of test.
Remarks: the ornamentation is quite variable. In the Middle Albian the overall size and degree of ornamentation increases gradually up-section (Carter and Hart, 1977; Hart, 1984), reaching an acme just below the Middle–Late Albian boundary.
Range: Middle Albian.

Eponides beisseli Schijfsma, 1946
Plate 7.9, Figs 4, 5 (× 65), Upper Chalk, Sidestrand, Norfolk, Lower Maastrichtian. Description: test free, asymmetrically biconvex, trochospiral; outline circular, occasionally lobate; chambers on dorsal side indistinct, only 6–7 chambers of final whorl visible; chambers triangular on ventral side, uninflated; sutures slightly depressed; aperture a distinct, narrow, slit along inner margin of final chamber, bordered by a thick, distinct, lip.
Range: Late Campanian and Early Maastrichtian.

Eponides concinna Brotzen, 1936
Plate 7.9, Figs 6, 7 (× 100), Upper Chalk, Kelveden, Essex, Santonian. Description: test free, plano-convex to biconvex; periphery circular to slightly lobate, sub-acute; on dorsal side there are three whorls of 7–10 uninflated, crescentric, curved, chambers; sutures indistinct on dorsal side but slightly depressed on ventral side; umbilicus variable, filled with a boss of clear calcite; aperture a narrow slit along inner margin of final chamber, bordered ventrally by a lip.
Remarks: a very variable species both in the number of chambers in the final whorl and in the presence and size of the umbilical boss (see varieties erected by Vasilenko (1961)).
Range: Santonian and Campanian.

Favusella washitensis (Carsey)
Plate 7.9, Figs 8–10 (× 70), Lower Chalk, Dover, Kent (Fig. 8), Gault Clay, Folkestone, Kent (Figs 9, 10), mid-Cenomanian (Fig. 8) and uppermost Middle Albian (Figs 9, 10). = *Globigerina washitensis* Carsey, 1926. Description: test free, low to high trochospiral coil, periphery lobate; chambers subspherical to spherical; 2–3 whorls, 4–5 chambers in the final whorl; sutures radial to slightly curved, depressed; chamber surfaces covered by a reticulate network of fine to coarse ridges, which produces a characteristic honeycomb pattern; aperture a low to moderate, interiomarginal umbilical-extraumbilical arch, bordered by a narrow lip.
Remarks: the surface ornamentation is quite diagnostic although there is great variation in the overall size and shape. The very strange stratigraphic distribution probably indicates a pronounced ecological control, and the validity of this species could, and has been, questioned by Masters (1977).
Range: this species occurs at two levels; one near to the Middle-Late Albian boundary, and the other in the mid-Cenomanian, immediately below the mid-Cenomanian non-sequence (Carter and Hart, 1977).

Frondicularia hastata Roemer, 1842
Plate 7.9, Fig. 11 (× 15), Speeton Clay, Speeton, Yorkshire, Lower Barremian (LB2). Description: test free, large, compressed, elongate, lanceolate; 8–16 chambers, increasing slowly in size as added; periphery smooth, rounded; sutures distinct, strongly angled at centre of the test, depressed; large proloculus, occasionally ornamented by one or two costae; aperture terminal, radiate, on a long neck.
Remarks: very similar to *F. midwayensis* Cushman, but the latter differs in having slightly raised sutures and a finely papillate surface.
Range: Latest Valanginian to mid-Barremian.

Cretaceous

Plate 7.9

Gavelinella baltica Brotzen, 1942
Plate 7.10, Fig 3–5 (× 50), Lower Chalk, Pitstone, Hertfordshire, Middle Cenomanian. Description: test free, trochospiral, biconvex, sides flattened, periphery rounded; 2–2½ whorls, 8–12 chambers in the final whorl; chambers all slightly inflated, increasing gradually in size as added except for the last 3–4 which expand more markedly; sutures distinct, depressed; aperture a low, interiomarginal slit, extending from near periphery to the umbilicus.
Remarks: can be confused with *G. intermedia* (Berthelin), the ancestral form, in the lastest Albian and earliest Cenomanian.
Range: Latest Albian to Late Cenomanian.

Gavelinella barremiana Bettenstaedt, 1952
Plate 7.10, Figs 1, 2 (× 60), Lower Atherfield Clay, Sevenoaks, Kent, Lower Aptian. Description: test free, trochospiral, biconvex, with sides flattened; periphery rounded; $2\frac{1}{2}$–$3\frac{1}{2}$ whorls, 10–12 chambers in final whorl, expanding uniformly in size; sutures distinct, flush; aperture a low interiomarginal slit extending from near the periphery to the umbilicus.
Remarks: there is very little variation in this species, which can usually be recognised by the large number of chambers in the final whorl.
Range: Barremian to Aptian.

Gavelinella brielensis Malapris-Bizouard, 1974
Plate 7.10, Figs 6–8 (× 60), Lower Atherfield Clay, Sevenoaks, Kent, Lower Aptian. Description: test free, trochospiral, flattened biconvex, periphery rounded; 2–2½ whorls, 9–10 chambers in final whorl, chambers only slightly inflated; sutures distinct, tending to become raised; aperture a low interiomarginal slit, extending from near periphery to umbilicus.
Remarks: *G. brielensis* seems to be (Malapris-Bizouard, 1974) the ancestral form of *G. intermedia* (Berthelin), which appears in the Albian.
Range: Early Aptian.

Gavelinella cenomanica (Brotzen)
Plate 7.10, Figs 9–11 (× 50), Lower Chalk, Pitstone, Hertfordshire, Lower Cenomanian. = *Cibicidoides (Cibicides) cenomanica* Brotzen, 1945. Description: test free, trochospiral, biconvex, periphery rounded to slight angled; 2–2½ whorls, 8–10 chambers in the final whorl, all expanding gradually in size; sutures distinct, slightly depressed; umbilicus is characterised by the presence of a more-or-less marked calcite rim, which forms a rim around the central depression; aperture a low, interiomarginal arch/slit, extending from near periphery to umbilicus.
Remarks: the prominence of the beaded, umbilical rim varies from very weak in the Albian, where this species evolves from *G. intermedia* (Berthelin), to the highly diagnostic form, more typical of the Cenomanian.
Range: Latest Albian to Late Cenomanian.

Cretaceous

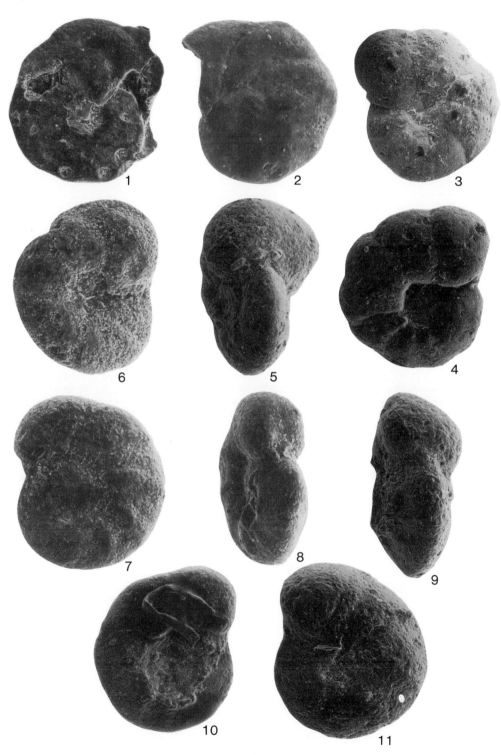

Plate 7.10

Gavelinella clementiana (d'Orbigny)
Plate 7.11, Figs 1–3 (× 65), Upper Chalk, Alum Bay, Ise of Wight, Upper Campanian. = *Rosalina clementiana* d'Orbigny, 1840. Description: test free, low trochospiral coil, periphery rounded; spiral side flattened, evolute, early whorls obscured by ornament; umbilical side weakly convex, involute, with small calcite boss, chamber distinct, 8–10 in final whorl; sutures initially raised, tending to become depressed; aperture a low interiomarginal slit, bordered by a distinct lip extending from periphery into umbilicus.
Remarks: a distinctive and characteristic species in the Late Campanian.
Range: Late Campanian.

Gavelinella cristata (Goel)
Plate 7.11, Figs 4–6 (× 100), Upper Chalk, Scratchells Bay, Isle of Wight, Lower Campanian. = *Pseudovalvulineria cristata* Goel, 1965. Description: test free, low trochospiral coil, periphery rounded; umbilicus filled with irregular calcite boss; 9–10 chambers in final whorl, distinct weakly inflated; sutures distinct, raised, thickened, becoming flush or depressed; aperture a low interomarginal slit, bordered by a distinct lip, extending from periphery towards umbilicus, covered by distinct, sub-triangular, chamber flaps.
Remarks: *G. cristata* may be distinguished from *G. clementiana* (d'Orbigny) by its convex spiral side and stronger ornamentation.
Range: Late Santonian and Early Campanian.

Gavelinella intermedia (Berthelin)
Plate 7.11, Figs 7–9 (× 100), Lower Chalk, Pitstone, Hertfordshire, Lower Cenomanian. = *Anomalina intermedia* Berthelin, 1880. Description: test free, periphery rounded, but may be slightly angled; dorsal side semi-involute with 1½–2 whorls visible, 10–12 chambers in final whorl; sutures distinct, depressed to slightly raised, slightly arcuate; aperture interiomarginal-equatorial, surrounded by a fairly wide lip.
Remarks: abundant in the Albian and Cenomanian of Britain. As used in this account, it does not include those forms with an umbilical boss (see Malapris, 1965; Michael, 1966).
Range: Albian and Cenomanian.

Gavelinella lorneiana (d'Orbigny)
Plate 7.11, Figs 10–12 (× 110), Upper Chalk, Alum Bay, Isle of Wight, Upper Campanian. = *Rosalina lorneiana* d'Orbigny, 1840. Description: test free, low trochospiral coil of 2½–3 whorls, outline circular, becoming lobate; periphery rounded to sub-acute with the occasional development of an imperforate peripheral band; chambers distinct, 8–10 in final whorl; sutures depressed; aperture a low interiomarginal slit bordered by an indistinct lip, expanding into umbilicus where it is covered by sub-triangular, imperforate, chamber flaps.
Remarks: the well-developed, imperforate chamber flaps and narrow umbilicus are highly distinctive.
Range: Turonian to early Late Campanian.

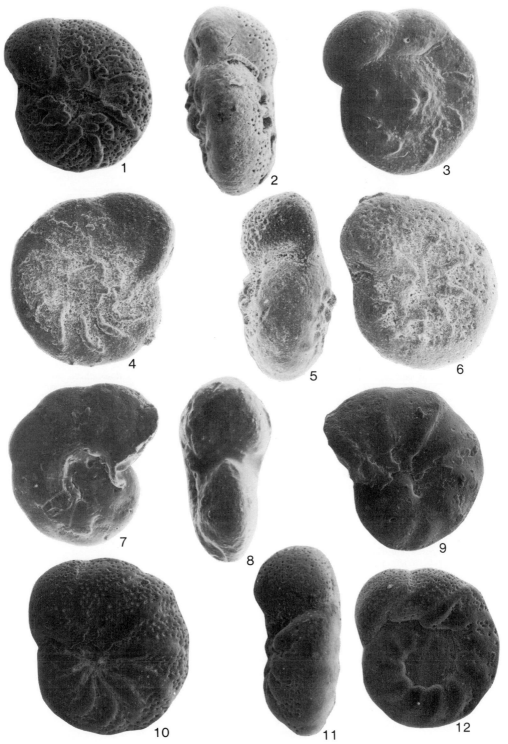

Plate 7.11

Gavelinella monterelensis (Marie)
Plate 7.12, Figs 1–3 (× 90), Upper Chalk, Alum Bay, Isle of Wight, Upper Campanian. = *Anomalina monterelensis* Marie, 1941. Description: test free, biconvex, low trochospiral coil of 2½–3 whorls; biumbonate, umbilical boss distinct, raised; spiral side boss distinct, broad, low; chambers numerous, 12–14 in final whorl; periphery circular, becoming weakly lobate, possessing indistinct imperforate keel; sutures distinct, depressed; aperture a moderately high interiomarginal arch, bordered by a thin lip, and extending into the umbilicus along the spiral suture partly covered by backward pointing, sub-triangular, chamber flaps.
Remarks: a distinctive, stratigraphically useful species.
Range: Late Campanian.

Gavelinella pertusa (Marsson)
Plate 7.12, Figs 4–6 (× 110), Upper Chalk, Overstrand, Norfolk, Upper Campanian. = *Discorbina pertusa* Marsson, 1878. Description: test free, low trochospiral coil, partially evolute, especially on spiral side, periphery rounded; chambers distinct, gradually increasing in size, 8–9 in final whorl; sutures flush to slightly depressed, indistinct, radial; test coarsely perforate; aperture an arcuate slit following the inner margin of the final chamber from the umbilicus round onto the periphery; umbilius broad, deep.
Remarks: the broad, open umbilicus is characteristic of this species.
Range: Coniacian to Maastrichtian.

Gavelinella sigmoicosta (Ten Dam)
Plate 7.12, Figs 7–9 (× 55), Speeton Clay, Speeton, Yorkshire, Upper Hauterivian (C1). = *Anomalina sigmoicosta* Ten Dam, 1948. Description: test free, plano-convex, periphery rounded, to sub-acute; ventral side convex, elevated, with deep umbilicus; 7–9 narrow, sigmoidal, depressed, chambers in final whorl; dorsal sutures strongly curved, raised, thickened; ventral sutures less distinct, slightly curved; aperture a narrow slit at base of apertural face, extending from umbilicus to half way towards the periphery.
Remarks: a distinctive and stratigraphically useful species.
Range: Late Hauterivian to Early Barremian.

Gavelinella stelligera (Marie)
Plate 7.12, Figs 10–12 (× 100), Upper Chalk, Scratchells Bay, Isle of Wight, Middle Campanian. = *Planulina stelligera* Marie, 1941. Description: test free, low trochospiral coil of 2½–3 whorls, periphery rounded to sub-acute; umbilical side has a broad shallow depression surrounding umbilicus, which is almost completely filled with chamber flaps; chambers distinct, 12–13 in final whorl; sutures indistinct, curved, flush; on spiral side chamber margins imperforate, thickened anterior to sutures, giving the appearance of a series of oblique, curved, ribs; aperture a narrow interiomarginal slit, bordered by an indistinct lip.
Remarks: the broad, curved, imperforate ribs and compressed test are characteristic of this species.
Range: atypical forms recorded from the Late Santonian, Early to Middle Campanian, becoming very rare in the early late Campanian.

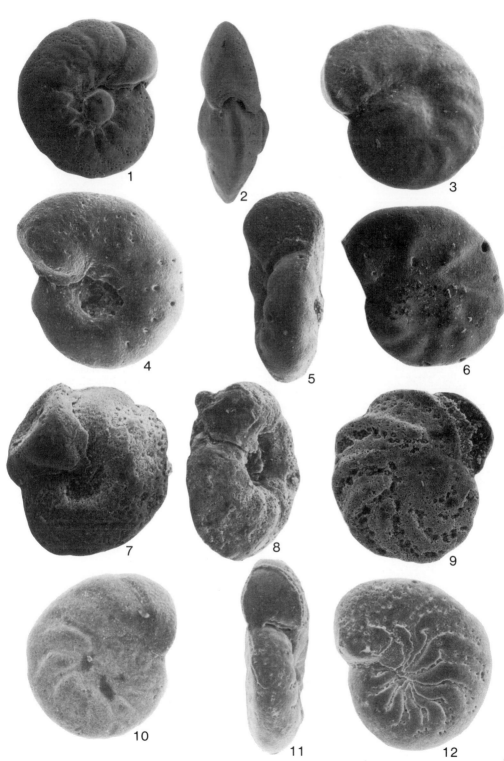

Plate 7.12

Gavelinella thalmanni (Brotzen)
Plate 7.13, Figs 1–3 (× 140), Upper Chalk, Euston, Suffolk (Fig 1, 2) and Scratchells Bay, Isle of Wight (Fig. 3), Upper Santonian (Figs 1, 2) and Middle Campanian (Fig. 3). = *Cibicides thalmanni* Brotzen, 1936. Description: test free, trochospiral, but appearing almost planispiral; chambers indistinct, not inflated, increasing gradually in size as added, 11–12 in final whorl; sutures flush, indistinct, straight, radial; ventral surface smooth, but dorsal surface with raised, rugose ribs, producing an irregular, nodose ornament; aperture an interiomarginal slit, extending from the periphery into the umbilical area of the ventral side.
Remarks: this stratigraphically useful species is easily distinguished from other members of the genus by the flattened nature of the ventral side, and the wide, shallow, dorsal umbilicus.
Range: Coniacian to early late Campanian.

Gavelinella usakensis (Vasilenko)
Plate 7.13, Figs 4–6 (× 100), Upper Chalk, Scratchells Bay, Isle of Wight, Middle Campanian, = *Anomalina (Pseudovalvulineria) clementiana* d'Orbigny var. *usakensis* Vasilenko, 1961. Description: test free, low to moderately high trochospiral coil, periphery rounded; spiral side convex, evolute, with earlier whorls obscured by thickened, smooth, calcite layer; umbilical side weakly convex with small umbilical boss; chambers distinct, 9–10 in final whorl, weakly inflated; sutures distinct, slightly depressed; aperture a low interiomarginal slit, bordered by a distinct lip, extending from periphery to umbilicus.
Remarks: this species is distinguished by its smooth, moderately convex, spiral side.
Range: Middle Campanian to early Late Campanian.

Globigerinelloides bentonensis (Morrow)
Plate 7.13, Figs 7–9 (× 100), Bed XIII, Gault Clay, Folkestone, Kent, Upper Albian. = *Anomalina bentonensis* Morrow, 1934. Description: test free, planispiral, bi-umbilicate, involute; two whorls, 6–9 chambers in the final whorl; sutures distinct, depressed, straight to slightly curved; aperture a low interiomarginal, umbilical-equatorial arch, with a prominent lip; relict apertural lips visible around the umbilicus on both sides.
Remarks: this is a characteristic species in the latest Albian of the British Isles. There is however a great deal of taxonomic confusion between this species and *G. cushmani* (Tappan), *G. caseyi* (Bolli, Loeblich and Tappan), and *G. eaglefordensis* (Moreman). After examination of topotype material Carter and Hart (1977) rejected *G. caseyi* and *G. eaglefordensis*, using *G. bentonensis* for the Late Albian and Cenomanian forms found in the British succession. Masters (1977) has erroneously suggested that *G. bentonensis* should be restricted to the Albian, even though Morrow's type material was of mid-Cenomanian age. However Masters does not extend its geographic distribution to the U.K. Both Masters and Price would refer this species to *G. cushmani*.
Range: Latest Albian and Cenomanian, although it is most abundant immediately below the Albian-Cenomanian boundary.

Globorotalites hiltermanni Kaever, 1961
Plate 7.13, Figs 10, 11 (× 90), Upper Chalk, Caister St. Edmunds, Norfolk, Upper Campanian. Description: test free, biconvex, trochospiral, 1½–2½ whorls, periphery acute, slightly keeled; chambers indistinct on spiral side; 7–9 chambers visible on umbilical side, indistinct, becoming slightly inflated; sutures indistinct, flush to weakly depressed; aperture narrow, elongate, interiomarginal slit, with distinct murus reflectus.
Remarks: the almost equally biconvex test, and closed umbilicus characterise this species.
Range: only recorded from the very latest Campanian.

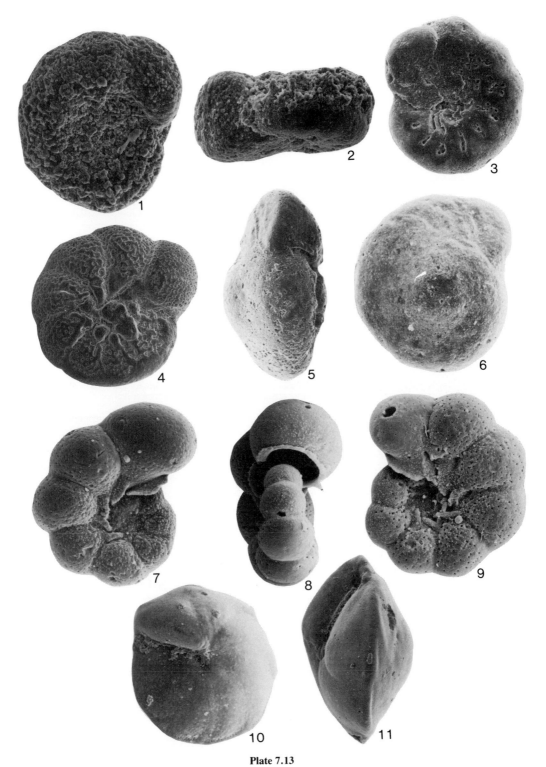

Plate 7.13

Globorotalites micheliniana (d'Orbigny, 1840)
Plate 7.14, Figs 1, 2 (× 85 and × 70), Upper Chalk, Catton Grove, Norfolk, Upper Campanian. = *Rotalina micheliniana* d'Orbigny, 1840. Description: test free, plano-convex, trochospiral; umbilical side involute, steeply conical, normally showing no pseudo-umbilicus; outline circular periphery, acute to sub-acute, slightly keeled; 6–7 chambers in final whorl, indistinct, conical; spiral side sutures indistinct; umbilical side sutures distinct, slightly depressed; aperture a narrow elongate, interiomarginal slit, which may be a slight lip.
Remarks: *G. micheliniana* was the first Late Cretaceous species of this genus to be described. Morphologically it shows a broad range of variation with respect to the height of the umbilical side and the width of the pseudo-umbilicus. Little of this variation (Goel, 1963) would appear to have any stratigraphic value. *G. cushmani* Goel – as used by Bailey (1978, unpublished thesis) – is probably an earlier member of the *G. micheliniana* plexus.
Range: Turonian to Campanian.

Globotruncana bulloides Vogler, 1941
Plate 7.14, Figs 3–5 (× 85), Upper Chalk, Sherringham (Fig. 3) and Keswick (Figs 4, 5), Norfolk, Upper Campanian. Description: test free, low trochospire, slightly biconvex, margins truncated with two distinct, widely-spaced, beaded, keels; chambers distinct, inflated both dorsally and ventrally, 6–7 in final whorl; sutures distinct, depressed, curved, masked by a lobate keel on the umbilical surface; aperture umbilical; umbilicas small with evidence for the development of a tegilla, although this is rarely preserved except in the Campanian.
Remarks: this species is distinguished by its small umbilicus, and widely spaced keels. There is a strong possibility that *G. paraventricosa* (Hofker) should be regarded as synonymous.
Range: Coniacian to Campanian.

Globotruncana contusa (Cushman)
Plate 7.14, Figs 6, 7 (× 75), North Sea Basin, Maastrichtian. = *Pulvinulina arca contusa* Cushman, 1926. Description: test free, trochospiral, spiro-convex with a deep umbilicus, outline sub-circular, weakly lobate; periphery truncated by two closely spaced, double keels; chambers indistinct, 5–7 in the final whorl, crescentric on spiral side, elliptical on umbilical side; sutures distinct, strongly curved, weakly raised and beaded on spiral side, while depressed, radial and slightly curved on the umbilical side; primary aperture interiomarginal-umbilical.
Remarks: this distinctive species is stratigraphically important in the North Sea Basin. Robaszynski *et al.* (1984) have suggested that this species, together with *G. fornicata* and *G. plummerae*, be placed in the new genus *Rosita*. As this has not yet gained wide acceptance these three species are listed here in the genus *Globotruncana*.
Range: Late Maastrichtian.

Globotruncana fornicata Plummer, 1931
Plate 7.14, Figs 8–10 (× 110), Upper Chalk, Alderford, Norfolk, Middle Campanian. Description: test free, trochospiral, asymmetrically biconvex; margins angular, truncated by two well-spaced keels; chambers distinct, 4–5 in final whorl; chambers elongate, crescentric on dorsal side, sub-rectangular on the umbilical side; umbilicus narrow, deep, covered by tegilla; primary aperture interiomarginal, umbilical.
Remarks: readily recognised by its moderately convex spiral side, and elongate, crescentric chambers. The initial chambers in the first whorl are also distinctly 'hedbergellid' in appearance.
Range: Latest Santonian and Campanian in the U.K., although the range in Southern Europe continues into the Maastrichtian.

Globotruncana linneiana (d'Orbigny)
Plate 7.14, Figs 11, 12 (× 85), Upper Chalk, Scratchells Bay, Isle of Wight (Fig. 11) and Shell/Esso 44/2–1, North Sea Basin (Fig. 12), Campanian. = *Rosalina linneiana*, d'Orbigny, 1839. Description: test free, low troschospiral, both sides almost flat, margins distinctly angular and truncated, with two sutures curved, raised on spiral side, slightly lobate to radial on umbilical side; primary aperture interiomarginal but preservation too poor to record development of a tegilla.
Remarks: this species, first described from Recent beach sands on Cuba, has had a much confused taxonomic history (Swiecicki, 1980, unpublished thesis). It is never common in the British succession.
Range: Santonian and Campanian.

Cretaceous

Plate 7.14

Globotruncana plummerae Gandolfi, 1955.
Plate 7.15, Figs 3–5 (× 110), Upper Chalk, Arminghall, Norfolk, Upper Campanian. Description: test free, low trochospiral coil, asymmetrically biconvex; periphery sub-rounded to sub-acute, truncated by prominent double keel; chambers distinct, 2–2½ whorls, 4–5 chambers in final whorl; chambers on dorsal side elongate, crescentic, inflated; chambers on umbilical side are elongate, sub-rectangular and weakly inflated; sutures depressed and curved; umbilicus narrow, deep; primary aperture interiomarginal, umbilical, covered by a tegilla.
Remarks: can be distinguished by its elongate, and inflated chambers, with those of the final whorl increasing rapidly in size as added.
Range: Late Campanian.

Globotruncanella havanensis (Voorwijk)
Plate 7.15, Figs 1, 2 (× 150), Shell/Esso 44/2–1, North Sea Basin (Fig. 1) and Upper Chalk, Trimingham, Norfolk (Fig. 2), Maastrichtian. = *Globotruncana havanensis* Voorwijk, 1937. Description: test free, low trochospiral coil; periphery acute to sub-acute, may possess imperforate, peripheral band; 4–5 distinct chambers in final whorl; sutures distinct, depressed, radial, slightly curved on umbilical side and strongly curved on spiral side; umbilicus narrow, moderately deep; aperture high umbilical-extraumbilical arch, opening into an umbilicus which may be covered by a delicate tegilla.
Remarks: a very distinctive species which possibly evolved from 'whiteinellid' stock in the latest Campanian. *G. cita* Bolli is a common synonym of this species.
Range: Maastrichtian.

Guembelitria cenomana (Keller)
Plate 7.15, Figs 6 (× 225), Bed XIII, Gault Clay, Folkestone, Kent, Upper Albian. = *Guembelina cenomana* Keller, 1935. Description: test free, triserial throughout, extremely small; chambers inflated and globular; sutures distinct, depressed; aperture an interiomarginal arch at base of final chamber, variable in outline but normally low.
Remarks: Masters (1977) has concluded, after examination of the holotype of *G. harrisi* Tappan, that it is a junior synonym of *G. cenomana*.
Range: Middle Albian to earliest Turonian.

Hedbergella brittonensis Loeblich and Tappan, 1961
Plate 7.15, Figs 7–9 (× 150), Lower Chalk, Pitstone, Hertfordshire, Cenomanian. Description: test free, high asymmetrical trochospire; 2–2½ whorls, 5½–7 chambers in final whorl; chambers inflated, globular, sutures radial, depressed; primary aperture umbilical-extraumbilical, bordered by lip.
Remarks: there is a problem in separating this species from the lower-spired *H. delrioensis* (Carsey), the higher-spired *Whiteinella paradubia* (Sigal), and the ancestral *H. infracretacea* (Glaessner). Higher in the Cenomanian and Turonian the aperture tends to become more umbilical in position, thereby changing the generic position to that of a *Whiteinella*.
Range: Cenomanian to Santonian.

Hedbergella delrioensis (Carsey)
Plate 7.15, Figs 10–12 (× 200), Lower Chalk, Pitstone, Hertfordshire, Cenomanian. = *Globigerina cretacea* d'Orbigny var *delrioenis* Carsey, 1926. Description: test free, trochospiral, biconvex, umbilicate, periphery rounded, gently lobate; chambers globular, sutures distinct, depressed; normally five chambers per whorl; aperture a simple interiomarginal, extraumbilical-umbilical arch, commonly bordered by a narrow lip.
Remarks: *H. delrioensis* is recognised by its low-medium spire, but in the Cenomanian its separation from the higher-spired *H. brittonensis* Loeblich and Tappan may sometimes be rather arbitrary.
Range: Middle Albian to Late Turonian.

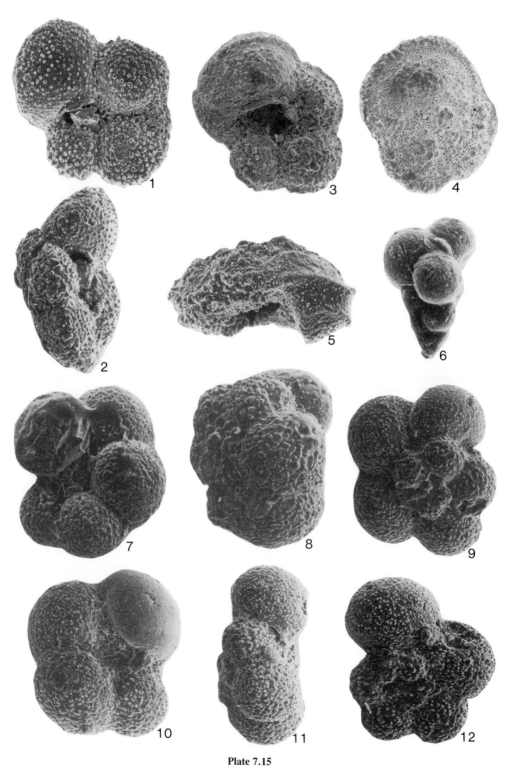

Plate 7.15

Hedbergella infracretacea (Glaessner)
Plate 7.16, Figs 1–3 (× 200), Bed XIII, Gault Clay, Folkestone, Kent, Upper Albian. = *Globigerina infracretacea* Glaessner, 1937. Description: test free, trochospiral, with moderately high spire, periphery rounded; five subspherical chambers per whorl, sutures radial, depressed; aperture extraumbilical, with imperforate lip.
Remarks: although Masters (1977) placed this species in the synonym of *H. delrioensis* (Carsey), both Price (1977b) and Carter and Hart (1977) have upheld their separate identity, while still recognising that they are both ends of a complete range of variability.
Range: Barremian to Late Albian in the North Sea Basin, but onshore the range is Middle and Late Albian.

Hedbergella planispira (Tappan)
Plate 7.16, Figs 4, 5 (× 200), Bed XIII, Gault Clay, Folkestone, Kent Upper Albian. = *Globigerina planispira* Tappan, 1940. Description: test free, small, very low trochospire, appearing almost planispiral in peripheral view; 2–2½ whorls, 6–7 chambers in final whorl, slowly increasing in size as added, chambers inflated, sutures distinct, depressed; aperture a low interiomarginal, extraumbilical-umbilical arch, bordered by an imperforate flap.
Remarks: small and very distinctive, although larger specimens are known from higher levels in the succession.
Range: Middle Albian to Late Cenomanian, with rare, atypical specimens, being found in the Turonian.

Hedbergella simplex (Morrow)
Plate 7.16, Figs 6–8 (× 150), Lower Chalk, Maiden Newton, Dorest, Upper Cenomanian. = *Hastigerinella simplex* Morrow, 1934. Description: test free, trochospiral, slightly asymmetrical, concavo-convex; 2 whorls, 4–6 chambers in final whorl; chambers globular, becoming radially elongated, sutures radial, very depressed; periphery rounded, clavate; aperture extraumbilical-umbilical, bordered by a well-developed lip.
Remarks: a distinctive species which includes *H. amabilis* Loeblich and Tappan and *H. simplicissima* (Magné and Sigal) in its synonymy.
Range: Cenomanian.

Heterohelix moremani (Cushman)
Plate 7.16, Fig. 9 (× 400), Bed XIII, Gault Clay, Folkestone, Kent, Upper Albian. = *Guembelina moremani* Cushman, 1938. Description: test free, slender, gradually tapering; biserial, 13–20 chambers, generally subequal in width, height, and thickness, slowly increasing in size; sutures depressed, aperture low-moderately high interiomarginal arch, bordered by narrow imperforate lip.
Remarks: no microspheric forms (i.e. those possessing an initial coil) have been recognised in the British succession.
Range: Middle Albian to Late Cenomanian (although Masters (1977) gives a Middle Albian to Maastrichtian range for this species).

Hoeglundina caracolla (Roemer)
Plate 7.16, Figs 10–12 (× 50), Speeton Clay, Speeton, Yorkshire, Barremian (LB4D). = *Gyroidina caracolla* Roemer, 1841. Description: test free, biconvex, trochoid, periphery acute, slightly keeled, 1½–2½ whorls, 6–10 chambers in final whorl; chambers triangular, depressed on umbilical side, curved and depressed on the spiral side; sutures prominent, raised, limbate; aperture oval, areal, with conspicuous lateromarginal apertures along the ventral periphery.
Remarks: there is a great deal of variation both in size and shape, although poor preservation hinders detailed study.
Range: Latest Ryazanian to Early Barremian; often occurs in flood abundance, particularly in the Late Hauterivian.

Plate 7.16

Hoeglundina carpenteri (Reuss)
Plate 7.17, Figs 1, 2 (× 65), Bed III, Gault Clay, Folkestone, Kent, Middle Albian. = *Rotalia carpenteri* Reuss, 1861.
Description: test free, biconvex, low trochospiral coil; periphery sharp, acute, crenulate but not developed into marked spines, 2–2½ whorls, 6–9 chambers in final whorl; chambers triangular on apertural side, curved on spiral side; all chambers slightly depressed between raised sutures; primary aperture a slit in the face of final chamber, with a slightly raised lip; lateromarginal chambers also present.
Remarks: the test ornament is highly variable ranging from smooth to more typically pustulose. It differs from *H. chapmani* (Ten Dam) in being larger and in having a crenulate periphery.
Range: Early and Middle Albian, with rare, (?) reworked, specimens being found in the earliest Late Albian.

Hoeglundina chapmani (Ten Dam)
Plate 7.17, Figs 3–5 (× 65), Bed IV, Gault Clay, Folkestone, Kent, Middle Albian. = *Epistomina chapmani* Ten Dam, 1948. Description: test free, biconvex to rarely plano-convex, low trochospiral coil; periphery acute, carinate, non-crenulate; 2–2½ whorls, eight or nine chambers in the final whorl; chambers triangular on apertural side, curved on spiral side; last one or two chambers may be depressed, otherwise smooth; sutures distinct, but flush with surface; primary aperture a slit in face of final chamber; lateromarginal apertures may also be present.
Remarks: this species, from the Aptian–Middle Albian interval of the British succession is clearly closely related to *E. caracolla* (Roemer) and *E. hechti* Bartenstein, Bettenstaedt and Bolli, both recorded from the Early Cretaceous of northeast England. It is possible that they may all be part of the same group, but as they occur at different stratigraphic levels in two separate basins of deposition such a conclusion would be difficult to substantiate.
Range: Early Aptian to earliest Late Albian.

Lagena hauteriviana Bartenstein and Brand, 1951
Plate 7.17, Fig. 6 (× 65), Speeton Clay, Speeton, Yorkshire, Hauterivian (C6). Description: test free, small oval to spherical, circular in transverse section; unilocular; central, tapering initial spine; aperture simple, terminal, at end of tubular neck.
Remarks: a distinctive species, though variation makes it difficult to identify at the sub-specific level. The length–breadth ratio above shows a continuous variation with *L. hauteriviana hauteriviana* at one extreme, and *L. hauteriviana cylindracea* Bartenstein and Brand at the other.
Range: Late Ryazanian to Early Barremian, but most abundant in the Hauterivian.

Lagena hispida Reuss, 1863
Plate 7.17, Figs 7, 8 (× 60), Speeton Clay, Speeton, Yorkshire, Barremian (LB4). Description: test free, small, with spherical, inflated chamber, and a long basal spine; surface is fine to coarsely hispid; aperture central, simple, on a long tubular neck.
Remarks: differs from *L. oxystoma* Reuss in having a coarsely hispid surface.
Range: Hauterivian to Early Barremian.

Lenticulina eichenbergi Bartenstein and Brand, 1951
Plate 7.17, Figs 9, 10 (× 25), Speeton Clay, Speeton, Yorkshire, Hauterivian (C7). Description: test free, biconvex, periphery acute, with delicate keel; up to 10 sub-triangular chambers in final whorl; sutures curved, limbate, ornamented by elevated pustules, sutures of later chambers tend to be unornamented; aperture radiate at peripheral angle.
Remarks: in the British succession *L. eichenbergi* and *L. guttata* (Ten Dam) have the same range, and are regarded as one, varying, population by Fletcher (1966, unpublished thesis), but Bartenstein (1977) gives two distinct ranges for these species in Germany.
Range: Valanginian to Late Aptian.

Lenticulina guttata (Ten Dam)
Plate 7.17, Figs 11, 12 (× 25), Speeton Clay, Speeton, Yorkshire, Hauterivian (C3). = *Planularia guttata* Ten Dam, 1946. Description: test free, biconvex, periphery acute, keeled; close coiled but with a tendency for the later chambers to uncoil; chambers subtriangular. 10–13 in the final whorl; sutures curved, limbate, raised, ornamented with numerous small, guttiform, pustules; aperture at peripheral angle, terminal, radiate.
Remarks: there is a great deal of variation in the degree of coiling, and also in the strength of the ornamentation (Fletcher, 1966, unpublished thesis).
Range: Late Valanginian to mid-Barremian.

Cretaceous

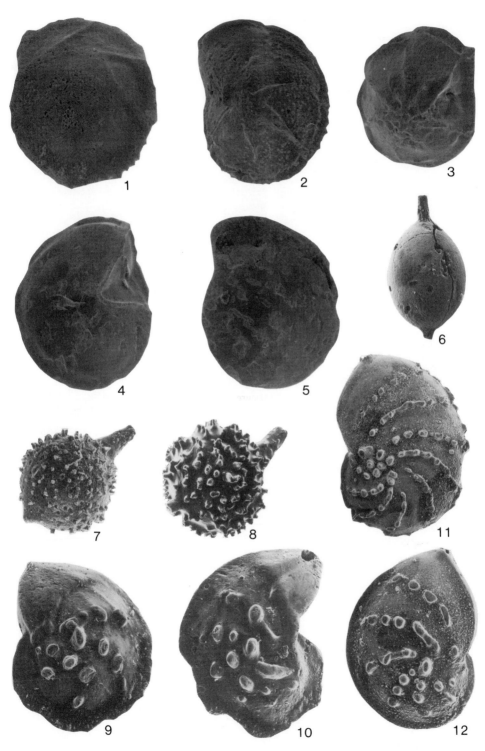

Plate 7.17

Lenticulina heiermanni Bettenstaedt, 1952
Plate 7.18, Fig. 1 (× 20), Speeton Clay, Speeton, Yorkshire, Barremian (LB5D). Description: test free, biconvex, periphery curved or slightly sub-angular, keeled; involute, 10–11 chambers in the final whorl; sutures limbate, raised, curving backwards to peripheral margin; clear, raised, calcite umbilical boss to which sutures are fused; aperture radiate, commonly on a small neck.
Remarks: there is very little variation, apart from the degree of elevation of the sutures above the surface of the test. The keel is easily broken giving a ragged outline.
Range: Late Hauterivian to Early Aptian, but most abundant in the Early Barremian.

Lenticulina muensteri (Roemer)
Plate 7.18, Fig. 2 (× 30), Speeton Clay, Speeton, Yorkshire, Hauterivian (C1B). = *Robulina muensteri* Roemer, 1839. Description: test free, biconvex, periphery smooth, acute, but not keeled; test involute 9–13 chambers in the final whorl; sutures limbate, flush with surface; large clear calcite boss, flush with surfaces; aperture radiate, normally at peripheral angle, but sometimes on a small extension.
Remarks: a distinctive species, but the range of variation gives rise to many morphotypes in the Jurassic and Lower Cretaceous. These have been well documented by Jendryka Fuglewicz (1975).
Range: This species ranges from the Ryazanian to Early Barremian in the Early Cretaceous.

Lenticulina ouachensis wisselmanni Bettenstaedt, 1952
Plate 7.18, Fig. 3 (× 25), Speeton Clay, Speeton, Yorkshire, Hauterivian (C7). Description: test free, biconvex, periphery acute, with well-developed keel; involute, but with a tendency to uncoil; 7–12 chambers in the final whorl; sutures limbate, raised, umbilical area surrounded by a raised rim; aperture radiate, on a small neck at the peripheral angle.
Remarks: extremely variable but can be distinguished by the characteristic high, sharp, rib, enclosing an oval umbilical area.
Range: Hauterivian.

Lenticulina saxonica Bartenstein and Brand, 1951
Plate 7.18, Fig. 4 (× 45), Speeton Clay, Speeton, Yorkshire, Barremian (LB4D). Description: test free, biconvex, periphery acute, with sharp keel; involute, 7–10 chambers in final whorl. sutures limbate, raised, curving towards aperture and meeting periphery at a low angle; raised sutures meet in the centre of the specimen; aperture terminal, radiate, on a small extension.
Remarks: there is little variation, but in some cases the raised sutures may fail to reach the centre. The keel is fragile and easily broken, giving a ragged outline to the test.
Range: Late Valanginian to Early Barremian.

'*Lenticulina*' *schloenbachi* (Reuss)
Plate 7.18, Fig. 5 (× 50), Speeton Clay, Speeton, Yorkshire, Hauterivian (C5K). = *Cristellaria schloenbachi* Reuss, 1863. Description: test free, elongate, compressed oval in cross-section; early portion close coiled, becoming uncoiled with 5–6 chambers in a rectilinear series; sutures indistinct in coiled portion, later distinct, oblique, and sometimes depressed; aperture radiate, at peripheral angle.
Remarks: similar to *Lenticulina (Astacolus) pachynota* (Ten Dam) but lacks the limbate, elevated sutures. The generic position of this species, in either *Astacolus* or *Lenticulina*, is a subject of debate.
Range: Late Ryazanian to Early Barremian.

'*Lenticulina*' *schreiteri* (Eichenberg)
Plate 7.18, Figs 6, 7 (× 40), Speeton Clay, Speeton, Yorkshire, Hauterivian (C2F). = *Elphidium schreiteri* Eichenberg, 1935. Description: test free, elongate, compressed; initially coiled, becoming uncoiled, with 3–8 low, broad, sub-triangular chambers in uncoiled part of test; surface of test with strong, reticulate ornament; aperture radiate, on a distinct neck.
Remarks: the ornament is quite variable, from strong and regular to very irregular and sometimes quite weak. To a lesser extent the degree of uncoiling is also quite variable causing problems of generic determination (*Lenticulina* vs. *Astacolus*). Taxonomically it is very close to *Vaginulinopsis reticulosa* (Ten Dam).
Range: Late Valanginian to Barremian.

Lingulogavelinella globosa (Brotzen)
Plate 7.18, Figs 8–10 (× 75), Plenus Marls, Bed 4, Betchworth Quarry, Surrey, Upper Cenomanian. = *Anomalinoides globosa* Brotzen, 1945. Description: test free, trochospiral, equally biconvex; periphery broadly rounded, lobate, spiral side centrally depressed; 2–2½ whorls, 6–8 inflated chambers in final whorl; sutures distinct, depressed, curved; aperture interiomarginal, slit-like, bordered by a lip; remnant lips of previous chambers forming star-shaped pattern around the umbilicus.
Remarks: *L. globosa* is a distinctive species, which appears to form the end of the Albian-Cenomanian 'lingulogavelinellid' lineage.
Range: Late Cenomanian to mid-Turonian.

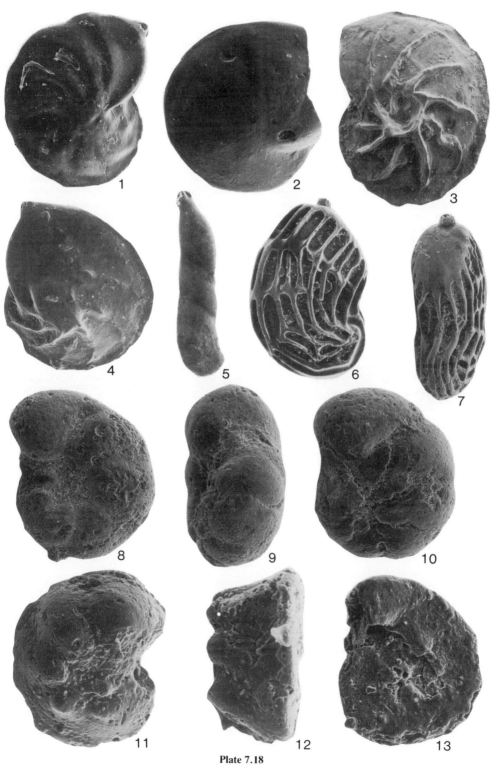
Plate 7.18

Lingulogavelinella jarzevae (Vasilenko)
Plate 7.18, Figs 11–13 (× 75), Glauconitic Marl, Folkestone, Kent, Lower Cenomanian. = *Cibicides (Cibicides) jarvezae* Vasilenko, 1954. Description: test free, plano-convex, periphery rounded to sub-acute; apertural side flat, with a star-shaped pattern made up of the relict apertural flaps; spiral side elevated, with 5–7 highly inflated chambers; sutures very depressed on spiral side, almost straight; aperture an interiomarginal slit extending from the periphery to umbilicus.
Remarks: this very distinctive species is morphologically very close to *Cibicides formosa* Brotzen, and is regarded as synonymous by some authors. The ranges of the two species do not, however, overlap and the use of a Campanian/Maastrichtian specific name is rejected for the present.
Range: Latest Albian to Early Cenomanian.

Lingulogavelinella sp. cf. *L. vombensis* (Brotzen)
Plate 7.19, Figs 1–3 (× 95), Upper Chalk, Helhoughton, Norfolk, Santonian. = *Pseudovalvulineria vombensis* Brotzen, 1942. Description: test free, low trochospiral, periphery rounded to sub-angular, early whorls on dorsal surface masked by a large calcareous boss, ventral side involute; sutures flush to slightly depressed on dorsal side, raised on ventral side; aperture a narrow, interiomarginal slit, covered by an apertural flap on the ventral surface; umbilicus filled with remnant apertural flaps, forming a radiate pattern.
Remarks: a distinctive species of great stratigraphic value. *L. vombensis* (Brotzen) is a Maastrichtian species and does not appear to be an extension of the present species' range.
Range: Coniacian to Santonian.

Loxostomum eleyi (Cushman)
Plate 7.19, Figs 4, 5 (× 160), Upper Chalk, Helhoughton, Norfolk, Santonian. = *Bolivinita eleyi* Cushman, 1927. Description: test free, elongate, compressed, biserial edge truncated, sub-angular; chambers flat, overlapping, reniform in outline; sutures distinct, flush to slighty raised, curved; aperture terminal, ovoid slit, surrounded by slightly raised lip.
Remarks: has a very useful range over a wide geographical area of Europe.
Range: Santonian to Middle Campanian; very rare in the Late Campanian.

Marginotruncana coronata (Bolli)
Plate 7.19, Figs 6, 7 (× 50), Middle Chalk, Beer, Devon, Turonian. = *Globotruncana lapparenti* Brotzen subsp. *coronata* Bolli, 1945. Description: test free, low trochospiral coil; 2–2½ whorls, 6–8 chambers in final whorl; chambers on umbilical side reniform, chambers on spiral side petaloid, chambers not inflated; sutures raised; chamber margins truncated, with two keels separated by an imperforate band; primary aperture extraumbilical-umbilical, may be covered with apertural flaps or a tegilla (especially in later forms).
Remarks: differs from *M. pseudolinneiana* (Pessagno) in having a more compressed profile, petaloid chambers on the spiral side, and in keels that are closer together.
Range: Late Turonian.

Marginotruncana pseudolinneiana Pessagno, 1967
Plate 7.19, Figs 8, 9 (× 75), Middle Chalk, Beer, Devon, Middle Turonian. Description: test free, very low trochospiral, both sides flat to slightly convex; periphery truncated, with two closely spaced, beaded keels; chambers elongate petaloid on spiral side, lobate on umbilical side, flat 5–7 in final whorl; sutures slightly depressed at the margins, otherwise raised; aperture extraumbilical-umbilical, low, interiomarginal arch, tegilla not recorded.
Remarks: there is a continuous morphological gradation from *M. pseudolinneiana* to *G. linneiana* (d'Orbigny).
Range: Turonian to Santonian.

Marginotruncana sigali (Reichel)
Plate 7.19, Fig. 10 (× 75), Middle Chalk, Beer, Devon, Middle Turonian. = *Globotruncana sigali* Reichel, 1950. Description: test free, trochospiral, single peripheral keel, formed by a very close double row of pustules; 2–3 whorls, 5–7 chambers in final whorl; chambers on umbilical side sub-rectangular, not inflated; chambers on spiral side subtrapezoidal, flat; sutures raised on umbilical side, 'U'-shaped, continuing along umbilical periphery; primary aperture extraumbilical-umbilical, umbilicus surrounded or covered by portici.
Remarks: a very distinctive, biconvex species; the oldest single keeled marginotruncanid in the British succession.
Range: Middle Turonian.

Marginulinopsis foeda (Reuss)
Plate 7.19, Figs 11, 12 (× 45), Speeton Clay, Speeton, Yorkshire, Hauterivian (C4C). = *Cristellaria foeda* Reuss, 1863. Description: test small, elongate, circular in cross-section; in microspheric form early chambers coiled, followed by 3–4 uncoiled chambers; chambers slightly inflated, sutures depressed; surface of test strongly hispid; aperture terminal, radiate, on a long tubular, lipped neck.
Remarks: there is considerable variation in the density and size of the hispid ornamentation; it is similar in shape to *M. gracilissima* (Reuss), but the latter species is completely smooth.
Range: Late Ryazanian to mid-Barremian.

Cretaceous

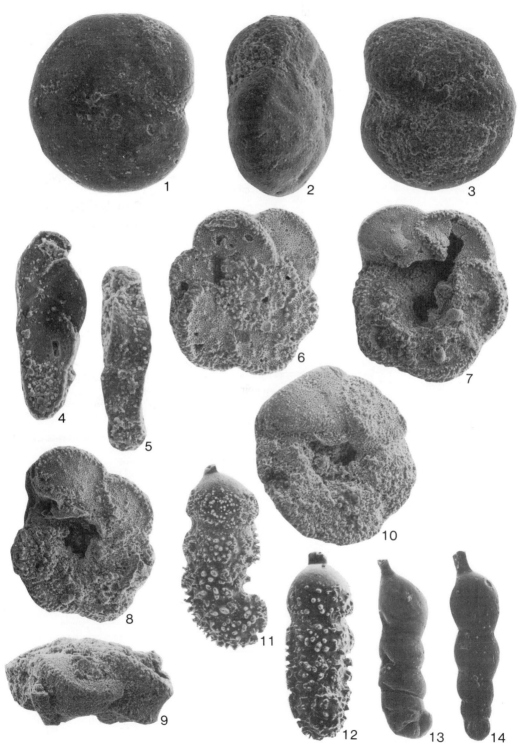

Plate 7.19

Marginulinopsis gracilissima (Reuss)
Plate 7.19, Figs 13, 14 (× 50), Speeton Clay, Speeton, Yorkshire, Barremian (LB3). = *Cristellaria gracillissima* Reuss, 1863. Description: test free, elongate, circular in cross-section; 4–5 coiled chambers followed by an uncoiled linear series of 3–4 chambers, the last of which is usually considerably inflated; sutures distinct, depressed, oblique in uncoiled portion of the test; aperture radiate, on a distinct neck.
Remarks: the overall shape of the test varies greatly. This species is very similar to *M. foeda* (Reuss) but differs in having a smooth test.
Range: Ryazanian to Early Barremian.

Neoflabellina praereticulata Hiltermann, 1952
Plate 7.20, Fig. 2 (× 60), Upper Chalk, Whitlingham, Norfolk, Upper Campanian. Description: test free, palmate, compressed, with angular to carinate edges; chambers initially coiled, uncoiling rapidly, chevron-shaped; sutures distinct, raised, slightly crenulated; aperture terminal, subcircular; test surface between sutures strongly ornamented by numerous short ridges, ornament on earliest chambers nodose.
Remarks: characterised by its markedly irregular, reticulate, ornament.
Range: Late Campanian to earliest Maastrichtian.

Neoflabellina reticulata (Reuss)
Plate 7.20, Fig. 3 (× 50), Upper Chalk, Trimingham, Norfolk, Lower Maastrichtian. = *Flabellina reticulata* Reuss, 1851. Description: test free, palmate to deltoid, sides flat, parallel, with angular to carinate edges; chambers initially coiled, rapidly uncoiling, chevron-shaped; sutures distinct, raised, crenulate; aperture terminal, sub-circular, on a short neck; test surface between sutures strongly ornamented by numerous ridges.
Remarks: differentiated from *N. praereticulata* Hiltermann by its more regular, reticulate, ornament, which extends even to the earliest chambers.
Range: Maastrichtian.

Neoflabellina rugosa (d'Orbigny)
Plate 7.20, Fig. 1 (× 50), Upper Chalk, Catton Grove, Norfolk, Upper Campanian. = *Flabellina rugosa* d'Orbigny, 1840. Description: test free, palmate, sides flattened, parallel, with angular to slightly carinate margins, chambers initially coiled, rapidly becoming uncoiled, chevron-shaped, narrow, increasing uniformly in size; sutures distinct, raised, aperture terminal, sub-circular on short neck; test surface between sutures ornamented by one or two rows of distinct raised papillae.
Remarks: the surface ornament makes this species distinctive, and many authors (e.g. Koch, 1977) have recognised several subspecies.
Range: Campanian.

Osangularia cordieriana (d'Orbigny)
Plate 7.20, Figs 4–6 (× 100), Upper Chalk, Scratchells Bay, Isle of Wight, Upper Campanian. = *Rotalina cordieriana* d'Orbigny, 1840. Description: test free, biconvex, biumbonate, periphery acute, slightly carinate; 3–4 whorls, 8–10 chambers in final whorl; umbilical side possessing pronounced, raised, umbilical boss of clear calcite; sutures distinct, slightly obscure on spiral side at first; aperture a narrow, oblique, slit along the base of final chamber on umbilical side, then forming a 'V'-shaped bend at an oblique angle up apertural face.
Remarks: characterised by its almost equally biconvex, biumbonate test and its slightly carinate margin.
Range: Campanian, but also rarely in Early Maastrichtian.

Osangularia navarroana (Cushman)
Plate 7.20, Figs 7, 8 (× 100), Upper Chalk, Trimingham, Norfolk, Lower Maastrichtian. = *Pulvinulinella navarrona* Cushman, 1938. Description: test free, equally biconvex; umbilical side possessing a moderately raised calcite boss; outline circular, periphery acute, bordered by a distinct keel; 2½ whorls, 10–12 chambers visible on umbilical side; sutures distinct, limbate, flush; aperture a narrow 'V'-shaped slit, occasionally separated into distinct interiomarginal and areal openings.
Remarks: the biconvex test and distinctly keeled margin serve to distinguish this species from the closely related *O. cordieriana* (d'Orbigny).
Range: Maastrichtian.

Osangularia whitei (Brotzen, 1936)
Plate 7.20, Figs 9, 10 (× 140), Upper Chalk, Euston, Suffolk, Santonian. = *Eponides whitei* Brotzen, 1936. Description: test free, trochospiral, equally biconvex, biumbonate, margin acutely angled, becoming carinate; chambers distinct, 7–8 in final whorl; sutures straight to slightly oblique, flush; aperture a narrow, oblique slit towards the inner margin of the final chamber with a very slightly raised lip.
Remarks: very similar to *O. cordieriana* (d'Orbigny), but can be distinguished by its almost horizontal periphery, as compared to the almost sigmoid periphery of *O. cordieriana*. Unfortunately this differentiation is impossible with some material of Late Santonian and Early Campanian age and it may prove necessary to discontinue the use of *whitei* and, instead, treat the whole range as being that of *cordieriana*.
Range: Mid-Coniacian to earliest Campanian.

Plate 7.20

Planularia crepidularis Roemer, 1842
Plate 7.20, Figs 11, 12 (× 50), Speeton Clay, Speeton, Yorkshire, Hauterivian (C3). Description: test free, compressed, sub-parallel sides; initially coiled, becoming uncoiled, 5–7 chambers in a linear series; sutures distinct, raised; periphery carinate, normally with three keels; aperture radiate, on a small neck.
Remarks: there is a wide range of variation, mainly in the degree of coiling and the intensity of the ornamentation. In the Jurassic some workers refer it to *Lenticulina (Planularia) tricarinella* Reuss.
Range: Jurassic to Early Aptain.

Pleurostomella barroisi Berthelin, 1880
Plate 7.20, Figs 13, 14 (× 75), Bed XIII, Gault Clay, Folkestone, Kent, Upper Albian. Description: test free, elongate, 7–8 chambers, cuneate, later approaching uniserial; sutures depressed, markedly oblique; aperture terminal with a projecting hood drawn out into sharp, pointed beak, bifid tooth on opposite side.
Remarks: both microspheric and megalospheric forms are known from the British succession. The distinct beak and bifid tooth separate this species from that which contains the microspheric *P. reussi* Berthelin and megalospheric *P. obtusa* Berthelin.
Range: Late Albian.

Praebulimina carseyae (Plummer)
Plate 7.20, Figs 15, 16 (× 60), Upper Chalk, Caister St. Edmunds, Norfolk, Upper Campanian. = *Buliminella carseyi* Plummer, 1931. Description: test free, ovate, initial end rounded; generally composed of four whorls, four chambers per whorl; sutures distinct; apertural face of final chamber elongate, often flattened; aperture variable, subterminal, extending along interiomarginal suture.
Remarks: can be identified by its step-like outline, quadriserial chamber arrangement, and inflated chambers. Early forms in the Early Campanian are smaller, slender, with less inflated chambers, while those in the Late Campanian are more typical.
Range: Campanian, and rarely in the earliest Maastrichtian.

Praebulimina laevis (Beissel)
Plate 7.21, Figs 1, 2 (× 75), Upper Chalk, Overstrand (Fig. 1) and Trimingham (Fig. 2), Norfolk, Lower Maastrichtian. = *Bulimina laevis*, Beissel, 1891. Description: test free, subfusiform, large; initial end rounded, then rapidly flaring, until final whorl; 3–4 whorls, four chambers to a whorl; chambers only slightly inflated, with sutures flush to only slightly depressed; apertural face sub-terminal, aperture variable, comma-shaped, occasionally bordered by a thin, indistinct, lip.
Remarks: Beissel (1891), indicating a wide range of morphology, may have included more than one species in the initial definition. As a result, this species has been regularly confused with *P. carseyae* (Plummer) and *P. obtusa* (d'Orbigny).
Range: Latest Campanian and Maastrichtian.

Praebulimina obtusa (d'Orbigny)
Plate 7.21, Figs 3 (× 60), 4 (× 50), Upper Chalk, Alum Bay, Isle of Wight, Upper Campanian. = *Bulimina obtusa* d'Orbigny, 1840. Description: test free, large, although two variants are known; variant one is bluntly pointed, rapidly flaring, 4–5 whorls, four slightly inflated chambers per whorl, sutures nearly flush; variant two is initially broadly rounded and then rapidly flaring, but almost parallel-sided in the last whorl; in both forms the loop-shaped aperture is set in a depression in the apertural face.
Remarks: the more inflated variant (Plate 7.21, Fig. 4) replaces the less inflated form (Plate 7.21 Fig. 3) in the lower levels of the middle Late Campanian.
Range: Late Campanian to Early Maastrichtian.

Praebulimina reussi (Morrow)
Plate 7.21, Figs 5, 6 (× 120), Upper Chalk, Wells, Norfolk (Fig. 5) and Scratchells Bay, Isle of Wight (Fig. 6), Middle Campanian. = *Bulimina reussi* Morrow, 1934. Description: test free, ovate, cross-section sub-circular; 4–5 whorls, chambers triserial throughout, becoming rapidly inflated, elongate, sutures distinct, slightly depressed; aperture variable, normally a narrow terminal slit at inner margin of final chamber.
Remarks: a very variable species that has been much confused in the literature.
Range: Santonian and Campanian.

Praeglobotruncana delrioensis (Plummer)
Plate 7.21, Figs 7, 8 (× 125), Lower Chalk, Eastbourne, Sussex, Middle Cenomanian. = *Globorotalia delrioensis* Plummer, 1931. Description: test free, biconvex, low trochospiral; 2–2½ whorls, 5–6 chambers in final whorl; chambers on umbilical side triangular in shape, gently inflated, sutures radial, depressed; chambers on spiral side gently inflated, sutures curved, slightly beaded; periphery lobate, margins of chambers marked by an accumulation of pustules; primary aperture extraumbilical-umbilical, bordered by a narrow lip.
Remarks: differs from *P. stephani* (Gandolfi) in having a lower trochospire, and a less developed marginal band of pustules.
Range: Early to Middle Cenomanian.

Cretaceous

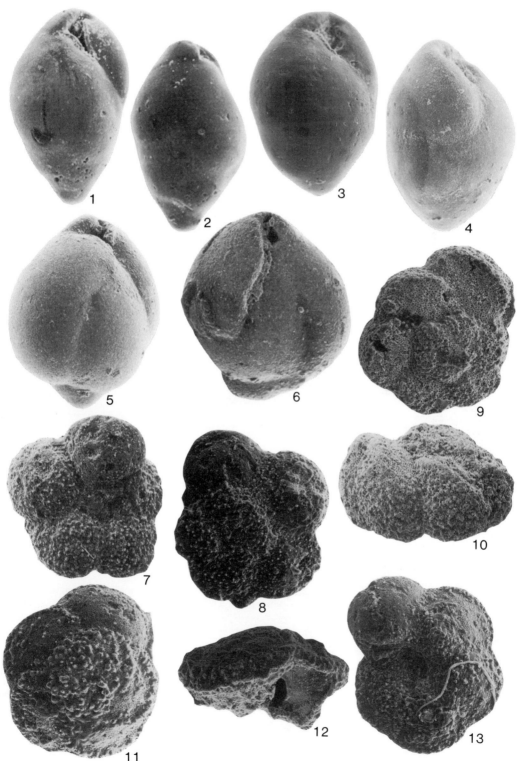

Plate 7.21

Praeglobotruncana helvetica (Bolli)
Plate 7.21, Figs 9, 10 (× 125), Middle Chalk, Beer, Devon, Turonian. = *Globotruncana helvetica* Bolli, 1945. Description: test free, plano-convex, 2–2½ whorls, 4–5½ chambers in the final whorl; chambers on umbilical side inflated, rugose, with radial, depressed, sutures; chambers on spiral side petaloid, with flat to concave surfaces, sutures raised; single keel usually forms edge to flat spiral surface; primary aperture extraumbilical-umbilical, bordered by a well-developed lip.
Remarks: this distinctive species normally has fewer chambers (4–5½ instead of 6–7) and is less ornamented than the typical, Tethyan, form.
Range: Early and Middle Turonian.

Praeglobotruncana stephani (Gandolfi)
Plate 7.21, Figs 11–13 (× 125), Lower Chalk, Buckland Newton, Dorest, Upper Cenomanian. = *Globotruncana stephani* Gandolfi, 1942. Description: test free, moderate trochospiral; 2–2½ whorls, 5–7 chambers in the final whorl; chambers on umbilical side triangular in shape, slightly inflated, sutures radial, depressed; chambers on spiral side petaloid, flat to gently inflated, sutures curved, distinctly beaded; peripheral margin beaded; primary aperture extraumbilical–umbilical, bordered by a well-developed lip.
Remarks: the higher spired, and more ornamented, *P. gibba* Klaus is commonly found near the Cenomanian–Turonian boundary.
Range: Middle Cenomanian to Early Turonian.

Pseudouvigerina cristata (Marsson)
Plate 7.22, Figs 1, 2 (× 100), Upper Chalk, Trimingham, Norfolk (Fig. 1) and Overstrand, Norfolk (Fig. 2), Lower Maastrichtian. = *Uvigerina cristata* Marsson, 1878. Description: test free, elongate, cross-section triangular; 4–5 whorls, triserial; chambers slightly inflated, the margins possessing distinct, widely spaced, double vertical costae; sutures distinct, depressed; aperture terminal at end of short neck, occasionally with a lip; internal narrow columnellar tooth plate present.
Remarks: a very distinctively ornamented species.
Range: Late Campanian and Maastrichtian.

Pullenia quaternaria (Reuss)
Plate 7.22, Fig. 3 (× 100), Upper Chalk, Caister St. Edmunds, Norfolk, Upper Campanian. = *Nonionina quaternaria* Reuss, 1851. Description: test free, planispiral, involute coil; 4–5 chambers in the final whorl; outline sub-circular, weakly lobate; periphery broadly rounded; chambers triangular, curved, final chamber covering the umbilicus on either side; sutures distinct, flush to weakly depressed; aperture a low interiomarginal slit, bordered by a thick, imperforate, band.
Remarks: *P. quaternaria* is characterised by its moderately compressed test, curved sutures and low apertural face.
Range: Middle Campanian to Maastrichtian.

Ramulina spandeli Paalzow, 1917
Plate 7.22, Fig. 4 (× 50), Speeton Clay, Speeton, Yorkshire, Hauterivian (C4). Description: test free, elongate, consisting of a single oviform chamber with long, stoloniferous tubes at either end; chamber and tube walls are coarsely hispid; aperture simple, rounded, at end of tubes.
Remarks: a distinctive species, even though it displays considerable variation in size and shape, and is always fragmentary.
Range: Hauterivian and Early Barremian.

Reussella kelleri Vasilenko, 1961
Plate 7.22, Figs 5, 6 (× 80), Upper Chalk, Newton-by-Castleacre, Norfolk, Santonian. Description: test free, subtriangular in outline, triserial throughout, edges sub-acute to angular; chambers triangular, strongly overlapping; sutures distinct, curved, often raised; margins spinose, occasionally developing small flanges which project from the edges; apertural slit, perpendicular to the inner margin of the final chamber, surrounded by a slightly raised lip.
Remarks: the first *Reussella* to appear in the British succession.
Range: Late Turonian to Early Campanian; extremely rare in the Middle Campanian.

Reussella szajnochae praecursor De Klasz and Knipscheer, 1954
Plate 7.22, Fig. 7 (× 100), Upper Chalk, Culver Cliff, Isle of Wight, Santonian. Description: test free, triangular, initial end pointed; test margins serrate, spinose; chambers triserial throughout, 5–6 whorls; aperture a slit-shaped opening extending up apertural face from mid-point of interiomarginal suture, with internal tooth plate, and external raised lip.
Remarks: this subspecies may be distinguished from *Reussella kelleri* Vasilenko by its smaller size, more spinose, generally non-carinate, test angles, and by its distinctly raised, limbate sutures.
Range: Santonian to Middle Campanian.

Cretaceous

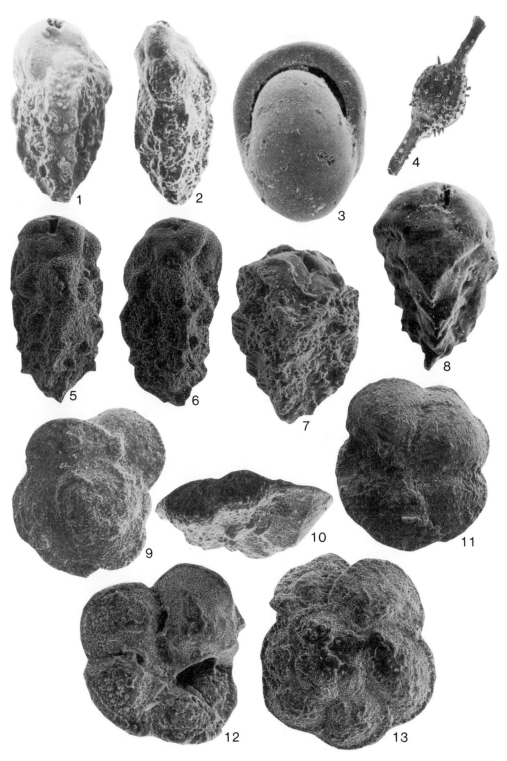

Plate 7.22

Reussella szajnochae szajnochae (Grzybowski)
Plate 7.22, Fig. 8 (× 100), Upper Chalk, Overstrand, Norfolk, Lower Maastrichtian. = *Verneuilina szajnochae* Grzybowski, 1896. Description: test free, triangular, robust; test margins sharp, serrate, spinose, non-carinate; triserial throughout, 4–6 whorls; sutures distinct, raised, limbate, carinate, projecting from test angles to form spines; aperture an elongate, slit-shaped opening, extending up apertural face from the mid-point of the interiomarginal suture, and bordered by a distinct lip; there is an internal tooth-plate.
Remarks: *R. pseudospinsosa* Troelsen from the Early Maastrichtian is a common synonym of this subspecies.
Range: Late Campanian, absent in the Early Maastrichtian but occurring rarely in the Late Maastrichtian.

Rotalipora appenninica (Renz)
Plate 7.22, Figs 9–11 (× 60), Lower Chalk, Eastbourne Sussex, Lower Cenomanian. = *Globotruncana appenninica* Renz, 1936. Description: test free, symmetrically biconvex, trochospiral; two whorls, 5–7 chambers in the final whorl; chambers on umbilical side triangular, sutures radial, depressed; chambers on spiral side petaloid, sutures distinct, raised; single keel; primary aperture extraumbilical-umbilical, secondary apertures umbilical initially, becoming sutural later.
Remarks: this species seems to appear later in the British succession than it does in the Tethyan localities, although Magniez-Jannin (personal communication) has recently found rare specimens in the upper part of Bed XIII, Gault Clay, at Folkestone.
Range: Latest Albian to Middle Cenomanian.

Rotalipora cushmani (Morrow)
Plate 7.22, Figs 1, 13 (× 100), Lower Chalk, Buckland Newton, Dorset, upper Cenomanian. = *Globorotalia cushmani* Morrow, 1934. Description: test free, biconvex, trochospiral; two whorls 4½–8 chambers in the final whorl; chambers on umbilical side triangular, strongly inflated, sutures radial, depressed; chambers on spiral side semi-circular, sutures raised; single keel; primary aperture extraumbilical-umbilical, secondary apertures sutural, bordered by well-developed lips.
Remarks: a most distinctive species, of wide-ranging stratigraphic value.
Range: Middle to Late Cenomanian.

Rotalipora greenhornensis (Morrow)
Plate 7.23, Figs 1–3 (× 100), Lower Chalk, Buckland Newton, Dorset, Upper Cenomanian. = *Globorotalia greenhornensis* Morrow, 1934. Description: test free, trochospiral, unequally biconvex; 2–2½ whorls, 7–10 chambers in final whorl, chambers on umbilical side trapezoidal, sutures curved, distinctly raised; chambers on spiral side crescentric, sutures raised; single keel; primary aperture extraumbilical-umbilical, secondary apertures umbilical, may be sutural in later chambers.
Remarks: differs from *R. brotzeni* (Sigal) in having a more pronounced asymmetrical test, and more chambers in the final whorl.
Range: Middle and Late Cenomanian.

Rotalipora reicheli Mornod, 1950
Plate 7.23, Figs 4–6 (× 100), Lower Chalk, Southerham, Sussex, Lower/Middle Cenomanian boundary. Description: test free, trochospiral, strongly asymmetrical; 2–2½ whorls, 6–8 chambers in the final whorl; chambers on umbilical side triangular, inflated, sutures depressed; chambers on spiral side crescentric, sutures raised, beaded; single keel; primary aperture extraumbilical-umbilical, secondary apertures umbilical, peri-umbilical ridges on all chambers forming a rim around the umbilicus.
Remarks: this species is very close to *R. deeckei* (Franke) but can be distinguished by its more flattened spiral side and depressed sutures on the umbilical side.
Range: occurs only near the Early/Middle Cenomanian boundary in the U.K.; rarely in the early Middle Cenomanian (Hart, 1979).

Rugoglobigerina rugosa (Plummer)
Plate 7.23, Figs 7–9 (× 100), Upper Chalk, Overstrand, Norfolk, Lower Maastrichtian. = *Globigerina rugosa* Plummer, 1927. Description: test free, low trochospiral coil of 2½–3 whorls, 4½–5 chambers in the final whorl; outline subcircular, periphery rounded; chambers distinct, subglobular, sutures depressed, radial; surface of each chamber ornamented with meridional pattern of discontinuous ridges or costellae; primary aperture interiomarginal umbilical.
Remarks: *R. rugosa* is recognised by its low trochospiral coil, rapidly expanding chambers, and moderately large umbilicus. As a species it shows a great deal of variability, which has resulted in a complex taxonomic history.
Range: Maastrichtian.

Pl. 7.23] **Cretaceous** 361

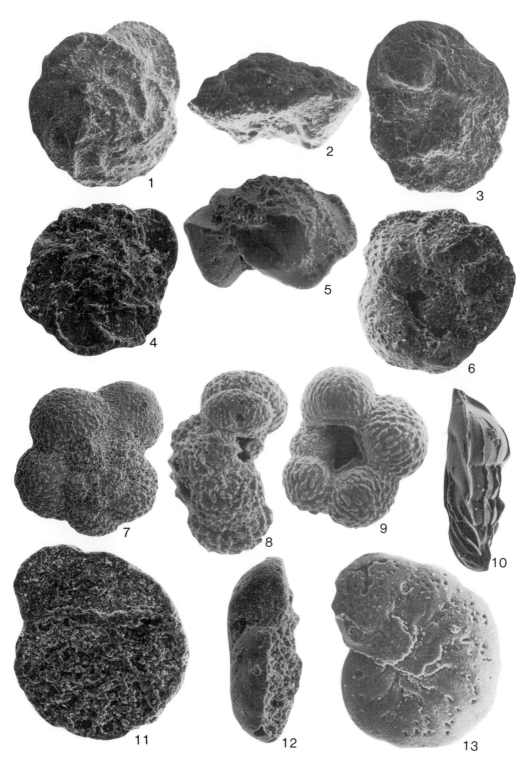

Plate 7.23

Saracenaria valanginiana Bartenstein and Brand, 1951
Plate 7.23, Fig. 10 (× 65), Speeton Clay, Speeton, Yorkshire, Ryazanian (D6AI). Description: test free, elongate, sub-triangular in cross-section; 2–3 close-coiled chambers followed by 3–5 slightly inflated chambers in a rectilinear series; sutures distinct, depressed; ornament comprises three keels at angles and distinct, fine, ribs; aperture radiate, at peripheral margin, on a small neck.
Remarks: a distinctive species that shows very little variation.
Range: Late Ryazanian and Early Valanginian.

Stensioeina exsculpta exsculpta (Reuss)
Plate 7.23, Figs 11–13 (× 100), Upper Chalk, Euston, Suffolk, Santonian. = *Rotalia exsculpta* Reuss, 1860. Description: test free, plano-convex to biconvex, margins acutely angled; chambers distinct on dorsal side, sub-rectangular, sutures distinct, sharply raised, forming elevated septal ridges; sutures on ventral side flush to slightly depressed; aperture an elongate slit along the inner, ventral, margin of the final chamber.
Remarks: this subspecies is distinguished from the '*S granulata* (Olbertz) lineage' in having a more sharply angled periphery, and in having a more distinctive septal ornament on the spiral side.
Range: Mid-Coniacian to Santonian; very rare in the earliest Campanian offshore.

Stensioeina exsculpta gracilis Brotzen, 1945
Plate 7.24, Figs 1–3 (× 110), Upper Chalk, Stiffkey, Norfolk, Middle Campanian. Description: test free, biconvex, trochospiral, periphery acute; $2\frac{1}{2}$–3 whorls, 10–12 chambers in final whorl; umbilical side weakly convex, involute; spiral side moderately to highly raised, evolute; sutures on umbilical side distinct, depressed; sutures on spiral side strongly elevated, forming a reticulate network with the spiral suture; aperture a low, interiomarginal opening, bordered by a distinct lip.
Remarks: the raised trochospiral coil, and sharply keeled margin, are characteristic of this subspecies.
Range: Santonian to Middle Campanian.

Stensioeina granulata granulata (Olbertz)
Plate 7.24, Figs 4–6 (× 100), Upper Chalk, Helhoughton, Norfolk, Santonian. = *Rotalia exsculpta granulata* Olbertz, 1942. Description: test free, low trochospiral coil, plano-convex, periphery rounded; chambers distinct on ventral side only, 9–10 in final whorl; sutures indistinct on dorsal side, elevated as low, broad, septal ridges on the ventral side; aperture a narrow, interiomarginal slit, following the ventral edge of the final chamber; umbilicus shallow, covered by a calcareous plate.
Remarks: following the work of Trümper (1968) *S. prae-exsculpta* (Keller) has been included in the synonymy of the present subspecies. The first appearance of this subspecies in the British succession is extremely important (Bailey, 1978, unpublished thesis; Bailey and Hart, 1979).
Range: Mid-Coniacian to earliest Santonian.

Stensioeina granulata polonica Witwicka, 1958
Plate 7.24, Figs 7–9 (× 100), Upper Chalk, Oldstairs Bay, Kent, Santonian. Description: test free, low troschospiral coil, periphery narrow, sub-angular to rounded; chambers distinct on both sides, nine in final whorl; sutures flush on dorsal side, slightly raised on ventral side, radial; aperture a narrow interiomarginal slit along the ventral side of the final chamber; umbilicus extremely shallow.
Remarks: intermediate between the *S. granulata* (Olberz) and *S. exsculpta* (Reuss) lineages, as suggested by Trümper (1968).
Range: Santonian.

Stensioeina pommerana Brotzen, 1936
Plate 7.24, Figs 10–12 (× 100), Upper Chalk, Alum Bay, Isle of Wight, Upper Campanian (Fig. 10), Caister St Edmunds, Norfolk, Upper Campanian (Fig. 11), and Overstrand, Norfolk, Lower Maastrichtian (Fig. 12). Description: test free, plano-convex, trochospiral, periphery acute; 2½–3 whorls; umbilical side involute, domed, umbilical region covered by large, irregular chamber flap; chambers on umbilical side distinct, weakly inflated, sutures depressed; all sutures on spiral side elevated, producing a highly irregular, reticulate network; aperture a low interiomarginal arch borderd by a thick lip.
Remarks: easily distinguished by its domed shape, angled periphery, and large well-developed umbilical chamber flap.
Range: rarely found in the Early Campanian; Middle Campanian to Maastrichtian.

Pl. 7.24] **Cretaceous** 363

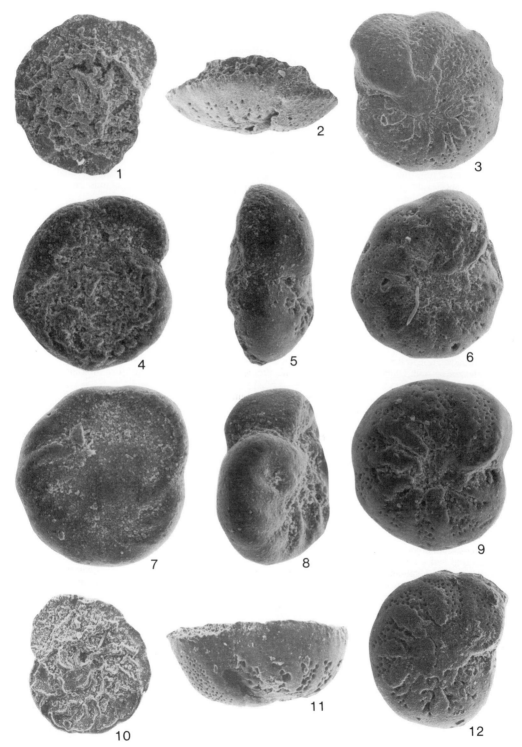

Plate 7.24

Vaginulina arguta Reuss, 1860
Plate 7.25, Figs 1, 2 (× 15), Speeton Clay, Speeton, Yorkshire, Hauterivian C2A. Description: test free, robust, compressed, triangular in outline; dorsal margin nearly straight; ventral margin may be slightly lobulate; spherical proloculus followed by up to 14 chambers; sutures oblique, limbate, raised; aperture simple, terminal, at dorsal angle, on a small neck.
Remarks: this species has been the subject of much confusion in the past, particularly with relation to *V. kochi* Roemer. Albers (1952) has shown that only smooth forms belong in *V. kochi* while those with raised sutures belong in the present species.
Range: Hauterivian to Early Barremian.

Vaginulina mediocarinata Ten Dam, 1950
Plate 7.25, Fig. 3, (× 15), Bed IX, Gault Clay, Folkestone, Kent, Upper Albian. Description: test free, harp-shaped, rectangular in cross-section; fine but sharp discontinuous, vertical striations cover the chamber surfaces; sutures depressed; rounded proloculus with or without costae; edges of test strongly costate; aperture terminal, radiate.
Remarks: this species differs from other ornamented Albian vaginulinids in its discontinuous surface striations that run the entire length of the test.
Range: Middle and Late Albian.

Vaginulina riedeli Bartenstein and Brand, 1951
Plate 7.25, Figs 4, 5 (× 20), Speeton Clay, Speeton, Yorkshire, Hauterivian (C8). Description: test free, elongate, robust, rectangular in cross-section; spherical proloculus followed by 4–7 slightly inflated chambers; peripheral angles of test acute, keeled; sutures oblique, slightly depressed; aperture radiate, at dorsal margin, on a small neck.
Remarks: closely related to *V. weigelti* Bettenstaedt, which may have evolved from *V. riedeli*.
Range: Hauterivian.

Vaginulinopsis humilis praecursoria Bartenstein and Brand, 1951
Plate 7.25, Fig. 6 (× 50), Speeton Clay, Speeton, Yorkshire, Valanginian (D3A). Description: test free, elongate, arcuate, oval in cross-section; initial coil followed by 4–6 chambers in a gently curving series; sutures in uncoiled part distinct, curved, raised; aperture radiate, at peripheral angle.
Remarks: this subspecies differs from *Lenticulina humilis* Reuss in having less prominent sutures, a less compressed test, and a rounder, not so tapering, proximal portion.
Range: Valanginian and Hauterivian. Due to difficulties caused by morphological variation occurrences in the Barremian are probably referable to *Lenticulina* (*Astacolus*) *neopachynota*.

Vaginulinopsis scalariformis Prothault, 1970
Plate 7.25, Fig. 7 (× 35), Upper Chalk, Quidhampton, Wiltshire, Santonian. Description: test free, large, compressed ovoid in cross-section; initial portion a tight planispiral coil, rapidly becoming uniserial; chambers distinct, low; sutures distinct, sub-horizontal; test marked by very characteristic transverse septal ridges; aperture radiate, terminally positioned on the peripheral dorsal angle of the final chamber.
Remarks: recorded only from southeast France and southeast England; this highly distinctive species appears to have some stratigraphic value.
Range: Earliest Santonian.

Valvulineria lenticula (Reuss)
Plate 7.25, Figs 8, 9 (× 120), Upper Chalk, Arminghall, Norfolk (Fig. 8) and Catton Grove, Norfolk, (Fig. 9), Upper Campanian. = *Rotalia lenticula* Reuss, 1845. Description: test free, trochospiral, biconvex, periphery rounded; chambers indistinct except in last whorl; sutures poorly visible, except in later stages, radial, curving slightly at the margins; aperture a narrow, slit-like opening along the inner margin of the final chamber; umbilical area covered by a distinct apertural flap.
Remarks: Harris and McNulty (1957) have described this species (and its variants) in great detail.
Range: Turonian to Early Maastrichtian.

Whiteinella aprica (Loeblich and Tappan)
Plate 7.25, Figs 10–12 (× 100), Middle Chalk, Beer, Devon, Turonian. = *Ticinella aprica* Loeblich and Tappan, 1961. Description: test free, trochospiral, biconvex to concavo-convex; 2–2½ whorls, 5–7 chambers in final whorl; chambers globular, pustulose, sutures radial, depressed; periphery lobate, with no keel; primary aperture tending to umbilical, bordered by a porticus.
Remarks: this species can be identified on the basis of its low spire.
Range: Turonian.

Whiteinella baltica Douglas and Rankin, 1969
Plate 7.25, Figs 13–15 (× 125), Upper Chalk, Quidhampton, Wiltshire, Santonian. Description: test free, trochospiral, biconvex to concavo-convex; 2–2½ whorls, 4–5 chambers in the final whorl; chambers globular, sutures depressed, radial to slightly curved; aperture umbilical, with evidence of a distinct flap in well-preserved specimens.
Remarks: this distinctive species, first described from Bornholm (Denmark), is very common in the British succession.
Range: Coniacian and Santonian.

Plate 7.25

Acknowledgements

The final text was prepared by MBH and typed by Mrs M. Luscott-Evans; the SEM photography was by MBH and AS (assisted by B. Lakey, R. Moate and C. Jocelyn—Polytechnic South West Electron Microscopy Unit); plates were prepared by MBH (assisted by Media Services, Polytechnic South West); and the diagrams and charts were prepared by MBH and J. Abraham (Polytechnic South West). Data were prepared by all contributors to the Cretaceous section, but thanks are also extended to D. J. Carter (ex-Imperial College), C. S. Harris (Trans-Manche Link, Dover), P. J. Bigg (Gearhart Geoconsultants Ltd), C. J. Wood (ex-British Geological Survey) and Prof. F. Robaszynski (Mons Polytechnic, Belgium), all of whom assisted in various ways during preparation. B. N. Fletcher's contribution is published with the approval of the Director, British Geological Survey.

REFERENCES

Adams, C. G., Knight, R. H. and Hodgkinson, R. L. 1973. An unusual agglutinating foraminifer from the Upper Cretaceous of England. *Palaeontology*, **16**, 637–644.

Albers, J. 1952. Taxonomie und Entwichlung einiger Arten von *Vaginulina* d'Orb, aus dem Barreme bei Hannover. *Mitt. geol. Staatsinst., Hamb.*, **21**, 75–112.

Arthur, M. A., Schlanger, S. O. and Jenkyns, H. C. (1987) the Cenomanian–Turonian Oceanic Anoxic Event, II. Palaeoceanographic controls on organic matter production and preservation. In Brooks, J. and Fleet, A. (eds) *Marine Petroleum Source Rocks, Spec. Publ. Geol. Soc. Lond.* **26**, 401–420.

Ascoli, P. 1976. Foraminiferal and ostracod biostratigraphy of the Mesozoic–Cenozoic, Scotian Shelf, Atlantic Canada. In 1st International Symposium on Benthonic Foraminifera of Continental Margins, Schafer, C. T. and Pelletier, B. R. (eds), *Maritime Sediments Spec. Publ.* No. 1, 653–771.

Ascoli, P., Poag, C. W. and Remane, J. 1984. Microfossil zonation across the Jurassic–Cretaceous boundary on the Atlantic Margin of North America. *Geol. Assoc. Canada, Special Paper* No. 27, 31–48.

Aubert, J. and Bartenstein, H. 1976. *Lenticulina (L.) nodosa*; additional observations in the worldwide Lower Cretaceous. *Bull. Centre Rech. Pau*, **10**, 1–33.

Bailey, H. W. 1975. A preliminary microfaunal investigation of the Lower Senonian at Beer, south-east Devon. *Proc. Ussher Soc.*, **3**, 280–285.

Bailey, H. W. and Hart, M. B. 1979. The correlation of the Early Senonian in Western Europe using Foraminiferida. *Aspekte der Kreide Europas*, IUGS, Series A, No. 6, 159–169.

Bailey, H. W., Gale, A. S., Mortimore, R. N., Swiecicki, A. and Wood, C. J. 1983. The Coniacian–Maastrichtian Stages of the United Kingdom, with particular reference to southern England. *Newsl. Stratigr.* **12**, 29–42.

Bailey, H. W., Gale, A. S., Mortimore, R. N., Swiecicki, A. and Wood, C. J. 1984. Biostratigraphical criteria for the reception of the Coniacian to Maastrichtian Stage boundaries in the Chalk of North West Europe. *Bull. Geol. Soc. Denmark*, **33**, 31–39.

Bandy, O. L. 1967, Cretaceous planktonic foraminiferal zonation. *Micropaleontology*, **13**, 1–31.

Barnard, T. 1958. Some Mesozoic adherent foraminifera from the Upper Cretaceous of England. *Palaeontology*, **1** 116–124.

Barnard, T. 1963a Polymorphinidae from the Upper Cretaceous of England. *Palaeontology*, **5** (for 1962), 712–726.

Barnard, T. 1963b The morphology and development of species of *Marssonella*, and *Pseudotextulariella* from the chalk of England. *Palaeontology*, **6**, 41–45.

Barnard, T. and Banner, F. T. 1953. Arenaceous Foraminifera from the Upper Cretaceous of England. *Q. Jl. geol. Soc. Lond.*, **109** 173–216.

Barnard, T. and Banner, F. T. 1980. The Ataxophragmiidae of England: Part 1, Albian-Cenomanian *Arenobulimina* and *Crenaverneuilina*. *Rev. Esp. Micropal.*, **12**, 383–430.

Barr, F. T. 1962. Upper Cretaceous planktonic foraminifera from the Isle of Wight, England, *Palaeontology*, **4**, 552–580.

Barr, F. T. 1966a, The foraminiferal genus *Bolivinoides* from the Upper Cretaceous of the British Isles. *Palaeontology*, **9**, 220–243.

Barr, F. T. 1966b. Upper Cretaceous foraminifera from the Ballydeenlea Chalk, Co. Kerry, Ireland. *Palaeontology*, **9**, 492–510.

Bartenstein, H. 1952, Taxonomische Revision und Nomenklator ze Franz E. Hecht, Standard—Gliederung der Nordwestdeutschen Unterkreide nach Foraminiferen (1938). Teil 1. Hauterive. *Senckenbergiana leth.*, **33**, 173; Teil 2. Barreme. *Senckenbergiana leth.*, **33**, 297.

Bartenstein, H. 1974. *Lenticulina (Lenticulina) nodosa* (Ruess, 1863) and its subspecies—world wide index foraminifera in the Lower Cretaceous. *Eclog. geol. Helv.*, **67**, 539–562.

Bartenstein, H. 1976a. Benthonic index foraminifera in the Lower Cretaceous of the northern hemisphere between East Canada and North West Germany. *Erdol Kohle*, **29** 254–256.

Bartenstein, H. 1976b. Practical applicability of a zonation with benthonic Foraminifera in the worldwide Lower Cretaceous. *Geol. Mijnb.*, **55** 83–86.

Bartenstein, H. 1977. Stratigraphic parallelism of the Lower Cretaceous in the northern hemisphere.

Newsl. Stratigr., **6**, 30–41.
Bartenstein, H. 1978a. Parallelisation of the Lower Cretaceous stages in North West Germany with index ammonites and index microfossils. *Erdol Kohle*, **31**, 65–67.
Bartenstein, H. 1978b. Phylogenetic sequences of Lower Cretaceous benthic Foraminifera and their use in biostratigraphy. *Geol. Mijnb.*, **57** 19–24.
Bartenstein, H. 1979. Worldwide zonation of the Lower Cretaceous using benthonic foraminifera. *Newsl. Stratigr.*, **7**, 142–154.
Bartenstein, H. and Bettenstaedt, F. 1962. Marine Unterkreide (Boreal und Tethys). In Simon, W. and Bartenstein, H. (Eds) *Leifossilien der Mikropaläontologie:* 225–297, pls 33–41, Berlin.
Bartenstein, H., Bettenstaedt, F. and Bolli, H. M. 1957, Die Foraminiferen der Unterkreide von Trinidad, B.W.I. Erster Teil: Cuche und Toco Formation. *Eclog. geol. Helv.*, **50**, 5–67.
Bartenstein, H. and Brand, E. 1951. Mikropaläontologische Untersuchungen zur Stratigraphie des nordwestdeutschen Valendis. *Abh. senckenb. naturforsch. Ges.*, **485**, 239–336.
Bartenstein, H. and Kaever, M. 1973. Die Unterkreide von Helgoland und ihre mikropaläontologische Gliederung. *Senckenberg, leth.*, **54**, 207–264.
Bartenstein, H. and Kovatecheva, T. 1982. A comparison of Aptian Foraminifera in Bulgaria and North West Germany. *Eclogae Geol. Helv.*, **75**, 621–667.
Beissel, L. 1891. Die Foraminifera de Aachener Kreide. *Abh. preuss. geol. Landesanst.*, part 3; 78 pp. 16 pls, Berlin.
Bellier, J.-P. 1971. Les Foraminifères Planctoniques du Turonien Type. *Rev. Micropalèont.*, **14**, 85–90.
Berthelin, G. 1880. Mémoire sur les Foraminifères de l'Etage Albien de Montcley (Doubs.). *Mém. Soc. geol. Fr.* (3), **1**, 1–84.
Bettenstaedt, F. 1952. Stratigraphische wichtige Foraminiferen Arten aus dem Barreme vorwiegend Nordwest-Deutschlands. *Senckenberg. leth.*, **33**, 263.
Birkelund, T., Hancock, J. M., Hart, M. B., Rawson, P. F., Remane, J., Robaszynski, F., Schmidt, F. and Surlyk, F. 1984. Cretaceous stage boundaries—Proposals. *Bull. Geol. Soc. Denmark*, **33**, 3–20.
Breistroffer, M. 1947. Sur les Zones d'Ammonites dans l'Albien de France et d'Angleterre. *Trav. Lab. géol. Univ. Grenoble*, **26**, 17–104.
Brotzen, F. 1934a, Vorläufiger Bericht über eine Foraminiferenfauna aud der schwedischen Schreibkreide. *Geol. För. Stockh. Förh.*, **56**, 77–80.
Brotzen, F. 1934b. Foraminiferen aus dem Senon Palästinas. *Z. dt. Ver. Palästinas*, (Leipzig), **57**, 28–72.
Brotzen, F. 1936. Foraminiferen aus dem Schwedischen untersten Senon von Eriksdal in Schonen. *Sver. geol. Unders. Afh.*, Ser. C, **30**, 1–206.
Brotzen, F. 1942. Die Foraminiferengattung *Gavelinella* nov. gen. unde die Systematik der Rotaliformes. *Sver. geol. Unders. Afh.*, Ser. C, **36**, (8), 1–60.
Brotzen, F. 1945. De Geologiska Resultaten fran Borrningarna vid Höllviken-Preliminär rapport Del 1: Kritan. *Sver. geol. Unders. Afh.*, Ser. C, **38** (7), 1–64.
Brotzen, F. 1948. The Swedish Paleocene and its foraminiferal fauna. *Sver. geol. Unders, Afh.*, Ser C, **42**, (2), 1–140.
Burnaby,T. P. 1962. The Palaeoecology of the foraminifera of the Chalk Marl., *Palaeontology*, **4**, 599–608.
Burrows, H. W., Sherborn, C. D. and Bailey, G. 1890. The Foraminifera of the Red Chalk of Yorkshire, Norfolk and Lincolnshire. *Jl. R. Microsc. Soc.*, **8**, 549–566.
Busnardo, R. 1965. Le stratotype du Barrémien. 1 Lithologie et Macrofaune. In Colloque sur le Crétacé inférieur. *Mém. Bur. Rech. Géol. minier.*, **34**, 101–116.
Butt, A. A. 1966. Foraminifera of the Type Turonian. *Micropaleontology*, **12**, 162–182.
Caron, M. 1966. Globotruncanidae du Crétacé supérieur du synclinical de la Gruyére (Préalpes Medianes, Suisse). *Revue Micropaléont.*, **9**, 68–93.
Carter, D. J. and Hart, M. B. 1977. Aspects of mid-Cretaceous stratigraphical micropalaeontology. *Bull. Br. Mus. nat. Hist (Geol.)*, **29**(1): 1–135.
Casey, R. 1961. The stratigraphical palaeontology of the Lower Greensand. *Palaeontology*, **3**, 487–621.
Casey, R. 1964. In: Dodson, M. H., Rex, D. C., Casey, R. and Allen P. Clauconite dates from the Upper Jurassic and Lower Cretaceous. In The Phanerozoic Time-Scale. *Q. Jl. geol. Soc. Lond.*, **120** S, 145–158.
Casey, R. 1973. The ammonite succession at the Jurassic–Cretaceous boundary in eastern England. In The Boreal Lower Cretaceous, Casey, R. and Rawson, P. F. (Eds), *Geol. Jl. Spec. Iss.* No, 5, 193–266.
Chapman, F. 1891–1898. The Foraminifera of the Gault of Folkestone. *Jl. R. microsc. Soc.*, (Part 1) 1891, 565–575; (Part 2) 1892, 319–330; (Part 3) 1892, 749–758; (Part 4) 1893, 579–595; (Part 5) 1893, 153–163; (Part 6) 1894, 421–427; (Part 7) 1894, 645–654; (Part 8), 1–14; (Part 9) 1896, 581–591; (Part 10) 1898, 1–49. The whole work has been reprinted by Antiquariaat Jumk (1970).
Cooper, M. R. 1977. Eustacy during the Cretaceous; Its implications and importance, *Palaeogeogr., Palaeoclimat., Palaeoecol.*, **22** 1–60.
Coquand, H. 1857. Position des *Ostrea columba* et *viauriculata* dans le groupe de la craie inférieur. *Bull. Soc. géol. Fr.*, (2) **14**, 745–766.
Dam, A. Ten, 1947. On foraminifera of the Netherlands No. 9. Sur quelques espèces nouvelles ou peu connues dans le Crétacé Inférieur (Albien) des Pays-Bas *Geol. Mijnb.*, **8**, 25–29.
Dam, A. Ten, 1948a, Foraminifera from the Middle Neocomian of the Netherlands. *J. Paleont.*, **22**, 175–192.
Dam, A. Ten, 1948b. Les espèces du genre *Epistomina* Terquem, 1883. *Rev. Inst. fr. Pètrole*, **3** 161–170.
Dam, A. Ten, 1950. Les Foraminifères de l'Albien des Pays-Bas. *Mém. soc. géol. Fr.*, n.s. **29**, (63), 1–66.
Deroo, G. 1966. Cytheracea (Ostracodes) du Maastrichtien de Maastricht (Pays-Bas) et des règions voisines: résultats stratigraphiques et paléontologiques de leur étude. *Meded. geol. Sticht.*, **C2**(2), 197 pp.

Dilley, F. C. 1969. The foraminiferal fauna of the Melton Carstone. *Proc. Yorks. geol. Soc.*, **37**, 321–322.

Douglas, R. G. 1969a. Upper Cretaceous planktonic foraminifera in northern California. 1 – Systematics. *Micropaleontology*, **15**, 151–209.

Douglas. R. G. 1969b. Upper Cretaceous biostratigraphy of northern California. *Proc. First Int. Conf. planktonic Microfossils, Geneva, 1967*, **2**, 126–152.

Douglas, R. G. and Rankin, C. 1969. Cretaceous planktonic foraminifera from Bornholm and their zoogeographic significance. *Lethaia*, **2**, 185–219.

Dumont, A. 1850. Rapport sur la carte géologique du royaume. *Bull Acad. r. Belg. Cl. Sci.*, **16**, 351–373.

Eicher, D. L. 1965. Foraminifera and biostratigraphy of the Graneros Shale. *J. Paleont.*, **39**, 875–909.

Eicher, D. L. 1966. Foraminifera from the Cretaceous Carlile Shale of Colorado. *Contr. Cushman Fdn. foramin. Res.*, **17**, 16–31.

Eicher, D. L. 1967. Foraminifera from Belle Fourche Shale and equivalents. in Wyoming and Montana. *J. Paleont.*, **41**, 167–188.

Eicher, D. L. 1969. Cenomanian and Turonian planktonic foraminifera from the western interior of the United States. *Proc. First Int. Conf. planktonic Microfossils, Geneva, 1967*, **2**, 163–174.

Eicher, D. L. and Worstell, P. 1970. Cenomanian and Turonian foraminifera from the Great Plains, United States. *Micropaleontology*, **16**, 269–324.

Fletcher, B. N. 1973. The Distribution of Lower Cretaceous (Berriasian-Barremian) Foraminifera in the Speeton Clay. In The Boreal Lower Cretaceous, Casey, R. and Rawson, P. F. (Eds.), *Geol. J. Spec. Iss. No. 5*, 161–168.

Frizzell, D. L. 1954. Handbook of Cretaceous Foraminifera of Texas. *Rep. Bur. econ. Geol. Univ. Tex.*, **22**, 1–232.

Fuchs, W. von, 1967. Die Foraminiferenfauna eines Kernes des höheren Mittel-Alb der Tiefbohrung DELFT 2 – Niederlands. *Jb. geol. Bundesanst. Wien*, **110**, 255–341.

Gawor-Biedowa, E. 1969. The genus *Arenobulimina* Cushman from the Upper Albian and Cenomanian of the Polish Lowlands. *Roczn. pol. Tow. geol.*, **39**, 73–104.

Gawor-Biedowa, E. 1972. The Albian, Cenomanian and Turonian Foraminifers of Poland and their Stratigraphic Importance. *Acta Palaeont, pol.*, **17**, 3–151.

Goel, R. K. 1965. Contributions à l'étude de Foraminiféres de Crétacé supérieur de la Baisse-Seine. *Bull. Bur. Rech. géol. min.*, **5**, 49–157.

Hallam, A., Hancock, J. M., LaBrecque, J. L., Lowrie, W and Channell, J. E. T. 1985. Jurassic to Paleogene; Part 1. Jurassic and Cretaceous geochronology and Jurassic to Paleogene magnetostratigraphy. In: Snelling, N. J. (ed). *The Chronology of the Geological Record*, Memoir 10, pp. 118–140. Geological Society of London.

Hancock, J. M. 1976. The Petrology of the Chalk. *Proc. Geol. Ass.*, **86**, 499–536.

Hancock, J. M. and Kauffman, E. G. 1979. The great transgressions of the Late Cretaceous. *Jl. geol. Soc. Lond.*, **136**, 175–186.

Haq, B. U., Hardebol, J. and Vail, P. R. 1987. Chronology of Fluctuating Sea levels since the Triassic. *Science* **235**, 1156–1167.

Harland, W. B., Cox, A. V., Llewellyn, P. G., Pickton, C. A. G., Smith, A. G. and Walters, R. 1982. *A geologic timescale*, Cambridge University Press, 131 pp.

Harris, R. W. and McNulty, C. L. Jr, 1975. Notes concerning a Senonian Valvulinerian. *J. Paleont.*, **3**, 865–868.

Hart, M. B. 1973. A correlation of the macrofaunal and microfaunal zonations of the Gault Clay in Southeast England. In: the Boreal Lower Cretaceous, Casey, R. and Rawson, P. F. (eds.), *Geol. J. Spec. Iss.* No. 5, 267–288.

Hart, M. B. 1975. Microfaunal analysis of the Membury Chalk succession. *Proc. Ussher. Soc.*, **3**, 271–279.

Hart, M. B. 1980a. A water depth model for the evolution of the planktonic Foraminiferida. *Nature*, **286**, 252–254.

Hart, M. B. 1980b. The recognition of mid-Cretaceous sea level changes by means of Foraminifera. *Cretaceous Res.*, **1**, 289–297.

Hart, M. B. 1982. Turonian foraminiferal biostratigraphy of southern England. *Mem. Mus. Nat. d'Histoire Nat.*, **46**, 203–207.

Hart, M. B. 1984. The Superfamily Robertinacea in the Lower Cretaceous of the U.K. and adjacent areas of NW Europe. *Benthos '83; 2nd Int. Symposium. Benthic Foraminifera* (Pau, April 1983), 289–298.

Hart, M. B. 1985. Oceanic Anoxic Event 2 on-shore and off-shore S.W. England. *Proc. Ussher. Soc.*, **6**, 183–190.

Hart, M. B. 1987. Cretaceous foraminifers from Deep Sea Drilling Project Site 612, Northwest Atlantic Ocean. In: Poag, C. W. and Watts, A. B. *et al.*, *Initial Reports of the Deep Sea Drilling Project*, **95**, 245–252, Washington, D.C.

Hart, M. B. and Bailey, H. W. 1979. The distribution of planktonic Foraminiferida in the mid-Cretaceous of N.W. Europe. *Aspekte der Kreide Europas*, IUGS., Ser. A, No. 6, 527–542.

Hart, M. B. and Ball, K. C. 1986. Late Cretaceous anoxic events, sea level changes and the evolution of the planktonic foraminifera. In Summerhayes, C. P. and Shackleton, N. J. (eds) *North Atlantic Paleoceanography*, Geol. Soc. Spec. Publ., **21**, 67–78.

Hart, M. B. and Bigg, P. J. 1981. Anoxic events in the Late Cretaceous Chalk seas of North-West Europe. In Neale, J. W. and Brasier, M. D. (eds), *Microfossils of Recent and Fossil Shelf Seas*. Horwood, Chichester 177–185.

Hart, M. B. and Swiecicki, A. 1987. Foraminifera of the Chalk Facies. In: Hart, M. B. (Ed.), *The Micropalaeontology of Carbonate Environments*, Horwood, Chichester, 121–137

Hart, M. B. and Leary, P. N. 1989 The stratigraphical and palaeogeographical setting of the late Cenomanian anoxic event. *Jl Geol. Soc. London*, **146**, 305–317

Hart, M. B., Manley, E. C. and Weaver, P. P. E. 1979. A biometric analysis of an *Orbitolina* fauna from

References

the Cretaceous succession at Wolborough, S. Devon. *Proc. Ussher. Soc.*, **4**, 317–326.

Hart, M. B. and Weaver, P. P. E. 1977. Turonian microbiostratigraphy of Beer, S.E. Devon. *Proc. Ussher Soc.*, **4**, 86–93.

Hilbrecht, H. and Hoefs, J. 1986. Geochemical and Palaeontological studies on the $\delta^{13}C$ anomaly in Boreal and North Tethyan Cenomanian–Turonian sediments in Germany and adjacent areas. *Palaeogeog., Palaeoclimatol., Palaeoecol*, **53**, 169–189.

Hinte, J. E. van, 1976. A Cretaceous time scale. *Bull. Am. Ass. Petrol. Geol.*, **60**, 498–516.

Jarvis, I., Carson, G., Hart, M. B., Leary, P. N. and Tocher, B. A. 1988a. The Cenomanian–Turonian (late Cretaceous) anoxic event in SW England: evidence fromm Hooken Cliffs near Beer, SE Devon, *Newsl. Stratigr.*, **18**, 147–164.

Jarvis, I., Carson, G. A., Cooper, M. K. E., Hart, M. B., Leary, P. N., Tocher, B. A., Horne, D. and Rosenfeld, A. 1988b. Microfossil assemblages and the Cenomanian–Turonian (late Cretaceous) Oceanic Anoxic Event. *Cretaceous Research*, **9**, 3–103.

Jendryka-Fuglewicz, B. 1975. Evolution of the Jurassic and Cretaceous smooth-walled *Lenticulina* (Foraminiferida) of Poland. *Acta palaeont. pol.*, **20**, 99–197.

Jefferies, R. P. S. 1962, The palaeoecology of the *Actinocamax pleuns* Subzone (Lowest Turonian) in the Anglo-Paris Basin. *Palaeontology*, **4**, 609–647.

Jefferies, R. P. S. 1963. The stratigraphy of the *Actinocamax plenus* Subzone (Turonian) in the Anglo-Paris Basin. *Proc. Geol. Ass.* **74**, 1–33.

Kauffman, E. G. 1970. Population sytematics, radiometrics and zonation – a new biostratigraphy. *Proc. N. Am. Paleont. Conv.*, **1**, 612–666.

Kemper, E. 1973a. The Valanginian and Hauterivian stages in northwest Germany. In The Boreal Lower Cretaceous, Casey, R. and Rawson, P. F. (eds), *Geol. J. Spec. Iss.* No. 5, 327–344.

Kemper, R. 1973b. The Aptian and Albian stages in northwest Germany. In The Boreal Lower Cretaceous, Casey, R. and Rawson, P. F. (eds), *Geol. J. Spec. Iss.* No. 5, 345–360.

Kennedy, W. J. 1969. The correlation of the Lower Chalk of south-east England. *Proc. Geol. Ass.* **80**, 459–560.

Kennedy, W. J. 1970. A correlation of the Uppermost Albian and the Cenomanian of south-west England. *Proc. Geol. Ass.*, **81**, 613–677.

Klaus, J. 1960a. Le 'Complexe schisteux intermédiaire' dans le synclinal de la Gruyère (Préalpes médianes). Stratigraphie et micropalèontologie, avec l'étude spéciale des Globotruncanides de l'Albian, du Cenomanian, et du Turonien. *Eclog. geol. Helv.*, **52**, 753–851.

Klaus, J. 1960b. Etude biométrique at statistique de espèces du genre *Praeglobotruncana* dans le Cénomanien de la Breggia. *Eclog. geol. Helv.*, **53**, 285–308.

Klaus, J. 1960c. Rotalipores et Thalmanninelles d'un niceau des Couches Rouges de l'Anticlinal d'Ai. *Eclog. geol. Helv.*, **53**, 704–709.

Koch, W. 1977. Biostratigraphie in der Oberkreide und Taxonomie von Foraminiferen. *Geol. Jb.*, Reihe **A 38**, 128pp.

Lecointre, G. 1959. Le Turonien dans sa région type, la Touraine. *C.r. Congr. Socs. Sav. Paris Sect. Sci.*, Colloque sur la Crétacé Supérieur Français (Dijon, 1959), 415–423.

Loeblich, A. R. and Tappan, H. 1961. Cretaceous planktonic foraminifera. 1 – Cenomanian. *Micropaleontology*, **7**, 257–304.

Loeblich, A. R. and Tappan, H. 1964. Protista 2, Sarcodina, chiefly 'Thecamoebians' and Foraminiferida. In Moore, R. C. (ed.), *Treatise on Invertebrate Paleontology*, Part C.1, 1–510.

Loeblich, A. R. and Tappan, H. 1974. Recent advances in the classification of Foraminiferida. In: Hedley, R. H. and Adams, C. G. (eds.), *Foraminifera*, **1**, 1–53.

Lott, G. K., Fletcher, B. N. and Wilkinson, I. P. 1986. The stratigraphy of the Lower Cretaceous Speeton Clay Formation in a cored borehole off the coast of north-east England. *Proc. Yorks. Geol. Soc.*, **46**, 39–56.

McGugan, A. 1957. Upper Cretaceous Foraminifera from Northern Ireland. *J. Paleont.*, **31**, 329–348.

McNeill, D. H. and Caldwell, W. G. E. 1981. Cretaceous Rocks and their Foraminifera in the Manitoba Escarpment. *Geol. Assoc. Canada, Special Paper* No. 21, 439pp.

Magniez-Jannin, F. 1975. Les Foraminifères de l'Albien de l'Aube: paléontologie, stratigraphie, écologie. *Cah. Paléontol.*, CNRS, 351 pp.

Magniez-Jannin, F. 1981. Decouverte de *Planomalina buxtorfi* (Gandolfi) et d'autres foraminiferes planctoniques inattendus dans l'Ablien Superieur d'Abbotscliff (Kent, Angleterre); consequences paléogeographiques et biostratigraphiques. *Geobios*, **14**, 91–97.

Malapris, M. 1965. Les Gavelinellidae et formes affines du gisement albien de Courcelles (Aube). *Revue Micropalèont.*, **8**, 131–150.

Malapris-Bizouard, M. 1974. Les premières Gavelinelles du Crétacé inférieur. *Bull. Inf. Gèol. Bass. Paris.* **40**, 9–23

Marie, P. 1938. Sur quelques foraminifères nouveaux ou peu connus du Crétacé du Bassin de Paris. *Bull. Soc. géol. Fr.*, (5), **8**, 91–104.

Marie, P. 1941. Les foraminifères de la Craie à *Belemnitella mucronata* du Bassin de Paris. *Mém. Mus. natn. Hist. nat. Paris*, n. ser., **12**, 1–296.

Marks, P. 1967. *Rotalipora* et *Globotruncana* dans la Craie de Théligny (Cénomanien: Dépt. de la Sarthe). *Proc. K. ned. Akad. Wet.*, **B70**, 264–275.

Masters, B. A. 1977. Mesozoic Planktonic Foraminifera; a worldwide review and analysis. In *Oceanic Micropalaeontology*, Ramsay, A.T.S. (Ed.), **1**, 301–731.

Michael, E. Die Evolution der Gavelinelliden (Foram) in der N.W.-deutschen Unterkreide. *Senckenberg, leth.* **47**, 411–459.

Mortimore, R. N. 1986. Stratigraphy of the Upper Cretaceous White Chalk of Sussex. *Proc. Geol. Ass.*, **91**, 97–139.

Neagu, T. 1965. Albian Foraminifera of the Rumanian Plain. *Micropaleonotology*, **11**, 1–38.

Neagu, T. 1969, Cenomanian Planktonic Foraminifera in the southern part of the Eastern Carpathians. *Roczn. pol. Tow. geol.*, **39**, 133–181.

Neale, J. W. 1974. Cretaceous. In: Rayner, D. H. and Hemingway, J. E. (eds.), *The Geology and Mineral Resources of Yorkshires*. Yorks. geol. Soc., 225–243.

Nyong, E. E. and Olsson, R. K. 1984. A palaeoslope model of Campanian to Lower Maestrichtian Foraminifera in the North American Basin and adjacent Continental margin. *Mar. Micropaleontol.*, **8**, 437–477.

Obradovich, J. D. and Cobban, W. A. 1975. Late Cretaceous Time Scale. In Caldwell, W. G. E. (ed.) *The Cretaceous System in the Western Interior of North America*, Geol. Assoc. Canada, Spec. Pap., **13**, 40–52.

Olsson, R. K. 1964. Late Cretaceous Planktonic Foraminifera from New Jersey and Delaware. *Micropaleontology*, **10**, 157–188.

Olsson, R. K. and Nyong, E. E. 1984. A Palaeoslope Model for Campanian–Lower Maestrichtian Foraminifera of New Jersey and Delaware. *J. foramin. Res.*, **14**, 50–68.

Orbigny, A. D. d'. 1842–1847. Pteropoda, Gasteropoda. *Paléontologie francaise, Terrains crétacés*, **2**, 456 pp., Atlas 236 pls, table, (1842). Brachipoda (includes Bivalvia). **4**, 390 pp., Atlas pls 490–599. (1847). Paris.

Pessagno, E. A. 1967. Upper Cretaceous planktonic foraminifera from the Western Gulf Coastal Plain. *Palaeontogr. am.*, **37**, 245–445.

Petters, S. W. 1977a. Upper Cretaceous planktonic Foraminifera from the subsurface of the Atlantic Coastal Plain of New Jersey. *J. foramin. Res.*, **7**, 165–187.

Petters, S. W. 1977b. *Bolivinoides* evolution and Upper Cretaceous biostratigraphy of the Atlantic Coastal Plain of New Jersey. *J. Paleont.*, **51**, 1023–1036.

Plummer, H. J. 1931. Some Cretaceous Foraminifera in Texas. *Publs. Bur. econ. Geol. Univ. Tex.*, **3101**, 109–203.

Price, R. J. 1967. Palaeoenvironmental interpretations in the Albian of western and southern Europe, as shown by the distribution of selected foraminifera. In *Proc. First Int. Symp. Benthonic Foraminifera of Continental Margins*, Schafter, C. T. and Pelletier, B. R. (eds) *Maritime Sediments, Spec. Publ. No. 1*, 625–648.

Price, R. J. 1977a. The stratigraphical zonation of the Albian sediments of north-west Europe, as based on foraminifera. *Proc. Geol. Ass.*, **88**, 65–91.

Price, R. J. 1977b. The evolutionary interpretation of the foraminiferida *Arenobulimina*, *Gavelinella*, and *Hedbergella* in the Albian of north-west Europe. *Palaeontology*, **20**, 503–527.

Rawson, P. F., Curry, D., Dilley, F. C., Hancock, J. M., Kennedy, W. J., Neale, J. W., Wood, C. J. and Worssam, B. C. 1978. A correlation of the Cretaceous rocks of the British Isles. Geol. Soc. Lond., Spec. Rpt., **9**, 70 pp.

Renevier, E. 1874. Tableau des terrains sédimentaires *Bull. Soc. vaud. Sci. mat.*, **13**, 218–252.

Rhys, G. H. 1975. A Proposed Standard Lithostratigraphic Nomenclature for the Southern North Sea. In Woodland, A. W. (ed.) *Petroleum and the Continental Shelf of N.W. Europe*. Vol. 1. Geology, 151–162, Academic Press, London.

Robaszynski, F., Amedro, F., Foucher, J. C., Gaspard, D., Magniez-Jannin, F., Manivit, H. and Sornay, J. 1980. Synthèse Biostratigraphique de l'Aptien au Santonian du Boulonnais a partir de sept groupes paléontologiques; Foraminifères, Nannoplancton, Dinoflagellés et Macrofaunas. *Rev. Micropaléont.*, **22**, 195–321.

Robaszynski, F. and Caron, M. 1979. Atlas of Mid-Cretaceous planktonic Foraminiferida (Boreal Sea and Tethys), *Cah. Micropaléont.*, Part 1, 1–185; Part 2, 1–181. (Published in both French and English).

Robaszynski, F., Caron, M., Gonzalez Donoso, J. M. and Wonders, A. A. H. 1984. Atlas of Late Cretaceous Globotruncanids. *Rev. Micropaléontol.*, **26**, 145–305.

Robinson, N. D. 1986 Lithostratigraphy of the Chalk Group of the North Downs, southeast England. *Proc. Geol. Ass.*, **97**, 141–170.

Sazonov, N. T. 1951. On some little-known ammonites of the Lower Cretaceous. *Byull. mosk. Obshch. Ispyt, Prir.*, **56**, 57–63 (in Russian).

Schlanger, S. O. and Jenkyns, H. C. 1976. Cretaceous anoxic events: causes and consequences. *Geol. Mijnb.*, **55**, 179–184.

Schlanger, S. O., Arthur, M. A., Jenkyns, H. C. and Scholle, P. A. (1987). The Cenomanian–Turonian Oceanic Event, I. Stratigraphy and distributions of organic-rich beds and the marine 13C excursion. In Brooks, J. and Fleet, A. (eds) *Marine Petroleum Source Rocks*. Spec. Publ. Geol. Soc. Lond. **26**, 371–399

Schmid, F. 1959. La definition des limites Santonien–Campanien et Campanien inférieur-supérieur en France et dans le nord-ouest de l'Allemagne. In *C.r. Congr. Socs. Sav. Paris, Sect. Sci.*, Colloque sur la Crétacé Supérieur Français (Dijon, 1959), 535–546.

Schmid, F. 1967. Die Oberkreide-Stufen Campan und Maastricht in Limburg (Sudniederlande, Nordostbelgien), Bei Aachen und in Nordwest-deutschland. *Ber. dt. Ges. geol. Wiss., A. Geol. Paläont.*, **12**, 471–478.

Schroeder,R., Simmons, M. D., Hart, M. B. and Williams, C. L. 1986. A Note on the Occurrence of *Orbitolina (Orbitolina) sefini* Henson, 1948 (Foraminiferida) in the Upper Greensand of S.W. England. *Cretaceous Research*, **7**, 381–387.

Sérone-Vivien, M. 1972. Contribution à l'etude de Sénonien en Aquitaine septentrionale, ses stratotypes: Coniacien, Santonien, Campanien. *Les Stratotypes Français 2*, CNRS, Paris, 195 pp.

Sliter, W. V. 1968, Upper Cretaceous Foraminifera from Southern California and North-West Baja California, Mexico. *Paleont. Contr. Univ. Kans.*, **49**, 1–141.

Stenestad, E. 1969. The genus *Heterohelix* Ehrenberg, 1843 (Foraminifera) from the Senonian of Denmark. *Proc. First Int. Conf. planktonic Microfossils, Geneva, 1967*, **2**, 614–661.

References

Terquem, O. 1883, Sur un nouveau genre de Foraminifères du Fuller's-earth de la Moselle. *Bull. Soc. géol. Fr.*,(3) **11**, 37–39.

Trümper, E. 1968. Variationsstatistische Untersuchungen an der Foraminiferen-Gattung *Stensioeina* Brotzen. *Geologie, 17Bh.*, 59–103.

Vasilenko, V. P. 1961. Upper Cretaceous foraminifers of the Mangyshlak Peninsula (descriptions, phylogenetic diagrams of some groups and stratigraphical analysis).). *Trudy vses. neft. nauchno-issled. geol. razv. Inst.*, **171**, 1–487. (in Russian).

Voigt, E. 1956. Zur Frage der Abgrenzung der Maastrict-Stufe. *Paläont. Z. Sonderh.*, **30**, 11–17.

Williams-Mitchell, E. 1948. The Zonal value of Foraminifera in the Chalk of England, *Proc. Geol. Ass.*, **59**, 91–112.

Ziegler, P. A. 1975. North Sea Basin History in the Tectonic Framework of North-Western Europe. In Woodland, A. W. (Ed.), *Petroleum and the Continental Shelf of N.W. Europe*, Vol. 1. Geology; 131–148.

8

Cretaceous of the North Sea

C. King, H. W. Bailey, C. A. Burton and A. D. King

8.1 INTRODUCTION

The North Sea coastline is an arbitrary boundary for biostratigraphical purposes, but the treatment here of the North Sea Cretaceous in a separate section is conditioned by several factors, such as the differing objectives of investigation (mainly for hydrocarbon exploration), and the different types of sample available (ditch cuttings and cores as opposed to outcrop samples). These have constrained the methodology of investigation, and the types of biostratigraphic data obtained. For these reasons, and due to the confidentiality of much of the offshore data, studies of the offshore and onshore areas have proceeded rather independently. In any future edition, it is anticipated that a fully integrated approach could be presented. Emphasis here is placed on the U.K. offshore area, but data have been obtained from all areas of the North Sea.

Cretaceous sediments are widely distributed throughout the North Sea (Figs 8.1, 8.2). They reach a maximum thickness of at least 2500 m, and in the centre of the North Sea are buried beneath 3000 m of Cenozoic sediments. Their distribution has been delineated by extensive seismic surveys, and by the drilling of hundreds of exploration and production wells for oil and gas over the past 25 years. Additional information has been provided by sea floor mapping and sampling programmes, and by shallow cored boreholes drilled by the British Geological Survey (B.G.S.) in connection with offshore mapping. Thus a fairly complete general picture of Cretaceous lithostratigraphy, facies distributions and depositional patterns has been established (Hancock, 1984).

Foraminiferids are present, often abundantly, in almost all marine Cretaceous sediments encountered offshore, and since the inception of offshore exploration they have been utilised extensively for dating, subdivision and correlation. Little information on foraminiferid assemblages from commercial boreholes has so far been published, although

Fig. 8.1 — Distribution of Early Cretaceous sediments in the North Sea and adjacent areas. Areas with over 1000 m of sediment preserved are encircled by dotted lines. '81/40' and '81/43' are British Geological Survey offshore cored boreholes.

Introduction

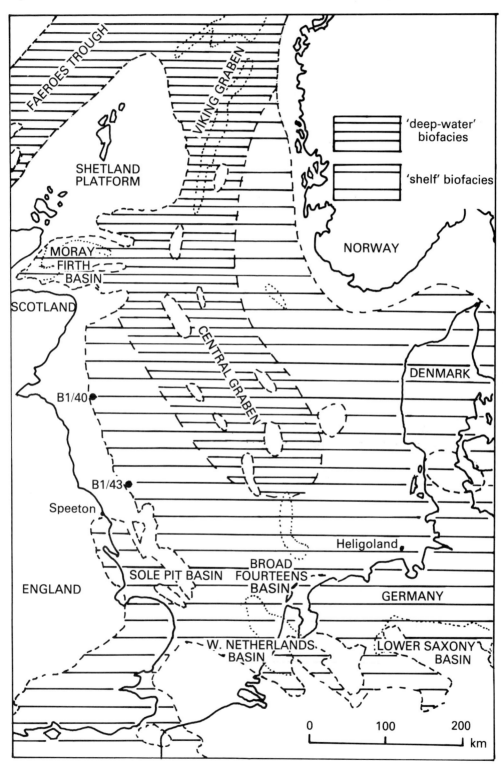

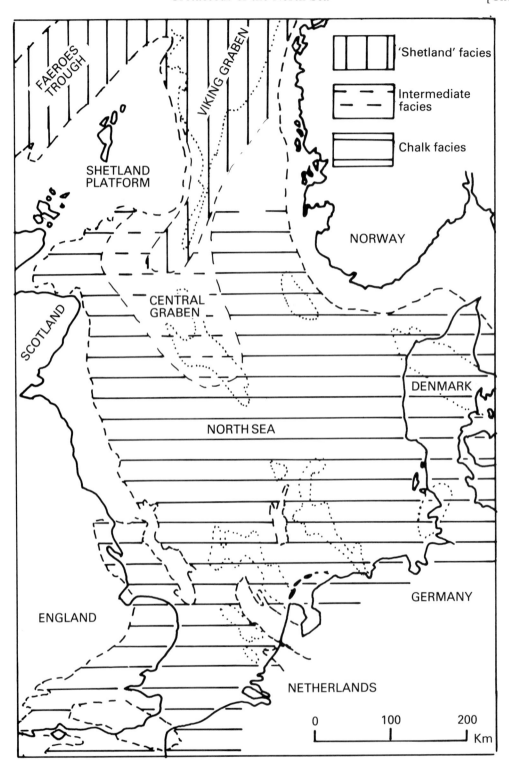

micropalaeontologial data have contributed crucially to the interpretation of North Sea Cretaceous stratigraphy.

The earliest significant information on North Sea Cretaceous foraminiferids was given by Bartenstein and Kaever (1973), who provided range-charts of Hauterivian to Albian foraminiferids from sea floor samples near Heligoland, dated by comparison with onshore northwest German microfaunas. Rasmussen (1974) provided the first data from hydrocarbon exploration wells, in the Danish offshore sector. Mid-Cretaceous foraminiferid assemblages from the Outer Moray Firth Basin were described in detail by Burnhill and Ramsay (1981), who erected zones based on planktonic foraminiferids for the Albian to Early Turonian sequence in this area. Crittenden (1982, 1984a, b, 1987) has discussed aspects of Early Cretaceous (Aptian and Albian) foraminiferid biostratigraphy in the southern North Sea, and has studied the geographic and stratigraphic distribution of the Albian species *Osangularia schloenbachi* in the North Sea and elsewhere (Crittenden, 1983).

The Early Cretaceous foraminiferid biostratigraphy of two B.G.S. offshore boreholes drilled near the western edge of the Cretaceous outcrop in the central North Sea (Fig. 8.1) has recently been published (Lott *et al.*, 1985, 1986). Late Cretaceous (Santonian–Maastrichtian) foraminiferid key species from two wells in the southern North Sea were tabulated in the first edition of this book (Hart *et al.*, 1981), and more information on the Late Cretaceous foraminiferid biostratigraphy of these sections is given by Ball (1984, 1985, 1986).

Rawson and Riley (1982) have commented on foraminiferid biostratigraphy and foraminiferid assemblage changes in the central and southern North Sea, in relation to Early Cretaceous regional eustatic and tectonic events. Almost no information has yet been published on the Cretaceous foraminiferids of the northern North Sea (Viking Graben and adjacent areas).

This chapter is a first attempt to synthesise the available data into a general account of Cretaceous foraminiferid biostratigraphy and its palaeoenvironmental controls, with range charts of the key index species. Its imperfections are clearly understood by the authors; it is intended to provide basic information, both for those unfamiliar with the North Sea, and also for those working on offshore sequences.

8.2. LOCATION OF IMPORTANT COLLECTIONS

Extensive collections of offshore Cretaceous foraminiferids are held by the stratigraphic laboratories of the major oil companies and biostratigraphic consultants, but these are largely of commercially confidential material, and are accessible only by special arrangement. Material from B.G.S. offshore boreholes is held by B.G.S, Keyworth (Nottingham). The material studied by Ball and Crittenden is in the Department of Geological Sciences, Polytechnic South West.

8.3 STRATIGRAPHIC DIVISIONS

8.3.1 Lithostratigraphy
Lithostratigraphic classification
Standard lithostratigraphic classifications for the North Sea Cretaceous sequences have been established for the U.K. offshore sector by Rhys (1974) and Deegan and Scull (1977), for the Netherlands sector by Nederlandse Aardolie Matschappij and Rijks Geologische Dienst (1980), for the Danish sector by Michelsen (1982) and Jensen *et al.* (1986),

Fig. 8.2 — Distribution of Late Cretaceous sediments in the North Sea and adjacent areas. Areas with over 1000 m of sediment preserved are encircled by dotted lines.

and for the Norwegian sector by Deegan and Scull (1977) and Hesjedal and Hamar (1983). The Norwegian, Danish and U.K. classifications are largely integrated one with another, but that for the Netherlands offshore sector is at present entirely separate (Fig. 8.3, 8.4). The scheme utilised in the southern North Sea (U.K. offshore) incorporates stratigraphic units defined in adjacent onshore areas (e.g. Spilsby Sandstone, Speeton Clay). Crittenden (1982) has advocated the use of the (more detailed) Netherlands stratigraphic subdivisions in the Early Cretaceous of the southern U.K. North Sea sector, to enable subdivision of the argillaceous sequence currently included within the Speeton Clay.

Early Cretaceous

Early Cretaceous marine sediments are represented mainly by claystones and marls, but sand bodies are present in some contexts. All are included in the Cromer Knoll Group (= Rijnland Group of the Netherlands offshore sector). Four major lithofacies can be differentiated:

(1) Shallow marine beach/barrier and deltaic sands. These include the Spilsby Sandstone of the southern U.K. sector and adjacent onshore areas, and the Vlieland Sandstone of the southern Netherlands offshore. They are restricted to the basin margins and to limited areas adjacent to uplifted fault blocks within the basin, and usually form wedge-shaped units prograding during basin-wide episodes of low eustatic sea level or during more localised tectonically controlled uplift. Foraminiferids are generally rare or absent.

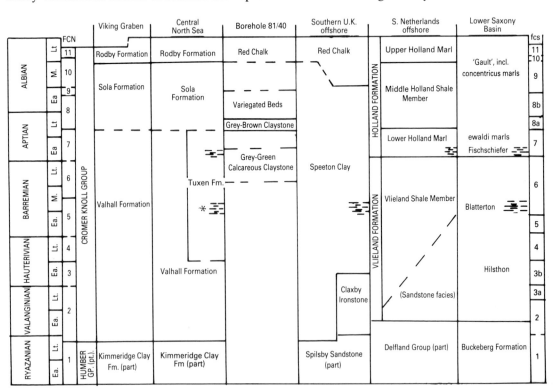

Fig. 8.3 — Lithostratigraphic classification of Early Cretaceous sediments in the North Sea and adjacent areas, related to the FCS and FCN Zones. The FCN Zones are applicable to the Viking Graben and to the Hidra Formation of the outer Moray Firth; the FCS Zones are applicable in other areas. Dashed lines indicate 'black-shale' facies; * = Munk Marl Bed and equivalents.

(2) Grey, grey–green, green, brown and red bioturbated claystones. These form the bulk of the Early Cretaceous sequence. They were deposited in a wide range of environments, from shallow shelf to bathyal, in normally to poorly oxygenated bottom waters, and usually contain abundant and diverse foraminiferid assemblages.

(3) 'Black-shale facies'. Thinly bedded and laminated dark brown to black organic-rich claystones are widespread (apparently almost basin-wide) at two levels, in the Early–Middle Barremian and the Early Aptian (see section 8.5). Thin and impersistent units of similar facies are recorded at other levels. These are interpreted as having been deposited beneath anoxic bottom-waters during episodes of stagnation, and are comparable with the 'black-shale facies' developed widely at this time in oceanic environments (Schlanger and Jenkyns, 1976).

(4) Calcareous clays (marls) and chalky limestones. These are usually white or light grey in colour, but some units are red or brown. They were deposited mainly during high eustatic sea levels, thus they are most widespread during the later Early Cretaceous, as worldwide sea levels became progressively higher. The "purest" calcareous facies are developed in areas starved of clastic sediments, particularly over 'highs' within the basin, often reflecting transgressive phases. They are usually greatly condensed (by comparison with time-equivalent clastic sequences). Episodes of sea-floor solution

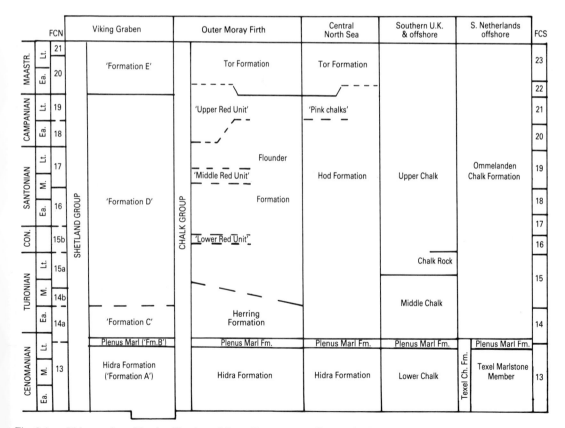

Fig. 8.4 — Lithostratigraphic classification of Late Cretaceous sediments in the North Sea and adjacent areas. CON. = Coniacian, MAASTR. = Maastrichtian.

and corrosion are indicated by the occurrence of dissolution seams and corroded microfossils (comparable to the Jurassic 'Ammonitico Rosso' facies of the Tethys, see Jenkyns, 1986). Foraminiferids are generally common to abundant.

(5) 'Deepwater sands'. These occur chiefly in the outer Moray Firth area, and include the Scapa Sand and similar units (Harker et al. 1987). They are fine- to coarse-grained, micaceous, often glauconitic sands, probably deposited by gravity-flow processes, and are interbedded with 'deepwater' clays. Foraminiferids are generally absent.

Late Cretaceous

The Late Cretaceous sediments of the North Sea can be subdivided into two main lithofacies: a 'shelf' and upper bathyal carbonate facies dominated by chalks, and a deeper-water (bathyal) dominantly argillaceous facies. These are classified lithostratigraphically as the Chalk Group and Shetland Group respectively (Deegan and Scull, 1977). In the North Sea, the Shetland Group is confined to the axial zone of the northern North Sea (the Viking Graben and adjacent areas), but very similar facies also extend widely over the Norwegian Sea and the southern margin of the Faeroes Trough. It grades laterally into the Chalk facies through an 'intermediate' zone in the Southern Viking Graben–outer Moray Firth area (Fig. 8.2). Chalks and chalky limestones occur in the Viking Graben in the Turonian and Maastrichtian, where they are regarded as Members of the Shetland Group. Fine-grained sandstones also occur locally in the Turonian and Campanian in this area.

Abundant foraminiferids are present throughout the Shetland Group. There is a strongly marked vertical differentiation of foraminiferid assemblages, discussed in detail below. Intervals dominated by planktonic foraminiferids and calcareous benthonic foraminiferids alternate with intervals containing only non-calcareous agglutinating foraminiferids. The latter indicate a bathyal environment of deposition.

The Chalk facies was deposited in predominantly low-energy carbonate shelf and upper bathyal environments. The superficially uniform lithology conceals a variety of subfacies (Kennedy, 1987), reflecting variable sedimentation rates and depths of deposition, partly dependent on eustatic sea-level changes.

In 'shelf' environments, foraminiferids are often abundant; calcareous benthonic foraminiferids predominate at most levels, but planktonic foraminiferids are abundant at some levels in the Cenomanian, Turonian, Coniacian, Santonian and Late Maastrichtian. The deeper-water chalks in the central North Sea contain less abundant and less diverse faunas, partly due to diagenetic recrystallisation, but partly reflecting the original distribution patterns. Significant lateral faunal variations can be identified, probably reflecting differing water depths, rates of deposition and degrees of oxygenation (see below), but study of these is still at an early stage.

Substantial volumes of these deeper-water chalks in the Central Graben have been redeposited by gravity-flow (Hatton, 1986), and these often contain reworked foraminiferids which may be significantly older than those in the adjacent chalks.

Chronostratigraphy

The 'standard' Cretaceous stages (except the Maastrichtian) are based on stratotypes outside the North Sea Basin. Identification of these stages in onshore sequences in the North Sea Basin has been established primarily through correlation based on ammonites, although recently foraminiferids, nannoplankton and inoceramid bivalves have also been utilised. The limits assigned to these stages in the North Sea Basin have therefore periodically changed following the progress of biostratigraphic research. The Early Cretaceous stages and ammonite zones quoted here follow Rawson et al. (1978), as modified for the Valanginian to Barremian by Kemper (1973a) and Kemper et al. (1981). The Late Cretaceous stages follow the proposals of Birkelund et al. (1984), and their calibration with the English onshore succession follows the recent revisions of Bailey et al. (1983, 1984). The substages of the Santonian, Campanian and Maastrichtian follow current German conventions (e.g. Koch, 1977).

It must be emphasized that this classification differs slightly but significantly from that used in some standard micropalaeontological texts (e.g. Bartenstein and Bettenstaedt, 1962; Hiltermann, 1962; Koch, 1977), and caution must be exercised when using range charts calibrated only to stages or substages. In particular, the Hauterivian/Barremian boundary has recently been amended; and the Cenomanian/Turonian boundary is placed above the Plenus Marl Formation (which was formerly included in the Early Turonian). The Late Turonian and Early Coniacian of Koch (1977) and earlier German authors are now regarded as Early and Middle Coniacian respectively (Seibertz, 1979).

The chronostratigraphic/biostratigraphic classification of the offshore Cretaceous succession has been established primarily by calibration of the ranges of benthonic foraminiferids and microplankton with onshore sections in the North Sea Basin.

Planktonic foraminiferids occur commonly at a number of levels from the Middle Barremian to the Late Maastrichtian. The assemblages in the Early Cretaceous lack most of the diagnostic species of the 'standard' international planktonic zonations (Caron, 1985). From the Cenomanian to Late Maastrichtian, however, some key index species occur, and permit some degree of inter-regional foraminiferid correlation, but benthonic foraminiferids provide most of the data for intrabasinal correlation.

A 'standard' (generally accepted) biostratigraphic subdivision and classification has not yet been proposed for the North Sea Cretaceous sequence. Although extensive studies have been carried out within the oil industry, these are essentially unpublished. The biostratigraphic zonation and correlations proposed here are based on offshore data available to the authors, and on comparisons with onshore sequences in England and Germany. They are evidently subject to modification and refinement.

8.4 DEPOSITIONAL HISTORY; PALAEOECOLOGY; FAUNAL ASSOCIATIONS

8.4.1 Geological history

The Cretaceous history of the North Sea Basin is summarised by Ziegler (1982) and Hancock (1984); Early Cretaceous events are discussed by Rawson and Riley (1982). Depositional patterns of basinal subsidence and sedimentation are due to the interaction of regional and local tectonism with eustatic sea-level changes. These are reflected in basin-wide sedimentary events. This is particularly striking in the Early Cretaceous, where lithological sequences and wireline geophysical log motifs are remarkably persistent laterally, although partly obscured by the rather fragmented lithostratigraphic nomenclature.

The anoxic 'black shale' stratified basin facies of the Late Jurassic Kimmeridge Clay Formation persisted in central North Sea areas until the Late Ryazanian, but was brought to an end by 'basin flushing and turnover' due to a terminal Late Ryazanian transgression (Rawson and Riley, 1982). The Early Cretaceous falls within the 'syn-rift' phase of North Sea development (Ziegler, 1982), characterised by subsidence interrupted by phases of gentle tectonism marked by block-faulting. These produced a rather complex pattern of sedimentation, in fault-controlled sub-basins, with facies shifts controlled both by the tectonism, and on a regional scale by eustatic sea-level changes. Generally rising sea levels eventually permitted the development of increasingly uniform facies in Aptian and Albian times. The mid-Cretaceous marks the change in the North Sea Basin from syn-rift to post-rift deposition. This event can be dated to the Turonian. Local fault-block movements mostly ended at this time, enabling steady subsidence and permitting sediments to drape over and eventually to completely blanket pre-existing structures. Subsequent Cretaceous depositional history is of intermittent but progressive eustatic sea-level rise, with minor regressive episodes. The highest sea levels were reached in the Campanian and Maastrichtian.

8.4.2 Palaeoecology and faunal associations
Preliminary comments
Very little work has been published on Cretaceous benthonic foraminiferid palaeoecology in the North Sea Basin. Classification of the foraminiferid assemblages into broadly defined biofacies, which reflect mainly depth and degree of oxygenation, is here attempted. Their boundaries are transitional, but they have been found to be valuable for the primary classification and environmental interpretation of the benthonic assemblages.

Early Cretaceous
Four major foraminiferid biofacies can be distinguished;

(1) A 'shelf' biofacies characterised by diverse nodosariids (including *Citharina*, *Nodosaria* and *Vaginulina*), epistominids (often large and ornate and occurring in 'flood' proportions at some levels), large calcareous-cemented agglutinants (including *Ammovertella*, *Arenobulimina*, *Haplophragmium* and *Tritaxia*) and the rotaliids *Gavelinella* and *Conorotalites*. This type of assemblage is widespread in the southern North Sea and adjacent onshore areas, including eastern England and north Germany.

(2) A deeper-water 'outer sublittoral–upper bathyal' biofacies, characterised by the occurrence of many of the taxa of assemblage 1, but with a more limited range of nodosariids, dominated by *Lenticulina* and *Pseudonodosaria*, without significant proportions of epistominids or large calcareous-cemented agglutinants. Rotaliids are rare until the Middle Barremian, but at higher levels *Gavelinella* and *Valvulineria* are common. Non-calcareous agglutinants are often present; *Falsogaudryinella* is common at some levels. This biofacies occurs mainly in the central North Sea and the Viking Graben.

(3) A 'restricted basin' biofacies dominated by (or composed exclusively of) non-calcareous agglutinating foraminiferids. Two subfacies can be discriminated: in shelf and upper bathyal environments the assemblage is dominated by *Ammobaculites*, *Glomospirella*, *Haplophragmoides*, *Reophax*, *Rhizammina* and (at some levels) *Textularia bettenstaedti*, as for instance in the Albian 'Zone 1' of Price (1977). In deeper-water environments a 'flysch-type' (Gradstein and Berggren, 1981) or '*Rhabdammina*-biofacies' (King, 1983) is present, including the taxa listed above, together with *Bathysiphon*, *Glomospira*, *Pelosina*, *Recurvoides* and *Rhabdammina*. This subfacies is widespread at several levels from the Hauterivian to the Middle Albian in the central North Sea and the Viking Graben.

(4) A 'carbonate biofacies' occurs in the 'clastic-starved' limestone facies occurring at several levels in the Valanginian, Hauterivian and Barremian over 'highs' within the basin in the central and northern North Sea. This includes elements of assemblage 2, together with rotaliids with tubular enrolled tests (spirillinids and involutinids) including *Aulotortus*, *Paalzowella*, *Patellina*, *Trocholina* and *Turrispirillina*. Water depths during deposition of this facies are uncertain, but are presumed to have been 'mid- to outer shelf'.

Late Cretaceous
(1) The Chalk facies can be characterised by

the relative abundance of the rotaliids *Bolivinoides, Gavelinella, Lingulogavelinella* and *Stensioeina*. Significant vertical and horizontal faunal variations can be observed, but their definition and interpretation has hardly been attempted so far. In the 'intermediate' argillaceous facies of the central North Sea, *Stensioeina* still occurs frequently.

(2) The calcareous benthonic assemblages of the 'Shetland' facies are typified by the frequency of nodosariids, the rotaliids *Eponides, Gyroidinoides, Nuttallides* and *Nuttallinella* and the agglutinant *Tritaxia* (*Pseudogaudryinella*). Genera found mainly or exclusively in this facies include *Allomorphina, Aragonia*, and *Semivulvulina*. The foraminiferid assemblages are largely distinct at the species level from those in the Chalk facies.

(3) A non-calcareous agglutinating foraminiferid assemblage of the 'Rhabdamminabiofacies' is present intermittently from the Turonian to the Maastrichtian in the 'Shetland' facies in the Viking Graben. Characteristic genera include *Ammodiscus, Bathysiphon, Haplophragmoides, Karreriella, Recurvoides* and *Trochamminoides*.

Foraminiferid assemblages and basin history
Study of successive foraminiferid assemblages in the Cretaceous of the North Sea Basin demonstrates the existence of synchronous, basin-wide, environmentally controlled changes in composition at the family and superfamily level. These changes can be simply but expressively analysed by dividing the assemblage at each level into three components:

(1) Non-calcareous agglutinants (e.g. *Glomospira, Haplophragmoides, Rhabdammina* and other elements of the 'Rhabdammina-biofacies', King, 1983).
(2) Calcareous benthonic foraminiferids, and agglutinants with calcareous cement (e.g. some species of *Gaudryina, Dorothia* and *Arenobulimina*).
(3) Planktonic foraminiferids.

For each sample studied, two ratios are calculated (similar techniques applied in the Cenozoic of the North Sea are discussed by King (Chapter 9)). These are here designated the NCA ratio (the percentage of individuals of non-calcareous agglutinants in the total benthonic assemblage) and the P ratio (the percentage of planktonic individuals in the total foraminiferid assemblage). If, for a single site, these ratios are plotted vertically, patterns emerge which are remarkably persistent laterally (Figs 8.5, 8.6). Assemblages with high P ratios alternate with assemblages with high NCA ratios; the shifts between successive assemblages are often relatively rapid, and can be identified in ditch cuttings with reasonable accuracy.

In the Early Cretaceous, NCA-dominated episodes are often developed throughout the Basin, except in shallow marine environments (e.g. in the Late Aptian–Early Albian) alternating with equally widespread plankton-rich episodes (e.g. in the late Early Aptian–early Late Aptian).

In the Late Cretaceous Chalk facies, NCA agglutinants are relatively rare; assemblage shifts between plankton-dominated and benthos-dominated phases can be differentiated, although they are still imperfectly documented in both onshore and offshore areas (see Hart and Bailey, 1979).

Within the Shetland facies of the northern North Sea, a well-defined and consistent sequence of assemblages can be defined (Fig. 8.6). Intervals with high P ratios (e.g. the Cenomanian, Middle–Late Turonian, Late Campanian, and Late Maastrichtian) alternate with intervals dominated by non-calcareous agglutinants. The calcareous microfaunas become progressively more restricted northwards, and in more northerly areas (Faeroes Basin and Norwegian Sea), from the Late Turonian to the mid-Late Maastrichtian the foraminiferid assemblages

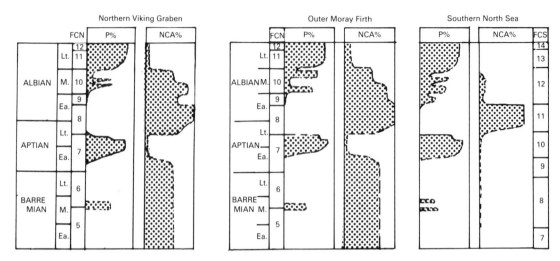

Fig. 8.5 — Foraminiferid assemblage patterns in the Early Cretaceous (Barremian – Albian) of the North Sea. 'P' = percentage planktonic foraminiferids in total assemblage, 'NCA' = percentage non-calcareous agglutinants in the total benthos. The vertical scale is arbitrary.

consist almost entirely of non-calcareous agglutinants.

Similar vertical changes are identifiable in the foraminiferid assemblages in the Cenozoic of the North Sea Basin; they are considered to reflect eustatic sea level changes and other events which influenced circulation patterns (Chapter 9).

High planktonic (P) percentages are interpreted as indicating episodes of high eustatic sea level, enabling the penetration of open oceanic circulation into the North Sea Basin. They are usually associated with diverse and abundant calcareous benthonic assemblages, reflecting the influence of open circulation leading to high oxygen levels at the sea floor, and can often be equated with transgressive events at the basin margin. Conversely, high NCA ratios reflect sluggish circulation and resultant stagnation (poor oxygenation) at the sea floor, often consequent on low eustatic sea levels, and they often correlate with regional regressive episodes at the basin margins (although local or regional tectonically induced 'ponding' may also play some part). As in the Cenozoic (Chapter 9), there is a marked negative correlation between high NCA and high P percentages, which reinforces this interpretation (Fig. 8.5). The Cretaceous history of the North Sea Basin can thus be divided into 'open' and 'restricted' phases. It is intended to expand on this concept elsewhere.

The definition of these successive assemblages is of value not only for evaluation of basin history, but in intrabasinal correlation. The apparent synchronism of the assemblage boundaries, at least in outer shelf and upper bathyal environments, gives them great importance as primary correlative events, independent of other techniques of biostratigraphic correlation. They have been used extensively in the definition of the foraminiferid zones introduced in the following section.

8.5 STRATIGRAPHIC DISTRIBUTION

8.5.1 Foraminiferid ranges
Early Cretaceous
The taxonomy and biostratigraphy of onshore Early Cretaceous foraminiferids in the North Sea Basin and adjacent areas are documented

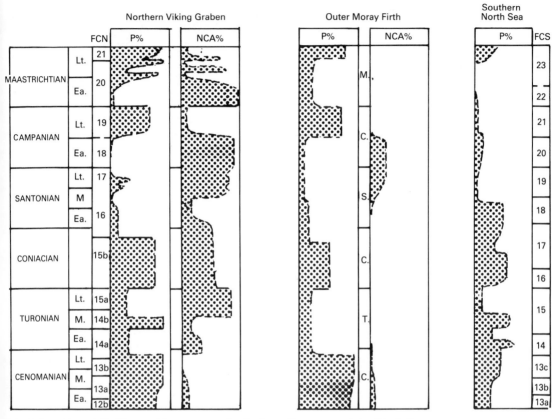

Fig. 8.6 — Foraminiferid assemblage patterns in the Late Cretaceous of the North Sea. Legend as for Fig. 8.5.

in numerous publications. These chiefly relate to northwest Germany, the Ryazanian to Barremian of the Speeton section (Yorkshire, England), and the Gault Clay and Upper Greensand (Middle and Late Albian) of southern England (see Chapter 7).

Key biostratigraphic range charts for Germany are given by Bartenstein and Bettenstaedt (1962), republished with revision of the ammonite zonation by Bartenstein (1978). Range charts for the U.K. are given by Fletcher (1973), Hart (1973), and by Hart et al. (1981) in the first edition of this Atlas. Range charts for two cored boreholes in the southern North Sea are given by Ball (in Lott et al., 1985) and Fletcher (in Lott et al., 1986), for the Early Cretaceous of Heligoland by Bartenstein and Kaever (1973), and for the Albian of the outer Moray Firth Basin by Burnhill and Ramsay (1981).

Although a large quantity of data now exist on foraminiferid ranges, foraminiferal zonations have been proposed only for the Albian (Carter and Hart, 1977; Price, 1977; Burnhill and Ramsay, 1981). The foraminiferid ranges have instead been calibrated to ammonite zones.

Late Cretaceous

Late Cretaceous foraminiferid range charts for southern England are given by Hart et al. (first and second editions of this Atlas), and for northwest Germany by Hiltermann (1962) and Koch (1977). Foraminiferid zonation schemes have been proposed for northwest Germany by Koch, for the Cenomanian of

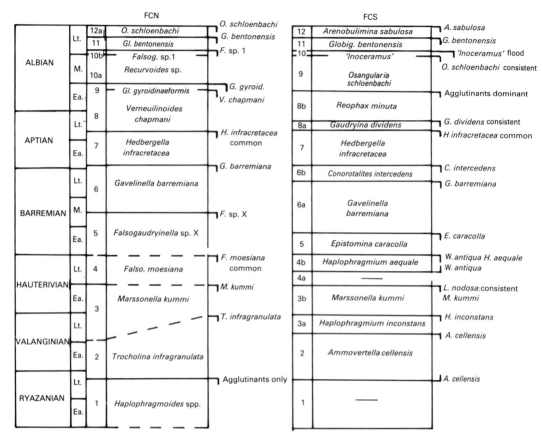

Fig. 8.7 — Foraminiferid Zones and datums in the Early Cretaceous of the North Sea.

southern England by Carter and Hart (1977), and for the Coniacian to Campanian of southern England by Bailey et al. (1983). Burnhill and Ramsay (1981) provide a range chart and a planktonic foraminiferid zonation for the Cenomanian and Early Turonian of the outer Moray Firth.

The benthonic foraminiferid assemblages in the Chalk Group of the southern North Sea are closely comparable to those recorded onshore in England, Germany and Denmark. All the taxa used in the scheme of Bailey et al. (1983) have been recorded in the North Sea, but as their zones are based partly on "first appearance datums" of taxa (FADs), which are not readily identifiable in ditch cuttings, it is preferred here to erect a new zonal scheme for the chalk facies of the North Sea.

8.5.2 Foraminiferid zonation

A scheme of foraminiferid zones is here proposed for the Cretaceous of the North Sea. Due to the existence of marked faunal differentiation (consequent mainly on depth-related biofacies) a single zonal scheme applicable to the whole area is not possible. Here, two zonal schemes are proposed, the first (prefixed FCS) for the 'shelf' facies widespread in the southern North Sea, including the Chalk facies of the Late Cretaceous, and the second (prefixed FCN) for the deeper-water outer sublittoral and bathyal facies prevalent in the central and north-

Stratigraphic Distribution

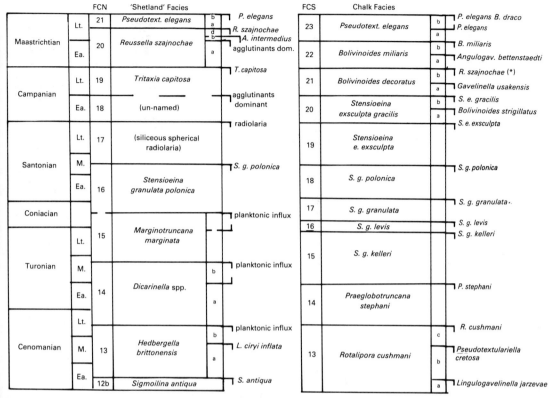

Fig. 8.8 — Foraminiferid Zones and datums in the Late Cretaceous of the North Sea.
* *Reussella szajnochae* also occurs rarely in the Late Maastrichtian.

ern North Sea, including the 'Shetland' facies of the Late Cretaceous. The zonal schemes, and their relationship to Cretaceous lithostratigraphy and chronology, are tabulated in Figs 8.7 and 8.8. The approximate areas within which these schemes are applicable are shown on Figs 8.1 and 8.2.

These schemes are derived from (and intended to be applied to) sections represented mainly by ditch cuttings, and therefore they emphasise the 'tops' (last appearance datums, LADs) of key species, and major assemblage changes. The constraints associated with the use of ditch cuttings samples from hydrocarbon exploration wells are discussed by King (1983) and King et al. (1981) and will not be repeated here. The boundaries of many of the zones are defined to correspond to the boundaries of the assemblages discussed in the previous section (see Figs 8.5 and 8.6 for a semi-quantitative presentation of these assemblages). It must be emphasised that accurate identification of the vertical assemblage changes and other zonal boundary criteria can be achieved only by study of relatively closely spaced samples (ideally at 10 m or closer intervals in the Early Cretaceous), and that attempts to identify them by utilising more widely spaced samples are likely to be misleading, due to contamination by caving and to the sometimes 'patchy' vertical ranges of key species.

Dating and correlation of the Late Cretaceous Zones in the Chalk facies is based on direct comparison of foraminiferid events and ranges with those recorded from the onshore sections in Germany and southern England, taking into account recent revisions in placement of stage and substage boundaries.

The relationship of the Late Cretaceous zones in the offshore Chalk facies to those proposed for onshore sections in southern England and Germany is tabulated in Fig. 8.9. Many of the 'offshore' zones are based on the index species utilised in the 'onshore' zonal schemes, but the definitions of the 'offshore' zones are different in all cases. This reflects the definition mainly by 'bases' (FADs) in the onshore sections, while for the offshore scheme mainly 'tops' (LADs) have been used. However, the faunal sequence and the ranges of key species are closely comparable throughout the southern North Sea and adjacent areas, and the 'offshore' scheme proposed here is equally applicable to onshore sequences in Chalk facies.

In the more argillaceous, deeper-water facies of the outer Moray Firth–northern Central Graben area, the species of *Bolivinoides* and gavelinellids utilised in zonation in the southern North Sea are rare, but *Stensioeina* is still common. The Cenomanian in this area, and in other parts of the central North Sea down to about latitude 56° 30′ N, is in an argillaceous facies (Hidra Formation). This is faunally dissimilar to the chalk facies of the southern North Sea, but is very similar to the Cenomanian of the northern North Sea, and is most appropriately classified using the 'FCN' zonal scheme. In the Early Cretaceous and the Cenomanian (up to the Plenus Marl Formation), the FCS scheme is applicable mainly in the southern North Sea (south of c. 55° N.) and in marginal areas of the central North Sea. From the Plenus Marl Formation upwards, the FCS zones are also applicable in the central North Sea and the outer Moray Firth, to c. 59° N.

8.5.3 Selection of key taxa
Early Cretaceous
Many taxa utilised as key species in range charts for onshore sequences in northwest Germany are geographically or environmentally restricted in distribution, and are not common in the southern North Sea. These include probable shallow marine taxa, such as *Triplasia*. Similarly, the zonal schemes for the Albian of southern England erected by Hart (1973) and Price (1977) cannot be utilised satisfactorily in the North Sea, as the Gault Clay and Upper Greensand facies to which they apply are relatively shallow marine facies restricted to marginal areas of the southern North Sea Basin. Many of their key taxa, including epistominids and nodosariids, occur only rarely in the deeper-water shelf facies of the southern North Sea. The key taxa selected here are based on comparisons of species recorded from the Speeton section (East Yorkshire, U.K.), sections in the U.K., Dutch and Danish offshore sectors, and the Heligoland and northwest German sections, utilising species which are relatively common and have consistent vertical ranges (Figs 8.10, 8.11).

Late Cretaceous
In the Chalk facies, above the Early Turonian, planktonic foraminiferids are abundant at some levels but comprise mainly long-ranging species. Therefore the foraminiferid zones are based largely on the well-documented, common and widespread benthonic genera *Bolivinoides*, *Gavelinella*, *Lingulogavelinella* and *Stensioeina*. *Neoflabellina* has been extensively used in northwest Germany for subdivision of the upper part of the Late Cretaceous, but is less common in most offshore sections, and has not been utilised in the current scheme. Foraminiferid ranges have been checked against published range-charts and samples from southern England and northwest Germany (Figs 8.12, 8.13).

8.5.4 FCS Zones
Zone FCS23 (*Pseudotextularia elegans* Zone). *Top*: defined at the LAD of *P. elegans*. This event coincides with the extinction of many other species of benthonic and planktonic foraminiferids. *Bolivinoides draco* (s.s.) is a

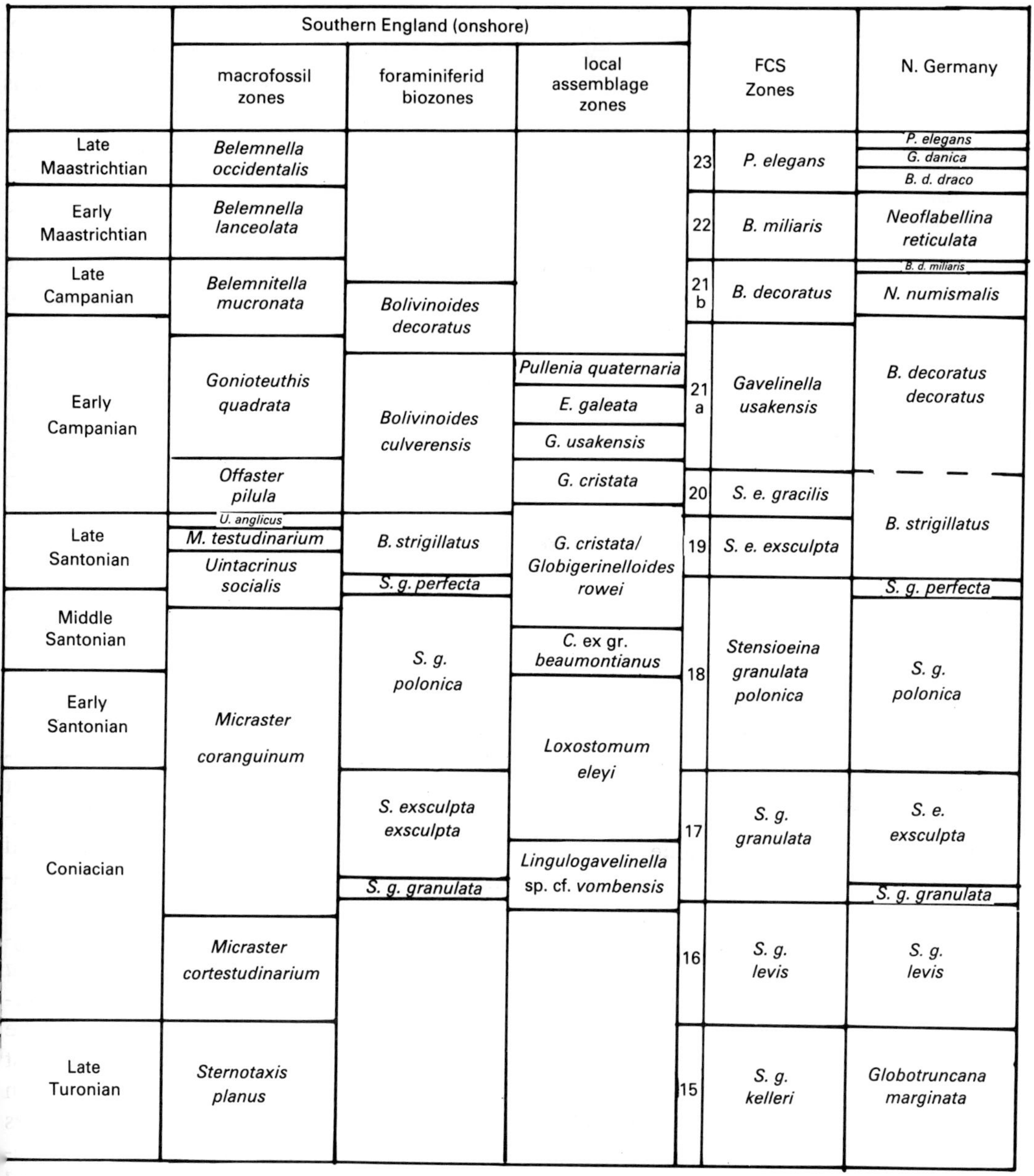

Stratigraphic Distribution

	Southern England (onshore)			FCS Zones		N. Germany
	macrofossil zones	foraminiferid biozones	local assemblage zones			
Late Maastrichtian	*Belemnella occidentalis*			23	*P. elegans*	*P. elegans* / *G. danica* / *B. d. draco*
Early Maastrichtian	*Belemnella lanceolata*			22	*B. miliaris*	*Neoflabellina reticulata*
Late Campanian	*Belemnitella mucronata*	*Bolivinoides decoratus*		21b	*B. decoratus*	*B. d. miliaris* / *N. numismalis*
Early Campanian	*Gonioteuthis quadrata*	*Bolivinoides culverensis*	*Pullenia quaternaria*	21a	*Gavelinella usakensis*	*B. decoratus decoratus*
			E. galeata			
			G. usakensis			
	Offaster pilula		*G. cristata*	20	*S. e. gracilis*	
Late Santonian	*U. anglicus* / *M. testudinarium*	*B. strigillatus*	*G. cristata / Globigerinelloides rowei*	19	*S. e. exsculpta*	*B. strigillatus*
	Uintacrinus socialis	*S. g. perfecta*				*S. g. perfecta*
Middle Santonian		*S. g. polonica*	*C. ex gr. beaumontianus*	18	*Stensioeina granulata polonica*	*S. g. polonica*
Early Santonian	*Micraster coranguinum*		*Loxostomum eleyi*			
Coniacian		*S. exsculpta exsculpta*		17	*S. g. granulata*	*S. e. exsculpta*
			Lingulogavelinella sp. cf. vombensis			
		S. g. granulata				*S. g. granulata*
	Micraster cortestudinarium			16	*S. g. levis*	*S. g. levis*
Late Turonian	*Sternotaxis planus*			15	*S. g. kelleri*	*Globotruncana marginata*

Fig. 8.9 — Relationship between macrofossil and microfossil Zones in the Late Cretaceous chalk facies of the southern North Sea and adjacent onshore areas. Data for southern England are after Bailey *et al.* (1983); data for north Germany are from Koch (1977).

useful supplementary index species, ranging throughout this Zone.
Subdivisions: two subzones can be differentiated:

Subzone FCS23b (*P. elegans* Subzone). Top defined as above.
Subzone FCS23a (*B. draco* Subzone). Top defined at the FAD of *P. elegans*.

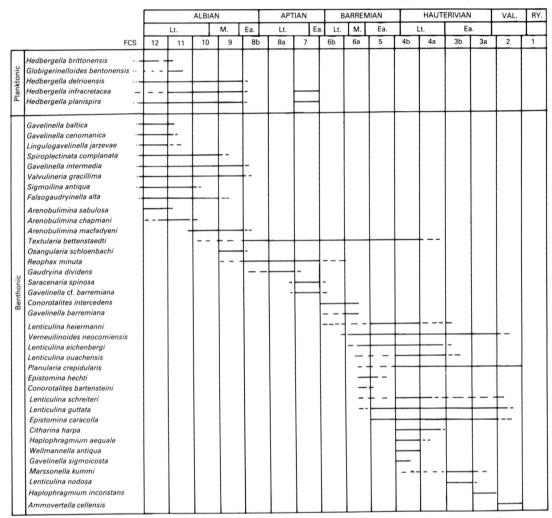

Fig. 8.10 — Ranges of key foraminiferids in the Early Cretaceous of the southern and central North Sea.

Assemblage: planktonic foraminiferids are common in the upper part of the Zone, including *Pseudotextularia*, *Rugoglobigerina* and globotruncanids; in FCS23a, planktonic foraminiferids are apparently rare, and benthonic foraminiferids are dominant.

Age: Late Maastrichtian. The major changes in the planktonic assemblage which define the top of this Zone (extinction of *P. elegans*, *Heterohelix*, globotruncanids, *Globigerinelloides*, etc.) are those which define the Maastrichtian/Danian boundary worldwide. Subzone FCS23b corresponds to the *P. elegans* Zone of Koch (1977), which is equivalent to the highest Late Cretaceous belemnite Zone (*casimirovensis* Zone, upper Late Maastrichtian) in Germany.

Lithostratigraphy: the top of this Zone corresponds to the top of the Tor Formation in the central North Sea. The Zone falls within the upper part of the Tor Formation, comprising white chalks and limestones. In this area, *Pseudotextularia elegans* and other planktonic foraminiferids occur near the top of the Formation, but at lower levels fossils are scarce and poorly preserved.

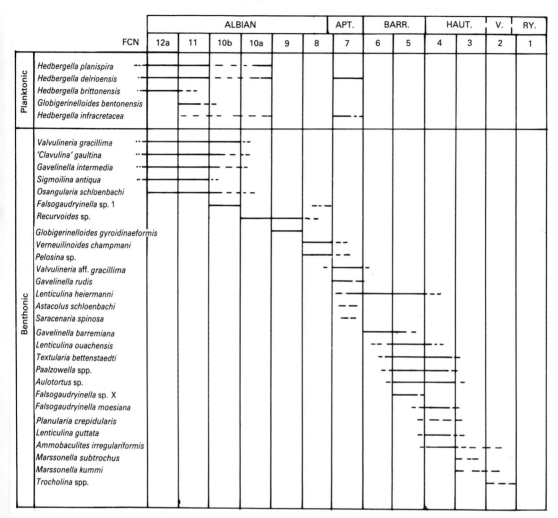

Fig. 8.11 — Ranges of key foraminiferids in the Early Cretaceous of the northern North Sea.

Zone FCS22 (*Bolivinoides miliaris* Zone).
Top: defined at the LAD of *B. miliaris*.
Subdivisions: two subzones can be distinguished in the southern North Sea:
  Subzone FCS22b (*B. miliaris* Subzone). Top defined as above.
  Subzone FCS22a (*Angulogavelinella bettenstaedti* Subzone). Top defined at the LAD of *A. bettenstaedti*. This species ranges down into the upper part of the Late Campanian. This Subzone can be identified only in the southern North Sea.

Assemblage: dominated by benthonic foraminiferids. *Rugoglobigerina* occurs consistently but in low abundance in the lower part of the Zone.
Age: Early Maastrichtian. The LAD of *B. miliaris* corresponds approximately to the top of the *lanceolata* belemnite Zone (top Early Maastrichtian) in northwest Germany (Koch, 1977). *A. bettenstaedti* 'tops' within the Early Maastrichtian in Germany (Hofker, 1957). This Zone cannot be clearly delimited in the outer Moray Firth and the central North Sea, where the chalks of the lower Tor Formation

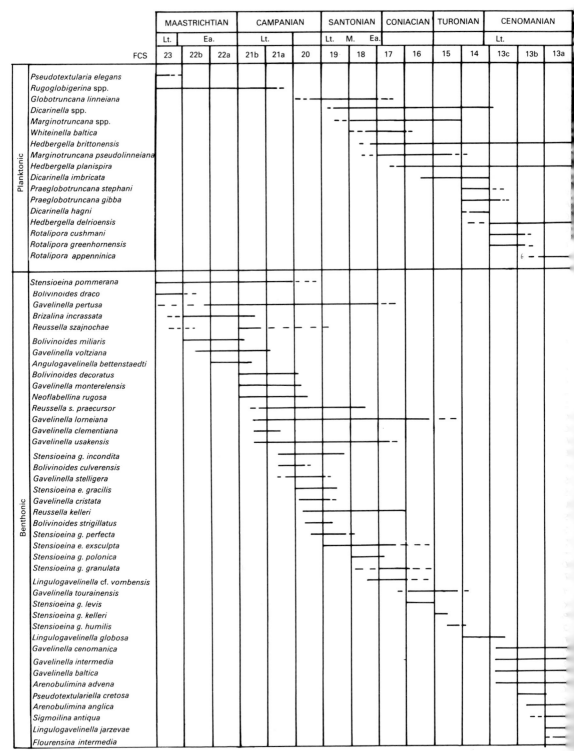

Fig. 8.12 — Ranges of key foraminiferids in the Late Cretaceous Chalk Group of the North Sea.

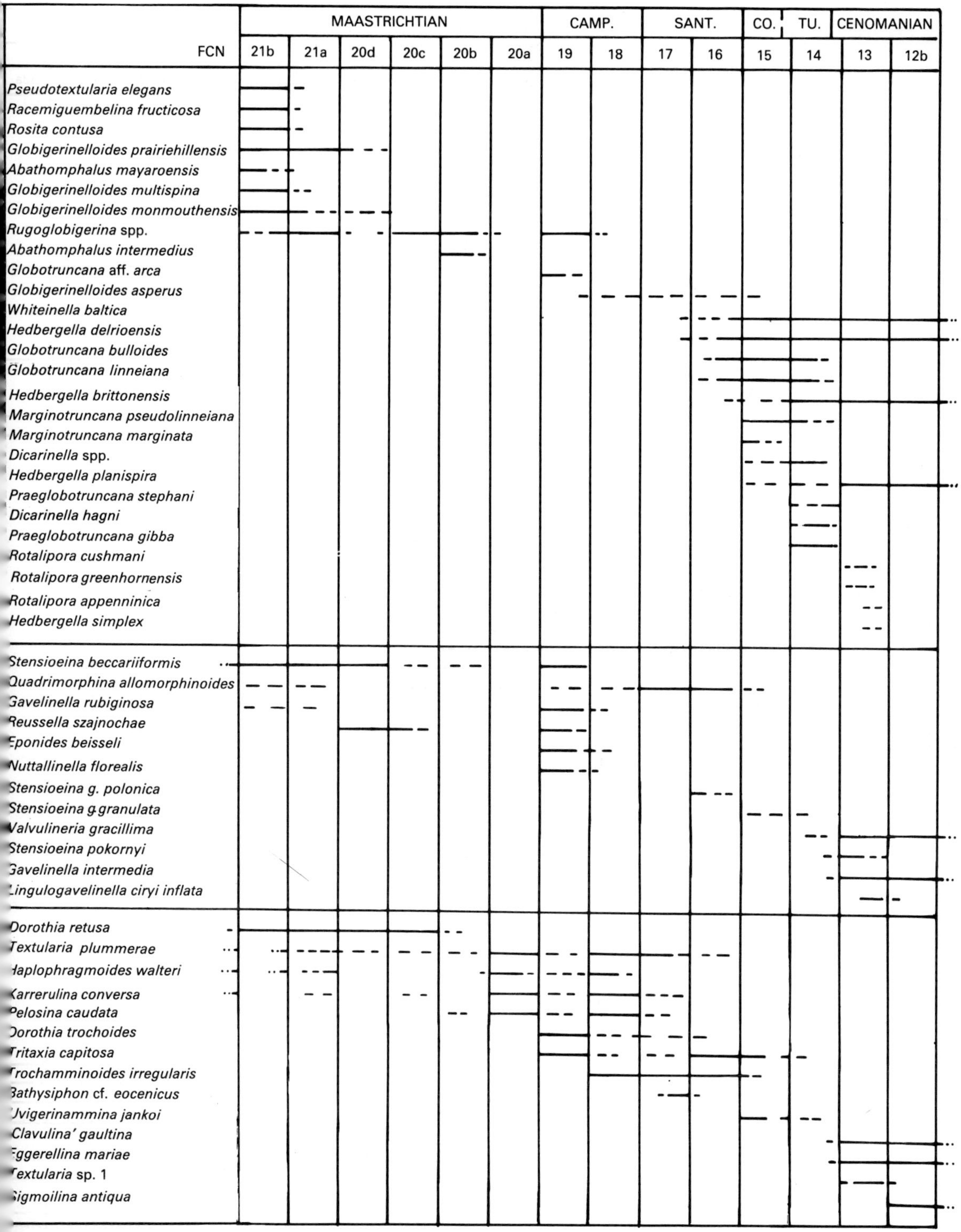

Fig. 8.13 — Ranges of key foraminiferids (including some long-ranging non-calcareous agglutinants) in the Late Cretaceous Shetland Group of the northern North Sea.

are usually poorly fossiliferous.
Lithostratigraphy: corresponds to the lower part of the Tor Formation in the central North Sea.

Zone FCS21 (*Bolivinoides decoratus* Zone).
Top: defined at the LAD (often a flood occurrence) of *Reussella szajnochae* (Swiecicki, in Bailey *et al.*, 1984). The LAD of consistent *B. decoratus* corresponds approximately to this event, but rare specimens of this species apparently occur sporadically in the lower part of the overlying Zone. *B. decoratus* is rare in the central North Sea, but here *R. szajnochae* and *Tritaxia capitosa* are common, and *T. capitosa* can be used as a supplementary index species (its LADs approximates to the top of the Zone). Their occurrence coincides with the more argillaceous facies at this level, indicating an approach to the 'Shetland' faunal and lithological facies.
Subdivisions: two subzones are recognisable in the southern North Sea:
    Subzone FCS21b (*B. decoratus* Subzone). Top defined as above.
    Subzone FCS21a (*Gavelinella usakensis* Subzone). Top defined at the LAD. of *G. usakensis*. This interval can be differentiated only in the southern North Sea; in the argillaceous facies of the outer Moray Firth and northern Central Graben, *G. usakensis* is absent. The LAD of *Stensioeina granulata incondita* provides a useful supplementary datum within the lower part of the Zone.
Assemblage: dominated by benthonic foraminiferids in the Chalk facies, where *Rugoglobigerina* is present in low abundance. In the outer Moray Firth and northern Central Graben, planktonic genera (*Rugoglobigerina, Heterohelix, Globigerinelloides*) are much more common, and dominate the microfauna at some levels.
Lithostratigraphy: the top of this Zone corresponds in the central North Sea to the top of the Flounder Formation (equivalent to the top Hod Formation in the Norwegian and Danish offshore sectors). The Flounder Formation comprises interbedded marly chalks, white chalks and marls. The upper part of the Flounder Formation, corresponding to Zones FCS21 and FCS20, is characterised by beds of pink/red marls, the 'Upper Red Marker' (see Fig. 8.4), and similar pink chalks are recognisable at the top of the Hod Formation.
Age: Late Campanian. In southern England the LAD of *R. szajnochae* corresponds closely to the Campanian/Maastrichtian boundary (Bailey *et al.*, 1984). The LAD of *B. decoratus* is also at this level in northwest Germany, but in eastern England it is apparently within the lower part of the Early Maastrichtian (Chapter 7).

Zone FCS20 (*Stensioeina exsculpta gracilis* Zone).
Top: defined at the LAD of *S. e. gracilis*.
Subdivisions: two Subzones are recognisable in the southern North Sea:
    Subzone FCS20b (*S. e. gracilis* Subzone). Top defined as above.
    Subzone FCS20a (*Bolivinoides strigillatus* Subzone). Top defined at the LAD of *B. strigillatus*.
Assemblage: dominated by benthonic foraminiferids. In the Chalk facies, planktonic forms (chiefly globotruncanids) are more common than at higher levels.
Lithostratigraphy: within the Hod Formation or Flounder Formation in the central North Sea.
Age: Early Campanian. The LAD of *S. e. gracilis* corresponds to the top of the Early Campanian in northwest Germany (Koch, 1977).

Zone FCS19 (*Stensioeina exsculpta exsculpta* Zone).
Top: defined at the LAD of *S. e. exsculpta*. This event corresponds approximately to the highest consistent occurrence of large spherical radiolaria, always calcified in the Chalk facies. These are not always present in the southern North Sea, but occur consistently in

the central North Sea (ranging down to the Late Cenomanian), and are used as supplementary index-fossils to define the top of this Zone.
Assemblage: dominated by benthonic foraminiferids. Globotruncanids occur consistently but in low numbers.
Lithostratigraphy: within the Flounder Formation or Hod Formation in the central North Sea. A unit of pink and red marls, the 'Middle Red Unit', is present in the outer Moray Firth and southern Viking Graben within the lower part of this zone (Fig. 8.4).
Age: Late to Middle Santonian. The LAD of *S. e. exsculpta* corresponds approximately to the top of the Late Santonian in northwest Germany (Koch, 1977), and is at a similar level in southern England.

Zone FCS18 (*Stensioeina granulata polonica* Zone).
Top: defined at the LAD of *S. g. polonica*.
Assemblage: in the upper part of the Zone benthonic foraminiferids are dominant. There is a downsection influx of planktonic foraminiferids (chiefly globotruncanids and hedbergellids) within the upper part of the Zone, and below this level planktonic taxa are abundant.
Lithostratigraphy: within the Flounder Formation or Hod Formation in the central North Sea.
Age: Middle to Early Santonian. The LAD of *S. g. polonica* falls within the upper Middle Santonian in northwest Germany (Koch, 1977), and is at a slightly higher level (basal Late Santonian) in southern England (Bailey *et al.*, 1983).

Zone FCS17 (*Stensioeina granulata granulata* Zone).
Top: defined at the LAD of *S. g. granulata*.
Assemblage: the composition of the assemblage in the upper part of the Zone is similar to that in the overlying Zone. There is a major downsection influx of planktonic foraminiferids (mainly *Globotruncana bulloides*, *Marginotruncana* spp. and hedbergellids) within the middle of the Zone; in the outer Moray Firth area they dominate the microfauna in the lower part of the Zone.
Lithostratigraphy: within the Flounder Formation or Hod Formation in the central North Sea.
Age: basal Santonian to Coniacian. The LAD of *S. g. granulata* is within the middle of the Early Santonian in northwest Germany (Koch, 1977).

Zone FCS16 (*Stensioeina granulata levis* Zone).
Top: defined at the LAD of *S. g. levis*. This taxon is not common in all parts of the southern North Sea. The LAD of *Gavelinella tourainensis* may be used as a supplementary datum level; it is probably just above the LAD of *S. g. levis*.
Assemblage: high proportions of planktonic foraminiferids are usual, as in the lower part of the overlying Zone, but they apparently decrease in abundance towards the base of the Zone.
Lithostratigraphy: within the lower part of the Flounder Formation or Hod Formation in the central North Sea. A unit of red-brown marls, the 'Lower Red Unit', is developed at the top of this Zone in the outer Moray Firth Basin and the south Viking Graben.
Age: Early Coniacian. The LAD of *S. g. levis* corresponds to the top of the Early Coniacian (Ober Turon of Koch, 1977) in northwest Germany. This subspecies is very rare in southern England, and its range is uncertain.

Zone FCS15 (*Stensioeina granulata kelleri* Zone).
Top: defined at the LAD of *S. g. kelleri*.
Assemblage: the top of this Zone corresponds to an abrupt downsection decrease in the relative proportion of planktonic foraminiferids. This event is well-defined in the 'marly' (intermediate) facies of the outer Moray Firth, and is also recognisable in the Chalk facies of the southern North Sea. In onshore sections in southern England, there

is a major downsection influx of planktonic taxa (dominantly *Marginotruncana* spp.) in the Middle Turonian, at a level equivalent to the lower part of the Zone. This event has not been definitely identified in the offshore area, where its recognition is impeded (in the sections studied by the authors) by poor sample quality and the condensation/sedimentary breaks usual at this level.

Lithostratigraphy: in the southern North Sea this Zone comprises the lower part of the Upper Chalk and the upper part of the Middle Chalk. In the central North Sea this Zone can be identified within the lower part of the Flounder Formation and perhaps in the upper part of the Herring Formation. The precise position of this formational boundary in terms of the foraminiferid Zones has not yet been established (see comments below).

Age: Late and Middle Turonian. The LAD of *S. g. kelleri* corresponds to the top of the Late Turonian (Mittel Turon of Koch, 1977) in northwest Germany. The index species is not recorded in onshore sequences in England or northern France.

Zone FCS14 (*Praeglobotruncana stephani* Zone).

Top: defined at the LAD of *P. stephani*. The LAD of *Lingulogavelinella globosa* corresponds approximately to this event, and *L. globosa* can be used as a supplementary index species. This Zone corresponds to the *Praeglobotruncana stephani* Zone of Burnhill and Ramsay (1981).

Assemblage: in southern England there is a planktonic foraminiferid 'abundance-level' in the middle of this Zone, followed by a decrease in the abundance of planktonic foraminiferids at the top of the Zone. These features are presumed to be regionally developed, but cannot yet be satisfactorily identified offshore, where planktonic foraminiferids are common and often dominate the microfauna throughout the Zone.

Lithostratigraphy: in the southern North Sea this Zone is represented within the lower part of the Middle Chalk and the Plenus Marl Formation. Its base is probably within the latter unit (see comments below). In the central North Sea and the outer Moray Firth it is identified in the Herring Formation and the Plenus Marl Formation. Burnhill and Ramsay (1981) record *Praeglobotruncana stephani*, *Dicarinella [Globotruncana] imbricata* and *Marginotruncana [Globotruncana] renzi* in the lower part of the Herring Formation, indicating a Middle to Early Turonian date (Zone FCS14). The upper part of the Herring Formation is usually poorly fossiliferous. *Dicarinella hagni* and *Praeglobotruncana gibba* have been recorded in the Herring Formation by the present authors, confirming an Early Turonian or basal Middle Turonian date. *Stensioeina granulata kelleri/humilis* is also recorded, which would indicate the presence of Zone FCS15, but it is possible that this specimen is caved from a higher level.

Age: basal Middle Turonian to Late Cenomanian. The LADs of *P. stephani* and *L. globosa* are in the lowest part of the *Terebratulina lata* macrofossil Zone (approximately Middle Turonian) in southern England (Chapter 7). N.B. The Cenomanian/Turonian boundary is here placed (following Birkelund, et al., 1984 and Kennedy, 1985) at a level which corresponds approximately to the base of the *Mytiloides labiatus* macrofossil Zone. This is well above the LAD of *Rotalipora*, which is the datum conventionally taken as the Cenomanian/Turonian boundary by micropalaeontologists (e.g. Bolli, et al., 1985).

Zone FCS13 (*Rotalipora cushmani* Zone).

Top: defined at the LAD of *R. cushmani*.

Subdivisions: three subzones can be differentiated:

Subzone FCS13c (*R. cushmani* Subzone). Top defined as above.

Subzone FCS13b (*Pseudotextulariella cretosa* Subzone). Top defined at the LAD of *P. cretosa*.

Subzone FCS13a (*Lingulogavelinella jarze-*

vae Subzone). Top defined at the LAD of *L. jarzevae*. This event corresponds in offshore areas approximately to the LAD of *Sigmoilina [Quinqueloculina] antiqua* (this species is recorded rarely in the overlying Subzone in southern England).

These Subzones are recognisable both in the Chalk facies and in the more argillaceous facies (Texel Marlstone) of the southern Netherlands offshore. As noted above, this Zone cannot be identified in the outer Moray Firth–Central Graben area. Here, from the top Cenomanian to the base of the Cretaceous, the FCN zonal scheme is used.

Assemblage: planktonic foraminiferids (hedbergellids) dominate the microfauna in the upper part of this Zone, but become less abundant downsection from about the middle of the Zone.

Lithostratigraphy: in southern England, *R. cushmani* has its highest occurrence within the lower part of the the Plenus Marl Formation, but offshore its LAD is usually at the top of the underlying Lower Chalk or Hidra Formation. This apparent discrepancy probably reflects the partial decalcification of the organic-rich Plenus Marl Formation, and its relatively thin development in most sections compared to sample spacing; here the top of Zone FCS13 is tentatively placed within the Plenus Marl formation. The Lower Chalk and the majority of the Texel Marlstone Member fall within this Zone.

Age: Late to Early Cenomanian. The LAD of *P. cretosa* in southern England is at the top of the *Turrilites acutus* ammonite Subzone, within the upper Middle Cenomanian. The LAD of *L. jarzevae* in the same area is within the *Mantelliceras saxbii* ammonite Subzone (middle Early Cenomanian). The top occurrence of consistent *Sigmoilina antiqua* in southern England is at approximately the same level (Chapter 7).

Zone FCS12 (*Arenobulimina sabulosa* Zone). *Top*: defined at the LAD of *A. sabulosa*. This event corresponds to the FAD of *Flourensina* spp. and of internally partitioned arenobuliminids.

Assemblage: dominated by planktonic foraminiferids (mainly hedbergellids) associated with calcareous benthonic foraminiferids and calcareous-cemented agglutinants.

Lithostratigraphy: in the U.K offshore the top of this Zone corresponds to the base of the Lower Chalk. It falls within the upper part of the Red Chalk in this area; in borehole 81/40 (southern U.K. offshore) the record of *Arenobulimina sabulosa* and *A. chapmani* in the upper part of the Red Chalk indicates this Zone (Lott *et al.*, 1985). It corresponds to the lowest part of the Texel Marlstone and the upper part of the Upper Holland Marl in the Netherlands offshore.

Age: Late Albian. The LAD of *Arenobulimina sabulosa* and the FAD of *Flourensina intermedia* occur in onshore sections at the Albian/Cenomanian boundary (Carter and Hart, 1977), but there is everywhere a stratigraphical break at this level, and it is possible that this correspondence may be fortuitous.

Zone FCS11 (*Globigerinelloides bentonensis* Zone).

Top: defined at the LAD of common *G. bentonensis*. Rare specimens occur in some areas in the overlying Zone, but in most areas the top of Zone FCS11 is in effect defined by the LAD of *G. bentonensis*. This event also marks a downsection increase in abundance of *Hedbergella planispira*. The FAD of *G. bentonensis* is probably within the lower part of the Zone.

Assemblage: dominated by planktonic foraminiferids (*Globigerinelloides* and hedbergellids), associated with calcareous benthonic foraminiferids and calcareous-cemented agglutinants.

Lithostratigraphy: this Zone lies within the Upper Holland Marl in the Netherlands offshore, and within the Red Chalk in parts of the U.K. offshore. In the cored borehole 81/40, *G. bentonensis* is recorded in the

lower part of the Red Chalk (Lott et al., 1985).
Age: Late Albian. An interval with abundant *G. bentonensis* (corresponding to the upper part of Zone FCS11) is identifiable in onshore sequences in southern England, northern France and in the subsurface of the Netherlands and Germany (Carter and Hart, 1977; Price, 1977). Its dating is complicated by the widespread condensation/hiatus at the Albian/Cenomanian boundary, but its top probably lies within the middle of the *dispar* ammonite Zone in southern England (Gallois and Morter, 1982). In the much thicker Late Albian–Early Cenomanian sequences in the southern Netherlands offshore, the LAD of *G. bentonensis* is well below the FAD of *Flourensina intermedia*, which is taken to mark the base of the Cenomanian. The FAD of *G. bentonensis* lies within the *auritus* ammonite Subzone (middle Late Albian) in southern England and northern France (Carter and Hart, 1977; Price, 1977; Sandman, pers. comm.). In the southern Netherlands offshore, *Arenobulimina macfadyeni* is recorded in the lower part of the Zone; in southern and eastern England this species ranges from Middle Albian into early Late Albian.

Zone FCS10 (*'Inoceramus'* Zone).
Top: defined at a downsection influx of abundant red-stained Inoceramid shell fragments ('prisms'). Although not a foraminiferid datum, this event is distinctive and geographically widespread. The LAD of consistent *Spiroplectinata complanata* and *Arenobulimina macfadyeni* are at approximately the same level, and there is an assemblage change at about the same level marked by the consistent appearance downsection of non-calcareous agglutinants.
Assemblage: there is a downsection change at approximately the top of this Zone, from a plankton-dominated assemblage to a benthos-dominated foraminiferid assemblage, including a consistent but relatively low proportion of non-calcareous agglutinants. *Hedbergella planispira* is common at some levels.
Lithostratigraphy: the top of this Zone is often marked by a downsection lithological change from grey to red claystones. Red claystones also occur in the underlying Zone. In the Netherlands offshore sector, this Zone corresponds to the lower part of the Upper Holland Marl, which includes beds of red claystone. It corresponds approximately to the red clays of the uppermost Speeton Clay (bed A3), but their microfauna is poorly known. In the southern U.K. offshore, this Zone (and the underlying Zones) are represented within an argillaceous sequence currently classified as Speeton Clay; Crittenden (1982) has suggested that the lithostratigraphic nomenclature used in the Netherlands offshore could be applied effectively in this area.
Age: Early Late Albian–?late Middle Albian. This and the underlying Zone can be correlated with the '*concentricus*-Schichten' of northwest Germany, characterised by the 'large numbers of *Inoceramus* prisms present in washed residues' (Kemper, 1973b), which include red claystones in some areas. They are approximately mid to late Middle Albian in age, but may range up into the Late Albian. In southern and eastern England, the highest level with abundant inoceramids is approximately at the top of the *orbignyi* ammonite Subzone (early Late Albian) (Gallois and Morter, 1982). *Spiroplectinata annectens* and *Sigmoilina antiqua* occur in this Zone in the southern Netherlands offshore; these species are common only from the base of the Late Albian in onshore sequences. It thus seems likely that part (at least) of this Zone is early Late Albian in age; the lower part may be Middle Albian. Accurate calibration with the ammonite Zones is not possible without additional onshore data.

Zone FCS9 (*Osangularia schloenbachi* Zone).
Top: defined at the LAD of consistent *O. schloenbachi*. Very rare specimens of this species are occasionally recorded in the Late Albian in the Netherlands offshore, but in

most sections *O. schloenbachi* is restricted to Zone FC9.

Assemblage: dominated by calcareous benthonic foraminiferids and calcareous-cemented agglutinants, with subordinate planktonic species (hedbergellids). Inoceramid 'prisms' are common, as in the Zone above.

Lithostratigraphy: the top of this Zone corresponds approximately to the top of the Middle Holland Shale (see Crittenden, 1987, Fig. 4). The base of the Zone is marked regionally by a highly glauconitic unit, which probably correlates with the Gault/Greensand Junction Beds and Carstone of southern and eastern England (late Early Albian and early Middle Albian).

Age: Middle Albian and latest Early Albian. The range of *Osangularia schloenbachi* in onshore successions is not satisfactorily established. It is commonest in the Netherlands and Germany, where accurate calibration to the ammonite zonation is not yet possible. Price (1977) gives it a continuous range from Middle Albian to Cenomanian, but this conflicts with data from eastern England, where it is recorded from the late Early Albian *tardefurcata* ammonite Zone (Rawson et al., 1978; Crittenden, 1983). No published records exist from the Gault Clay, but it has recently been recorded from the *loricatus* ammonite Zone (Middle Albian) of East Anglia (Sandman, pers. comm.). Available data from Germany suggest a similar range (late Early Albian to Middle Albian). It is recorded by Hecht (1938) from the *regularis* (= *tardefurcata*) ammonite Zone (as *Planulina* D5), and this appears to be its level of greatest abundance, as in England. Late Albian records are mentioned above.

Zone FCS8 (*Reophax minuta* Zone).

Top: defined at a downsection change from a foraminiferid assemblage dominated by calcareous benthonic foraminiferids and calcareous-cemented agglutinants to one composed dominantly or exclusively of non-calcareous agglutinants. The LAD of consistent *R. minuta* and the LAD of *Textularia bettenstaedti* are both at this level.

Subdivisions: two Subzones can be recognised:

Subzone FCS8b (*R. minuta* Subzone). Top defined as above. The assemblage in this Subzone usually consists almost entirely of non-calcareous agglutinants.

Subzone FCS8a (*Gaudryina dividens* Subzone). Top taken at the LAD of *G. dividens*. This event corresponds to an assemblage change defined by the incoming downsection of a significant proportion of calcareous benthonic foraminiferids and calcareous agglutinants.

Assemblage: see comments above. Planktonic foraminiferids are very rare or absent.

Lithostratigraphy: within the Middle Holland Shale in the Netherlands offshore sector. In borehole 81/40 (Lott et al., 1985), Subzone FCS8b is probably represented by the thin unit of 'Variegated Beds' with a wholly non-calcareous agglutinating assemblage. The underlying 'Grey–Brown Mudstones' contain a mixed calcareous and non-calcareous benthonic foraminiferid assemblage, including *Gaudryina dividens*, and can be assigned to Subzone FCS8a.

Age: Early Albian and Late Aptian. The event taken to define the top of Zone FCS8 marks a major change in the foraminiferid assemblage which can be identified at outcrop in the Lower Saxony Basin, northwest Germany. Here it corresponds to the mid-*tardefurcata* Zone transgression (Kemper, 1973b), which marks a change from a restricted environment to a more open marine environment. The quantitative change from agglutinant-dominated to dominantly calcareous benthonic assemblages at this level is illustrated by Price (1977, Fig. 3). The equivalent transgressive event in southern England is probably marked by the contact of the Lower Greensand with the condensed Gault–Lower Greensand Junction beds (Rawson et al., 1978), dated as within the *tardefurcata* Zone. In northwest Germany, *R. minuta* has its top consistent occurrence in the *regularis* Subzone of the *tardefurcata* Zone (Bartenstein and Bettenstaedt, 1962; Bartenstein, 1978).

The LAD of *Gaudryina dividens* in northwest Germany corresponds approximately to the top of the *nutfieldensis* ammonite Zone (mid-Late Aptian), (Bartenstein, 1978). This calibration is confirmed independently in borehole 81/40, where the 'Grey–Brown Mudstone', which can be assigned to Subzone FCS8a, contains ostracods diagnostic for the *nutfieldensis* Zone (Wilkinson *in* Lott *et al.*, 1985). Thus the Albian/Aptian boundary lies within Subzone FCS8b.

The widespread occurrence of volcanic ash beds in Subzone FCS8a forms a distinctive marker horizon. They are recorded in the Lower Saxony Basin (Zimmerle, 1979), in borehole 81/40 (Lott *et al.*, 1985), in many parts of the Central Graben and outer Moray Firth (L. Riley, pers. comm.) and occur also in the *nutfieldensis* Zone in southern England.

Zone FCS7 (*Hedbergella infracretacea* Zone). *Top*: defined at the LAD of abundant *H. infracretacea*. This event marks the reappearance downsection of a planktonic-dominated microfauna (often more than 90% planktonic foraminiferids), comprising *Hedbergella infracretacea* and *H. delrioensis*.
Subdivisions: two subzones can be differentiated:
   Subzone FCS7b (*H. infracretacea* Subzone): top defined as above.
   Subzone FCS7a (un-named): top defined at the FAD of abundant *H. infracretacea*.
Assemblage: in the upper part of the Zone (Subzone FCS7b) the assemblage is dominated by planktonic foraminiferids (*Hedbergella* spp.), which often occur in 'flood' proportions, and may form up to 90% of the assemblage. Two successive lithological and microfaunal units can often be differentiated within this interval, characterised by the relative abundance and distinctive preservation of the foraminiferids: an upper calcareous claystone unit with green, grey or grey-brown stained microfossils (unit FCS7b2), and a lower unit of red marls with brick red microfossils (unit FCS7b1). The main 'flood' horizon of planktonic foraminiferids is associated with the lower unit. These units can be differentiated over a wide area, from the southern North Sea to the Viking Graben.

The lower part of the Zone contains an assemblage dominated by calcareous benthonic foraminiferids and non-calcareous agglutinants. Planktonic foraminiferids are often common in ditch cuttings, but are interpreted as caved from the overlying Subzone.
Lithostratigraphy: in the Netherlands offshore, this Zone corresponds to the lowest part of the Middle Holland Shale and the Lower Holland Marl. The sediments corresponding to the lower part of the Zone are usually grey claystones, including a poorly fossiliferous 'black-shale' facies corresponding to the Fischschiefer of the Lower Saxony Basin.
In borehole 81/40 (Ball, in Lott *et al.*, 1985), the highest occurrence of *H. infracretacea*, defining the top of the Zone, is in the lower part of the 'Grey-Brown Mudstone'; this interval corresponds to the 'green *Hedbergella*' unit (FCS7b2). The underlying 'Red Mudstone' contains the main *H. infracretacea* flood occurrence, and corresponds to the 'red *Hedbergella*' unit (FCS7b1). In Heligoland the *H. infracretacea* 'flood horizon' is developed in a 'gelblich–hellbraune' marl, so rich in planktonic foraminiferids as to be called a 'Globigerinenkalk', overlying laminated shales of 'Fischschiefer' facies (Bartenstein and Kaever, 1973).
Age: early Late Aptian and Early Aptian. In the Lower Saxony Basin the *H. infracretacea* Subzone (FCS8b) corresponds to a grey and red calcareous claystone unit rich in planktonics (the '*ewaldi*-mergel' and similar facies), which include the mid-*nutfieldensis* to mid?-*deshayesi* ammonite Zones (Kemper, 1973b; Bertram and Kemper, 1982). In borehole 81/40, the 'Red Mudstone' is dated on the basis of the ostracod assemblages (Wilkinson, in Lott *et al.*, 1985) as Late Aptian (*nutfieldensis* and *martinioides* ammonite Zones). At

outcrop in the Lower Saxony Basin, the Fischschiefer is dated as Early Aptian, approximately early *deshayesi* and *forbesi* ammonite Zones (Kemper, 1973b). Thus the top of the Zone probably lies within the lower part of the *nutfieldensis* Zone (mid-Late Aptian), and the top of Subzone FCS7a lies within the middle of the Early Aptian.

Zone FCS6 (*Conorotalites intercedens* Zone).
Top: defined at the LAD of *C. intercedens*.
Subdivisions: divided into two Subzones:
 Subzone FCS6b (*C. intercedens* Subzone). Top defined as above.
 Subzone FCS6a (*Gavelinella barremiana* Subzone). Top defined at the LAD of *G. barremiana*. *Epistomina hechti* has its highest occurrence within this interval, and can be used as a supplementary index species. The FAD of *G. barremiana* is within this subzone. The lower part of the Subzone is usually represented by a poorly fossiliferous 'black-shale' facies.
Assemblage: dominated by calcareous benthonic taxa. Small planktonic foraminiferids (*Hedbergella* sp.) are abundant at one or more levels within Subzone FCS6a, and may prove to delimit a regional marker horizon. This is apparently the earliest occurrence of planktonic foraminiferids in the North Sea Basin.
Lithostratigraphy: in the Netherlands offshore, the top of this Zone corresponds to the top of the Vlieland Shale. In the lower part of the Zone is a widespread 'black-shale' unit, which corresponds to the 'Blatterton' of the Lower Saxony Basin, and is also represented at Speeton (Rawson and Mutterlose, 1983) and in the Heligoland section (Bartenstein and Kaever, 1973).
Age: basal Aptian to late Early Barremian. The top of this Zone corresponds approximately to the base of the Fischschiefer (top *Prodeshayesites fissicostatus* ammonite Zone, Kemper, 1973b). The highest part of this Zone is therefore dated as basal Aptian. The 'Blatterton' facies is dated as late Early Barremian to early Middle Barremian in northwest Germany and at Speeton (mid-*elegans* to mid-*Hoplocrioceras fissicostatus* ammonite Zones, Rawson and Mutterlose, 1983).

Zone FCS5 (*Epistomina caracolla* Zone).
Top: defined at the LAD of *E. caracolla*. This species is often common, but the susceptibility of epistominids to diagenetic dissolution can make the Zone difficult to identify.
Assemblage: dominated by calcareous benthonic foraminiferids and calcareous-cemented agglutinants. Epistominids are often abundant.
Lithostratigraphy: within the Vlieland Shale in the Netherlands offshore.
Age: Early Barremian. In northwest Germany and at Speeton the LAD of *E. caracolla* is immediately beneath the base of the Blatterton (upper part of the Early Barremian).

Zone FCS4 (*Haplophragmium aequale* Zone).
Top: defined at the LAD of *H. aequale*. This event corresponds approximately to the LAD of *Gavelinella sigmoicosta*, which can be used as an accessory marker for the top of the Zone. The LAD of consistent *Wellmannella antiqua* is at the same level, but small specimens occur sporadically in the lower part of the overlying Zone.
Subdivisions: divided into two Subzones:
 Subzone FCS4b (*G. sigmoicosta* Subzone). Top defined as above. *G. sigmoicosta* is restricted to this Subzone.
 Subzone FCS4a (unnamed). Top defined at the FAD of *W. antiqua*, which corresponds approximately to the FAD of *H. aequale*.
Assemblage: composed of calcareous benthonic foraminiferids and calcareous-cemented agglutinants.
Lithostratigraphy: within the Vlieland Formation in the Netherlands offshore sector.
Age: basal Barremian and Late Hauterivian. The LADs of *Gavelinella (Conorotalites) sigmoicosta* and *Haplophragmium aequale* are

within the *rarocinctum* Zone (basal ammonite zone of the Barremian) at Speeton (Fletcher, 1973) and probably at a similar level in northwest Germany (Bartenstein, 1978), making allowances for the revision of the Barremian/Hauterivian boundary in northwest Europe by Kemper *et al.* (1981).

Zone FCS3 (*Marssonella kummi* Zone).
Top: defined at the LAD of consistent *M. kummi*; rare specimens are recorded in the overlying Subzone. The LAD of *Lenticulina nodosa* (s.s.) is at about the same level, and this species serves as a supplementary index-fossil.
Subdivisions: divided into two Subzones:
  Subzone FCS3b (*Marssonella kummi* Subzone). Top defined as above.
  Subzone FCS3a (*Haplophragmium inconstans* Subzone). Top defined at the LAD of *H. inconstans*.
Assemblage: composed of calcareous benthonic foraminiferids and calcareous-cemented agglutinants.
Lithostratigraphy: within the Vlieland Shale in the Netherlands offshore sector.
Age: Early Hauterivian and latest Valanginian. The LAD of consistent *Marssonella kummi* and the LAD of *Lenticulina nodosa* are within the *regale* ammonite Zone (uppermost Early Hauterivian) at Speeton (Fletcher, 1973) and at about the top of the *regale* Zone in northwest Germany (Bartenstein, 1978), where available ammonite information is less detailed. The LAD of *Haplophragmium inconstans* corresponds approximately to the top Valanginian in northwest Germany (Bartenstein, 1978).

Zone FCS2 (*Ammovertella cellensis* Zone).
Top: defined at the LAD of *A. cellensis*.
Assemblage: composed of calcareous benthonic foraminiferids and calcareous-cemented agglutinants.
Lithostratigraphy: this Zone includes the lowest part of the Cromer Knoll Group. In parts of the southern North Sea it rests on non-marine sediments of the Delfland Group.

Age: Late to Early Valanginian. In Germany the LAD of *A. cellensis* is within the Late Valanginian.

Zone FCS1 (*Haplophragmoides* spp. Zone).
Top: defined at the FAD of *Ammovertella cellensis*.
Assemblage: a limited assemblage of poorly preserved non-calcareous agglutinants (mainly *Haplophragmoides* spp.) is present in the dysaerobic/anaerobic facies of the Kimmeridge Clay Formation in the central North Sea Basin. In more proximal areas, where other marine argillaceous facies are present (e.g. in parts of the Norwegian and Danish offshore areas), a mixed assemblage of agglutinants and calcareous benthonic foraminiferids is recorded.
Lithostratigraphy: this Zone is only recognised in parts of the offshore area, where it is represented by the highest part of the Kimmeridge Clay Formation or by marine claystones of 'normal marine' facies (basal Speeton Clay, Bream Formation). Earliest Cretaceous sediments in most onshore sections and in parts of the southern North Sea are represented by non-marine sediments (Wealden facies), or by shallow marine sands (e.g. Spilsby Sandstone) with sparse foraminiferid faunas.
Age: Ryazanian.

8.5.5 FCN Zones
Zone FCN21 (*Pseudotextularia elegans* Zone).
Top: defined at the major break in planktonic assemblages at the Cretaceous/Palaeocene boundary, marked by the LAD of *P. elegans*, *Heterohelix* spp., *Globigerinelloides* spp. and *Rugoglobigerina* spp.
Subdivisions: two Subzones are recognised:
  Subzone FCN21b (*P. elegans* Subzone): top defined as above. *P. elegans* is probably restricted to this interval. The microfauna is dominated by planktonic foraminiferids, including *Rosita contusa* and the taxa mentioned above.
  Subzone FCN21a (*Rugoglobigerina* spp. Subzone): top defined at a downsection

increase in abundance of *Rugoglobigerina* spp., from less than 5% of the planktonic foraminiferid assemblage to > 50% in some areas.

Assemblage: dominated by planktonic foraminiferids, which comprise 80–90% of the microfauna in the upper part of the Zone. The diversity of the planktonic foraminiferid assemblage is higher in this interval than at any other level in the Cretaceous of the North Sea Basin. It includes the genera *Abathomphalus*, *Globigerinelloides*, *Globotruncanella*, *Heterohelix*, *Pseudotextularia*, *Racemiguembelina*, *Rosita* and *Rugoglobigerina*. In the lower part of the Zone, the proportion of planktonic foraminiferids decreases to less than 30% of the total assemblage. A low proportion of non-calcareous agglutinants (< 10%) is present throughout.

Lithostratigraphy: this Zone lies within the upper part of 'Formation E' of the Shetland Group, comprising calcareous and non-calcareous claystones with thin limestone and chalk beds in some areas.

Age: Late Maastrichtian. The major break in the planktonic foraminiferid assemblage at the top of this Zone is that which defines the Maastrichtian/Danian boundary. The top of the Zone corresponds to the top of Zone FCS23. The planktonic taxa in Subzone FCN21b include *Abathomphalus mayaroensis*, the index species of the Late Maastrichtian *A. mayaroensis* Zone (Caron, 1985).

Zone FCN20 (*Reussella szajnochae* Zone).

Top: defined at the LAD of *R. szajnochae*. In the northern North Sea, this species ranges from the top of Zone FCN20 to the base of Zone FCN19, but within this interval it is apparently absent in Subzones FCN20b and 20a.

Subdivisions: in the northern Viking Graben and adjacent areas, four Subzones can be distinguished:

Subzone FCN20d (*R. szajnochae* Subzone): top defined as above. The top of this Subzone is marked by a downsection influx of planktonic foraminiferids, which dominate the microfauna in the upper part of the Subzone. In the lower part of the Subzone, non-calcareous agglutinants are dominant.

Subzone FCN20c: top defined by a downsection influx ('flood') of *Rugoglobigerina* spp. At this level the proportion of planktonic foraminiferids increases downsection from 10% to > 90%.

Subzone FCN20b (*Abathomphalus intermedius* Subzone). Top defined at the LAD of *A. intermedius*, which is (in the North Sea Basin) apparently restricted to this Subzone. The microfauna is dominated by planktonic foraminiferids, but they form a lower proportion of the foraminiferids than in the overlying Subzone.

Subzone FCN20a. Top defined at a downsection increase in the proportion of non-calcareous agglutinants, which comprise 90–100% of the foraminiferid assemblage throughout the Subzone.

Assemblage: see details under Subzones.

Lithostratigraphy: this Zone lies within the lower part of 'Formation E'.

Age: Late to Early Maastrichtian. *A. intermedius* ranges from the middle of the *G. gansseri* Zone (approximately basal Late Maastrichtian in northwest European usage) to the top of the *mayaroensis* Zone (Late Maastrichtian) (Robaszynski et al., 1984). Thus Subzone FCN20b is Late Maastrichtian in age. No restricted Early Maastrichtian microfossils have been recorded from the lower part of the Zone, but the top of the Early Maastrichtian is here tentatively placed at the top of Subzone FCN20a.

Zone FCN19 (*Tritaxia capitosa* Zone).

Top: defined at the LAD of *T. capitosa*. *Reussella szajnochae* reappears downsection at the top of the Zone, and this event is a supplementary marker for the top of the Zone.

Assemblage: the top of this zone is marked by a downsection change in overall assemblage composition from dominantly agglutinating foraminiferids to dominantly calcareous

benthonic foraminiferids and planktonic foraminiferids. This event is well marked in most parts of the northern North Sea, but in the most northerly areas calcareous foraminiferids become rarer, and the top of the Zone can be difficult to define.
Lithostratigraphy: the top of this Zone corresponds to the top of 'Formation D', defined by a downsection change from non-calcareous claystones to marls and calcareous claystones. In the upper part of the Formation the claystones are often red in colour, corresponding to the 'Upper Red Unit' of the Flounder Formation, and the foraminiferids are often red-stained.
Age: Late Campanian. The top of this Zone can be correlated with the 'top *Reussella szajnochae* event' defining the top of Zone FCS21 in the southern North Sea.

Zone FCN18 (unnamed).
Top: defined at a downsection assemblage change from dominantly calcareous foraminiferids to dominantly or exclusively non-calcareous agglutinants.
Assemblage: dominantly or exclusively non-calcareous agglutinants.
Lithostratigraphy: within 'Formation D'; this interval comprises mainly non-calcareous claystones with thin limestone beds.
Age: approximately Early Campanian, determined by stratigraphical context. No short-ranging taxa have been recorded.

Zone FCN17 ('*Radiolaria*' Zone).
Top: defined at the downsection appearance of large siliceous spherical reticulate radiolaria (?'*Cenosphaera*' sp.). These occur sporadically from this level down to the Late Cenomanian. Pyritised 'lensoid' radiolaria/diatoms also occur consistently in this Zone, and are a valuable index fossil, particularly in more northerly areas where the spherical radiolaria may be rare or absent, but must be differentiated from the small pyritised *Coscinodiscus* spp. which occur more widely in the Late Cretaceous.
Assemblage: the microfauna comprises mainly non-calcareous agglutinants, but at some levels planktonic and calcareous benthonic foraminiferids are recorded.
Lithostratigraphy: within 'Formation D', comprising mainly non-calcareous claystones. A wedge of 'deepwater' sands is present within the upper part of this Zone in the northern Viking Graben, passing distally into interlaminated siltstones and claystones.
Age: Late to Middle Santonian. The downsection appearance of radiolaria at the top of this interval is equated with the corresponding event (top of Zone FCS19) dated as top Santonian in the southern North Sea.

Zone FCN16 (*Stensioeina granulata polonica* Zone).
Top: defined at the LAD of *Stensioeina granulata polonica*.
Assemblage: there is a change in foraminiferid assemblage composition at about the top of this Zone, from the dominantly non-calcareous agglutinating assemblage in the overlying interval, to a mixed calcareous and non-calcareous assemblage, associated with common Inoceramid shell debris. Planktonic foraminiferids (long-ranging globotruncanids and *Whiteinella* spp.) reappear consistently downsection just below the top of the Zone.
Lithostratigraphy: within 'Formation D'.
Age: Middle to Early Santonian. *S. g. polonica* ranges in the southern North Sea from 'mid'-Middle Santonian to basal Santonian, and this interval is dated accordingly. The top of the Zone probably corresponds approximately to the top of Zone FCS18.

Zone FCN15 (*Marginotruncana marginata* Zone).
Top: defined at a downsection influx of abundant planktonic foraminiferids, including common *Marginotruncana* spp. The LAD of *M. marginata* is within the upper part of the Zone. The LAD of the agglutinant *Uvigerinammina jankoi* corresponds approximately to the top of this Zone.
Subdivisions: two Subzones can be defined:
  Subzone FCN15b: top defined as above.

Subzone FCN15a: top taken at a downsection assemblage change to dominantly non-calcareous agglutinants and calcareous benthonic taxa.
Assemblage: planktonic foraminiferids dominate the microfauna in Subzone FCN15b, associated with calcareous benthonic foraminiferids and non-calcareous agglutinants. In Subzone FCN15a, the microfauna is dominated by non-calcareous agglutinants, with subordinate calcareous benthonic and planktonic foraminiferids.
Lithostratigraphy: within 'Formation D'.
Age: Coniacian and ?Late Turonian. The planktonic foraminiferids in FCN15b are mainly long-ranging taxa, but indicate a Coniacian or Late Turonian date. The occurrence of rare specimens of *Stensioeina granulata* (*S. g. granulata* or *S. g. levis*) indicates a Coniacian date for this interval. The age of Subzone FCN15a has not been directly determined, but in context a Late Turonian date seems most probable.

Zone FCN14 (*Dicarinella* spp. Zone).
Top: defined at a downsection influx of abundant planktonic foraminiferids, dominantly hedbergellids and *Dicarinella* spp.
Subdivisions: divided into two Subzones:
  Subzone FCN14b: top defined as above.
  Subzone FCN14a: top defined at a downsection assemblage change from dominantly planktonic foraminiferids to dominantly calcareous benthonic foraminiferids.
Assemblage: in FCN14b, dominated by planktonic foraminiferids. Subzone FCN14a contains an assemblage in which planktonic foraminiferids are common at some levels, but non-calcareous agglutinants are usually dominant.
Lithostratigraphy: Subzone FCN14b encompasses the lowest part of 'Formation D' of the Shetland Group. The top of Subzone FCN14a corresponds approximately to the top of 'Formation C', characterised by interbedded claystones, chalky limestones and calcareous glauconitic sandstones. Formation B (= Plenus Marl Formation) falls within the lowest part of this Zone.
Age: Middle Turonian to Late Cenomanian. The planktonic foraminiferids in Subzone FCN14b include *Dicarinella imbricata* (Early Turonian to Coniacian) and *D. hagni* (Early Turonian to basal Middle Turonian). *Marginotruncana* is apparently absent, suggesting a date older than mid-Middle Turonian. Occasional specimens of *Stensioeina granulata kelleri/humilis* and *S. g. humilis/interiecta* are recorded. The co-occurrence of these taxa indicates a probable Middle Turonian date. Subzone FCN14a contains only long-ranging taxa, but the Plenus Marl Formation at its base can be dated regionally as Late Cenomanian.

Zone FCN13 (*Hedbergella brittonensis* Zone).
Top: defined at a downsection influx of abundant planktonic foraminiferids (*Hedbergella* spp.). This Zone corresponds approximately to the *Rotalipora cushmani* Zone of Burnhill and Ramsay (1981).
Assemblage: dominated (> 90%) by planktonic foraminiferids (*Hedbergella delrioensis* and *H. brittonensis*), with subordinate calcareous benthonic foraminiferids and with small proportions of non-calcareous agglutinants.
Subdivisions: divided into two Subzones:
  Subzone FCN13b (*H. brittonensis* Subzone). Top defined as above. *Stensioeina pokornyi* is a supplementary index species, almost restricted to this Subzone, although its LAD is probably just above the top of the Zone.
  Subzone FCN13a (*Lingulogavelinella ciryi inflata* Subzone). Top defined at the LAD of *L. c. inflata*.
Lithostratigraphy: this Zone corresponds approximately to the Hidra Formation (= 'Formation A' of Deegan and Scull, 1977).
Age: Late to Early Cenomanian. *Rotalipora cushmani* and *R. greenhornensis* occur consistently in the upper part of this Zone in the

outer Moray Firth, but become progressively rarer in more northerly locations in the Viking Graben. They indicate a Late to Middle Cenomanian date. The top of the Zone approximately corresponds to the top of Zone FCS13. In the lower part of the Zone, occasional records of *R. appenninica* probably indicate an Early Cenomanian date.

Zone FCN12 (*Sigmoilina antiqua* Zone).
Top: defined at the LAD of *S. antiqua*. This event also corresponds approximately to the LAD of consistent common *Hedbergella planispira*. This Zone corresponds approximately to the *Hedbergella brittonensis* Zone of Burnhill and Ramsay (1981).
Subdivisions: divided into two Subzones:
Subzone FCN12b (*S. antiqua* Subzone). Top defined as above.
Subzone FCN12a (*Osangularia schloenbachi* Subzone). Top defined at the LAD of *O. schloenbachi*.
Assemblage: the microfauna is dominated by planktonic foraminiferids (*Hedbergella* spp.), which comprise ~ 90% of the microfauna. The benthonic foraminiferids are predominantly calcareous taxa and calcareous-cemented agglutinants, but non-calcareous agglutinants comprise up to 10% of the benthonic assemblage.
Lithostratigraphy: the boundary between the Hidra Formation and Rodby Formation is defined at a downsection change from limestone/marl to calcareous claystone. As interpreted in the outer Moray Firth area by Burnhill and Ramsay (1981), this boundary lies just above the LAD of *O. schloenbachi* (within the basal part of FCN12b). A similar lithological break at approximately the same level can be traced northwards throughout the Viking Graben, but in some areas a stratigraphically older boundary is more prominent, at about the top of Zone FCN11, and this has sometimes been taken as the top Rodby Formation. There is a 'black-shale' unit in the uppermost part of the Rodby Formation, which is widely identifiable, and often prominent due to the occurrence of black-stained foraminiferids.
Age: Early Cenomanian to Late Albian. The LAD of consistent *S. antiqua* is within the Early Cenomanian in southern England (Hart *et al*, 1981). The LAD of *O. schloenbachi* is normally within the Middle Albian in the southern North Sea, although very rare specimens are recorded in the Late Albian. Its vertical range is clearly longer in the northern North Sea; unpublished palynological evidence suggests that here the top Albian lies at (or slightly above) the LAD of *O. schloenbachi*. This accords with the lithostratigraphic evidence, as the Rodby/Hidra boundary seems likely to correspond to the upward increase in carbonate content marking the Gault/Chalk and Upper Holland Marl/Texel Marlstone boundaries in the southern North Sea Basin, which both approximate to the base Cenomanian.

Zone FCN11 (*Globigerinelloides bentonensis* Zone).
Top: defined at the LAD of *G. bentonensis*. This Zone corresponds to the upper part of the *G. bentonensis* Zone of Burnhill and Ramsay (1981). *G. bentonensis* occurs commonly in the outer Moray Firth and adjacent areas, but is rarer in the Viking Graben, although still present in most sections. The FAD of *G. bentonensis* is within this Zone.
Assemblage: dominated by planktonic foraminiferids (*Hedbergella* spp. and *G. bentonensis*), which comprise 90% of the assemblage in the upper part of the Zone. *Hedbergella planispira* is abundant through most of the Zone, and is the dominant planktonic foraminiferid below the FAD of *G. bentonensis*.
Lithostratigraphy: this Zone is present within the Rodby Formation.
Age: Late Albian. The LAD of *G. bentonensis* is equated with the corresponding event in the southern North Sea, within the Late Albian, which defines the top of Zone FCS11. The FAD of *G. bentonensis* falls

within the upper part of the *cristatum* ammonite Zone (middle Late Albian) in onshore sequences.

Zone FCN10 (*Falsogaudryinella* sp. 1 Zone).
Top: defined at the LAD of *F.* sp. 1 (= *Uvigerinammina* sp. 1 of Burnhill and Ramsay, 1981). This Zone corresponds to the lower part of the *G. bentonensis* Zone of Burnhill and Ramsay.
Subdivisions: divided into two Subzones:
  Subzone FCN10b (*F.* sp. 1 Subzone). Top defined as above. This Subzone corresponds to the total range of *F.* sp. 1 in the Albian, although it is also recorded at a lower level, in the Late Aptian (in the lower part of Zone FCN8).
  Subzone FCN10a (*Recurvoides* sp. Subzone). Top taken at a down-section increase in relative abundance and diversity of non-calcareous agglutinants, which are often red-stained.
Assemblage: the top of this Zone corresponds to a down-section assemblage change from dominantly planktonic and calcareous benthonic foraminiferids to dominantly non-calcareous agglutinants. Calcareous benthonic foraminiferids occur at some levels, and influxes of planktonic foraminiferids (*Hedbergella planispira*) are seen at one or more levels in the central North Sea. In general, agglutinants become increasingly dominant northwards.
Lithostratigraphy: a down-section lithological change from calcareous claystone to non-calcareous claystone, which corresponds approximately to the top of this Zone, is usually taken to define the top of the Sola Formation. As in the case of the Rodby Formation, definition of this boundary in the absence of biostratigraphic data is often inconsistent.
Age: Late to Middle Albian. Rare, poorly preserved specimens of *Falsogaudryinella* sp. 1 are recorded in the southern North Sea (Netherlands offshore), where they occur consistently in the highest part of Zone FCS10. This level is dated as probably early Late Albian. The '*Recurvoides* assemblage' of Subzone FCN10a, which often occurs in red claystones, and in the outer Moray Firth is accompanied by common red-stained Inoceramid 'prisms', is closely analogous to the non-calcareous agglutinating foraminiferid assemblage characterising Zones FCS10 ('*Inoceramus* Zone') and FCS9 in the Netherlands offshore, and is therefore assigned a Middle Albian date.

Zone FCN9 ('*Globigerinelloides*' *gyroidinaeformis* Zone).
Top: defined at the LAD of '*G.*' *gyroidinaeformis*. This species often occurs abundantly, but is usually restricted to a very thin interval. It is not entirely restricted to this Zone, as it is recorded rarely in the lower part of Zone FCN8. This Zone corresponds to the *G. gyroidinaeformis* Zone of Burnhill and Ramsay (1981).
Assemblage: dominated by calcareous benthonic foraminiferids.
Lithostratigraphy: this Zone falls within the Sola Formation as usually defined. It often corresponds to a more calcareous interval within the Formation (the 'Sola Marl' of L. Riley, pers. comm.)
Age: Middle Albian. As noted by Burnhill and Ramsay (1981), '*G.*' *gyroidinaeformis* has not been recorded from onshore sequences in the North Sea area. It occurs rarely in the southern North Sea, but apparently only in Subzone FCS8a. In southern France, it is recorded mainly from the Middle Albian (Moullade, 1966).

Zone FCN8 (*Verneuilinoides chapmani* Zone).
Top: defined at the LAD of *V. chapmani*.
Assemblage: at the top of the Zone there is a down-section change to an exclusively non-calcareous agglutinating assemblage, which differs from the '*Recurvoides* assemblage' of the Middle Albian by its greater diversity, and is dominated by *Bathysiphon, Glomospira, Glomospirella, Haplophragmoides*,

Pelosina and *Rhabdammina*. Calcareous benthonic foraminiferids often reappear downsection in the lowest part of the Zone.
Lithostratigraphy: this Zone lies within the lower part of the Sola Formation. A very widespread and distinct lithological change, marked by the downsection appearance of a proportion of coarser clastic material, mainly silt and fine sand grade, in the claystones, associated with sphaerosiderite, is recognisable from the central North Sea to the northern Viking Graben, and corresponds to the top of the Zone. Significant 'deepwater sands' occur in this Zone in the outer Moray Firth area.
Age: Early Albian and Late Aptian. The faunal change marking the top of this Zone is comparable to the event defining the top of the *Reophax minuta* Zone (FCS8) in the southern North Sea, which is dated as late Early Albian. This is confirmed by the record of rare and poorly preserved specimens of *V. chapmani* in the upper part of the *R. minuta* Zone in the southern North Sea. Unpublished palynological evidence indicates that the top Aptian falls within the upper half of this Zone.

Zone FCN7 (*Hedbergella infracretacea* Zone).
Top: defined at a downsection influx of abundant planktonic foraminiferids (*Hedbergella infracretacea* and *H. delrioensis*).
Assemblage: the assemblage in the upper part of the Zone is dominated (80–90%) by planktonic foraminiferids, which occur in 'flood' abundance in some areas, as in the southern North Sea. They are accompanied by calcareous benthonic foraminiferids and calcareous-cemented agglutinants. Successive 'green' and 'red' units can be identified in some areas, as in the corresponding Zone FCS7 in the southern North Sea. In the lower part of the Zone, calcareous benthonic taxa predominate.
Lithostratigraphy: The top of this Zone corresponds approximately to the top of the Valhall Formation. In the lower part of the Zone, a 'black shale' unit is present in the outer Moray Firth area. This can probably be correlated with the Fischschiefer of the southern North Sea.
Age: early Late Aptian and Early Aptian. The '*H. infracretacea* influx' can be correlated with the corresponding event in the southern North Sea, which defines the top of Zone FCS7, within the early Late Aptian. The Fischschiefer is Early Aptian in age in its type-area.

Zone FCN6 (*Gavelinella barremiana* Zone).
Top: defined at the LAD of *G. barremiana*.
Assemblage: dominated by calcareous benthonic foraminiferids, associated with non-calcareous agglutinants. Small planktonic foraminiferids (*Hedbergella* sp.) are common at one level.
Lithostratigraphy: this Zone falls within the Valhall Formation, and is represented mainly by claystones.
Age: Late Barremian and late Middle Barremian. In the southern North Sea, the LAD of consistent *G. barremiana*, presumed to equate with the top of this Zone, is at the top of the Barremian in northwest Germany.

Zone FCN5 (*Falsogaudryinella* sp. X Zone).
Top: defined at the LAD of *F.* sp. X (see 'Index Species').
Subdivisions: two Subzones can be differentiated:
  Subzone FCN5b: top defined as above. This Subzone corresponds to the overlap of the ranges of *F.* sp. X and *Gavelinella barremiana*. In the northern North Sea the LAD of *Textularia bettenstaedti* is within the lower part of this Subzone, and forms a useful datum level (in the southern North Sea this species ranges up to the Albian).
  Subzone FCN5a: top defined at the FAD of *G. barremiana*. The FAD of *F.* sp. X is within or at the base of this Subzone.
Assemblage: usually dominated by non-calcareous agglutinants, although significant proportions of calcareous benthonic foraminiferids occur in some areas.

Lithostratigraphy: this Zone falls within the Valhall Formation.
Age: Middle to Early Barremian. The FAD of *G. barremiana* is at the base of the Middle Barremian in the southern North Sea and adjacent onshore areas. A Middle Barremian date for the upper part of this Zone is confirmed by unpublished palynological data.

Zone FCN4 (*Falsogaudryinella moesiana* Zone).
Top: defined at the LAD of consistent *F. moesiana*; occasional specimens are recorded in the overlying Zone. *F. moesiana* occurs commonly in both argillaceous and carbonate facies.
Assemblage: comprises calcareous benthonic foraminiferids and non-calcareous agglutinants.
Lithostratigraphy: this Zone falls within the Valhall Formation. In condensed sequences (e.g. over structural 'highs'), this and underlying Zones are often represented by white or red micritic limestones, frequently with very abundant microfaunas (see above).
Age: Late Hauterivian. This dating is derived mainly from palynological evidence; the Barremian/Hauterivian boundary is not yet accurately defined with respect to the microfaunal zonal boundaries, but occasional specimens of *Gavelinella (Conorotalites) sigmoicosta* are recorded in this Zone, confirming a Late Hauterivian date.

Zone FCN3 (*Marssonella kummi* Zone).
Top: taken at the FAD of *Falsogaudryinella moesiana*. This event corresponds approximately to the LAD of *Marssonella kummi*.
Assemblage: comprises variable proportions of calcareous benthonic foraminiferids and non-calcareous agglutinants.
Lithostratigraphy: this Zone falls within the Valhall Formation.
Age: Early Hauterivian. The LAD of *M. kummi* defines the top of Subzone FCS3b in the southern North Sea, which approximates to the top of the Early Hauterivian.

From this Zone, down to the base of the Valhall Formation, the foraminiferid assemblage often consists almost exclusively of long-ranging taxa (polymorphinids, nodosariids and non-calcareous agglutinants), and is difficult both to subdivide and to correlate with sequences in the southern North Sea.

Zone FCN2 (*Trocholina infragranulata* Zone).
Top: taken at the LAD of *Trocholina infragranulata*. This species is common only in carbonate facies.
Assemblage: dominated by calcareous benthonic foraminiferids.
Lithostratigraphy: this Zone forms the basal part of the Valhall Formation. It is often represented by a white or light grey pyritic marl or limestone unit (the 'basal Valhall Limestone', Utvik Formation or Leek Formation), but in the most highly condensed sequences, this 'basal Valhall' unit may include much younger Zones.
Age: Early Hauterivian?–Valanginian–?Late Ryazanian. The validity of this Zone is uncertain; it may be in part a lateral facies of Zone FCN3. It is clearly defined only in the carbonate facies. Dating is difficult, as only facies-dependent or long-ranging taxa are recorded.

Zone FCN1 (*Haplophragmoides* spp. Zone).
Top: taken at a downsection assemblage change to a low-diversity, exclusively non-calcareous agglutinating assemblage.
Assemblage: comprises only a few species of non-calcareous agglutinants (mainly *Haplophragmoides* and *Trochammina* spp.), usually crushed and difficult to identify.
Lithostratigraphy: the major microfaunal assemblage change defining the top of this Zone corresponds to the boundary between the Valhall Formation and the Kimmeridge Clay Formation. This boundary is interpreted as an upward change from a stagnant anoxic seafloor environment to one with open marine circulation (Rawson and Riley, 1982), and is sharply defined throughout the central and northern North Sea.

Age: Late Ryazanian–?Early Ryazanian. This Zone corresponds approximately to the similarly named Zone (FCS1) of the southern North Sea. Palynological evidence (Rawson and Riley, 1982), reinforced by occasional unpublished ammonite records from cored sections, indicates that the top of this Zone lies within the upper part of the Late Ryazanian. The base of the Zone is left undefined; Early Ryazanian and latest Jurassic (Late Volgian) sediments in Kimmeridge Clay facies in the northern North Sea are almost devoid of benthonic microfossils.

8.6 Index Species

The majority of the species included on the range charts in this chapter (Figs 8.12 and 8.13) are described and illustrated in Chapter 7 (Cretaceous), and others in Chapter 9 (Cenozoic of the North Sea). The remaining species are here tabulated, with brief taxonomic notes. Many Early Cretaceous benthonic species are also illustrated by Bartenstein and Bettenstaedt (1962). It should be noted that the vertical distribution of species in the 'deep-water' facies (FCN Zones) often differs from their range in the 'shelf' biofacies (FCS Zones). Thus some species which occur throughout the basin (e.g. *Falsogaudryinella moesiana*) are not utilised here as index species in both biofacies, and others (e.g. *Textularia bettenstaedti*) appear on both range-charts but with different vertical ranges.

SUBORDER TEXTULARIINA

Ammobaculites irregulariformis Bartenstein and Brand, 1951.
Plate 8.1, Fig. 1 (× 22). Late Valanginian, north Germany (after Bartenstein and Brand). Description: test rather coarsely agglutinated, comprising a small coiled stage followed by a longer uncoiled stage, which is rather thin and usually irregularly curved; chambers are short and not greatly inflated.
Range: Valanginian–Hauterivian in the North Sea. Restricted to the Valanginian in north Germany.

'Clavulina' gaultina Morozova.
Plate 8.1, Fig. 2 (× 24). Albian, U.S.S.R. (after Morozova, see Ellis and Messina, 1940). = *Clavulina gaultina* Morozova, 1948, ?= *Clavulina gabonica* Le Calvez, de Klasz and Brum, 1971. Description: test elongated, moderately finely agglutinated; an arrow-shaped pyramidal triserial section of 7–8 chamber rows with acute angles is followed by a uniserial series of five somewhat inflated chambers, with depressed sutures.
Remarks: if the uniserial stage is broken off, the triserial section resembles a small *Tritaxia*, and can be confused with the triserial stage of *Tritaxia capitosa* (Cushman).
Range: Middle Albian to Early Cenomanian in the North Sea (mainly in the 'deepwater' biofacies of the central and northern North Sea). Not recorded from onshore sections in the North Sea Basin, but widespread elsewhere.

Dorothia retusa (Cushman).
Plate 8.1, Fig. 3, 4 (× 36). 'Late Cretaceous' (probably Palaeocene), Trinidad (after Cushman, 1937). = *Gaudryina retusa* Cushman, 1926 (see Cushman, 1937, p.85). Description: test large (to 2 mm), expanding rapidly from the apex; cross-section subcircular, chambers somewhat inflated; a final biserial stage of 4–6 chambers.
Remarks: Some records of *D. bulletta* (Carsey) may be referrable to this species (e.g. Hanzlikova, 1972).
Range: Late Maastrichtian in the 'Shetland facies' of the northern North Sea.

Dorothia trochoides (Marsson).
Plate 8.1, Fig. 7 (× 64). Campanian, Czechoslovakia (after Hanzlikova, 1972). = *Gaudryina crassa* var. *trochoides* Marsson, 1878 (see Cushman, 1937, p.79). Description: test broadly conical, cross-section circular, chambers not inflated, apertural face flattened.
Range: Campanian in the central and northern North Sea (in the 'Shetland' facies and argillaceous chalks).

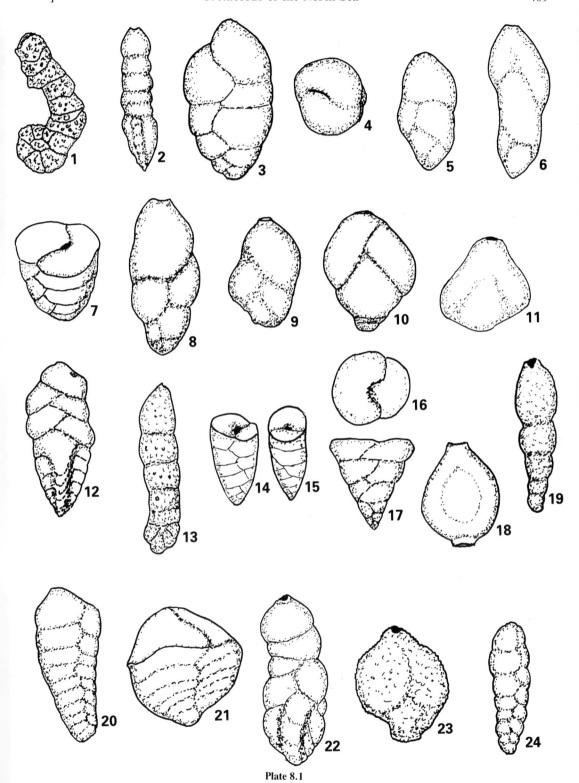

Plate 8.1

Falsogaudryinella alta (Magniez-Jannin).
Plate 8.1, Figs 5, 6 (× 90), Albian, France (after Bartenstein, 1977). = *Uvigerinammina alta* Magniez-Jannin, 1975.
Description: test varying from short and oval to relatively elongated, widest at mid-growth, cross-section rounded, chambers somewhat inflated.
Remarks: The taxonomy of *Falsogaudryinella* ('*Uvigerinammina*' auctt. of the Early Cretaceous) is still incompletely understood; there is a great deal of vertical and horizontal morphological variation, and several of the taxa seen in the North Sea are apparently unrecorded from onshore sections. This genus is nevertheless of considerable biostratigraphic importance, and the most common species are described here. A wide range of *Falsogaudryinella* specimens is illustrated from the Barremian to Albian of DSDP Site 549 (North Atlantic) by Magniez and Sigal (1985).
Range: Middle and Late Albian in the North Sea.

Falsogaudryinella moesiana (Neagu).
Plate 8.1, Figs 8, 9 (× 120), Albian, Rumania (after Bartenstein, 1977). = *Uvigerinammina moesiana* Neagu, 1965.
Description: test varies in shape from short and ovoid to more elongated; cross-section rounded-trigonal; chambers moderately inflated.
Remarks: differs from *F. alta* by its more trigonal cross-section and more inflated chambers, and is typically rather more elongated. The types are from the Albian, but in the North Sea an apparently identical form characterises the Late Hauterivian.
Range: Restricted to the Late Hauterivian (common) and Early Barremian (rare) in the central and northern North Sea 'deep-water' biofacies. In the 'shelf' biofacies of the southern North Sea it also occurs in the Aptian and Albian.

Falsogaudryinella sp. 1.
Plate 8.1, Fig. 10 (× 80), Albian, North Sea (after Burnhill and Ramsay, 1981). = *Uvigerinammina* sp. 1 Burnhill and Ramsay. Description: test very finely agglutinated, elongate-oval, expanding rapidly from apex, widest at the last whorl, sutures sharply defined.
Remarks: this species differs from all others referred to *Falsogaudryinella* by its 'glossy' surface texture and well-defined sutures, but it has the characteristic elliptical slit aperture of *Falsogaudryinella*.
Range: latest Aptian–basal Albian (rare), and early Late Albian (common), in the 'deep-water' biofacies of the central and northern North Sea, Faeroes Trough and Norwegian Sea.

Falsogaudryinella sp. X
Plate 8.1, Fig. 11 (× 100). Early Barremian, North Sea. Description: test small, short, with a very short initial stage and a very swollen last whorl tapering rapidly towards the aperture.
Remarks: this is the smallest species of *Falsogaudryinella*, and can be confused with juvenile *F. moesiana*, but is a distinctive and widespread form, restricted to the Early and early Middle Barremian in the 'deep-water' biofacies of the central and northern North Sea and Norwegian Sea.

Gaudryina dividens Grabert, 1959.
Plate 8.1, Fig. 12 (× 90). Late Aptian, north Germany (after Bartenstein and Bettenstaedt, 1962). Description: test finely agglutinated; initial triserial stage pyramidal, with deeply incut chamber margins; followed sometimes by a biserial stage, which may be longer than the triserial stage. Remarks: this species occurs commonly in the 'shelf' biofacies of the southern North Sea and adjacent onshore areas, but is very rare in the 'deep-water biofacies' of the central and northern North Sea. In the latter area, juvenile or poorly preserved specimens of *Clavulina gaultina* have sometimes been mistakenly identified as *G. dividens*.
Range: Late Aptian to Early Albian.

Haplophragmium inconstans erectum Bartenstein and Brand, 1951.
Plate 8.1, Fig. 13 (× 20), Valanginian, north Germany (after Bartenstein and Brand). Description: test rather-coarsely agglutinated, large and robust; short coiled stage of 3–4 chambers followed by a straight uncoiled uniserial stage of 3–6 slightly inflated chambers.
Range: Late Valanginian to Early Hauterivian in the 'shelf' biofacies of the North Sea and adjacent onshore areas.

Marssonella kummi Zedler, 1961.
Plate 8.1, Figs 14, 15 (× 55), Valanginian, north Germany (after Bartenstein and Bettenstaedt, 1962. = *Marssonella* sp. 1 (aff. *oxycona* Reuss) of Bartenstein and Brand, 1951. Description: test finely agglutinated, elongate, conical, becoming parallel-sided in later stages; chambers flattened, sutures flush, apertural face rather flattened.
Range: Valanginian to Early Hauterivian in the 'deep-water' biofacies of the central and northern North Sea; Valanginian to Early Barremian in the 'shelf' biofacies.

Marssonella subtrochus Bartenstein, 1962.
Plate 1, Figs 16, 17 (× 45). Middle Barremian, Trinidad (after Bartenstein and Bettenstaedt). = *Valvulina* D11 of Hecht, 1938, *Marssonella* cf. *trochus* (d'Orbigny) of Bartenstein and Bettenstaedt, 1962. Description: test finely agglutinated, conical, with final chambers more widely expanded to give a 'flaring' profile, often exaggerated by post-mortem distortion; apertural face flattened.

Pelosina caudata (Montanaro-Gallitelli)
Plate 8.1, Fig. 18 (× 45), Santonian, Czechoslovakia (after Hanzlikova, 1972). = *Saccammina caudata* Montanaro-Gallitelli, 1958. Description: test finely agglutinated, apparently subspherical (but usually flattened by post-mortem compression), somewhat pear-shaped, with short tubular apertures at base and top.
Remarks: this species is almost identical morphologically to detached chambers of *Hormosina ovulum* (Grzybowski).
Range: Santonian?, Campanian and Maastrichtian in the 'Shetland' biofacies. Apparently ranges from Early Cretaceous to Palaeocene elsewhere.

Reophax minuta Tappan, 1940.
Plate 8.1, Fig. 19 (× 35), Early Albian, north Germany (after Bartenstein and Bettenstaedt, 1962). Description: test moderately finely agglutinated, elongated, slightly curved, with six or more convex elliptical chambers (usually flattened by post-mortem compression), rapidly increasing in size.
Range: Middle Barremian to Middle Albian (mainly Late Aptian to Early Albian) in the 'shelf' biofacies of the southern North Sea and adjacent onshore areas.

Spiroplectinata complanata (Reuss).
= *Proroporus complanatus* Reuss, 1860. Description: test thin, elongated; a long flattened biserial stage is followed by a very short uniserial stage of inflated chambers.
Remarks: this species is associated in the Middle and Late Albian with *S. annectens* (Parker and Jones) and *S. bettenstaedti* Grabert, which form the culmination of lineages deriving from *Gaudryina dividens* (see description and illustrations in Bartenstein and Bettenstaedt, 1962).
Range: Middle and Late Albian of the 'shelf' biofacies.

Textularia bettenstaedti Bartenstein and Oertli, 1977.
Plate 8.1, Fig. 20 (× 55), Early Albian, north Germany (after Bartenstein and Oertli). = *Textularia foeda* of authors, non *T. foeda* Reuss, 1846. Description: test moderately finely agglutinated, elongated, narrow, often slightly twisted or curved; with numerous short and slightly inflated chambers; characteristically with dark mineral grains along sutures.
Remarks: *T. foeda* is a similar but morphologically distinct Late Cretaceous species.
Range: Late Hauterivian to Middle Albian (mainly Late Aptian to Early Albian) in the 'shelf' biofacies of offshore and onshore areas; but Valanginian to Middle Barremian in the 'deep-water' biofacies of the central and northern North Sea.

Textularia sp. 1 Burnhill and Ramsay, 1981.
Plate 8.1, Fig. 21 (× 85), Cenomanian, North Sea (after Burnhill and Ramsay). Description: test finely agglutinated, expanding rapidly from the apex, becoming parallel-sided in the final stages, with a diamond-shaped cross-section; chambers flattened, margins acutely rounded; apertural face rather flattened.
Range: Cenomanian of the 'Shetland' biofacies in the central and northern North Sea.

Tritaxia capitosa (Cushman)
Plate 8.1, Fig. 22 (× 45), Maastrichtian, Czechoslovakia (after Hanzlikova, 1972). = *Gaudryinella capitosa* Cushman, 1933, *Pseudogaudryinella capitosa* (Cushman). Description: test wall moderately finely agglutinated, elongated; early triserial and biserial stages rounded triangular in cross-section, final uniserial stage of 2–4 inflated chambers.
Remarks: this species is referred in some unpublished reports to *Tritaxia dubia* (Reuss), but the latter species has a triangular cross-section throughout, and its uniserial stage is reduced to a terminal chamber.
Range: Coniacian to Late Campanian in the 'Shetland' facies of the Viking Graben and adjacent areas (with abundance peaks in the Early Santonian and Late Campanian), and in Late Campanian argillaceous chalks and marls in the Central graben and outer Moray Firth. Rare specimens are recorded in the Chalk facies of the southern North Sea.

Uvigerinammina jankoi Majzon, 1943
Plate 8.1, Fig. 23 (× 90), Turonian?, Poland (after Bartenstein, 1977). Description: test moderately coarsely agglutinated, ovate, triserial, final chambers biserial, chambers inflated; terminal rounded aperture with a short neck.
Range: Turonian and Coniacian of the 'Shetland' biofacies.

Verneuilinoides chapmani (ten Dam).
Plate 8.1, Fig. 24 (× 35), Hauterivian, Netherlands (after ten Dam, 1946). = *Verneuilina chapmani* ten Dam, 1946. Description: test elongated, moderately finely agglutinated, slightly twisted; chambers rather inflated.
Remarks: it is uncertain if *V. chapmani* is the correct name for this species, as the types from the Netherlands are from a different stratigraphic level to the specimens from the North Sea. It is similar to *V. subfiliformis* Bartenstein, but is more robust and less elongated, with fewer chambers. *V. neocomiensis* (Mjatliuk) is smaller and has less inflated chambers. It is similar to *V. borealis* Tappan and *V. tailleuri* Tappan (from the Early Cretaceous of Canada).
Range: Late Aptian and Early Albian in the 'deep-water' facies of the central and northern North Sea. Rare and poorly preserved specimens at the same level in the 'shelf' biofacies.

SUBORDER LAGENINA

Astacolus schloenbachi (Reuss).
Plate 8.2, Figs 1, 2 (× 70), Late Aptian, north Germany (after Bartenstein and Bettenstaedt, 1962). = *Cristellaria schloenbachi* Reuss, 1863. Description: test compressed, curved, with a rounded triangular cross-section; chambers narrow, S-shaped.
Range: Aptian in the 'deep-water' facies of the central and northern North Sea. This species has a more extended range (up to the Middle Albian) in the 'shelf' biofacies.

Lenticulina nodosa (Reuss).
Plate 8.2, Fig. 3 (× 40). Early Hauterivian, north Germany (after Bartenstein, 1974). = *Robulina nodosa* Reuss, 1863. Description: test compressed, with low sutural ribs, weak in umbilical area, but thickening towards the periphery, and swelling over the periphery to give a 'nodose' profile.
Remarks: the typical form (*L. nodosa nodosa* of Bartenstein, 1974) must be differentiated from similar taxa occurring at higher levels in the Early Cretaceous.
Range: typically Early Hauterivian of the 'shelf' biofacies.

Saracenaria spinosa Eichenberg, 1935.
Plate 8.2, Fig. 4 (× 35). Aptian, north Germany (after Bartenstein and Bettenstaedt, 1962). Description: test curved, cross-section triangular, with a deeply concave ventral side bordered by acute keels from which the chamber ends are prolonged as short spines.
Range: Aptian, occurring in both the 'deep-water' and 'shelf' biofacies.

SUBORDER GLOBIGERININA

For descriptions and illustrations of the following species, standard texts on planktonic foraminiferids should be consulted (e.g. Caron, in Bolli *et al.*, 1985; Pessagno, 1967; Robaszynski *et al.*, 1984).

Abathomphalus intermedius (Bolli, 1951)
Dicarinella imbricata (Mornod, 1949/1950)
Globigerinelloides multispina (Lalicker, 1948)
Globigerinelloides prairiehillensis Pessagno, 1967
Globotruncana rugosa (Marie, 1941)
Globotruncanella monmouthensis (Olsson, 1960)
Hedbergella simplex (Morrow, 1934)
Marginotruncana marginata (Reuss, 1845)
Praeglobotruncana gibba Klaus, 1960
Pseudotextularia elegans (Rzehak, 1891)
Racemiguembelina fructicosa (Egger, 1899)

'*Globigerinelloides*' *gyroidinaeformis* Moullade, 1966.
Plate 8.2, Figs 5,6 (× 60). Albian, France (after Moullade). Description: test usually slightly trochospiral, chambers globular, inflated; 4–5 chambers in the last whorl.
Remarks: this species is probably not a planktonic foraminiferid; the slightly trochospiral test seems to preclude assignment to *Globigerinelloides*, and it occurs abundantly but unaccompanied by any other planktonic species.
Range: Early/Middle Albian boundary, very widespread in the 'deep-water' biofacies. Not yet recorded from onshore sections in the North Sea Basin (see Burnhill and Ramsay, 1981).

SUBORDER ROTALIINA

Angulogavelinella bettenstaedti Hofker, 1957
Plate 8.2, Figs 7,8 (× 55). Maastrichtian, Czechoslovakia (after Hanzlikova, 1972). Description: test unequally biconvex, spiral side more convex than umbilical side; periphery acute, keeled; 8–10 chambers in the last whorl, with raised sutures.
Range: Late Campanian to Early Maastrichtian in the Chalk facies.

Conorotalites bartensteini (Bettenstaedt)
Plate 8.2, Fig. 9 (× 70). Middle Barremian, north Germany (after Bartenstein and Bettenstaedt, 1962). = *Globorotalites bartensteini* Bettenstaedt, 1952. Description: test planoconvex, spiral side flat, umbilical side highly convex; cross-section roughly rectangular; periphery acute.
Remarks: this species is the earliest member of the lineage *C. bartensteini–C. intercedens–C. aptiensis* (see illustrations and descriptions in Bartenstein and Bettenstaedt, 1962). Range: Early to Middle Barremian.

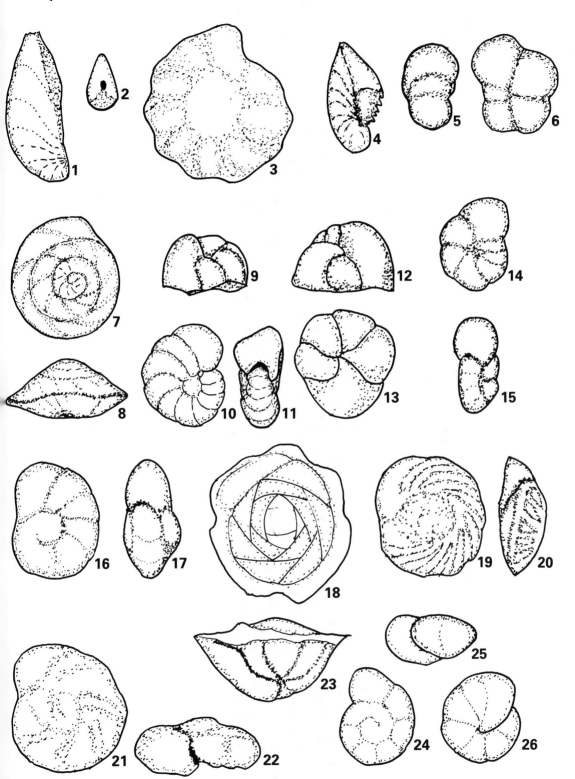

Plate 8.2

Conorotalites intercedens (Bettenstaedt).
Plate 8.2, Figs 12, 13 (× 70). Middle Barremian, north Germany (after Bartenstein and Bettenstaedt, 1962). = *Globorotalites intercedens* Bettenstaedt, 1952. Description: similar to *C. bartensteini*, but umbilical side higher, with semicircular cross-section.
Remarks: *C. aptiensis* (Bettenstaedt) continues the morphological trend, with a subconical profile and a less flattened spiral side.
Range: Middle Barremian to Early Aptian.

Gavelinella rudis (Reuss).
Plate 8.2, Figs 14, 15 (× 65). Middle Albian, Rumania (after Neagu, 1965). = *Rosalina rudis* Reuss, 1863. Description: test unequally biconvex, with spiral side rather flattened, and moderately deep umbilicus on the umbilical side; about seven chambers in the last whorl, becoming increasingly inflated.
Remarks; this species is more inflated than *G. intermedia* (Berthelin), and does not have the low umbilical boss on the spiral side characterising that species.
Range: Aptian and (rarely) Early and Middle Albian.

Gavelinopsis tourainensis Butt, 1966.
Plate 8.2, Figs 16, 17 (× 70). Turonian, France (after Butt). Description: test unequally convex to plano-convex; umbilical side flattened, with a prominent umbilical knob, spiral side with a very prominent umbilical boss; about ten chambers in the last whorl, only the last 4–5 distinct.
Range: Turonian and Coniacian of the 'chalk' biofacies. A similar form is widespread within the mid-Cenomanian in the outer Moray Firth, probably *Gavelinopsis berthelini* (Ten Dam).

Lingulogavelinella ciryi inflata Malapris-Bizouard, 1967.
Plate 8.2, Figs 10, 11 (× 130). Late Albian, France (after Malapris-Bizouard). Description: test inflated, cross-section plano-convex with a flattened spiral side and convex umbilical side; periphery rounded; 9–11 chambers in the last whorl.
Remarks: *L. ciryi ciryi* Malapris-Bizouard (Early to Late Albian) is thinner and has a biconvex cross-section. *L. globosa* (Brotzen) is more inflated, with fewer chambers in the last whorl.
Range: Early and Middle Cenomanian.

Nuttallinella florealis (White).
Plate 8.2, Figs 18, 23 (× 55). Maastrichtian, Czechoslovakia (after Hanzlikova, 1972). = *Gyroidina florealis* White, 1928. Description: test plano-convex, umbilical side convex and inflated; periphery acute, keeled and flanged; flange is wide and rather undulating.
Range: Campanian in the 'Shetland' facies.

Osangularia schloenbachi (Reuss).
Plate 8.2, Figs 19, 20 (× 75). Albian, DSDP Site 258 (after Crittenden, 1983). = *Rotalia schloenbachi* Reuss, 1863. Description: test low trochospiral, biconvex: periphery subacute to acute, with a thick peripheral keel; 8–12 chambers in the last whorl, strongly curved, with raised and thickened sutures.
Remarks: see Crittenden, 1983 for a detailed discussion and synonymy of this species.
Range: Early to Middle Albian, very rarely in the Late Albian, in the 'shelf' biofacies. Mainly in the Middle and Late Albian in the 'deep-water' biofacies. Very widespread worldwide.

Stensioeina granulata humilis Koch, 1977.
Description: test small, compressed, periphery rounded, spiral side low convex; a network of irregular and ill-defined raised ridges on the spiral side.
Remarks: the subspecies of *S. granulata* are important biostratigraphic markers in both the 'Chalk' and 'Shetland' biofacies. Refer to Koch (1977) and Chapter 7 for detailed descriptions and illustrations. *S. g. granulata* (s.s.) differs from the subspecies described here in the presence of a raised granular ornament all over the spiral side.
Range: Middle Turonian.

Stensioeina granulata levis Koch, 1977.
Description: test moderately compressed, biconvex; with weak and ill-defined sutural ribs on the spiral side.
Range: Early Coniacian.

Stensioeina granulata kelleri Koch, 1977.
Description: test compressed, plano-convex; spiral side flattened, with weak sutural ribs.
Range: Late Turonian.

Stensioeina pokornyi Scheibnerova, 1963
Plate 8.2, Figs 21, 22 (× 115). Turonian, north Germany (after Koch, 1977).
Description: test small, compressed; periphery rounded; with weak sutural ribs and an irregular low umbilical boss on the spiral side.
Remarks: interpretation of this species follows Koch (1977) and Burnhill and Ramsay (1982). The illustrations of Scheibnerova suggest that the North Sea Basin taxon may in fact be a different species.
Range: Late Cenomanian to ?Early Turonian in the 'Shetland facies'; in the Early Turonian Chalk facies in north Germany.

Valvulineria gracillima ten Dam, 1947.
Plate 8.2, Figs 24–26 (× 75). Albian, Netherlands (after ten Dam). Description: test small, low trochospiral, inflated, periphery rounded, about seven chambers in the last whorl.
Remarks: this variable species or species-group ranges from Middle Albian to Cenomanian in the North Sea. Detailed study would probably identify biostratigraphically useful subspecies. A related species, common in the late Early Aptian and early Late Aptian (Zone FCN7), is separated as *V.* aff. *gracillima* on the range-chart. It is more inflated, and has fewer chambers (4–5) in the last whorl. It may be referable to *Valvulineria [Gyroidina] kasahstanica* Myatliuk, 1949, from the Aptian and Albian of Kazakhstan. The closely allied species *V. parva* Khan, common in shallow marine facies of the Albian in onshore sections, has a more convex umbilical side, a flatter spiral side, and a more lobate periphery (see Magniez-Jannin, 1975).

ACKNOWLEDGMENTS

Thanks are due to a number of colleagues including Phil Copestake, Brian Fletcher, Colin Harris, Les Riley and Ross Sandman for information and discussion on North Sea and onshore Cretaceous lithostratigraphy and biostratigraphy, and for provision of offprints or preprints of their published work.

REFERENCES

Bailey, H. W. and Hart, M. 1979. Correlation of the Early Senonian in Western Europe using Foraminifera. *Aspekte der kreide Europas.* I.U.G.S. Series A, No. 6, 159–169.

Bailey, H. W., Gale, A. S., Mortimore, R. N., Swiecicki, A. and Wood, C. J. 1983. The Coniacian–Maastrichtian Stages of the United Kingdon, with reference to southern England. *Newsl. Stratigr.*, 12, 29–42.

Bailey, H. W., Gale, A. S., Mortimore, R. N., Swiecicki, A. and Wood, C. J. 1984. Biostratigraphical criteria for the recognition of the Coniacian to Maastrichtian stage boundaries in the Chalk of north-west Europe, with particular reference to southern England. *Bull. geol. Soc. Denmark*, 33 31–39.

Ball, K. C. 1984. The foraminifer *Vaginulinopsis scalariformis* — a new record from the Santonian (Cretaceous) of the southern North Sea Basin. *Rep. Br. Geol. Surv.*, 16, 10–12.

Ball, K. C. 1985. A Foraminiferal Biostratigraphy of the Upper Cretaceous of the Southern North Sea (U.K. Sector). Unpublished Ph.D. Thesis. Plymouth Polytechnic/C.N.A.A.

Ball, K. C. 1986. Unusually large Late Campanian–Early Maastrichtian foraminifera from the Southern North Sea Basin. *J. micropalaeontol.*, 5, 11-17.

Bartenstein, H. 1974. *Lenticulina (Lenticulina) nodosa* (Reuss 1863) and its subspecies — worldwide index foraminifera in the Lower Cretaceous. *Eclogae geol. Helv.*, 67, 539–562, 2 pls.

Bartenstein, H. 1977. *Falsogaudryinella* n.g. (Foraminifera) in the Lower Cretaceous. *N. Jb. Geol. Paläont. Mh.* 1977, H.7, 385–401.

Bartenstein, H. 1978. Paleontological zonation. Parallelisation of the Lower Cretaceous stages in North West Germany with index ammonites and index microfossils. *Erdol Kohle, Erdgas, Petrochemie*, 31, 65-67.

Bartenstein, H. and Bettenstaedt, F. 1962. Marine Unterkreide (Boreal und Tethys). In Simon, W., and Bartenstein, H. (eds.) *Leitfossilien der Mikropaläntologie*, 225–297, pls 33-41. Borntraeger, Berlin.

Bartenstein, H. and Brand, E. 1951. Mikropaläontologische Untersuchungen zur Stratigraphie des nordwestdeutschen Valendis. *Abh. senckenb. naturf. Ges.*, 485, 239-336, 25 pls.

Bartenstein, H. and Kaever, M. 1973. Die Unterkreide von Helgoland und ihre mikropaläntologische gliederung. *Senckenberg. leth.*, 54, 107–164.

Bartenstein, H. and Oertli, H. J. 1977. *Textularia bettenstaedti* n. sp., approved benthonic index foraminifer in the Central European Lower Cretaceous. *N. Jb. Geol. Paläont. Mh.*, 1977, 15–24.

Bertram, H. and Kemper, E. 1982. Die Foraminiferen des spaten Apt und fruhen Alb Nordwestdeutschlands. *Geol. Jb.*, A65, 481–497.

Birkelund, T, Hancock, J. M., Hart, M. B., Rawson, P. F., Remane, J., Robaszynski, F., Schmid, F., and Surlyk, F. 1984. Cretaceous stage boundaries — proposals. *Bull. Geol. Soc. Denmark*, 33, 3–20.

Bolli, H. M., Saunders, J. B. and Perch-Nielsen, K. (eds) 1985. *Plankton Stratigraphy*. Cambridge University Press.

Burnhill, T. J. and Ramsay, W. V. 1981. Mid-Cretaceous Palaeontology and Stratigraphy, Central North Sea. In *Petroleum Geology of the north-west European Continental Shelf*. Institute of Petroleum, London, 245–254.

Butt, A. 1966. Foraminifera of the type Turonian. *Micropaleontology*, 12, 162–182.

Caron, M. 1985. Cretaceous planktic foraminifera. In Bolli, H. M., Saunders, J. B. and Perch-Nielsen, K. (eds.) *Plankton Stratigraphy*. Cambridge University Press, 17–86.

Carter, D. J. and Hart, M. B. 1977. Aspects of mid-Cretaceous stratigraphical micropalaeontology. *Bull. Br. Mus. nat. Hist. (Geol.)*, 29, 1–135, 4 pl.

Crittenden, S. 1982. Lower Cretaceous lithostratigraphy NE of the Sole Pit area in the U.K. southern North Sea. *Journ. Petroleum Geol.*, 5, 191-201.

Crittenden, S. 1983. *Osangularia schloenbachi* (Reuss 1863): an index foraminiferid species from the Middle Albian to Late Aptian of the southern North

Sea. *Neues Jb. Geol. Paläont. Abh.*, **167**, 40–64.

Crittenden, S. 1984a. A preliminary account of Aptian benthic foraminifera from the southern North Sea (U.K. sector). In *Benthos '83, Second International Symposium on Benthic Foraminifera. Bull. Centres Rech. Explor.-Prod. Elf-Aquitaine*, Mem. **6**, 185–190.

Crittenden, S. 1984b. A note on the Early Cretaceous biostratigraphy (Foraminifera) of borehole 49/24-1 (Shell-Esso) in the southern North Sea. *J. micropalaeontol.*, **3**, 1–10.

Crittenden, S. 1987. Aptian lithostratigraphy and biostratigraphy (foraminifera) of block 49 in the southern North Sea (U.K. sector). *J. micropalaeontol.*, **6**, 11–20.

Cushman, J. A. 1937. A Monograph of the Foraminiferal Family Valvulinidae. *Spec. Pub. Cushman Lab. Foraminiferal Res.*, **8**, 1–210, 24 pls.

Dam, A. ten. 1946. Arenaceous foraminifera and Lagenidae from the Neocomian (Lower Cretaceous) of the Netherlands. *J. Paleont.*, **20**, 570–577, pls 87, 88.

Dam, A. ten. 1947. On foraminifera of the Netherlands: No. 9 — Sur quelques espèces nouvelles ou peu connues dans le Crétacé inférieur (Albien) des Pays-Bas. *Geol. en Mijnb.*, n.s., **9**, 25–29.

Deegan, C. E. and Scull, B. J., 1977. A standard lithostratigraphic nomenclature for the central and northern North Sea. *Rep. Inst. Geol. Sci.*, 77/25, 1–36.

Ellis, B. F. and Messina, A. R. 1940 (and later supplements). *Catalogue of Foraminifera*. American Museum of Natural History, New York.

Fletcher, B. N. 1973. The distribution of Lower Cretaceous (Berriasian–Barremian) foraminifera in the Speeton Clay. In Casey, R. and Rawson, P. F. (eds.) The Boreal Lower Cretaceous. *Geol. Jl. Spec. Issue*, **5**, 161–168.

Gallois, R. W. and Morter, A. A. 1982. The stratigraphy of the Gault of East Anglia. *Proc. Geol. Ass.*, **93**, 351–368.

Gradstein, F. M. and Berggren, W. A. 1981. Flysch-type agglutinating foraminifera and the Maastrichtian to Paleogene history of the Labrador and North Seas. *Mar. Micropaleontol.*, **6**, 211–268, 9 pls.

Hancock, J. M. 1984. Cretaceous. In Glennie, K. W. (ed). *Introduction to the Petroleum Geology of the North Sea*. Blackwell, Oxford, 133–150.

Hanzlikova, E. 1972. Carpathian Upper Cretaceous Foraminiferida of Moravia (Turonian–Maastrichtian). *Rozpr. Ustr. ust. geol.* **39**. 160 pp., 40 pl.

Harker, S. D., Gustav, S. H. and Riley, L. A. 1987. Triassic to Cenomanian stratigraphy of the Witch Ground Graben. In Brooks, J. and Glennie, K. (eds) *Petroleum Geology of North West Europe*. Graham & Trotman, London, 809–818.

Hart, M. B. 1973. A correlation of the macrofaunal and microfaunal zonations of the Gault Clay in southeast England. In Casey, R. and Rawson, P. F. (eds) The Boreal Lower Cretaceous. *Geol. Jl. Spec. Issue*, **5**, 267–288.

Hart, M. B., Bailey, H. W., Fletcher, B., Price, R., and Swiecicki, A. 1981. Cretaceous. In Jenkins, D. G. and Murray, J. (eds) *Stratigraphical Atlas of Fossil Foraminifera*, (1st ed). Ellis Horwood, Chichester, 149–27.

Hart, M. B. and Bailey, H. W. 1979. The distribution of Planktonic Foraminiferids in the Mid-Cretaceous of NW Europe. *Aspekte der Kreide Europas*. I.U.G.S Series A, No. 6, 527–542.

Hatton, I. 1986. Geometry of allochthonous Chalk Group members, Central Trough, North Sea. *Mar. Petroleum Geol.*, **3**, 79–98.

Hecht, F. E. 1938. Standard-Gliederung der nordwestdeutschen Unterkreide nach Foraminiferen. *Abh. Senckenb. naturforsch. Ges.*, **443**, 1–42.

Hesjedal, A. and Hamar, G. P., 1983. Late Cretaceous stratigraphy and tectonics of the south-southeastern Norwegian offshore. In J. P. H. Kaasschieter and T. J. A. Reyers (eds) *Petroleum Geology of the southeastern North Sea and the adjacent onshore areas*. *Geologie Mijn.*, **62**, 135–144.

Hiltermann, H. 1962. Oberkreide des nördlichen Mitteleuropa. In *Leitfossilien der Mikropaläontologie*, 299–338, 10 pls, Berlin (Borntraeger).

Hofker, J. 1957. Foraminiferen der Oberkreide von Nordwestdeutschland und Holland. *Beih. geol. Jb.*, **27**, 464 pp.

Jenkyns, H. C. 1986. Pelagic environments. In Reading, H. G. (ed). *Sedimentary Environments and Facies* (2nd edition), 343–398. Blackwell, Oxford.

Jensen, T. F., Holm, L., Frandsen, N. and Michelsen, 0. 1986. Jurassic–Lower Cretaceous lithostratigraphic nomenclature for the Danish Central Trough. *Dansk geol. Unders*, Ser. A, 65 pp.

Kemper, E. 1973a. The Valanginian and Hauterivian stages in northwest Germany. In Casey, R. and Rawson P. F. (eds) The Boreal Lower Cretaceous. *Geol. Jl. Spec. Issue*, **5**, 327–344.

Kemper, E. 1973b. The Aptian and Albian stages in Northwest Germany. In Casey, R. and Rawson, P. F. (eds) The Boreal Lower Cretaceous. *Geol. Jl. Spec. Issue*, **5**, 345–360.

Kemper E., Rawson, P. F. and Thieuloy, J. P. 1981. Ammonites of Tethyan ancestry in the early Lower Cretaceous of north-west Europe. *Palaeontology*, **24**, 251–311, pls 34-47.

Kennedy, W. J. 1985. Integrated biostratigraphy of the Albian to basal Santonian. In R. A. Reyment and P. Bengtson (compilers). *Mid-Cretaceous Events: report on results obtained 1974-1983 by IGCP Project No. 58*, 91–108. Publications from the Palaeontological Institution of the University of Uppsala, Special Volume 5.

Kennedy, W. J. 1987. Late Cretaceous and Early Paleocene Chalk Group sedimentation in the Greater Ekofisk area, North Sea Central Graben. *Bull. Centres Rech. Explor.-Prod. Elf-Aquitaine*, **11**, 91–126, 11 pl.

King, C. 1983. Cainozoic micropalaeontical biostratigraphy of the North Sea. *Rep. Inst. Geol. Sci.*, 82/7, 1–40, 6 pl.

King, C., Bailey, H. W., King, A. D., Meyrick, R. W.

and Roveda, V. L. 1981. North Sea Cenozoic. In Jenkins, D. G. and Murray, J. W. (eds) *Stratigraphical Atlas of Fossil Foraminifera*, (1st ed). Ellis Horwood, Chichester, 294–298.

Koch, W. 1977. Biostratigraphie in der Oberkreide und Taxonomie von Foraminiferen. *Geol. Jb.*, A **38**, 11–123, 17 pl.

Lott, G. K., Ball., K. C. and Wilkinson, I. P. 1985. Mid Cretaceous stratigraphy of a cored borehole in the western part of the central North Sea Basin. *Proc. Yorks. Geol. Soc.*, **45**, 235–248.

Lott, G. K., Fletcher, B. N. and Wilkinson, I. P. 1986. The stratigraphy of the Lower Cretaceous Speeton Clay Formation in a cored borehole off the coast of north-east England. *Proc. Yorks. Geol. Soc.*, **46**, 39–56.

Malapris-Bizouard, M. 1967. Les Lingulogavelinelles de l'Albien inférieur et moyen de l'Aube. *Rev. Micropaleont.*, **10**, 128–150.

Magniez, F. and Sigal, J. 1985. Barremian and Albian Foraminifera, Site 549, Leg 80. In Graciansky, P. C. de, Poag, C. W. *et al.*, *Init. Repts. DSDP*, **80**. U.S. Govt. Printing Office, Washington, 601–628, 9 pl.

Magniez-Jannin, F. 1975. Les Foraminifères de l'Albien de l'Aube: paléontologie, stratigraphie, écologie. *Cah. Paléontol., CNRS*. 351 pp.

Michelsen, O. 1982. Geology of the Danish Central Graben. *Danm. Geol. Unders.* B, **8**, 1–133.

Moullade, M. 1966. Étude stratigraphique et micropaléontologique du Cretacé inférieur de la fosse vocontienne. *Lyons Univ. Fac. Sci. Lab. Géol. Doc.* **15**, 369 pp.

Neagu, T. 1965. Albian foraminifera of the Rumanian Plain. *Micropaleontology* **11**, 1–38, 10 pl.

Nederlandse Aardolie Matschappij B. V. and Rijks Geologische Dienst. 1980. Stratigraphical nomenclature of the Netherlands. *Kon. Ned. Geol. Mijn. Gen. Verh.*, **32**, 1–77.

Pessagno, E. A. 1967. Upper Cretaceous Planktonic Foraminifera from the Western Gulf Coastal Plain. *Palaeontogr. Am.*, **5**, No. 37, 245–445, pl. 48–101.

Price, R. J. 1977. The stratigraphical zonation of the Albian sediments of north-west Europe, as based on foraminifera. *Proc. Geol. Ass.*, **88**, 65–91.

Rasmussen, L. B. 1974. Some geological results from the first five Danish exploration wells in the North Sea. *Danm. geol. Unders.* 3 Rk., **42**, 46 pp.

Rawson, P. F., Curry, D., Dilley, F. C., Hancock, J. M., Kennedy, W. J., Neale, J. W., Wood, C. J. and Worssam, B. C. 1978. A correlation of Cretaceous rocks in the British Isles. *Geol. Soc. Lond., Special Report No.* **9**, 70 pp.

Rawson, P. F. and Mutterlose, J. 1983. Stratigraphy of the Lower B and basal Cement Beds (Barremian) of the Speeton Clay, Yorkshire, England. *Proc. Geol. Ass.*, **94**, 133–146.

Rawson, P. F. and Riley, L. A. 1982. Latest Jurassic–Early Cretaceous events and the 'Late Cimmerian unconformity' in the North Sea area. *Bull. Amer. Assoc. Petr. Geol.*, **66**, 2628–2648.

Rhys, G. H. (compiler). 1974. A proposed standard lithostratigraphic nomenclature for the southern North Sea and an outline structural nomenclature for the whole of the (U.K.) North Sea. *Rep. Inst. Geol. Sci.*, **74/18**, 1–14.

Robaszynski, F., Caron, M., Gonzalez Donoso, J. M. and Wonders, A. A. H. 1984. Atlas of Late Cretaceous Globotruncanids. *Rev. Micropaléont.*, **26**, 145–305. 54 pl.

Schlanger, S. O. and Jenkyns, H. C. 1976. Cretaceous oceanic anoxic events: causes and consequences. *Geol. Mijnb.*, **55**, 435–449.

Seibertz, E. 1979. Probleme der Turon-Gliederung Nordeuropas (Oberkreide) im uberregionalen Vergleich. *Newsl. Stratigr.*, **7**, 166–170.

Ziegler, P. A. 1982. *Geological Atlas of Western and Central Europe*. Shell International Petroleum Maatschappij B.V.

Zimmerle, W. 1979. Lower Cretaceous tuffs in Northwest Germany and their geotectonic significance. *Aspekte der Kreide Europas*. I.U.G.S. Series A, No. 6, 385–402. Stuttgart.

9

Cenozoic of the North Sea

C. King

9.1 INTRODUCTION AND HISTORY OF INVESTIGATIONS

The onshore Cenozoic sediments of northwest Europe have been studied since the earliest days of geology, and their foraminiferid faunas are now mostly well documented. When offshore exploration for hydrocarbons began in the North Sea in the 1960s, it was discovered that these onshore sediments represent the fringes of a major intracontinental Cenozoic basin, the North Sea Basin, filled by up to 3200 m of mainly marine sediments. Subdivision and correlation of these sediments has been achieved largely by the application of microfaunal (foraminiferid) biostratigraphy, but currently with increasing emphasis on the use of dinoflagellates and calcareous nannoplankton.

In the first edition of this 'Atlas' a brief summary of the Cenozoic foraminiferid biostratigraphy of the North Sea was given (King et al., 1981). Subsequently a more detailed account was published (King, 1983) with the introduction of zonal schemes based on benthonic foraminiferids and planktonic microfossils (mainly foraminiferids, but including also larger radiolaria, diatoms and ?tintinnids) and with range charts for 122 foraminiferid species.

Details of other publications on North Sea Cenozoic foraminiferids (up to 1982) are given in King (1983). Information on the microfaunas of the Netherlands offshore sector is incorporated in Letsch and Sissingh (1983). Microfaunal biostratigraphy of wells in the southern Norwegian sector is summarised by Moe (1983), and the Early Eocene–Late Palaeocene foraminiferid biostratigraphy of a cored borehole in the U.K. offshore sector is described briefly by Hughes in Lott et al. (1983).

Feyling-Hanssen and his colleagues (e.g. Feyling-Hanssen, 1982) have established the foraminiferid biostratigraphy of several site investigation boreholes penetrating Quaternary sequences in the central and northern North Sea. Quaternary microfaunas have also been obtained from other shallow boreholes and seabed cores in various parts of the North Sea, but these are outside the scope of the present study.

Gradstein et al. (in press) propose a microfaunal zonal scheme for the central

North Sea based on computer analysis of data from 29 wells, which is closely similar to the scheme of King (1983). Twenty concurrent-range zones are proposed.

It must be emphasised that rigorous biostratigraphic and taxonomic study of the North Sea benthonic foraminiferid assemblages is still at a relatively early stage, and many features of their taxonomy, stratigraphical distribution and environmental controls remain to be elucidated. The planktonic foraminiferids are discussed only briefly here; it is believed that available data are insufficient to update previous studies adequately, except for parts of the Eocene and the late Neogene.

Although data from many wells have been incorporated into this study, details of vertical distribution have often been derived from relatively few wells in which good sample quality, unusually good preservation or exceptionally complete sequences have been available. Such data are still accumulating; therefore the present account is still very much a progress report, and is essentially a revision, extension and updating of King (1983), with emphasis on the benthonic foraminifera.

The scope of this book is restricted primarily to Britain, but this chapter is based on data from all areas of the North Sea, including the Netherlands, German and Norwegian offshore sectors. It must be emphasised that the terms 'offshore' and 'onshore' in this text refer to present-day geography, and are not used in a palaeogeographic context.

9.2 LOCATION OF IMPORTANT COLLECTIONS

Collections of Cenozoic foraminiferids from the North Sea are held by the stratigraphical laboratories of the major oil companies drilling in the area, and by biostratigraphic consultants, but access to these is normally possible only by special arrangement. The British Geological Survey, Keyworth, holds material from offshore boreholes and seabed cores drilled as part of its offshore mapping programme.

9.3 STRATIGRAPHIC DIVISIONS

Lithostratigraphy

Formal lithostratigraphic units have been defined for the central and northern North Sea (U.K. and Norwegian sectors) by Deegan and Scull (1977), for the Danish sector by Michelsen (1982) (mainly numbered units), and for the Netherlands sector by the Nederlanse Aardolie Matschappij (NAM) and the Rijks Geologische Dienst (RGD) (1980). No formal scheme has been proposed for the southern U.K. sector, but here there is a tendency to utilise either the names of adjacent onshore stratigraphic units, or Deegan and Scull's terminology, depending upon the context. The lithostratigraphic schemes are most refined in the Palaeocene and Early Eocene, where study has been concentrated due to the occurrence of important oil and gas reservoirs, and additional subdivisions have been proposed in some areas (e.g. Knox et al. 1981, and Stewart, 1987). Later Eocene and younger sequences are still inadequately documented.

The co-existence of several different lithostratigraphic classifications for the North Sea Cenozoic, limited by arbitrary geographical boundaries, is due to the independent growth of information in each area, but it is becoming increasingly unsatisfactory. It is hoped that a unified terminology will ultimately be adopted for the whole area (and including the adjacent onshore areas). The major units utilised in the U.K and Norwegian offshore sectors are tabulated in Fig. 9.9.

Chronostratigraphy

Calibration of the offshore North Sea sequence with the standard Cenozoic chronostratigraphic/biostratigraphic scale is based mainly on indirect second- or third-order correlations, via onshore sections in the North Sea Basin and offshore North Atlantic sequences, as discussed below. Provisional correlations were proposed by King (1983), and a revised version is given here (Figs 9.4–9.7). Dating of the adjacent onshore sequences in terms of the standard planktonic microfossil zones has itself been established only recently, and is still rather imprecise at some levels, particularly in the Neogene. Therefore calibration of the offshore sequence is liable to further modification, as more information becomes available.

The calibration is discussed in detail below: the basic technique adopted is to establish microfaunal datums in the offshore sequence (mainly zone boundaries) which are correlated to onshore or North Atlantic microfaunal sequences. The available nannofossil or planktonic foraminiferid data from these latter sequences are then used to correlate them with the international standard biostratigraphic zones, and hence to standard stages and systems. Dinoflagellates can be used at some levels to provide additional data, but the dinoflagellate biostratigraphy of onshore sequences is still imperfectly established.

9.4 DEPOSITIONAL HISTORY; PALAEOECOLOGY; FAUNAL ASSOCIATIONS

Depositional history

The North Sea Basin originated during the Permian, as an intracontinental basin generated by lithospheric stretching and subsidence, consequent on the inception and development of the North Atlantic–Arctic rift system. Its history comprises a pre-rift phase, a Mesozoic syn-rift phase characterised by extensional block-faulting, and a post-rift phase (mid-Cretaceous to Recent) (see Ziegler, 1982 for a summary of North Sea Basin history in its northwest European context). The post-rift phase is marked in central areas of the Basin by generally continuous subsidence and sedimentation, mostly without significant differential local tectonic control, except for a tectonic episode in the mid-Palaeocene and long-continued halokinesis in the southern North Sea and adjacent onshore areas. Sediment thicknesses are greatest in the axial areas of the Central Graben and the Viking Graben, and these areas were probably the deepest parts of the Basin throughout its history (Fig. 9.1).

The Early Palaeocene (Danian) is represented mainly by carbonates (chalks and bioclastic limestones), representing a continuation of Late Cretaceous depositional patterns. Following mid-Palaeocene uplift of the basin margins, during the 'Laramide' tectonic episode, associated with reactivation of the graben bounding faults, these carbonates were replaced by a clastic depositional regime. Late Palaeocene and Early Eocene sequences in the outer Moray Firth, northern Central Graben and Viking Graben include large volumes of coarse clastic sediments derived largely from the uplifted Shetland Platform, deposited in deltaic and submarine-fan accumulations. A major episode of igneous activity in the latest Palaeocene and basal Eocene in the Rockall–Faeroes–northwest Scotland–East Greenland area, associated with reactivation of the Rockall–Faeroes Rift, led to the deposition of a series of volcanic ash layers over a wide area of northwest Europe, including the North Sea.

Later Cenozoic sediments are predominantly clays and silts, with marginal–marine and shallow marine sand bodies developing

Fig. 9.1 — Cenozoic sediments of the North Sea Basin and adjacent areas (after Ziegler, 1982). Thickness in metres.

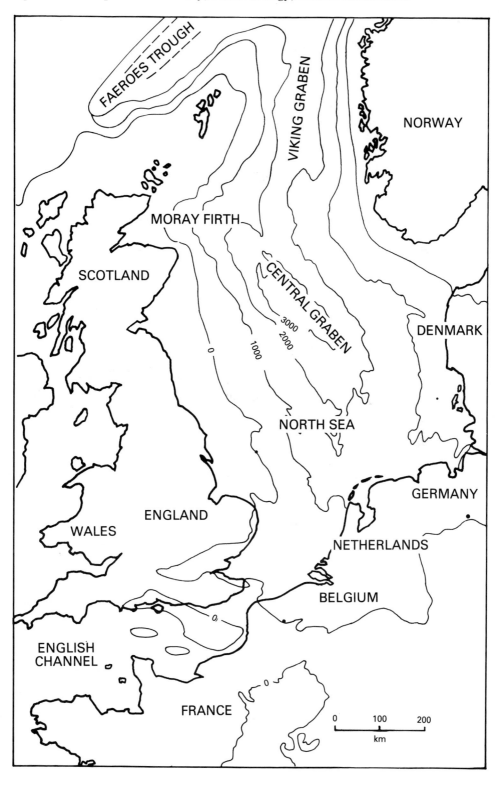

towards the margins of the Basin during regressive phases. Water depths fluctuated continually, largely in response to the relationship between subsidence rates and eustatic sea-level changes, and this is reflected in the foraminiferid biofacies, as discussed below. The microfaunas provide good control on calculation of sedimentation rates and determination of depositional environments.

The North Sea Basin was connected to oceanic areas during the Cenozoic mainly via its northern end, through the Faeroes Trough to the North Atlantic Ocean, with a shallower intermittent connection to the Atlantic through the English Channel, and an eastward indirect connection to the Tethys through north Germany and Poland. The response of the microfaunas to the periodic development of open marine circulation through these connections, at times of high sea level, can be well documented (see below).

Palaeoecology and faunal associations
Palaeoecology of benthonic foraminiferids

King (1983) distinguished three depth-related benthonic foraminiferid biofacies within the North Sea Cenozoic; these can be recognised also in the onshore sequences:

(1) 'Inner sublittoral biofacies'. Characterised by the frequency of elphidiids, polymorphinids, miliolids, attached cibicidids and rotaliids (including larger rotaliids, such as *Nummulites*, in the Eocene). Depths of 0–50 metres are likely for this biofacies, by comparison with data in Murray and Wright (1974).

(2) 'Outer sublittoral-epibathyal biofacies'. This can be subdivided into two depth-related subfacies:

  (2a) Characterised by significant proportions of large nodosariids, valvulineriids, bolivinids, large buliminids, and (in the Neogene) cassidulinids. This biofacies corresponds approximately to the 'Midway type' assemblage of Berggren and Aubert (1975), and indicates water depths of approximately 50–200 m.

  (2b) Characterised by the association of pleurostomellids, stilostomellids, small gyroidinids, *Pullenia*, and *Oridorsalis*; specimens are generally small, and large nodosariids are much rarer than in biofacies 2a. This corresponds approximately to the 'Velasco-type' assemblage of Berggren and Aubert (1975); comparison with similar assemblages elsewhere, and its context in the North Sea, suggests depths of > 200 m.

These subfacies can only be differentiated from Late Palaeocene to Early Miocene.

(3) '*Rhabdammina* biofacies' (non-calcareous bathyal biofacies): dominated by or composed exclusively of non-calcareous 'flysch-type' agglutinating foraminiferids, characterised by the dominance of Astrorhizacea, Ammodiscacea and cyclamminids, with characteristic genera including *Ammodiscus, Glomospira, Bathysiphon, Psammosiphonella* and *Reticulophragmium*. Calcareous foraminiferids may be present, but are usually small, and often affected by postmortem dissolution. This biofacies, where present, always occurs in the deepest parts of the basin, and laterally and vertically replaces biofacies 2b. Depths of deposition of at least 200 m are indicated (see discussion below).

If these biofacies are mapped out for a single 'time slice' they are seen to occupy concentric belts centred on the axis of the North Sea Basin. This reinforces their interpretation as largely depth-controlled assemblages, but does not necessarily imply exact correspondence of biofacies boundaries to specific water depths.

Depth ranges for individual Cenozoic deepwater benthonic species are given by van Morkhoven et al. (1986), but these are usually very broad. Once the relatively broad depth-related biofacies have been defined, foraminiferid distribution within each biofacies probably reflects other environmental factors.

Analysis of Late Pliocene and Pleistocene assemblages is complicated by the probability of transport of microfossils from shallow shelf environments into deeper water by ice-rafting; this may explain the commonly observed association of abundant *Ammonia*, *Elphidium* and *Haynesina* with *Bulimina* and *Uvigerina*. In this same interval arctic, boreal and 'lusitanian' assemblages can be identified by comparison with Recent species distributions, enabling glacial and interglacial episodes to be identified (Feyling-Hanssen, 1982).

Relationship of foraminiferid assemblages to basin history

Non-calcareous agglutinating foraminiferid assemblages

The ratio of agglutinating to calcareous benthonic foraminiferids is often used as an environmental index. However, agglutinants fall readily into two groups inhabiting differing habitats, and (as fossils) forming two distinct assemblages, based on the presence or absence of calcium carbonate as a significant part of the cement in the test wall. This division into two groups roughly follows taxonomic lines, but cuts across some families and even genera. The Valvulinidae and most Textulariidae, Eggerellidae and Dorothiidae fall into the first group, which can be called the CA (calcareous agglutinant) group. These occur in association with calcareous benthonic foraminiferids in 'normal' marine environments (well-oxygenated, with normal salinity, and saturated with respect to $CaCO_3$). The second group (NCA, or non-calcareous agglutinants), includes the majority of the Astrorhizacea, Ammodiscacea, Ammosphaeroidinidae, Cyclammininae, Haplophragmoididae and Trochamminidae. These often dominate assemblages in which calcareous foraminiferids are rare or absent.

NCA assemblages can occur in many environments, from intertidal and lagoonal to abyssal, but in the North Sea they are predominantly deep-water, and fall into the '*Rhabdammina*-biofacies'. The widespread development of 'flysh-type' ('*Rhabdammina*-biofacies') NCA foraminiferid assemblages from Late Palaeocene to mid-Miocene is one of the prominent features of the Cenozoic history of the North Sea Basin. The environmental controls on these assemblages have been discussed by Moorkens (1976) and by Gradstein and Berggren (1981). Restricted water circulation, leading to low oxygen levels and high CO_2 levels at the seafloor, produces a slightly acid reducing environment in which calcium carbonate cannot be precipitated. Thus the benthonic foraminiferal assemblage consists almost exclusively of NCAs. Conditions which are less extreme permit the existence of some 'tolerant' calcareous taxa, but the low $CaCO_3$ content of the sediment often leads to post-mortem dissolution of carbonate. Hence the tests of calcareous foraminiferids occurring in this biofacies are often preserved as pyrite or siderite moulds. The general rarity or absence of planktonic foraminiferids in sediments containing this type of assemblage is a result of this seafloor dissolution of calcareous tests.

A significant factor leading to restricted circulation on this scale is basin compartmentalisation—the development of silled basins whose surface and near-surface waters circulate normally and are well-oxygenated, but with restricted circulation developing below the level of the sill. Such conditions in the North Sea Basin are liable to be accentuated during phases of low eustatic sea level, particularly if sea levels fall low enough to prevent free circulation through the 'sill' separating the North Sea Basin and the Norwegian Sea from the North Atlantic. In Fig. 9.2, the relative extent of NCA assemblages in the North Sea Basin is compared to the proposed worldwide eustatic sea level cycles (TPl to Q2) of Vail *et al.* (1977).

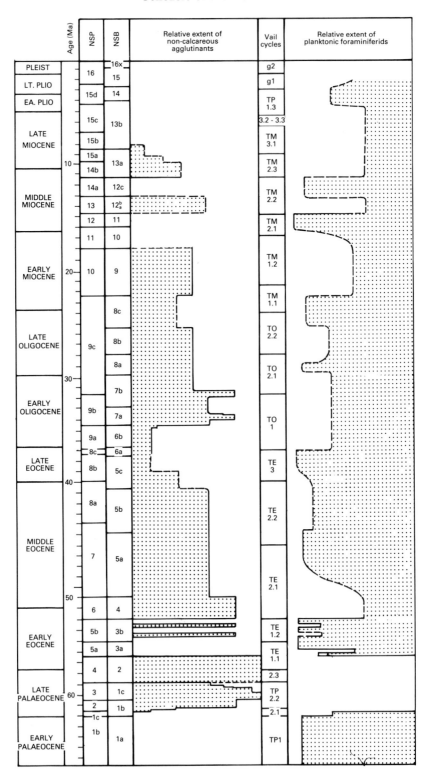

'NCA' assemblages first appear in the Cenozoic in the basal Late Palaeocene, coincident with Vail's low sea level phase defining the base of cycle TP2.1. They extend very widely in the Late Palaeocene (Zone NSA1). The NCA assemblage defining Zone NSA2 is not of '*Rhabdammina*-biofacies' type, and co-exists with rare calcareous benthonic taxa indicating a shallow shelf environment. This episode correlates with the Late Palaeocene sea-level falls of TP2.3 and TE1.1. The '*Rhabdammina*-biofacies' reappears in the Early Eocene, and is particularly widespread in the late Early Eocene and early Middle Eocene, coincident with the major eustatic sea-level fall in the late Early Eocene (base of cycle TE2.1). In the late Middle Eocene, NCA assemblages are rather more restricted geographically, with a notable contraction in the upper part of Zone NSB5. In the Late Eocene and earliest Oligocene their exact distribution is uncertain, due to difficulties in dating and widespread unconformity, but continued contraction is likely, perhaps reflecting the rising sea levels inferred by Vail *et al*. NCA assemblages become very extensive in the mid-Oligocene, but are not obviously related to eustatic sea-level rise. They contracted significantly in the mid-Oligocene, perhaps related to the sea-level fall defining the base of cycle TE2.1, but persist in the Central Graben until the Early Miocene. The general disappearance of diverse NCA assemblages in the early Middle Miocene (dated as ±NN4) may correlate with the major Mid-Miocene high sea-level phase (base of cycle TM2.2). A limited NCA assemblage is again present in the Central Graben in the Late Miocene (±NN8–9) which can be correlated with the successive sea-level falls defining the bases of cycles TM2.3 and TM 3.1.

Planktonic foraminiferids
The North Sea Basin was connected to the Atlantic Ocean during the Cenozoic mainly *via* the Faeroes Trough. A more indirect connection to the Tethys existed at some periods, via eastern Europe and the U.S.S.R. (Ziegler, 1982). Faunal evidence indicates that the northern connection, via the Faeroes Trough, was the most important, and that it provided the main pathway for planktonic faunas, except perhaps in the Late Eocene and Early Oligocene, when the abundant planktonics in the southern North Sea may have arrived via the eastern 'corridor'.

The marked vertical changes in abundance and geographical spread of plankton in the North Sea Basin are thus envisaged as controlled primarily by eustatic sea-level changes; high sea levels permitted influx of North Atlantic water masses with accompanying plankton, whereas low sea levels restricted the entry of plankton. If the relative geographical extent of plankton-rich foraminiferid assemblages in the North Sea Basin (from central basin towards more marginal areas) is plotted against the Cenozoic timescale, a generally inverse relationship to the distribution of NCA assemblages is revealed (Fig. 9.2), and a partial correlation can be made with the high sea level episodes of Vail *et al.*

The disappearance of planktonic forms in the basal Late Palaeocene (from the base of NSP3) coincides with the sea-level fall defining the base of cycle TP2.1. They reappear in the Early Eocene during a major transgressive episode which appears to have no clear equivalence to Vail's cycles. The major fall in planktonic abundance at the base of Zone NSP6 corresponds well to the base of cycle TE2.1, and their reappearance in abundance in the later Middle and Late

Fig. 9.2 — Relative geographical spread (from basin centre towards margin) of non-calcareous agglutinating foraminiferid assemblages, and of foraminiferid assemblages with common planktonic foraminiferids, related to cycles of relative sea-level change (cycles after Vail, 1977, recalibrated to the chronology of Berggren *et al.*, 1985). Each cycle comprises an initial sea-level fall followed by a sea-level rise.

Eocene fits well with the progressive sea-level rises in upper TE2.1 and upper TE3. There is, however, no obvious response to Vail's major mid-Oligocene sea-level fall (base TO2.1).

An important reduction in planktonic abundance in the Early Miocene (NSP9) probably corresponds to the base of cycle TM1.2. The subsequent very wide spread of plankton-rich assemblages in the late Early Miocene and Middle Miocene (NSP12 to NSP14a) corresponds to the highest Neogene sea levels (TM2.1 and TM2.2). Planktonic abundance falls in the Late Miocene, but picks up again in the mid-Pliocene, corresponding to the high sea levels in the TP1 and TP2 cycles.

These correlations must be regarded as tentative, as local and basinal tectonic events clearly played some part in controlling faunal distributions. The inverse correlation between the distribution of NCA benthonic assemblages and plankton-rich microfaunas (Fig. 9.12) is nevertheless very striking, indicating that the Cenozoic history of the North Sea Basin can be clearly subdivided into successive 'open' and 'restricted' episodes. These are more appropriately discussed in detail elsewhere.

Major changes in the calcareous benthonic assemblage.
The basal Late Palaeocene is marked by the entry and spread throughout the Basin of a 'Midway-type' assemblage, including many species which are widespread Palaeocene index fossils, such as *Bulimina midwayensis* and *Stensioeina beccariiformis*. This assemblage rapidly becomes attenuated and disappears later in the Palaeocene. In the latest Palaeocene and basal Eocene, calcareous benthonic assemblages are virtually confined to inner shelf settings. Following successive Early Eocene transgressions, outer shelf and bathyal habitats were repopulated by species from outside the Basin. Many are Palaeocene survivors, but distinctive new species occur, including the *Vaginulinopsis decorata* plexus and *Turrilina brevispira*. Eocene microfaunas have many similarities to those of the North Atlantic and central Europe.

There is a major turnover within the Early Oligocene, at the base of Zone NSB7, consequent on Late Eocene/basal Oligocene regression and regional uplift, followed by the 'Rupelian transgression'. Many species and genera enter the Basin for the first time at this level, including *Cassidulina*, *Rotaliatina*, *Sphaeroidina* and *Valvulineria*. There is a major extinction in the Early Miocene, followed by renewed diversification in the late Early Miocene and Middle Miocene. Affinities of Oligocene and Miocene faunas are predominantly with central Europe, and similarity to Atlantic faunas is by now limited, but the *Asterigerina guerichi* plexus appears to be a distinctively 'boreal' group, recorded also from the Norwegian Sea, Svalbard, the Canadian Atlantic Shelf and the Canadian Arctic.

From the Early Pliocene, cassidulinids and *Elphidium* spp. become common. There is a marked faunal change beginning in the mid-Late Pliocene (from NSB14b, intensifying at the base of Zone NSB15), characterised by the entry of arctic taxa including *Elphidium groenlandicum*, *C. bartletti*, *Islandiella* spp., and *Buccella* spp., coincident with the first appearance of ice-rafted detritus. The Late Pliocene and Pleistocene assemblages are dominated by elphidiids, cassidulinids, and *Islandiella*; alternation of glacial and interglacial episodes is evidenced by the presence of warmer-water taxa (*Uvigerina peregrina*, *Sigmoilopsis celata*, *Bulimina gibba*) at some levels. These assemblages are very similar to Late Pliocene and Pleistocene microfaunas from arctic and sub-arctic areas.

9.5 STRATIGRAPHIC DISTRIBUTION

Foraminiferid biozones
A dual microfaunal zonal scheme was introduced for the North Sea Cenozoic by King (1983). This comprises a parallel series of benthonic (NSB) Zones, based on benthonic

foraminiferids, and planktonic (NSP) Zones, based on planktonic foraminiferids and other planktonic or presumed planktonic microfossils (radiolaria, diatoms and *Bolboforma*) which are recovered in the > 125 μm-size fraction of samples processed for microfaunal analysis. The zonal boundaries are based on selected biostratigraphic events, chiefly 'tops' (LADs, last appearance datums) of individual species, but including events marked by significant microfaunal 'turnover'. Ranges of species are derived largely from ditch cuttings.

The problems involved in determining microfaunal ranges from ditch cuttings have been outlined by King et al. (1981) and King (1983). Emphasis must be placed on 'tops' as they are often the only objectively recognisable events. Nevertheless, other aspects of the vertical distribution of individual taxa and assemblages, including acme levels and FADs (first appearance datums) have been taken into account as far as possible in this study. The erection of 'assemblage-zones' is not practicable, due to the major lateral biofacies changes within the area studied.

The philosophy of erecting a single series of benthonic zones for an area as large as the North Sea Basin, with a wide range of depositional environments, is clearly debatable, and such a scheme involves certain compromises. Nevertheless, in outer sublittoral and upper bathyal environments, the biostratigraphic events defining the NSP and NSB zones are remarkably consistent throughout the Basin, and can also be utilised readily outside the North Sea in the Faeroes Trough and the Norwegian Sea. This probably reflects significant control of both benthonic and planktonic microfaunas within this depth-range by eustatic sea-level changes, as discussed above. In shallow water (inner sublittoral) environments, the NSB zonal scheme is still partially identifiable.

In the restricted deep-water '*Rhabdammina*-biofacies' of the central North Sea, a combination of the dominance of agglutinants and the partial dissolution of calcareous benthonic and planktonic tests limits the use of both the NSB and NSP zonations, especially in the Eocene and Oligocene. Here a separate zonal scheme is proposed, based exclusively on NCA foraminiferids, which can be utilised in restricted deepwater environments where calcareous foraminiferids are rare or absent.

A computer-generated zonal scheme for the central North Sea has recently been proposed by Gradstein et al. (in press), which resembles parts of the present scheme. The limited database utilised by these authors has however produced only a low-resolution scheme, and their biostratigraphic conclusions are not always supported by the larger body of data presented here. Their zonation is compared with the present zonations in Fig 9.10.

Zonations

The zonal schemes are summarised in Fig. 9.3. The NSB and NSP zones were defined by King (1983); minor modifications in definition, and some new subdivisions, are introduced here, and some zones are renamed to take account of taxonomic revisions. The NSA zones are introduced here; they are based on non-calcareous agglutinating foraminiferids, and are intended for use mainly in the non-calcareous '*Rhabdammina*-biofacies' which characterises the central axial areas of the North Sea Basin through the Palaeocene to Miocene interval.

Those NSB zones which were originally defined wholly or partly by NCA (NSB9, NSB4, NSB2 and NSB1) are here redefined in terms of calcareous benthonic taxa (although their boundaries remain unchanged).

The zones are 'interval-zones' in the sense of Hedberg (1976), defined mostly either as the intervals between the 'tops/exits' (LADs) of individual species, or with boundaries defined by changes in the total benthonic or planktonic assemblage. It should be emphasised that the zonal schemes were designed to be useful for biostratigraphers working in the oil industry. They are

Ma	NSP Zones			NSB Zones			NSA Zones	
0	16b	Neo. pachyderma (S)	N.p.	16x	Nonion labradoricum	M.p.C.g.		
	16a	N. pachyderma (D)		15b	Cibicides grossus			
				15a	Cibicidoides pachyderma			
	15d	Globorotalia puncticulata		14b	M. pseudotepida			
				14a				
	15c	Neogloboquadrina atlantica	N. atlasntica		Cib. limbatosuturalis			
				13b	Uvigerina venusta saxonica	U. venusta		
	15b	N. atlantica (D)						
	15a	N. acostaensis					12	(unnamed)
10	14b	(Bolboforma metzmacheri)	B. metz.	13a	Uvigerina pigmea			
	14a	(Bolboforma spiralis)		12c	Uvigerina sp. A			
				12b	Elphidium antoninum	U. pig.		
	13	(Bolboforma clodiusi)		12a	U. semiornata saprophila		11	Martinottiella bradyana
	12	Sphaeroid. disjuncta		11	Asterigerina g. staeschei			
	11	Globorotalia praescitula		10	Uvigerina tenuipustulata			
20	10	(Diatom sp. 4 King 1983)		9	Plectofrondicularia seminuda		10	Spirosigmoilinella sp. A
				8c	Bolivina antiqua		9	Ammodiscus sp. B
	9c	(Diatom sp. 3 King 1983)	Diatom sp. 3	8b	Elphidium subnodosum	B. antiqua		
30				8a	Asterigerina g. guerichi		8	Karreriella chilostoma
				7b	Rotaliatina bulimoides			
	9b	(unnamed)		7a	Cassidulina carapitana		7	Cribrostomoides scitulus
	9a	Globorotalia danvillensis		6b	Uvigerina germanica	U. ger.	6b	Karrerulina conversa
	8c	Globigerinatheka index		6a	Cibicidoides truncanus			
	8b	(unnamed)	G. index	5c	Planulina costata		6a	Amm. macrospira
40	8a	Truncorotaloides spp.		5b	Lenticulina gutticostata			
						P. costata	5	Spiroplectammina aff. spectabilis
	7	Pseudohastigerina spp.		5a	Neoeponides karsteni			
50	6	(Cenosphaera sp.)		4	(unnamed)		4b	Reticulo. amplectens
							4a	Textularia plummerae
	5b	Pseudohastigerina wilcoxensis	S. lin.	3b	Bulimina sp. A	G. hilt.	3	(unnamed)
	5a	Subbotina gr. linaperta		3a	Gaudryina hiltermanni			
	4	(Coscinodiscus sp. 1)		2	Unnamed)		2	Verneuilinoides subeocaenus
60	3	(unnamed)		1c	Bulimina trigonalis		1b	T. ruthvenmurrayi
	2	('Cenodiscus' sp.)		1b	Stensioeina beccariiformis	S. spectabilis	1a	S. spectabilis
	1c	Globorotalia chapmani	G. pseu.					
	1b	Globorotalia pseudobulloides		1a	Tappanina selmensis			
	1a	Globoconusa daubjergensis						

intended to be applicable on a routine basis to sections in which only ditch cuttings are available, and also to be applied throughout the North Sea Basin, as far as practicable. If only restricted areas are considered, or if core samples are available, considerably more detailed schemes can be constructed for local correlations, but because of lateral facies changes, they are not usually applicable to basin-wide correlations.

The datums used to define the NSB and NSP zones and subzones are tabulated in Figs 9.4–9.7. The following comments are intended as supplementary to these figures, and not as a complete description of the zones. Information on the non-foraminiferid component of the NSP zones is included where appropriate. Reference should be made to King (1983) for further details; the data given here are essentially revisions and emendations of that publication. The zones are reviewed here from top to base of the succession, following the order they are normally encountered in offshore borehole studies. Ranges of benthonic foraminiferids (excluding non-calcareous agglutinants) are given in Figs 9.11–9.13. Ranges of non-calcareous agglutinants are given in Fig. 9.14.

Benthonic foraminiferid Zones (NSB Zones)
Zones NSB17 and NSB16 These Zones were based largely on sequences in the central and southern North Sea. Zone NSB16, defined by the occurrence of common *Elphidiella hannai*, cannot usually be differentiated from Zone NSB17 in sections north of 57° N. *E. hannai* is confined to relatively shallow marine environments, and its occurrence now appears for this reason to be significantly diachronous. Subzone NSB16a (*Elphidium oregonense* subzone) is even more restricted in its occurrence, for similar reasons; *E. oregonense* has recently been recorded rarely in the northern North Sea and the Norwegian Sea, but well below the LAD of *Cibicides grossus* (the event defining the top of Zone NSB15). The LAD of *Cibicides grossus* can be identified throughout the North Sea, but its synchronism within this area now also seems suspect. It is probable that in the southern North Sea, where Zone NSB15 is overlain by inner shelf sediments with *Elphidiella hannai*, *C. grossus* became extinct earlier than in the deeper-water environments seen further north. This conclusion is also supported by other biostratigraphic and palaeomagnetic data (see below).

Due to these complications, it is proposed here to introduce a new Zone for the northern North Sea and other areas where Zones NSB16 and NSB17 cannot be identified. This is designated Zone NSB16x (*Nonion labradoricum* Zone). Its top is defined by the LAD of *Nonion labradoricum*, and its base is defined by the LAD of *Cibicides grossus*. The index species is common and widely distributed, and is probably restricted to this zone, which may extend into Holocene sediments.

Borehole sections through the younger Pliocene and Pleistocene of the North Sea have been studied by Feyling-Hanssen and his associates. They recognise a unit with *Cibicides grossus* in the central North Sea, which corresponds to the *C. grossus* zone (NSB15). Overlying units (e.g. Zones 2–9 of Feyling-Hanssen, 1982) contain *Nonion labradoricum*, and can be assigned to the *N. labradoricum* Zone (NSB16x).

Zone NSB15 Subzone NSB15a (*Cibicides pseudoungerianus* Subzone of King 1983), is here renamed the *Cibicidoides pachyderma* Subzone, following taxonomic revision by Van Morkhoven *et al.* (1986).

Fig. 9.3 — Microfaunal zonal schemes for the Cenozoic of the North Sea. NSP = Planktonic zones (non-foraminiferid taxa in brackets). NSB = Benthonic foraminiferid zones. NSA = Non-calcareous agglutinating foraminiferid zones.

Zone NSB14 Here divided into two subzones:
Subzone NSB14b (*Monspeliensina pseudotepida* Subzone). The top of this Subzone is coincident with the top of Zone NSB14.
Subzone NSB14a (*Cibicidoides limbatosuturalis* Subzone). The top of this Subzone is defined by the LAD of *C. limbatosuturalis*. This event is identifiable widely in the central and southern North Sea, but in the northern North Sea Zone NSB14 is usually relatively thin, and the two subzones cannot readily be separated.

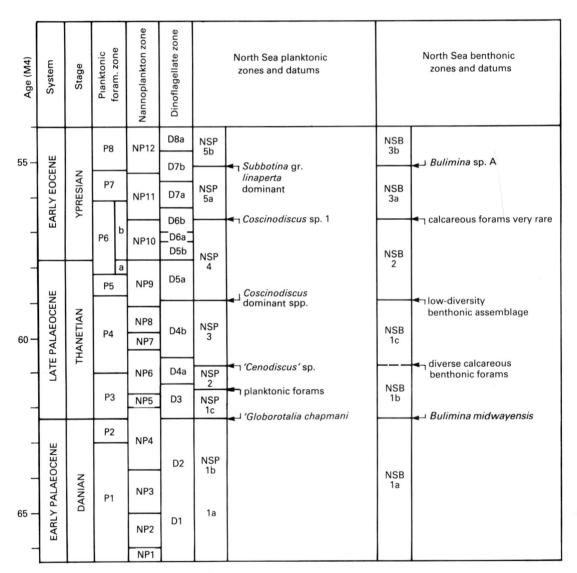

Fig. 9.4 — NSB and NSP zones and microfaunal marker events in the Palaeocene of the North Sea. Geochronology, stage nomenclature, nannoplankton (NP) and planktonic foraminiferid (P) zones are from Berggren *et al.* (1985). Dinoflagellate zones are from Costa and Manum (1988).

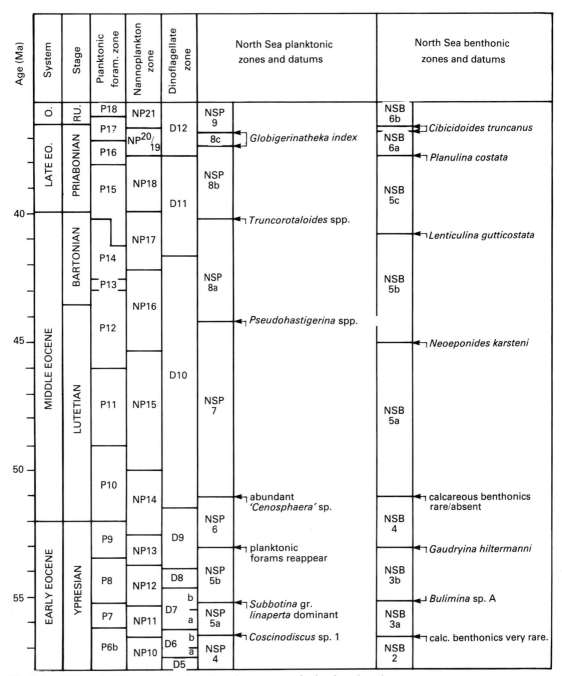

Fig. 9.5 — NSB and NSP zones microfaunal marker events and microfaunal marker events in the Eocene of the North Sea. Explanation as for Fig. 9.4.

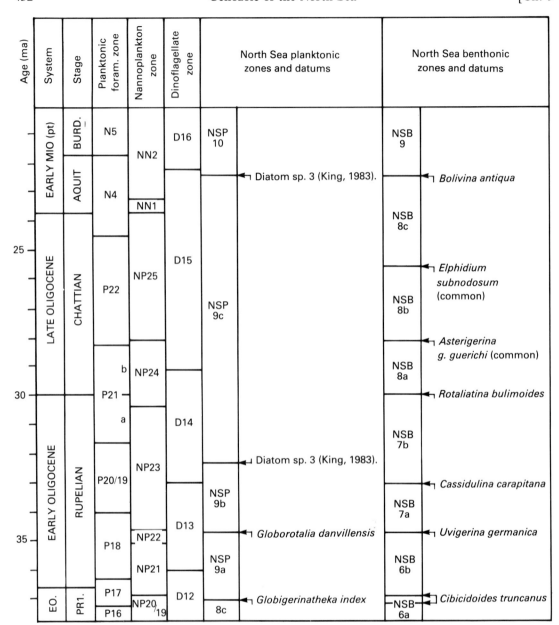

Fig. 9.6 — NSB and NSP zones and microfaunal marker events in the Oligocene and Early Miocene of the North Sea. Explanation as for Fig. 9.4. AQUIT. = Aquitanian, BURD. = Burdigalian.

[Sec. 9.5]   **Stratigraphic Distribution**   433

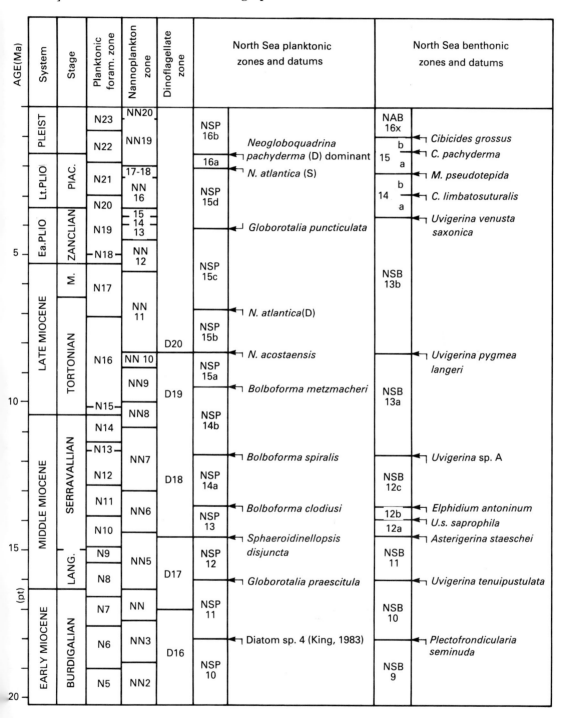

Fig. 9.7 — NSB and NSP zones and microfaunal marker events in the Miocene, Pliocene and Pleistocene of the North Sea. Explanation as for Fig.9.4. S = sinistral, D = dextral, LANG. = LANGHIAN, M. = MESSINIAN, PIAC. = PIACENZIAN.

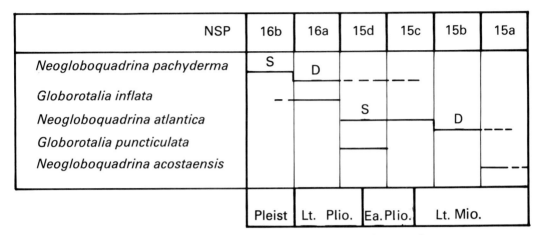

Fig. 9.8 — Ranges of key planktonic foraminiferids in the Late Miocene to Pleistocene of the North Sea. S = sinistral, D = dextral.

Zone NSB13 Subzone NSB13b (*U. venusta* Subzone of King, 1983) is here renamed the *U. v. saxonica* Subzone, following taxonomic re-assessment.

Zone NSB12 This zone was divided by King (1983) into two subzones; here a third subzone is introduced:
Subzone NSB12c (= Subzone NSB12b, *Uvigerina* sp. cf. *hemmooriensis* Subzone, NSB12b, of King, 1983) is here renamed the *Uvigerina* sp. A Subzone (see taxonomic section below).
Subzone NSB12b (*Elphidium antoninum* Subzone) is new; its top is defined at the LAD of *E. antoninum*.
Subzone NSB12a (*Uvigerina semiornata* Subzone of King 1983) is here renamed the *Uvigerina semiornata saprophila* Subzone, due to taxonomic re-assessment.

Zone NSB11 (*Asterigerina staeschei* Zone). Top defined at LAD of *A. staeschei*.

Zone NSB10 (*Uvigerina tenuipustulata* Zone). Top defined at LAD of *U. tenuipustulata*.

Zone NSB9 The *Plectofrondicularia seminuda–Silicosigmoilina* sp. Zone of King (1983), here renamed simply the *P. seminuda* Zone. Top defined at the LAD of *P. seminuda*.

Zone NSB8 The name and the definition of this Zone are both revised here. Both index taxa proposed by King (1983), *Valvulineria mexicana grammensis* and *Fursenkoina schriebersiana*, range above the top of the zone, although their consistent occurrence in association is characteristic of Subzone NSB8c. Zone NSB8 is here renamed the *Bolivina antiqua* Zone; the top of the Zone is here redefined at the highest occurrence (LAD) of *Bolivina antiqua*. This event corresponds closely to the boundary as originally defined; *Almaena osnabrugensis* is a supplementary index species which has its LAD at approximately the same level, but is common only in shallow-water environments.
Subzone NSB8c (*V. m. grammensis–F. schriebersiana* Subzone of King, 1983) is here renamed the *Bolivina antiqua* Subzone.

Zone NSB7 Subzone NSB7b is here subdivided into units 7b2 and 7bi. The top of unit 7bi is defined by the highest occurrence of *Cibicidoides mexicanus*; *Frondicularia budensis* and *Vaginulinopsis* sp. B (*V.* sp. aff. *decorata* of King, 1983) also have their highest occurrence at this level.

Sec. 9.5]   **Stratigraphic Distribution**   435

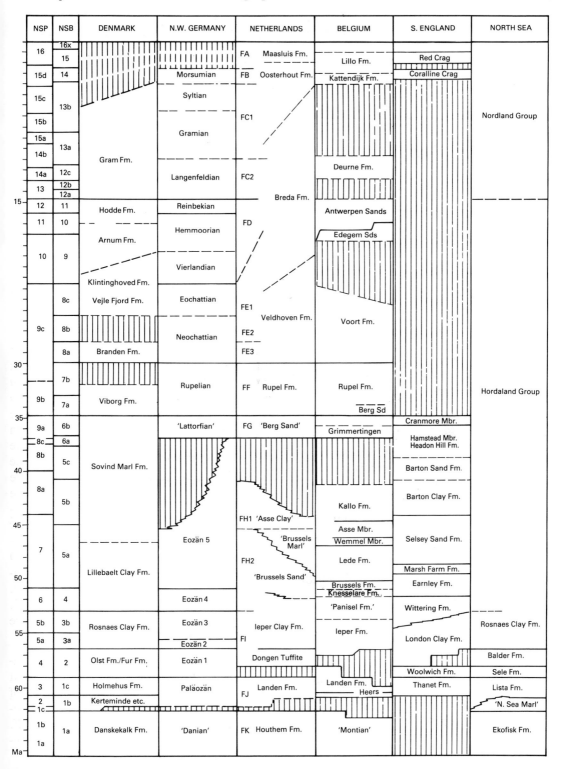

Fig. 9.9 — Correlation of North Sea Basin Cenozoic sedimentary sequences.

436   **Cenozoic of the North Sea**   [Ch. 9

NSP	NSB	NSA	Gradstein et al	Verdenius & van Hinte
14	13	12	*C. teretis*	
13	12		*Globorotalia praescitula-G. zealandica*	*Spirosigmoilinella – Martinottiella communis*
12	11	11		
11	10			
10	9	10		
9c	8	9		*Spirosigmoilinella – Karreriella*
		8		
9b	7	7	*Rotaliatina bulimoides*	
9a	6b	6b		
8	6a	6a	*Gl. index*	
	5c			
	5b	5	*Reticulophragmium amplectens*	*Spiroplectammina spectabilis – Cyclammina amplectens*
7	5a			
6	4	4b		
		4a		
5	3	3	*Subbotina patagonica*	
	2	2	*Coscinodiscus* spp.	
3,4	1c	1b	*Tr. ruthvenmurrayi -R. paupera*	
2	1b	1a		
1c				
1b	1a		*Subbotina pseudobulloides*	
1a				

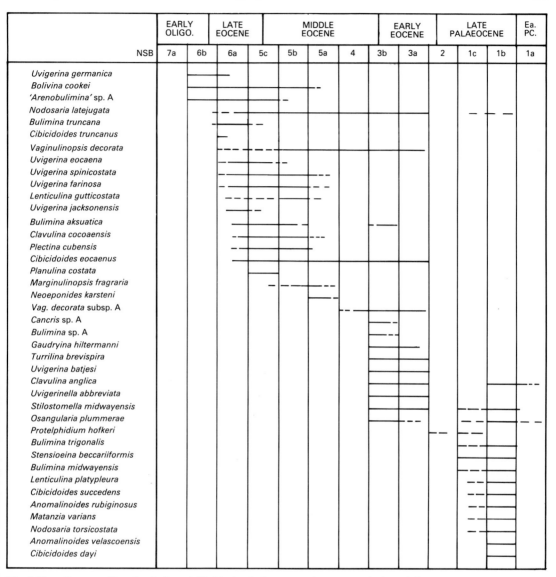

Fig. 9.11 — Ranges of benthonic foraminiferids (excluding non-calcareous agglutinants) in the Palaeocene and Eocene of the North Sea.

Zone NSB6 The *Angulogerina germanica* Zone of King, 1983, here renamed the *Uvigerina germanica* Zone. Subzone NSB6a (*Cibicidoides granulosus* Subzone of King, 1983) is here renamed the *Cibicidoides truncanus* Subzone, following taxonomic revision by van Morkhoven et al. (1986). Subzone NSB6a is here subdivided into units 6a2 and 6a1. The junction between these units is defined at the FAD of *Cibicidoides truncanus*; thus unit 6a2 is defined by the total range of *C. truncanus*.

Fig. 9.10 — Microfaunal zonal schemes of Gradstein et al. (in press) for the central North Sea, and Verdenius and van Hinte (1983) for the Norwegian Sea, compared with the zonal schemes utilised here.

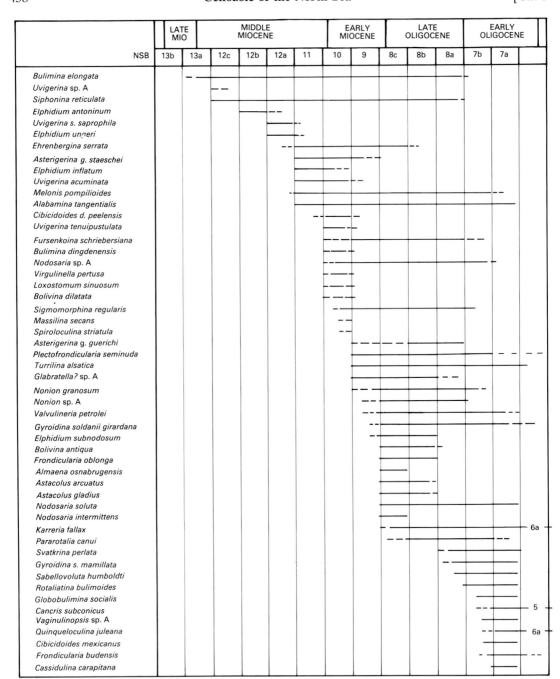

Fig. 9.12 — Ranges of benthonic foraminiferids (excluding non-calcareous agglutinants) in the Early Oligocene to Middle Miocene of the North Sea.

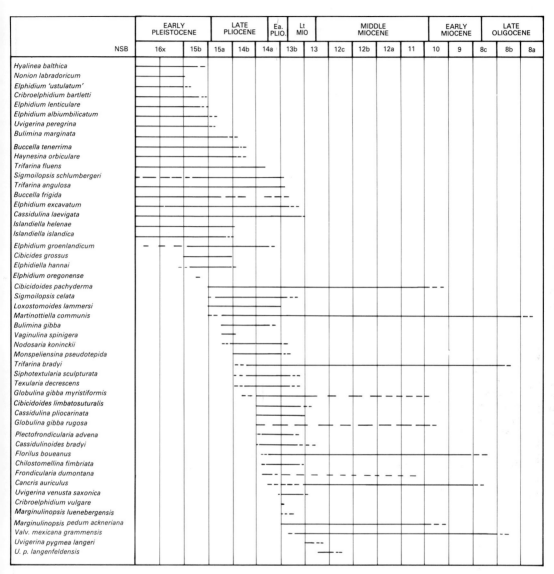

Fig. 9.13 — Ranges of benthonic foraminiferids (excluding non-calcareous agglutinants) in the Late Oligocene to Pleistocene of the North Sea.

Zone NSB5 The *Planulina palmerae* Zone of King (1983), here renamed the *Planulina costata* Zone, following taxonomic revision by Van Morkhoven *et al.* (1986). This zone is subdivided here into three Subzones:

Subzone NSB5c (*Planulina costata* Subzone): defined as the interval between the LAD of *Planulina costata* and the LAD of consistent *Lenticulina gutticostata*.

Subzone NSB5b (*Lenticulina gutticostata* Subzone): defined as the interval between the LAD of consistent *L. gutticostata* and the LAD of *Neoeponides karsteni*.

Subzone NSB5a (*Neoeponides karsteni* Subzone): top defined at the LAD of *N. karsteni*. *N. karsteni* probably does not range be-

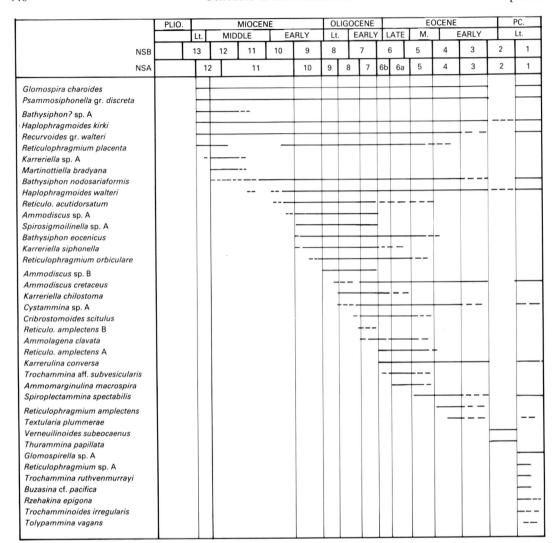

Fig. 9.14 — Ranges of non-calcareous agglutinating foraminiferids in the Cenozoic of the North Sea.

low this Zone. These Subzones are recognisable only in mid- to outer sublittoral environments.

The base of this Zone (except in the central axial areas of the North Sea Basin), is defined at a major faunal turnover, marked by the downsection replacement of a dominantly calcareous benthonic foraminiferid assemblage by a dominantly agglutinating assemblage.

Zone NSB4 (unnamed). The *Cyclammina amplectens* Zone of King (1983). The *Reticulophragmium [C.] amplectens* Zone is now transferred to the NSA zones (see below). The zone is now defined as the interval between the FAD of *Neoeponides karsteni* and the LAD of *Gaudryina hiltermanni*. No suitable index species is available; the zone is characterised by the absence or scarcity of calcareous benthonic and calcareous-cemented agglutinating taxa, compared with the intervals above and beneath. Low-diversity calcareous assemblages occur in this interval in inner sublittoral environments, where it corresponds approximately to the '*Planulina burlingtonensis tendami* Zone' of Letsch and Sissingh (1983).

Zone NSB3 Here subdivided into two subzones:
Subzone NSB3b (*Bulimina* sp. A Subzone). Top defined at the LAD of *Gaudryina hiltermanni*. The top of this Subzone marks the downsection re-occurrence of common calcareous foraminiferids beneath the largely non-calcareous foraminiferid assemblage in Zone NSB4. Species restricted to this subzone include *Bulimina* sp. A and *Cancris* sp. A.
Subzone NSB3a (*Gaudryina hiltermanni* Subzone): top defined by the FAD of *Bulimina* sp. A and *Cancris* sp. A. *G. hiltermanni* ranges throughout Zone NSB3, but there are apparently no common species restricted to Subzone NSB3a. *Turrilina brevispira* is common.

Zone NSB2 This zone is characterised by the virtual absence of calcareous foraminiferids, except in inner sublittoral environments. The benthic microfauna is normally restricted to a few species of poorly preserved agglutinants. *Protelphidium hofkeri* is often the only calcareous foraminiferid recorded in this Zone; it is widespread but generally rare.

Zone NSB1 This zone was defined by King (1983) on the basis of NCA foraminiferid assemblages. It is here redefined in terms of calcareous benthonics, renamed the *Stensioeina beccariiformis* Zone, and divided into three subzones:
Subzone NSB1c (*Bulimina trigonalis* Subzone). The top of this Subzone is not always well-defined: calcareous foraminiferids are often sparse, but typically include *B. trigonalis*, *B. midwayensis*, *Stensioeina beccariiformis* and *Stilostomella midwayensis*. The microfauna is usually dominated by non-calcareous agglutinants, except in shallow marine environments.
Subzone NSB1b (*Stensioeina beccariformis* Subzone). The top of this Subzone is defined by a marked downsection increase in diversity and dominance of calcareous benthonics. This event can be recognised throughout the North Sea, and in central basinal areas is sharply defined, as calcareous taxa are virtually absent in the overlying interval. Species restricted to this subzone include *Cibicidoides dayi* and *Anomalinoides velascoensis*, but these occur only in bathyal environments. Most other species recorded in this Subzone probably range up into NSB1c. *Bulimina midwayensis* and *Stensioeina beccariformis* are usually common.
Subzone NSB1a (*Tappanina selmensis* Subzone of King, 1983). This corresponds to the chalks of the Ekofisk Formation. The definition of the top of the Subzone is here amended to the FAD of *Bulimina midwayensis*.

Planktonic Zones (NSP Zones)
Range-charts for planktonic foraminiferids are given in King (1983). These are not repeated here, but a range-chart of the key index species and coiling variants in Zones NSP15 and NSP16 is given here (Fig. 9.8).

The ranges of many planktonic foraminiferid species in the Cenozoic of the North Sea Basin are shorter than in the adjacent North Atlantic Ocean; this reflects periodic interruptions in oceanic water circulation into the North Sea, due to tectonic or eustatic sea-level changes, as discussed above. Some important planktonic datums (e.g. the FAD of *Pseudohastigerina*) are thus younger in the North Sea Basin than in oceanic areas.

Zone NSP16 This Zone is here subdivided into two Subzones:
Subzone NSP16b (sinistral *N. pachyderma* Subzone)
Subzone NSP16a (dextral *N. pachyderma* Subzone)

The boundary between them is defined at the downsection coiling change from predominantly sinistral to predominantly dextral in *Neogloboquadrina pachyderma*.

Zone NSP15 This Zone is here subdivided into four Subzones:
Subzone NSP15d (*Globorotalia puncticulata* Subzone). Top defined at the LAD of *Neogloboquadrina atlantica*. This corresponds approximately to the LAD of *G. puncticulata*.
Subzone NSP15c (sinistral *Neogloboquadrina atlantica* Subzone). Top defined at the FAD of *G. puncticulata*. Assemblages are dominated by long-ranging taxa and sinistral *N. atlantica*.
Subzone NSP15b (dextral *N. atlantica* Subzone). Top defined by the downsection coiling change from predominantly sinistral to predominantly dextral in *N. atlantica*.
Subzone NSP15a (*Neogloboquadrina acostaensis* Subzone). Top defined by the LAD of *N. acostaensis* (= *N.* gr. *acostaensis/humerosa* of King, 1983).

Zone NSP14–NSP10 Detailed analysis of the planktonic foraminiferids in this interval still remains to be carried out; refinements in the subdivision are anticipated when this is completed. Zone NSP12, the *Sphaeroidinellopsis subdehiscens* Zone of King (1983), is here renamed the *S. disjuncta* Zone, following taxonomic reassessment of the index species.

Zone NSP9 This is here subdivided into three Subzones:
Subzone NSP9c (Diatom sp. 3 Subzone). Top defined at the LAD of Diatom sp.3 (King, 1983).
Subzone NSP9b (unnamed). Top defined at the FAD of Diatom sp. 3. This datum level is not yet accurately calibrated to the NSB or NSA Zones.
Subzone NSP9a (*Globorotalia danvillensis* Subzone). Top defined at the LAD of *Globorotalia danvillensis* (Howe and Wallace). This species is not common in offshore sequences, but occurs more consistently in onshore areas. The Subzone is defined to comprise the planktonic equivalent to Subzone NSB6b.

Zone NSP8 Divided here into three Subzones:
Subzone NSP8c (*Globigerinatheka index* Subzone). Top defined at the LAD of *G. index*. This Subzone corresponds to the total range of *G. index*.
Subzone NSP8b (unnamed Subzone). Top defined at the FAD of *G. index*. This Subzone corresponds to the interval between the FAD of *G. index* and the LAD of *Truncorotaloides* spp.
Subzone NSP8a (*Truncorotaloides* spp. Subzone). Top defined at the LAD of *Truncorotaloides* spp., including *T.? pseudodubius* (Bandy).

Zone NSP7 Top defined at the LAD of consistent *Pseudohastigerina* spp. Rare specimens of *Pseudohastigerina* may occur above this level, in Subzone NSP8a, but although *P. micra* ranges into the Early Oligocene in onshore sections, it seems very rare or absent above NSP8a in the North Sea.

Zone NSP6 Defined at the LAD of abundant large spherical radiolaria (*Cenosphaera* sp.).

Zone NSP5 The *Globigerina* ex gr. *linaperta* Zone of King, 1983, here renamed the *Subbotina* gr. *linaperta* Zone. Here subdivided into two subzones:
Subzone NSP5b (*Pseudohastigerina wilcoxensis* Subzone). Top defined at the downsection reappearance of planktonic foraminiferids beneath the generally calcareous plankton-free interval of Zone NSP6. *P. wilcoxensis* is frequent, but ranges above this zone. The FAD of *P. wilcoxensis* is within the lower part of this subzone.
Subzone NSP5a (*Subbotina* gr. *linaperta* Subzone). Top defined at a change in the planktonic foraminiferid assemblage; below this level *Subbotina* gr. *linaperta* comprises > 80% of the assemblage, while above it is

much less common. Large spherical finely reticulate radiolaria are often abundant.

Zone NSP4 The top of this Zone is defined at the boundary between the overlying interval with common planktonic foraminiferids (Zone NSP5), and the underlying essentially non-calcareous interval, with planktonic microfossils represented mainly by abundant large pyritised or siliceous diatoms (including *Coscinodiscus* spp. 1 and 2). This junction is well-defined throughout the North Sea Basin.

It may be possible to divide this Zone into two Subzones, as in the lower part of the Zone a more diverse diatom assemblage is recorded in some areas, with several morphologically distinctive forms. However, the chronological significance of this assemblage is still under study.

Zone NSP3 The top of this zone is defined at a major basin-wide biostratigraphic event, which corresponds to the boundary between the Sele Formation and the Lista Formation. This marks the base of the '*Coscinodiscus* assemblage' of Zone NSP4. In Zone NSP3, planktonic foraminiferids are absent; occasional small pyritised diatoms are recorded.

Zone NSP2 Top defined at the LAD of large reticulate spherical/lenticular radiolaria, (often referred to the genus *Cenodiscus*, but the lenticular shape is here believed to be due to compression of originally spherical tests). These occur commonly to abundantly throughout Zone NSP2. No planktonic foraminiferids are recorded.

Zone NSP1 Top defined at the 'reappearance' downsection of planktonic foraminiferids, which are apparently entirely absent in Zones NSP2, NSP3 and NSP4. Divided here into three Subzones:
Subzone NSP1c (*Globorotalia chapmani* Subzone) (formerly the upper part of Subzone NSP1b of King, 1983). Top defined as indicated above. Characterised by the association of *Globorotalia pseudobulloides*, *G. compressa*, *Subbotina triloculinoides*, *Globigerina* spp., and with widespread occurrence of *G. chapmani*. Very rare specimens of *G. uncinata* have also been recorded.
Subzone NSP1b (*Globorotalia pseudobulloides* Subzone). Top defined at the FAD of *G. chapmani*.
Subzone NSP1a (*Globoconusa daubjergensis* Subzone). Top defined at the LAD of *G. daubjergensis*.

More detailed microfaunal subdivisions of the Early Palaeocene chalks of the Ekofisk Formation have been proposed (e.g. Stewart, 1987), but this part of the succession is often poorly fossiliferous or with only poorly preserved planktonic foraminiferids.

Agglutinating foraminiferid Zones (NSA Zones)
The zones defined here are based mainly on non-calcareous agglutinants of the '*Rhabdammina*-biofacies' (King, 1983) which characterise restricted bathyal environments in the axial areas of the North Sea Basin (Gradstein and Berggren, 1981). Although most of the taxa in this biofacies are relatively long-ranging, a number of entry and exit events have been identified which appear to be regionally consistent. These zones can be compared with the zonal scheme proposed for the Eocene to Miocene of the Norwegian Sea (Verdenius and van Hinte, 1983). Satisfactory comparison with the ranges of these taxa in other areas is difficult, due to inconsistency in identification within this group. Accurate calibration of these zones with the NSB zones is not completely established, due to the scarcity of calcareous fossils in areas where non-calcareous agglutinating assemblages are developed, especially in the Oligocene and Eocene. Calibration has been established mainly from sections on the fringes of the '*Rhabdammina*-biofacies', where 'mixed' CA and NCA assemblages occur. The ranges of selected NCA are tabulated on Fig. 9.14.

Zone NSA12 (un-named Zone). This is the youngest significant assemblage of non-calcareous agglutinants in the North Sea, and is characterised mainly by long-ranging species. It corresponds approximately to Subzone NSB13a.

Zone NSA11 (*Martinottiella bradyana* Zone). Top defined at the LAD of the index species. The top of this zone corresponds approximately to the top of Zone NSB12.

Zone NSA10 (*Spirosigmoilinella* sp. A Zone). Top defined at the LAD of the index species. This zone marks a downsection diversification of the NCA assemblage in the central North Sea, with the highest occurrence of *Ammodiscus* sp. A. The top of this zone corresponds closely to the top of Zone NSB9.

Zone NSA9 (*Ammodiscus* sp. B Zone). Top defined at the LAD of the index species. In the central North Sea, this event marks a further significant downsection increase in abundance and diversity of the agglutinating assemblage, which corresponds approximately to the top of Zone NSB9.

Zone NSA8 (*Karreriella chilostoma* Zone). Top defined at the LAD of the index species, which corresponds approximately to the middle of Subzone NSB8a.

Zone NSA7 (*Cribrostomoides scitulus* Zone). Top defined at the LAD of consistent *C. scitulus*; only rare specimens are present above this level. The top of this zone corresponds to a level within the middle of Zone NSB7. The top of the Zone is defined at the FAD of *Spirosigmoilinella* sp. A.

Zone NSA6 (*Karrerulina conversa* Zone). *K. conversa* is common, but ranges up to the lowest part of the overlying zone. It is divided into two subzones:
Subzone NSA6b (*K. conversa* Subzone)
Subzone NSA6a (*Ammomarginulina macrospira* Subzone). Top defined at the LAD of *A. macrospira*. This unit is given only the status of a subzone, as (in the sections studied by the present author) it is uncertain if the LAD of *A. macrospira* is sufficiently well-defined to be utilised as a zonal marker event.

The top of Zone NSA6 presumably corresponds to the top of Zone NSB6; the NSA6b/6a subzone junction cannot yet be accurately correlated with the calcareous microfaunal sequence.

Zone NSA5 (*Spiroplectammina* aff. *spectabilis* Zone). Top defined at the LAD of *S*. aff. *spectabilis*. This event probably lies within Subzone NSB5c, but its exact correlation is still uncertain. *S*. aff. *spectabilis* occurs consistently through the upper part of the Zone; the lower part is usually poorly fossiliferous, and contains only a rather sparse and restricted assemblage of agglutinants.

Zone NSA4 (*Reticulophragmium amplectens* Zone) (= *Cyclammina amplectens* Zone of King, 1983). Top defined at the highest consistent occurrence of the index species (see the description of the index species for its discrimination from similar taxa occurring at other levels). This event corresponds to the top of Zone NSB4. Divided into two subzones:
Subzone NSA4b (*R. amplectens* Subzone). Top defined as above.
Subzone NSA4a (*Textularia plummerae* Subzone). Top defined at the highest occurrence of consistent *T. plummerae*.
The validity of this subdivision is uncertain, as forms superficially similar to *T. plummerae* occur at higher levels, but it is certainly most common in Subzone NSA4a. The base of the zone probably marks the FAD of *Reticulophragmium amplectens*.

Zone NSA3 (unnamed Zone). The top of this Zone is tentatively defined at the FAD of *Reticulophragmium amplectens*. It corresponds to Zone NSB3, and is characterised by a low-diversity assemblage, composed almost exclusively of long-ranging taxa.

Zone NSA2 (*Verneuilinoides subeocaenus* Zone). The boundaries of this Zone correspond to those of Zone NSB2. '*Rhabdammina*-biofacies' agglutinants are very rare or absent, and the sparse and generally poorly preserved NCA assemblage comprises mainly indeterminate crushed *Haplophragmoides/Trochammina*. *V. subeocaenus* and *Thurammina papillata* are widespread, especially in the southern North Sea, and are probably restricted to this zone. A more diverse agglutinating foraminiferid assemblage occurs in some areas in the lowest part of the Zone.

Zone NSA1 (*Spiroplectammina spectabilis* Zone). This is equivalent to Zone NSB1b. Two subzones are defined here:
Subzone NSA1b (*Trochammina ruthvenmurrayi* Subzone). Top defined by the reappearance downsection of a diverse and abundant agglutinating assemblage of the '*Rhabdammina*-biofacies'. This event is recognisable basin-wide. The highest occurrence of 'typical' *Spiroplectammina spectabilis* occurs at this level. *T. ruthvenmurrayi* and *Reticulophragmium* sp. A are probably restricted to this subzone.
Subzone NSA1a (*S. spectabilis* Subzone). Top defined at the FAD of *T. ruthvenmurrayi*. Characterised by the abundance of *S. spectabilis* at some levels, with a lower diversity assemblage than in NSA1b. The base is defined by the earliest appearance of the '*Rhabdammina*-biofacies' in the Cenozoic of the North Sea. This corresponds approximately to the base of Subzone NSB1b. NCAs are very rare in the underlying Ekofisk Formation, and this is not included in the zonal scheme.

Relationship of microfaunal zones to named lithostratigraphic units
The chalks of the Ekofisk Formation correspond to Subzones NSP1a and b, and NSB1a. The overlying marl unit (variously referred to as the North Sea Marl, 'Danian Marl', or 'Maureen equivalent') corresponds approximately to Subzones NSP1c, NSB1b and NSA1b. Subzone NSP2 is a thin Zone which corresponds to the transition from this marl unit to the overlying non-calcareous claystones (Lista Formation). In the Central Graben, the Maureen Formation falls within NSP1c and NSP2. The top of NSP2 corresponds approximately to the 'Thanetian ash marker' (Phase 1 pyroclastic sediments of Knox and Morton, 1983). The Lista Formation is equivalent to NSP3, NSB1c and NSA1b. The boundary between the Lista Formation and the overlying Sele Formation corresponds to the base of NSP4, NSB2 and NSA2. All these Zones are approximately co-extensive, and correspond to the Sele and Balder Formations. The lowest unit of the Hordaland Group is usually a red claystone which corresponds to NSP5, NSB3 and NSA2. No detailed lithostratigraphic classification has yet been formalised for the overlying sediments.

Calibration of the North Sea zonal schemes with the standard stratigraphic scale
Calibration of the North Sea microfaunal zones with the chrono- and biostratigraphic scales of Hailwood *et al.* (1979) and Hardenbol and Berggren (1978) was attempted by King (1983). Sufficient data have subsequently been accumulated to enable significant improvement of the precision of the calibration.

The chronostratigraphic and biostratigraphic scales used here are those of Berggren *et al.* (1985a) (with the substitution of Thanetian for Selandian) and Berggren *et al.* (1985b). These are based on comprehensive syntheses of the available data, and incorporate significant input from magnetostratigraphic studies. Few first-order correlations with these scales are as yet possible; the majority of the calibrations are made via dinoflagellate, planktonic foraminiferid and nannoplankton data from onshore sequences in the North Sea Basin, which usually involve second- or third-order correlations. However, most of the correlations are considered to be accurate to about one nannoplankton zone, and at some levels are considerably more refined. The proposed correlations are summarised in Figs 9.4–9.7. These are derived from data summarised in King (1983), and from additional data discussed below.

Pleistocene, Pliocene and Late Miocene
The *Elphidium oregonense* Subzone (NSB16a) can be identified in the Netherlands (onshore), where it correlates with the Praetiglian pollen stage. This is dated via palaeomagnetic correlations at ±2.00 Ma (as summarised in Jenkins *et al.*, 1985, Fig. 2). This correlation implies significant diachronism of the LAD of *Cibicides grossus* (top NSB15) which is well above the FAD of sinistral *N. pachyderma* (1.6–1.75 Ma) in the northern North Sea (see below, and discussion of NSB zones above).

Planktonic foraminiferid events in the Late Miocene to Pleistocene interval are here revised, following the work of Weaver and Clement (1986) on North Atlantic planktonic foraminiferid datums and their calibration with the palaeomagnetic scale. Several of these datums can be identified in the North Sea, and four of them have been used to define NSP Zone or Subzone boundaries:

(a) FAD of sinistral *Neogloboquadrina pachyderma* (base NSP16b): dated as 1.65–1.75 Ma in the North Atlantic.
(b) FAD of *Globorotalia inflata* (close to base NSP16a): dated as 2.1 Ma in the North Atlantic.
(c) LAD of *N. atlantica* (base NSP16a): dated as 2.3 Ma in the North Atlantic.
(d) FAD of *Globorotalia puncticulata* (base NSP15c): dated as 4.16 Ma in the North Atlantic.
(e) Sinistral to dextral coiling transition in *N. atlantica* (base NSP15b): dated as 6.5–7 Ma in the North Atlantic.

These correlations imply a significantly younger chronological age for these Zones than previously deduced (King, 1983), and hence for the onshore units correlated with them (by second-order correlation via the NSB zones) (see Fig. 9.9). They indicate that the Pliocene/Pleistocene boundary corresponds to the base of Subzone NSP16b, the Late/Early Pliocene boundary lies within Subzone NSP15c and the Early Pliocene/Late Miocene boundary lies within Subzone NSP15b.

Middle and Early Miocene
The interval from c.8 Ma to c.15 Ma in the North Sea is still very imprecisely calibrated with the international biostratigraphic scale. Unpublished biostratigraphic data suggests that the Serravallian/Tortonian boundary approximately corresponds to the Langenfeldian/Gramian boundary in northwest Germany (itself probably approximating to the NSB13a/13b boundary).

Following recent taxonomic revision, the index species of Zone NSP12, originally identified as *Sphaeroidinellopsis subdehiscens*, is here redetermined as *S. disjuncta* (Finlay). This species ranges in oceanic environments through Zones N9 to N16 (Bolli and Saunders, 1985). This redetermination removes the discrepancy between nannofossil and planktonic foraminiferid dating for this level discussed by King (1983, p.19). The LAD of *Globorotalia praescitula* in the North Sea (defining the top of Zone NSP11) is probably older than its extinction level in open oceanic environments, and does not necessarily correspond to the top of Zone N11 as suggested by King (1983).

The correlation of the Early Miocene in the North Sea Basin is still uncertain. This is partly due to the unavailability of detailed stratigraphic data for this interval in onshore sequences. The LAD of *Plectofrondicularia seminuda*, which defines the top of Zone NSB9, is within the lower Hemmoorian (Behrendorfian) in Germany (±NN3) (Spiegler, 1974); this is the only currently available calibration point for this event.

Oligocene

The NSB9/NSB8 boundary is not readily identifiable in published onshore sequences, as only generalised range charts are available. Data from northwest Germany (Spiegler, 1974) indicates that several foraminiferid taxa which do not range above Zone NSB8, including *Almaena osnabrugensis* and *Bolivina antiqua*, are restricted to Neochattian and older sediments. This suggests that the top of NSB8 may correspond to the Vierlandian/Chattian boundary. This boundary cannot be accurately placed within the standard biostratigraphic scale, but is probably within the interval NP25-NN2. The occurrence of *Globigerinoides* gr. *trilobus* within Subzone NSB8c in the outer Moray Firth area indicates a Miocene date (N4 or younger) for at least part of this Subzone. The Oligocene/Miocene boundary is therefore tentatively placed within NSB8c.

The basal part of the Boom Clay/Rupel Clay, with *Cassidulina carapitana* (= Subzone NSB7a), is assigned to the lower part of Zone NP23 in Belgium (Steurbaut, 1986) and Germany (von Benedek and Müller, 1976). In the Piepenhagen section, near Bunde, north Germany, the 'basal Rupel transgression' (= base NSB7) lies within the very short zone NP22 (von Benedek and Müller, 1976), and this is probably also true in Denmark.

Onshore stratigraphic units assigned to Subzone NSB6b (Grimmertingen Sands, Silberberg Beds, and Upper Hamstead Beds) are all correlated with NP21–NP20 (King, 1983; Steurbaut, 1986), and thus this Subzone is essentially Early Oligocene in age.

Eocene

Unpublished data from Denmark (Heilmann-Clausen, King and Thomsen, in preparation) indicates that the ranges of *Cibicidoides truncanus* (defining NSB6a) and *Globigerinatheka index* (defining Subzone NSP8c) lie within NP19/20, and are thus younger than originally deduced from indirect evidence (King, 1983). The Danish data also indicate that the LAD of *Planulina costata* (top NSB5) corresponds approximately to the NP18/NP19 boundary.

The LAD of *Truncorotaloides* spp. (defining top NSP8a) corresponds to the top Middle Eocene in the standard biostratigraphic scale (top P14 in 1969 definition, see Berggren et al., 1985a).

The LAD of consistent *Pseudohastigerina* (top NSP7), (excluding Oligocene records), and the slightly older LAD of *Neoeponides karsteni* (top NSB5a), both fall within the lower part of NP16 in Belgium (King, in press, b) and, so far as can be determined from fragmentary data, at the same level in northwest Germany and Denmark.

Calibration of Zones NSP7-5 and NSB5a-3 to the NP and P Zones is based on data in King (1983) and on more recent foraminiferid and nannoplankton data from Denmark (Heilmann-Clausen et al., 1985; King, in preparation), and Belgium (Hooyberghs, 1983; Steurbaut, in press; King, in press, b).

The continuing difficulty in placing the 'ash series' (Balder and Sele Formations and their onshore equivalents, corresponding to Zone NSB2) in an inter-regional biostratigraphic context is due to the complete absence of calcareous plankton in these units. However, they contain several distinctive dinoflagellate assemblages, which can be partially identified outside the North Sea Basin.

The NSB2/NSB3 junction (= NSP4/NSP5 junction) corresponds approximately to the very short dinoflagellate Zone D6a (*Wetzeliella astra* Zone) and falls at the top of the Balder Formation. The *W. astra* Zone was correlated approximately with the NP9/NP10 boundary by study of DSDP sites in the North Atlantic (Costa and Müller, 1978). This correlation has been questioned by Knox (1984) on the basis of available North Atlantic nannoplankton dates and volcanic ash correlations between the North Sea Basin and the Goban Spur (DSDP Leg 80). He proposes that the NP9/NP10 boundary (and hence the Palaeocene/ Eocene boundary) falls at the base of the Sele Formation. However, a new review of the data (King, in press, a), suggests that the NP9/NP10 boundary corresponds approximately to the Sele Formation/Balder Formation boundary, and this correlation is followed here.

Palaeocene
The base of the Sele Formation (= base of NSB2/NSP4) corresponds to the base of dinoflagellate Zone D5 (*Apectodinium hyperacanthum* Zone). In onshore sequences, nannoplankton diagnostic of Zone NP9 have been recorded from beds falling within the lowest part of the *A. hyperacanthum* Zone. The upper part of the underlying dinoflagellate Zone D4 (*Alisocysta margarita* Zone) can be correlated with NP8. It therefore seems probable that the NP8/NP9 boundary corresponds approximately to the NSB1/NSB2 boundary (the base of the Sele Formation).

Zone NSP3 can be identified in the Holmehus Clay Formation of Denmark (King, unpublished data), which can be correlated using dinoflagellates with the lower part of the Thanet Beds of England (Heilmann-Clausen, 1985). This level has recently been assigned to NP6 or NP7 (Siesser *et al.*, 1987). Zone NSP2 can be identified in the underlying 'Palaeocene grey clay' and 'Kerteminde Marl' of Denmark, whose base is dated as upper NP4 or lower NP5 (Thomsen and Heilmann-Clausen, 1983). *Globorotalia chapmani* (which ranges from upper P3 to P6) is recorded in NSP2 in the North Sea, and also in Denmark. Rare specimens of *G. angulata* (P3 to lower P4) and *G. uncinata* (P2 to lower P3) have been recorded from this interval in the North Sea. This indicates that NSP1c corresponds approximately to P3.

Correlation with other areas
Onshore areas of the North Sea Basin
Correlations with onshore sequences in the North Sea Basin were summarised by King (1983). A modified version of the correlation is given here (Fig. 9.9), incorporating some additional information. It is proposed that the North Sea zonations utilised here should be extended to onshore areas where middle to outer shelf and bathyal marine facies exist.

Letsch and Sissingh (1983) have introduced a foraminiferid zonal scheme for the Netherlands (onshore and offshore), but this does not represent a significant advance over the scheme of Doppert (1975). The stratigraphic ranges given for many taxa (in their Fig. 15) are certainly too long, probably a reflection of the effect of caving in ditch cuttings.

Norwegian Sea
The North Sea Cenozoic zonal schemes can be readily applied, almost without modification, to the Cenozoic sequences proved by hydrocarbon exploration in the Norwegian Sea, although no detailed biostratigraphic data have yet been released from this area. As in the North Sea, outer sublittoral–epibathyal biofacies and 'Rhabdammina-biofacies' can be differentiated.

Early Palaeocene is often poorly developed, but Late Palaeocene agglutinating assemblages (Zone NSA1) are widespread. The 'large *Coscinodiscus* assemblage' (Zone NSP4) is present in the Late Palaeocene–basal Eocene 'ash-series', overlain by Early Eocene red clays (Rosnaes Formation equivalent) with *Subbotina* gr. *linaperta* (Zone NSP5). Later Eocene sediments contain dominantly or exclusively NCA assemblages in most sites drilled, with *Reticulophragmium amplectens* (Zone NSA4) followed by *Spiroplectammina* aff. *spectabilis* (Zone NSA5). Late Eocene has not been definitely identified microfaunally.

Early Oligocene sequences are again dominated by agglutinants; records of *Cassidulina carapitana* and *Rotaliatina bulimoides* confirm the presence of Subzones NSB7a and 7b. Late Oligocene is less clearly defined, but its presence is indicated by records of Diatom sp. 3 (King, 1983) at higher levels. Diatom sp.4 and Diatom sp.5 (King, 1983) are widespread in an overlying interval, indicating the Early Miocene Zone NSP10. This is followed by an interval with occasional *Sphaeroidinellopsis disjuncta* and rare *Asterigerina staeschei*, indicating the presence of the Middle Miocene Zones NSP12 and NSB10/11. Higher levels in the Middle and Late Miocene are indicated by the successive occurrence of *Bolboforma spiralis* (Subzone NSP14a) and *B. metzmacheri* (NSP14b), associated with *Martinottiella bradyana* (Zone NSA11). The later Late Miocene and Early Pliocene are not well-defined biostratigraphically. Very thick Late Pliocene–Pleistocene glaciomarine sequences with *Cibicides grossus* (NSB15) are overlain by a unit with common *Nonion labradoricum* (Zone NSB16x).

The calcareous benthonic assemblage in the Early Eocene of DSDP Site 338: Core 32–34 can be readily assigned to Zone NSB3. The presence of *Cancris 'subconicus'* (probably *C.* sp. A) probably indicates Subzone NSB3b (Talwani, *et al.*, 1976, p. 193). In the overlying Early Eocene to Miocene of DSDP Norwegian Sea sites (Verdenius and van Hinte, 1983), non-calcareous agglutinating foraminiferid assemblages of the '*Rhabdammina*-biofacies' are present. Zones established for this sequence can be compared with the North Sea NSA Zones (Fig. 9.10).

The Eocene '*Spiroplectammina spectabilis–Cyclammina amplectens* Zone' of Verdenius and van Hinte corresponds to NSA4 and NSA5. The overlying '*Spirosigmoilinella–Karreriella siphonella* Zone' includes several distinct successive assemblages. The earliest occurrence of *Spirosigmoilinella* sp. at the base of this Zone correlates with its FAD at the base of Zone NSA7 in the North Sea. *Cribrostomoides scitulus* (*Budashevaella* aff. *multicamerata* of Verdenius and van Hinte) is recorded only in the lowest part of this Zone at Site 345 (Cores 25.4 to 26.5), suggesting correlation with Zone NSA7. *Karreriella chilostoma* (*K. siphonella* of Verdenius and van Hinte) is apparently restricted to the lower half of the Zone. Its highest occurrence (in Core 27.2 at Site 348) may correlate with its LAD in the North Sea, defining the top of Zone NSA8. *K. siphonella* is recorded again at higher levels, but these specimens are probably another species.

The '*Spirosigmoilinella–Martinottiella communis* Zone' probably corresponds approximately to Zones NSA10 and NSA11; *M. communis* of Verdenius and van Hinte is here identified as *M. bradyana*, the index of Zone NSA11. *Bolboforma clodiusi* is recorded by Powell (1986) from Core 29 at Site 341, within the upper part of this Zone. This confirms the presence of Zone NSP13. Abundant *B. metzmacheri* is recorded from the overlying Core 28, indicating Zone NSP14.

Dextral specimens of *Neogloboquadrina atlantica* occur in Cores 27.4 to 26 (top) at Site 341 (Talwani, *et al.*, 1976), indicating Zone NSP15b. Sinistral specimens of this species occur from Core 25.2 upwards, defining the base of Subzone NSP15c. The highest occurrence of *N. atlantica* is in Core 20.1, defining the top of Zone NSP15d. No equivalent of Zones NSP16a or NSB15 appear to be present at this site, suggesting an unconformity at this level. The overlying sequence contains *Nonion labradoricum* (= NSB16x) and sinistral *Neogloboquadrina pachyderma* (= NSP16b).

Faeroes Trough

On the southern edge of the Faeroes Trough, many of the North Sea zones can be identified in Cenozoic marine sequences penetrated by wells drilled for hydrocarbon exploration. Planktonic foraminiferid assemblages are generally more diverse than in the North Sea, reflecting proximity to the Atlantic Ocean. Early Palaeocene sediments contain planktonic foraminiferid assemblages similar to those in Zone NSP1; the 'mid'-Palaeocene radiolarian unit (Zone NSP2) is recognisable. Later Palaeocene sediments contain agglutinating assemblages referable to Zone NSA1. *Coscinodiscus* sp.1 occurs in the 'ash-series', indicating Zone NSP4. In deep-water Eocene sequences, agglutinating Zones NSA4 and NSA5 can be identified; in shallower marine environments the occurrence of *Neoeponides karsteni*, *Planulina costata* and *Cibicidoides truncanus* enables identification of Subzones NSB5a, NSB5b and NSB6a. Oligocene to Pliocene sediments are represented in the areas drilled mainly by fragmentary and incomplete sequences, and their microfaunas are not yet studied in detail. Late Pliocene glaciomarine sediments with *Cibicides grossus* can be assigned to Zone NSB15.

Hughes and Jenkins (1981) record Middle/Early Miocene assemblages with *Asterigerina staeschei* and *Elphidium inflatum* (probably equivalent to Zone NSB10) from a BGS borehole north of Scotland.

Svalbard

A small outcrop of Oligocene marine sediments is reported from Svalbard (Spitzbergen) by Feyling-Hanssen and Ulleberg (1984). Two microfaunal zones are identified — the lower (TA, *Bolivina* cf. *antiqua* Zone) contains an assemblage including *B.* cf. *antiqua*, *Nonion granosum*, and *Turrilina alsatica*, suggesting correlation with Zones NSB8 or NSB7. The overlying zone (TB, *Asterigerina guerichi* Zone) with *T. alsatica*, *A. guerichi*, *Rotaliatina bulimoides* and *Gyroidina soldanii mamillata*, suggests a similar age and may correlate with the Subzone NSB8a/NSB7 transition.

Labrador Sea–Grand Banks–Scotian Shelf (Canadian offshore)

Cenozoic microfaunal biostratigraphy of this area has been studied in varying degrees of detail (e.g. Gradstein and Agterberg, 1982; Miller, et al., 1982). Many of the benthonic taxa recorded here also occur in the North Sea, including the stratigraphically important taxa *Bulimina midwayensis*, *Stensioeina [Gavelinella] beccariiformis*, *Spiroplectammina spectabilis* (Palaeocene), *Reticulophragmium [Cyclammina] amplectens*, *Cibicidoides truncanus*, *Turrilina brevispira* (Eocene), *Turrilina alsatica* (Oligocene), and *Asterigerina guerichi* (probably *A. g. staeschei*) (Miocene). More detailed comparisons with North Sea assemblages are difficult to make, due to the absence of detailed information on foraminiferid ranges.

East Greenland and Baffin Island

Feyling-Hanssen (1985) has identified a stratigraphic unit with *Cibicides grossus* in marine late Cenozoic sections at Qivituq Peninsula (East Greenland) and Clyde Foreland (Baffin Island), which can be correlated with North Sea Zone NSB15. In these sections *Nonion labradoricum* is restricted to the interval above the LAD of *C. grossus*, indicating that the *Nonion labradoricum* Zone (NSB16x) can be identified here also (equivalent to Zones D–G of Feyling-Hanssen at Qivituq Peninsula, and his *Cassidulina teretis* and *Islandiella islandica* Zones at Clyde Foreland).

Canadian Arctic

Cenozoic microfaunas have been recorded from oil exploration wells in the Beaufort Sea–Mackenzie Delta area (Canada, Northwest Territories) (e.g. Young and McNeil, 1984; Dixon et al., 1984). The lowest assemblage ('*Haplophragmoides* spp. assemblage') is composed almost entirely of NCA foraminiferids. It is somewhat similar in aspect to Oligocene and Eocene 'Rhabdammina-biofacies' assemblages from the North Sea (although less diverse), and includes *Reticulophragmium [Cyclammina] amplectens*, *Trochammina* (aff.) *subvesicularis*, and probably *Spirosigmoilinella* sp.A (*Rzehakina epigona* of Dixon et al.).

The overlying '*Cibicides* spp. assemblage' apparently includes several distinct successive microfaunas. *Gyroidina soldanii mamillata* (*Rotaliatina* cf. *R. mexicanus* of Young and McNeil, Plate 9.5, Fig. 8) is illustrated from the lower part of this unit, occurring with *Asterigerina guerichi* (s.l.), *Turrilina alsatica*, and *Valvulineria petrolei*, indicating the probable presence of the Oligocene Zones NSB7 and NSB8. The immediately overlying sediments are apparently barren, but the highest part of the '*Cibicides* spp. assemblage' (Kopanoar M-13 well, 1341–1494 m) contains common *C. grossus* and other taxa which indicate correlation with Zone NSB15. The highest microfaunal assemblage, the '*Elphidium* spp. assemblage', with *E. ustulatum*, may correspond approximately to NSB17.

General comments
A sequence of faunal assemblages can be identified in the Norwegian Sea and the Shetlands–Faeroes Basin which correlates closely with the North Sea Zones. Further afield in Arctic and sub-Arctic areas, considerable similarity of the Cenozoic benthonic foraminiferid assemblages is still evident. This degree of regional homogeneity indicates that the vertical faunal distributions probably reflect the influence of palaeo-oceanographic events which affected wide areas almost simultaneously. Some of these events may be circulation changes due to plate-tectonic events associated with the development of the North Atlantic–Arctic rift system; others probably reflect eustatic sea-level changes, as discussed above.

9.6 INDEX SPECIES

Introduction
The range charts (Figs 9.11–9.14) illustrate the stratigraphic distribution of selected benthonic foraminiferids in the North Sea Cenozoic. The ranges of the calcareous benthonic taxa and agglutinants (excluding NCAs) are calibrated against the NSB Zones, and the ranges of non-calcareous agglutinants (Fig. 9.14) against the NSA Zones. Range charts for planktonic foraminiferids are not given, except for key Late Miocene–Pleistocene taxa, as insufficient data is available to update the range charts given in King (1983). The selection is biased in favour of short-ranging and (as far as possible) geographically widespread taxa, but a selection of long-ranging NCAs is included, as they tend to dominate the NCA assemblages.

Descriptions of taxa are confined mainly to specific or subspecific characteristics; for generic characteristics, reference should be made to standard texts. The taxonomic treatment of NCAs mainly follows Gradstein and Berggren (1981) and Kaminski, *et al*. (in press). Significant unpublished taxonomic data on North Sea Palaeogene agglutinants has also been supplied by Mike Charnock and Bob Wynn Jones, and this is acknowledged with thanks. Ranges refer only to the North Sea, except where indicated, and are given in terms of the correlation adopted here.

SUBORDER TEXTULARIINA

'Non-calcareous' species are indicated by 'NCA'.

Ammodiscus cretaceus (Reuss)
Plate 9.1, Fig. 3 (× 32), Middle Eocene, Labrador Sea (after Miller, *et al.*, 1982). = *Operculina cretacea* Reuss, 1845. Probable synonyms: *Ammodiscus incertus* (d'Orbigny) of authors (*non Operculina incerta* d'Orbigny, 1839); *A. glabratus* Cushman and Jarvis, 1928; *A. peruvianus* Berry, 1928 of Gradstein and Berggren (1981); *A. siliceus* (Terquem) of Verdenius and van Hinte (1983) (*non Involutina silicea* Terquem, 1862). Description: test very fine grained, tightly coiled, with numerous (8–12) depressed whorls (NCA).
Remarks: specimens here referred to this species occur from Late Cretaceous to Oligocene, although assigned in the literature to several different species. Much of the apparent morphological variation (in outline and degree of compression) is due to postmortem deformation. *A. siliceus* is a moderately coarsely agglutinated Jurassic species, with fewer whorls than *A. cretaceus*, according to the illustration in Loeblich and Tappan (1964). See Kaminski *et al.* (in press) for synonymy.
Range: Late Cretaceous (Santonian) to Early Oligocene. The types are from the Planermergel (Turonian/Coniacian) of Czechoslovakia.

Ammodiscus sp. A.
Plate 9.1, Fig. 4 (× 27), Zone NSA9, Late Oligocene, North Sea. Description: test moderately coarse-grained, coiling slightly irregular, with 4–5 whorls, which are apparently compressed (NCA).
Remarks: similar to *A. pennyi* Cushman and Jarvis, 1928, but with a larger number of whorls. Not definitely identified with any described species.
Range: Early Oligocene to Early Miocene.

Ammodiscus sp. B.
Plate 9.1, Fig. 5 (× 36), Zone NSA9, Late Oligocene, North Sea. Description: test small, fine-grained, thin, numerous whorls (seven or more), tightly and somewhat irregularly coiled; whorls depressed; last whorl tending to be constricted in several places (NCA).
Range: Early Oligocene to Late Oligocene.

Ammolagena clavata (Jones and Parker)
Plate 9.1, Fig. 6 (× 32), Eocene, Labrador Sea (after Miller *et al.*, 1982). = *Trochammina irregularis* (d'Orbigny) var. *clavata* Jones and Parker, 1860. Description: test attached, oval, bulbous, fine-grained, with an elongated neck (NCA).
Remarks: occurs attached to other agglutinating foraminiferids and to sand grains.
Range: Late Palaeocene to Early Oligocene.

Ammomarginulina macrospira Bykova, 1953
Plate 9.1, Fig. 7 (× 130), Eocene, U.S.S.R. (after Bykova). = *Ammobaculites* aff. *polythalamus* Loeblich of Berggren and Gradstein (1981). Description: test small, moderately coarse-grained, comprising a relatively large initial coil with about six chambers, followed by a short uncoiled stage with rather irregular chambers; sutures may be emphasized by concentrations of dark mineral grains (NCA).
Remarks: an important index-fossil for the Late Eocene in the '*Rhabdammina*-biofacies'.
Range: Middle?–Late Eocene (from Maastrichtian according to Gradstein and Berggren, 1981).

'*Arenobulimina*' sp. A.
Plate 9.1, Fig. 8 (× 63), Zone NSB6, Late Eocene, North Sea (after King, 1983). = *Arenobulimina* sp. of King (1983). Description: test moderately fine grained, with some dark mineral grains incorporated into the test wall; trochospiral, test expanding regularly, last whorl forming over half of the test; chambers slightly inflated, sutures slightly depressed; 3–4 chambers in the last whorl; aperture small, rounded, interiomarginal.
Range: Middle Eocene?, Late Eocene to basal Early Oligocene.

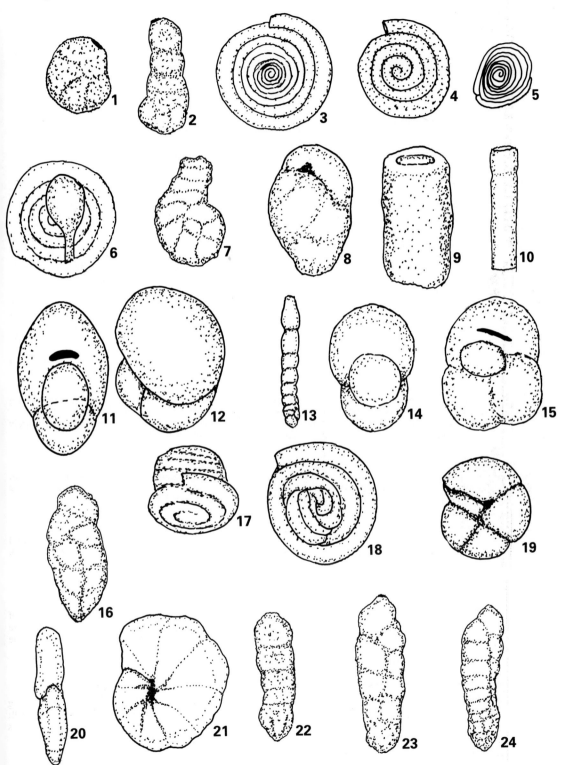

Plate 9.1

Bathysiphon eocenicus Cushman and Hanna, 1927
Plate 9.1, Fig. 9 (× 38), Middle Eocene, California (after Cushman and Hanna). Description: test large (up to 1 mm in diameter), fine-grained, tubular, thick walled, probably with widely spaced constrictions; test wall white, with a friable surface, often speckled with dark minerals, and sometimes incorporating scattered siliceous sponge spicules. Occurs as fragments which may be several centimetres in length (NCA).
Remarks: this is much larger and thicker walled than *B. nodosariaformis*, and has a distinctive test surface. Several other specific names may be applicable to this morphotype, including *Bathysiphon taurinensis* Sacco, 1893. It is included by Gradstein and Berggren (1981, p.11, Fig. 4 only) in their *Bathysiphon discreta* variety 'B'. *B. eocenica* of Verdenius and van Hinte (1983) is much coarser-grained, and is probably quite a different species.
Range: Late Cretaceous (Campanian) to Oligocene.

Bathysiphon nodosariaformis Subbotina, 1950
Plate 9.1, Fig. 10 (× 26), Eocene, North Sea (after Gradstein and Berggren, 1981). = *Bathysiphon discreta* var. A (Gradstein and Berggren, 1981), *Bathysiphon* sp. (Kaminski *et al.*, in press). Description: test tubular, smooth-surfaced, very fine-grained ('glassy' when silicified), moderately thick-walled, occasionally 'wrinkled' and with widely spaced constrictions which may be elongated into 'necks'; occurs mainly as approximately straight fragments (NCA).
Remarks: *B. nodosariaformis* appears to be the closest match both morphologically and stratigraphically for this common and long-ranging 'species'. *B. alexanderi* Cushman, 1933 (Late Cretaceous) may be synonymous, but the degree of variation in the development of constrictions is of uncertain taxonomic significance. *B. nodosariaformis* is synonymised with *B. eocenica* by Verdenius and van Hinte (1983), but the two are quite distinct morphologically. This species is referred to as *Bathysiphon discreta* variety 'A' by Gradstein and Berggren (1981, Figs 5, 6, *non* Fig. 4) but it does not appear to be related to *Psammosiphonella/Rhabdammina/Bathysiphon discreta*. It is probably identified by Verdenius and van Hinte (1983) (together with *B. eocenicus* as here interpreted) as the Recent species *B. filiformis* Sars, but the latter species is characterised by the incorporation of common sponge spicules into its wall; these are entirely absent from *B. nodosariaformis*. See the synonymy in Kaminski *et al.* (in press). Attribution to *Bathysiphon* is uncertain, depending on whether the incorporation of sponge spicules in the test is regarded as the distinctive characteristic of the genus; the generic and specific classification of fossil single-chambered tubular agglutinants is currently rather confused, partly due to their occurrence mainly as fragments.
Range: Late Cretaceous to Miocene.

Bathysiphon? sp. A.
Description: test fine-grained, small (0.25–0.3 mm diameter), tubular (generally partially crushed), irregularly rectilinear, with regular closely-spaced constrictions (NCA).
Remarks: the exact affinities of this form are uncertain, but it is a useful index species for the Middle and Late Miocene in agglutinant-dominated assemblages.

Buzasina cf. *pacifica* (Krasheninnikov)
Plate 9.1, Figs 11, 12 (× 64), Maastrichtian, Labrador Sea (after Gradstein and Berggren, 1981). = cf. *Labrospira pacifica* Krasheninnikov, 1973. Description: test moderately fine-grained, wall smooth; planispiral, compressed, with a subrounded arched periphery; three chambers in the last whorl, the last chamber is very large (NCA).
Range: Late Palaeocene (from Maastrichtian elsewhere).

Clavulina anglica (Cushman)
Plate 9.1, Fig. 22 (× 26). Early Eocene, England (after Cushman, 1937a). = *Pseudoclavulina anglica* Cushman, 1936. Description: test elongate, robust, rather coarsely agglutinated; the initial triserial stage is large, trigonal in cross-section, followed by a uniserial stage with a circular cross-section, comprising up to six short chambers, later chambers becoming inflated.
Remarks: synonymised with *Clavulinoides aspera* (Cushman) by some authors, but the latter species (and genus) are characterised by the triangular cross-section of the uniserial stage.
Range: Early Palaeocene to Early Eocene (and Middle Eocene in onshore sequences).

Clavulina cocoaensis Cushman, 1936
Plate 9.1, Fig. 13 (× 26). Middle Eocene, Alabama (U.S.A.) (after Cushman, 1937a). Description: test thin, elongate, moderately coarsely agglutinating; the initial triserial stage is rounded in cross-section, followed by an elongated uniserial stage of six or more chambers, becoming increasingly elongated in the adult stage.
Remarks: thinner and less robust than *C. anglica*, and with a proportionately smaller triserial section.
Range: Middle to Late Eocene.

Cribrostomoides scitulus (Brady)
Plate 9.2, Figs 9, 10 (× 38), Recent, North Atlantic (after Brady, 1884). = *Haplophragmium scitulum* Brady, 1881, *Labrospira scitula* (Brady), *Budashevaella* sp. aff. *multicamerata* (Budasheva) of Verdenius and van Hinte (1983). Description: test planispiral, compressed, rather smooth; periphery broadly rounded; last whorl becoming more evolute, tending to become slightly trochospiral; chambers depressed, about 9–10 chambers in the last whorl, with sutures well-defined and radial (NCA).
Remarks: this species resembles a globose *Haplophragmoides*, but the last part of the final whorl becomes slightly trochospiral. As in many of the other NCAs in the North Sea Cenozoic, the aperture is rarely visible, so assignment to *Cribrostomoides* is somewhat uncertain.
Range: Middle Eocene to Early Oligocene.

Cystammina sp.A.
Plate 9.1, Figs 14, 15 (× 140), Eocene, North Sea (after Gradstein and Berggren, 1981). = *Praecystammina globigerinaeformis* Krashenninikov of Gradstein and Berggren (1981), *non* Krashenninikov, 1973. Description: test subglobular, fine-grained, probably streptospiral; three subspherical chambers in the last whorl; aperture areal, with a lip (NCA).
Range: Late Palaeocene to Early Oligocene.

Gaudryina hiltermanni Meisl, 1959
Plate 9.1, Fig. 16 (×29), Zone NSB3, Early Eocene, Germany (after Staesche and Hiltermann, 1940). = *Clavulina* aff. *szaboi* Hantken of Staesche and Hiltermann. Description: test large, moderately coarse-grained; a short triserial stage is followed by a long biserial stage with a sharp-edged triangular cross-section.
Range: Early Eocene.

Glomospira charoides (Jones and Parker)
Plate 9.1 Fig. 17 (× 60). Palaeocene, Tasman Sea (after Webb, 1975). = *Trochammina squamata* var. *charoides* Jones and Parker, 1860. Description: test fine-grained, approximately spherical in shape; coiling of early whorls is fairly regular, but the last whorl may be much larger, encircling test equatorially or in a plane at 90° to previous whorls (NCA).
Remarks: as interpreted here, *G. charoides* includes *G. charoides corona* Cushman and Jarvis, 1928 and *G.* 'gordialis' of Gradstein and Berggren (1981) (*non* Jones and Parker, 1860), which can be seen to intergrade within single populations. See Kaminski *et al*. (in press).
Range: Early Cretaceous to Late Miocene.

Glomospirella sp. A
Plate 9.1, Fig. 18 (× 96), Early Eocene, Labrador Sea (after Miller *et al*., 1982). = *Glomospirella* sp. of Miller *et al*., 1982, and Gradstein, *et al*., (in press). Description: test fine-grained; early whorls coiled streptospirally, similar to *Glomospira*; the last four whorls are planispiral and compressed, similar to *Ammodiscus* (NCA).
Range: Late Palaeocene.

Haplophragmoides kirki Wickenden, 1932
Plate 9.1, Fig. 19 (× 53), Early Miocene, Norwegian Sea (after Verdenius and van Hinte, 1983). Description: test small, very fine-grained, compressed, involute, periphery broadly rounded; chambers moderately inflated; typically $4\frac{1}{2}$–5 chambers in the last whorl (NCA).
Remarks: specimens are usually partly or wholly crushed. *H. burrowsi* Haynes, 1958 (Late Palaeocene, England) is probably a junior synonym.
Range: Late Cretaceous to Late Miocene.

Haplophragmoides walteri (Grzybowski)
Plate 9.1, Figs 20, 21 (× 53), Oligocene, Norwegian Sea (after Verdenius and van Hinte, 1983). = *Trochammina walteri* Grzybowski, 1898. Description: test very fine-grained, compressed (usually affected by postmortem distortion), involute; periphery acute, narrowly rounded; 8–12 chambers in the last whorl (NCA).
Remarks: *H. compressa* Leroy (see Verdenius and van Hinte, 1983), which also occurs in the Cenozoic of the North Sea, is similar but less compressed, more evolute, and with fewer chambers; when crushed it may be difficult to separate from *H. walteri*. See Kaminski *et al*. (in press).
Range: Late Cretaceous (Santonian) to Early Miocene.

Karreriella chilostoma (Reuss)
Plate 9.2, Figs 1, 2 (× 38), Zone NSB7, Early Oligocene, Germany (after Cushman, 1937b). = *Textularia chilostoma* Reuss, 1852. Description: test small, short, fine-grained; in the megalospheric form a short initial trochospiral stage is followed by an elongated biserial stage of around six chambers; in the microspheric form a larger trochospiral stage is followed by a short biserial stage. The chambers are inflated, somewhat elongated longitudinally; the aperture forms a narrow slit at the base of the apertural face (NCA).
Remarks: synonymised with *K. siphonella* by Kaasschieter (1961) and other authors, but the latter species is coarser-grained, with a calcareous cement, and has a different aperture and different chamber shapes (see below).
Range: Early Oligocene to early Late Oligocene.

Karreriella siphonella (Reuss)
Plate 9.2, Fig. 3 (× 21), Zone NSB7, Early Oligocene, Germany (after Cushman, 1937b). = *Gaudryina siphonella* Reuss, 1851. Description: test moderately finely agglutinated, with calcareous cement; chambers inflated and subglobular, test often somewhat twisted; the megalospheric form is elongated, with a short trochospiral stage and a longer biserial stage; the microspheric form is short and tapering, trochospiral almost to the final chamber; the aperture is circular or elliptical, with a short neck, and is placed above the base of the apertural face. Remarks: The species identified as *K. siphonella* by Verdenius and van Hinte (1983) is very fine-grained, forms part of a non-calcareous foraminiferid assemblage, and is probably *K. chilostoma*.
Range: Early Oligocene to Early Miocene.

Karreriella sp.A.
Description: test moderately fine-grained, chambers not greatly inflated; the megalospheric form is elongated, cylindrical, with a circular cross-section in the biserial stage, comprising approximately half of the test length; the microspheric form is short, expanding rapidly from the apex, with a long trochospiral stage, becoming finally triserial. The aperture is circular, with a short neck, in a re-entrant on the apertural face.
Remarks: this species differs from *K. chilostoma* by its coarser grain size, its partly calcareous cement, its aperture and its circular cross-section. It differs from *K. siphonella* by its cylindrical shape in the megalospheric form.
Range: Middle to Late Miocene.

Karrerulina conversa (Grzybowski)
Plate 9.1, Fig. 23 (× 64), Eocene, Labrador Sea (after Gradstein and Berggren, 1981), Fig. 24 (× 90). Palaeocene, Tasman Sea (after Webb, 1975). = *Gaudryina conversa* Grzybowski, 1901, *Karreriella conversa* (Grzybowski), *Plectina conversa* (Grzybowski), *Karreriella apicularis* (Cushman, 1911). Description: test elongated, often slightly twisted, moderately fine-grained, initially trochospiral, later chambers becoming biserial to semi-uniserial in the microspheric form; the chambers are moderately inflated; aperture terminal (NCA).
Range: Late Cretaceous (Santonian) to Early Oligocene.

Martinottiella bradyana (Cushman)
Plate 9.2, Fig. 6 (× 26), Zone NSA11, Middle/Late Miocene, Norwegian Sea (after Verdenius and van Hinte, 1983). = *Listerella bradyana* Cushman, 1936. Description: test fine-grained; an initial thickened trochospiral stage is followed by an elongated cylindrical uniserial stage of up to ten chambers, with slightly depressed sutures; the apertural face is rounded, the aperture is central, terminal, forming a short slit with a slight neck (NCA).
Remarks: this species is apparently identical with *M. bradyana*, described originally from the Miocene to ?Recent of the Pacific. It is recorded from the Miocene of the Norwegian Sea by Verdenius and van Hinte (1983) as *M. communis*, but the latter species is more robust, more coarsely agglutinated, with fewer chambers in the uniserial section and a partially calcareous cement. *M. bradyana* is a valuable index-fossil for the Middle Miocene to basal Late Miocene in the '*Rhabdammina*-biofacies'.
Range: Middle to Late Miocene.

Martinottiella communis (d'Orbigny)
Plate 9.2, Figs 4, 5 (× 45), Early Miocene, Belgium (after de Meuter, 1980). = *Clavulina communis* d'Orbigny, 1846. Description: test moderately coarse-grained; an initial trochospiral stage comprising about one quarter of the test length is followed by a cylindrical uniserial stage of c.5–6 chambers.
Remarks: the Oligocene records include specimens which should perhaps be assigned to *M. muensteri* (Cushman, 1936).
Range: Late Oligocene to Late Pliocene, ?Pleistocene.

Matanzia varians (Glaessner)
Plate 9.2, Figs 7, 8 (× 33), Subzone NSB1b, Late Palaeocene, Netherlands (after ten Dam, 1944). = *Textulariella varians* Glaessner, 1937, *Hagenowella paleocenica* Hofker, 1949. Description: test moderately fine-grained; microspheric form expanding rapidly, megalospheric form more elongated; initially triserial, becoming biserial in the final stages; cross-section approximately circular, aperture interiomarginal, loop-shaped. Later chambers partially divided by longitudinal partitions (often visible through the chamber wall).
Remarks: see Kaminski *et al.* (in press).
Range: Late Palaeocene.

Plectina cubensis Cushman and Bermudez, 1936
Plate 9.2, Figs 11, 12 (× 25), Eocene, Cuba (after Cushman, 1937b). Description: test moderately coarse-grained, short, initial trochospiral stage, last few chambers biserial, highly inflated, subglobular; aperture circular, areal, terminal.
Range: Middle and Late Eocene.

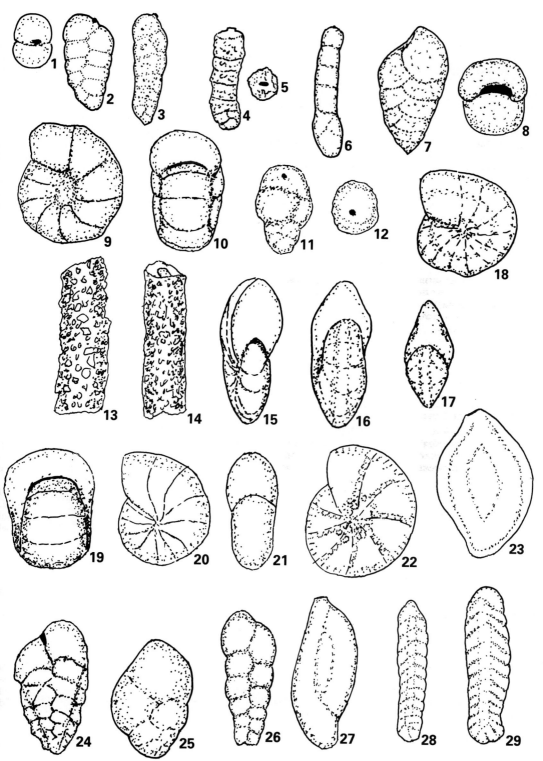

Plate 9.2

Psammosiphonella gr. *discreta* (Brady)
Plate 9.2, Figs 13, 14 (× 34), Eocene ?, Labrador Sea (after Gradstein and Berggren, 1981). = gr. *Rhabdammina discreta* Brady, 1884, *Bathysiphon discreta* var. B Gradstein and Berggren, 1981, *Rhabdammina* ex gr. *discreta* Kaminski, *et al.* (in press). Description: test coarsely agglutinating, elongated, tubular, apparently open at both ends, with rather widely spaced constrictions (NCA).
Remarks: the grain size of the coarser particles agglutinated (mainly quartz grains) varies considerably, and in some specimens the test wall comprises only very coarse quartz grains with very little visible cement. These variations have been given formal taxonomic status by some authors, but in the North Sea Cenozoic they seem to reflect the relative availability of coarse clastic grains: it is noticeable that the coarsest-grained specimens occur in association with coarse sand units. Probable synonyms include *Rhabdammina cylindrica* Glaessner (a very coarse-grained form) and *R. eocenica* Cushman and Hanna. *Bathysiphon* is typically fine-grained and thick-walled. and *Rhabdammina* is probably best restricted to branching tubular forms; *Psammosiphonella* seems the most appropriate genus for this form. See Kaminski, *et al.* (in press).
Range: Late Cretaceous to Miocene.

Recurvoides gr. *walteri* (Grzybowski)
= *Haplophragmium walteri* Grzybowski, 1898. Description: test streptospirally coiled, moderately coarse-grained, variable in shape; some specimens are almost spherical; c.4–5 chambers in the last whorl; aperture basal to areal (NCA).
Remarks: the rather broad interpretation of this taxon follows Gradstein and Berggren (1981).
Range: Late Cretaceous to Miocene.

Reticulophragmium acutidorsatum (Hantken)
Plate 9.2, Fig. 15 (× 23), Eocene, Labrador Sea (after Gradstein and Berggren, 1981). = *Haplophragmium acutidorsatum* Hantken, 1868, *Cyclammina acutidorsata* (Hantken). Description: test highly compressed, disc-shaped, alveolar structure complex; periphery acute; 10–16 chambers in the last whorl (usually 12–13) (NCA).
Remarks: differentiated from *R. amplectens* by its profile and its lower number of chambers (although it can be confused with the '*amplectens* variants' occurring in the Middle Eocene to Oligocene discussed below). This species is synonymised with *R.(C.) placenta* by Gradstein and Berggren (1981), but is regarded here as separable morphologically. The species of *Reticulophragmium* in the North Sea Basin are differentiated mainly on gross morphology, but more detailed study of alveolar structure would certainly enhance the definition and stratigraphic utility of the species of this genus.
Range: Middle Eocene to Early Miocene.

Reticulophragmium amplectens (Grzybowski)
Plate 9.2, Figs 16, 17, 18 (× 56), Eocene, North Sea (after Gradstein and Berggren, 1981). = *Cyclammina amplectens* Grzybowski, 1898. Description: 'disc-shaped; the test is thick in the centre and tapers sharply towards a narrow periphery. The umbilical centre is depressed' (Gradstein and Berggren, 1981) (NCA).
Remarks: the form regarded here as 'typical' *R. amplectens* (but not necessarily identical with Grzybowski's species, as topotype material has not been studied) has 13–20 chambers in the last whorl, and a characteristic 'ogival' cross-section, with a slightly swollen area around the depressed umbilicus. It is restricted to Zone NSA4 (Early–Middle Eocene). In the later Middle Eocene and Late Eocene a similar form occurs (here referred to as *R. amplectens* subsp. A), with a slightly lower number of chambers, but with an '*acutidorsata*' cross-section and apparently a less complex alveolar structure. In the Early Oligocene, a similar form is frequent (*R. amplectens* subsp. B), with an '*acutidorsata*' profile, but with 16–19 chambers (compared to 10–16 chambers in *R. acutidorsata*). Although these 'subspecies' are referred here to *R. amplectens*, they may be genetically more closely related to *R. acutidorsata*. Further study of this group is necessary, particularly of the internal structure.

Reticulophragmium orbiculare (Brady)
Plate 9.2, Fig. 19 (× 38), Oligocene, North Sea (after Gradstein and Berggren, 1981). = *Cyclammina orbicularis* Brady 1881, *Cyclammina rotundidorsata* (Hantken) of Gradstein and Berggren (1981), non *Haplophragmium rotundidorsatum* Hantken, 1875. Description: test globose, chambers depressed, periphery broadly rounded, 10–12 chambers in the last whorl (NCA).
Range: Late Eocene to Late Oligocene, ? Early Miocene.

Reticulophragmium placenta (Reuss)
Plate 9.2, Fig. 20 (× 19), Eocene, Labrador Sea (after Gradstein and Berggren, 1981). Fig. 21 (× 32). Eocene, Norwegian Sea (after Verdenius and van Hinte, 1983). =*Nonionina placenta* Reuss, 1851, *Cyclammina placenta* (Reuss). Description: test compressed, periphery narrowly to broadly rounded, 10–16 chambers in the last whorl (NCA).
Remarks: this broadly interpreted species is distinguished from *R. acutidorsatum* by its less compressed profile and more rounded periphery. It includes specimens referred to *Cyclammina cancellata* Brady by Gradstein and Berggren (1981).
Range: Middle Eocene to Miocene.

Reticulophragmium sp. A.
Plate 9.2, Fig. 22 (× 38), Zone NSA1b, Late Palaeocene, North Sea. = '*R.* cf. *garcilassoi* (Palaeocene morphotype)'

of Kaminski *et al.* (in press), *R. paupera* of Gradstein *et al.* (in press). Description: test compressed, disc-shaped, approximately eight chambers in the last whorl; the alveolar structure is represented by primary alveoles only, confined to areas adjacent to the sutures and periphery (NCA).
Remarks: this is the oldest *Reticulophragmium* recorded in the North Sea Basin. It is differentiated from all others by its 'primitive' alveolar structure and its low number of chambers. It is included as a primitive variant of *R. amplectens* by some workers, but is morphologically quite distinct from that species. No satisfactory name can be given to this species. It is recorded from the Palaeocene of Trinidad by Kaminski *et al.* (in press) as *R.* cf. *garcilassoi*, but the types of *R. garcilassoi* (Frizzell) have evenly distributed alveoles. It is figured as *R. paupera* (no author given) from the North Sea by Gradstein *et al.* (in press), but if this intended to refer to *Cyclammina paupera* Chapman (Miocene of Australia), it is unlikely to be the same species.
Range: Late Palaeocene.

Rzehakina epigona (Rzehak)
Plate 9.2, Fig. 23 (× 75), Maastrichtian/Palaeocene, Labrador Sea (after Gradstein and Berggren, 1981). = *Silicina epigona* Rzehak, 1895. Description: test fine-grained, highly compressed, involute, with a concave 'umbilical' area; periphery bluntly keeled; two chambers in the last whorl, flattened and compressed (NCA).
Range: Late Cretaceous (Campanian) to Late Palaeocene.

Sabellovoluta humboldti (Reuss)
Plate 9.1, Fig. 1 (× 20), Zone NSB7, Early Oligocene, North Sea (after King, 1983); Fig. 2 (× 15), Zone NSB7, Early Oligocene, Germany (after Bartenstein, 1952). = *Spirolina humboldti* Reuss, 1851, *Ammobaculites humboldti* (Reuss). Description: test large, moderately coarse-grained, compressed; initial planispiral coil of 4–6 chambers followed by an uncoiled series of 3–4 moderately inflated chambers.
Range: Late Eocene ?, Early Oligocene to basal Late Oligocene.

Siphotextularia sculpturata (Cushman and ten Dam)
Plate 9.2, Fig. 24 (× 32), Late Pliocene, Netherlands (after Cushman and ten Dam). = *Textularia sculpturata* Cushman and ten Dam, 1947. Description: test moderately coarsely agglutinated, subquadrangular in cross-section, expanding regularly from the apex, or tending to be more parallel-sided; chambers short, wide, with flattened, deeply depressed sutures.
Range: Late Miocene?, Early to Late Pliocene.

Spiroplectammina spectabilis (Grzybowski)
Plate 9.2, Fig. 28 (× 32) (*S.* aff. *spectabilis*), Middle Eocene, Norwegian Sea (after Verdenius and van Hinte, 1983); Fig. 29 (× 38), Zone NSA1, Late Paleocene, North Sea (after Moe, 1983). = *Spiroplecta spectabilis* Grzybowski, 1898, *Bolivinopsis spectabilis* (Grzybowski). Description: test fine-grained, elongated, flattened to rhomboid cross-section; large initial coil (in microspheric form) followed by a biserial section of 15+ chambers. The chambers have depressed sutures and a flattened peripheral 'flange' (NCA).
Remarks: a distinctive species, ranging from Campanian to Late Eocene in the North Sea Basin. Eocene specimens, here referred to as *S.* aff. *spectabilis*, are smaller, thinner and more compressed than those at older levels.

Spirosigmoilinella sp. A
Plate 9.2, Fig. 27 (× 68), Late Oligocene/Early Miocene, North Sea (after Gradstein and Berggren, 1981). = *Silicosigmoilina* sp. of King (1983), '*Rzehakina*' sp. of Gradstein and Berggren (1981), *Spirosigmoilinella compressa* Matsunaga of Miller *et al.* (1982), *S.* sp. of Verdenius and van Hinte (1983). Description: test small, compressed, fine-grained, ovoid-elongate, shape rather variable; sutures indistinct; the last chamber is often prolonged into an apertural neck (NCA).
Remarks: identification of this species with *S. compressa* is considered still uncertain.
Range: Early Oligocene to Early Miocene.

Textularia decrescens Cushman and ten Dam, 1947
Plate 9.2, Fig. 25 (×30), Pliocene, Netherlands (after Cushman and ten Dam, 1947). Description: test moderately coarsely agglutinated, short and very broad, periphery rounded; few wide chambers, increasing rapidly in size.
Remarks: differs from similar species, such as *T. gramen* d'Orbigny, by its more compressed cross-section.
Range: Late Miocene?, Early to Late Pliocene.

Textularia plummerae Lalicker, 1935
Plate 9.2, Fig. 26 (× 48), Eocene, Labrador Sea (after Gradstein and Berggren, 1981). Description: test moderately coarse-grained, elongated, with c. 10–12 moderately inflated chambers (NCA).
Remarks: this is said to differ from *Spiroplectammina navarroana* Cushman only by the absence of an initial planispiral coil, but apparently the types of both species have the initial chambers broken off (Kaminski, *et al.*, in press; Charnock and Jones, pers. comm.). The two species are therefore probably synonymous. The species illustrated by Verdenius and van Hinte (1983, Plate 7, Fig. 1) from the Eocene of the Norwegian Sea as *Dorothia principiensis* Cushman and Bermudez is probably referrable to *T. plummerae*. The specimen they illustrate as *S. navarroana* (Plate 7, Figs 6, 7) is probably also misidentified.
Range: Late Cretaceous to Early Eocene.

Thurammina papillata Brady, 1879
Plate 9.3, Fig. 1 (× 75), Recent, Atlantic (after Brady). Description: test moderately fine-grained, thick-walled, irregular in shape, roughly spherical to ovoid; test wall with a variable number of large blunt hollow projections, each with a terminal aperture (NCA).
Range: Late Palaeocene to Early Eocene.

Tolypammina vagans (Brady)
Plate 9.3, Fig. 2 (× 12), Recent, Atlantic (after Brady). = *Hyperammina vagans* Brady, 1884. Description: test attached, unilocular, elongated, fine-grained, tubular, uncoiled, winding; small rounded proloculus (NCA).
Remarks: occurs attached to quartz grains.
Range: Late Palaeocene.

Trochammina ruthvenmurrayi Cushman and Renz, 1946
Plate 9.3, Figs 3, 4 (× 49), Subzone NSA1b, Late Palaeocene, North Sea (after Gradstein and Berggren, 1981). = *Trochammina* aff. *albertensis* Wickenden of Gradstein and Berggren (1981). Description: test moderately fine-grained, biconvex, compressed, circular, with a subangular periphery; varies considerably in shape, from biconvex to subconical, with a flattened umbilical side; around seven chambers in the last whorl (NCA). Remarks: similar in shape to *Globotruncana contusa*.
Range: Late Palaeocene.

Trochammina aff. *subvesicularis* Hanzlikova, 1955
Plate 9.3, Figs 5, 6 (× 60). Zone NSA6, Late Eocene, North Sea (after Moe, 1983). Description: test moderately fine-grained. spiral side flattened, periphery subrounded, umbilical side convex, 4–5 chambers in the last whorl (NCA).
Remarks: *T. subvesicularis* (s.s.) has a more convex umbilical side, giving a subconical profile.
Range: Middle and Late Eocene.

Trochamminoides irregularis White, 1928
Plate 9.3, Fig. 7 (× 45), Palaeocene, Tasman Sea (after Webb, 1975). Description: test fine-grained, coiled in several different planes; the last whorl is usually roughly planispiral, with about eight rather inflated chambers (NCA).
Range: Late Cretaceous (Campanian) to Late Palaeocene.

Verneuilinoides subeocaenus (Wick)
Plate 9.3, Fig. 8 (× 45), Zone NSA2, Late Palaeocene/Early Eocene, Germany (after Staesche and Hiltermann, 1940). = *Verneuilina subeocaena* Wick, 1950, *V*. sp. indet. Staesche and Hiltermann. Description: test expanding rapidly from apex, cross-section rounded triangular, chambers inflated (NCA).
Remarks: a characteristic species of the 'ash-series' and coeval units in onshore sequences, which occurs sparsely but widely in Zone NSA2 in the central and southern North Sea.
Range: Late Palaeocene to Early Eocene.

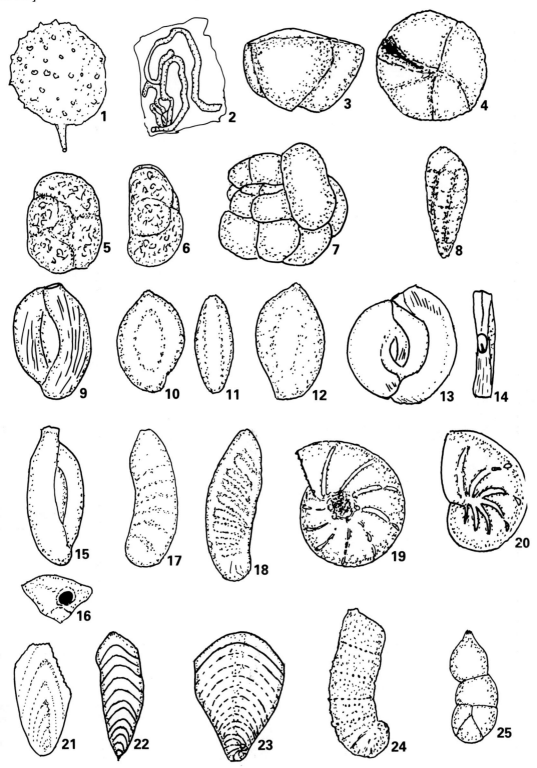

SUBORDER MILIOLINA

Massilina secans d'Orbigny, 1846
Plate 9.3, Fig. 9 (× 18), Zone NSB10, Middle Miocene, Germany (after Spiegler, 1974). Description: test large, highly compressed, circular–oval profile, either smooth or with weak oblique striae (*M. secans obliquestriata* (Halkyard)).
Remarks: occurs in association with *Spiroloculina striatula* in the upper part of Zone NSB10 in the outer Moray Firth and southern Viking Graben, and at a similar level in Germany and the Netherlands.
Range: Early and ? Middle Miocene.

Quinqueloculina juleana d'Orbigny, 1846
Plate 9.3, Fig. 15, 16 (× 55), Middle Eocene, England (after Murray and Wright, 1974), Description: test elongate-oval, chambers keeled, aperture with a short neck.
Range: Late Eocene to Early Oligocene. Also Early and Middle Eocene in onshore sequences.

Sigmoilopsis celata (Costa)
Plate 9.3, Fig. 12 (× 39), Pliocene, Italy (after AGIP Mineraria, 1957). = *Sigmoilina celata* Costa. Description: similar to *S. schlumbergeri*, but more flattened, larger, and more coarse-grained, with a rounded periphery, the last chamber relatively larger and sutures indistinct.
Range: Late Miocene ?, Early to Late Pliocene.

Sigmoilopsis schlumbergeri (Silvestri)
Plate 9.3, Figs 10, 11 (× 45), Zone NSB15, Late Pliocene, North Sea (after King, 1983). = *Sigmoilina schlumbergeri* Silvestri, 1904. Description: test ovate, compressed, moderately coarse-grained, periphery angular.
Range: Early Pliocene to Recent.

Spiroloculina striatula ten Dam and Reinhold, 1942.
Plate 9.3,. Figs 13, 14 (× 11), Zone NSB10, Early Miocene, Netherlands (after ten Dam and Reinhold). Description: test large, flattened, roughly circular in profile, smooth except for week oblique striae which are variably developed; periphery flattened; two chambers in the last whorl.
Remarks: see comments under *Massilina secans*.
Range: Early and ? Middle Miocene.

SUBORDER LAGENINA

Astacolus arcuatus (Philippi)
Plate 9.3, Fig. 17 (× 15), Zone NSB8, Late Oligocene, Germany (after Grossheide and Trunko, 1965). = *Marginulina arcuata* Philippi, 1843. Description: test large, compressed, initial planispiral coil followed by an evenly curved uncoiled stage; smooth except for slightly thickened test over sutures.
Range: Late Oligocene to ?Early Miocene.

Astacolus gladius (Philippi)
Plate 9.3., Fig. 18 (× 10), Zone NSB8, Late Oligocene, Germany (after Grossheide and Trunko, 1965). = *Marginulina gladius* Philippi, 1843. Description: test large, compressed; small initial coil followed by an elongated, irregularly curved, uncoiled stage; sutures limbate, with wide rounded sutural ribs, fading towards the periphery.
Remarks: differs from *A. arcuatus* by the presence of sutural ribs.
Range: Late Oligocene to ? basal Early Miocene.

Frondicularia budensis (Hantken)
Plate 9.3, Fig. 21 (× 23), Late Eocene, Germany (after Kiesel, 1970). = *Flabellina budensis* Hantken, 1875. Description: test small, smooth, compressed, elliptical; periphery rounded; small initial coil followed by a laterally expanding uniserial stage, with arcuately arched chambers.
Range: Late Eocene? to Early Oligocene.

Frondicularia dumontana Reuss, 1861
Plate 9.3, Fig. 22 (× 15), Middle Miocene, Belgium (after De Meuter, 1980). Description: similar to *F. oblonga*, but more elongated, expanding less rapidly, and with sutures tending to be more sharply angled; smooth, or with numerous thin longitudinal striae on all or part of the test.
Range: Middle Miocene to basal Late Pliocene.

Frondicularia oblonga Roemer, 1838
Plate 9.3, Fig. 23 (× 8), Zone NSB8, Late Oligocene, Germany (after Batjes, 1958). Description: test large, robust, very variable in shape, typically broad and flattened, with evenly curved sutures and rounded to truncated periphery; smooth, or with slightly limbate sutures. Remarks: *Frondicularia ovata* (Roemer) and *F. obliqua* (Roemer) are probable synonyms, but the synonymy given by Batjes (1958) is probably too extensive.
Range: Late Oligocene to ?basal Early Miocene.

Lenticulina gutticostata Guembel, 1868
Plate 9.3, Fig. 19 (× 34), Middle Eocene, northern Germany (after Kiesel, 1970). Description: test compressed, periphery keeled, sutures recurved; ornament rather variable, comprising sutural ribs which may break up into rows of tubercles, ending in thickened knobs at their umbilical end, which tend to fuse into an irregular ring around the umbilicus.
Range: Middle Eocene to ?Late Eocene.

Lenticulina platypleura (Jones)
Plate 9.3, Fig. 20 (× 20), Subzone NSB1b, Late Palaeocene, Denmark (after King, 1983). = *Cristellaria platypleura* Jones, 1852, *Astacolus platypleura* (Jones), *Lenticulina multiformis* Franke, 1911. Description: test compressed, varying in shape from lenticuline to astacoline; periphery acute; sutures limbate, with raised sutural ribs of variable strength, crossed by short irregular longitudinal ribs.
Remarks: very variable in shape and ornamentation; classification in *Lenticulina* is here preferred, as an initial lenticuline coil is usually present.
Range: Late Palaeocene.

Marginulinopsis fragraria (Guembel)
Plate 9.3, Fig. 24 (× 30), Middle Eocene, Germany (after Staesche and Hiltermann, 1940). = *Marginulina fragraria* Guembel, 1868, *Cristellaria* Te51 Staesche and Hiltermann. Description: test curved, compressed: periphery acute, with thin keel, becoming narrowly rounded and unkeeled in adult; sutures slightly depressed, low sutural ribs ornamented by c.10–15 small tubercles, initially coarse, but becoming smaller and eventually fading, which may grade into short longitudinal ribs: chamber surface granulated; the aperture has a moderately long neck.
Remarks: similar to *Vaginulinopsis decorata*, but with more delicate sutural ornament, only very short longitudinal ribs (if present) and no keel in the adult stage.
Range: Middle Eocene.

Marginulinopsis luenebergensis (Clodius)
Plate 9.4, Fig. 1 (× 20), Late Miocene, Germany (after Langer, 1969). = *Cristellaria luenebergensis* Clodius, 1922. Description: test with small initial planispiral coil, remaining part of test uncoiled and elongated; oval cross-section; short slightly inflated chambers, smooth except for one or two short spines on initial coil; periphery keeled on initial coil, otherwise rounded.
Range: Late Miocene to Early Pliocene.

Marginulinopsis pedum ackneriana (Neugeboren)
Plate 9.3, Fig. 25 (× 34), Middle Miocene, Netherlands (after ten Dam and Reinhold, 1942). = *Marginulina ackneriana* Neugeboren, 1851. Description: test with an inflated initial planispiral coil, followed by an elongated uncoiled stage with chambers moderately inflated, and periphery rounded. Remarks: differs from *M. luenebergensis* by its inflated chambers and by the absence of keel or spines on the initial coil.
Range: Middle Miocene to basal Early Pliocene.

Nodosaria intermittens Roemer, 1838
Plate 9.4, Fig. 7 (× 13), Zone NSB8, Late Oligocene, Germany (after Grossheide and Trunko, 1965). Description: test very variable in shape, straight or curved; short, rather oblique rounded longitudinal ribs, strongest over sutures and fading over centre of chambers. Remarks: as in many nodosariids, generic classification is somewhat ambiguous; this species is often assigned to *Dentalina*.
Range: Late Oligocene to basal Early Miocene.

Nodosaria koninckii (Reuss)
Plate 9.4, Fig. 2 (× 8), Zone NSB 14?, Pliocene, North Sea (after King, 1983). = *Dentalina koninckii* Reuss, 1861. Description: test large, slightly curved, with numerous thin longitudinal ribs, continuous across sutures, often irregularly bifurcating; sutures slightly depressed, becoming oblique on later chambers.
Range: Early Pliocene to Late Pliocene.

Nodosaria latejugata Guembel, 1868
Plate 9.4, Fig. 3 (× 45), Zone NSB3, Early Eocene, Germany (after Staesche and Hiltermann, 1940). Description: test large, chambers inflated, sutures depressed; few thick longitudinal ribs, crossing sutures. Remarks: *N. minor* Hantken, similar morphologically but with sutures only slightly depressed, is probably a dimorphic form of *N. latejugata*.
Range: Late Palaeocene to Late Eocene.

Nodosaria soluta (Reuss)
Plate 9.4, Fig. 4 (×30), Late Pliocene, Italy (after AGIP Mineraria, 1957). = *Dentalina soluta* Reuss, 1851, *Nodosaria radicula* (Linne) of authors. Description: test slightly curved; early chambers globular, later chambers becoming pear-shaped, slowly expanding in size; characteristically with fine 'granular' ornament at base of later chambers.
Remarks: *Dentalina/Nodosaria globifera* Reuss has a similar range and morphology and may be synonymous. The generic classification of this and similar species is somewhat ambiguous.
Range: Early Oligocene to basal Early Miocene.

Nodosaria torsicostata ten Dam, 1944
Plate 9.4, Fig. 5 (× 17), Early Palaeocene, Greenland (after Hansen, 1970, as *N. latejugata*). = *N. latejugata* (pars) of authors. Description: similar to *N. latejugata*, but the ribs are thinner and more numerous, and the sutures more constricted; the ribs tend to run obliquely across the sutures.
Remarks; this species is often confused with *N. latejugata*, but it is a morphologically distinctive Late Palaeocene index species in the North Sea Basin and adjacent areas.

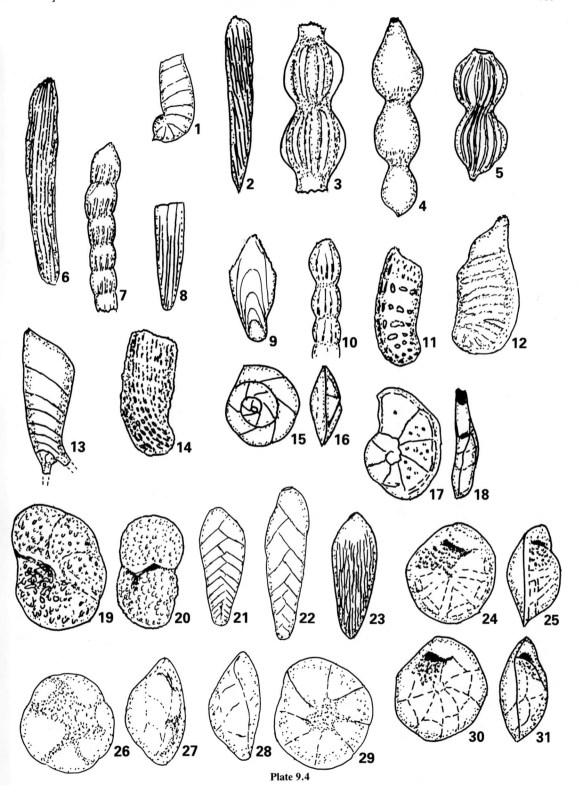

Plate 9.4

Nodosaria sp. A.
Plate 9.4, Fig. 6 (× 15), Zone NSB8, Late Oligocene, Germany (after Grossheide and Trunko, 1965). = *Nodosaria vertebralis* (Batsch) of authors, non *Nautilus (Orthoceras) vertebralis* Batsch, 1781. Description: test large, early part of test approximately straight, later becoming slightly curved; numerous platy flat-topped longitudinal ribs, continuous across sutures, tending to bifurcate irregularly. Remarks: this species is similar to *N. koninckii*, but its chambers are more cylindrical, the ribs are flat-topped and the sutures not oblique. Although it is currently referred to *N. vertebralis*, topotype specimens of the latter species, from the Pliocene or Recent of the Mediterranean, illustrated by Cushman (1931), are much more coarsely ribbed and are clearly a different species.
Range: Early Oligocene to Early Miocene.

Plectofrondicularia advena (Cushman)
Plate 9.4, Fig. 9 (× 60), Early Pliocene, Germany (after Langer, 1969). = *Frondicularia advena* Cushman, 1923. Description: test compressed, smooth, initial end narrow, initial planispiral coil very reduced, followed by a slowly expanding elliptical section, with arcuately angled chambers; sutures slightly depressed.
Remarks: originally described from Recent sediments of the North Atlantic Ocean.
Range: Late Miocene to early Late Pliocene.

Plectofrondicularia seminuda (Reuss)
Plate 9.4, Fig. 8 (× 30), Zone NSB9, Early Miocene, North Sea (after King, 1983). = *Frondicularia seminuda* Reuss, 1851. Description: test small, elongated, compressed; periphery rounded; with a variable number of longitudinal ribs (between three and 12) on early chambers, fading on later chambers.
Range: Early Oligocene to Early Miocene.

Stilostomella midwayensis (Cushman and Todd)
Plate 9.4, Fig. 10 (× 26)), Late Palaeocene, Tunisia (after Aubert and Berggren, 1976). = *Ellipsonodosaria midwayensis* Cushman and Todd, 1946, *Nodosaria spinulosa* (Montagu) of authors, non *Nautilus spinulosus* Montagu. Description: test large for the genus, elongated elliptical chambers with distinct necks; ornament very variable, from large backward-pointing spines to short ribs with spiny ends.
Range: Late Palaeocene to Early Eocene.

Vaginulina spinigera Brady, 1881
Plate 9.4, Fig. 13 (× 13). Recent, North Atlantic (after Brady, 1884). Description: test smooth, elongate, slowly increasing in width towards the aperture; sutures somewhat limbate; two long spines (usually broken) extending at divergent angles from near the initial end. Remarks: originally described from Recent sediments of the Atlantic and Pacific. In the North Sea it is widespread (but not very common) within a thin interval near the base of Zone NSB15.
Range: Late Pliocene.

Vaginulinopsis decorata (Reuss).
Plate 9.4, Fig. 11 (× 15), Zone NSB3, Early Eocene, North Sea (after King, 1983). = *Cristellaria decorata* Reuss, 1855, *Marginulina decorata* (Reuss). Description: test large, compressed, periphery subrounded to keeled; small initial planispiral coil, followed by a slightly curved uncoiled section; chambers short and wide; sutures limbate; ornament very variable, comprising coarse sutural tubercles, often elongated spirally or longitudinally into short ribs; the longitudinal ribs become more continuous towards the periphery; the chamber surface is granular. Remarks: there is a morphological and chronological gradation from an Early Eocene form with smooth sutural ribs (tuberculate only in the first 2–3 chambers), referred here to *V. decorata* subsp. A, through an Early and Middle Eocene form with thick tuberculate sutural ribs, to a Middle and Late Eocene form without sutural ribs, but with 7–10 longitudinally elongated tubercles along each suture. More study could delineate useful chronological subspecies within this very variable taxon. *Vaginulinopsis [Marginulina] wetherellii* (Jones) (= *Marginulina enbornensis* Bowen) from the Early Eocene of England, probably falls within the range of variation of *V. decorata*.
Range: Early Eocene to Late Eocene.

Vaginulinopsis decorata subsp.A
Plate 9.4, Fig. 12 (× 23), Zone NSB3, Early Eocene, North Sea. Description: similar to *V. decorata*, but with smooth sutural ribs, except on the initial coil, where irregular tuberculation is present.
Range: Early Eocene to ? Middle Eocene.

Vaginulinopsis sp.B.
Plate 9.4, Fig. 14 (× 19), Zone NSB7, Early Oligocene, North Sea (after King, 1983). = *Vaginulinopsis* sp. aff. *decorata* (Reuss) of King, 1983. Description: test compressed, periphery rounded; small initial coil followed by uncoiled stage of c. 8–10 chambers: chambers short and wide, sutures not prominent, ornamented by c. 10–15 rows of interrupted longitudinal ribs or longitudinally elongated tubercles, tending to become more continuous and sharper-edged towards the periphery, tending to fade on the last chambers.
Remarks: a widespread species in the lower part of the Early Oligocene, usually recorded as *Marginulina decorata*, but distinguished by its more compressed cross-section and its predominantly longitudinal ornament.

SUBORDER ROTALIINA

Alabamina tangentialis (Clodius)
Plate 9.4, Figs 15, 16 (× 40), Zone NSB7, Early Oligocene, Germany (after Batjes, 1958). = *Pulvinulina tangentialis* Clodius, 1922. Description: test highly compressed, biconvex, periphery acute; umbilical side more convex than spiral side; sutures slightly limbate; 5–6 chambers in last whorl.
Remarks: the common Middle Eocene to Early Oligocene species *A. wollerstorfii* (Franke) is larger, less compressed, with a more convex umbonal side and a more rounded periphery.
Range: Early Oligocene to Middle Miocene.

Almaena osnabrugensis (Roemer)
Plate 9.4, Figs 17, 18 (× 38), Zone NSB8, Late Oligocene, Germany (after Batjes, 1958). = *Planulina osnabrugensis* Roemer, 1838. Description: test coarsely perforate, compressed, flat sided; periphery acute, keeled, becoming flattened and double-keeled in last half-whorl; around eight chambers in last whorl, sutures raised, limbate, thickened on umbilical side in last half whorl and projected backwards on periphery to form second interrupted 'keel'.
Remarks: this distinctive species is restricted to Subzone NSB8c.
Range: Late Oligocene to ?basal Early Miocene.

Anomalinoides rubiginosus (Cushman)
Plate 9.4, Figs 19, 20 (× 50), Early Palaeocene, Tunisia (after Aubert and Berggren, 1974). = *Anomalina rubiginosa* Cushman, 1926, *Anomalinoides/Gavelinella [Cibicides]danica* (Brotzen, 1948). Description: test coarsely perforate, biconvex; periphery rounded; 7–8 chambers in last whorl, early part of last whorl coarsely and irregularly pitted, with irregular raised ridges, last 2–3 chambers becoming inflated, with depressed sutures,
Remarks: see van Morkhoven *et al.* (1986) for discussion and synonymy. A closely related species, very coarsely perforate and pitted, tending to develop thickened sutural ribs, is frequent in the Early and Middle Eocene. It is usually referred to *Anomalina grosserugosa* (Guembel) but is probably more correctly assigned to *Anomalinoides capitatus* (Guembel).
Range: Late Cretaceous (Campanian) to Late Palaeocene, ?Early Eocene.

Asterigerina guerichi guerichi (Franke)
Plate 9.4, Figs 30, 31 (× 56), Zone NSB8, Late Oligocene, Belgium (after Batjes, 1958). = *Discorbina guerichi* Franke, 1912. Description: test approximately circular, biconvex, lensoid cross-section; umbilical side and spiral side approximately equally convex; eight chambers in last whorl.
Range: Late Oligocene to Early Miocene.

Asterigerina guerichi staeschei ten Dam and Reinhold, 1941
Plate 9.4, Figs 24, 25 (× 56), Zone NSB11, Middle Miocene, Germany (after Batjes, 1958). Description: test subcircular to rounded quadrangular, biconvex; umbilical side much more convex than spiral side; 6–7 chambers in last whorl.
Remarks: similar to *A. g. guerichi*, but distinguished by its shape and cross-section, by the presence of granulation around the umbilicus spread more widely over the umbonal area, and by its smaller supplementary chambers. These two subspecies are probably chronological variants; morphologically intermediate specimens occur in the Late Oligocene and Early Miocene.
Range: Early to Middle Miocene.

Bolivina antiqua d'Orbigny, 1846
Plate 9.4, Fig. 22 (× 78). Zone NSB8, Late Oligocene, Netherlands (after Doppert, 1980). Description: test smooth, elongate, periphery rounded, compressed; sutures flush.
Range: Late Oligocene to basal Late Miocene.

Bolivina cookei Cushman, 1922
Plate 9.4, Fig. 23 (× 85), Zone NSB6, Late Eocene, North Sea (after King, 1983). Description: test elongate, periphery acute; numerous thin longitudinal ribs, rather irregular and often bifurcating, fading towards the aperture.
Range: Middle Eocene to Early Oligocene.

Bolivina dilatata Reuss, 1850
Plate 9.4, Fig. 21 (× 68), Zone NSB11; Middle Miocene, Germany (after Batjes, 1958). Description: test elongated, compressed, tapering; periphery subacute to subrounded; chambers narrow, elongated, smooth except for weak central rib in early stages.
Range: Early to Middle Miocene.

Buccella frigida (Cushman)
Plate 9.4, Figs 26, 27 (× 78), Zone NSB15, Late Pliocene/Early Pleistocene, North Sea (after King, 1983). = *Pulvinulina frigidus* Cushman, 1922. Description: test small, average maximum diameter 0.35 mm, unequally biconvex, with a higher spiral side; periphery subrounded; approximately six chambers in last whorl; umbilical area and sutures on umbilical side filled by granular calcite.
Remarks: a similar form, but apparently morphologically separable, occurs in the Late Miocene.
Range: Early Pliocene?. Late Pliocene to Recent.

Buccella tenerrima (Bandy)
Plate 9.4, Figs 28, 29 (× 50), Zone NSB16, Pleistocene, Norway (after Feyling-Hanssen *et al.*, 1971). = *Rotalia tenerrima* Bandy, 1950. Description: test larger than *B. frigida*, maximum average diameter 0.6 mm, biconvex to planoconvex, with low conical spiral side and less convex umbilical side, periphery subangular; 8–10 chambers in last whorl; relatively coarse granular calcite filling umbilicus and extending narrowly along sutures.
Range: Late Pliocene to Pleistocene. Living in arctic areas.

Bulimina aksuatica Morozova, 1939
Plate 9.5, Fig. 1 (× 60), Eocene, U.S.S.R. (after Morozova). = *Bulimina truncana* var. *aksuatica* Morozova. Description: test small, tapering evenly, triangular in cross-section, chambers not inflated; sutures ill-defined; numerous fine longitudinal ribs, crossing sutures without interruption.
Range: Early Eocene to Late Eocene.

Bulimina dingdenensis Batjes, 1958
Plate 9.5, Fig. 2 (× 64), Zone NSB11, Middle Miocene, Germany (after Batjes, 1958). Description: test small, 'pyramidal', subtriangular cross-section; chambers inflated, rounded to triangular; sutures deeply depressed; chambers ornamented by short hollow spines/tubercles.
Range: Early and Middle Miocene

Bulimina elongata d'Orbigny, 1846
Plate 9.5, Fig. 3 (× 56), Zone NSB12, Middle Miocene, North Sea, (after King, 1983). Description: test small, smooth, elongated, maximum width usually midway between base and apex; chambers not greatly inflated. Remarks: this species tends to be confused with elongate variants of the *Bulimina marginata/aculeata* group, but is regarded here as morphologically and stratigraphically distinct. As differentiated here, *B. elongata* is a small, entirely smooth, elongated taxon which is restricted to Subzone NSB13a and older levels; *B. marginata* and associated species (including weakly spinose elongate variants often referred to *B. elongata*) do not occur below Zone NSB14 (see also de Meuter, 1980).
Range: Late Oligocene to basal Late Miocene.

Bulimina gibba Fornasini, 1902
Plate 9.5, Fig. 4 (× 45), Zone NSB15, Late Pliocene, North Sea (after King, 1983). Description: test elongate, tapering; earlier part of the test tending to be triangular in cross-section; early chambers variably spinose; later chambers rounded, smooth.
Remarks: similar to *B. aculeata* d'Orbigny, which is less elongate and has more inflated chambers. Illustrated by King, 1983 (Plate 2, Fig. 18), as *B. marginata* (pars).
Range: Early Pliocene?; Late Pliocene to Recent.

Bulimina marginata d'Orbigny, 1826
Plate 9.5, Fig. 5 (×45), Pleistocene, Gulf of Mexico (after van Morkhoven, *et al.*, 1986). Description: test expanding rapidly from apex; chambers inflated; basal margins of chambers undercut, with numerous short spines.
Remarks: very variable, both in shape and ornamentation. Differs from *B. gibba* by its angular chamber margins and less elongate test.
Range: Late Pliocene to Recent.

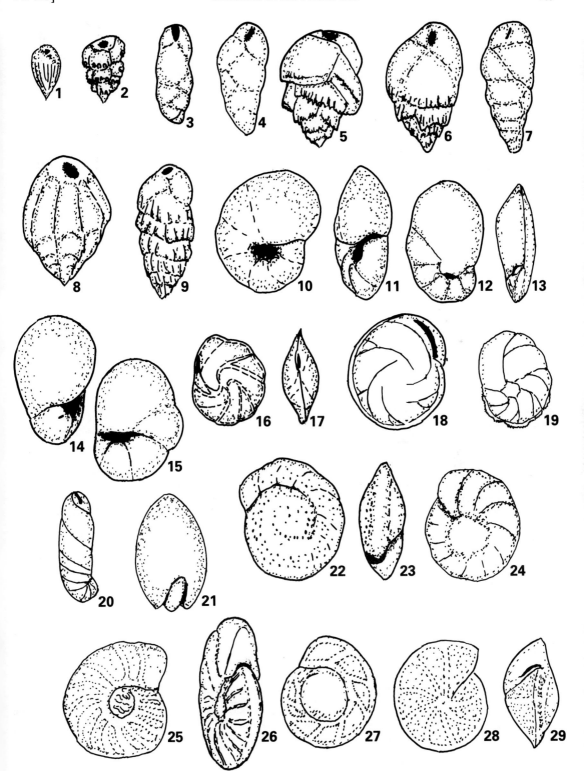

Plate 9.5

Bulimina midwayensis Cushman and Parker, 1936.
Plate 9.5, Fig. 6 (× 70), Subzone NSB1b, Late Palaeocene, North Sea (after King, 1983). Description: test triangular in profile, rounded in cross-section, expanding rapidly; chambers inflated, lower part of chambers with weak ribs ending in short blunt overhanging spines. Remarks: a geographically widespread index species for the Palaeocene. A similar form occurs in the Early Eocene (Subzone NSB3b) in the North Sea Basin.
Range: Late Palaeocene.

Bulimina trigonalis ten Dam, 1944
Plate 9.5, Fig. 7 (× 60), Late Palaeocene, Tunisia (after Aubert and Berggren, 1975). Description: test rather small, (c.0.3 mm in length), smooth, elongated, tapering, early part trigonal in cross-section, often slightly twisted; later part of test becoming rounded; chambers moderately inflated.
Remarks: very similar and perhaps identical to *B. thanetensis* Haynes (Late Palaeocene, England).
Range: Late Palaeocene.

Bulimina truncana Guembel, 1868
Plate 9.5, Fig. 8 (× 68), Zone NSB6, Late Eocene, North Sea (after King, 1983). Description: test large, stout, rapidly expanding; sutures flush; strong longitudinal ribs, extending continuously from initial end to last whorl.
Range: Late Eocene.

Bulimina sp. A.
Plate 9.5, Fig. 9 (× 56), Subzone NSB3b, Early Eocene, North Sea (after King, 1983). = *Bulimina* sp. nov. King, 1983. Description: test elongated, tapering, rounded in cross-section; chambers overhanging; sutures deep, with 8–10 prominent ribs per chamber, projecting slightly below base of chambers.
Remarks: not comparable to any described Eocene species, but similar to some Neogene taxa, such as *B. calabra* Seguenza.
Range: Early Eocene.

Cancris auriculus (Fichtel and Moll)
Plate 9.5, Figs 12, 13 (× 47), Zone NSB11, Middle Miocene, Germany (after Batjes, 1958). = *Nautilus auricula* Fichtel and Moll, 1803. Description: test highly compressed; periphery acute, keeled; chambers in the last whorl expanding rapidly in size.
Range: Early Miocene to Late Miocene, ?Early Pliocene.

Cancris subconicus (Terquem)
Plate 9.5, Figs 14, 15 (× 80). Late Eocene, North Sea (after Hughes, 1981). = *Rotalina subconica* Terquem, 1882, *Cancris turgidus* Cushman and Todd, 1942. Description: test rather coarsely perforate, compressed; periphery subrounded, somewhat lobate; chambers inflated, last chamber large and highly inflated.
Range: Late Eocene to Early Oligocene; Middle Eocene to Late Oligocene in onshore sequences.

Cancris sp. A.
Plate 9.5, Figs 10, 11 (× 112), Subzone NSB3b, Early Eocene, North Sea (after King, 1983, as *C. auriculus primitivus*). Description: test rather coarsely perforate, compressed; periphery bluntly subangular; around seven chambers in the last whorl, expanding rather slowly except for the final chamber, which forms almost one-quarter of the last whorl. Remarks: referred to *C. auriculus primitivus* Cushman and Todd by King (1983), but now regarded as a distinct species which cannot be matched satisfactorily with any named taxon.
Range: restricted to the Early Eocene, Subzone NSB3b, in the North Sea, and recorded at the same level in Denmark and northwest Germany.

Cassidulina carapitana Hedberg, 1937
Plate 9.5, Figs 16, 17 (× 45), Subzone NSB7a, Early Oligocene, Belgium (after Batjes, 1958). Description: test compressed, lenticular, periphery acute; rather inflated umbonal area; 9–12 chambers in last whorl; sutures sharply flexed around umbonal area.
Remarks: this species can be confused with *C. laevigata*, especially if poorly preserved, but its inflated umbonal area and flexed sutures are distinctive.
Range: Early Oligocene (Subzone NSB7a).

Cassidulina laevigata d'Orbigny, 1826
Plate 9.5, Fig. 18 (× 60), Pleistocene, North Sea. Description: test compressed; periphery acute, may develop a thin keel; 10–14 chambers in last whorl.
Remarks; see comments under *Islandiella helenae*.
Range: Late Miocene to Recent.

Cassidulina pliocarinata van Voorthuysen, 1950
Plate 9.5, Fig. 19 (× 53), Pliocene, Netherlands (after van Voorthuysen, 1950). Description: test compressed; thin peripheral keel with irregular 'ragged' edge, becoming more regular on the last chambers; 10–14 chambers in the last whorl.
Remarks: can be confused with specimens of *C. laevigata* which have a damaged periphery.
Range: Late Miocene to Late Pliocene.

Cassidulinoides bradyi (Brady)
Plate 9.5, Fig. 20 (× 56), Recent, North Atlantic (after Brady, 1884). = *Cassidulina bradyi* Brady, 1884. Description: test compressed, elongate–ovate; early part enrolled biserial, later chambers uncoiled; sutures flush to slightly depressed.
Range: the types are from Recent sediments, but in the North Sea this species is apparently restricted to the Late Miocene and Early Pliocene.

Chilostomellina fimbriata Cushman, 1926
Plate 9.5, Fig. 21 (× 72), Zone NSB13, Late Miocene, North Sea (after King, 1983). Description: test subspherical to elongate ovoid, planispiral; chambers increasing rapidly in size, last chamber very large and almost completely enveloping previous chambers.
Range: Late Miocene to basal Late Pliocene.

Cibicides grossus (ten Dam and Reinhold)
Plate 9.6, Figs 12, 13 (× 30), Zone NSB15, Late Pliocene/Early Pleistocene, North Sea (after King, 1983). = *Cibicides lobatulus* var. *grossa* ten Dam and Reinhold, 1941, *Cibicides rotundatus* Stschedrina, 1964. Description: test large, rather coarsely perforate, inflated, periphery rounded; spiral side rather flattened, with the last 3–4 chambers becoming more inflated; umbilical side highly convex; 8–10 chambers in last whorl.
Remarks: an index species for the Late Pliocene and earliest Pleistocene in circum-Arctic areas, recorded from the North Sea Basin, the Norwegian Sea, Greenland, the Canadian Arctic and Siberia.

Cibicidoides dayi (White)
Plate 9.5, Figs 25, 26 (× 45), Early Palaeocene, Tunisia (after Aubert and Berggren, 1975). = *Planulina dayi* White, 1928. Description: test compressed biconvex, periphery subacute; 12–17 chambers in last whorl; depressed spiral suture; spiral side with irregular raised ridges on umbonal area; umbilical side smooth, mainly imperforate, with umbilicus covered by smooth area.
Remarks: see the discussion in van Morkhoven *et al.* (1986).
Range: Late Cretaceous to Late Palaeocene.

Cibicidoides dutemplei peelensis (ten Dam and Reinhold)
Plate 9.6, Figs 3, 4 (× 45), Early/Middle Miocene, Belgium (after Batjes, 1958). = *Cibicides peelensis* ten Dam and Reinhold, 1942, *Heterolepa dutemplei peelensis* (ten Dam and Reinhold). Description: test biconvex, umbilical side tending to be more convex than spiral side, periphery acute; moderately coarsely perforate, except over sutures and on periphery; sutures depressed, radial on umbilical side, limbate and oblique on spiral side.
Remarks: *C. dutemplei dutemplei* (d'Orbigny) has strongly curved sutures on the umbilical side, and a generally planoconvex profile. *C. d. peelensis* is differentiated from *C. mexicanus* by its more compressed profile and chambers more elongated spirally.
Range: Early and Middle Miocene.

Cibicidoides eocaenus (Guembel)
Plate 9.6, Figs 1, 2 (×30), Middle Eocene, Atlantic Ocean (after Tjalsma and Lohmann, 1983). = *Rotalia eocaena* Guembel, 1868, *Heterolepa eocaena* (Guembel), *Cibicides perlucida* Nuttall, 1932, *C. tuxpamensis* Cole, 1928. Description: test biconvex, coursely perforate, except for non-perforate spiral suture; periphery angular, spiral side moderately convex to almost flattened; sutures oblique; umbilical side highly convex, with glassy umbilical area; a large glassy umbonal knob in some microspheric forms. Remarks: see van Morkhoven *et al*. (1986) for discussion of the synonymy of this and allied species.
Range: Early Eocene to Late Eocene.

Cibicidoides limbatosuturalis (van Voorthuysen)
Plate 9.6, Figs 8, 9 (× 56), Pliocene, Netherlands (after van Voorthuysen). = *Cibicides cookei* var. *limbatosuturalis* van Voorthuysen, 1950. Description: test biconvex; periphery acute, may be keeled; umbilical side evenly convex, sutures radial and limbate; spiral side less convex, umbonal area inflated but last whorl flattened to slightly concave, sutures oblique; around eight chambers in the last whorl.
Remarks: a distinctive short-ranging species, rather similar to the '*dutemplei*' group, but without coarse perforation and with a characteristic profile on the spiral side.
Range: Late Miocene to basal Late Pliocene (Subzone NSB14a).

Cibicidoides mexicanus (Nuttall)
Plate 9.5, Figs 27–29 (× 47), Zone NSB7, Early Oligocene, Belgium (after Batjes, 1958). = *Cibicides mexicana* Nuttall, 1932, *Heterolepa dutemplei praecincta* (Franzenau, 1884). Description: test biconvex to planoconvex, rather coarsely perforate; spiral side flattened to slightly concave, sutures oblique, with early whorls covered by a thickened area of glassy calcite; umbilical side very convex, sutures curved, defined by slightly limbate non-perforate bands.
Remarks: distinguished from *C. eocaenus* by its less inflated chambers, its finer and more even perforation, and its flattened last whorl on the spiral side. *C. dutemplei, C. mexicanus* and *C. eocaenus* are related species often assigned to the genus *Heterolepa*. The taxonomic interpretation used here follows van Morkhoven *et al*. (1986).
Range: Early Oligocene.

Cibicidoides pachyderma (Rzehak)
Plate 9.5, Figs 22–24 (× 42), Subzone NSB15a, Late Pliocene, North Sea (after King, 1983). = *Truncatulina pachyderma* Rzehak, 1886, *Cibicides pseudoungerianus* Cushman, 1931. Description: test biconvex, periphery acute, keeled; spiral side coarsely perforate, umbilical side very finely perforate, with slightly swollen umbonal area; 10–12 chambers in the last whorl. Remarks: see van Morkhoven *et al*. (1986) for discussion of taxonomy and synonymy.
Range: Middle Miocene to basal Pleistocene.

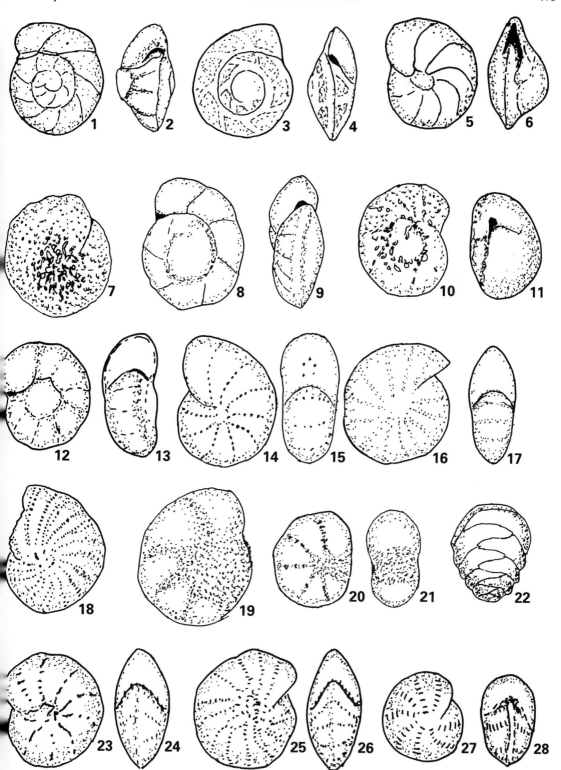

Plate 9.6

Cibicidoides succedens (Brotzen)
Plate 9.6, Figs 5, 6 (× 75), Zone NSB1, Late Palaeocene, Sweden (after Brotzen, 1948) = *Cibicides succedens* Brotzen, 1948. Description: test planoconvex or slightly biconvex; spiral side almost flat, umbilical side strongly convex, swelling into a prominent central boss; periphery acute, may be keeled; sutures limbate, raised and strongly curved on the spiral side.
Range: Late Palaeocene.

Cibicidoides truncanus (Guembel)
Plate 9.6, Fig. 7 (× 38), Zone NSB6, Late Eocene, North Sea. = *Rotalia truncana* Guembel, 1868, *Cibicidoides granulosus* (Bykova) of authors non *Pseudoparrella granulosa* Bykova, 1952. Description: test biconvex, coarsely perforate, umbilical side more convex than spiral side; periphery sharp, with a broad 'frilled' keel; coarsely perforate; umbilical side covered by a pattern of raised ridges, forming a 'rosette-like' ornament; 9–10 chambers in the last whorl.
Remarks: a worldwide Middle–Late Eocene index species (van Morkhoven *et al.*, 1986).
Range: Late Eocene.

Cibicidoides velascoensis (Cushman)
Plate 9.6, Figs 10, 11 (× 70), Late Palaeocene, south Atlantic (after van Morkhoven *et al.*, 1986). = *Anomalina velascoensis* Cushman, 1925. Description: test planoconvex, periphery rounded; spiral side relatively flattened, coarsely perforate, sutures indistinct, with a thickened central boss; umbilical side very convex, rounded, smooth; 8–9 chambers in last whorl.
Range: Campanian to Late Palaeocene (see van Morkhoven *et al.*, 1986).

Cribroelphidium bartletti (Cushman)
Plate 9.6, Figs 14, 15 (× 50), Pleistocene, Norway (after Feyling-Hanssen *et al.*, 1971). = *Elphidium bartletti* Cushman, 1933. Description: periphery broadly rounded; sutures radial, slightly depressed, with numerous very short septal bridges, giving a 'beaded' appearance; sutural fissures reduced to a series of pores; 10–12 chambers in the last whorl; aperture comprises several large pores on the apertural face.
Remarks: the multiple areal aperture precludes assignment to *Elphidium*.
Range: Pleistocene. Living in arctic and subarctic areas.

Cribroelphidium vulgare Voloshinova, 1952
Description: test compressed, periphery broadly rounded, chambers moderately inflated; sutures depressed, numerous short septal bridges, with septal fissures reduced to pores; aperture multiple, formed by areal and basal pores.
Remarks: this species has been synonymised with *C. bartletti* by most authors (e.g. Feyling-Hanssen, *et al.*, 1971). The specimens from the North Sea referred to this species are smaller (average maximum diameter 0.4 mm, compared to 0.8–0.9 mm for *C. bartletti*), more involute, and less compressed than *C. bartletti*. They are also stratigraphically older than *C. bartletti*, occurring in the Early Pliocene. The type material of *C. vulgare* is from the Late Miocene of Sakhalin Island, U.S.S.R.
Range: Early Pliocene. A very short-ranging species, but widespread in the topmost part of Zone NSB13 in the central North Sea.

Ehrenbergina serrata Reuss, 1850
Plate 9.6, Fig. 22 (× 70), Zone NSB8, Late Oligocene, Germany (after Spiegler, 1973). Description: test enrolled biserial in juvenile, becoming uncoiled; uncoiled section smooth, compressed, periphery acute and serrate.
Remarks: the evolution of this species and the related species *E. variabilis* Trunko is detailed by Spiegler (1973).
Range: Late Oligocene to Middle Miocene.

Elphidiella hannai (Cushman and Grant)
Plate 9.6, Figs 16, 17 (× 38), Zone NSB16, Pleistocene, North Sea (after King, 1983). = *Elphidium hannai* Cushman and Grant, 1927. Description: test compressed, biconvex; periphery subrounded to subangular; sutures limbate, flush, with a double row of septal pits; up to 15 chambers in the last whorl.
Remarks: this species is similar to *E. arctica* (Parker and Jones), which has a broadly rounded periphery, a flat-sided

cross section, and only 10–12 chambers in the last whorl. *E. arctica* is apparently restricted to later Pleistocene and Recent sediments, while *E. hannai* is characteristic of the Early Pleistocene in northwest Europe.
Range: this species occurs abundantly in the 'Early Pleistocene' of the southern North Sea (Zone NSB16); here specimens are small (average maximum diameter 0.5 mm, with only 8–10 chambers in the last whorl); in the northern North Sea it is common in Zone NSB15, where it is much larger (up to 10 mm in diameter) and has around 15 chambers in the last whorl.

Elphidium albiumbilicatum (Weiss)
Plate 9.6, Fig. 19 (× 150), Quaternary, Norwegian Sea (after Kihle and Lofaldi, n.d.) = *Nonion pauciloculum albiumbilicatum* Weiss, 1954, *Nonion depressulus* (Walker and Jacob) forma *asterotuberculatus* van Voorthuysen, 1957. Description: periphery rounded; sutures depressed, obscured by granular ornament which extends over the umbilical area; 7–9 chambers in the last whorl.
Range: Pleistocene and Recent.

Elphidium antoninum (d'Orbigny)
Plate 9.6, Figs 20, 21 (× 75), Late Miocene, Netherlands (after Doppert, 1980). = *Polystomella antonina* d'Orbigny, 1846. Description: test small, compressed, periphery rounded; sutures depressed, with about 10 septal bridges, obscured by granular ornament; 7–10 chambers in the last whorl.
Range: Middle Miocene.

Elphidium excavatum (Terquem)
Plate 9.6, Figs 23, 24 (× 56), Zone NSB15, Pleistocene, North Sea (after King, 1983). = *Polystomella excavata* Terquem, 1876, *Elphidium incertum clavatum* Cushman, *E. selseyense* Heron-Allen. Description: periphery rounded, slightly lobate; sutures slightly curved, depressed, with elongated sutural fissures, partly obscured by granular ornament, crossed by a few septal bridges; 7–9 chambers in the last whorl. Remarks: a highly variable species with a complex taxonomic history. 'Formae' defined by Feyling-Hanssen (1972) reflect different climatic environments, and can be used to identify glacial/interglacial episodes. *E. excavatum* 'forma' *clavata* Cushman has a yellowish-brown translucent test, and is common in arctic (glacial) sequences in the North Sea Basin and elsewhere in northwest Europe.
Range: Early Pliocene to Recent.

Elphidium groenlandicum Cushman, 1923
Plate 9.6, Figs 25, 26 (× 26), Zone NSB15, Late Pliocene/Early Pleistocene, North Sea (after King, 1983). Description: test large, compressed, with a subacute periphery; sutures slightly curved, slightly depressed, with numerous septal bridges; umbonal area slightly inflated; 12–15 chambers in the last whorl.
Remarks: This species has its highest consistent occurrence in the middle of Subzone NSB15b in the North Sea, but rare specimens occur at higher levels, especially in the northern Viking Graben, and it is also recorded from younger Pleistocene sediments in the North Sea Basin (e.g. by Feyling-Hanssen et al., 1971).
Range: Early Pliocene?, Late Pliocene and Pleistocene. Living in arctic areas.

Elphidium inflatum (Reuss)
Plate 9.6, Figs 27, 28 (× 50), Early/Middle Miocene, Belgium (after Batjes, 1958). = *Polystomella inflata* Reuss, 1861. Description: test slightly compressed, test wall with 'granular' texture; periphery broadly rounded, thick rounded keel; numerous septal bridges; 9–10 chambers in the last whorl.
Range: Early to Middle Miocene.

Elphidium oregonense Cushman and Grant, 1927
Plate 9.6, Fig. 18 (× 26), Zone NSB17, Pleistocene, North Sea (after King, 1983). Description: test large for the genus, up to 1.8 mm in diameter, compressed, rather flat-sided, periphery rounded; sutures depressed, numerous septal bridges; many chambers (around 20 in the last whorl). Remarks: a distinctive large species occurring commonly in a thin interval in the Early Pleistocene in the southern North Sea and adjacent areas (Subzone NSB16a), and recorded rarely in the northern North Sea.
Range: Early Pleistocene.

Elphidium subarcticum Cushman, 1944
Plate 9.7, Fig. 1 (× 62), Quaternary, Norwegian Sea (after Kihle and Løfaldi, n.d.). Description: periphery rounded, lobate; sutures slightly depressed, with numerous short septal bridges, and widely spread very fine granular ornament; 8–10 chambers in the last whorl, slightly inflated.
Range: Pleistocene to Recent.

Elphidium subnodosum (Roemer)
Plate 9.7, Figs 2, 3 (× 22), Subzone NSB8b, Late Oligocene, Belgium (after Batjes, 1958) = *Robulina subnodosa* Roemer, 1838. Description: test large, biconvex; periphery acute, keeled in adult; raised umbonal boss; numerous chambers (8–20 per whorl, increasing in number in last whorl); sutures vary from short fissures with few septal bridges to longer fissures.
Remarks: 'typical' large specimens are restricted to the Late Oligocene (Subzone NSB8b); smaller specimens, doubtfully referred to this species, with a morphology similar to juvenile *E. subnodosum*, occur from Early Oligocene to Early Miocene.

Elphidium ungeri (Reuss)
Plate 9.7, Figs 4, 5 (× 39), Zone NSB11, Middle Miocene, Germany (after Langer, 1969) = *Polystomella ungeri* Reuss, 1850. Description: test small, compressed; periphery arched, narrowly subrounded to subangular, with a smooth raised band or rounded keel; 10–12 chambers in the last whorl; sutures slightly depressed, with numerous elongated septal bridges.
Remarks: similar to *E. inflatum* but more compressed, with a narrower periphery and a larger numbers of chambers in the last whorl.
Range: Early? and Middle Miocene.

Elphidium 'ustulatum Todd, 1957'
Plate 9.7, Fig. 6 (× 68), Quaternary, Norwegian Sea (after Kihle and Lofaldi, n.d.). Description: test compressed; sutures curved, with short sutural fissures (partly filled by granular ornament), which do not extend to the periphery or the umbonal area; about 9–10 chambers in the last whorl. Remarks: topotypes of *E. ustulatum*, from the 'Pliocene' (actually Miocene?) of Carter Creek, Alaska, kindly donated by Dave McNeil, are quite distinct from this form, which requires renaming.
Range: Pleistocene.

Florilus boueanus (d'Orbigny)
Plate 9.7, Figs 7, 8 (× 42), Zone NSB10, Early Miocene, North Sea (after King, 1983). = *Nonionina boueana* d'Orbigny, 1846. Description: test moderately large, evolute to rather involute, compressed; periphery subangular to subrounded; chambers increase rapidly in height in last whorl; 10–15 chambers in last whorl; umbilicus filled by granular calcite, extending variably into sutures; sutures radial to curved, depressed.
Remarks: a variable long-ranging species, within which two main subgroups can be identified; one is compressed, with numerous chambers and curved sutures, the other (including *Nonion dingdeni* Cushman) has fewer chambers, a less compressed last whorl and radial to slightly curved sutures. The exact taxonomic status of these subgroups is uncertain (see De Meuter, 1980). Similar but smaller forms, only doubtfully specifically distinct, occur in the Eocene in onshore sequences (*Florilus commune* d'Orbigny sp.).
Range: Late Oligocene to Early Pliocene.

Fursenkoina schreibersiana (Czjzek)
Plate 9.7, Fig. 9 (× 38), Zone NSB8, Late Oligocene, North Sea (after King, 1983). = *Virgulina schreibersiana* Czjzek, 1848. Description: test large for the genus, elongated, expanding slightly from the apex, compressed in cross-section; periphery rounded; chambers are not greatly inflated.
Range: Early Oligocene?; Late Oligocene to Early Miocene.

Glabratella ? sp. A.
Plate 9.7, Figs 10, 11 (× 135), Zone NSB8, Late Oligocene, North Sea (after King, 1983). Description: test small, compact, trochospiral, rather coarsely perforate, granular, often brown in colour; chambers globular, 4–5 chambers in the last whorl; umbilicus obscured by coarse granulation.
Remarks: the exact affinities of this species are uncertain; it rather resembles a small globigerinid in shape and texture. The generic assignment is uncertain, as this species occurs associated with '*Rhabdammina*-biofacies' agglutinants; whereas *Glabratella* occurs mainly in shallow marine environments.
Range: Late Oligocene to Early Miocene.

Globobulimina socialis (Bornemann)
Plate 9.7, Fig. 14 (× 49), Zone NSB7, Early Oligocene, Germany (after Cushman and Parker, 1947). = *Bulimina socialis* Bornemann, 1855. Description: test ovate, subglobular, chambers inflated, sutures depressed; last chamber forms most of the test; aperture loop-shaped. Remarks: several similar species occur in the Cenozoic of the North Sea. *G. auriculata* (Bailey) (Pliocene to Recent), is more elongated, with less inflated chambers, a rather pointed apex and a terminal aperture. *Praeglobobulimina ovata* (d'Orbigny) (Late Palaeocene to Early Oligocene) is also more elongated, with more numerous chambers.
Range: Early Oligocene.

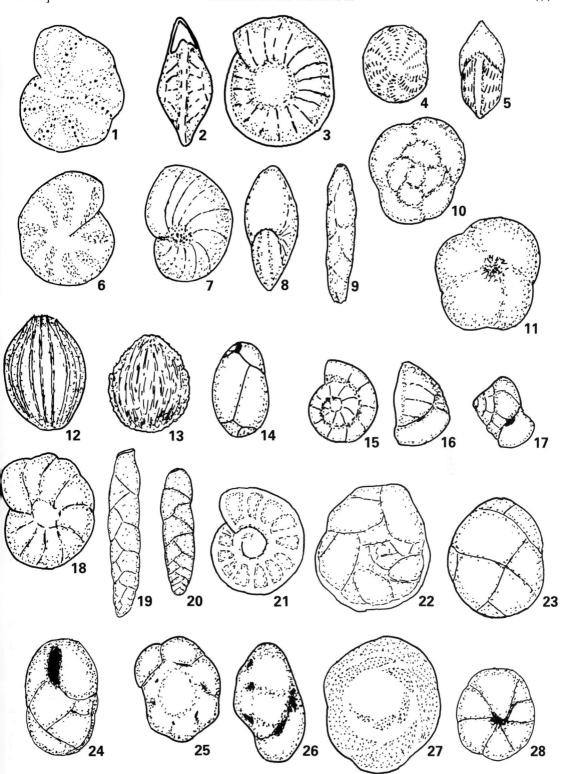

Plate 9.7

Globulina gibba myristiformis (Williamson)
Plate 9.7, Fig. 12 (× 95), Recent, Great Britain (after Murray, 1971). = *Polymorphina myristiformis* Williamson, 1858. Description: test globular; prominent sharp continuous longitudinal ribs, occasionally interrupted or bifurcating. Range: Middle Miocene to Late Pliocene. Also occurs in Recent sediments in the North Sea and adjoining waters (Murray, 1971).

Globulina gibba rugosa d'Orbigny, 1846
Plate 9.7, Fig. 13 (× 38), Pliocene, North Sea. Description: test globular, subspherical; numerous closely spaced short irregular longitudinal ribs, with rather spinose ends, grading to short blunt spines.
Range: Early Miocene to Late Pliocene.

Gyroidina soldanii girardana (Reuss)
Plate 9.7, Figs 15, 16 (× 45), Zone NSB7, Early Oligocene, Belgium (after Batjes, 1958) = *Rotalina girardana* Reuss, 1851. Description: *G. soldanii soldanii* has a biconvex to planoconvex lenticular test with a flattened or slightly convex spiral side and 8–10 chambers in the last whorl. *G. s. girardana* has a more conical profile, with a higher, more compressed last whorl, and a depressed spiral suture in the last whorl.
Range: ?Late Eocene; Early Oligocene to basal Early Miocene.

Gyroidina soldanii mamillata (Andreae)
Plate 9.7, Fig. 17 (× 60), Zone NSB7, Early Oligocene, Belgium (after Batjes, 1958). = *Rotalia mamillata* Andreae, 1884. Description: similar to *G. s. girardana*, but with a high conical spiral side and a rounded periphery.
Remarks: the height of the spiral side is rather variable. Specimens with a very high spiral side approach *Rotaliatina bulimoides* in morphology, and there is possibly a genetic relationship between these taxa.
Range: Early Oligocene to early Late Oligocene.

Haynesina orbiculare (Brady)
Plate 9.7, Fig. 18 (× 47), Zone NSB15, Late Pliocene/Early Pleistocene, North Sea (after King, 1983). = *Nonionina orbicularis* Brady, 1881, *Protelphidium orbiculare* (Brady). Description: test compressed, periphery broadly rounded; sutures slightly depressed, with sutural slits of variable length; 10–12 chambers in the last whorl.
Range: Late Pliocene to Pleistocene. Living in arctic areas.

Hyalinea balthica (Schroeter)
Plate 9.7, Fig. 21 (× 50), Recent, North Atlantic (after Brady, 1884). = *Nautilus balthicus* Schroeter, 1783. Description: test small, compressed; flattened sides, periphery subrounded to subacute; sutures radial, limbate; thickened peripheral keel; secondary apertures under umbilical flaps; 9–12 chambers in the last whorl.
Range: Pleistocene to Recent.

Islandiella helenae Feyling-Hanssen and Buzas, 1976
Plate 9.7, Fig. 22 (× 54), Recent, Alaska (after Feyling-Hanssen and Buzas, 1976). = *Cassidulina teretis* Tappan of authors *non* Tappan, 1951. Description: test large, translucent to hyaline, compressed; periphery subacute; ten chambers in the last whorl, alternately large and small. Remarks: differentiated from *Cassidulina teretis* by its wall structure and apertural characteristics. The latter are not easily observable, but the radial wall structure of *Islandiella* gives it a characteristic 'glossy' and often 'brilliant' appearance, compared with the rather dull 'frosted' granular texture of most *Cassidulina* species (see Feyling-Hanssen and Buzas, 1976). *I. norcrossi* (Cushman), which also occurs in the Pleistocene of the North Sea, is similar in shape, but smaller, and its chambers are of approximately equal size.
Range: Late Pliocene to Pleistocene. Living in arctic areas.

Islandiella islandica (Nørvang)
Plate 9.7, Figs 23, 24 (× 60), Pleistocene, Baffin Island (after Feyling-Hanssen, 1980). = *Cassidulina islandica* Nørvang, 1945, *Cassidulina crassa* d'Orbigny (*pars*) of authors (including King, 1983) *non* d'Orbigny, 1839.

Description: test large, up to 1.0 mm in diameter, glassy, inflated, periphery broadly rounded; sutures slightly depressed; aperture large, elliptical to subtriangular. Remarks: similar to *Cassidulina reniforme* Nørvang (= *Cassidulina crassa pars* of authors *non* d'Orbigny, 1839), but distinguished by its size (*C. reniforme* has a mean diameter of c.0.25 mm), its apertural characteristics and its wall structure (see Sejrup and Guilbault, 1980).
Range: Late Pliocene to Recent.

Karreria fallax Rzehak, 1891
Description: test attached in life, initial trochospiral section followed by 2–3 uniserial chambers; one side of test (originally attached to substrate) flattened or concave, with a sharp periphery, the other side is convex and inflated. Remarks: see the illustration in Aubert and Berggren (1976, plate 12, Fig.5).
Range: Late Eocene to Late Oligocene, ?basal Early Miocene. Maastrichtian to Eocene in onshore sequences.

Loxostomoides lammersi (Cushman and ten Dam)
Plate 9.7, Fig. 19 (× 40), Pliocene, Netherlands (after Cushman and ten Dam) = *Loxostomum lammersi* Cushman and ten Dam, 1947. Description: test 'glassy', finely perforate, elongated, largely parallel-sided; periphery compressed, broadly rounded; chambers rather elongated, sutures distinct,
Range: Late Miocene to Late Pliocene.

Loxostomum sinuosum (Cushman)
Plate 9.7, Fig. 20 (× 64), Zone NSB11, Middle Miocene, Germany (after Batjes, 1958). = *Loxostoma sinuosa* Cushman, 1936. Description: test elongated, compressed, periphery rounded; later chambers become lobate, with rather sinuous sutures.
Range: Early and ?Middle Miocene.

Melonis pompilioides (Fichtel and Moll)
Plate 9.8, Figs 5, 9 (× 70), Middle Miocene, Austria (after van Morkhoven *et al.*, 1986). = *Nautilus pompilioides* Fichtel and Moll, 1798, *Nonion pompilioides* (Fichtel and Moll), *Nonionina soldanii* d'Orbigny, 1846. Description: test compressed, involute, chambers depressed, periphery broadly rounded, sutures limbate, umbilicus deep, approximately 9–11 chambers in the last whorl.
Remarks: For the complex taxonomy of this species, see van Morkhoven *et al.*, 1986. *M. barleeanum* (Williamson) (Pliocene to Recent in the North Sea) and the closely related or identical species *M. affine* (d'Orbigny) (Eocene to Miocene) are very similar to *M. pompilioides*, but with compressed and more evolute whorls.
Range: Early Oligocene to Middle Miocene (ranges up to Pleistocene in oceanic environments).

Monspeliensina pseudotepida (van Voorthuysen)
Plate 9.7, Figs 25 (× 44), 26 (× 50), Zone NSB14, Late Pliocene, North Sea (after King, 1983). = *Streblus beccarii* var. *pseudotepidus* van Voorthuysen, 1950. Description: test trochospiral, biconvex, periphery subrounded, outline lobate; chambers inflated, sutures depressed; 6–7 chambers in the last whorl; accessory apertures on the spiral side.
Remarks: there appear to be significant morphological differences between the specimens of *M. pseudotepida* from the North Sea Basin and those referred to this species from the Atlantic coasts of France and southern England (compare Hughes and Jenkins, 1981, plate 9.2, Figs 14).
Range: Early to Late Pliocene.

Neoponides karsteni (Reuss)
Plate 9.7, Figs 27 (× 75), 28 (× 43), Zone NSB5, Middle Eocene, North Sea (after King, 1983). = *Rotalia karsteni* Reuss, 1855, *Eponides schreibersi* d'Orbigny of authors *non* *Neoeponides schreibersii* (d'Orbigny). Description: test compressed, biconvex, periphery subacute, with a thin keel; chambers narrow; sutures on spiral side strongly oblique, sutures on umbilical side radial, deepening towards the umbilicus to produce lobate chamber ends projecting slightly into the deep narrow umbilicus; 5–6 chambers in the last whorl.
Range: Middle Eocene.

Nonion granosum (d'Orbigny)
Plate 9.8, Figs 3, 4 (× 45), Zone NSB8, Late Oligocene, Belgium. = *Nonionina granosa* d'Orbigny, 1846. Description: test compressed, periphery varying from broadly to narrowly rounded, sutures curved to almost radial, umbilicus filled by granular secondary calcite, approximately ten chambers in the last whorl,
Remarks: this variable species is here taken to include *Nonion roemeri* Cushman and *N. granulosum* ten Dam and Reinhold.
Range: Early Oligocene to Middle Miocene.

Nonion labradoricum (Dawson)
Plate 9.8, Figs 6, 7 (× 45), Pleistocene, Sweden (after Knudsen, 1982). = *Nonionina labradorica* Dawson, 1860. Description: test compressed, periphery subangular, apertural face rounded triangular; around ten chambers in the last whorl, increasing rapidly in size.
Remarks: restricted to Zone NSB16x; recorded only in the northern North Sea and more northerly areas.
Range: Pleistocene. Living in arctic areas.

Nonion sp. A
Plate 9.8, Figs 1, 2 (× 93), Zone NSB8, Late Oligocene, North Sea (after King, 1983). = *Elphidium* cf. *latidorsatum* (Reuss) of King, 1983. Description: test small; chambers depressed, moderately inflated; periphery broadly rounded; 4–5 chambers in the last whorl; sutures radial, deeply incised; umbilicus deep, may be partly or wholly filled by granular calcite, which extends over the apertural face and adjoining areas; where the aperture is visible it forms a basal slit.
Remarks: the affinities of this species are uncertain. In gross morphology it resembles *Elphidium latidorsatum* (Reuss) (see Murray and Wright, 1974, Plate 12, Fig. 10), but lacks the septal bridges of *Elphidium*. It is here assigned to *Nonion*; there is considerable similarity to *N. parvulum* (Grzybowski), as figured by Murray and Wright, 1974, Plate 13, Figs 14–15, but their specimen may be incorrectly referred to Grzybowski's species. It occurs commonly in probable bathyal environments in the central North Sea, often associated with non-calcareous agglutinants of the '*Rhabdammina*-biofacies'.
Range: Late Oligocene to Early Miocene.

Osangularia plummerae Brotzen, 1940
Plate 9.8, Fig. 8 (× 60), Late Palaeocene, Tunisia (after Aubert and Berggren, 1976). = *Parrella expansa* Toulmin, 1941, *Osangularia expansa* (Toulmin). Description: test compressed, low biconvex to planoconvex; spiral side almost flat, may be slightly concave: thin peripheral keel; chambers on spiral side depressed, sutures limbate; 7–10 chambers in the last whorl.
Range: Early Palaeocene to Early Eocene.

Pararotalia canui (Cushman)
Plate 9.8, Figs 10, 11 (× 60), Zone NSB7, Early Oligocene, Belgium (after Batjes, 1958). = *Rotalia canui* Cushman, 1928. Description: test small, biconvex; periphery rounded to subacute, keeled, often with a short peripheral spine projecting from each chamber; chambers inflated, 5–6 chambers in last whorl; umbilical area with a large central knob surrounded by a deep groove.
Range: Early to Late Oligocene.

Planulina costata (Hantken)
Plate 9.8, Figs 12, 13 (× 32), Late Eocene, Gulf of Mexico (after van Morkhoven, *et al.*, 1986). = *Truncatulina costata* Hantken, 1875, *Planulina palmerae* van Bellen, 1941. Description: test highly compressed; spiral side flat or slightly depressed, umbilical side low convex; periphery keeled, 10–12 chambers in last whorl; sutures limbate, strongly curved. Remarks: see van Morkhoven *et al.* (1986) for description and synonymy.
Range: Late Eocene. Also occurs rarely in the Early Oligocene (Subzone NSB7, not included on the range chart).

Protelphidium hofkeri Haynes, 1956
Plate 9.8, Fig. 14 (× 75), Zone NSB1, Late Palaeocene, England (after Banner and Culver, 1978). Description: test compressed; periphery broadly rounded, chambers tending to be slightly lobate; 8–9 chambers in last whorl; sutures depressed; umbilicus filled by granular calcite which extends along sutures.
Range: Late Palaeocene to Early Eocene.

Rotaliatina bulimoides (Reuss)
Plate 9.8, Fig. 15 (× 90), Zone NSB7, Early Oligocene, Belgium (after Batjes, 1958). = *Rotalina bulimoides* Reuss, 1851. Description: test trochospiral, very variable in shape, high conical to elongate, tightly coiled; approximately six chambers in the last whorl.
Remarks: see comments under *Gyroidina soldanii mamillata*.
Range: Early Oligocene.

Sigmomorphina regularis (Roemer)
Plate 9.8, Fig. 16 (× 48), Zone NSB8, Late Oligocene, Netherlands (after Doppert, 1980). = *Polymorphina regularis* Roemer, 1838. Description: test large, variable in shape, flattened to low biconvex, profile subcircular to oval–elongate, chambers slightly inflated.
Remarks: *S. schwageri* (Karrer) is a similar species, characterised by a low raised central longitudinal ridge, occurring in the Early and Middle Miocene in the North Sea Basin.
Range: Early Oligocene to Early Miocene.

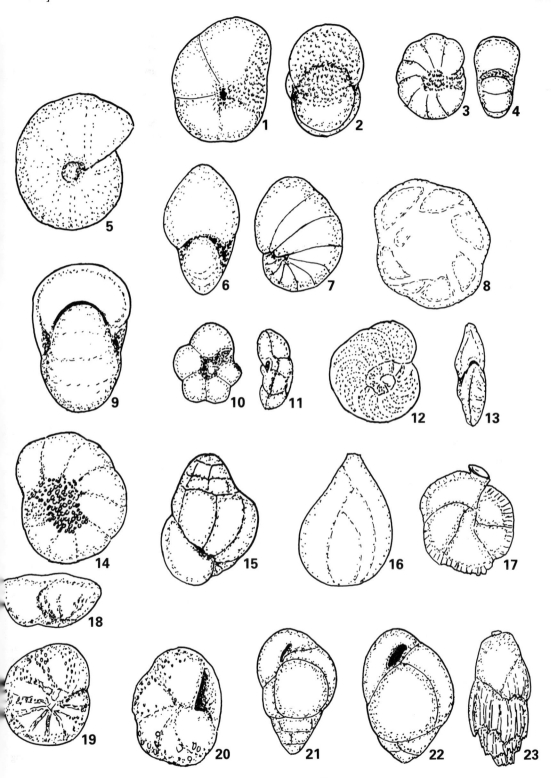

Plate 9.8

Siphonina reticulata Czjzek, 1848
Plate 9.8, Fig. 17 (× 60), Early Pliocene, Italy (after AGIP Mineraria, 1957). = *Siphonina fimbriata* Reuss, 1850. Description: test compressed, spiral and umbilical sides equally biconvex, with flattened 'frilled' periphery; four chambers in the last whorl.
Range: Late Oligocene to Middle Miocene.

Stensioeina beccariiformis (White)
Plate 9.8, Figs 18, 19 (× 72), Subzone NSB1b, Late Palaeocene, Denmark (after van Morkhoven et al., 1986). = *Rotalia beccariiformis* White, 1928, *Rotalia parvula* ten Dam, 1944. Description: test compressed, periphery broadly rounded; spiral side relatively flat, smooth except for scattered coarse perforations; umbilical side coarsely perforate, with limbate sutures which divide into irregular raised ridges around the umbilicus.
Remarks: see discussion of taxonomy in van Morkhoven et al. (1986).
Range: Late Cretaceous (Santonian) to Late Palaeocene.

Svatkrina perlata (Andreae)
Plate 9.8, Fig. 20 (× 126), Zone NSB7, Early Oligocene, North Sea (after Hughes, 1981). = *Pulvinulina perlata* Andreae, 1884, *Alabamina perlata* (Andreae). Description: test small, biconvex, finely perforate, with scattered larger pores on spiral side; periphery rounded, often surrounded by small tubercles; six chambers in the last whorl.
Range: Early to Late Oligocene.

Turrilina alsatica Andreae, 1884
Plate 9.8, Fig. 21 (× 100), Zone NSB8, Late Oligocene, North Sea (after King, 1983). Description: test trochospiral, elongated conical, tapering; chambers moderately inflated, 3–4 chambers in the last whorl, which comprises about half of the test.
Remarks: this species is probably a descendant of *T. brevispira*; specimens intermediate in morphology occur in the Middle and Late Eocene. It is most common in the Early Oligocene and basal Late Oligocene, but typical specimens occur in the central North Sea as high as the top of Zone NSB9 (Early Miocene).
Range: Late Eocene?; Early Oligocene to Early Miocene.

Turrilina brevispira ten Dam, 1944
Plate 9.8, Fig. 22 (× 210), Zone NSB3, Early Eocene, Denmark (after King, 1983). Description: test small, trochospiral; chambers inflated, three chambers in the last whorl, which comprises about two-thirds of the test.
Remarks: see discussion in van Morkhoven et al. (1986).
Range: Early Eocene. As noted above, specimens which may be referable to this species also occur in the Middle and Late Eocene.

Uvigerina acuminata Hosius, 1895
Plate 9.8, Fig. 23 (× 42), Zone NSB10, Early Miocene, North Sea (after King, 1983). Description: test moderately elongated; chambers not inflated, ornamented by ribs of variable length which end in short blunt spines at the lower end of the chambers.
Remarks; this species differs from *U. semiornata* in shape and degree of inflation of chambers, and has more prominent ribs (although the strength of ribbing is very variable).
Range: Early and Middle Miocene.

Trifarina angulosa (Williamson)
Plate 9.9, Fig. 1 (× 68), Pleistocene, Norway (after Feyling-Hanssen et al., 1971) = *Uvigerina angulosa* Williamson, 1858, *Angulogerina angulosa* (Williamson). Description: test moderately large, cross-section rounded triangular; chambers flat-sided, with a prominent central longitudinal rib forming a sharp angle, flanked by a few shorter ribs.
Range: Early Pliocene to Recent.

Trifarina bradyi Cushman, 1923
Plate 9.9, Fig. 2 (× 90), Zone NSB10, Early Miocene, Belgium (after De Meuter, 1980). Description: test small, smooth, ovate–elongate, often somewhat twisted; triangular in cross-section, keeled at the angles along the whole length of the test.
Range: Late Oligocene to Late Pliocene.

Trifarina fluens (Todd)
Plate 9.9, Fig. 3 (× 112), Pleistocene, Norway (after Feyling-Hanssen, et al., 1971) = *Angulogerina fluens* Todd, 1947. Description: test moderately large, cross-section roughly circular, chambers moderately inflated, ornamented by a few rather irregular thin ribs.
Range: Pleistocene. Living in arctic areas.

Uvigerina germanica (Cushman and Edwards)
Plate 9.9, Fig. 4 (× 110), Late Eocene/Early Oligocene, Germany (after Cushman and Edwards). = *Angulogerina germanica* Cushman and Edwards, 1938. Description: test small, ovate, maximum width near the centre of the test,

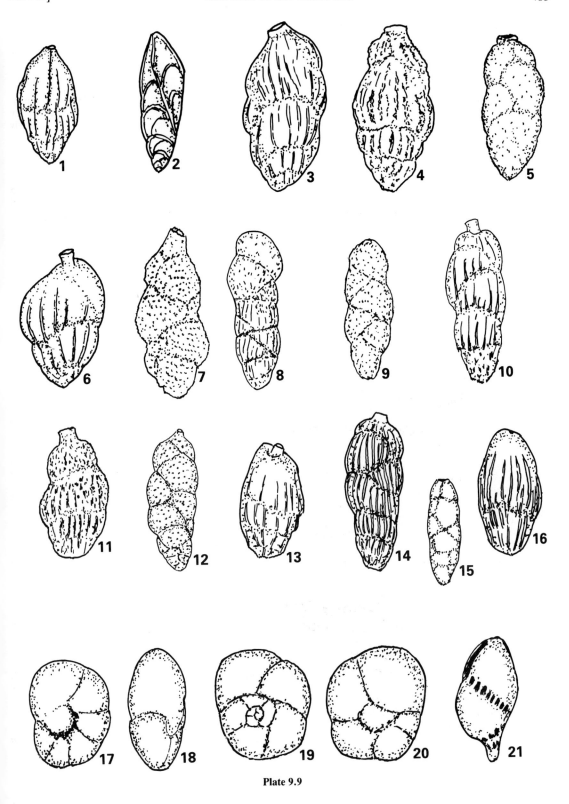

Plate 9.9

cross-section roughly circular, chambers inflated, sutures rather depressed; numerous thin longitudinal ribs on earlier chambers, which become broken up into irregular tubercles and tiny spines on later chambers.
Remarks: this species seems better referred to *Uvigerina* rather than to *Angulogerina* (= *Trifarina*).
Range: Late Eocene to basal Early Oligocene.

Uvigerina batjesi Kaasschieter, 1961
Plate 9.9, Fig. 5 (× 95), Zone NSB3, Early Eocene, England (after Murray and Wright, 1974). Description: test small, elongate; mainly triserial, final section biserial to uniserial; chambers moderately inflated, finely hispid; sutures depressed.
Remarks: this species may be synonymous with *U. garzaensis* Cushman and Siegfus and *U. elongata* Cole, but without direct comparison of specimens this cannot be proven. 'These small hispid species are very similar' (Boersma, 1984).
Range: Early Eocene.

Uvigerina eocaena Guembel, 1870
Plate 9.9, Fig. 6 (× 56), Late Eocene, Italy (after Braga et al., 1975). Description: test large, short, expanding rapidly from the apex, sutures indistinct; few thick longitudinal ribs (two to four per chamber) partly crossing sutures.
Range: Middle? and Late Eocene.

Uvigerina farinosa Hantken, 1875
Plate 9.9, Fig. 7 (× 150), Late Eocene, France (after Boersma, 1984). Description: test small, elongated; chambers moderately inflated, densely ornamented by tiny blunt spines; tending to become biserial/uniserial, with a long apertural neck.
Remarks: similar to *U. batjesi*, but more elongated and more densely ornamented.
Range: Middle and Late Eocene. Early Eocene to Late Oligocene in oceanic areas (Boersma, 1984).

Uvigerina jacksonensis Cushman 1925
Plate 9.9, Fig. 13 (× 43), Zone NSB6, Late Eocene, North Sea (after King, 1983). Description: test large, stout, compact; chambers moderately inflated, ornamented by 5–7 coarse platy ribs per chamber, which may overhang previous chambers.
Range: Late Eocene. Early Eocene to Late Oligocene in oceanic areas (Boersma, 1984)

Uvigerina peregrina Cushman, 1923
Description: similar to *U. venusta*, but larger, with less inflated chambers and fewer longitudinal ribs, and tending to become biserial at an earlier stage.
Range: Pleistocene to Recent.

Uvigerina pygmea langenfeldensis von Daniels and Spiegler, 1977.
Plate 9.9, Fig. 8 (× 60), 'Late' (Middle) Miocene, Germany (after von Daniels and Spiegler). Description: test small, elongated; chambers moderately inflated, ornamented by densely packed very short blunt spines.
Remarks: differentiated from *U. tenuipustulata* by its much coarser ornament and its less globular chambers. *U. p. langenfeldensis*, *U. p. langeri* and *U. venusta* form a chronological series characterised by a progressive change in ornament from wholly spiny to wholly ribbed.
Range: Middle?–Late Miocene.

Uvigerina pygmea langeri von Daniels and Spiegler, 1977
Plate 9.9, Fig. 9 (× 55), Late Miocene, Germany (after von Daniels and Spiegler). Description: test small, elongated; chambers rather inflated; earlier chambers ornamented by thin ribs, which become weaker in the adult stage; later chambers with closely spaced small blunt spines.
Range: Late Miocene.

Uvigerina semiornata saprophila von Daniels and Spiegler, 1977
Plate 9.9, Fig. 10 (× 62), Middle Miocene, Germany (after von Daniels and Spiegler). Description: test elongated, chambers somewhat inflated, ornamented by rather widely spaced thin ribs which tend to end in short spines at the lower end.
Range: Middle Miocene.

Uvigerina spinicostata Cushman and Jarvis, 1929
Plate 9.9, Fig. 11 (× 110), Zone NSB6, 'Early Oligocene' (Late Eocene), Denmark (after Boersma, 1984). Description: test elongate, becoming biserial, rarely uniserial; ornamented by platy ribs (8–12 per chamber), continuous from chamber to chamber, breaking up into coarse spines on final chambers.
Range: Middle Eocene to basal Early Oligocene. Early Eocene to Late Oligocene in oceanic areas (Boersma, 1984).

Uvigerina tenuipustulata van Voorthuysen, 1950
Plate 9.9, Fig. 12 (× 45), Early/Middle Miocene, Belgium (after Batjes, 1958). Description: test small, elongated; chambers inflated, ornamented by numerous very small short blunt spines.
Range: Early and Middle Miocene.

Uvigerina venusta saxonica von Daniels and Spiegler, 1977.
Plate 9.9, Fig. 14 (× 60), Late Miocene, Germany (after von Daniels and Spiegler). Description: test elongated, chambers slightly to moderately inflated, numerous longitudinal costae (14–20 per chamber), sutures depressed, test

triserial throughout.
Range: Late Miocene to Early Pliocene.

Uvigerina sp. A
Plate 9.9, Fig. 16 (× 33), Subzone NSB12c, Middle Miocene, North Sea (after King, 1983). = *U.* sp. cf. *hemmooriensis* von Daniels and Spiegler of King, 1983. Description: test large, broadly fusiform, chambers ornamented with around 6–8 thick rounded ribs, tending to cross sutures; tending to become uniserial.
Remarks: this large and distinctive species characterises a thin interval within the upper part of the Middle Miocene in the southern and central North Sea. It is not recorded from onshore sections. It is difficult to match exactly with any of the large number of described Miocene species.

Uvigerinella abbreviata (Terquem)
Plate 9.9, Fig. 15 (× 60), Middle Eocene, Belgium (after Kaasschieter, 1961). = *Uvigerina abbreviata* Terquem, 1882. Description: test elongated, triserial, smooth, parallel-sided except near the apex, rounded cross-section, chambers slightly inflated; sutures slightly depressed; aperture terminal, without neck.
Range: Early Eocene (also Middle Eocene in onshore sequences).

Valvulineria mexicana grammensis Langer, 1963
Plate 9.9, Figs 17, 18 (× 75), Zone NSB8, Late Oligocene, North Sea (after King, 1983). Description: test compressed, profile subcircular to oval, periphery narrowly rounded, six chambers in the last whorl.
Range: Late Oligocene to Late Miocene.

Valvulineria petrolei (Andreae)
Plate 9.9, Figs 19, 20 (× 85), Zone NSB9, Early Miocene, North Sea (after King, 1983). = *Pulvinulina petrolei* Andreae, 1884. Description: test compressed, profile rounded quadrangular to pentagonal, periphery rounded; spiral side rather flattened, umbilical side convex; five chambers in last whorl, increasing rapidly in size.
Range: Early Oligocene to Early Miocene.

Virgulinella pertusa (Reuss)
Plate 9.9, Fig. 21 (× 53), Zone NSB10, Early/Middle Miocene, North Sea (after King, 1983). = *Virgulina pertusa* Reuss, 1861. Description: test variable in shape, elongated, tapering; long retral processes. Remarks: a very distinctive short-ranging species.
Range: Early to Middle Miocene.

ACKNOWLEDGEMENTS

The author wishes to thank colleagues at Paleoservices Ltd. for discussions on various aspects of North Sea Cenozoic stratigraphy. Mike Charnock, Rolf Feyling-Hanssen, Graham Jenkins, Bob Wynn Jones, Dave McNeil, Etienne Steurbaut, Phil Weaver and others kindly provided preprints, offprints and other data which have enabled incorporation of up-to-date information into the text.

REFERENCES

AGIP Mineraria 1957. *Foraminiferi Padani (Terziario e Quaternario). Atlante iconografico e distributione stratigraphica.* AGIP, Milano, 52 pls.

Aubert, J. and Berggren, W. A. 1976. Paleocene benthic foraminiferal biostratigraphy and paleoecology of Tunisia. *Bull. Centre Rech. Pau. — S.N.P.A.*, **10**, 379–469, 12 pls.

Aubry, M. P. 1983. Biostratigraphie du Paléogène Epicontinental de l'Europe du Nord-Ouest. Etude fondée sur les nannofossiles calcaires. *Docum. Lab. Géol. Lyon*, **87**, 1–317, 8 pls.

Banner, F. T. and Culver, S. J. 1978. Quaternary *Haynesina* n.gen. and Paleogene *Protelphidium* Haynes; their morphology, affinities and distribution. *J. Foraminiferal Res.*, **8**, 177–207, 10 pls.

Bartenstein, H. 1952. Taxonomische Bemerkungen zu den *Ammobaculites*, *Haplophragmium*, *Lituola* und verwandten Gattungen (For.). *Senckenbergiana leth.*, **33**, 297–343.

Batjes, D. A. J. 1958. Foraminifera of the Oligocene of Belgium. *Mem. Inst. R. Sci. Nat. Belg.*, **14**, 1–188, 13 pls.

Berggren, W. A. and Aubert, J. 1975. Paleocene benthonic foraminiferal biostratigraphy, paleobiogeography and paleoecology of Atlantic–Tethyan regions. Midway-type fauna. *Palaeogeogr., Palaeoclimatol., Palaeoecol.*, **18**, 73–192.

Berggren, W. A., Kent, D. V. and Flynn, J. J. 1985a. Jurassic to Paleogene. Part 2. Paleogene geochronology and chronostratigraphy. In Snelling, N. J. (ed.) *The chronology of the geological record*. Geol. Soc. Memoir **10**, 141–195.

Berggren, W. A., Kent, D. V. and van Couvering, J. A. 1985b. The Neogene. Part 2. Neogene geochronology and chronostratigraphy. In Snelling, N. J. (ed.) *The chronology of the geological record*. Geol. Soc. Memoir **10**, 211–260.

Boersma, A. 1984. *Handbook of common Tertiary*

Uvigerina. Microclimates Press, New York. 207 pp., 73 pls.

Bolli, H. M. and Saunders, J. B. 1985. Oligocene to Holocene low latitude planktic foraminifera. In Bolli, H. M., Saunders, J. B. and Perch-Nielsen, K. (eds.) *Plankton Stratigraphy*. Cambridge University Press, 55–262.

Brady, H. B. 1884. Report on the Foraminifera dredged by H.M.S. Challenger during the years 1873–1876. *Rept. Scientific Results Explor. Voyage. H.M.S. Challenger, Zoology*, **9**, 814 pp., 116 pls.

Braga, G., De Biase, R., Grunig, A., and Proto Decima, F. 1975. Foraminiferi Bentonici del Paleocene et Eocene della Sezione di Possagno. *Schweiz. Pal. Abhand.*, **96**, 85–199, 6 pls.

Brotzen, F. 1948. The Swedish Paleocene and its foraminiferal fauna. *Sver. Geol. Unders.*, **42**, 1–140, 19 pls.

Brown, S. and Downie, C. 1984. Dinoflagellate biostratigraphy of Late Paleocene and Early Eocene sediments from holes 552, 553A and 555 leg 81 Deep Sea Drilling Project (Rockall Plateau). *Initial Rep. Deep Sea Drill. Proj.*, **81**, 565–579, 9 pls.

Bykova, N. K. 1953. (Material on the paleoecology of Paleogene Suzakian foraminifera from the Fergana Valley) (in Russian). Leningrad, *Vses. Neft. Nauchno-Issted. Geol.-Razved. Inst. (V.N.I.G.R.I.)*, Trudy, 1953, n.s., vypusk, **73**.

Costa, L. and Downie, C. 1976. The distribution of the dinoflagellate *Wetzeliella* in the Palaeogene of north-western Europe. *Palaeontology*, **19**, 591–614.

Costa, L. I., and Downie, C., 1979. Cenozoic dinocyst stratigraphy of sites 403 to 406 (Rockall Plateau), DSDP Leg 48. In Montadert, L. and Roberts, D. G. (eds) *Initial Rep. Deep Sea Drill. Proj.*, **48**, 513–529, 3 pls.

Costa, L. I. and Manum, S. 1988. Regional dinoflagellate zonation for the NW-European Teriary. In The North West European Tertiary Basin. *Geol. Jb.*, 100.

Costa, L. I. and Mueller, C. 1978. Correlation of Cenozoic dinoflagellate and nannoplankton zones from the north-east Atlantic and north-west Europe. *Newsl. Stratigr.*, **7**, 65–72.

Cushman, J. A. 1931. Notes on the Foraminifera described by Batsch in 1791. *Contrib. Cushman Lab. Foraminiferal Res.*, **7**, 63–71, pls 8, 9.

Cushman, J. A. 1937a. A monograph of the foraminiferal family Verneuilinidae. *Spec. Pub. Cushman Lab. Foraminiferal Res.*, **7**, 1–157, 20 pls.

Cushman, J. A. 1937b. A monograph of the foraminiferal family Valvulinidae. *Spec. Pub. Cushman Lab. Foraminiferal Res.*, **8**, 1–210, 24 pls.

Cushman, J. A. and Edwards, P. 1938. Notes on the Oligocene species of *Uvigerina* and *Angulogerina*. *Contrib. Cushman Lab. Foraminiferal Res.*, **14**, 74–89, pls. 13–15.

Cushman, J. A. and Hanna, G. D. 1927. Foraminifera from the Eocene near Coalinga, California. *Proc. Acad. Sci. California*, ser.4, **16**.

Cushman, J. A. and Parker, F. 1947. *Bulimina* and related foraminiferal genera. *Prof Pap. U.S. Geol. Surv.*, **210-D**, 55–176, pls 15–30.

Cushman, J. A. and ten Dam, A. 1947. Some new species of foraminifera from the Lower Pliocene of the Netherlands. *Contrib. Cushman Lab. Foraminiferal Res.*, **23**, 57–59, pls. 14.

De Meuter, F. J. C. 1980. Benthonic Foraminifera from the Miocene of Belgium. *Aardkundige Meded.*, **1**, 1–170, 8 pls.

Deegan, C. E. and Scull, B. J. 1977. A proposed standard lithostratigraphic nomenclature for the central and northern North Sea. *Rep. Inst. Geol. Sci.*, 77/25.

Dixon, J., McNeil, D. H., Dietrich, J. R., Bujak, J. P. and Davies, E. H. 1984. Geology and biostratigraphy of the Dome Gulf et al. Hunt Kopanoar M-13 well, Beaufort Sea. *Geological Survey of Canada, Paper* **82–13**, 1–28, 6 pls.

Doppert, J. C. W. 1975. Foraminiferenzonering van het Nederlandse Onder-kwartair en Tertiar. In Zagwijn, W. H. and van Staalduinen, C. J. (eds) *Toelichting bij de Geologische overzichstkaarten van Nederland*. Rijks Geologische Dienst, Haarlem, pp. 114–118.

Doppert, J. C. W. 1980. Lithostratigraphy and biostratigraphy of marine Neogene deposits in the Netherlands. *Meded. Rijks Geol. Dienst*, **32**, 257–311, 21 pls.

Edwards, L. E., Goodman, D. K., and Witmer, R. J. 1984. Lower Tertiary (Pamunkey Group) Dinoflagellate Biostratigraphy, Potomac River area, Virginia and Maryland. In Frederiksen, N. O. and Krafft, K. (eds) *Cretaceous and Tertiary Stratigraphy, Paleontology and Structure, Southwestern Maryland and Northeastern Virginia*. (American Association of Stratigraphic Palynologists Field Trip Volume and Guidebook) 137–152

Feyling-Hanssen, R. W. 1972. The foraminifera *Elphidium excavatum* (Terquem) and its variant forms. *Micropaleontology*, **18**, 337–354.

Feyling-Hanssen, R. W. 1980. Microbiostratigraphy of young Cenozoic marine deposits of the Qivituq Peninsula, Baffin Island. *Mar. Micropaleont.*, **5**, 153–184, pl.6.

Feyling-Hanssen, R. W. 1982. Foraminiferal zonation of a boring in Quaternary deposits of the northern North Sea. *Bull. Geol. Soc. Denmark*, **31**, 29–47.

Feyling-Hanssen, R. W. 1985. Graensen mellem Tertiar og Kvartaer i Nordsoen og i Arktis, fastlagt og korreleret ved hjaelp of benthoniske foraminiferer. *Dansk Geol. Foren., Arsskrift for 1985*, 19–33, 2 pls.

Feyling-Hanssen, R. W. and Buzas, M. A. 1976. Emendation of *Cassidulina* and *Islandiella helenae*, new species. *J. Foraminiferal Res.*, **6**, 154–158.

Feyling-Hanssen, R. W. and Ulleberg, K. 1984. A Tertiary–Quaternary section at Sarsbukta, Spitsbergen, Svalbard, and its foraminifera. *Polar Research* (n.s.), **2**, 77–106, 4 pls.

Feyling-Hanssen, R. W., Jorgensen, J. A., Knudsen, K. L. and Andersen, A. J. 1971. Late Quaternary foraminifera from Vendsyssel, Denmark and Sandnes, Norway. *Bull. Geol. Soc. Denmark*, **21**, 67–317, 26 pls.

Gradstein, F. M. and Agterberg, F. P. 1982. Models of Cenozoic foraminiferal stratigraphy — Northwestern

References

Atlantic Margin. In Cubitt, J. M. and Reyment, R. A. (eds) *Quantitative stratigraphic correlation*. John Wiley, pp. 119–166.

Gradstein, F. M. and Berggren, W. A. 1981. Flysch-type agglutinating foraminifera and the Maastrichtian to Paleogene history of the Labrador and North Seas. *Mar. Micropaleont.*, **6**, 211–268, 9 pls.

Gradstein, F. M., Kaminski, M. A. and Berggren, W. A. (in press). Cenozoic foraminiferal biostratigraphy of the central North Sea. In *Proceedings of the Second International Workshop on Agglutinating Foraminiferids*, Vienna, June 1986.

Grossheide, K. and Trunko, L. 1965. Die Foraminiferen des Doberges bei Bunde und von Astrup mit beitragen zur Geologie dieser Profile (Oligozän, NW-Deutschland). *Beih. Geol. Jb.*, **60**, 1–213, 19 pls.

Hailwood, E. A., Bock, W., Costa, L., Dupeuple, P. A., Muller, C. and Schnitker, D. 1979. Chronology and biostratigraphy of Northeast Atlantic sediments, D.S.D.P. Leg 48. In Montadert, L. and Roberts, D. G. (eds) *Initial Rep. Deep Sea Drill. Proj.*, **48**, 1119–1141.

Hansen, H. J. 1970. Danian Foraminifera from Nugssuag, West Greenland with special reference to species occurring in Denmark. *Meddel. om Gronland*, **193**, 1–132, 14 pls.

Hardenbol, J. and Berggren, W. A. 1978. A new Paleogene numerical time scale. In *Contributions to the Geologic Time Scale*. American Association of Petroleum Geologists, Tulsa, pp. 213–234.

Hedberg, H. B. 1976. *International Stratigraphic Guide*. John Wiley, New York. 200 pp.

Heilmann-Clausen, C. 1985. Dinoflagellate stratigraphy of the uppermost Danian to Ypresian in the Viborg borehole, central Jylland, Denmark. *Danm Geol. Unders. Ser. A*, **7**, 1–69, 15 pls

Heilmann-Clausen, E., Nielsen, O. B., Gersner, F. 1985. Lithostratigraphy and depositional environments in the Upper Paleocene and Eocene of Denmark. *Bull. Geol. Soc. Denmark*, **33**, 287–323.

Hooyberghs, H. J. F. 1983. Contribution to the study of planktonic foraminifera in the Belgian Tertiary. *Aardkundige Meded.*, **2**, 1–131, 23 pls.

Hughes, M. J. 1981. Contribution to the Oligocene and Eocene microfaunas of the southern North Sea. In Neale, J. W. and Brasier, M. D. (eds) *Microfossils from recent and fossil shelf seas*. Ellis Horwood, Chichester, pp. 186–294.

Hughes, M. J. and Jenkins, D. G. 1981. Neogene. In Jenkins, D. G. and Murray, J. W. (eds) *Stratigraphical Atlas of Fossil Foraminifera*. Ellis Horwood, Chichester, pp. 268–285, pls 91–93.

Jenkins, D. G., Bowen, D. G. Adams, C. G., Shackleton, N. J. and Brassell, S. C. 1985. The Neogene. Part 1. *In* Snelling, N. J. (ed), *The chronology of the geological record*. Geol. Soc. Memoir **10**, 199–210.

Kaasschieter, J. P. H. 1961. Foraminifera of the Eocene of Belgium. *Mem. Inst. R. Sci. Nat. Belg.*, **147**, 1–271, 16 pl.

Kaminski, M. A., Gradstein, F. M., Berggren, W. A., Geroch, S. and Beckman, J. P. (in press). Flysch-type agglutinating foraminiferid assemblages from Trinidad: taxonomy, stratigraphy and paleobathymetry. In *Proceedings of the Second International Workshop on Agglutinating Foraminiferids*, Vienna, June 1986.

Kiesel, Y. 1970. Die Foraminiferenfauna der paläozänen und eozänen Schichtfolge der Deutschen Demokratischen Republik. *Paläont. Abh.*, **A4**, 163–394. 27 pls.

Kihle, R. and Lofaldi, M. (undated). *Atlas of Foraminifera from unconsolidated sediments on the Norwegian Continental Shelf*. N.T.N.F. publication 35. Continental Shelf Division, I.K.U., Oslo.

King, C. 1983. Cainozoic micropalaeontological biostratigraphy of the North Sea. *Rep. Inst. Geol. Sci.*, **82/7**, 1–40, 6 pl.

King, C. (in press, a). Classification and correlation of the basal London Clay Formation and Oldhaven Beds in southern England and adjacent areas. *Tertiary Res.*

King, C. (in press, b). Eocene stratigraphy of the Knokke borehole (Belgium). *Toelichtende Verhand. Geol. Kaart en Mijnkaart Belgie.*

King, C., Bailey, H. W., King, A. D., Meyrick, R. W. and Roveda, V. L. 1981. North Sea Cainozoic. In Jenkins, D. G. and Murray, J. W. (eds) *Stratigraphical Atlas of Fossil Foraminifera*. Ellis Horwood, Chichester, pp. 294–298.

Knox, R. W. O'B. 1984. Nannoplankton zonation and the Paleocene/Eocene boundary beds of NW Europe: an indirect correlation by means of volcanic ash layers. *J. Geol. Soc. London*, **141**, 993–999.

Knox, R. W. O'B. and Morton, A. C. 1983. Stratigraphical distribution of Early Palaeogene pyroclastic deposits in the North Sea Basin. *Proc. Yorks. Geol. Soc.*, **44**, 355–363.

Knox, R. W. O'B., Morton, A. C. and Harland, R. 1981. Stratigraphical Relationship of Palaeocene Sands in the U.K. Sector of the Central North Sea. In *Petroleum Geology of the Continental Shelf of North-West Europe*. Institute of Petroleum, London, pp. 267–281.

Knudsen, K. L. 1982. Foraminifers. In E. Olausson (ed) The Pleistocene/Holocene boundary in South-Western Sweden. *Sver. Geol. Unders. C.*, **794**, 148–177.

Langer, W. 1969. Beitrag zur kenntnis einiger Foraminiferen aus dem mittleren und oberen Miozän des Nordsee-Beckens. *Neues Jahrb. Geol. Paläontol.*, **133**, 23–78, 4 pls.

Letsch, W. J. and Sissingh, W. 1983. Tertiary Stratigraphy of the Netherlands. *Geol. Mijnbouw*, 62, 305–318.

Loeblich, A. R. and Tappan, H. 1964. *Treatise on invertebrate paleontology. Part C: Protista 2*. 2 vols. 900 pp., 5311 figs. Geological Society of America, New York.

Lott, G. K., Knox, R. W. O'B, Harland, R. and Hughes, M. J. 1983. The stratigraphy of Palaeogene sediments in a cored borehole off the coast of north-east Yorkshire. *Rep. Inst. Geol. Sci.*, **83/9**, 1–10.

Michelsen, O. 1982. Geology of the Danish Central Graben. *Danm. Geol. Unders. Ser. B*, **8**, 1-133.

Miller, K. G., Gradstein, F. M. and Berggren, W. A. 1982. Late Cretaceous to Early Tertiary agglutinated benthic foraminifera in the Labrador Sea. *Micropaleontology*, **28**, 1–30, pls 1–3.

Moe, A. 1983. Tertiary biostratigraphy of the southernmost part of the Norwegian sector of the Central Trough. In Norwegian Petroleum Directorate Paper **32** (Norwegian Petroleum Directorate, Stavanger), 23–33.

Moorkens, T. 1976. Paläokologische Bedeutung einiger Vergesellschaften von sandschaligen Foraminiferen aus dem europaischen Alttertiar und ihre Beziehung zu Muttergestein. Compendium 75/76 *Erdol und Kohle*, 77–95.

Morozova, V. G. 1939. (The stratigraphy of the Upper Cretaceous and Lower Tertiary deposits in the Emba oil-bearing district according to the foraminiferal fauna) (in Russian). *Bull. Soc. Nat. Moscou, Sect. Geol.*, **17**.

Murray, J. W. 1971. *An Atlas of British Recent Foraminiferids*. 244 pp., 96 pls. Heinemann, London.

Murray, J. W. 1984. Paleogene and Neogene Benthic Foraminiferids from Rockall Plateau. In Roberts, D. G. Schnitker, D. *et al.* (eds) *Initial Reports Deep Sea Drill. Proj.* **81**, 503–534, 5 pls.

Murray, J. W. and Wright, C. A. 1974. Palaeogene foraminiferida and palaeoecology, Hampshire and Paris Basins and English Channel. *Spec. Pap. Palaeontol.*, **14**, 1–171, 20 pls.

Nederlandse Aardolie Matschappij B. V. and Rijks Geologische Dienst 1980. Stratigraphic nomenclature of the Netherlands. *Kon. Ned. Geol. Mijn. Gen. Verh.*, **32**, 1–77.

Powell, A. 1986. A new species of *Bolboforma (Incertae Sedis)* from the Miocene of the Vøring Plateau, northern Norway. *J. micropalaeont.*, **5** 71–74, 1 pls.

Rhys, G. H. 1974. A proposed standard lithostratigraphic nomenclature for the southern North Sea and an outline structural nomenclature for the whole of the (U.K.) North Sea. *Rep. Inst. Geol. Sci.*, **74/18**, 1–14.

Sejrup, H.-P. and Guilbault, J. P. 1980. *Cassidulina reniforme* and *C. obtusa* (Foraminifera), taxonomy, distribution and ecology. *Sarsia*, **65**, 79–85.

Siesser, W. G., Ward, D. J. and Lord, A. R. 1987. Calcareous nannoplankton from the Thanetian stage stratotype. *J. micropalaeont.* **6**, 85–102.

Spiegler, D. 1973. Die Entwicklung von *Ehrenbergina* (Foram.) im höheren Tertiär NW-Deutschlands. *Geol. Jahrb.*, **A6**, 3–23, 3 pls.

Spiegler, D. 1974. Biostratigraphie des tertiärs zwischen Elbe und Weser/Aller (Benthische Foraminiferen. Oligo-Miozän). *Geol. Jahrb.*, **A16**, 27–69, 2 pls.

Staesche, K. and Hiltermann, H. 1940. Mikrofaunen aus dem Tertiär Nordwestdeutschlands. *Abh. Reichsanst. Bodenforsch.*, N.F., **201**, 1–26, 53 pls.

Steurbaut, E. 1986. Late Middle Eocene to Middle Oligocene calcareous nannoplankton from the Kallo well, some boreholes and exposures in Belgium and a description of the Ruisbroek Sand Member. *Meded. Werkgr. Tert. Kwart. Geol.*, **23**, 49–83, 2 pls.

Steurbaut, E. (in press). Tertiary calcareous nannoplankton from the Knokke Well (NW Belgium). *Toelichtende Verhand. Geol. Kaart en Mijnkaart Belgie*.

Stewart, L. J. 1987. A revised stratigraphic interpretation of the Early Palaeogene of the Central North Sea. In *Petroleum Geology of North-West Europe*. Graham and Trotman, London, Vol. 1, 557–576.

Talwani, M., Udintsev, G. *et al.* 1976. Sites 338–343. In *Initial Rep. Deep Sea Drill. Proj.*, **38**, 151–387.

Ten Dam, A. 1944. Die stratigraphische gliederung des niederlandischen Paläozäns und Eozäns nach Foraminiferen (mit ausnahme von Sud-Limburg). *Meded. Geol. Sticht*, C-V-3, 1–142, 6 pls.

Ten Dam, A. and Reinhold, T. 1942. Die stratigraphische gliederung des niederlandischen Oligo-Miozäns nach foraminiferen (mit ausnahme von Sud-Limburg). *Meded. Geol. Sticht.*, C-V-2, 1–106, 10 pls.

Thomsen, E. and Heilmann-Clausen, C. 1983. The Danian–Selandian boundary at Svejstrup, with remarks on the biostratigraphy of the boundary in western Denmark. *Bull. Soc. Geol. Denmark*, **33**, 341–362.

Tjalsma, R. C. and Lohmann, G. P. 1983. Paleocene–Eocene bathyal and abyssal benthic foraminifera from the Atlantic Ocean. *Micropaleontology, Spec. Publ.*, **4**, 1–90, 22 pls.

Townsend, H. A. and Hailwood, E. A. 1985. Magnetostratigraphic correlation of Palaeogene sediments in the Hampshire and London Basins, southern U.K. *J. Geol. Soc. London*, **142**, 957–982.

Vail, P. R. 1977. Seismic stratigraphy and global changes of sea level. In Seismic stratigraphy: applications to hydrocarbon exploration. *Mem. American Association of Petroleum Geologists*, Tulsa, **26**, 49–212.

Van Morkhoven, F. C. P. M., Berggren, W. A., and Edwards, A. S. 1986. Cenozoic cosmopolitan deepwater benthic foraminifera. *Bull. Centres Rech. Explor-Prod. Elf-Aquitaine, Mém.* **11**, 1–421, 126 pls.

Van Voorthuysen, J. 1950. The quantitative distribution of the Pleistocene, Pliocene and Miocene foraminifera of boring Zaandam (Netherlands). *Meded. Geol. Sticht. (n.s.)*, **4**, 51–72, 4 pls.

Verdenius, J. G. and van Hinte, J. 1983. Central Norwegian–Greenland Sea: Tertiary arenaceous foraminifera, biostratigraphy and environment. In Verdenius, J. G., van Hinte, J. E., and Fortuin, A. R. (eds) *Proceedings of the First Workshop on Arenaceous Foraminifera* 7–9 September, 1981. I.K.U., Trondheim, 173–223, 7 pl.

Von Benedek, P. N. and Muller, C. 1976. Grenze Unter-Mittel Oligozän am Doberg bei Bunde (Westfalen) 1. Phyto- und Nannoplankton. *N. Jb. Geol. Paläont. Mh. H3*, 129–144.

Von Daniels, C. H. and Spiegler, D. 1977. *Uvigerina* (Foram.) im Neogen Nordwestdeutschlands. *Geol. Jahrb., Ser. A*, 40, 3–59, 6 pls.

Weaver, P. P. E. and Clement, B. M. 1986. Synchroneity of Pliocene planktonic foraminiferid datums in the North Atlantic. *Mar. Micropaleont.*, **10**, 295–307.

Webb, P. N. 1975. Paleocene Foraminifera from DSDP Site 283, South Tasman Basin. In Kennett, T. P., Montz, R. E. *et al.* (eds) *Initial Rep. Deep Sea Drill. Proj.*, **29**, 833–843, 3 pl.

Young, F. G., and McNeil, D. H. 1984. Cenozoic stratigraphy of the MacKenzie Delta, Northwest Territories. *Geological Survey of Canada, Bull.* **336**, 1–63, 6 pl.

Ziegler, P. A. 1982. *Geological Atlas of Western and Central Europe*. Shell International Petroleum Maatschappij B. V.

10

Palaeogene

J. W. Murray, D. Curry, J. R. Haynes, C. King

10.1 INTRODUCTION

Marine and marginal marine Palaeogene sediments on land are found only in southern England. They are preserved in two main areas, the Hampshire Basin centred on the Isle of Wight, and the London Basin comprising the lower valley and estuary of the River Thames. Most of the Palaeogene sediments are clastic sands and clays with occasional limestones. A major unconformity separates them from the underlying late Cretaceous chalk. The unconformity marks a period of uplift, slight deformation, and erosion of the chalk under subaerial and submarine conditions.

The onshore deposits represent a marginal development of a more complete succession in the North Sea and to a lesser extent in the English Channel (Smith and Curry, 1975). The oldest Palaeogene deposit, the Thanet Formation of late Palaeocene age, is found only in the area of the Thames estuary. It clearly represents a transgressive event from the adjacent North Sea Basin, and in contrast, the Reading Formation and the London Clay Formation were laid down throughout the area of the Hampshire and London Basins. Nevertheless, it is believed that inversion of the Mesozoic Weald Basin began to separate the two areas at some period in early Eocene times (Ziegler, 1975). The resultant shoal formed a partial to total barrier between the two basins for the remainder of the Palaeogene. This is reflected in the faunas of the Hampshire Basin which are mainly of northern affinity but from time to time show southern elements thought to have been introduced via the western English Channel. Depositional environments ranged from continental fluvial and lacustrine, through brackish marginal marine lagoons and deltas, to normal marine. The successions resulting from these varying environmental conditions, with faunas limited by environmental and geographic changes, are complex. Correlation within England is not always easy and that between England and mainland Europe is difficult.

The division of responsibilities between the contributors is as follows: planktonic and larger benthic species, D. Curry; Thanetian benthic species, J. R. Haynes; range charts for London Clay Formation of London

Basin, C. King; Eocene–Oligocene benthic species and remainder of text, J. W. Murray.

10.2 COLLECTIONS OF PALAEOGENE FORAMINIFERA

British Museum (Natural History)
Adams:
  Bracklesham Group, Whitecliff Bay and Selsey (Adams, 1962).
Barr and Berggren:
  Thanet Formation, Thanet (Barr and Berggren, 1964)
Bhatia:
  Solent Group, Headon Hill, Bembridge Limestone and Bouldnor Formations, Isle of Wight (Bhatia, 1955).
Bowen:
  London Clay Formation, Alum and Whitecliff Bays; Barton Clay Formation, Barton (Bowen, 1954).
Burrows and Holland:
  Thanet Formation (Burrows and Holland, 1897). Bracklesham Group, Bracklesham (named but undescribed).
Burton:
  Barton Clay Formation, Barton (named but undescribed).
Earland and Edwards:
  Bracklesham Group, Bracklesham; Barton Clay Formation, Barton (named but undescribed).
Haynes:
  type Thanetian (Haynes, 1956–1958).
Heron-Allen:
  shore shands from Selsey with reworked Bracklesham Group material.
Murray and Wright:
  London Clay Formation to Bouldnor Formation (Bembridge Marls), Whitecliff Bay; London Clay Formation to Barton Clay Formation, Alum Bay, London Clay Formation, Swanwick; Bracklesham Group, Bracklesham; Barton Clay Formation, Barton; Solent Formation (M. Headon Beds), Headon Hill and Colwell Bay; Bouldnor Formation (U. Hamstead Beds), Hamstead (Murray and Wright, 1974).
Sherborn and Chapman:
  London Clay Formation, Piccadilly, London (Sherborn and Chapman, 1886, 1889).
Venables:
  London Clay Formation, Bognor (Venables, 1962).

10.3 STRATIGRAPHIC DIVISIONS

The history of stratigraphic studies on the Palaeogene of southern England extends back to the early part of the last century. A summary of information in the mid 1970s was drawn up by Curry *et al.* (1978). They advised against the use of stage names because the European successions on which they are based are of varying facies and incomplete.

Curry *et al.* (1978) used two geochronometric scales on their correlation charts. They favoured that by Odin, *et al.*, (1978) based on dates derived from sedimentary glauconite against that of Berggren, *et al.* (1978) based mostly on high temperature minerals, as linked to mammal faunas. Since that time three alternative geochronometric scales have been introduced: Curry and Odin (1982), Harland, *et al.*, (1982) and Berggren, *et al.*, (1985). Furthermore, the boundaries between epochs are not always coincident in the various schemes (see Fig. 10.1).

Correlation between English successions is based on a great variety of biostratigraphic evidence (see Curry, *et al.*, 1978). Correlation with the standard nannoplankton (NP) and planktonic foraminiferal zones was at that time based on few observations. However, the recent work of Aubry (1983, 1986) on nannofossils and of Townsend and Hailwood (1985) on magnetostratigraphy, has provided better correlation within Britain and also with standard NP zones and magnetostratigraphic

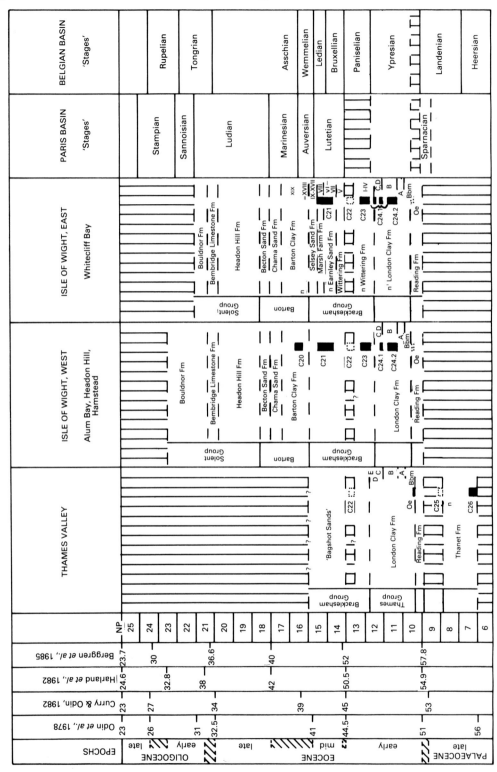

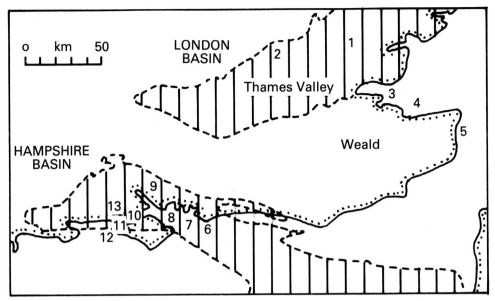

Fig. 10.2 — Distribution of Palaeogene strata (shaded). 1, Ongar, Essex; 2, Hadley Wood, Middlesex; 3, Sheppey, Kent; 4, Reculver, Kent; 5, Pegwell, Kent; 6, Selsey, West Sussex; 7, East Wittering, West Sussex; 8, Whitecliff Bay, Isle of Wight; 9, Chilling, West Sussex; 10, Hamstead, Isle of Wight; 11, Colwell Bay, Isle of Wight; 12, Alum Bay, Isle of Wight; 13, Barton, Hampshire.

scale (Aubry, et al., 1986). These new results are incorporated in Fig. 10.1.

In view of the continuing development of the Palaeogene time scales and of our understanding of the correlation of the English succession with them, all data are plotted with reference to lithostratigraphic successions (Figs. 10.3–12). Recent changes in terminology include those of King (1981) for the London Clay Formation, Daley and Insole (1984) and Insole and Daley (1985) for the late Eocene to early Oligocene Solent Group, and Edwards and Freshney (1987) for the Reading Formation to Headon Formation. In this work, the base of the Barton Clay Formation is taken at the bottom of Fisher Bed XVIII at Whitecliff Bay (Edwards and Freshney, 1987, took the base at the bottom of Fisher Bed XVII, a sandy lenticle of limited lateral extent).

Fig. 10.1 — Stratigraphic summary. Epochs: oblique shading indicates uncertainty in placing the boundary. Thames Valley: Thanet and Reading Formations after Curry et al. (1978), Thames Group after King (1981). Isle of Wight: Reading Formation after Knox (1984), London Clay formation, Bracklesham Group and Barton Group after Edwards and Freshney (1987) plus London Clay Formation Divisions A–D from King (1981). Solent Group after Curry et al. (1978) and Insole and Daley (1985). Bbm = Basement Bed Member. I–XIX, bed numbers after Fisher (1862). Calcareous nannoplankton (n) after Aubry (1986); n' = occurrence at Lower Swanwick. Magnetostratigraphy after Townsend and Hailwood (1985) and Aubry et al. (1986), normal phases numbered C20–C26; Oe = Oldhaven event, normal event C22 is considered to be missing and its probable position shown by a dotted rectangle. Vertical ornament indicates a gap in the sedimentary record. Approximate positions of the Paris and Belgian basin 'stages' are given to assist the interpretation of European range data.

Useful references to individual formations and localities are listed below:

London Basin
General, field guides (Pitcher, *et al.*, 1967; Blezard, *et al.*, 1967). Thanet Formation (Burrows and Holland, 1897). London Clay formation (Wrigley, 1924, 1940; King, 1981, 1984).

Hampshire Basin
General, Field Guides to the Isle of Wight (Daley and Insole, 1984) and Southampton area (Curry and Wisden, 1958), White (1915, 1921). London clay Formation, Whitecliff Bay, Alum Bay, (White, 1921; King, 1981).
Lower Swanwick (Curry and King, 1965).
Bracklesham Group, Whitecliff Bay, Alum Bay (White, 1921; Plint, 1982, 1983a, b, 1984), Selsey (Curry, *et al.*, 1977).
Barton Clay Formation, (Hooker, 1986), Whitecliff Bay, Alum Bay (White, 1921), Barton (Burton, 1929, 1933).
Solent Group, Whitecliff Bay, (White, 1921; Stinton, 1971; Insole and Daley, 1985), Headon Hill and Hamstead, (White, 1921; Insole and Daley, 1985).

10.4 PUBLICATIONS ON FORAMINIFERA

The following list is not exhaustive; it includes only those papers of stratigraphic value.

Thanet Formation:
Burrows and Holland (1897), Haynes (1954, 1955, 1956, 1958a, b, c), Haynes and El-Naggar (1964), Wood and Haynes (1957), Barr and Berggren (1964, 1965), Berggren (1965), El-Naggar (1967), Brönnimann *et al.*, (1968).

London Clay Formation:
Sherborn and Chapman (1886, 1889), Chapman and Sherborn (1889), Davis (1928), Kaasschieter (1961), Bignot (1962), Venables (1962), Brönnimann *et al.*, (1968), Wright (1972a, b), Murray and Wright (1974).

Bracklesham Group:
Wrigley and Davis (1937), Curry (1937, 1962), Kaasschieter (1961), Blondeau and Curry (1963), Murray and Wright (1974).

Barton Beds:
Curry (1937), Bowen (1955), Kaasschieter (1961), Murray and Wright (1974).

Solent Group:
Bhatia (1955, 1957), Brönnimann *et al.* (1968), Vella (1969), Murray and Wright (1974).

General papers dealing with correlation using Foraminifera:
Curry (1965, 1967), Curry *et al.* (1969, 1978), Brönnimann *et al.* (1968).

10.5 PALAEOECOLOGY OF THE FORAMINIFERA

Palaeoecological interpretations are normally based on a comparison of the fossil assemblages with data for modern living assemblages. In doing this it is assumed that individual genera have not changed their ecological preferences through geological time. However, this assumption is not always valid and may on occasion be shown to be wrong because of inconsistency with other data. For instance, certain genera of the Nodosariacea, which in modern seas occupy normal marine shelf and slope habitats, in former times clearly also occupied shallow water including somewhat brackish environments (cf. Larsen and Jørgensen, 1977). Further problems arise through postmortem modification of assemblages (see Murray, 1976b and 1984 for reviews).

There is now a vast quantity of information available on modern foraminifera and this has been summarised in Murray (1973) and Boltovskoy and Wright (1976). The following generalisations can be made concerning the distribution of shelf genera:

Genera occurring in water of normal salinity (~ 35‰)
  Asterigerina
  Cancris
  Discorbis
  Globulina
  Gyroidina
  Melonis
  Nonionella
  Pullenia
  Rosalina
  Textularia
  Uvigerina

Genera occurring in seawater with a salinity of 32 to 35‰ (i.e. slight brackish tolerance). *indicates a somewhat greater brackish tolerance.

  Bolivina
  Brizalina
  Buccella*
  Bulimina
  Cibicides
  Globobulimina
  Nonion*
  Quinqueloculina

Genera common in brackish environments

  Ammobaculites
  Elphidium
  Protelphidium

Palaeoecological interpretations of British Palaeogene foraminiferal assemblages have been made by Bhatia (1955, 1957), Haynes (1958c), Wright (1972a), and Murray and Wright (1974). From these sources it can be seen that almost all the Palaeogene deposits of southern England that have yielded foraminiferal assemblages have been deposited in shallow shelf seas (< 100 m deep) or marginal marine environments. The following environmental associations of dominant genera occur at different levels within the sequence:

Brackish marsh
  Simple agglutinated genera; very low diversity. (Care must be taken not to confuse these with those shelf assemblages which have lost their calcareous component as a result of postmortem solution).

Brackish lagoon
  Protelphidium
  Pararotalia (only P. curryi)
  Ammobaculites
  Buccella

Intertidal lagoon, salinity > 32‰
  Rosalina araucana
  Quinqueloculina reicheli
  Turrilina acicula

Normal marine lagoon
  Quinqueloculina
  Cibicides

Shallow shelf of normal or near-normal salinity
  Cibicides
  Melonis
  Globulina
  Cancris
  Elphidium
  Buccella
  Quinqueloculina
  Asterigerina
  Brizalina
  Nonion
  Nummulites
  Alabamina
  Anomalinoides

As southeastern England was far removed from oceanic conditions during Palaeogene times environments were not generally favourable for planktonic foraminifera. For the most part these comprise small, immature tests brought into the depositional area from time to time through transport by water currents (Murray, 1976a). However, planktonic foraminifera are abundant at certain levels in the London Clay Formation of the London Basin where they are accompanied by other organisms suggesting water depths of > 200 m.

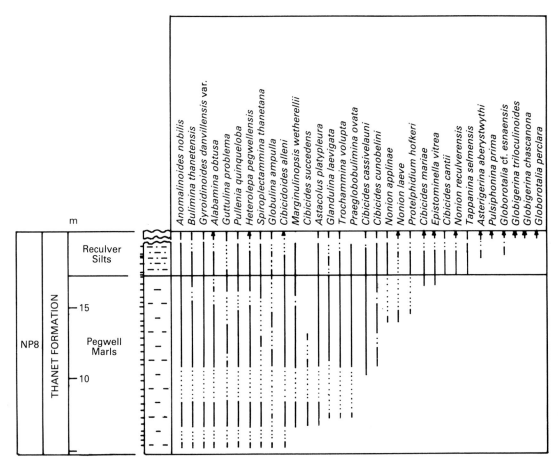

Fig. 10.3 — Range chart for the Thanet Formation (based on Haynes, 1956, 1958c). The main part of the succession is that exposed at Pegwell. At the top, additional data are given on occurrences in the 4 m of Reculver silts at Reculver, but this part of the succession is not to scale. The ticks to the left of the column show the position of the samples examined. Nannoplankton zone after Aubry (1985, 1986). Heights refer to level above the base of the Thanet Formation.

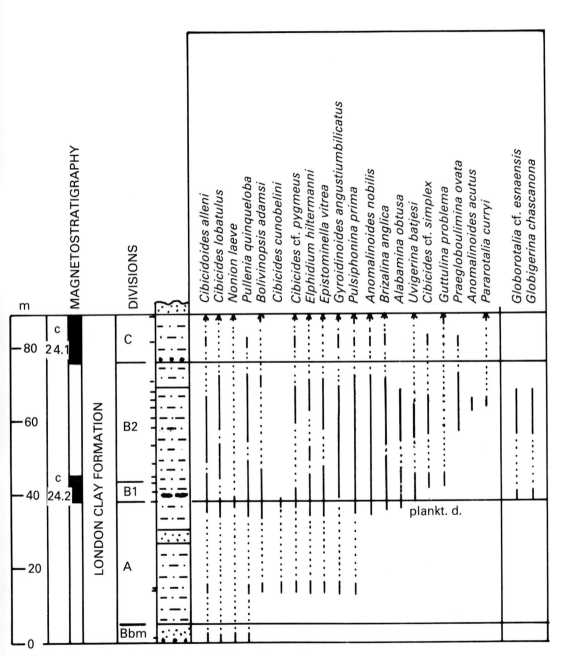

Fig. 10.4 — Range chart for the London Clay Formation, Whitecliff Bay, Isle of Wight. Magnetostratigraphy from Aubry et al. (1986); normal polarity in black. plankt. d. = planktonic datum. Bbm: Basement Bed Member.

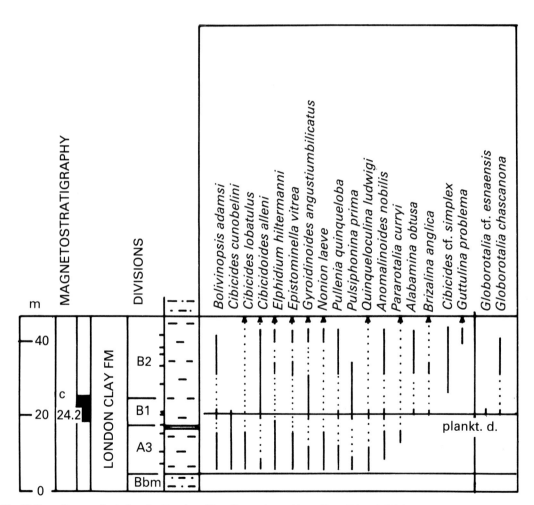

Fig. 10.5 — Range chart for the London Clay Formation, Alum Bay, Isle of Wight. Plankt. d: planktonic datum. Bbm: Basement Bed Member.

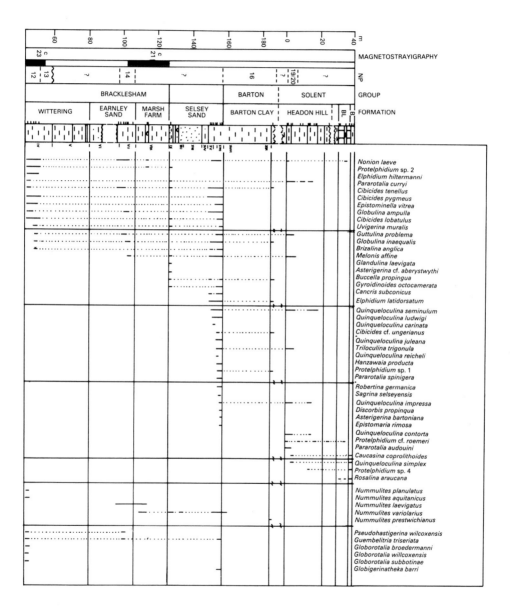

Fig. 10.6 — Range chart for the Bracklesham to Solent Groups, Whitecliff Bay, Isle of Wight. BL: Bembridge Limestone Formation; B = Bouldnor Formation. Roman numerals after Fisher (1862).

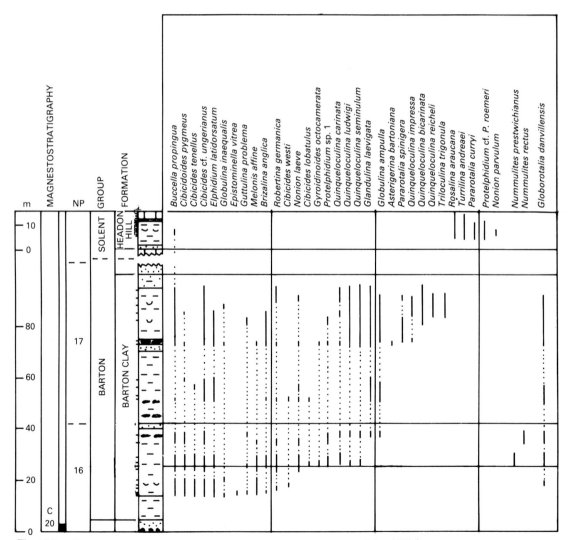

Fig. 10.7 — Range chart for the Bracklesham to Solent Groups, Alum Bay, Isle of Wight.

Sec. 10.6] Foraminiferal Biostratigraphy

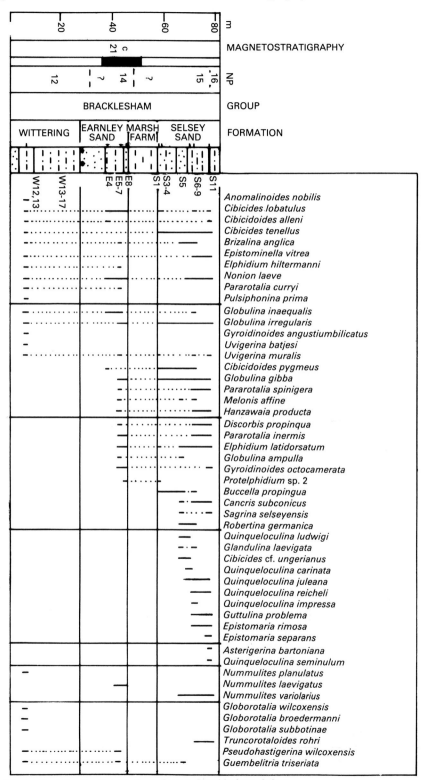

Fig. 10.8 — Range chart for the Bracklesham Group, Bracklesham–Selsey.

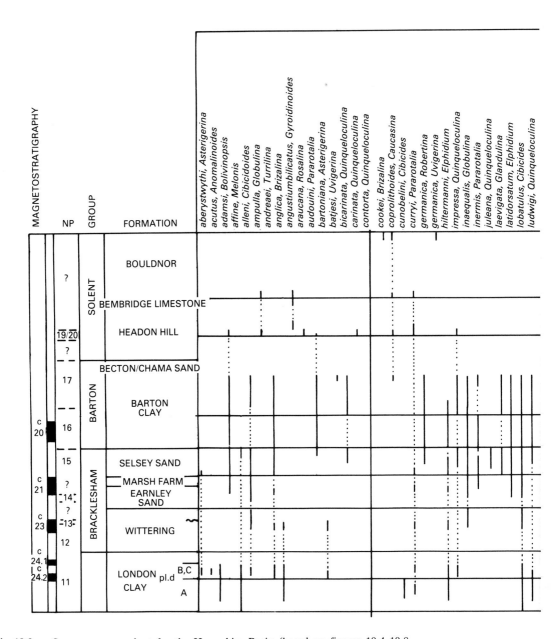

Fig. 10.9 — Summary range chart for the Hampshire Basin (based on figures 10.4–10.8 and data in Murray and Wright, 1974, and Wright, 1972). Pl.d: planktonic datum.

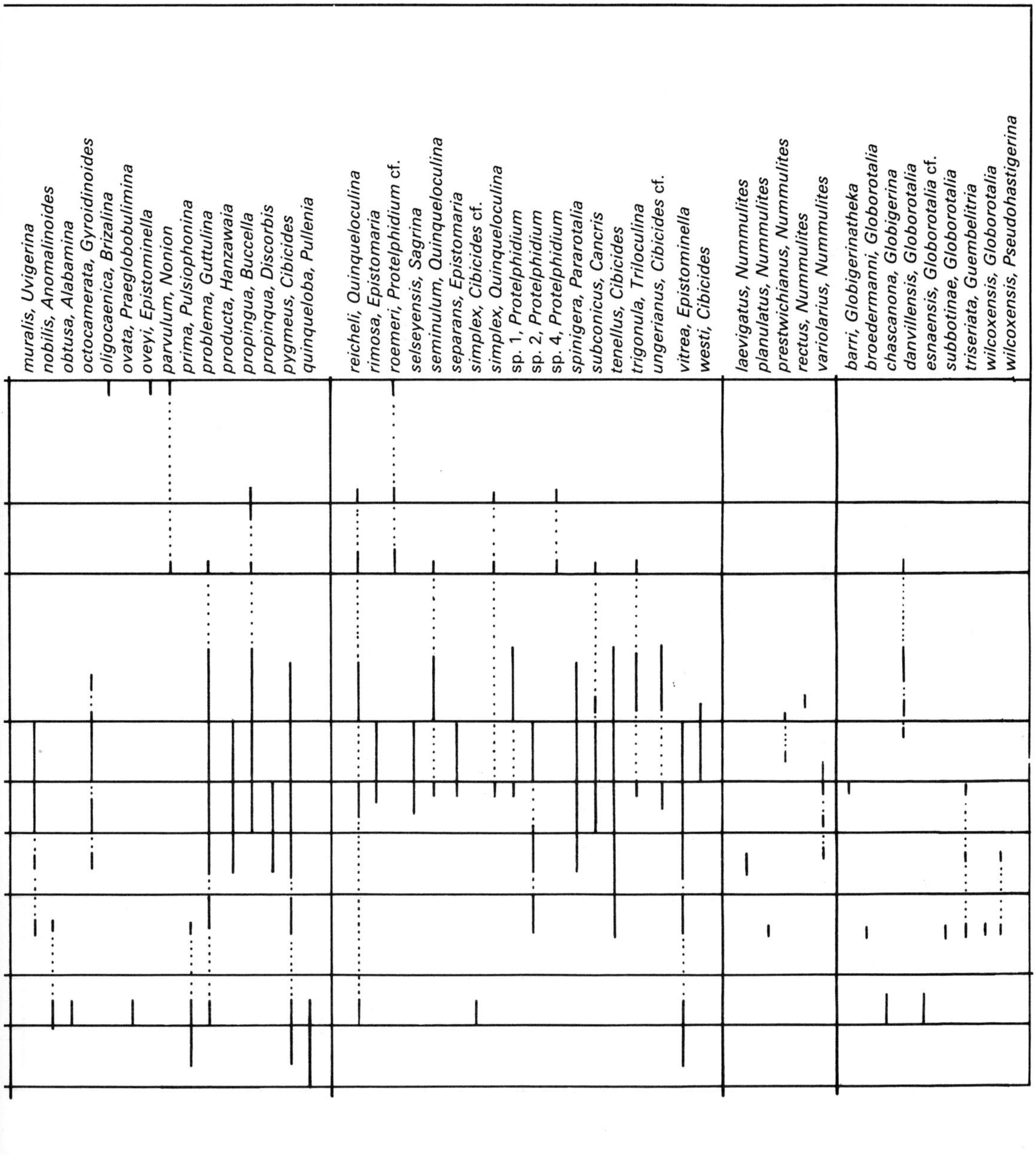
muralis, Uvigerina
nobilis, Anomalinoides
obtusa, Alabamina
octocamerata, Gyroidinoides
oligocaenica, Brizalina
ovata, Praeglobobulimina
oveyi, Epistominella
parvulum, Nonion
prima, Pulsiophonina
problema, Guttulina
producta, Hanzawaia
propingua, Buccella
propinqua, Discorbis
pygmeus, Cibicides
quinqueloba, Pullenia
reicheli, Quinqueloculina
rimosa, Epistomaria
roemeri, Protelphidium cf.
selseyensis, Sagrina
seminulum, Quinqueloculina
separans, Epistomaria
simplex, Cibicides cf.
simplex, Quinqueloculina
sp. 1, Protelphidium
sp. 2, Protelphidium
sp. 4, Protelphidium
spinigera, Pararotalia
subconicus, Cancris
tenellus, Cibicides
trigonula, Triloculina
ungerianus, Cibicides cf.
vitrea, Epistominella
westi, Cibicides
laevigatus, Nummulites
planulatus, Nummulites
prestwichianus, Nummulites
rectus, Nummulites
variolarius, Nummulites
barri, Globigerinatheka
broedermanni, Globorotalia
chascanona, Globigerina
danvillensis, Globorotalia
esnaensis, Globorotalia cf.
subbotinae, Globorotalia
triseriata, Guembelitria
wilcoxensis, Globorotalia
wilcoxensis, Pseudohastigerina

MAGNETOSTRATIGRAPHY											
NP											8
FORMATION	THANET	READING	LONDON CLAY								
DIVISION	P	R	Bbm	A1	A2	A3	B	C	D	E	

Magnetostratigraphy markers: Oe, 24.2 c, 24.1 c, c

Globulina ampulla
Gyroidinoides danvillensis var.
Guttulina problema
Spiroplectammina thanetana
Bulimina thanetensis
Heterolepa pegwellensis
Cibicidoides alleni
Anomalinoides nobilis
Alabamina obtusa
Pullenia quinqueloba
Cibicides succedens
Glandulina laevigata
Astacolus platypleura
Trochammina? volupta
Praeglobobulimina ovata
Cibicides cassivelauni
Cibicides cunobelini
Protelphidium hofkeri
Nonion applinae
Nonion laeve
Cibicides mariae
Epistominella vitrea
Cibicides cantii
Tappanina selmensis
Nonion reculverensis
Asterigerina aberystwythi
Pulsiphonina prima
Elphidium hiltermanni
Cibicides lobatulus
Pararotalia curryi
Brizalina anglica
Gyroidinoides angustiumbilicatus
Uvigerina batjesi
Anomalinoides acutus
Bolivinopsis adamsi
Turrilina brevispira
Anomalinoides aff. aragonensis
Cibicides westi
Clavulina anglica
Nodosaria latejugata
Marginulinopsis wetherellii
Gaudryina hiltermanni
Globigerina triloculinoides
Globorotalia perclara
Globorotalia cf. esnaensis
Globigerina chascanona
Globigerina patagonica
Globorotalia reissi
Pseudohastigerina wilcoxensis

One unique association of benthic foraminifera is that of Selsey Formation, Fisher Bed XVII at Whitecliff Bay and its equivalent (Fisher Bed 21, Selsey). The fauna includes species not found at any other level in the British Palaeogene: *Articulina* spp., *Arenagula kerfornei*, *Dendritina elegans*, *Epistomaria rimosa*, *Fasciolites fusiformis*, *Linderina brugesi*, *Orbitolites complanatus*, *Rotalia* spp., etc. This is interpreted as the product of a normal marine to hypersaline embayment, of warm temperature, and with a vegetational cover over part of the sedimentary substrate.

10.6 FORAMINIFERAL BIOSTRATIGRAPHY

No attempt has been made to erect a zonal scheme.

Reworked Cretaceous species occur at various levels and are easily recognised. Reworking of Cretaceous and probably Palaeocene planktonic species into the type Thanetian deposits has caused difficulties in dating the deposits (El-Naggar, 1967; Brönnimann, *et al.*, 1968). The presence of planktonic forms at particular stratigraphic levels (regardless of the species identified) has been used for correlation (Vella, 1969; Wright 1972b) on the grounds that such an incursion is likely to represent a simultaneous change in water mass movement throughout the depositional basin.

The influence of environment on the benthic species is very marked. They occur in deposits interpreted as normal marine inner to mid shelf, slightly brackish inner shelf, brackish lagoon, and hypersaline lagoon. The introduction of new species and the local extinction of species is invariably due to environmental (including biogeographical) change. Thus, all species are represented by partial ranges or a succession of partial ranges. Therefore, the stratigraphic distribution of species within a succession is mainly of local value for the purposes of correlation. The summary range charts (Figs 10.9–10.12) list those species most useful for stratigraphic correlation in the Hampshire Basin and London Basin.

10.7 SPECIES ENTRIES

For each species the entry is arranged as follows: name used, primary synonym (common synonym where appropriate), description, remarks, palaeoecology. PB = Paris Basin. Range based on Le Calvez (1970) = (L) or Murray and Wright (1974) = (MW). BB = Belgian Basin. Range based on Kaasschieter (1961) = (K), Batjes (1958) = (B), Blondeau (1972) = (Bl) or Le Calvez (1970) = (L). The ranges observed in the Paris and Belgian Basins are expressed in terms of local 'stages' mainly because this enables the entry to be short. The correlation of these 'stages' with the local lithostratigraphic successions and with that of southern England is shown in Fig. 10.1. TR = total range observed in Britain. The known range elsewhere may be longer. e = early, m = middle, l = late.

It should be noted that most of the species described are somewhat variable in size. The measurements given refer to the figured specimen.

The species are arranged alphabetically by genera within the suborders Textulariina and Miliolina. The forms with hyaline walls (suborder Rotaliina of Loeblich and Tappan, 1964) are arranged alphabetically with the exception of the Glandulinidae and Polymorphinidae which are placed at the end.

Fig. 10.10 — Summary range chart for the London Basin (based on Fig. 10.3 and data provided by D. Curry for the Reading basement bed and Mr C. King for the London Clay Formation). P: Pegwell Member; R: Reculver Member; Oe: Oldhaven Event.

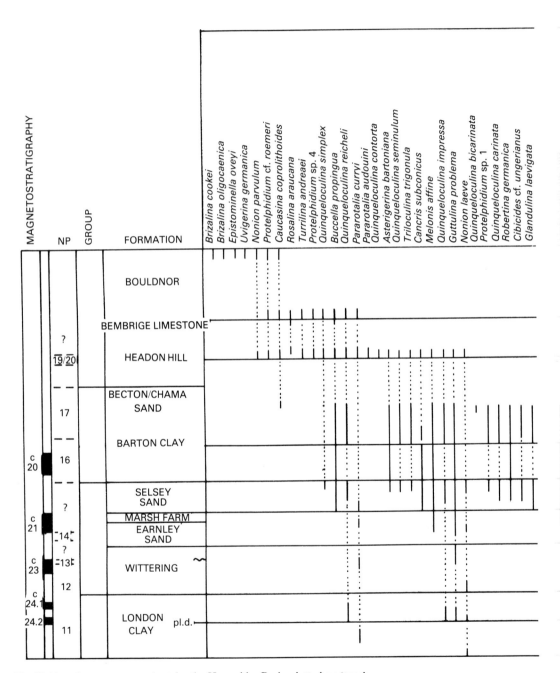

Fig. 10.11 — Summary range chart for the Hampshire Basin plotted on 'tops'.

Foraminiferal Biostratigraphy

A biostratigraphic range chart showing the stratigraphic ranges of foraminiferal species including:

Pararotalia inermis, *Elphidium latidorsatum*, *Globulina inequalis*, *Cibicides tenellus*, *Globulina ampulla*, *Brizalina anglica*, *Quinqueloculina ludwigi*, *Cibicides lobatulus*, *Pararotalia spinigera*, *Cibicides pygmeus*, *Gyroidinoides octocamerata*, *Cibicides westi*, *Elphidium hiltermanni*, *Epistomaria separans*, *Epistomaria rimosa*, *Sagrina selseyensis*, *Hanzawaia producta*, *Uvigerina muralis*, *Protelphidium sp. 2*, *Epistominella vitrea*, *Quinqueloculina juleana*, *Discorbis propinqua*, *Cibicidoides alleni*, *Asterigerina aberystwythi*, *Anomalinoides nobilis*, *Uvigerina batjesi*, *Pulsiphonina prima*, *Gyroidinoides angustiumbilicatus*, *Cibicides cf. simplex*, *Alabamina obtusa*, *Praeglobobulimina ovata*, *Bolivinopsis adamsi*, *Anomalinoides acutus*, *Cibicides cunobelini*, *Nummulites rectus*, *Nummulites prestwichianus*, *Nummulites variolarius*, *Nummulites laevigatus*, *Nummulites planulatus*, *Globorotalia danvillensis*, *Globigerinatheka barri*, *Guembelitria triseriata*, *Pseudohastigerina wilcoxensis*, *Globorotalia broedermanni*, *Globorotalia wilcoxensis*, *Globigerina chascanona*, *Globorotalia cf. esnaensis*.

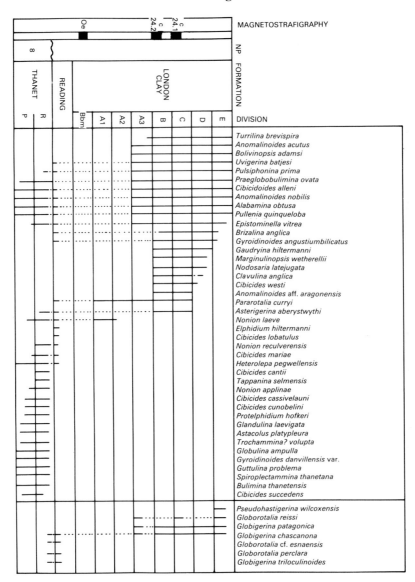

Fig. 10.12 — Summary range chart for the London Basin plotted on 'tops'.

10.7.1 SMALL BENTHIC SPECIES

SUBORDER TEXTULARIINA
Bolivinopsis adamsi (Lalicker)
Plate 10.1, Fig. 1 (× 66), length 600 μm; Fig. 2 (× 74); London Clay Formation, Alum Bay, Isle of Wight. Figs 10.4, 10.5, 10.9–12 = *Spiroplectammina adamsi* Lalicker, 1935. Test initially planispiral then biserial, flaring, compressed; axial portion thick.
Remarks: PB no record. BB Ypresian-Paniselian (K). TR e Eoc. Palaeoecology: normal marine to slightly brackish, inner to mid shelf, fine sediment substrate.

Clavulina anglica (Cushman)
Plate 10.1, Fig. 3 (× 28), length 1.44 mm; Fig. 4, (× 35); London Clay Formation, Hadley Wood, Middlesex. Figs 10.10, 10.12 = *Pseudoclavulina anglica* Cushman, 1936. Description: early portion of test triserial, triangular in section with flat or slightly concave sides; later portion uniserial; aperture terminal, round.
Remarks: confined to the London Basin. PB Lutetian (L). BB Bruxellian-Wemmelian (L). TR e Eoc. Palaeoecology: normal marine shelf.

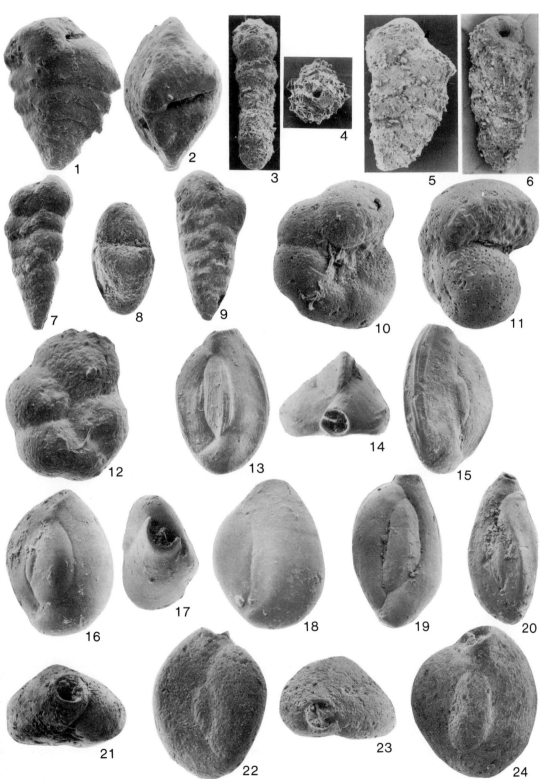

Plate 10.1

Gaudryina hiltermanni (Meisl)
Plate 10.1, Figs 5, 6 (× 24), length 1.66 mm; London Clay Formation, Sheppey, Kent. Figs 10.10, 10.12. = *Gaudryina (Pseudogaudryina) hiltermanni* Meisl, 1959. Description: test triangular in cross-section with slightly concave faces; aperture subterminal, round to oval.
Remarks: confined to the London Basin. PB, BB no record. TR e Eoc. Palaeoecology: normal marine, shelf.

Spiroplectammina thanetana (Lalicker)
Plate 10.1, Fig. 7 (× 40); Fig. 8 (× 80); Fig. 9 (× 40); length 1 mm; Thanet Formation, Reculver Silts Member, Reculver, Kent. Fig. 10.3. = *Spiroplectammina thanetana* Lalicker, 1955. Description: compressed rhomboid in section, periphery subacute; chambers low, up to 26 in microspheric (Fig. 7) and 16 in megalospheric (Fig. 9) generations; angle of taper ~ 40° with 3–4 chambers in initial spire.
Remarks: differs from *S. palaeocenica* (Cushman) in having more chambers. PB Thanetian (Rouvillois, 1960). BB no record. TR 1 Palaeoc. Palaeoecology: normal marine, inner to mid shelf.

Trochammina? volupta (Haynes)
Plate 10.1, Figs 10–12 (× 74), diam. 500 μm, Thanet Formation, Upper Pegwell Marls Member, Pegwell, Kent. Figs 10.3, 10.10, 10.12. = *Gyroidinoides voluptus* Haynes, 1956. Description: globose with rounded periphery, 6–7 chambers in each whorl; aperture long and low, extending from umbilicus to periphery; wall of detrital grains, white, perforate.
Remarks: unusual wall structure and perforation makes generic assignment doubtful. PB, BB. Thanetian (Curry) TR 1 Palaeoc. Palaeoecology: normal marine inner shelf.

SUBORDER MILIOLINA
Quinqueloculina bicarinata d'Orbigny, 1878
Plate 10.1, Fig. 13 (× 33); Fig. 14 (× 40); Fig. 15 (× 33); length 1.2 mm; Barton Clay formation, Alum Bay, Isle of Wight. Figs 10.7, 10.9, 10.11. Description: test oval, subtriangular in cross-section; wall smooth except at periphery where each chamber is ornamented with two keels; aperture round with a small tooth.
Remarks: the double peripheral keels together with the oval outline distinguish this species from all others. PB Lutetian–Marinesian (MW) BB Wemmelian (K) TR 1 Eoc, Palaeoecology: slightly brackish (32‰) to normal marine, inner shelf.

Quinqueloculina carinata d'Orbigny, 1826
Plate 10.1, Figs 16–18 (× 100), length 400 μm; Barton Clay Formation, Middle Barton Beds (F), Barton, Hampshire. Fig. 10.6–9, 10.11. Description: test oval, subtriangular in cross-section, with subangular chamber margins; wall smooth; aperture rounded with a tooth.
Remarks: this species resembles *Q. seminulum* but differs in being smaller and in having a small tooth in the aperture. PB Cuisian-Ludian (L, MW) BB Ledian-Asschian (K) TR m-1 Eoc. Palaeoecology: slightly brackish (32‰) to normal marine, inner shelf.

Quinqueloculina contorta d'Orbigny, 1846
Plate 10.1, Fig. 19 (× 45); Fig. 20 (× 60); Fig. 21 (× 45); length 800 μm; Solent Group, Headon Hill Formation, (Middle Headon Beds), Whitecliff Bay, Isle of Wight. Figs 10.6, 10.9, 10.11. Description: test elongate, chambers subquadrangular in section giving a truncate periphery; aperture round with a tooth.
Remarks: this species is distinctive because of the truncate periphery but the forms with a rounded periphery grade into *Q. ludwigi*. PB Lutetian-Sannoisian (L) BB no record. TR 1 Eoc. Palaeoecology: normal marine, lagoonal.

Quinqueloculina impressa Reuss, 1851
Plate 10.1, Figs 22–24 (× 59), length 680 μm. Solent Group, Headon Hill Formation, (Middle Headon Beds), Whitecliff Bay, Isle of Wight. Figs 10.6–9, 10.11. Description: test oval, periphery rounded, surface smooth; aperture round with a small tooth.
Remarks: differs from other species of *Quinqueloculina* with a rounded periphery (*Q. ludwigi*, *Q. reicheli* and *Q. simplex*) in being oval rather than elongate. PB no record. BB Ledian-Asschian (K) Tongrian (B) TR e-1 Eoc. Palaeoecology: slightly brackish (32‰) to normal marine, inner shelf.

Quinqueloculina juleana (d'Orbigny, 1846)
Plate 10.2, Fig. 1 (× 50); Fig, 2 (× 75); Fig. 3 (× 50); length 800 μm; Bracklesham Group, Selsey Sand Formation (Fisher Bed 21), Selsey, West Sussex. Figs 10.6, 10.8, 10.9, 10.11. Description: test elongate, chamber peripheries with keels, aperture round, with a tooth, situated on a short neck.
Remarks: the strongly developed keels, elongate test and aperture on a neck distinguish this species from all others. PB Auversian (L) BB Ledian-Asschian (K) Tongrian-Rupelian (L, B) TR m Eoc. Palaeoecology: slightly brackish (32‰) to normal marine, inner shelf, on muddy sand substrates.

Quinqueloculina ludwigi Reuss, 1866
Plate 10.2, Fig. 4 (× 118); Fig. 5 (× 177); Fig. 6 (× 118); length 340 μm; Bracklesham Group, Selsey Sand Formation, Whitecliff Bay, Isle of Wight. Figs 10.5–9, 10.11. Description: test elongate oval, smooth, chambers rounded in section; aperture circular, with a bifid tooth, on a short neck.

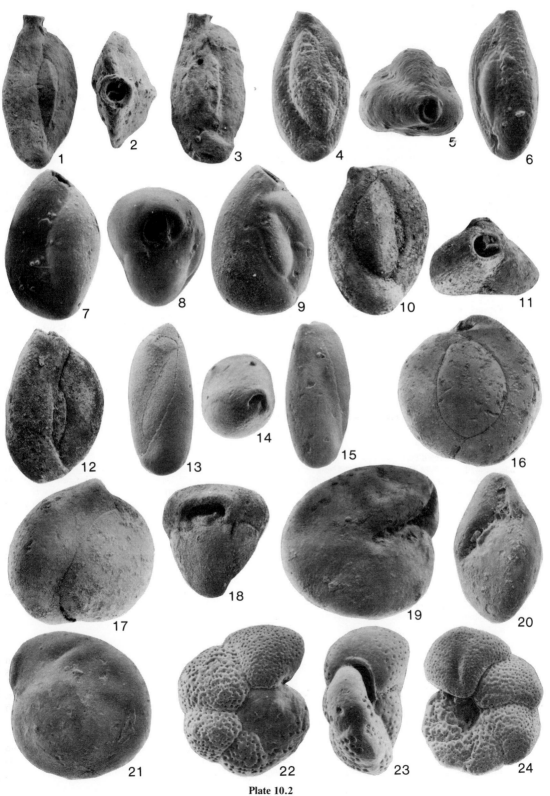
Plate 10.2

Remarks: differs from *Q. simplex* in having chambers with a circular cross section and therefore depressed sutures. PB Lutetian-Sannoisian (L), BB Ledian-Asschian (K) Tongrian (L) Rupelian (B), TR e-1 Eoc. Palaeoecology: slightly brackish (32‰) to normal marine, inner shelf, on muddy substrates.

Quinqueloculina reicheli Le Calvez, 1966
Plate 10.2, Fig. 7 (× 118); Fig. 8 (× 135); Fig. 9 (× 119); length 340 μm; Bracklesham Group, Selsey Sand Formation, Whitecliff Bay, Isle of Wight. Figs 10.6–9, 10.11. = *Quinqueloculina (Scutuloris) reicheli* Le Calvez, 1966. Description: test oval, rounded in section, smooth; aperture the obliquely open end of the last chamber, sometimes with a flap.
Remarks: the oblique aperture is distinctive. PB Lutetian-Sannoisian (L, MW) BB Tongrian-Rupelian (L), TR e Eoc–e Olig. Palaeoecology: slightly brackish (32‰) to hypersaline, intertidal lagoonal, to inner shelf.

Quinqueloculina seminulum (Linné)
Plate 10.2, Figs 10–12 (× 25), length 1.6 mm, Solent Group, Headon Hill Formation, (Middle Headon Beds), Whitecliff Bay. Figs 10.6–9, 10.11. = *Serpula seminulum* Linné, 1758. Description: test oval, subtriangular in section with subangular chamber margins, smooth; aperture an oval opening with an elongate tooth.
Remarks: differs from *Q. carinata* in being larger and in having a big tooth in the elongate aperture. PB Sannoisian (L). BB Paniselian-Asschian (K) Tongrian-Rupelian (L, B) TR m-1 Eoc.; Rec. Palaeoecology: modern forms are normal marine, inner shelf inhabitants.

Quinqueloculina simplex Terquem, 1882
Plate 10.2, Fig. 13 (× 125); Fig. 14 (× 140); Fig. 15 (× 115); length 320 μm; Solent Group, Headon Hill Formation, (Middle Headon Beds), Whitecliff Bay, Isle of Wight. Figs 10.6, 10.9, 10.11. = *Quinqueloculina simplex* Terquem, emend. Le Calvez, 1947. Description: test very elongate, rounded in cross-section, smooth: aperture round with a small tooth.
Remarks: the very elongate form and small aperture are distinctive. PB Lutetian-Sannoisian (L, MW) BB Tongrian–Rupelian (B as *Scutuloris oblonga*) TR m Eoc.-e Olig. Palaeoecology: slightly brackish (32‰) to hypersaline, intertidal lagoonal to inner shelf.

Triloculina trigonula (Lamarck)
Plate 10.2, Figs 16–18 (× 87), length 460 μm, Solent Group, Headon Hill Formation, (Middle Headon Beds), Whitecliff Bay, Isle of Wight. Figs 10.6, 10.7, 10.9, 10.11. = *Miliolites trigonula* Lamarck, 1804. Description: test triloculine, of rounded triangular cross-section; aperture a terminal arch with a small tooth.
Remarks: this is the only common species of *Triloculina* in the British Palaeogene. PB Cuisian-Sannoisian (L, MW) BB Paniselian-Asschian (K) TR m-1 Eoc; Rec. Palaeoecology: normal marine, inner shelf.

SUBORDER ROTALIINA
Alabamina obtusa (Burrows and Holland)
Plate 10.2, Figs 19–21 (× 80), diam. 500 μm. London Clay Formation, Alum Bay, Isle of Wight. Figs 10.3–5, 10.9–12. = *Pulvinulina exigua* (Brady) var. *obtusa* Burrows and Holland, 1897. Description: test nearly biconvex, umbilical side higher than spiral side; generally 5 chambers in final whorl; periphery subrounded; apertural face infolded, aperture an interiomarginal slit; diam. 0.2–0.6 mm.
Remarks: differs from *A. wilcoxensis* Toulmin in having a subrounded rather than subangular periphery, and in being more nearly biconvex rather than planoconvex. PB Thanetian (Curry) Cuisian (MW) BB Thanetian (Curry) Ypresian (K) TR 1 Palaeoc.-e Eoc. Palaeoecology: slightly brackish, inner shelf, on muddy substrates.

Anomalinoides acutus (Plummer)
Plate 10.3, Figs 1–3 (× 100), diam. 400 μm, London Clay Formation, Whitecliff Bay, Isle of Wight. Figs 10.4, 10.9. = *Anomalina ammonoides* (Reuss) var. *acuta* Plummer, 1926. Description: test trochospiral, very compressed; spiral side moderately involute with a thickened central boss; umbilical side involute with central boss surrounded by thickened ends of sutures; periphery acute; 13–15 narrow chambers in the last whorl; aperture an arch extending over the periphery and onto the umbilical side.
Remarks: differs from *A. nobilis* in its acute periphery and numerous chambers. PB Thanetian (L) Cuisian (MW) BB Ypresian-Asschian (K) TR e Eoc. Palaeoecology: normal marine.

Anomalinoides aff. *A. aragonensis* Nuttall
Plate 10.2, Figs 22–24 (× 44), diam. 0.9 mm, London Clay Formation, Hadley Wood, Middlesex. Figs 10.10, 10.12. = *Anomalina dorri* Cole var. *aragonensis* Nuttall, 1930. Description: test trochospiral; spiral side flat, umbilical side convex, periphery rounded; aperture an interiomarginal slit; wall coarsely perforate.
Remarks: The large size of the test and the coarse wall pores are distinctive. PB Thanetian (L) BB Paniselian-Wemmelian (K) TR e Eoc. Palaeoecology: normal marine.

Anomalinoides nobilis Brotzen, 1948
Plate 10.3, Figs 4–6 (× 130), diam. 300 μm, London Clay Formation, Whitecliff Bay, Isle of Wight. Figs 10.3–5, 10.8–12. Description: test low trochospiral, almost planispiral; spiral side partly involute with a shallow umbilicus; 7–9 chambers in final whorl; periphery rounded; aperture extending from periphery onto spiral side.

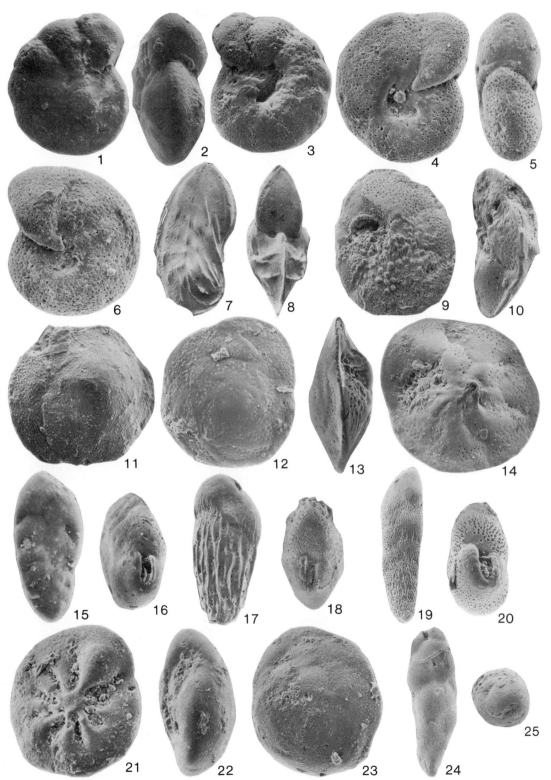

Plate 10.3

Remarks: this species superficially resembles *Melonis affine* but the latter is planispiral and involute on both sides. PB Cuisian (MW) BB no record, TR 1 Palaeoc.-e Eoc. Palaeoecology: normal marine to slightly brackish, inner-mid shelf, mainly on fine sediment substrates.

Astacolus platypleura (Jones)
Plate 10.3, Figs 7, 8 (× 22), length 1.8 mm, Thanet Formation, Lower, Pegwell Marls Member, Pegwell, Kent. Figs 10.9, 10.10. = *Cristellaria platypleura*, Jones, 1852. Common synonym: *Lenticulina multiformis* Franke. Description: compressed, carinate; up to 13 chambers in close-coiled microspheric and up to ten chambers in megalospheric generation, becoming high and astacoline; sutures limbate and raised with variable development of spiral cross-ornament.
Remarks: PB, BB no record. TR 1 Palaeoc. Palaeoecology: normal marine, mid and outer shelf, on silt and mud.

Asterigerina aberystwythi Haynes, 1956
Plate 10.3, Figs 9–11 (× 130), diam. 300 µm. Thanet Formation, Reculver Silts Member, Reculver, Kent. Figs 10.3, 10.6, 10.9–12. Description: biconvex or subconoidal with spiral side raised and umbilical side flattened; periphery acute, 4–5 chambers visible on umbilical side, supplementary chamberlets small; aperture an arch at base of apertural face with tuberculate ornament below.
Remarks: differs from *A. bartoniana* in being smaller, having fewer chambers and in being flattened on the umbilical side. PB, BB no record. TR 1 Palaeoc. Palaeoecology: normal marine-?brackish, inner shelf.

Asterigerina bartoniana (Ten Dam)
Plate 10.3, Figs 12–14 (× 130), diam 300 µm. Barton Clay Formation, Middle Barton Beds (D), Barton, Hampshire. Figs 10.6–9, 10.11. = *Rotalia granulosa* Ten Dam, 1944., renamed *Rotalia bartoniana*, Ten Dam, 1947. Description: test trochospiral, biconvex, umbilical side more convex than spiral side; 6–10 chambers in last whorl; umbilicus filled by calcite plug; granular ornament on umbilical side.
Remarks: see *A. aberystwythi*. PB Cuisian-Auversian (L. MW) BB Ypresian-Asschian (K) e Tongrian (B) TR m-1 Eoc. Palaeoecology: normal marine, inner shelf.

Brizalina anglica (Cushman)
Plate 10.3, Fig. 15 (× 110); Fig. 16 (× 220); length 360 µm; London Clay Formation, Alum Bay, Isle of Wight. Figs 10.4–12. = *Bolivina anglica* Cushman, 1936. Description: test biserial, elongate, about three times as long as broad; compressed, periphery rounded, aperture loop-shaped.
Remarks: this species is relatively broader than *B. oligocaenica*. PB Thanetian (L) BB Ypresian-Asschian (K). TR e-1 Eoc. Palaeoecology: normal marine, shelf.

Brizalina cookei (Cushman)
Plate 10.3, Fig. 17 (× 155); Fig. 18 (× 260); length 260 μm; Solent Group, Bouldnor Formation, (Upper Hamstead Beds, Cerithium Bed), Hamstead, Isle of Wight. Figs 10.9, 10.11. = *Bolivina cookei* Cushman, 1922. Description: test biserial, elongate, periphery subacute; numerous fine costae extending along most of the test length; last chambers smooth.
Remarks: the presence of costae distinguishes this from other species of *Brizalina*. PB Sannoisian (MW) BB Wemmelian-Asschian (K) Rupelian (L) TR m Olig. Palaeoecology: normal marine, shelf, mud substrate.

Brizalina oligocaenica (Spandel)
Plate 10.3, Fig. 19 (× 100); Fig. 20 (× 300); length 400 μm; Solent Group, Bouldnor Formation, (Upper Hamstead Beds, Cerithium Bed), Hamstead, Isle of Wight. Figs 10.9, 10.11. = *Bolivina oligocänica* Spandel, 1909. Description: test biserial, elongate, about four times as long as broad; periphery rounded; numerous chambers; sutures straight, oblique; aperture loop-shaped.
Remarks: this species is narrower and more elongate than *B. anglica*. PB Stampian (MW) BB Rupelian (L) TR m Olig. Palaeoecology: normal marine shelf, mud substrate.

Buccella propingua (Reuss)
Plate 10.3, Figs 21–23 (× 200), diam. 200 μm; Barton Clay Formation, Middle Barton Beds (D), Barton, Hampshire. Figs 10.6–10.11. = *Rotalia propingua* Reuss, 1856. Common synonym: *Ammonia propingua* (Reuss). Description: test trochospiral, biconvex, periphery subacute; sutures on spiral side oblique, those on umbilical side sigmoid, deeply incised in their inner part and ornamented with pustules; primary aperture an interiomarginal slit obscured by pustules.
Remarks: PB Lutetian-Auversian (Curry) Marinesian-Sannoisian (L, MW) BB Bruxellian-Asschian (K) 1 Tongrian–Rupelian (B) TR m Eoc.-1 Olig. Palaeoecology: slightly brackish inner to mid shelf, generally on fine sediment substrates.

Bulimina thanetensis Cushman and Parker, 1947
Plate 10.3, Fig. 24 (× 100); Fig. 25 (× 125); length 400 μm; Thanet Formation, Lower Pegwell Marls Member, Pegwell, Kent. Figs 10.3, 10.10, 10.12. Description: elongate, initial part trigonal with marked spiral suture, adult rounded and parallel sided; chambers up to 30 in microspheric and up to 20 in megalospheric generations, last four making up about half the test; aperture comma shaped with toothplate (simple trough with slight flange).
Remarks: *B. trigonalis* Ten Dam is smaller, *B. simplex* Ten Dam has a quadrangular aperture, *B. paleocenica* Brotzen is sharply triangular and *B. rosenkrantzi* Brotzen has an enlarged final whorl. PB, BB no record. TR 1 Palaeoc. Palaeoecology: marginal marine to mid shelf, probably tolerant of muddy bottoms with lowered oxygen levels.

Cancris subconicus (Terquem)
Plate 10.4, Figs 1–3 (× 130), diam. 300 μm, Bracklesham Group, Selsey Sand Formation, Whitecliff Bay, Isle of Wight. Figs 10.6, 10.8, 10.9, 10.11. = *Rotalina subconica* Terquem, 1882. Common synonym: *Valvulineria subconica* (Terquem). Description: test trochospiral, ovate, periphery rounded; last chamber more elongate than the rest.
Remarks: PB Cuisian-Marinesian (L, MW) BB Ypresian-Asschian (K) TR m-1 Eoc. Palaeoecology: normal marine, inner to mid shelf, varied substrates.

Caucasina coprolithoides (Andreae)
Plate 10.4, Fig. 4 (× 145), length 280 μm, Solent Group, (Bembridge Marls), Whitecliff Bay, Isle of Wight. Figs 10.6, 10.9, 10.11. = *Bulimina coprolithoides* Andreae, 1884. (syn. *Buliminella carteri* Bhatia, 1955). Description: initial portion trochospiral, but main part of test triserial; aperture an elongate loop at the inner margin of the last chamber.
Remarks: juvenile forms of this species were named *Buliminella carteri* by Bhatia (1955). The initial trochospiral stage distinguishes *C. coprolithoides* from *Bulimina* species. PB Sannoisian (L, MW) BB Tongrian-Rupelian (L) TR 1 Eoc.–m Olig. Palaeoecology: slightly brackish, normal marine, slightly hypersaline, marginal lagoons, mainly muddy substrates.

Cibicides cantii Haynes, 1957
Plate 10.4, Figs 5–7 (× 135), diam. 300 μm, Thanet Formation, Reculver Silts Member, Reculver, Kent. Figs 10.3, 10.10, 10.12. Description: planoconvex, periphery acute; chambers initially lobate becoming arcuate 6–8 in final whorl, wall coarsely perforate on spiral side; aperture extends from the periphery onto the spiral side.
Remarks: a small species. PB, BB no record. TR 1 Palaeoc. Palaeoecology: normal marine, ?brackish, inner shelf.

Cibicides cassivelauni Haynes, 1957
Plate 10.4, Figs 8–10 (× 73), diam. 550 μm, Thanet Formation, Upper Pegwell Marls Member, Pegwell, Kent. Figs 10.10, 10.12. Description: planoconvex with subangular periphery; 7–8 chambers in final whorl becoming arcuate on spiral side by end of second whorl; wall coarsely perforate on spiral side but on umbilical side only so on last few chambers; aperture extending from periphery onto spiral side.
Remarks: differs from *C. lobatulus* in lacking general development of large pores on umbilical side. PB, BB no record TR 1 Palaeoc. Palaeoecology: normal marine, inner-mid shelf.

Cibicides (Cibicidina) cunobelini Haynes, 1957
Plate 10.4, Figs 11–13 (× 80), diam. 500 μm, Thanet Formation, Reculver Silts Member, Reculver, Kent. Figs 10.3–5, 10.9–12. Description: equally biconvex or umbilical side more convex, periphery subrounded; 9–10 chambers in outer whorl of microspheric and 7–8 in megalospheric generation; sutures limbate on spiral side and swell into small bosses towards umbilicus but in later part of test are replaced by lappets along base of last few chambers; aperture extending from periphery onto spiral side.
Remarks: differs from *C. (C.) mariae* in being larger and in its convex spiral side. PB Thanetian-Sparnacian (L) Cuisian (MW) BB no record. TR 1 Palaeoec. Palaeoecology: brackish-normal marine, inner shelf, on fine sediment.

Cibicides lobatulus (Walker and Jacob)
Plate 10.4, Figs 14–16 (× 100), diam. 360 μm, Bracklesham Group, Selsey Sand Formation, Selsey, West Sussex. Figs 10.4–12. = *Nautilus lobatulus* Walker and Jacob, 1798. Description: test trochospiral, spiral side flat to concave, evolute, umbilical side convex, involute; 7–9 chambers in the outer whorl; periphery acute; aperture extending from the periphery along 2–3 chambers on spiral side.
Remarks: distinguished from *C. cantii* by its larger number of chambers and the absence of the coarse pores on the spiral side. *C. pseudoungerianus* generally has a more convex spiral side. PB Thanetian-Sannoisian (L, MW) BB Ypresian–Asschian (K) Tongrian-Rupelian (B) TR e-1 Eoc., Rec. Palaeoecology: marine, inner to mid shelf, often in areas disturbed by currents. *C. lobatulus* lives clinging or attached to firm substrates such as shells.

Cibicides (Cibicidina) mariae (Jones)
Plate 10.4, Figs 17–19 (× 160), diam. 250 μm, Thanet Formation, Reculver Silts Member, Reculver, Kent. Figs 10.10, 10.12. = *Rosalina mariae* Jones, 1852. Common synonym: *Cibicides newmanae* (Plummer). Description: concavo-convex, partially involute with well developed lappets; 8–9 chambers in outer whorl; suture limbate on spiral side, aperture extending from periphery onto spiral side.
Remarks: see *C. (C.) cunobelini* PB, BB no record. TR 1 Palaeoc. Palaeoecology: normal marine-?brackish, inner shelf.

Cibcides cf. *simplex* Brotzen, 1948
Plate 10.4, Figs 20–22 (× 165), diam. 240 μm, London Clay Formation, Whitecliff Bay, Isle of Wight. Figs 10.4, 10.5, 10.9, 10.11. Description: test trochospiral, spiral side flat to convex, partially involute; umbilical side convex, involute; periphery rounded; 6–10 chambers in outer whorl; aperture extending from periphery along 2–3 chambers on the spiral side.
Remarks: PB, BB no record TR e Eoc. Palaeoecology: brackish, inner shelf, muddy sand substrates.

Palaeogene

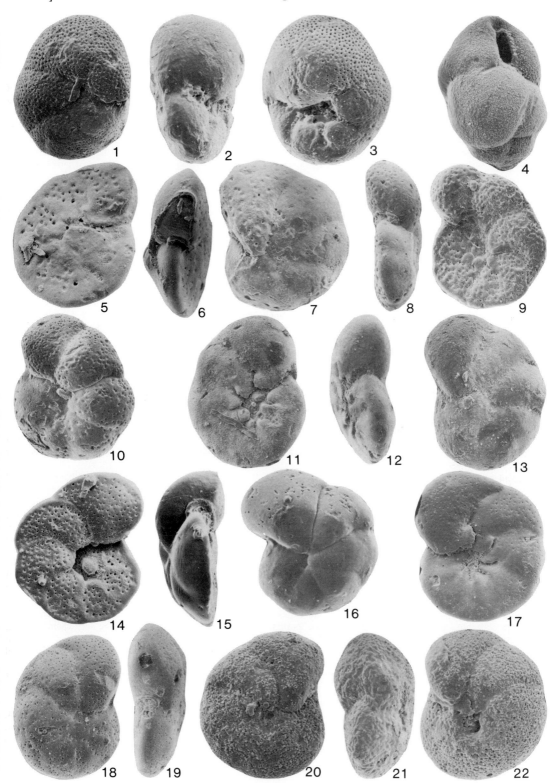

Plate 10.4

Cibicides (Cibicidina) succedens Brotzen, 1948
Plate 10.5, Figs 1–3 (× 90), diam. 440 μm, Thanet Formation, Lower Pegwell Marls Member, Pegwell, Kent. Figs 10.3, 10.10, 10.12. Description: planoconvex or with spiral side slightly raised, bi-umbonate; periphery subangular; 8–10 chambers in final whorl; suture limbate on spiral side.
Remarks: PB Thanetian (L) BB no record. TR 1 Palaeoc. Palaeoecology: marginal marine to mid shelf, possibly tolerant of poorly aerated mud bottom.

Cibicides tenellus (Reuss)
Plate 10.5, Figs 4–6, (× 145), diam. 280 μm, Bracklesham Group, Selsey Sand Formation, Whitecliff Bay, Isle of Wight. Figs 10.6–9, 10.11. = *Truncatulina tenella* Reuss, 1865. Description: test trochospiral, spiral side flat, evolute, umbilical side strongly convex, involute; 8–12 chambers in final whorl, periphery acute; aperture extending from the periphery onto the spiral side.
Remarks: *C. cantii*, *C. cassivelauni*, and *C. lobatulus* generally have fewer chambers and are less strongly convex on the umbilical side. PB Auversian-Marinesian (MW), BB Ledian-Asschian (K) e Tongrian (B) TR e-1 Eoc. Palaeoecology: marine, inner to mid shelf.

Cibicides cf. *ungerianus* (d'Orbigny)
Plate 10.5, Figs 7–9 (× 100), diam. 400 μm, Barton Clay Formation, Middle Barton Beds (F), Barton, Hampshire. Figs 10.6–9, 10.11. = *Rotalina ungeriana* d'Orbigny, 1846. Description: test trochospiral, spiral side slightly convex, partly involute, umbilical side convex, involute, with a small open umbilicus; periphery rounded; 8–11 chambers in outer whorl; aperture extends from the periphery along 2–3 chambers on the spiral side; wall coarsely perforate; thickened bosses on the early portion of the spiral side.
Remarks: the distinctive feature is the thickening of the early portion of the spiral side with bosses. The Eocene forms have a more rounded periphery than the forms from the Neogene. PB Marinesian (MW). BB Paniselian-Asschian (K) Rupelian (B) TR m-1 Eoc. Palaeoecology: normal marine, inner or mid shelf.

Cibicides westi Howe, 1939
Plate 10.5, Figs 10–12 (× 135), diam. 300 μm, London Clay Formation, Ongar, Essex. Figs 10.7, 10.9–11. Description: test trochospiral; spiral side flat, evolute; umbilical side strongly convex, involute; periphery acute; aperture a peripheral arch extending onto the spiral side.
Remarks: differs from *C. tenellus* in the absence of coarse pores on the umbilical side and in having relatively few coarse pores on the spiral side. PB Ypresian-Bartonian (L) Cuisian-Marinesian (MW) BB Ypresian-Wemmelian (K) TR m-1 Eoc. Palaeoecology: normal marine.

Cibicidoides alleni (Plummer)
Plate 10.5, Figs 13–15 (× 110), diam. 360 μm, London Clay Formation, Whitecliff Bay, Isle of Wight. Figs 10.3–5, 10.8–12. = *Truncatulina alleni* Plummer, 1927. Common synonyms: *Cibicidoides proprius* Brotzen, 1948; *Cibicides proprius* (Brotzen). Description: test trochospiral, flattened, biconvex, spiral side evolute with thickened early part, umbilical side involute with shallow umbilicus; periphery subacute; 8–11 chambers in outer whorl; aperture a peripheral arch extending onto the spiral side; wall coarsely perforate.
Remarks: a less inflated biconvex form than *C. pygmeus*. PB Thanetian-Lutetian (L, MW) BB Ypresian-Paniselian (K). TR 1 Palaeoc.-e Eoc. Palaeoecology: brackish to normal marine, inner to mid shelf, varied substrates.

Cibicidoides pygmeus (Hankten)
Plate 10.5, Figs 16–18 (× 135), diam. 300 μm, Bracklesham Group, Selsey Sand Formation, Whitecliff Bay, Isle of Wight. Figs 10.4, 10.6–9, 10.11. = *Pulvinulina pygmea* Hantken, 1875. Common synonym: *Cibicides pygmeus*. Description: test trochospiral, biconvex, spiral side evolute, umbilical side involute; 8–10 chambers in outer whorl; periphery subacute, aperture extending, from the periphery along the umbilical face of the last chamber; wall coarsely perforate.
Remarks: the biconvexity is more inflated than in *C. alleni*. PB no record. BB Wemmelian-Asschian (K) Rupelian (B), TR e-1 Eoc. Palaeoecology: marine, inner to mid shelf.

Discorbis propinqua (Terquem)
Plate 10.5, Figs 19–21 (× 100), diam. 400 μm, Bracklesham Group, Selsey Sand Formation, Selsey, West Sussex. Figs 10.6–9, 10.11. = *Rosalina propinqua* Terquem, 1882 = *Discorbis propinqua* (Terquem) emend. Le Calvez, 1949. Description: test trochospiral, planoconvex, with flat umbilical side, periphery keeled; on the umbilical side the chambers have non-perforate extensions which fuse to form a star-shaped umbilical boss; sutures slightly depressed on spiral side, deeply depressed on umbilical side.
Remarks: PB Cuisian-Marinesian (L, MW) BB Bruxellian-Wemmelian (K), TR m Eoc. Palaeoecology: normal marine, inner shelf, varied substrates.

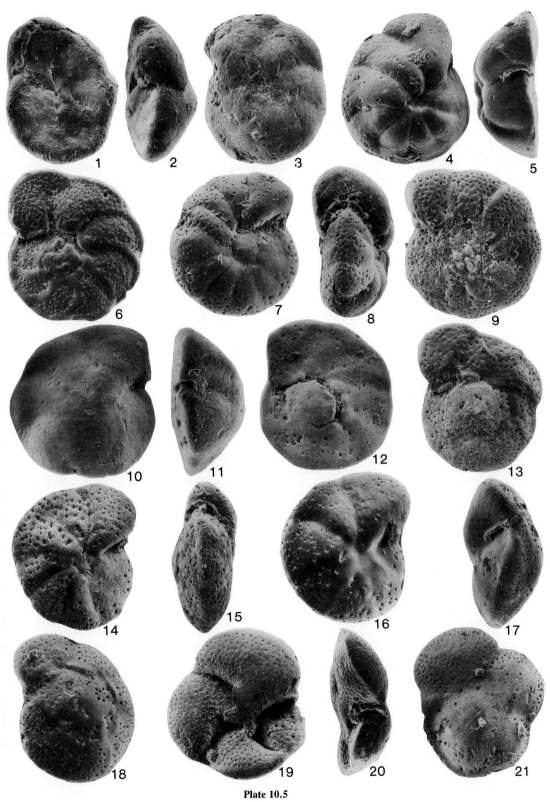

Plate 10.5

Elphidium hiltermanni Hagn, 1952
Plate 10.6, Figs 1, 2 (× 155), diam. 260 µm, Bracklesham Group, Wittering Formation, Whitecliff Bay, Isle of Wight. Figs 10.4–6, 10.8–12. Description: test planispiral, somewhat compressed, periphery rounded; shallow umbilici; 6–7 chambers in last whorl; sutures depressed, retral processes short; aperture interiomarginal, of pores hidden by pustules which form rows along the apertural face and extend onto the adjacent part of the earlier test.
Remarks: resembles *E. latidorsatum* but has a much less inflated test. PB Cuisian-Sannoisian (L, MW), BB Bruxellian–Wemmelian (K) Rupelian (L), TR e-1 Eoc. Palaeoecology: brackish to normal marine, estuarine to inner shelf, varied substrates.

Elphidium latidorsatum (Reuss)
Plate 10.6, Figs 3, 4 (× 135), diam. 300 µm, Barton Clay Formation, Whitecliff Bay, Isle of Wight. Figs 10.6–9, 10.11. = *Polystomella latidorsata* Reuss, 1864. Description: test planispiral, involute, inflated, periphery rounded; sutures nearly flush with numerous small retral processes between which there are pore-like openings; aperture of pores hidden by pustular ornament which covers the apertural face and the adjacent areas of older test.
Remarks: differs from *E. hiltermanni* in having a very inflated test. PB Auversian-Sannoisian (L, MW) BB Ypresian–Wemmelian (K) TR m-1 Eoc. Palaeoecology: marine, inner to mid shelf.

Epistomaria rimosa (Parker and Jones)
Plate 10.6, Figs 5–7 (× 135), diam. 300 µm, Bracklesham Group, Selsey Sand Formation, Selsey, West Sussex. Figs 10.6–9, 10.11. = *Discorbina rimosa* Parker and Jones, 1865. Description: test trochospiral, elongate in outline, periphery rounded, spiral side gently convex with deep sutures; umbilical side involute, internal walls give appearance of supplementary chambers around the umbilicus; primary aperture an interiomarginal slit, but secondary apertures are present along periphery, on the umbilical face, and in the sutures.
Remarks: PB Cuisian-Marinesian (L, MW) BB no record TR m Eoc. Palaeoecology: normal marine, shelf, varied substrates.

Epistomaria separans Le Calvez, 1949
Plate 10.6, Figs 8, 9 (× 105), diam. 380 µm, Bracklesham Group, Selsey Sand Formation, Selsey, West Sussex. Figs 10.8, 10.9, 10.11. Description: test trochospiral, with convex spiral side, flat umbilical side and rounded periphery; sutures between chambers of final whorl deeply incised, giving the appearance of separation of the chambers; primary aperture an interiomarginal slit, secondary apertures in peripheral parts of incised sutures.
Remarks: PB Lutetian-Marinesian (L, MW) BB no record, TR m Eoc. Palaeoecology: normal marine shelf.

Epistominella oveyi (Bhatia)
Plate 10.6, Figs 10–12 (× 285), diam. 140 µm, Solent Group, Bouldnor Formation, (Upper Hamstead Beds, Cerithium Bed), Hamstead. Figs 10.9, 10.11. = *Pseudoparrella oveyi* Bhatia, 1955. Description: test trochospiral, 6–8 chambers in final whorl; planoconvex, with conical spiral side, periphery subrounded; sutures on spiral side slightly swept back, those on umbilical side radial; aperture comma-shaped.
Remarks: microspheric forms are more conical than the megalospheric forms. Differs from *E. vitrea* in being planoconvex, and in having less swept-back sutures on the spiral side. *E. oveyi* resembles and may represent juvenile *Caucasina coprolithoides*. PB Marinesian (L) BB Paniselian-Wemmelian (K) Rupelian (B), TR Olig. Palaeoecology: brackish to marine, shallow shelf, muddy substrate.

Epistominella vitrea Parker, 1953
Plate 10.6, Figs 13–15 (× 285), diam. 140 µm, London Clay Formation, Alum Bay, Isle of Wight. Figs 10.3–12. Description: test trochospiral, 5–7 chambers in final whorl; biconvex, periphery subrounded, sutures on spiral side backward curving, those on umbilical side radial; aperture comma-shaped.
Remarks: Haynes (1956) has recognised slight morphological differences between micro and megalospheric generations. See *E. oveyi* for differences. PB Cuisian-Lutetian (MW) BB no record. TR 1 Palaeoc.-1 Eoc. Palaeoecology: slightly brackish to normal marine, inner to mid shelf, on muddy substrates.

Gyroidinoides angustiumbilicatus (Ten Dam)
Plate 10.6, Figs 16–18 (× 145), diam. 280 µm, London Clay Formation, Whitecliff Bay, Isle of Wight. Figs 10.4, 10.5, 10.8–12. = *Gyroidina augustiumbilicata* Ten Dam, 1944. Description: test trochospiral planoconvex, with strongly convex umbilical side and flat to slightly convex spiral side, periphery subacute; sutures radial on both sides; umbilical side with deep umbilicus; aperture an interiomarginal slit.
Remarks: differs from *G. octocamerata* in having a more highly convex umbilical side and in the orientation of the sutures. PB, BB no record. TR e Eoc. Palaeoecology: brackish to normal marine, shelf, muddy substrates.

Gyroidinoides danvillensis var. *gyroidinoides* (Bandy)
Plate 10.6, Figs 19–21 (× 135), diam. 300 µm. Thanet Formation, Upper Pegwell Marls Member, Pegwell, Kent. Figs 10.10, 10.12. = *Valvulineria danvillensis* var. *gyroidinoides* Bandy, 1949. Description: subglobose, umbilical side highest, periphery rounded; 8–9 chambers in final whorl; apertural face quadrangular, aperture interiomarginal.
Remarks: differs from *G. danvillensis* s.s. (Howe and Wallace) in having 8–9 chambers per whorl rather than 6. PB, BB no record. TR 1 Palaeoc. Palaeoecology: normal marine, inner to mid shelf, probably tolerant of poorly aerated muddy bottoms.

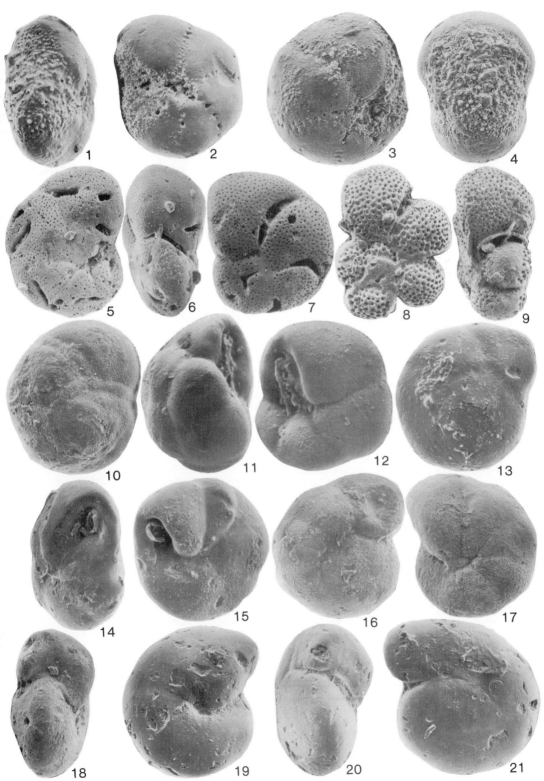

Plate 10.6

Gyroidinoides octocamerata (Cushman and Hanna)
Plate 10.7, Figs 1–3 (× 200), diam. 200 μm, Bracklesham Group, Selsey Sand Formation, West Sussex. Figs 10.6–9, 10.11. = *Gyroidina soldanii* d'Orbigny var. *octocamerata* Cushman and Hanna, 1927. Description: test trochospiral, planoconvex, with high convex spiral side, periphery rounded; sutures oblique on spiral side, radial on umbilical side; umbilical side with depressed umbilicus; aperture an interiomarginal slit.
Remarks: see *G. angustiumbilicatus*. PB Cuisian-Stampian (L), BB Ypresian-Asschian (K), TR m-l Eoc. Palaeoecology: normal marine, inner to mid shelf.

Hanzawaia producta (Terquem)
Plate 10.7, Figs 4–6 (× 105), diam. 380 μm, Bracklesham Group, Selsey Sand Formation, Selsey, West Sussex. Figs 10.6–9, 10.11. = *Truncatulina producta* Terquem,, 1882. = *Cibicides productus* (Terquem) emend. Le Calvez, 1949. Description: test trochospiral, planoconvex, with flat, fairly involute spiral side and convex involute umbilical side; periphery angular with keel; 6–7 chambers in outer whorl, the later ones having flaps on the spiral side; aperture extending from periphery on to spiral side.
Remarks: the sharp keel and the concave surface next to it on the umbilical side distinguishes this species from others of '*Cibicides*' type. PB Cuisian-Marinesian (L, MW) BB Ypresian-Asschian (K), TR e-l Eoc. Palaeoecology: normal marine, shelf.

Heterolepa pegwellensis (Haynes)
Plate 10.7, Figs 7–9 (× 160), diam. 250 μm, Thanet Formation, Upper Pegwell Marls Member, Pegwell, Kent. Figs 10.3, 10.10, 10.12. = *Hollandina pegwellensis* Haynes, 1956. Description: biconvex or highest in spiral side, periphery acute and warped; up to 27 chambers in microspheric (7 : 9 : 9 : - or 7 : 9 : 8 : - in successive whorls), becoming twice as long as high by 20th, up to 22 in megalospheric generation (7 : 8 : 7 : - or 7 : 7 : 7 : - in successive whorls), becoming twice as long as high by 13th; sutures flush, markedly swept back on spiral side; coarse pores on spiral side; aperture an interiomarginal slit.
Remarks: a very small species (< 300 μm). PB, BB no record. TR 1 Palaeoc. Palaeoecology: normal marine, inner to mid shelf, probably tolerant of poorly aerated bottoms.

Marginulinopsis wetherellii (Jones)
Plate 10.7, Figs 10, 11 (× 21), length 1.9 mm, London Clay Formation, Hadley Wood, Middlesex. Figs 10.3, 10.10, 10.12. = *Marginulina wetherellii* Jones, 1854. Common synonym: *Marginulina enbornensis* Bowen. Description: test initially planispiral then uncoiling, laterally compressed; aperture radiate on a neck, terminal; test commonly ornamented with longitudinal costae or tubercles.
Remarks: PB no record, BB Ypresian (K), TR 1 Palaeoc.– e Eoc. Palaeoecology: normal marine.

Melonis affine (Reuss)
Plate 10.7, Figs 12, 13 (× 125), diam. 320 μm, Barton Clay Formation, Middle Barton Beds (D), Barton, Hampshire. Figs 10.6–9, 10.11. = *Nonionina affinis* Reuss, 1851. Common synonym: *Nonion affine*. Description: test planispiral, involute, compressed, biumbilicate, periphery rounded; around ten chambers in last whorl; aperture an interiomarginal arch.
Remarks: PB Lutetian-Sannoisian (L, MW), BB Ypresian-Asschian (K). Tongrian-Rupelian (B), TR m-l Eoc. Palaeoecology: marine, inner to mid shelf.

Nodosaria latejugata Gümbel, 1868
Plate 10.7, Fig. 14 (× 15), length 2–6 mm, London Clay Formation, Hadley Wood, Middlesex. Figs 10.10, 10.12. Description: test uniserial, chambers inflated, with stout longitudinal ribs that cross the depressed sutures; circular in cross section.
Remarks: PB no record, BB Paniselian (K), TR e Eoc. Palaeoecology: normal marine.

Nonion applinae Howe and Wallace, 1932
Plate 10.7, Figs 15, 16 (× 135), diam. 300 μm, Thanet Formation, Reculver Silts Member, Reculver, Kent. Figs 10.10, 10.12. Description: compressed, periphery rounded, apertural face high, oval; 7–9 chambers; sutures radial; umbilici small, granulate.
Remarks: PB, BB no record. TR 1 Palaeoc. Palaeoecology: normal marine-?brackish, inner shelf.

Nonion laeve (d'Orbigny)
Plate 10.7, Figs 17, 18 (× 100), diam. 400 μm, Bracklesham Group, Selsey Sand Formation, Selsey, West Sussex. Figs 10.3–8, 10.10–12. = *Nonionina laevis* d'Orbigny, 1826. Common synonym: *Elphidium laeve*. Description: test planispiral, compressed; periphery subangular to rounded; 10–14 chambers in last whorl; umbilici with umbilical boss; sutures depressed, ornamented with tubercules; aperture an interiomarginal slit obscured by tubercles.
Remarks: this form shows varation in the number of chambers, depth of depression of the sutures and the number and size of the umbilical bosses. PB Sparnacian-Ludian (L, MW). BB Ypresian-Asschian (K), TR 1 Palaeoc.-e Olig. Palaeoecology: brackish to normal marine, marginal marine to inner shelf.

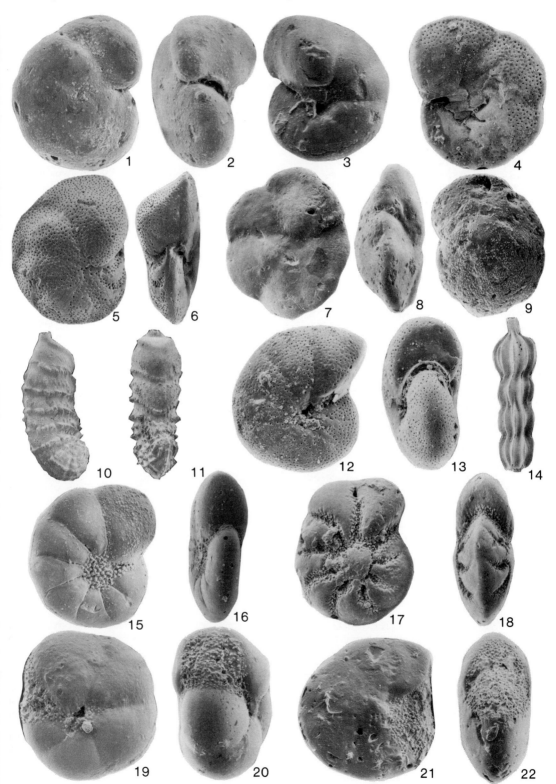

Plate 10.7

Nonion parvulum (Grzybowski)
Plate 10.7, Figs 19, 20 (× 160), diam. 250 μm, Solent Group, Bouldnor Formation, (Upper Hamstead Beds, Cerithium Bed), Hamstead. Figs 10.7, 10.9, 10.11. = *Anomalina parvula* Grzybowski, 1896. Description: test planispiral, involute, periphery rounded; sutures depressed; five chambers in last whorl; aperture an interiomarginal slit, normally obscured by pustulose material.
Remarks: superficially similar to *Elphidium hiltermanni* but with fewer chambers and an absence of retral processes. PB Sannoisian (L, MW) BB no record. TR 1 Eoc.-m Olig. Palaeoecology: brackish to normal marine, lagoonal, mainly muddy substrates.

Nonion reculverensis Haynes, 1956
Plate 10.7, Figs 21, 22 (× 135), diam. 300 μm, Thanet Formation, Reculver Silts Member, Reculver, Kent. Figs 10.10, 10.12. Description: planispiral, periphery entire, subangular, bi-umbonate; eight chambers in outer whorl; sutures limbate, swept back, aperture obscured by granular pustules.
Remarks: a small species. Some individuals show incipient retral processes across the sutures,. PB, BB no record. TR 1 Palaeoc. Palaeoecology: normal marine-?brackish, inner shelf.

Pararotalia audouini (d'Orbigny)
Plate 10.8, Figs 1–3 (× 57), diam. 700 μm. Solent Group, Headon Hill Formation, (Middle Headon Beds), Whitecliff Bay, Isle of Wight. Figs 10.6, 10.9, 10.11. = *Rotalia audouini* d'Orbigny, 1826. Common synonym. *Pararotalia subinermis* Bhatia, 1955. Description: test trochospiral, unequally biconvex; 6–8 chambers in final whorl; periphery sharply keeled and with granular ornament; keel produced into spine-like outgrowths in latest chambers; umbilical side very convex with deep umbilicus largely filled with a calcite boss.
Remarks: differs from *P. inermis* (Terquem) in having a more angled outline in equatorial view, and from *P. spinigera* (Le Calvez) in having a sharper periphery and only short spines. PB Lutetian-Marinesian (L, MW) BB present but range not distinguished from other *Pararotalia* spp. (K) TR 1 Eoc. Palaeoecology: normal marine, inner shelf, varied substrates.

Pararotalia curryi Loeblich and Tappan, 1957
Plate 10.8, Figs 4–6 (× 135), diam. 300 μm, Solent Group, Headon Hill Formation, (Middle Headon Beds), Whitecliff Bay, Isle of Wight. Figs 10.4–12. Common synonym: *Rotalia canui* Cushman of some authors. Description: test trochospiral, biconvex, periphery subacute; spiral side convex; umbilical side convex with umbilical plug; 4–6 chambers in outer whorl, sometimes with a short peripheral spine.
Remarks: this species resembles *P. spinigera* but the latter is planoconvex and has more pronounced peripheral spines. PB Cuisian-Sannoisian (L, MW) BB Tongrian-Rupelian (L). TR e Eoc.-e. Olig. Palaeoecology: mainly brackish but extending into normal marine, inner shelf.

Pararotalia inermis (Terquem)
Plate 10.8, Figs 7–9 (× 57), diam. 700 μm, Bracklesham Group, Selsey Sand Formation, Selsey, West Sussex. Figs 10.8, 10.9, 10.11. = *Rotalina inermis* Terquem, 1882. = *Pararotalia inermis* (Terquem) emend. Loeblich and Tappan, 1957. Description: test trochospiral, biconvex, periphery acute and keeled with occasional short spines; umbilical side with deep umbilicus around an umbilical plug, each chamber with a boss adjacent to the umbilicus; aperture on apertural face, with lip.
Remarks: differs from *P. audouini* in having a more rounded outline, being more equally biconvex, and in having a boss on the umbilical portion of each chamber. *P. spinigera* is planoconvex and much smaller. PB Lutetian-Marinesian (L, MW) BB no record, TR m-1 Eoc. Palaeoecology: normal marine, inner shelf, varied substrates.

Pararotalia spinigera (Le Calvez)
Plate 10.8, Figs 10–12 (× 115), diam. 350 μm, Bracklesham Group, Selsey Sand Formation, Selsey, West Sussex. Figs 10.6–9, 10.11. = *Globorotalia spinigera* (Terquem) Le Calvez, 1949 (not *Rosalina spinigera* Terquem) = *Pararotalia spinigera* (Le Calvez) emend. Loeblich and Tappan, 1957. Description: test trochospiral, planoconvex, spiral side convex, umbilical side flat, periphery angled; umbilical side with umbilicus and plug; 5–6 chambers in outer whorl, the older ones bearing a short peripheral spine; aperture on apertural face, bearing a lip.
Remarks: this species resembles *P. curryi* but the latter is biconvex and has only feebly developed peripheral spines. PB Sparnacian-Ludian (L, MW), BB present but range not distinguished from that of other *Pararotalia* spp. TR e-1 Eoc. Palaeoecology: normal marine, inner shelf, varied substrates.

Praeglobobulimina ovata (d'Orbigny)
Plate 10.8, Fig. 13 (× 120), length 340 μm, London Clay Formation, Whitecliff Bay, Isle of Wight. Figs 10.3, 10.4, 10.9–12. = *Bulimina ovata* d'Orbigny, 1846. = *Praeglobobulimina ovata* (d'Orbigny) emend. Haynes, 1954. Description: test triserial, elongate, with strongly overlapping chambers; aperture an elongate loop perpendicular to the basal margin of the apertural face.
Remarks: PB no record. BB Wemmelian-Asschian (K) TR 1 Palaeoc.-e Eoc. Palaeoecology: slightly brackish, inner shelf, mainly sandy clay substrates.

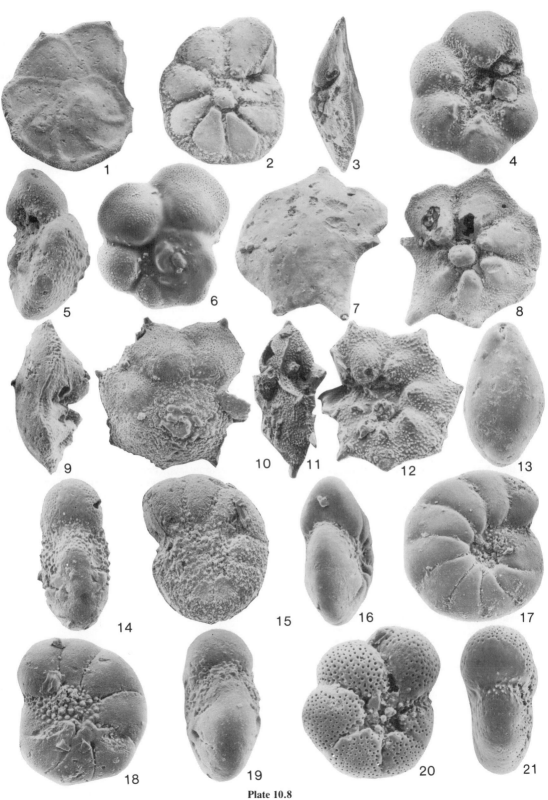

Plate 10.8

Protelphidium hofkeri Haynes, 1956
Plate 10.8, Figs 14, 15 (× 165), diam. 240 μm, Thanet Formation, Reculver Silts Member, Reculver, Kent. Figs 10.10, 10.12. Description: planispiral, periphery broadly rounded, lobate; 8–9 chambers in outer whorl; sutures backward curving, impressed and excavated; umbilici filled with granular calcite that extends along sutures; aperture a low slit.
Remarks: this is the most pustulate of all the *Protelphidium* spp. BB Thanetian (L) BB no record. TR 1 Palaeoc. Palaeoecology: normal marine-?brackish, inner shelf.

Protelphidium sp. 1
Plate 10.8, Figs 16, 17 (× 135), diam. 300 μm, Barton Clay Formation, Middle Barton Beds (F) Barton, Hampshire. Figs 10.6, 10.7, 10.9, 10.11. = *Protelphidium* sp. 1 of Murray and Wright, 1974. Description: test planispiral, biconvex; periphery subacute; 8–10 chambers in outer whorl; umbilici with a small amount of tubercular ornament; aperture an interiomarginal row of pores.
Remarks: test more compressed and biconvex than in other *Protelphidium* species. PB Marinesian (MW), BB no record. TR m-1 Eoc. Palaeoecology: slightly brackish to normal marine, inner to mid shelf.

Protelphidium sp. 2
Plate 10.8, Figs 18, 19 (× 165), diam. 240 μm, Bracklesham Group, Wittering Formation, Whitecliff Bay, Isle of Wight. Figs 10.6, 10.8, 10.9, 10.11. = *Protelphidium* sp. 2 of Murray and Wright, 1974. Description: test planispiral, biconvex; periphery rounded; 7–9 chambers in the outer whorl; sutures depressed and ornamented with tubercles except over the periphery; tubercules also ornament the umbilici; aperture an interiomarginal row of pores.
Remarks: this species has a more biconvex test than *P.* cf. *roemeri* or *Protelphidium* sp. 1 and is less coarsely perforate than *Protelphidium* sp. 4. PB, BB no record. TR e-m Eoc. Palaeoecology: brackish (20–30‰), estuarine.

Protelphidium sp. 4
Plate 10.8, Figs 20, 21 (× 250), diam. 160 μm, Solent Group, Headon Hill Formation, (Middle Headon Beds), Colwell Bay, Isle of Wight. Figs 10.6, 10.9, 10.11. = *Protelphidium* sp. 4 of Murray and Wright (1974). Description: test planispiral, chambers inflated, periphery rounded and lobulate; 6–7 chambers in the outer whorl; sutures deeply depressed; umbilicus ornamented with tubercules; aperture an interiomarginal row of pores; wall pores coarse except on apertural face.
Remarks: this is a small species with inflated chambers and a relatively coarsely perforate wall. PB, BB no record. TR 1 Eoc.-e Olig. Palaeoecology: brackish to normal marine, estuarine and lagoonal.

Protelphidium cf. *P. roemeri* (Cushman)
Plate 10.9, Figs 1, 2 (× 160), diam. 250 μm, Solent Group, Bouldnor Formation, (Bembridge Marl), Whitecliff Bay, Isle of Wight. Figs 10.6, 10.7, 10.9, 10.11. = *Nonion roemeri* Cushman, 1936. Description: test planispiral, with fairly flat sides; periphery rounded and lobulate; 8–10 chambers in the outer whorl; tubercular ornament extends from the umbilici into the umbilical ends of the depressed sutures; aperture an interiomarginal row of pores.
Remarks: the flat-sided form and rounded periphery distinguishes this from other *Protelphidium* species. PB Sannoisian (MW) BB 1 Olig. TR 1 Eoc.-m Olig. Palaeoecology: brackish, lagoonal.

Pullenia quinqueloba (Reuss)
Plate 10.9, Figs 3, 4 (× 135), diam. 300 μm. London Clay Formation, Alum Bay, Isle of Wight. Figs 10.3–5, 10.9, 10.10, 10.12. = *Nonionina quinqueloba* Reuss, 1851. Description: test planispiral, involute, compressed with subrounded periphery; 5–6 chambers in the last whorl; aperture an interiomarginal slit.
Remarks: superficially similar to *Nonion parvulum* but lacking the pustulose ornament around the aperture. PB Auversian-Marinesian (L). BB Ypresian-Asschian (K) Rupelian (B). TR 1 Palaeoc.-e Eoc.; Rec. Palaeoecology: normal marine, mid shelf, muddy substrate.

Pulsiphonina prima (Plummer)
Plate 10.9, Figs 5–7 (× 200), diam. 200 μm, London Clay Formation, Alum Bay, Isle of Wight. Figs 10.3–5, 10.8–12. = *Siphonina prima* Plummer, 1927. Description: test trochospiral compressed, biconvex, periphery acute; 4–5 chambers in last whorl; aperture an arch close to the periphery.
Remarks: PB Thanetian-Cuisian (L, MW) BB Ypresian-Bruxellian (K). TR 1 Palaeoc.-e Eoc. Palaeoecology slightly brackish to normal marine, inner shelf, mainly muddy substrates.

Robertina germanica Cushman and Parker, 1938
Plate 10.9, Fig. 8 (× 110), length 360 μm, Barton Clay Formation, Middle Barton Beds (C), Barton, Hampshire. Figs 10.6–9, 10.11. Description: test high trochospiral, elongate; chambers divided internally; aperture loop shaped; wall aragonitic.
Remarks: PB no record. BB Ledian-Asschian (K). TR m-1 Eoc. Palaeoecology: normal marine shelf.

Palaeogene

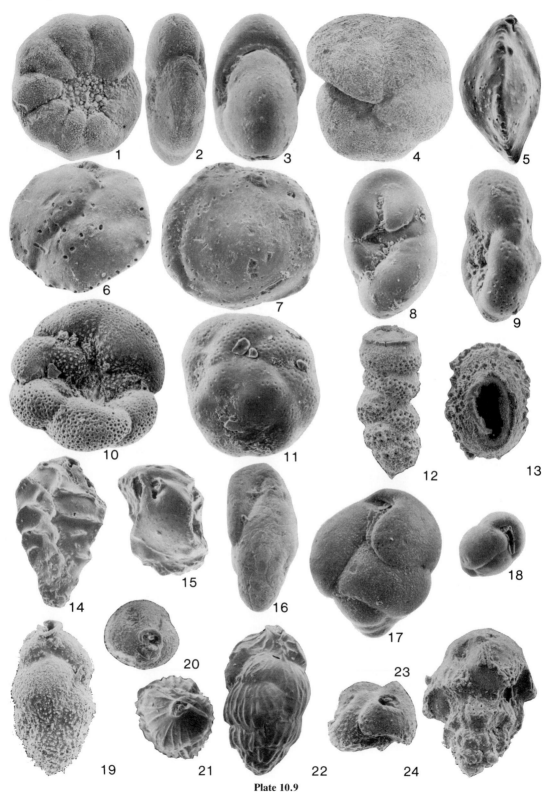

Plate 10.9

Rosalina araucana d'Orbigny, 1839
Plate 10.9, Figs 9–11 (× 155), diam. 260 μm, Solent Group, Headon Hill Formation, (Upper Headon Beds), Headon Hill, Isle of Wight. Figs 10.6, 10.7, 10.9, 10.11. Common synonyms: *Discorbis araucanus, Valvulinera araucana*. Description: test trochospiral, planoconvex, spiral side gently convex; umbilical side has depressed, star-shaped umbilicus; 8 chambers in last whorl, final chamber large; aperture interiomarginal, umbilical.
Remarks: PB no record. BB Tongrian-Rupelian (B, as *Discorbis* sp.). TR 1 Eoc.-e Olig. Palaeoecology: brackish to normal marine, interitidal to shallow lagoonal on muddy substrates.

Sagrina selseyensis (Heron-Allen and Earland)
Plate 10.9, Fig. 12 (× 100), final chamber missing; Fig. 13 (× 160); length 400 μm; Bracklesham Group, Selsey Sand Formation, Selsey, West Sussex. Figs 10.6, 10.8, 10.9, 10.11. = *Bigenerina selseyensis* Heron-Allen and Earland, 1909. Description: test initially biserial, (6–8 chambers), then uniserial with 2–5 chambers of oval cross section; aperture oval, terminal, with a raised lip.
Remarks: BB Ledian-Asschian (K). TR m-l Eoc. Palaeoecology: normal marine, inner shelf.

Tappanina selmensis (Cushman)
Plate 10.9, Fig. 14 (× 100); Fig. 15 (× 120); length 400 μm; Thanet Formation, Reculver Silts Member, Reculver, Kent. Figs 10.10, 10.12. = *Bolivinita selmensis* Cushman, 1933. Description: test biserial with concave faces; aperture a narrow arch at base of the final chamber.
Remarks: PB Thanetian (L). TR 1. Palaeoc. Palaeoecology: normal marine.

Turrilina andreaei Cushman, 1928
Plate 10.9, Fig. 16 (× 155), length 260 μm, Solent Group, Headon Hill Formation, (Middle Headon Beds), Headon Hill, Isle of Wight. Figs 10.7, 10.9, 10.11. = *Bulimina acicula* Andreae, 1884 (non *Bulimina acicula* Costa, 1856). Description: test a high trochospiral of around six whorls with many chambers; sutures depressed; aperture a small slit in a depressed area of the apertural face.
Remarks: PB Sannoisian (L, MW). BB Tongrian (L, as *T. acicula*) TR 1 Eoc.-e Olig. Palaeoecology: slightly brackish, normal marine, to hypersaline lagoons.

Turrilina brevispira Ten Dam, 1944
Plate 10.9, Figs 17, 18, Fig. 17 (× 200); length 200 μm, London Clay Formation, Sheppey, Kent. Figs 10.10, 10.12. Description: a high trochospiral with three inflated chambers in the final whorl.
Remarks: PB Lutetian (MW). BB Ypresian (K). TR e Eoc. Palaeoecology: normal marine.

Uvigerina batjesi Kaasschieter, 1961
Plate 10.9, Figs 19, 20 (× 145), length 280 μm, London Clay Formation, Whitecliff Bay, Isle of Wight. Figs 10.4, 10.8–12. Description: test initially triserial, becoming uniserial, almost circular in section; chambers inflated, sutures depressed in later part; aperture terminal, with a slight neck and lip.
Remarks: the circular cross-section and absence of costae distinguish this species from *U. germanica* and *U. muralis*. PB no record. BB Ypresian-Paniselian (K). TR e Eoc. Palaeoecology: slightly brackish, inner to mid shelf, muddy substrates.

Uvigerina germanica (Cushman and Edwards)
Plate 10.9, Figs 21, 22 (× 135), length 300 μm, Solent Group, Bouldnor Formation, (Upper Hamstead Beds, Cerithium Bed), Hamstead, Isle of Wight. Figs 10.9, 10.11. = *Angulogerina germanica* Cushman and Edwards, 1938. Description: test initially triserial becoming uniserial, triangular in section; chambers inflated, earlier ones ornamented with longitudinal costae, later ones smooth; aperture terminal, large, elliptical, with a lip.
Remarks: the presence of costae is distinctive. PB Sannoisian (L). BB Rupelian (B). TR m Olig. Palaeoecology: normal marine, shelf, muddy substrate.

Uvigerina muralis Terquem, 1882
Plate 10.9, Figs 23, 24 (× 200), length 200 μm, Bracklesham Group, Selsey Sand Formation, Selsey, West Sussex. Figs 10.6, 10.8, 10.9, 10.11. Common synonym: *Angulogerina muralis* (Terquem). Description: test initially triserial, becoming uniserial; sides angular, formed of irregular chambers with truncated margins; sutures deep and wide; aperture small on a short neck.
Remarks: the triangular cross section is distinctive. PB Sparnacian-Marinesian (L, MW). BB Paniselian-Asschian (K). TR m-l Eoc. Palaeoecology: normal marine, inner shelf, muddy substrate.

GLANDULINIDAE AND POLYMORPHINIDAE

Glandulina laevigata (d'Orbigny)
Fig. 10.13a, b (× 50), length 420 μm, Barton Clay Formation, Middle Barton Beds (C), Barton, Hampshire. Figs 10.3, 10.6–12. = *Nodosaria (Glandulina) laevigata* d'Orbigny, 1826. Description: test fusiform, pointed at both ends, circular in section; early chambers biserial in microspheric forms, but mainly or entirely uniserial; sutures flush; aperture terminal, radiate.

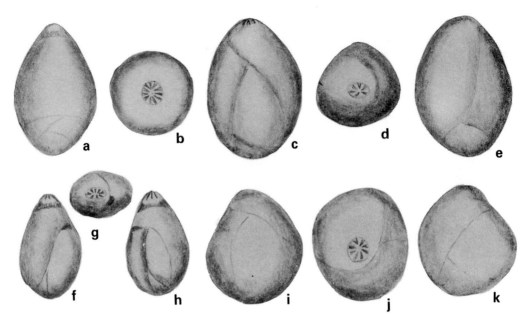

Fig. 10.13. a, b. *Glandulina laevigata* (d'Orbigny); c, d, e. *Globulina ampulla* (Jones); f, g, h. *Globulina inaequalis* Reuss; i, j, k. *Guttulina problema* d'Orbigny.

Remarks: PB Marinesian (L, MW). BB Paniselian-Asschian (K) Rupelian (B). TR 1 Palaeoc.-1 Eoc. Palaeoecology: normal marine, inner to mid shelf.

Globulina ampulla (Jones)
Fig. 10.13c-e (× 50), length 460 μm, Barton Clay Formation, Lower Barton Beds (A2), Barton, Hampshire. Figs 10.3, 10.6–12. = *Polymorphina ampulla* Jones, 1852. Description: test oval, pointed at both ends, subcircular in cross-section; few chambers, added in planes 144° apart; sutures very slight depressed; aperture radiate.
Remarks: differs from *G. gibba* in being oval rather than globular; differs from *G. inaequalis* and *G. rotundata* in both profile and cross section. PB Thanetian-Auversian (L, MW). BB no record. TR 1 Palaeoc. Palaeoecology: normal marine, inner to mid shelf.

Globulina inaequalis (Reuss, 1850)
Fig. 10.13f–h (× 50), length, 300 μm, Bracklesham Group, Selsey Sand Formation, Whitecliff Bay, Isle of Wight. Figs 10.6–9, 10.11. Description: test oval, broader at the base, compressed oval in cross section; sutures more or less flush; aperture radiate.
Remarks: differs from *G. gibba*, *G. ampulla* and *G. rotundata* in being compressed rather than round in cross-section. PB Cuisian-Sannoisian (L, MW). BB Ypresian-Asschian (K). Tongrian-Rupelian (B as part of *G. gibba*) Rupelian (L). TR e-1 Eoc. Palaeoecology: normal marine, inner to mid shelf.

Guttulina problema d'Orbigny, 1826
Fig. 10.13i–k (× 50), length 450 μm, Barton Clay Formation, Upper Barton Beds, (H), Barton, Hampshire. Figs 10.3–12. Description: test somewhat elongate, subtriangular in cross section; chambers arranged in a quinqueloculine series; sutures depressed; aperture radiate.
Remarks: PB Thanetian-Sannoisian (L, MW). BB Ypresian-Asschian (K). Tongrian-Rupelian (B). TR 1 Palaeoc.-1 Eoc. Palaeoecology: normal marine, inner to mid shelf.

Planktonic species

Globigerina chascanona Loeblich and Tappan, 1957
Plate 10.10, Figs 1–3 (× 120), diam. 270 μm, London Clay Formation, Leca Works, Ongar, Essex. (= *Globorotalia esnaensis* MW). Description: test tightly coiled, rather high-spired; $4\frac{1}{2}$–5 chambers in last whorl, increasing slowly; surface pitted, with spines especially on the umbilical side; aperture an umbilical arch, with a narrow lip.
Remarks: a variable species, widespread in the London Clay. PB Ypresian (Varengeville, as *intermedia*, Bignot, 1963), BB Ypresian-Paniselian (K, as cf. *varianta*). TR l Palaeoc.-e Eoc.

Globigerina triloculinoides Plummer, 1926
Plate 10.10, Figs 4, 5 (× 120), diam. 210 μm, Thanet Formation, Reculver, Kent. Figs 10.3, 10.10, 10.12. Description: test a low trochospiral, spiral side almost flat; $3\frac{1}{2}$ globular chambers in final whorl, rapidly increasing in size; surface with rather wide-spaced pore-pits; aperture a low arch, with a thick lip.
Remarks: rare in the Thanet Formation, and possibly derived. PB, BB not recorded. Danian of Denmark (Auctt) TR Palaeoc.

Globigerina patagonica Todd and Kniker, 1952
Plate 10.10, Figs 6–8, 10–12 (× 120), diam. 330 μm, London Clay Formation, Leca Works, Ongar, Essex. Figs 10.10, 10.12. Description: test low-spired, typically $3\frac{1}{2}$ subglobular chambers in the final whorl; surface with well-defined pore-pits on all chambers; aperture a slightly twisted umbilical arch.
Remarks: Specimens with three chambers per whorl resemble *triangularis* White. PB no record, BB Ypresian (K, as *triloculinoides*), NW Germany (Berggren, 1969). TR e Eoc.

Globigerinatheka barri Brönnimann, 1952
Plate 10.10, Fig. 9 (× 500); Fig. 13 (× 120); diam. 280 μm; Bracklesham Group, Selsey Sand Formation, Amusium Bed, (Chilling, Hampshire) (Curry *et al.*, 1968)). Description: test almost spherical, initially trochospiral, but later chambers overlapping, the last covering nearly half of the test; multiple apertures are tiny, marginal in position, and have a thickened lip (Fig. 9).
Remarks: PB, BB no record. TR m-l Eoc.

Globorotalia broedermanni Cushman and Bermudez, 1949
Plate 10.10, Figs 14–16 (× 120), diam. 240 μm, Bracklesham Group, Wittering Formation, (Bed W12), East Wittering, West Sussex. Figs 10.6, 10.8, 10.9, 10.11. Description: test a low trochospiral, appoaching biconical, with $4\frac{1}{2}$–5 radially compressed and somewhat angular chambers in the last whorl; periphery in side view subangular, surface hispid.
Remarks: PB Cuisian (L), BB no record. TR e-m Eoc.

Globorotalia danvillensis (Howe and Wallace)
Plate 10.10, Figs 17–19 (× 120), diam. 250 μm, Barton Clay Formation, Bed A2, Barton, Hampshire. Figs 10.7, 10.9, 10.11. = *Globigerina danvillensis* Howe and Wallace, 1932. = *G.* cf. *angustiumbilicata*, Brönnimann *et al.* (L, MW). Description: test low trochospiral, with $4\frac{1}{2}$–$5\frac{1}{2}$ almost spherical chambers in last whorl; surface progressively more spiny in earlier-formed chambers; aperture semicircular, without marked lip.
Remarks: The ornament is very characteristic. PB. Auversian, Marinesian (L), Stampian (? derived), BB Tongrian (Vieux-Joncs) (Curry). TR l Eoc.-m Olig.

Globorotalia cf. *esnaensis* (Leroy)
Plate 10.10, Figs 26, 27 (× 120), diam. 210 μm, Thanet Formation, Reculver, Kent. Figs 10.3–5, 10.9–12. = *Globigerina esnaensis* Leroy, 1953. Description: test subspherical, last whorl of $3\frac{1}{2}$–4 rounded chambers which, viewed from the umbilical area, have a semi-oval to quadrate profile; surface pitted; aperture slightly extraumbilical, with lip.
Remarks: The characteristic quadrate profile seen in specimens from the early Eocene of Germany is only feebly shown in the British material, hence the use of the "cf." prefix. PB no record, BB Ypresian (Moorkens, 1968)? TR l Palaeoc.-e Eoc.

Globorotalia perclara Loeblich and Tappan, 1957
Plate 10.10, Figs 28–30 (× 120), diam. 210 μm, Thanet Formation, Reculver, Kent. Figs 10.3, 10.10, 10.12. Description: test trochospiral, almost planoconvex, last whorl of 5–6 subspherical chambers; sutures radial; periphery lobed; umbilicus open; aperture interiomarginal; ornament of pustules, becoming stronger on earlier chambers.
Remarks: PB Thanetian (Curry), BB no record. TR m Palaeoc.-e Eoc.

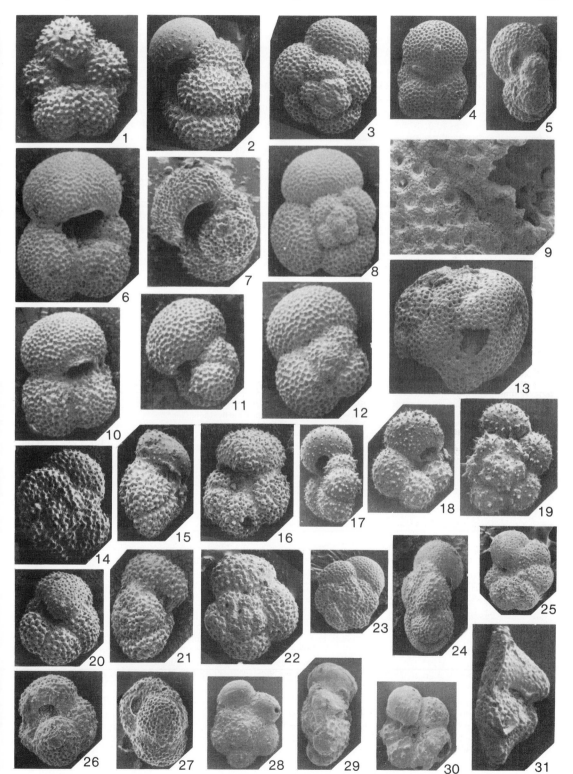

Plate 10.10

Globorotalia reissi Loeblich and Tappan, 1957
Plate 10.10, Figs 23–25 (× 120), diam. 180 µm, London Clay Formation, Leca Works, Ongar, Essex. Figs. 10.10, 10.12. Description: test low trochospiral, last whorl of six slightly compressed chambers; umbilicus shallow, small; aperture small, interiomarginal; ornament slight.
Remarks: specimens from Sheppey have a distinct keel, and were referred to *pseudoscitula* Glaessner (Brönnimann *et al.*, 1968). PB Ypresian (Varengeville), BB Paniselian (Brönnimann, *et al.*, 1968). TR e Eoc.

Globorotalia wilcoxensis Cushman and Ponton, 1932
Plate 10.10, Figs 20–22 (× 120), diam. 240 µm, Bracklesham Group, Wittering Formation, (Bed W12) East Wittering, Sussex. Figs 10.6, 10.8, 10.9, 10.11. Description: test trochospiral, umbilical side very convex, final whorl of four subangular chambers, surface hispid, periphery lobate.
Remarks: PB (L, as *pseudotopilensis* Subbotina), BB no record. TR e Eoc.

Globorotalia subbotinae Morozova, 1939
(a) Plate 10.11, Fig. 1; (b) Plate 10.10, Fig. 31; (× 120); diam. 290 µm. (a) Bracklesham Group, Wittering Formation, (Bed W12), East Wittering, Sussex; (b) Bracklesham Group, English Channel, 50° 11.5′N, 1°31′ W. Figs 10.6, 10.8, 10.9, 10.11. Description: test subconical, last whorl of four angulate chambers, with a strong peripheral keel; surface very rugose, umbilicus small; aperture small, interiomarginal.
Remarks: Fig. 31 has an unusually strong keel and might be ascribed to *marginodentata* Subbotina. PB Cuisian (L), BB Paniselian? (K, fide Berggren, 1969), TR e Eoc.

Truncorotaloides rohri Brönnimann and Bermudez, 1953
Plate 10.11, Figs 2–4 (× 120), diam. 300 µm. Bracklesham Group, Selsey Sand Formation, Selsey, West Sussex. Fig. 10.8. Description: test low trochospiral, with 4–5 subquadrangular chambers in last whorl; surface covered with close-spaced, fine spines.
Remarks: as is generally the case in the Anglo-Paris area, specimens typically do not display sutural apertures. PB Lutetian (L) BB Bruxellian–Wemmelian (Brönnimann *et al.*, 1968). TR m Eoc.

Guembelitria triseriata (Terquem)
Plate 10.11, Figs 5, 6 (× 120), length 220 µm, Bracklesham Group, Wittering Formation, (Bed W12) East Wittering, Sussex. Figs 10.6, 10.8, 10.9, 10.11. = *Textilaria triseriata* Terquem, 1882. Description: test elongate, triserial, chambers globular, smooth; aperture an interiomarginal arch with a twisted and thickened lip.
Remarks: PB Cuisian-Marinesian (L), BB no record. TR Eoc. (not earliest).

Pseudohastigerina wilcoxensis (Cushman and Ponton)
Plate 10.11, Figs 7–9 (× 120), diam. 290 µm, Bracklesham Group, Wittering Formation, (Bed W12), East Wittering, Sussex. Figs 10.6, 10.8–12. = *Nonion wilcoxensis* Cushman and Ponton, 1932. Description: planispiral, last whorl of six subglobular chambers, rapidly increasing; aperture a semicircular arch, with a marked lip; surface smooth except for the presence of pores.
Remarks: PB no record, BB Paniselian (K, as *micra*) TR e Eoc.

Larger benthic species

Nummulites aquitanicus Benoist, 1888
Plate 10.11, Figs 20, 21 form B (× 6), diam. 7 mm, Bracklesham Group, Wittering Formation, (Bed IV), Whitecliff Bay, Isle of Wight. Fig. 10.6. Description: test biconical, septal filaments meandriform, pillars present both along and between filaments; spiral lamina of about eight turns (B), chambers high.
Remarks: PB no record, BB one doubtful record. TR 1 Eoc.

Nummulites laevigatus (Bruguière)
Plate 10.11, Figs 12–16, form B, Figs 12, 13, 16, (× 3); form A, Figs 14, 15 (× 6), diam. (B) 15 mm, (A) 3.5 mm; Bracklesham Group, Earnley Formation, (Bed E6), Bracklesham, West Sussex. = *Camerina laevigata* Bruguière, 1792. Figs 10.6, 10.8, 10.9, 10.11. Description: test biconvex, septal filaments reticulate, pillars present, mostly along filaments; spiral lamina of 15 (B) or five (A) turns, chambers rather high, septa oblique.
Remarks: an important marker throughout the Anglo-Paris-Belgian area. PB Lutetian (B1), BB Lutetian (B1) TR m Eoc.

Nummulites planulatus (Lamarck)
Plate 10.11, Figs 10, 11 form B (× 6), diam. 7 mm. Bracklesham Group, Wittering Formation, (Bed IV), Whitecliff Bay, Isle of Wight. = *Lenticulites planulata* Lamarck, 1804. Figs 10.6, 10.8, 10.9, 10.11. Description: test lenticular, septal filaments meandriform, pillars absent; spiral lamina of about seven turns (B), chambers high, septa upright.
Remarks: an important marker throughout the Anglo-Parisian-Belgian area. PB Cuisian (B1), BB Paniselian (B1), TR e Eoc.

Palaeogene

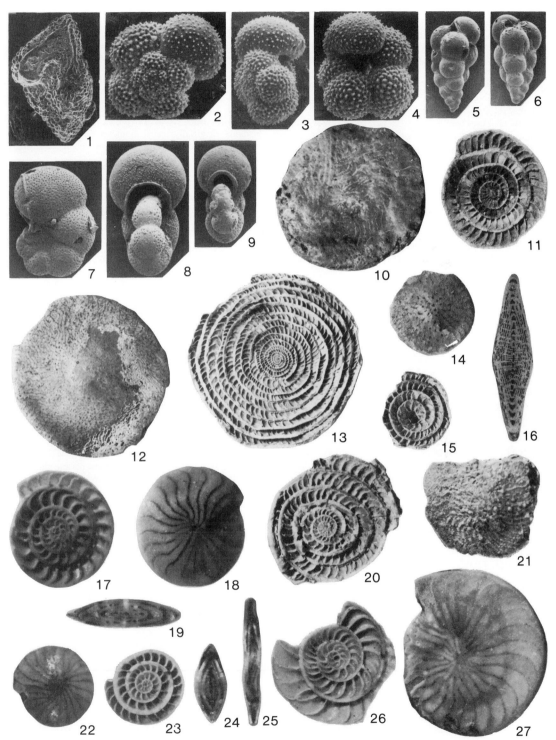

Plate 10.11

Nummulites prestwichianus (Jones)
Plate 10.11, Figs 25–27 form A (× 12), diam. 3.8 mm. Barton Clay Formation, base bed, Alum Bay, Isle of Wight. Figs 10.7, 10.9, 10.11. = *Nummulina planulata* var. *prestwichiana* Jones, 1862. Description: test discoidal, septal filaments radial to sinuate, spiral lamina of four turns (A), chamber height increasing rapidly; A and B generations indistinguishable externally.
Remarks: apart from one record in the English Channel, known only from the Hampshire Basin and southern USSR (Nemkov, 1968). TR m/1 Eoc. boundary.

Nummulites rectus Curry, 1937
Plate 10.11, Figs 17–19 form A (× 12), diam. 2.7 mm, Barton Clay Formation, (10 m above base), Alum Bay, Isle of Wight. Description: test lenticular, A and B forms indistinguishable externally, ratio diameter to thickness about four; septal filaments sinuate, spiral lamina of 4–5 turns (A), chamber height increasing rather slowly; septa strongly curved.
Remarks: as for *N. prestwichianus*.

Nummulites variolarius (Lamarck)
Plate 10.11, Figs 22–24 form A (× 12), diam. 1.9 mm; Bracklesham Group, Selsey Sand Formation, (Bed XVII), Whitecliff Bay, Isle of Wight. Figs 10.6, 10.8, 10.9, 10.11. = *Lenticulites variolaria* Lamarck, 1804. Description: test lenticular, A and B forms almost indistinguishable externally, ratio diameter to thickness about 2.8, septal filaments radial, curving near margin, spiral lamina of four turns (A), septa oblique.
Remarks: an important marker in the Anglo-Paris-Belgian area. Range overlaps slightly those of *N. laevigatus* and *N. prestwichianus*. PB Lutetian-Auversian (B1), BB Bruxellian-Ledian, Wemmelian? (B1). Tr m Eoc.

ACKNOWLEDGEMENTS

Dr C. G. Adams, British Museum (Natural History) kindly allowed access to the collections of Foraminifera. Mr J. Jones and Mr B. Evans printed the figures of plates 10.1–9 from negatives taken by J. W. Murray. For plates 10.10 and 10.11, electron microscope facilities were provided by Professor T. Barnard; the planktonic species illustrations were prepared by Mr M. Gay and those of the nummulites by Mr M. Gray (University College London). Additional negatives of Thanetian benthic species were provided by Professor J. R. Haynes. Mrs G. Wright and Miss J. Eggins typed the manuscript. We wish to thank them all.

REFERENCES

Adams, C. G. 1962. *Alveolina* from the Eocene of England. *Micropaleontology*, **8**, 45–54.

Aubry, M. P. 1985. Biostratigraphie du Paléogène épicontinental de l'Europe du nord-ouest. Étude fondéé sur les nannofossiles. *Docum. Lab. Géol. Lyon*, **89**, 1–317.

Aubry, M. P. 1986. Palaeogene calcareous nannoplankton biostratigraphy of northwestern Europe. *Palaeogeography, Palaeoclimatol, Palaeoecol.*, **55**, 267–334.

Aubry, M. P., Hailwood, E. A., and Townsend, H. A. 1986. Magnetic and calcareous-nannofossil stratigraphy of the lower Paleogene formations of the Hampshire and London basins. *J. geol. Soc. Lond.*, **143**, 729–735.

Barr, F. T. and Berggren, W. A. 1964. Lower Tertiary planktonic Foraminifera from the Thanet Formation of England. *Int. geol. Congr.*, 22nd ser., 1964(3), 118–136.

Barr, F. T. and Berggren, W. A. 1965. Planktonic Foraminifera from the Thanet Formation (Paleocene) of Kent, England. *Stockh. Contr. Geol.*, **13**, 9–26.

Batjes, D. A. J. 1958. Foraminifera of the Oligocene of Belgium. *Mem. Inst. r. Sci. nat. Belg.*, **143**, 1–188.

Berggren, W. A. 1965. Further comments on planktonic Foraminifera of the type Thanetian. *Contr. Cushman Fdn.*, **16**, 125–127.

Berggren, W. A. 1969. Paleogene biostratigraphy and planktonic foraminifera of Northern Europe. *In* Brönnimann, P. and Renz, H. H. (Eds.) *Proc. first Intern. Conf. planktonic Microfossils*, Geneva 1967. E. J. Brill, Leiden, 121–159.

Berggren, W. A., Kent, D. V. and Flynn, J. J. 1985. Paleogene chronology and chronostratigraphy. *Mem. geol. Soc. Lond.*, **10**, 141–195.

Berggren, W. A., McKenna, M. C., Hardenbol, J. and Obradovich, J. D., 1978. Revised Paleogene polarity time scale. *Jl. Geol.*, **86**, 67–81.

Bhatia, S. B. 1955. The foraminiferal fauna of the Late Palaeogene sediments of the Isle of Wight, England, *J. Paleont.*, **29**, 665–693.

Bhatia, S. B. 1957. The paleoecology of the late Palaeogene sediments of the Isle of Wight, England. *Contr. Cushman, Fdn.*, **8**, 11–28.

Bignot, G. 1962. Etude micropaléontologique de la formation de Varengeville du gisement Eocène du Cap d'Ailly (Seine-Maritime). *Rev. Micropaléont.*, **5**, 167–184.

Bignot, G. 1963. Foraminifères planctoniques et foraminifères remaniés dans la Formation de Varengeville. *Bull. Soc. géol. Normandie*, **53**, 1–12.

References

Blezard, R. G., Bromley, R. G., Hancock, J. M., Hester, S. W., Hey, R. W. and Kirkaldy, J. F. 1967. The London Region (North of the Thames). *Geol. Assoc. Guide.* **30A**, 1–34.

Blondeau, A. 1972. *Les Nummulites.* Vuibert, Paris, 255 pp.

Blondeau, A. and Curry, D. 1963. Sur la présence de *Nummulites variolarius* (Lmk) dans les diverses zones du Lutétien des bassins de Paris, de Bruxelles et du Hampshire. *Bull. Soc. géol. Fr.,* ser. 7, **5**, 275–277.

Boltovskoy, E. and Wright, R, 1976. *Recent Foraminifera.* Dr. W. Junk, The Hague, 515 pp.

Bowen, R. N. C. 1954. Foraminifera from the London Clay. *Proc. Geol. Assoc.,* **65**, 125–174.

Bowen, R. N. C. 1955. Smaller foraminifera from the Upper Eocene of Barton. *Micropaleontology,* **3**, 53–60.

Brönnimann, P., Curry, D., Pomerol, C. and Szöts, E. 1968. Contribution à la connaissance des Foraminifères planctoniques de l'Eocène (incluant le Paléocène) du Bassin anglo-franco-belge. *Mém. Bur. Rech. géol. min.,* **58**, 101–108.

Burrows, H. and Holland, R. 1897. The Foraminifera of the Thanet Beds of Pegwell Bay. *Proc. Geol. Assoc.* **15**, 19–52.

Burton, E. St. J. 1929. The horizons of Bryozoa (Polyzoa) in the Upper Eocene Beds of Hampshire. *Q. Jl. geol. Soc. Lond.,* **85**, 223–239.

Burton, E. St. J. 1933. Faunal horizons of the Barton Beds in Hampshire. *Proc. Geol. Assoc.,* **44**, 131–167.

Chapman, F. and Sherborn, C. D. 1889. The Foraminifera from the London Clay of Sheppey. *Geol. Mag.,* Dec. III, **6**, 497–499.

Curry, D. 1937. The English Bartonian nummulites. *Proc. Geol. Assoc.,* **48**, 229–246.

Curry, D. 1962. Sur la découverte de *Nummulites variolarius* (Lamarck) dans le Lutétian des bassins de Paris et du Hampshire. *C. r. Somm. Séanc.Soc, géol. Fr.,* **9**, 247.

Curry, D. 1965. The Palaeogene Beds of south-east England. *Proc. Geol. Assoc.,* **76**, 151–174.

Curry, D. 1967. Problems of correlation in the Anglo-Paris-Belgian Basin. *Proc. Geol. Assoc.,* **77**, 437–467 (for 1966).

Curry, D., Adams, C. G., Boulter, M. C., Dilley, F. C., Eames, F. E., Funnell, B. M. and Wells, M. K. 1978. A correlation of Tertiary rocks in the British Isles. *Geol. Soc. Lond., Special Report,* **12**, 72 pp.

Curry, D., Gulinck, M., and Pomerol, C. 1969. Le Paléocène et l'Eocène, dans les Bassins de Paris, de Belgique et d'Angleterre. *Mém. Bur. Rech. géol. min.,* **69**, 361–369.

Curry, D., Hodson, F. and West, I. M. 1968. The Eocene succession in the Fawley tranmission tunnel. *Proc. Geol. Assoc.,* **79**, 179–206.

Curry, D., King, A. D., King, C. and Stinton, F. C. 1977. The Bracklesham Beds (Eocene) of Bracklesham Bay and Selsey, Sussex. *Proc. Geol. Assoc.,* **88**, 243–254.

Curry, D. and King, C. 1965. The Eocene succession at Lower Swanwick Brickyard, Hampshire, *Proc. Geol. Assoc.,* **76**, 29–35.

Curry, D. and Odin, G. S. 1982. Dating of the Palaeogene. In: Odin, G. S. (ed.) *Numerical dating in stratigraphy,* John Wiley, Chichester, pp. 607–630.

Curry, D. and Wisden, D. E. 1958. Geology of the Southampton area including the coast sections at Barton, Hants., and Bracklesham, Sussex. *Geol. Assoc. Guide,* **14**, 1–16.

Daley, B. and Insole, A. 1984. The Isle of Wight. *Geol. Assoc. Guide,* **25**, 1–34.

Davis, A. G. 1928. The geology of the City and South London Railway, Clapham-Morden extension. *Proc. Geol. Assoc.,* **39**, 339–352.

Edwards, R. A. and Freshney, E. C. 1987. Lithostratigraphical classification of the Hampshire Basin Palaeogene deposits (Reading Formation to Headon Formation). *Tertiary Research,* **8**, 43–73.

El-Naggar, Z. R. 1967. Planktonic Foraminifera in the Thanet Sands of England, and the position of the Thanetian in Paleocene stratigraphy. *J. Paleont.,* **41**, 575–586.

Fisher, O. 1862. On the Bracklesham Beds of the Isle of Wight Basin. *Quart. J. geol. Soc. Lond.,* **18**, 65–94.

Harland, W. B., Cox, A. V., Llewellyn, P. E., Pickton, C. A., Smith, A. G. and Walters, R. 1982. *A geologic time scale.* Cambridge University Press. 131 pp.

Haynes, J. 1954. Taxonomic position of some British Paleocene Buliminidae. *Contr. Cushman Fdn.,* **5**, 185–191.

Haynes, J. 1955. Pelagic Foraminfera in the Thanet Beds and the use of Thanetian as a stage name. *Micropaleontology,* **1**, 189.

Haynes, J. 1956. Certain smaller British Paleocene Foraminifera Part 1, Nonionidae, Chilostomellidae, Epistominidae, Discorbidae, Amphisteginidae, Globigerinidae, Globorotaliidae, and Gümbelinidae, *Contr. Cushman Fdn.,* **9**, 79–101.

Haynes, J. 1958a. Certain smaller British Paleocene Foraminifera Part III. *Contr. Cushman Fdn.,* **9**, 4–16.

Haynes, J. 1958b. Certain smaller British Paleocene Foraminifera Part IV. Arenacea, Lagenidea, Buliminidea and Chilostomellidae. *Contr. Cushman Fdn.,* **9**, 58–77.

Haynes, J. 1958c. Certain smaller British Palaeocene Foraminifera Part V. Distribution. *Contr. Cushman Fdn.,* **9**, 83–92.

Haynes, J. and El-Naggar, Z. R. 1964. Reworked Upper Cretaceous and Danian planktonic foraminifera in the type Thanetian of England. *Micropaleontology,* **10**, 345–356.

Hooker, J. J. 1986. Mammals from the Bartonian (middle/late Eocene) of the Hampshire Basin, southern England. *Bull. Brit. Mus. (Nat. Hist.) Geol.,* **39**, 191–478.

Insole, A. and Daley, B. 1985,. A revision of the lithostratigraphical nomenclature of the Late Eocene and Early Oligocene of the Hampshire Basin, Southern England. *Tertiary Res.,* **7**, 67–100.

Kaasschieter, J. P. H. 1961. Foraminifera of the Eocene of Belgium. *Mem. Inst. r. Sci. nat. Belg.,* **147**, 1–271.

King, C.1981. The stratigraphy of the London Clay and

associated deposits. *Tertiary Res. Spec. Pap.*, **6**, 1–158.

King, C. 1984. The stratigraphy of the London Clay Formation and Virginia Water Formation in the coastal sections of the Isle of Sheppey (Kent, England). *Tertiary Res.*, **5**, 121–160.

Knox, R. W. O'B. 1984. Nannoplankton zonation and the Paleocene/Eocene boundary beds of NW Europe: an indirect correlation by means of volcanic ash layers. *J. geol. Soc. Lond.*, **141**, 993–999.

Larsen, A. R. and Jørgensen, N. O. 1977. Palaeobathymetry of the Lower Selandian of Denmark on the basis of Foraminifera. *Bull. geol. Soc., Denmark*, **26**, 175–184.

Le Calvez, Y. 1970. Contribution à l'étude des foraminifères paléogènes du Bassin de Paris. *Cah. Paléont.*, 326 pp.

Loeblich, A. R. Jun., and Tappan, H. 1964. Sarcodina chiefly 'Thecamoebians' and Foraminiferida. In Moore, R. C. (ed.). *Treatise on invertebrate paleontology*. New York, Geol. Soc. Amer., pt. C, Protista **2**, 900 pp.

Moorkens, T. 1968. Quelques foraminifères planctoniques de l'Yprésien de la Belgique et du Nord de la France. *Bull. Bur. Rech. géol. min.*, No. **58**, 109–129.

Murray, J. W. 1973. *Distribution and ecology of living benthic foraminiferids*. Heinemann Educational Books, London, 288 pp.

Murray, J. W. 1976a. A method of determining proximity of marginal seas to an ocean. *Mar. Geol.*, **22**, 103–119.

Murray, J. W. 1976b. Comparative studies of living and dead benthic foraminiferal distributions. In Hedley, R. H. and Adams, C. G. (Eds.) *Foraminifera*, **2**, 45–109.

Murray, J. W. 1984. Benthic foraminifera: some relationships between ecological observations and palaeoecological interpretations. In Oertli, H. J. (ed.) *Benthos '83*, 465–469.

Murray, J. W. and Wright, C. A. 1974. Palaeogene Foraminiferida and palaeoecology, Hampshire and Paris Basins and the English Channel. *Spec. Pap. Palaeontology*, **14**, 1–171.

Nemkov, G. 1968. Nummulites de l'U.R.S.S., leur évolution, systématique et distribution stratigraphique. *Mém. Bur. Rech. géol. min.*, **58**, 71–78.

Odin, G. S. and Curry, D. 1985. The Palaeogene time-scale: radiometric dating versus magnetostratigraphic approach. *J. geol. Soc. Lond.*, **142**, 1179–1188.

Odin, G. S., Curry, D. and Hunziker, J. C. 1978. Radiometric dates from NW European glauconites and the Palaeogene time-scale. *J. geol. Soc. Lond.*, **135**, 481–497.

Pitcher, W. S., Peake, N. B., Carreck, J. N., Kirkaldy, J. F. and Hancock, J. M. 1967. The London Region (South of the Thames). *Geol. Assoc. Guide*, **30B** 1–32.

Plint, A. G. 1982. Eocene sedimentation and tectonics in the Hampshire Basin. *J. geol. Soc. Lond.*, **139**, 249–254.

Plint, A. G. 1983a. Liquefaction, fluidization and erosional structures associated with bituminous sands of the Bracklesham Formation (Middle Eocene) of Dorset, England. *Sedimentology*, **30**, 525–535.

Plint, A. G. 1983b. Facies, environments and sedimentary cycles in the Middle Eocene Bracklesham Formation of the Hampshire Basin: evidence for global sea-level changes? *Sedimentology*, **30**, 625–653.

Plint, A. G. 1984. A regressive coastal sequence from the Upper Eocene of Hampshire, Southern England. *Sedimentology*, **31**, 213–225.

Rouvillois, A. 1960. Le Thanétien du Bassin de Paris. *Mém. Mus. nat. Hist.* n.s. ser. C, **8**, 1–91.

Sherborn, C. D. and Chapman, F. 1886. On some microzoa from the London Clay exposed in the drainage works, Piccadilly, London, 1885. *Jl. R. microsc. Soc.*, **6**, 737–767.

Sherborn, C. D. and Chapman, F. 1889. Additional note on the foraminifera of the London Clay exposed in the drainage works, Piccadilly, London, 1885. *Jl. R. microsc. Soc.*, 483–488.

Smith, A. J. and Curry, D. 1975. The structure and geological evolution of the English Channel. *Phil. Trans. R. Soc.*, A **279**, 3–20.

Stinton, F. C. 1971. Easter field meeting in the Isle of Wight. *Proc. Geol. Assoc.*, **82**, 403–410.

Townsend, H. A. and Hailwood, E. A. 1985. Magnetostratigraphic correlation of Palaeogene sediments in the Hampshire and London Basins, southern U.K. *J. geol. Soc. Lond.*, **142**, 957–982.

Vella, P. 1969. Correlation of base of Middle Headon Beds between Whitecliff Bay and Colwell Bay, Isle of Wight. *Geol. Mag.*, **106**, 606–608.

Venables, E. M. 1962. The London Clay of Bognor Regis. *Proc. Geol. Assoc.*, **73**, 245–271.

White, H. J. O. 1915. The geology of the country near Lymington and Portsmouth. *Mem. geol. Surv. G.B.*, Sheets 330–331, 78 pp.

White, H. J. O. 1921. A short account of the geology of the Isle of Wight. *Mem. geol. Surv. G.B.*, 219 pp.

Wood, A. and Haynes, J. R. 1957. Certain smaller British Paleocene Foraminifera Part II. *Cibicides* and its allies. *Contr. Cushman Fdn.*, **8**, 45–53.

Wright, C. A. 1972a. Foraminiferids from the London Clay at Lower Swanwick and their palaeoecological interpretation. *Proc. Geol. Assoc.*, **83**, 337–347.

Wright, C. A. 1972b. The recognition of a planktonic foraminiferid datum in the London Clay of the Hampshire Basin. *Proc. Geol. Assoc.*, **83**, 413–419.

Wrigley, A. 1924. Faunal divisions of the London Clay, illustrated by some exposures near London. *Proc. Geol. Assoc.*, **35**, 245–259.

Wrigley, A. 1940. The faunal succession in the London Clay, illustrated in some new exposures near London. *Proc. Geol. Assoc.*, **51**, 230–245.

Wrigley, A. and Davis, A. G. 1937. The occurrence of *Nummulites planulatus* in England, with a revised correlation of the strata containing it. *Proc. Geol. Assoc.*, **48**, 203–228.

Ziegler, P. A. 1975. Geologic evolution of North Sea and its tectonic framework *Bull. Am. Ass. Petrol. Geol.*, **59**, 1073–1097.

11

Neogene

D. G. Jenkins, C. King, M. Hughes

11.1 Introduction

Miocene and Pliocene marine sediments within the area of the U.K. and its surrounding continental shelves were deposited mainly within three sedimentary basins (Fig. 11.1).

(1) The North Sea Basin

This major intracontinental basin contains a thick sequence of Neogene marine sediments. The succession and microfaunas are summarised in Chapter 9, as part of a description of the entire Cainozoic sequence in the North Sea. Neogene sediments of the North Sea Basin extend onto onshore areas in Britain only in a small area of East Anglia, where they are represented by the Coralline Crag and 'Trimley Sands', and possibly also by the Lenham Beds of Kent.

The latest Miocene/earliest Pliocene 'Trimley Sands' (Balson, in press) are represented only by reworked pebbles and cobbles, mainly of phosphorite-cemented sandstone (the 'Suffolk box-stones'), now forming a component of the lag deposit at the base of the Coralline Crag and the Pleistocene Red Crag. Mollusc moulds are well-documented, but no determinable foraminifera have been recorded.

The Coralline Crag comprises fine to coarse-grained bioclastic carbonates (calcarenites and calcirudites) with quartz and glauconite as subordinate components, partly affected by aragonitic dissolution, and partly irregularly calcite-cemented. It was deposited in low and high energy shallow marine environments (less than 100 m in depth). The Coralline Crag has been regarded as Pliocene since the nineteenth century: its dating has been given more precision by recent studies on planktonic foraminifera and nannoplankton (Jenkins and Houghton, 1987, Jenkins et al., 1988), which establish the date as Late Pliocene (upper N19/N21, NN16?).

The foraminifera of the Coralline Crag were first described by Jones et al. (1866–1897). Amongst the other authors whose work was included in Jones' monograph were Burrows and Holland, who contributed the distribution lists (Jones et al., 1866–1897, pp. 374–394) and the description of a number of new species. Carter (1951, 1957) initially examined the distribution of the foraminifera

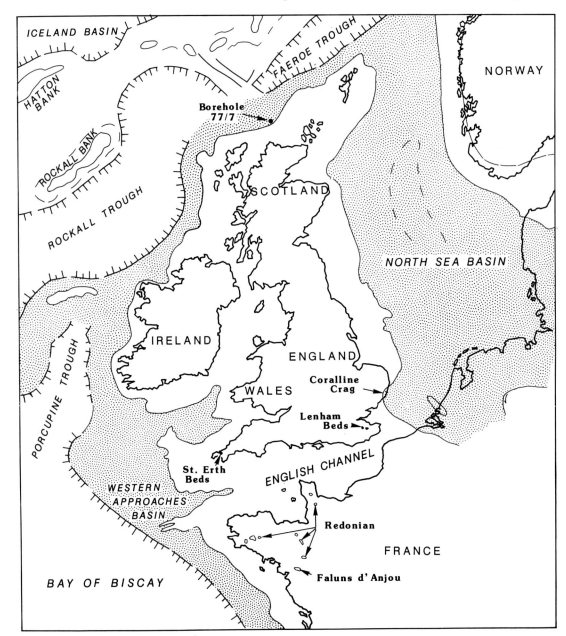

Fig. 11.1 — Distribution of Neogene (Miocene and Pliocene) sediments in the British Isles and surrounding areas (after Ziegler, 1982).

by size in order to determine the palaeoenvironment, and later discussed the genera *Alliatina* and *Alliatinella*, using the present day distribution of the former as further evidence for the palaeoenvironment of these beds. Wilkinson (1980) in a study of the ostracods, stratigraphy and palaeoenvironment briefly discusses the foraminifera.

Doppert (1975) provided an extensive list of foraminifera from a range of localities and stratigraphic levels. The most important recent study is by Hodgson and Funnell (1987), discussing foraminiferal palaeoecology and illustrating a number of species by scanning electron microscope (SEM) photographs.

The Lenham Beds are limonite-cemented sands and sandstones which occur only as displaced and disturbed units filling solution-hollows in the North Downs of Kent, south east England. They are entirely decalcified, but contain mollusc moulds indicating a Middle or Late Miocene date; no foraminifera are recorded. They may represent a remnant of an Atlantic transgression (A. Janssen, pers. comm.), and are perhaps not related to the North Sea Basin.

(2) The continental shelf of the South-Western Approaches and the Western Approaches Basin

This area forms a small part of the wedge of marine Neogene sediments underlying the entire European Atlantic margin, which thicken towards the shelf-edge. The Western Approaches Basin is a narrow prolongation of this wedge into the western end of the English Channel. The former existence of a marine environment in the Late Pliocene is evidenced by the small outlier of the St. Erth Beds in Cornwall.

The Neogene stratigraphy of the Western Approaches Basin is summarised by Evans and Hughes (1984), based on information mainly from sea-bed samples and shallow-cored boreholes. In the axis of the western part of the Basin there is apparently a continuous marine sequence of calcareous clays from the Oligocene through into the Early Miocene, but in more marginal locations the Early Miocene oversteps unconformably onto Eocene and older sediments. The Neogene sequence is divided into three Formations. The Jones Formation (formerly 'Globigerina Silts') comprises Early and Middle Miocene glauconitic argillaceous calcilutites and calcareous claystones, with abundant planktonic foraminifera, deposited in outer shelf to upper bathyal environments. The overlying Cockburn Formation comprises Middle Miocene silty calcarenites, deposited in a higher-energy inner sublittoral environment (depths of 40–80 m?). This is overlain unconformably by the Little Sole Formation (Pliocene–Pleistocene), comprising argillaceous sands deposited in a marine shelf environment. Benthic foraminifera from these units are described by Curry et al., (1962), Murray in Curry et al., (1965), Hughes in Warrington and Owens, (1977).

The St. Erth Beds are a small outlier of fossiliferous marine Neogene clays overlying unfossiliferous sands, resting unconformably on the Devonian. They attracted attention in the mid-nineteenth century because of their marine molluscs, but their age and stratigraphical relationships have long been uncertain. The history of their study, and of attempts to date them, have been summarised by Mitchell, (1973) and Jenkins, (1982). Recent studies of the nannofossils and the rare but surprisingly diverse planktonic foraminifera in the St. Erth Beds permits a Late Pliocene date to be conclusively established (Jenkins et al., 1986). They can be considered as an 'outlier' of the Western Approaches Basin, owing their high altitude to post-Pliocene warping or high sea level stand. Benthic foraminifera from the St. Erth Beds have been studied by Millett (1886 to 1902), Macfadyen (1942), Funnell (in Mitchell, 1965), and Margerel (in Mitchell, 1973).

Small outliers of Miocene and Pliocene shallow marine sediments are present in western France (Faluns d'Anjou, Redonian), and these can also be regarded as remnants of much more extensive sediments probably deposited during phases of high eustatic sea levels. Redonian foraminifera are described by Margerel (1970, 1971, 1972).

(3) The Faeroes Trough

Neogene sediments in this area are not well

known. Information on the stratigraphy and microfauna derived from oil and gas exploration wells is given in Chapter 9. The sediments on the margins of the Trough have been penetrated by several shallow-cored stratigraphic boreholes drilled by the British Geological Survey (Institute of Geological Sciences, 1978) where Miocene glauconitic sands with marine microfaunas have been proved. Their microfauna is of North Sea Basin affinities, and some species are illustrated here (Plate 11.1).

11.2 Planktonic foraminifera

Two onshore Pliocene localities in England have recently yielded planktonic foraminifera, namely the St. Erth Beds and the Coralline Crag (Jenkins 1982; Jenkins et al., 1986, 1988; Jenkins and Houghton, 1987).

The St. Erth Beds have yielded 15 species which were described and illustrated in Jenkins et al. (1986); the species ranges suggest that the age of the fauna is Late Pliocene G. inflata Zone of Weaver and Clement (1986), Jenkins et al. (1986). The age can be narrowed down to 2.1–1.9 Ma on the presence and absence of key species. The fauna has yielded specimens of Globorotalia inflata (d'Orbigny) which appear in the northeast Atlantic at 2.1 Ma (Weaver and Clement, 1986), and the absence of Globorotalia truncatulinoides (d'Orbigny) which appeared at 1.9 Ma (Berggren et al., 1967, 1985) is regarded as significant (Jenkins et al., 1986). Other stratigraphically important species in the St. Erth Beds include Globorotalia praehirsuta Blow, Globorotalia tosaensis Takayangi and Saito, Pulleniatina primalis Banner and Blow, and dextrally coiled Neogloboquadrina pachyderma (Ehrenberg).

King (1983) recorded Neogloboquadrina atlantica (Berggren) from the Coralline Crag and eight additional species have been recorded and illustrated by Jenkins et al. (1988). The age of the Coralline Crag is Early–Late Pliocene G. puncticulata Zone between 4.1 Ma and 3.2 Ma based on the overlapping ranges of Globorotalia puncticu-

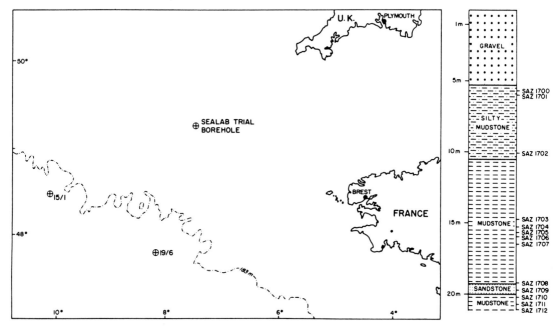

Fig. 11.2 — Map to show the position of Sealab Trial Borehole and two dredge samples (15/1 and 19/6)

lata (Deshays) which appeared at 4.1 Ma and the extinction of *N. atlantica* at 2.3 Ma in the northeast Atlantic (Weaver and Clement, 1986) (see Jenkins and Houghton, 1987). Offshore, King has made brief comments on the planktonic foraminifera from the North Sea basin (see Chapter 9 of this book). On the seabed of the Western Approaches of the English Channel, Neogene deposits have been recorded and planktonic foraminifera were listed by Curry (in Curry *et al.*, 1962) and by Murray (in Curry *et al.*, 1965). Jenkins (1977) described and illustrated 35 species and subspecies from the Sealab Trial borehole in the English Channel, 100 km southwest of the Isles of Scilly at latitude 49° 10.73' N, longitude 7° 27.86' W in a water depth of 137–140 m (Fig. 11.2). The age of the fauna is the Early Miocene *G. trilobus trilobus* Zone and 18 species have been chosen as diagnostic in the 20.06 m of cored Early Miocene silty mudstone, mudstone and thin sandstone. Jenkins (1977) suggested that the *G. trilobus* Zone in Europe is probably within the Burdigalian Stage and Iaccarino (1985) has also placed this zone in the same Stage in the Mediterranean area. Dredge samples from the *Sarsia* in the same area have yielded faunas from the Early Miocene *P. glomerosa curva* Zone at latitude 47°48' N, longitude 8° 09.50' W and the Middle Miocene *G. mayeri mayeri* Zone at latitude 48° 27' N, longitude 10° 07' W.

In the South Western Approaches, Evans and Hughes (1984) have recorded some species of planktonic foraminifera from the Early Miocene (Borehole 81/9, latitude: 49° 08.5' N; longitude: 6° 0.9' W) and also Middle Miocene faunas (Borehole 81/3A, latitude 48° 59.5' N; longitude 7° 47.8' W).

11.3 LOCATION OF COLLECTIONS OF IMPORTANCE

Bristol University: samples collected by the Bristol University group from the South Western Approaches to the English Channel.

Southampton University: Murray (in Curry *et al.*, 1965)

British Museum (Natural History), London: Coralline Crag, Jones *et al.*, 1866–1897; Carter, 1951; Jenkins *et al.*, 1986, 1988; St. Erth Beds, Millett (1886a, b, 1895, 1897, 1898, 1902).

Nantes University, Institut des Sciénces de la Nature, France: St. Erth, Mitchell material studied by Margerel.

British Geological Survey, Keyworth: offshore material reported on in survey publications plus Evans and Hughes (1984), Jenkins (1977).

11.4 STRATIGRAPHIC DIVISIONS

The ages and relationships of the Miocene and Pliocene lithostratigraphic units are summarised in Fig. 11.3, and compared to the eustatic sea-level cycles of Vail (1977). The biostratigraphic dating is derived mainly from Evans and Hughes (1984), Jenkins *et al.* (1986, 1988) and Jenkins and Houghton (1987); the chronostratigraphic scale is from Vail (1977), slightly modified. It must be emphasized that the dating (and particularly the chronological limits) of most of these units is not precisely established, but, as suggested by Evans and Hughes (1984), the limits of the marine units appear to correspond to high eustatic sea level episodes. Periods with low eustatic sea levels are generally unrepresented by sediments, except in the more axial parts of the Western Approaches basin and the North Sea Basin.

Correlation with sequences in Belgium and the Netherlands

Correlation of the North Sea sequences with those in continental Europe is discussed by King (1983), and tabulated in Chapter 9 of this 'Atlas'. The Coralline Crag can be correlated with the Late Pliocene benthic Zone NSB14 and the planktonic Zone NSP15 in

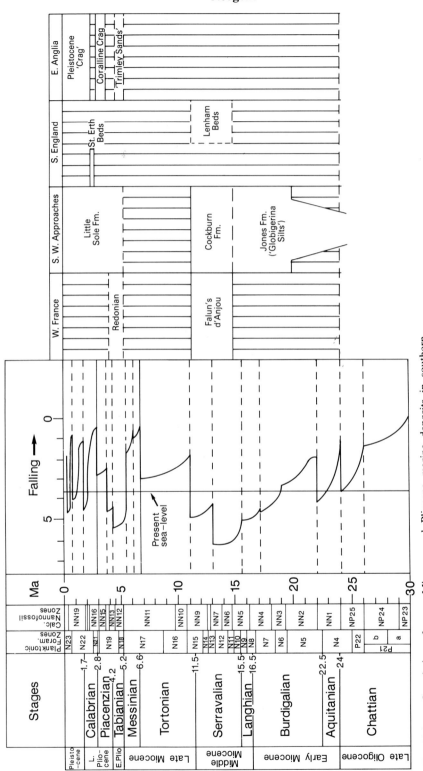

Fig. 11.3 — Correlation of some Miocene and Pliocene marine deposits in southern England and adjacent areas, related to the sea-level curve of Vail (1977). The Pliocene–Pleistocene boundary has recently been redated at c.1.6 Ma.

the North Sea. Similar benthic microfaunas occur in the FB Zone (lower Oosterhout Formation) of the Netherlands and the BFN5 Zone (Luchtbal Sands) of Belgium. Further discussion on the correlation of the Coralline Crag is given by Hodgson and Funnell (1987).

11.5 DEPOSITIONAL HISTORY, PALAEOECOLOGY, FAUNAL ASSOCIATIONS

Depositional history

The depositional history of the North Sea and the Faeroes Trough are discussed in Chapter 9 of this book. The depositional history of the sequence in the Western Approaches, and possible controls on sedimentation in this area, are discussed above.

Palaeoecology and faunal associations (benthic)

Coralline Crag

The benthic foraminiferal associations of the Coralline Crag have recently been discussed in detail by Hodgson and Funnell (1987), following earlier work by Carter (1951, 1957), and related to the lithofacies of Balson (1981a, b). For details the Hodgson and Funnell paper should be referred to; they distinguish three main biofacies:

(A) A *Cibicides* and *Cassidulina*-dominated facies (mainly *Cibicides lobatulus* and *Cassidulina laevigata*), occurring in Balson's Facies A (Silty Sands), interpreted as a moderate to low energy environment in water depths of 50–100 m

(B) A *Pararotalia*-dominated facies (*P. serrata*) associated with Balson's Facies B (Sandwave facies), interpreted as a sandwave regime at water depths of less than 50 m.

(C) A *Cibicides–Textularia* dominated facies (mainly *C. lobatulus*, *C. pseudoungerianus*, and *T. sagittula*), associated with Balson's Facies C (Skeletal Sand Facies), deposited on a shoal swept by high energy currents at depths of less than 50 m.

Hodgson and Funnell refute the conclusions of Carter (1951) that the foraminiferal species in the Coralline Crag can be divided into two groups, one related in abundance to clastic grains of similar size, indicating that they have been transported as sedimentary particles, and the other unrelated to the grain size of the sediment, thus presumed to be a biocoenosis. They indicate that the relationship is likely to be much more complex, due to the different density and shape of foraminiferal tests as compared to quartz grains.

S.W. Approaches

Murray (in Curry et al., 1965) listed the benthic species occurring in the Miocene *Globigerina* Silts (Jones Formation) of the western English Channel. The most frequent species present in the Aquitanian were *Brizalina scalprata miocenica* (Macfadyen), *Florilus grateloupi* (d'Orbigny) and *Trifarina bradyi* Cushman which were present at every station placed in that stage. The first two named were present also at all stations recognised as Vindobonian (= Langhian) together with *Cibicides dutemplei* (d'Orbigny) and *Elphidium crispum* (Linné). Different subordinate species were recorded for each stage but it was noted that the two faunas were very similar.

The only detailed study so far was carried out on samples from the Jones Formation ('*Globigerina* Silts') of the Sealab Trial Borehole (Fig. 11.2).

Six assemblages have been recognised from the Sealab Trial borehole (Hughes, 1977, unpublished thesis, see Fig. 11.4) and dated as Burdigalian (Jenkins, 1977). The benthic assemblages, for convenience named after the dominant species, can be related to the planktonic subzones in the Trial borehole.

Subzone	Assemblages
Globorotalia praescitula	Gyroidina aff. parvus
	Bolivina hebes–Bulimina alazanesis
	Bulimina elongata
	Bolivina hebes–Bulimina alazanensis
Globorotalia semivera	Cibicides ungerianus
	Astrononion perfossum
	Pararotalia sp.

The benthic assemblages are facies controlled and a return to a pre-existing palaeoenvironmental condition may result in the return of a previous assemblage as shown by the B. hebes–B. alazanensis assemblage. The three assemblages in the *semivera* Subzone represent a gradually increasing water depth from inner to outer shelf. The lithology is a glauconitic sandstone with extensive recrystallisation of the fossils. The rocks of the *praescitula* Subzone are a silty, greenish grey mudstone. B. alazanensis, a living species, with a frequency of 10–15% over a range of levels in the borehole succession, is considered to be significant in determining the palaeobathymetry. The species is found at the present time only in water depths greater than 400 m (Pflum and Frerichs, 1976) and a middle bathyal regime is postulated for this assemblage. B. alazanensis has not been recorded from elsewhere in the *Globigerina* Silts, but neither have rocks of Burdigalian age except on the continental slope (Jenkins, 1977). The two remaining assemblages are regarded as outer shelf. It would appear that a marked change in water depth and lithology coincide with the Subzone boundary in this borehole succession. A number of benthic species are restricted to the Miocene; three of these, *Astrononion perfossum, Unicosiphonia zsigmondyi* and *Virgulinella pertusa*, are present in the Trial borehole, whilst the last named has been recorded from a number of other western English Channel stations (Hughes in Warrington and Owens, 1977).

Curry (in Curry et al., 1962) postulated that the Miocene sea in the western English Channel was a gulf with no connection with the Miocene basin situated in the region of the present North Sea. Murray (in Curry et al., 1965) agreed with this and suggested a water depth slightly greater than the present day 128 m. A greater water depth is indicated for the early part of the *praescitula* Subzone, at least in the area of the Trial borehole (see above), but Hughes (1977 unpublished thesis) agrees with both Curry and Murray in considering that the English Channel and the North Sea were not connected during the Miocene.

St. Erth Beds

The total geographical isolation of the St. Erth Beds has resulted in considerable importance being attached to the stratigraphical and palaeoenvironmental interpretations based on the foraminifera, as well as other groups of fossils. Margerel (in Mitchell, 1973) recorded over 100 species, but noted that there were only five which he would consider as characteristic of the formation: *Quinqueloculina cliarensis* (Heron-Allen and Earland), *Q. seminulum, Monspeliensina pseudotepida, Ammonia* sp. and *Faujasina subrotunda*. However it should be noted that the form of *Buccella frigida* (referred to as *B.* aff. *frigida* herein) present in this formation is recognisably different from the Pleistocene species. Funnell (in Mitchell, 1965) determined the palaeoenvironment by analysing the number of species in terms of faunal dominance and faunal diversity. The result from the former study suggested a water depth of not more than 20 m, whilst the diversity factor indicated 60–100 m. It was noted that the conditions around Cornwall were very different from the Gulf of Mexico where the criteria for the interpretation of the water depth were originally established. Funnell came to the conlusion that the fauna originated in shallow water of less than twenty metres, but was deposited in depths of up to 100 m. Margerel had nothing to add to this conclusion.

Faeroes Trough

Two Miocene benthic assemblages from northwest Scottish waters are present in British Geological Survey borehole 77/7 (see IGS 1978, p. 23 for lithostratigraphical log). The younger assemblage dominated by *Cibicides peelensis* Ten Dam and *Ehrenbergina healyi* is placed in the Middle Miocene and the older assemblage with high frequencies of *Asterigerina staeschei* and *Elphidium inflatum* are considered as Early Miocene. *Uvigerina semiornata* and *Florilus boueanus* (d'Orbigny) are common to both assemblages, the latter replacing *F. dingdeni* recorded in the western English Channel. These are typical middle shelf assemblages.

General comments

There are clear differences between the 'Atlantic' foraminiferal assemblages of the Western Approaches Basin and St. Erth, and the 'North Sea Basin' assemblages of the Coralline Crag and the Faeroes Trough. Although a significant proportion of species occur in both 'provinces', many are restricted to one or other area. These differences appear most marked in the Miocene; 'Atlantic' species include *Brizalina scalprata, Bulimina alazanensis* and *Florilus grateloupi*, and 'North Sea Basin' species include *Asterigerina staeschei* and *Elphidium inflatum*. These features reflect the relative isolation of the North Sea Basin at this time, with, as noted above, probably no marine connection through the English Channel. 'Atlantic'-type assemblages are known to characterise the North Atlantic continental shelf from the area west of Ireland down to the Aquitaine Basin. Pliocene assemblages are less easy to compare, due to differences in facies and lack of information on deeper-shelf Pliocene foraminifera in 'Atlantic' areas. The limited evidence suggests more faunal mixing between the two areas, although the English Channel is believed to have been still closed. Differences are also seen in the vertical distribution of species occurring in both areas, e.g. *Monspeliensina pseudotepida* occurs at St. Erth in younger Pliocene sediments than in the North Sea Basin (see King, Chapter 9). Clearly, much more work is necessary to form a clear picture of these distribution patterns and their significance.

Palaeoecology of the planktonic foraminifera

The 35 species and subspecies in the Early Miocene Sealab sequence of the English Channel suggest that there were oceanic conditions with normal salinity prevailing at this time. It had been suggested by Curry *et al.* (1965) that the Miocene foraminifera of the Western Approaches indicated a sea temperature similar to the present day sea-surface temperature of 15–18°C. The presence of *Globigerinoides sacculifer* in the Sealab samples indicated a higher range of temperatures, because it lives in the present day oceans in waters of 15–30°C. Confirmation of this higher estimate has come from an oxygen isotope analysis of the tests of *G. sacculifer*; there was a rise of temperature from 15.5°–16°C in the two lowermost samples to 22°C in the highest sample in the *G. trilobus* Zone (Jenkins and Shackleton, 1979).

There is a species diversity gradient of planktonic foraminifera from about 22 species in the tropics to one in the freezing polar seas. Species diversity changes in the Sealab borehole tend to parallel changes in palaeotemperature (Jenkins and Shackleton, 1979).

From a study of the St. Erth species which are living today Jenkins (1982) postulated that the original sea-water was in the upper range of 10–18°C. Further work by Jenkins *et al.*, 1987, yielded specimens of *Globorotalia tosaensis* and *Pulleniatina primalis* which suggests that the sea-water temperature in the Later Pliocene was subtropical.

The low diversity of nine species in the

Coralline Crag tends to suggest a water temperature range of 10–18°C in the southern North Sea between 4.1 and 2.3 Ma (Jenkins et al., 1988), but confirmation of this will come from the study of North Atlantic faunas now in progress.

11.6 Index Species

A considerable amount of literature on Miocene foraminifera has been published by Continental workers since the early classic work of d'Orbigny (1846) on the Vienna basin, revised by Marks (1951), which was followed by Reuss (1851, 1856, 1861, 1863a, b), Hosius (1892) and Clodius (1922) in Germany, and Hantken (1875) in Hungary. The first attempts at biostratigraphical zonation based on species ranges and assemblages was made by Staesche and Hiltermann (1940) in Germany, and Ten Dam and Reinhold (1941a, b, c, 1942) in the Netherlands, to be followed by Indans (1958, 1962, 1965) and Langer (1963a, b, 1969) in West Germany and Batjes (1958) in Belgium. More recently Spiegler (1974) has published a comprehensive review of 532 species ranges through the Oligocene and Miocene of northwest Germany. In addition detailed analysis of the ranges of the species and sub-species of *Asterigerina* (Gramann, 1964), *Ehrenbergina* (Spiegler, 1973) and *Uvigerina* (Daniels and Spiegler, 1977) have been carried out. Miocene foraminiferal biostratigraphy is engaging workers also in Rumania (Gheorghian, 1971, 1974, 1975, Popescu, 1975) and Poland (Odrzywolska-Bienkowa, 1977). A review of the Aquitaine area has been completed recently (Poignant and Pujol, 1976, 1978).

The proceedings of the various meetings of the Congrés du Néogène Mediterranéan and the reports of the IGCP Project 124, The Northwest European Tertiary Basin, also contain biostratigraphical papers based on foraminifera. Many of the more recent works listed above also contain palaeoenvironmental interpretations and all are pertinent to the study of British off-shore Miocene assemblages.

The foraminifera of the Pliocene of the Netherlands and Belgium are comparable to those found in the Coralline Crag. A biostratigraphical framework for the Netherlands Plio-Pleistocene was first established by Ten Dam and Reinhold (1941d). Subsequently the results of further work were made available (Van Voorthuysen 1958, Toering and Van Voorthuysen 1973). Recently the separate studies in Belgium (De Mueter and Laga, 1977) and in the Netherlands (Doppert, 1975) have been unified, resulting in a biostratigraphy based on benthic foraminifera of the Neogene of these two areas (Doppert et al., 1979). The benthic species have been studied by M. J. Hughes, C. King and the planktonic species by D. G. Jenkins.

Benthic foraminifera are recorded from the Miocene of three offshore boreholes and on-land sections in the Crags and St. Erth Beds (Fig. 11.5).

Thirty-five species and subspecies of planktonic foraminifera were recorded from 13 samples taken from the Sealab Trial borehole (Jenkins, 1977) and 18 have been chosen as diagnostic in the 20.06 m of cored Lower Miocene silty mudstone, mudstone and thin sandstone (Figs 11.2 and 11.5). A record of some of the species in two Middle Miocene dredge samples is also shown in Fig. 11.5; the *P. glomerosa curva* Zone sample from Lat. 47° 48' N, Long. 8° 09.50' W and the *G. mayeri mayeri* Zone sample from Lat. 48° 27' N, Long. 10° 07' W.

The classification used is that of Loeblich and Tappan (1984). The illustrated specimens have been deposited in the collections of the the British Geological Survey, Keyworth.

Sec. 11.6] **Index Species** 547

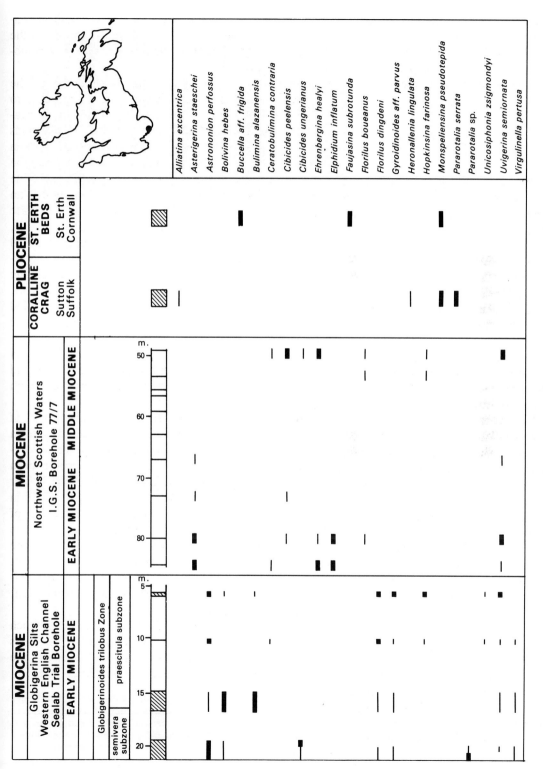

Fig. 11.4 — Range chart of Neogene benthic foraminifera.

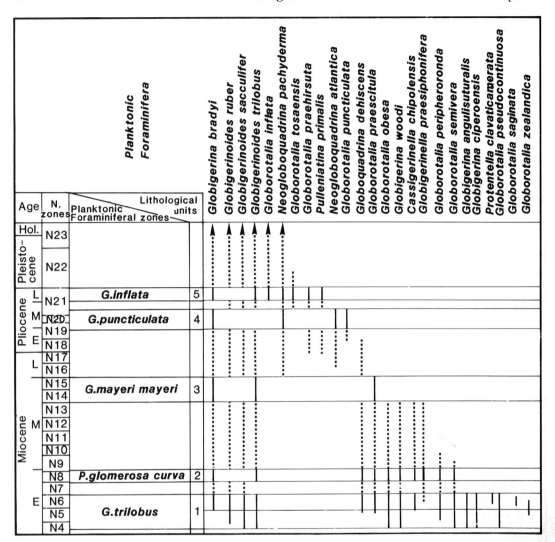

Fig. 11.5 — Range table of Neogene planktonic foraminifera; lithological units refer to 1: Sealab Trial Borehole (see Fig. 11.2); 2 and 3: *Globigerina* silts; 4: Coralline Crag; 5: St. Erth Beds.

Benthic species

Alliatina excentrica (di Napoli Alliata)
Plate 11.1, Fig. 1 (× 75), Fig. 2 (× 90) (both MPK 2655), Pliocene, Coralline Crag; Sudbury Park Pit, Suffolk, colln. I. P. Wilkinson. Fig. 11.4. = *Cushmanella excentrica* di Napoli Alliata, 1952. Description: dextrally trochospiral; up to ten chambers in final whorl, increasing rapidly in height; small accessory chambers at umbilical margins on both sides of test; apertures, slit at base, and oval perforation sometimes closed by a thin calcareous plate, in the centre of the apertural face; a groove extends across the apertural face from the ventral umbilicus to the areal aperture.

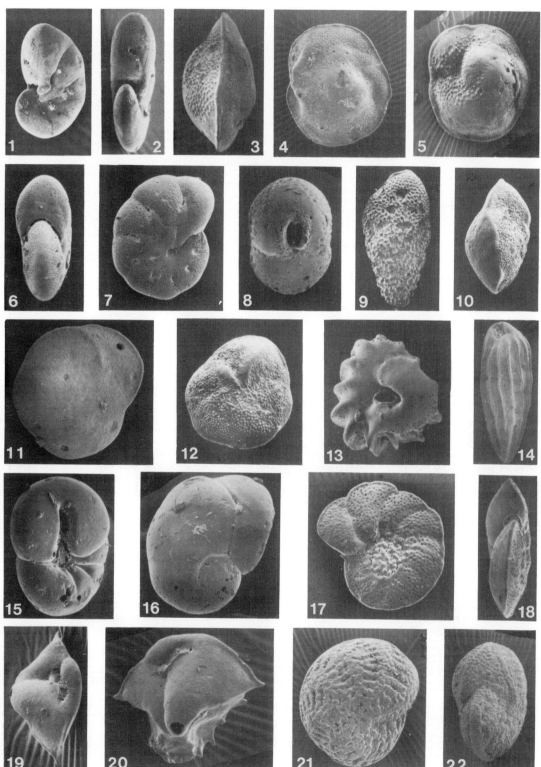

Plate 11.1

Remarks: differs from *Alliatinella gedgravensis* Carter in being less strongly trochospiral, the areal aperture is central rather than laterally offset, less strongly arched apertural groove.

Asterigerina staeschei Ten Dam and Reinhold, 1941
Plate 11.1, Fig. 3 (MPK 2640) (× 105), Fig. 4 (MPK 2641) (× 80), Fig. 5 (MPK 2642) (× 70), Lower Miocene; Figs 3, 4 IGS borehole 77/7 at 80 m (Sample CSC 2533), Fig. 5 IGS borehole 77/9 at 74.4 m (Sample CSC 2525); Fig. 11.4. Description: biserially trochospiral; biconvex; umbilical side higher than spiral side; keeled; umbilical chambers large, nearly triangular, together forming a dome higher than peripheral chambers which are narrow; strong granular ornament around aperture.
Remarks: differs from *A. guerichi* (Franke) in the greater convexity of the umbilical side, in the well-developed keel and the greater development of ornament around the aperture.

Astrononion perfossus (Clodius)
Plate 11.1, Fig. 6 (MPK 2643) (× 125) Fig. 7 (MPK 2644) (× 135), Lower Miocene; Sealab Trial borehole at 5.48 m (Sample CSB 2704); Fig. 11.4. = *Nonionina perfossa* Clodius, 1922 (syn. *Nonion nanum* Van Voorthuysen, 1950). Description: involute planispiral; compressed; nine chambers in final whorl; periphery rounded; small deep umbilici; distinctive opening at peripheral extremity of infilling at inner ends of each suture; apertural face imperforate.

Bolivina hebes Macfadyen, 1930
Plate 11.1, Fig. 8 (MPK 2645) (× 230), Fig. 9 (MPK 2646) (× 85), Lower Miocene; Sealab Trial borehole, Fig. 8 at 6.00 m (Sample CSB 2705), Fig. 9 at 10.06 m (Sample SAZ 1702); Fig. 11.4. Description: spatulate; biserial; broad, gently tapering with rounded initial end in side view; broad in end view; 6–8 pairs of chambers; basal margin of chambers strongly dentate; chambers surface strongly reticulate; aperture oval with collar; internal toothplate visible.

Buccella aff. *frigida* (Cushman)
Plate 11.1, Fig. 10 (MPK 2649) (× 135), Fig. 11 (MPK 2647) (× 160), Fig. 12 (MPK 2648) (× 150), Pliocene, St. Erth Beds; St. Erth, Cornwall (Sample SAA 2107); Fig. 11.4. = *Pulvinulina frigida* Cushman, 1922. Description: trochospiral; 6–7 chambers in final whorl; periphery sharply angled with slight keel; umbilical side covered with granular ornament.
Remarks: differs from Pleistocene *B. frigida* in the extension of the granular ornament to the entire ventral surface and in its sharper periphery.

Bulimina alazanensis Cushman, 1927
Plate 11.1, Fig. 13 (MPK 2650) (× 300), Fig. 14 (MPK 2651) (× 90), Lower Miocene; Sealab Trial borehole, Fig. 13 at 14.70 m (Sample CSB 2707), Fig. 14 at 10.06 m (Sample SAZ 1702); Fig. 11.4. Description: triserial; test small,

tapering, with 12 strong longitudinal costae; aperture deeply recessed with prominent elevated rim of fixed border and toothplate.
Remarks: differs from *B. rostrata* Brady in presence of notches in the costae where they cross the sutures and the absence of a strong apical spine.

Ceratobulimina aff. *contraria* (Reuss)
Plate 11.1, Fig. 15 (MPK 2653) (× 130), Fig. 16 (MPK 2652) (× 120), Lower Miocene; Sealab Trial borehole at 5.98 m (Sample SAZ 1710); Fig. 11.4. = *Rotalina contraria* Reuss, 1851. Description: dextrally trochospiral; 7–8 rapidly enlarging chambers in the final whorl; deeply umbilicate; aperture situated in deep broad groove extending from umbilicus up the apertural face.
Remarks: differs from the Oligocene type in the lower height and relatively greater width of the apertural face (see Langer 1969, p. 62).

Cibicides ungerianus (d'Orbigny)
Plate 11.1, Fig. 17 (MPK 2730) (× 60), Fig. 18 (MPK 2731) (× 65), Lower Miocene; Sealab Trial borehole, Fig. 17 at 6.00 m (Sample CSB 2705), Fig. 18 at 5.98 m (Sample SAZ 1701); Fig. 11.4. = *Rotalina ungeriana* d'Orbigny, 1846. Description: trochospiral; compressed; 12–14 chambers in final whorl; periphery acute, sometimes slightly keeled; sutures curved; early whorls on evolute spiral side covered with coarse granulations; coarsely perforate; aperture, at base of final chamber on the umbilical side and extending along the spiral suture.
Remarks: differs from *Heterolepa dutemplei* (d'Orbigny) in the greater compression and the dorsal granulation.

Ehrenbergina healyi Finlay, 1947
Plate 11.1, Fig. 19 (× 100), Fig. 20 (× 125) (both MPK 2654); Lower Miocene; IGS borehole 77/7 at 80 m (Sample CSC 2533); Fig. 11.4. Description: biserial; initial end enrolled with short spines; ventral side of later chambers strongly inflated; margin serrated; aperture extending from base of final chamber up the apertural face parallel to dorsal margin, with distinct lip on ventral edge.
Remarks: differs from *E. serrata* Reuss in the strong inflation of the ventral side.

Elphidium inflatum (Reuss)
Plate 11.1, Fig. 21 MPK 2657), Fig. 22 (MPK 2658) (both × 70), Lower Miocene; IGS borehole 77/9 at 73.4 m (Sample CSC 2525); Fig. 11.4. = *Polystomella inflata* Reuss, 1861. Description: involute, planispiral; slightly compressed; ten chambers in final whorl; 15–20 thick sutural bridges; sutural pores appear as slits; apertural face low; aperture, indistinct pores at base of final chamber.
Remarks: the only specimens available are somewhat corroded, the type description refers to a smooth wall. The preservational features, however, may be common in high latitude specimens.

Faujasina subrotunda Ten Dam and Reinhold, 1941
Plate 11.2, Fig. 1 (MPK 2659) (× 75), Fig. 2 (MPK 2660) (× 70), Fig. 3 (MPK 2661) (× 80); Pliocene, St Erth Beds, St Erth, Cornwall (Sample SAA 2107); Fig. 11.4. Description: trochospiral; plano-convex; periphery bluntly keeled; all chambers visible on flat spiral side, only those of the final whorl on convex side; 10–12 chambers in final whorl; sutural bridges well developed and numerous; surface ornament finely granular; apertural face broad with granular ornament; aperture, pores at base of final chamber.
Remarks: differs from *F. carinata* d'Orbigny and *F. compressa* Margerel in its greater thickness.

Florilus dingdeni (Cushman)
Plate 11.2, Fig. 4 (MPK 2662) (× 45), Fig. 5 (MPK 2663) (× 60); Lower Miocene; Sealab Trial borehole at 10.06 m (Sample SAZ 1702); Fig. 11.4. = *Nonion dingdeni* Cushman, 1936. Description: evolute; planispiral; compressed; 10–11 chambers in final whorl, increasing rapidly in height; periphery narrowly rounded; sutures depressed; wide umbilici infilled with granular calcite which extends into the sutural depressions.
Remarks: differs from *F. boueanus* (d'Orbigny) in that the granular infilling extends into the sutural depressions.

Gyroidinoides aff. *parvus* (Cushman and Renz)
Plate 11.2, Fig. 6 (MPK 2664) (× 150). Fig. 7 (PK 2665) (× 170), Fig. 8 (MPK 2666) (× 140). Lower Miocene; Sealab Trial borehole; Figs 6 and 8 at 15.25 m (Sample SAZ 1704), Fig. 7 at 5.89 m (Sample SAZ 1701); Fig. 11.4. = *Gyroidina parva* Cushman and Renz, 1941. Description: small, trochospiral; spiral side flat or slightly domed; on spiral side final whorl lower than previous one at the margin; $5\frac{1}{2}$–6 chambers in final whorl; umbilical area depressed; umbilicus closed; periphery rounded; outline lobate; outline of apertural face rounded; aperture extends from umbilicus to periphery with lip on final chamber extending for about three-quarters of distance from the umbilicus.
Remarks: differs from type in its lesser height of test, the less marked lowering of the final whorl and the wider apertural face and from other northern European Miocene species in its smallness and rounded periphery.

Heronallenia lingulata (Burrows and Holland)
Plate 11.2, Fig. 9, 10 (MPK 2667) (both × 170), Pliocene, Coralline Crag; Sudbury Park Pit, Suffolk; colln. I. P. Wilkinson; Fig. 11.4. = *Discorbina lingulata* Burrows and Holland, 1896. Description: trochospiral, concavo-convex; compressed; five chambers in final whorl rapidly increasing in height; periphery bluntly carinate; each chamber surface on spiral side bears a raised boss which together with the limbate sutures present a distinctive ornament with high relief.

Hopkinsina farinosa (Hantken)
Plate 11.2, Fig. 11 (MPK 2668) (× 70), Lower Miocene; Sealab Trial borehole at 5.48 m (Sample CSB 2704); Fig. 11.4. = *Uvigerina farinosa* Hantken, 1875. Description: elongate; triserial becoming loosely biserial; biserial portion half or more of test length in adult specimens; sutures depressed becoming wide and deep in biserial part; ornament finely postulate, with fine spines along the sutures frequently bridging the wide sutures of the biserial part; aperture terminal on neck with lip; internal toothplate clearly visible.
Remarks: differs from *Uvigerina tenuipustulata* Van Voorthuysen and *Angulogerina gracilis* (Reuss) in the greater development and looseness of the biserial part and also from the latter in the coarser ornament.

Monspeliensina pseudotepida (Van Voorthuysen)
Plate 11.2, Fig. 12 (MPK 2670) (× 70), Fig. 13 (MPK 2671) (× 70), Fig. 14 (MPK 2672) (× 60), Pliocene, St. Erth Beds; St. Erth, Cornwall (Sample SAA 2107); Fig. 11.4. = *Streblus beccarii* (Linné) *pseudotepidus* Van Voorthuysen, 1950. Description: trochospiral; biconvex; 6–7 chambers in final whorl; periphery broadly rounded; umbilicus deep and narrow; sutures on umbilical side deeply excavated, in well-preserved specimens umbilicus and inner end of sutures covered by plates extending back from subsequent chambers; aperture at base of final chamber midway between umbilicus and periphery; accessory apertures on spiral side of intersection of septal and spiral sutures.
Remarks: differs from *Ammonia* spp. in the presence of the accessory apertures on the spiral side.

Pararotalia serrata (Ten Dam and Reinhold)
Plate 11.2, Fig. 15 (MPK 2673) (× 85), Fig. 16 (MPK 2674) (× 70, Fig. 17 (MPK 2675) (× 80); Pliocene, Coralline Crag; Sudbury Park Pit, Suffolk; colln. I. P. Wilkinson; Fig. 11.4. = *Rotalia serrata* Ten Dam and Reinhold, 1941. Description: trochospiral; 8–9 chambers in final whorl; periphery sharply angled with short stout spines at the centre of the periphery of each chamber; outline stellate; chambers ventrally very convex with deep sutures and umbilicus, the latter filled with a calcite boss.

Unicosiphonia zsigmondyi (Hantken)
Plate 11.2, Fig. 18 (MPK 2677) (× 165), Fig. 19 (MPK 2676) (× 22); Lower Miocene; Sealab Trial borehole, Fig. 18 at 5.48 m (Sample CSB 2704), Fig. 19 at 10.06 m (Sample SAZ 1702); Fig. 11.4. = *Nodosaria (Dentalina) zsigmondyi* Hantken, 1868 (syn. *Rectobolivina marentinensis* Ruscelli, 1952). Description: large; straight to slightly arcuate; biserial becoming uniserial; biserial part restricted to first two pairs of chambers; circular in section; chambers with basal crenulations extending back over suture; wall smooth; aperture terminal with internal toothplate.
Remarks: frequently broken but individual chambers may be recognised by crenulations and very prominent internal columella process; under reflected light wall appears translucent with perforations clearly visible.

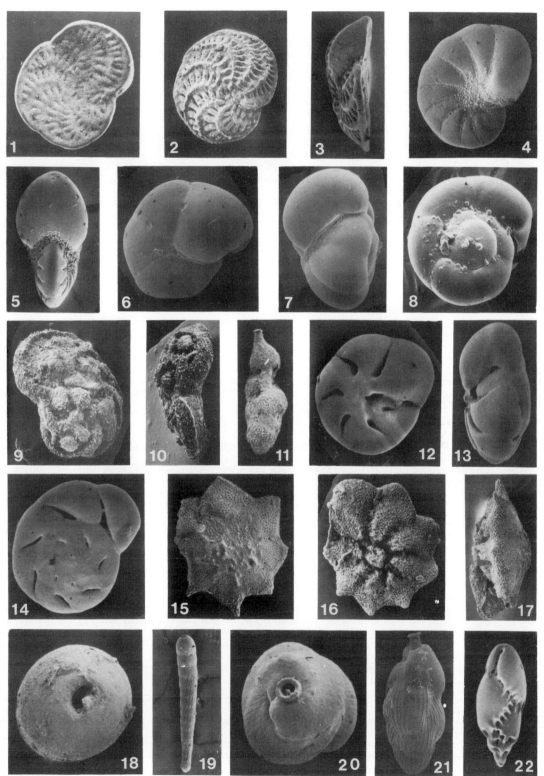

Plate 11.2

Uvigerina semiornata d'Orbigny, 1846
Plate 11.2, Fig. 10 (MPK 2669) (× 80), Fig. 21 (MPK 2656) (× 60), Lower Miocene; Sealab Trial borehole, Fig. 20 at 10.05 m (Sample CSB 2706), Fig. 21 at 10.06m (Sample SAZ 1702); Fig. 11.4. Description: elongate; triserial; rounded in section; costae prominent on early chambers and lower part of later chambers, upper part of final chamber smooth; aperture terminal on a stout neck with phialine lip and prominent internal toothplate.
Remarks: the various subspecies described from the north German Miocene do not appear to have differentiated from the central type in the two areas investigated by this writer.

Virgulinella pertusa (Reuss)
Plate 11.2, Fig. 22 (MPK 2678) (× 50), Lower Miocene; Sealab Trial borehole at 16.50 m. (Sample SAZ 1707); Fig. 11.4. = *Virgulina pertusa* Reuss, 1861. Description: elongate: triserial later biserial; rounded in section; two morphological types, either uniformly tapering test or an abrupt change from narrow early whorls to wide parallel sided chambers; very long digitate retral processes developed across sutures and onto previous chambers, early chambers may be completely obscured; surface smooth; aperture long and narrow extending from the base of final chamber in the direction of growth.
Remarks: very fragile but fragments of the distinctive retral processes serve to record its presence.

Planktonic Species

Cassigerinella chipolensis (Cushman and Ponton)
Plate 11.3, Fig. 1 (× 125). Fig. 11.5. = *Cassidulina chipolensis* Cushman and Ponton, 1932. Description: very small (c. 0.19 mm diameter), biserially arranged chambers continuing to spiral in same plane, biumbilicate, periphery broadly rounded.
Remarks: designated a planktonic taxon mainly because of its widespread geographic distribution; the genus became extinct in the Middle Miocene.
Range: Britain, Lower Miocene *G. trilobus* Zone to Middle Miocene *P. glomerosa curva* Zone; France, Upper Aquitanian (Jenkins, 1966) and in the Upper Oligocene of the Aquitaine Basin (Pujol, 1970). Total range: Late Eocene to Middle Miocene.

Globigerina bradyi Wiesner, 1931
Plate 11.3, Fig. 2 (× 125), Fig. 11.5 (syn *Globigerinoides parva* Hornibrook, 1961). Description: very small test (c. 0.16 mm diameter), high spired and lipped umbilical aperture and rare supplementary apertures on the spiral side as in the figured specimen,
Remarks: related to *Globigerinita glutinata* (Egger) which has a lower spired test and an umbilical bulla; living at present in cooler waters of 0–10°C.
Range: Britain, Lower Miocene *G. trilobus* Zone to Middle Miocene *G. mayeri mayeri* Zone; France, Upper Oligocene to Lower Miocene in the Aquitaine Basin (Pujol, 1970). Total range: Upper Oligocene to Recent.

Globigerina ciperoensis angulisuturalis Bolli, 1957
Plate 11.3, Fig. 3 (× 125), Fig. 11.5. Description: five chambers in the final whorl with relatively deeply cut sutures on the umbilical side.
Remarks: differs from *G. ciperoensis ciperoensis* and *G. ciperoensis angustiumbilicata* in having incised sutures.
Range: Britain, Lower Miocene *G. trilobus* Zone; France, type Lower Aquitainian (Jenkins, 1966), Upper Oligocene to Lower Miocene in the Aquitaine Basin (Pujol, 1970, Bizon *et al.*, 1972). Total range: Upper Oligocene–Lower Miocene.

Globigerina ciperoensis ciperoensis Bolli, 1954
Plate 11.3, Fig. 4 (× 75), Fig. 11.5. Description: five chambers in the final whorl increasing slowly in size and with an umbilical aperture.
Remarks: it lacks the lipped aperture of *G. ciperoensis angustiumbilicata*.
Range: Britain, Lower Miocene *G. trilobus* Zone; France, Lower Miocene in the Aquitaine Basin (Pujol, 1970). Total range: Upper Oligocene to Lower Miocene.

Globigerina woodi Jenkins, 1960
Plate 11.3, Figs 5, 6 (× 75). Fig. 11.5. Description: 3–4 chambers in the final whorl increasing slowly in size; a large rimmed aperture and coarse wall.
Remarks: it differs from *G. bulloides* in having a higher arched aperture and coarser wall structure and from its descendant *G. decoraperta* in having a much lower spired test.
Range: Britain, Lower Miocene *G. trilobus* Zone to the Middle Miocene *P. glomerosa curva* Zone; France, type Aquitainian–Burdigalian (Jenkins, 1966) and the Lower to Upper Miocene in the Aquitaine and Mediterranean Basins (Bizon *et al.*, 1972).

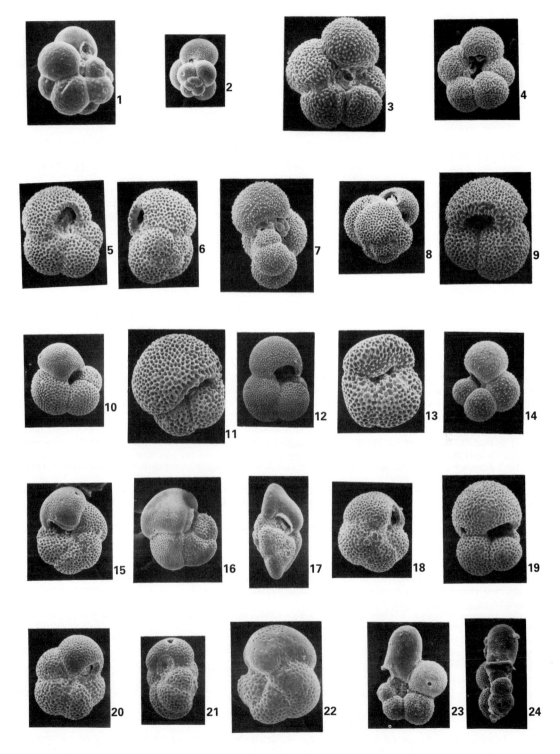

Plate 11.3

Globigerinella praesiphonifera (Blow)
Plate 11.3, Fig. 7 (× 75). Fig. 11.5. = *Hastigerina (H). siphonifera praesiphonifera* Blow, 1969. Description: five chambers in final whorl, nearly planispirally coiled; a finely pitted wall with small spine bases.
Remarks: differs from its trochospirally coiled ancestor *Globorotalia obesa* in having five chambers in the final whorl instead of four, and from its descendant *G. aequilateralis* which has a planispirally coiled adult test.
Range: Britain, Upper *G. trilobus* Zone of the Lower Miocene. Total range; Lower to Middle Miocene (Blow, 1969).

Globigerinoides ruber (d'Orbigny)
Plate 11.3, Figs 8, 9, Fig. 8 (× 50), Fig. 9, (× 75). Fig. 11.5. = *Globigerina rubra* d'Orbigny, 1839. Description: small rounded apertures positioned opposite the sutures, normally on a high spired test as in Fig. 8.
Remarks: differs from other Miocene–Recent *Globigerinoides* in the position of the apertures and in having a high spiral test. Living at present in the upper 50 m of oceanic water within a temperature range of 10–30°C.
Range: Britain, Lower Miocene *G. trilobus* Zone to Middle Miocene *P. glomerosa curva* Zone; France, Lower Miocene to Middle Pliocene in the Aquitaine Basin (Pujol, 1970). Total range: Lower Miocene–Recent.

Globigerinoides sacculifer (Brady)
Plate 11.3, Fig. 10 (× 50). Fig.11.5. = *Globigerina sacculifer* Brady, 1877. Description: three to four chambers in the final whorl with a sac-like final chamber.
Remarks: the distinctive final chamber distinguishes it from other species of the genus and its relatively thin wall from *Sphaeroidinella dehiscens*; it lives today in the upper 50 m of oceanic water within a temperature range of 14.5–30°C with maximum numbers found in the 24–30°C range.
Range: Britain, Lower Miocene *G. trilobus* Zone; France, Mediterranean and Aquitaine Basins from the Lower Miocene to Pleistocene (Pujol, 1970). Total range: Lower Miocene to Recent.

Globigerinoides trilobus (Reuss)
Plate 11.3, Figs 11, 12, Fig. 11 (× 75), Fig. 12 (× 37.5). Fig. 11.5. = *Globigerina triloba* Reuss, 1850. Description: three chambers in the final whorl with a low arched umbilical aperture and one supplementary aperture on the spiral side; final chamber can be enlarged (as illustrated in Fig. 11) during the evolutionary development towards its descendant *G. bisphericus* Todd.
Remarks: distinguished from its immediate ancestor *Globigerina woodi connecta* in the possession of a supplementary aperture on the spiral side; *G. trilobus* is related to *G. sacculifer* but is distinguished by not having the sac-like chamber.
Range: Britain, Lower Miocene *G. trilobus* Zone to Middle Miocene *G. mayeri mayeri* Zone; France, type Upper Burdigalian (Jenkins, 1966) and in the Lower to Middle Miocene in the Aquitaine Basin (Pujol, 1970). Total range: Lower Miocene to Recent.

Globoquadrina dehiscens (Chapman, Parr and Collins)
Plate 11.3, Fig. 13 (× 75). Fig. 11.5. = *Globorotalia dehiscens* Chapman, Parr and Collins, 1934. Description: normally four chambers in the final whorl, quadrate in umbilical view, with a low arched aperture at base of final chamber.
Remarks: distinguished from *G. altispira* which is high spired with five chambers in the final whorl and from *G. globosa* which has more inflated chambers and is less quadrate in outline.
Range: Britain, Lower Miocene *G. trilobus* Zone to Middle Miocene *G. mayeri mayeri* Zone; France, type Aquitainian–Burdigalian (Jenkins, 1966) and in the Aquitaine Basin from Lower to Upper Miocene. Total range: Lower Miocene to Lower Pliocene.

Globorotalia obesa Bolli, 1957
Plate 11.3, Fig. 14 (× 75), Fig. 11.5. Description: four chambers in the final whorl increasing slowly in size, low-arched aperture, and a hispid wall.
Remarks: it is distinguished from its immediate descendant *G. praesiphonifera* in only having four chambers in the final whorl and has a low trochospiral test.
Range: Britain, Lower Miocene *G. trilobus* Zone to Middle Miocene *P. glomerosa curva* Zone; France, type Burdigalian (Jenkins, 1966) and from the Upper Oligocene to Upper Miocene in the Aquitaine Basin (Pujol, 1970). Total range: Lower to Upper Miocene.

Globorotalia peripheroronda Blow and Banner 1966
Plate 11.3, Fig. 15, (× 75). Fig. 11.5. Description: 5–6 chambers in the final whorl, rounded periphery with strongly recurved sutures on the spiral side and a low-arched lipped aperture.
Remarks: Britain, Lower Miocene *G. trilobus* Zone to Middle Miocene *P. glomerosa curva* Zone; France, from the Lower to Middle Miocene in the Aquitaine Basin (Pujol, 1970). Total range: Lower to Middle Miocene.

Globorotalia praescitula Blow, 1959
Plate 11.3, Figs 16, 17, (× 100). Fig. 11.5. Description: $4-4\frac{1}{2}$ chambers in the final whorl with strong, recurved sutures on spiral side, acutely angled rounded periphery with a low arched lipped aperture.
Remarks: distinguished from most descendants such as *G. archaeomenardii*, *G. miozea* and *G. praemenardii* in not possessing a keeled or incipient keeled periphery, and from *G. scitula* in having a more coarsely pitted wall, a more rapid increase in chamber size and in having a more bi-convex shape in side view.
Range: Britain, Lower Miocene *G. trilobus* Zone to Middle Miocene *G. mayeri mayeri* Zone; France, Lower to Middle Miocene in the Aquitaine Basin (Pujol, 1970). Total range: Lower to Middle Miocene.

Globorotalia pseudocontinuosa Jenkins, 1967
Plate 11.3, Fig. 18 (× 75). Fig. 11.5. Description: four chambers in final whorl, coarsely ornamented wall and a large comma-shaped aperture.
Remarks: distinguished from *G. semivera* which has five chambers in the final whorl and from *G. obesa* which has a low arched aperture and a pustular wall.
Range: Britain, Lower Miocene *G. trilobus* Zone; France, Upper Burdigalian and in one Helvetian (?) sample. Total range: Lower Miocene.

Globorotalia saginata Jenkins, 1966
Plate 11.3, Fig. 19 (× 100). Fig. 11.5. Description: $3\frac{1}{2}$ chambers in the final whorl, chambers increase rapidly in size, low arched aperture.
Remarks: distinguished from *G. obesa* by having a more rapid increase in chamber size and in having a more coarsely ornamented wall structure.
Range: Britain, Lower Miocene *G. trilobus* Zone; France, type Aquitainian–Burdigalian (Jenkins, 1966). Total range: Middle Oligocene to Middle Miocene.

Globorotalia semivera (Hornibrook)
Plate 11.3, Figs 20, 21 (× 75). Fig. 11.5. = *Globigerina semivera* Hornibrook, 1961. Description: five chambers in the final whorl, large open aperture, radial sutures and coarse wall.
Remarks: differs from *G. pseudocontinuosa* in having five chambers in the final whorl, and from *G. peripheroronda* in having radial sutures and a larger more open aperture.
Range: Britain, Lower Miocene *G. trilobus* Zone to Middle Miocene *P. glomerosa curva* Zone; France, type Aquitainian–Burdigalian (Jenkins, 1966), and as *G. acrostoma* by Pujol (1970) from the Lower-Middle Miocene in the Aquitaine Basin. Total range: Upper Oligocene–Middle Miocene.

Globorotalia zealandica Hornibrook, 1958
Plate 11.3, Fig. 22 (× 100). Fig. 11.5. Description: four chambers in the final whorl, with a flattened spiral and convex umbilical side, high arched aperture and a coarse wall.
Remarks: differs from *G. pseudocontinuosa* in having a more rapid increase in chamber size in the final whorl and in having recurved sutures on the spiral side.
Range: Britain, Lower Miocene *G. trilobus* Zone. Total range: Lower Miocene.

Protentella clavaticamerata Jenkins, 1977
Plate 11.3, Figs 23, 24 (× 100) Fig. 11.5. Description: small test (holotype is 0.26 mm diameter), planispirally coiled, final chambers increasing fairly rapidly in size and becoming radially elongate, aperture centrally placed, lipped at base of final chamber.
Remarks: the smaller planispirally coiled, involute test distinguishes it from *G. praesiphonifera*.
Range: Britain, Upper *G. trilobus* Zone, Lower Miocene. Total range: upper Lower Miocene.

Globorotalia inflata (d'Orbigny)
Plate 11.4, Figs 1, 2 (× 75), Fig. 11.5 = *Globigerina inflata* d'Orbigny, 1839. Description: 3.6–3.25 chambers in the final whorl (Malmgren and Kennett, 1981), rounded quadrate umbilical view outline with large aperture.
Remarks: differs from its immediate ancestor *Globorotalia puncticulata* (Deshays) which has 4.0–3.6 chambers in the final whorl and has a smaller aperture (Plate 11.4, Figs 6–8); it lives in the sub-Arctic and sub-Antarctic to the subtropical faunal province with a temperature range of 5–24°C.
Range: in the north Atlantic *G. inflata* made its first appearance in the Late Pliocene at 2.1 Ma (Weaver and Clement, 1986); it occurs in the St. Erth Beds. Total range: Late Pliocene to Recent.

Globorotalia praehirsuta Blow, 1969
Plate 11.4, Figs 3, 4, 5 (× 165), Fig. 11.5. Description: relatively small species with unkeeled periphery and with 4.5 chambers in the final whorl.
Remarks: similar in size to *Globorotalia scitula* (Brady) but distinguished from this dextrally coiled species by having less curved sutures on the spiral side and having more pustules on the umbilical side.
Range: Britain in the St. Erth Beds; Blow (1969) recorded its range from the Late Miocene (Zone N18) to the Pleistocene (Zone N22), but in the north Atlantic its extinction is within the Late Pliocene (Poore, 1978). Total range: Late Miocene to Pleistocene.

Globorotalia puncticulata (Deshays)
Plate 11.4, Figs, 6, 7, 8, (× 80). Fig. 11.5. = *Globigerina puncticulata* Deshays, 1832. Description: 4.0–3.6 chambers in the final whorl, increasing slowly in size.
Remarks: distinguished from its descendant species *G. inflata* in having more chambers in the final whorl, a smaller aperture and a less rounded periphery.
Range: Britain, found in the Coralline Crag; in the northeast Atlantic it ranges from Early Pliocene 4.16 Ma and its extinction is at 2.2–2.4 Ma (Weaver and Clement, 1986; Weaver, 1986). Total range: Early Pliocene to early Late Pliocene.

Globorotalia tosaensis Takayanagi and Saito, 1962
Plate 11.4, Figs 9, 10, 11 (× 75). Fig. 11.5. Description: has a rounded umbilical outline and lacks a keeled periphery.
Remarks: it is distinguished from its descendant species *G. truncatulinoides* (d'Orbigny) in having a rounded umbilical outline and in not possessing a keel.
Range: Britain, found in the St. Erth Beds. Total range: Late Pliocene to Early Pleistocene (Blow, 1969; Bolli and Saunders in Bolli *et al.*, 1985).

Neogloboquadrina atlantica (Berggren)
Plate 11.4, Figs 12, 13, 14, (× 100), 15 (× 70). Fig. 11.5. = *Globigerina atlantica* Berggren, 1972. Description: large globigerine *Neogloboquadrina* species; coarse wall structure and with globular chambers which increase rapidly in size with an open umbilicus; 3.5–4 chambers in the final whorl.
Remarks: distinguished from *N. humerosa* (Takayanagi and Saito) which has 5–6 chambers in the final whorl which increase in size slowly, and from *N. pachyderma* which is much smaller (Plate 11.4, Figs 16, 17, 18) and has a distinct lipped aperture.
Range: Britain, Coralline Crag, also in the Red Crag but these could be reworked; northeast Atlantic: Late Miocene to Late Pliocene, i.e. c. 7.0–2.3 Ma (Weaver and Clement, 1986). Total range: Late Miocene to Late Pliocene.

Neogloboquadrina pachyderma (Ehrenberg)
Plate 11.4, Figs 16, 17, 18 (× 155). Fig. 11.5. = *Aristerospira pachyderma* Ehrenberg, 1861. Description: relatively small with 4.0 to 4.5 chambers in the final whorl; low arched aperture with a lip and a coarse wall.
Remarks: it differs from *N. humerosa* which has 5–6 chambers in the final whorl and from *N. atlantica* which has a globigerine shape and has a much larger test; in the northern hemisphere it ranges from the Arctic to transitional faunal provinces, i.e. in surface waters with temperatures from 0 to 18°C.
Range: Britain, found in the Coralline Crag and St. Erth Beds where it is dextrally coiled; in the north Atlantic it ranges from Late Miocene to Recent (Poore, 1978; Weaver, 1986). Total range: Late Miocene to Recent (Jenkins, 1967).

Pulleniatina primalis Banner and Blow, 1967
Plate 11.4, Fig. 19, 20 (× 70). Fig. 11.5. Description: streptospirally coiled, thick walled with 3.5 to four chambers in the final whorl and a large low-arched aperture.
Remarks: it is distinguished from its immediate descendant *P. obliquiloculata* (Parker and Jones) which has fewer chambers in the final whorl and has a much more rounded and larger test.
Range: Britain, in St. Erth Beds; Indo-Pacific: Late Miocene to Late Pliocene (Parker, 1967); Saito *et al.* (1981) claim that it ranges into the Pleistocene but they may have misidentified juvenile *P. primalis* as *P. obliquiloculata* (Jenkins *et al.*, 1986). Total range: Late Miocene to Late Pliocene.

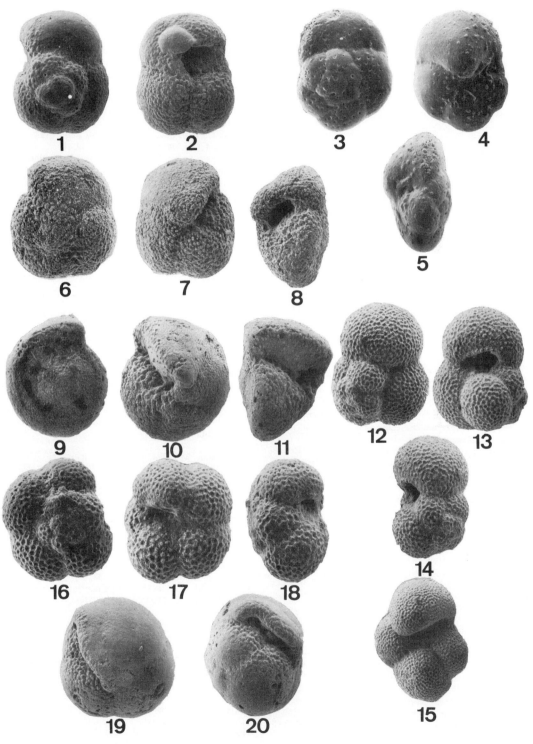

Plate 11.4

ACKNOWLEDGEMENTS

Mrs. C. Whale typed the manuscript, Mr. A. Lloyd drafted the figures and Dr. J. Whittaker (British Museum of Natural History) provided facilities for SEM of Plate 11.4. We wish to thank them all.

REFERENCES

Balson, P. S. 1981a. The sedimentology and palaeoecology of the Coralline Crag (Pliocene) of Suffolk. Unpublished Ph. D. thesis, London., 364 pp.

Balson, P. S. 1981b. Facies-related distribution of bryozoans of the Coralline Crag (Pliocene) of eastern England. In: Larwood, G. P. and Nielsen, C. N. (eds) *Recent and Fossil Bryozoa*. Olsen & Olsen. pp. 1–6.

Batjes, D. A. J. 1958. Foraminifera of the Oligocene of Belgium. *Mém. Inst. r. Sci. Nat. Belg.*, **143**.

Berggren, W. A., Phillips, J. D., Bertels, A. and Wall, D. 1967. Late Pliocene–Pleistocene stratigraphy in deep-sea cores from the south central North Atlantic. *Nature*, **216**, 253–254.

Berggren, W. A., Kent, D. V. and van Couvering, J. A. 1985. The Neogene: Part 2, Neogene geochronology and chronostratigraphy, In: Snelling, N. (ed.), Chronology and the geological record. *Geol. Soc. Lond.* Mem. no. **10**, 211–260.

Bizon, G., Bizon, J., Aubert, J. and Oertli, H. 1972. Atlas des principaux foraminifères planctoniques du bassin méditerranéen. Oligocène á Quaternaire. Paris: *Editions Technip*, 1–316.

Blow, W. H. 1969. Late Middle Eocene to Recent planktonic foraminiferal biostratigraphy. In Brönnimann, P., and Renz, H. H (eds) *Proceedings of the first international conference in planktonic microfossils*. Leiden: E. J. Brill, **1**, 199–421.

Bolli, H. M. and Saunders, J. B. 1985. Oligocene to Holocene low latitude planktonic foraminifera. In Bolli, H. M., Saunders, J. B. and Perch-Nielsen, K. (eds) *Plankton stratigraphy*, pp. 155–262, Cambridge University Press.

Cambridge, P. G. 1977. Whatever happened to the Boytonian? A review of the marine Plio-Pleistocene of the southern North Sea basin. *Bull. geol. Soc. Norfolk*, **29**, 23–45.

Carter, D. J. 1951. Indigenous and exotic foraminifera in the Coralline Crag of Sutton, Suffolk. *Geol. Mag.*, **88**, 236–248.

Carter, D. J. 1957. The distribution of the foraminifera *Alliatina excentrica* (di Napoli Alliata) and the new genus *Alliatinella*. *Palaeontology*, **1**, 76–86.

Clodius, G. 1922. Die Foraminiferen des obermiozänen Glimmertons in Norddeutschland mit besonderer Berücksichtung der Aufschlüsse in Mecklenburg. *Arch. Ver. Freunde Naturg. Mecklenb.*, **75**, 76–145.

Curry, D., Martini, E., Smith, A. J. and Whittard, W. F. 1962. The geology of the western approaches of the English Channel. I. Chalky rocks from the upper reaches of the continental slope. *Phil. Trans. R. Soc.*, **B245**, 267–290.

Curry, D., Murray, J. W. and Whittard, W. F. 1965. The geology of the western approches of the English Channel. III. The *Globigerina* Silts and associated rocks. *Colston Pap.*, **17**, 239–264.

Curry D., Adams, C. G. Boulter, M. C., Dilley, F. C., Eames, F. E., Funnell, B. M. and Wells, M. K. 1978. A correlation of Tertiary rocks in the British Isles. *Geol. Soc. Lond., Special Report*, **12**, 72pp.

Daniels, C. H. von and Spiegler, D. 1977. *Uvigerinen* (Foram.) im Neogen Nordwestdeutschlands (Das Nordwestdeutsche Tertiärbecken, Beitrag Nr. 23). *Geol. Jb.*, **A40**, 3–59.

De Meuter, F. J. and Laga, P. G. 1977. Lithostratigraphy and biostratigraphy based on benthonic foraminifera of the Neogene deposits of northen Belgium. *Bull. Soc. belge. Géol. Paléont. Hydrol.*, **85**, 133–152.

Doppert, J. W. C. 1975. Foraminiferenzonering van het Nederlandse Onder-Kwartair en Tertiair. 114–118. in *Toelichting bij geologische overzichtskaarten van Nederland*. Zagwijn, W. H. (ed.). Haarlem, Netherlands.

Doppert, J. W. C., Laga, P. G. and De Meuter, F. J. 1979. Correlation of the biostratigraphy of marine Neogene deposits, based on benthonic foraminifera, established in Belgium and the Netherlands. *Meded. Rijks. geol. Dienst*, **31-1**.

Evans, C. D. R. and Hughes, M. J. 1984. The Neogene succession of the South Western Approaches, Great Britain. *Journ. Geol. Soc. Lond.*, **141**, 315–326.

Gheorghian, M. 1971. Sur quelques affleurements de dépots ottnangiens de Roumanie et sur leur contenu microfaunique. *Mem. Inst. Geol. Bucarest*, **14**, 103–121.

Gheorghian, M. 1974. Considerations of the genus *Hidina* (order Foraminiferida, Eichwald 1830). *Dari Seama Sedint. Inst. Geol.*, **60/3**, 23–31.

Gheorghian, M. 1975. Asupra biostratigrafiei depozitelor Miocene din Romania (Stadiul 1974) [On biostratigraphy of Miocene deposits in Rumania (Stage 1974)]. *Dari Seama Sedint. Inst. Geol.*, **61/4**, 85–104 (in Rumanian).

Gramann, F. 1964. Die Arten der Foraminiferen-Gattung *Asterigerina* d'Orb. im Tertiär NW-Deutschlands. *Paläont. Z.*, **38**, 207–222.

Hantken, M. von, 1875. Die Fauna der *Clavulina szaboi* Schichten. I. Foraminiferen. *Mitt. Jb. K. ung. geol. Ant.*, **4**, 1–39.

Harmer, F. W. 1898. The Pliocene deposits of the east of England: the Lenham Beds and the Coralline Crag. *Q. Jl. geol. Soc. Lond.*, **54**, 308–356.

Harmer, F. W. 1900. On a proposed new classification of the Pliocene deposits of the east of England. *Rep. Br. Ass. Advmt Sci.*, Dover 1899, 751–753.

Harmer, F. W. 1902. A sketch of the Later Tertiary history of East Anglia. *Proc. Geol. Asso.*, **17**, 416–479.

Hinsch, W., Kaever, M. and Martini, E. 1978. Die Fossilführung des Erdfalls von Nieheim (SE-Westfalen) und seine Bedeutung für Paläeogeog-

raphie im Campan und Miozän. *Paläont. Z.*, **52**, 219–245.
Hodgson, G. E. and Funnell, B. M. 1987. Foraminiferal biofacies of the early Pliocene Coralline Crag. In Hart, M. B. (ed.) *Micropalaeontology of Carbonate Environments*. Ellis Horwood, pp. 44–73.
Hosius, A. 1892. Beiträge zur Kenntnis der Foraminiferen-Fauna des Miocens. *Verh. naturh. Ver. preuss. Rheinl.*, **49**, 148–197.
Iaccarino, S. 1985. Mediterranean Miocene and Pliocene planktonic foraminifera. In: Bolli, H. M., Saunders, J. B. and Perch-Nielsen, K., (eds), *Plankton Stratigraphy*, pp. 283–314, Cambridge University Press.
Indans, J. 1958. Mikrofaunistische Korrelationen im marinen Tertiär der Niederrheinischen Bucht. *Fortschr. Geol. Rheinld Westf.*, **1**, 223–238.
Indans, J. 1962. Foraminiferen-Fauna aus dem Moizän des Niederrheingebietes. *Fortschr. Geol. Rheinld Westf.*, **6**, 19–82.
Indans, J. 1965. Mikrofaunistisches Normalprofil durch das marine Tertiär der Neiderrheinsichen Bucht. *Fortschr.-Ber. Nordrhein-West*, **1484**.
Institute of Geological Sciences. 1978. IGS Boreholes 1977. *Rep. Inst. Geol. Sci.*, **78/21**.
Jenkins, D. G. 1966. Planktonic foraminifera from type Aquitainian–Burdigalian of France. *Contr. Cushman Found.*, **17**, 1–15.
Jenkins, D. G. 1967. Planktonic foraminiferal zones and new taxa from the Lower Miocene to Pleistocene of New Zealand. *New Zealand Journ. Geol. Geophys.*, **10**, 1064–1078.
Jenkins, D. G. 1977. Lower Miocene planktonic foraminifera from a borehole in the English Channel. *Micropaleontology*, **23**, 297–318.
Jenkins, D. G. 1982. The age and palaeoecology of the St. Erth Beds, southern England, based on planktonic foraminifera. *Geol. Mag.* **119**, 201–205.
Jenkins, D. G. and Houghton, S. D. 1987. Age, correlation and paleoecology of the St. Erth Beds and the Coralline Crag of England. *Meded. Werkgr. Tert. Kwart. Geol.* **24**, 147–156.
Jenkins, D. G. and Shackleton, N. 1979. Parallel changes in species diversity and palaeotemperature in the Lower Miocene. *Nature*, **278**, 50–51.
Jenkins, D. G., Whittaker, J. E. and Carlton, R. 1986. On the age and correlation of the St. Erth Beds, S.W. England, based on planktonic foraminifera. *Journ. Micropal.* **5**(2), 93–105.
Jenkins, D. G., Curry, D., Funnell, B. J. and Whittaker, J. E. 1988. Planktonic foraminifera from the Pliocene Coralline Crag of Suffolk, Eastern England. *Journ. Micropal.*, **7**(1), 1–10.
Jones, T. R., Parker, W. K. and Brady, H. B. 1866–1897. *A monograph of the foraminifera of the Crag*. Palaeontographical Society, London.
King, C. 1983. Cainozoic micropalaeontological biostratigraphy of the North Sea. *Rep. Inst. Geol. Sci.*, **82/7**, 1–40.
Langer, W. 1963s. Einige wenig bekannte Foraminiferen aus dem mittleren und oberen Moizän des Nordseebeckens. *Neues Jb. Geol. Paläont. Abh.*, **117**, 169–184.
Langer, W. 1963b. Bemerkungen zur Stratigraphie nach Foraminiferen im mittleren und oberen Miozän von N-und NW-Deutschland. *Neues Jb. Geol. Paläont. Mh.*, 543–558.
Langer, W. 1969. Beitrag zur Kenntnis einiger Foraminiferen aus dem mittleren und oberen Miozän des Nordsee-Beckens. *Neues Jb. Geol. Paläont. Abh.*, **133**, 23–78.
Macfadyen, W. A. 1942. A post-glacial microfauna from Swansea Docks. *Geol. Mag.*, **79**, 133–146.
Malmgren, B. A. and Kennett, J. P. 1981. Phyletic gradualism in a Late Cenozoic planktonic foraminiferal lineage; D.S.D.P. Site 284, southwest Pacific. *Paleobiology*, **7**, 230–240.
Margerel, J. P. 1970. Les foraminifères des marnes à 'Nassa Prismatica' du Boscq d'Aubigny. *Bull. Soc. belge Géol Paléont. Hydrol.*, **79**, 133–156.
Margerel, J. P. 1971. Le genre *Faujasina* d'Orbigny dans le Plio-Pléistocène du Bassin Nordique Européen. *Revue Micropaléont.*, **14**, 113–120.
Margerel, J. P. 1972. Les foraminifères du Néogène de l'Ouest de la France intéret paléoécologique, paléogéographique et stratigraphique. *Bull. Soc. géol. Fr.*, (7), **14**, 121–126.
Marks, P. 1951. A revision of the smaller foraminifera from the Miocene of the Vienna Basin. *Contr. Cushman Fdn., foramin. Res.*, **2**, 33–73.
Millett, F. W. 1886a. Notes on the fossil foraminifera of the St Erth Clay pits. *Trans. R. geol. Soc. Corn.*, **10**, 213–216.
Millett, F. W. 1886b. Additional notes on the foraminifera of St Erth Clay. *Trans. R. geol. Soc. Corn.*, **10**, 221–226.
Millett, F. W., 1895. The foraminifera of the Pliocene beds of St Erth. *Trans. R. geol. Soc. Corn.*, **11**, 655–661.
Millett, F. W. 1897. The foraminifera of the Pliocene beds of St Erth in relation to those of other deposits. *Trans. R. geol. Soc. Corn.*, **12**, 43–46.
Millett, F. W. 1898. Additions to the list of foraminifera from the St Erth Clay. *Trans. R. geol. Soc. Corn.*, **12**, 174–176.
Millett, F. W. 1902. Notes on the *Faujasinae* of the Tertiary beds of St Erth. *Trans. R. geol. Soc. Corn.*, **12**, 719–720.
Mitchell, G. F. 1965. The St. Erth Beds – an alternative explanation. *Proc. Geol. Ass.* **76**, 345–366.
Mitchell, G. F. 1973. The Late Pliocene marine formation of St Erth, Cornwall. *Phil. Trans. R. Soc.*, **B226**, 1–37.
Odrzywolska-Bienkowa, E. 1977. Wybrane profile Miocenu Opolszczyzny w swietle baden mikropaleontologicznych [Selected Miocene profiles in the Opole region in the light of micropalaeontological investigations]. *Przegl. Geol.*, **1**, 12–16 (in Polish).
Orbigny, A. d', 1846. *Foraminifères fossiles du bassin tertiaire de Vienne (Autriche)*. Paris, Gide et Comp.
Parker, F. L. 1967. Late Tertiary biostratigraphy (planktonic foraminifera) of tropical Atlantic deep-sea sections. *Revista Esp. Micropaleont.*, **5**, 253–289.
Pflum, C. E. and Frerichs, W. E. 1976. Gulf of Mexico deep-water foraminifera. *Spec. Publs. Cushman Fdn.*, **14**.
Poignant, A. and Pujol, C. 1976. Nouvelles données

micropaléontologiques (foraminifères planctoniques et petits foraminifères benthiques) sur le stratotype de l'Aquitanien. *Géobios*, **9**, 607–663.

Poignant, A. and Pujol, C. 1978. Nouvelles données micropaléontologiques (foraminifères planctoniques et petits foraminifères benthiques) sur le stratotype bordelais du Burdigalien. *Géobios*, **11**, 655–712.

Poore, R. Z., 1978. Oligocene through Quaternary planktonic foraminiferal biostratigraphy of the North Atlantic: D.S.D.P. Leg 49. *Init. Repts. Deep Sea Drilling Project*, **49**, 447–517.

Popescu, G. 1975. Etudes des foraminifères due Miocène inférieur et moyen du nord-ouest de la Transylvanie. *Mem. Inst. Geol. Bucarest*, **23**.

Prestwich, J. 1871. On the structure of the Crag-beds of Suffolk and Norfolk with some observatons on their organic remains. I. The Coralline Crag of Suffolk. *Q. Jl. geol. Soc. Lond.*, **27**, 115–146.

Pujol, C. 1970. Contribution à l'étude des Foraminiferes planktoniques néogènes dans le Bassin Aquitaine. *Bull. Inst. Géol. Bassin Aquitaine*, **9**, 201–219.

Reuss, A. E. 1851. Uber die fossilen Foraminiferen und Entomostraceen der Septarienthone der Umgegend von Berlin. *Z. dt. geol. Ges.*, **3**, 49–92.

Reuss, A. E. 1856. Beiträge zur Charakteristik der Tertiärschichten des nördlichen und mittleren Deutschlands. *Sber. Akad. Wiss. Wien.*, **18**, 197–273.

Reuss, A. E. 1861. Beiträge zur Kenntnis der teriären Foraminiferenfauna. *Sber. Akad. Wiss. Wien.*, **42**, 355–370.

Reuss, A. E. 1863a. Les foraminifères du Crag d'Anvers. *Bull. Acad. r. Sci. Belg., Cl. Sci.* (2), **15**, 137–162.

Reuss, A. E. 1863b. Beiträge zur Kenntnis der tertiären Foraminiferen-Fauna. *Sber. Akad. Wiss. Wien.*, **48**, 36–71.

Saito, T., Thompson, R. P. and Berger, D. 1981. *Systematic index of Recent and Pleistocene planktonic foraminifera*. 190 pp. University of Tokyo Press.

Spiegler, D. 1973. Die Entwicklung von *Ehrenbergina* (Foram.) im höheren Tertiär NW-Deutschlands. *Geol. Jb.*, **A6**, 3–23.

Spiegler, D. 1974. Biostratigraphie des Tertiärs zwischen Elbe und Weser/Aller (Bentische Foraminiferen, Oligo-Miozän). (Das Nordwestdeutsche Tertiärbecken) Beitrag Nr. 2. *Geol. Jb.*, **A16**, 27–69.

Staesche, K. and Hiltermann, H. 1940. Mikrofaunen aus dem Tertiär Nordwestdeutschlands. *Abh. Reichstelle Bodenforsche.*, NF 201, 1–26.

Ten Dam, A. and Reinhold, T. 1941a. The genus *Darbyella* and its species. *Geologie Mijnb.*, **3**, 108–111.

Ten Dam, A. and Reinhold, T. 1941b. Nonionidae as Tertiary index foraminifera. *Geologie Mijnb.*, **3**, 209–212.

Ten Dam, A. and Reinhold, T. 1941c. Asterigerinen als index-Foraminiferen fuer das Nordwest-Europaeische Tertiaer. *Geologie Mijnb.*, **3**, 220–223.

Ten Dam, A. and Reinhold, T. 1941d. Die Stratigraphische Gliederung des Niederländischen Plio-Plistozäns nach Foraminiferen. *Meded. Geol. Sticht.* **C-V-1**.

Ten Dam, A. and Reinhold, T. 1942. Die Stratigraphische Gliederung des Neiderländischen Oligo-Miozäns nach Foraminiferen (mit ausnahme von S. Limburg). *Meded. Geol. Sticht.* **C-V-2**.

Toering, K. and Van Voorthuysen, J. H. 1973. Some notes about a comparison between the Lower Pliocene foraminiferal faunae of the south-western and north-eastern parts of the North Sea basin. *Revue Micropaléont.*, **16**, 50–58.

Vail, P. R. 1977. Seismic stratigraphy and global changes of sea level. In Seismic stratigraphy: applications to hydrocarbon exploration. *Mem. American association of Petroleum Geologists*, Tulsa, **26**, 49–212.

Van Voorthuysen, J. H. 1958. Les foraminifères Mio-Pliocenes et Quaternaires due Kruisschans. *Mém. Inst. r. Sci. nat. Belg.*, **142**.

Warrington, G. and Owens, B. (Compilers), 1977. Micropalaeontological biostratigraphy of offshore samples from south-west Britain. *Rep. Inst. Geol. Sci.*, **77/7**.

Weaver, P. P. E. 1986. Late Miocene to Recent planktonic foraminifers from the North Atlantic: Deep Sea Drilling Project Leg 94. *Init. Reps. Deep Sea Drilling Project*, **94**, 703–727.

Weaver, P. P. E. and Clement, B. M. 1986. Synchroneity of Pliocene planktonic foraminiferal datums in the North Atlantic. *Marine Micropal.* **10**, 295–307.

Wilkinson, I. P. 1980. Coralline Crag ostracods and their environmental and stratigraphic significance. *Proc. Geol. Ass.*, **91**, 291–306.

Ziegler, P. A. 1982. *Geological Atlas of Western and Central Europe*. Shell International Petroleum Maatschappij B.V.

12

Quaternary

B. M. Funnell

12.1 INTRODUCTION

The earliest systematic descriptions of British Quaternary foraminifera were made by Jones (1865), and Jones et al. (1866–1897). The first of these references listed foraminifera from the Hoxnian Nar Valley Clay, whilst the multi-authored Palaeontographical Society Monograph on the 'Crag Foraminifera' eventually illustrated many species from the Pliocene and early Pleistocene deposits of East Anglia. It was commenced before, and completed after, the voyage of H.M.S. Challenger and H. B. Brady's monograph of the Challenger collections of Recent foraminifera.

There followed a lapse of many years before any further studies of British Quaternary foraminifera were made. Macfadyen (1932, 1933, 1942, 1952) then returned to the subject, publishing a number of papers, firstly on miscellaneous Crag foraminifera, and subsequently on Post-Glacial (i.e. Flandrian) assemblages. His papers post-date the early work of Cushman on worldwide collections of Recent foraminifera, but still rely heavily on the taxonomy of late nineteenth and early twentieth century U.K. investigators of Recent species. In 1941 Ten Dam and Reinhold published the first account of early Quaternary foraminifera from the Netherlands using Cushman's classification and nomenclature. Van Voorthuysen (1949, 1955) continued and developed Ten Dam and Reinhold's work in the Netherlands, introducing extensive use of quantitative methods, until his retirement in 1972.

12.2 LOCATION OF IMPORTANT COLLECTIONS

Some slides of materials examined by Jones et al. (1866–1897) are still preserved at the British Museum (Natural History) and the Geological Museum, South Kensington, but are of historical interest only. Hypotypes of species investigated by Funnell (unpublished thesis) are lodged at the Sedgwick Museum, Cambridge.

12.3 STRATIGRAPHIC DIVISIONS

The Quaternary period in the British Isles has been divided into stages based on pollen assemblage biozones. Fifteen stages have so far been recognised in the Pleistocene epoch, whilst the Holocene epoch comprises one stage only. In general the stages represent alternating cool and temperate conditions in the early Quaternary, and alternating glacial and interglacial conditions in the late Quaternary. The earliest evidence for periglacial ground-ice conditions (ice wedges) is found in the Baventian Stage. The earliest evidence of glacial deposits is found in the Anglian Stage.

Since 1952 the base of the Quaternary in the British Isles has been taken at the base of the Red Crag Formation, and this practice is continued in this account. Thus defined it probably corresponds to the base of the Reuverian 'C' (pollen) stage of the Netherlands succession. The immediately following cool Praetiglian (pollen) Stage of the Netherlands sequence, in which the Pacific foraminiferal species *Elphidium oregonense* appears, is not recognisable in the British Isles, where *E. oregonense* has not so far been found. The Praetiglian, now commonly taken as the base of the Quaternary in the Netherlands, seems likely to be equivalent in age to, or to slightly post-date, the Butley Crag towards the top of the Red Crag member of the Red Crag Formation.

The occurrence of *Neogloboquadrina atlantica* in at least the earlier, normally magnetised part of the Red Crag Formation (Funnell 1987) strongly indicates that these deposits pre-date both the 2.3 Ma extinction level of *N. atlantica* in the North Atlantic and the end of the Gauss period of normal polarity at 2.4 Ma. By the Netherlands convention of setting the Praetiglian as the base of the Quaternary at least the majority of the Red Crag Formation would be confirmed as Pliocene. The next normally magnetised foraminifera-containing marine sediments in the British Isles, (which might equate with the Olduvai event 1.91–1.76 Ma), are those of the Baventian Stage. By the international convention, which sets the base of the Quaternary at the top of the Olduvai event all, or almost all, of the Norwich Crag Formation would also revert to the Pliocene, leaving the Quaternary foraminifera of the British Isles (thus restricted) essentially identical in species composition with those of Recent North Sea and North Atlantic assemblages. The upper limit of the Quaternary extends to the present day. Table 12.1 shows the full sequence of Quaternary stages so far recognised in the British Isles, together with a selected list of some geological formations to which they correspond, and an indication of references to Quaternary foraminifera of the relevant age.

12.4 BRIEF SUMMARY OF FAUNAL ASSOCIATIONS AND FACIES

Most Quaternary foraminiferal associations in the British Isles are of intertidal or inner sublittoral facies. Cool and warmer water associations tend to occur repetitively in successive cool and warmer stages. Deeper water associations occur in continental shelf, especially North Sea, boreholes (Gregory and Bridge, 1979; Gregory and Harland, 1978; Harland, *et al.*, 1978; Hughes *et al.*, 1977; King, 1983, Chapter 9, this volume), and in the Netherlands. Cold, late Quaternary deeper water associations are found above sea-level as a result of Post-Glacial isostatic uplift.

Climatic and water depth influences exert a dominating control on British Quaternary foraminiferal assemblages. Appearances of new species by evolution or immigration, and disappearances by extinction or emigration are subordinate. Therefore true stratigraphic index species are few in number and most fine stratigraphic discrimination necessarily depends on other criteria such as pollen analysis or ^{14}C dating. Palaeoenvironmental

Table 12.1 Stratigraphical subdivisions of the Quaternary: (f = freshwater, m = marine)

Epochs	Stages	Formations	Members	References
Holocene	Flandrian			Adams and Haynes, 1965; Coles and Funnell, 1981; Culver and Banner, 1978; Haynes *et al.*, 1977; Lees, 1975; Macfadyen, 1933, 1942, 1952, 1955; Murray and Hawkins, 1976; Peacock *et al.*, 1978
Pleistocene	Devensian		(glacial)	Haynes *et al.*, 1977; Jansen *et al.*, 1979; Lord 1980; Peacock *et al.*, 1977, 1978.
	Ipswichian			
	Wolstonian		(glacial)	
	Hoxnian			Fisher *et al.*, 1969; Jones, 1865; Lord and Robinson, 1978; van Voorthuysen, 1955; West *et al.*, 1984
	Anglian		(glacial)	Macfadyen, 1932; Bristow and Gregory, 1983
	Cromerian	Cromer	Bacton (f) Mundesley (m) West Runton (f)	West *et al.*, 1984
	Beestonian Pastonian Pre-Pastonian	Forest Bed	Runton (f/m) Paston (m) Sheringham (f) Sidestrand	
	Bramertonian Baventian Antian Thurnian	Norwich Crag	Chillesford Norwich Crag	Funnell, 1961, 1980, 1983a, b, 1987; Funnell and West, 1962, 1977; Funnell *et al.*, 1979; King, 1983; West *et al.*, 1980; Beck *et al.*, 1972; Funnell, 1961, 1983b, 1987; Funnell and Booth, 1983
	Ludhamian		Ludham Crag	
	Pre-Ludhamian	Red Crag	Red Crag	Funnell and West, 1977; King, 1983

interpretation however, based exclusively on living species, is well developed.

Characteristic faunal associations include:

(a) Polymorphinid–textulariid-containing assemblages, reminiscent of the late Neogene (Pliocene) faunas of the Coralline Crag, characterise the Red Crag Formation (Red and Ludham Crag members). These contain strong Lusitanian faunal elements, appear to indicate water depths of about 50 m, and are typically deposited in sand-wave sedimented deposits. As well as polymorphinid and textulariid species, a variety of cibicidid and rosalinid species are usually present, and larger and more heavily calcified species of *Elphidium*; the species *Pararotalia serrata* is also typical.

(b) *Elphidiella*-dominated associations are typical of all the post-Ludhamian/pre-Anglian stages. Temperate (interglacial) stages contain higher percentages of *Ammonia beccarii*, which in particular dominates the 500–250 μm fraction of the intertidal assemblages of the Chillesford Crag. Cool (glacial) stages contain the highest percentages of *Elphidiella*, together with significant percentages of such cold-water species of *Elphidium* as *E. frigidum* and *Protelphidium orbiculare*. Typically *A. beccarii* is almost or totally unrepresented in the cold stages.

(c) Post-Anglian associations are very similar to modern assemblages from comparable environments. Textulariid genera, such as *Trochammina* and *Jadammina*, dominate high intertidal salt-marsh assemblages, *Protelphidium* and *Elphidium* species, together in temperate (interglacial) periods with *A. beccarii*, are typical of lower intertidal tidal flats. The pattern is clear in assemblages from the Hoxnian, Ipswichian and Flandrian temperate (interglacial) stages. Deeper water associations, usually, but not always of Devensian (Late-Glacial) age, are dominated by or contain high proportions of *Elphidium clavatum*, *Bulimina* spp., *Cassidulina* spp. and even *Uvigerina* spp.

12.5 INDEX SPECIES

SUBORDER ROTALIINA

Ammonia beccarii (Linné)
Plate 12.1, Figs 1–3 (× 50), = *Nautilus beccarii* Linné, 1758. Description: 9–11 chambers in final whorl; rounded periphery; margins of umbilical radial sutures thickened and crenulate; often an umbilical boss.
Remarks: confined to temperate (interglacial) stages in the North Sea Basin.

Cibicides lobatulus (Walker and Jacob)
Plate 12.1, Figs 4–6 (× 60), = *Nautilus lobatulus* Walker and Jacob, 1798. Description: 6–9 chambers in final whorl; periphery usually acute, margin usually lobate; spiral side flat, slightly convex or irregular, umbilical side convex.
Remarks: less common in cold (glacial) stages in the North Sea basin, but the only species of *Cibicides* which is tolerant of arctic conditions.

Cibicides subhaidingerii Parr, 1950
Plate 12.1, Figs 7–9 (× 70). Description: circa nine chambers in final whorl; periphery rounded, margin smooth; spiral side slightly convex, umbilical side strongly convex, coarsely perforate.
Remarks: although this species is now apparently restricted to the Pacific, it occurs in temperate (interglacial) stages in the early Pleistocene of the North Sea basin.

Elphidiella hannai (Cushman and Grant)
Plate 12.1, Figs 13–14 (× 65), = *Elphidium hannai* Cushman and Grant, 1927. Description: 9–14 chambers in final whorl; periphery rounded, margin smooth sometimes becoming slightly lobate; planispiral biconvex; bifurcating canal system opens as pairs of pores along radial sutures.
Remarks: the most common and ubiquitous species in littoral and inner sublittoral environments of the early Pleistocene of the North Sea basin. Not known from the British Isles area after the Anglian stage, it is confined to the northern Pacific at the present day.

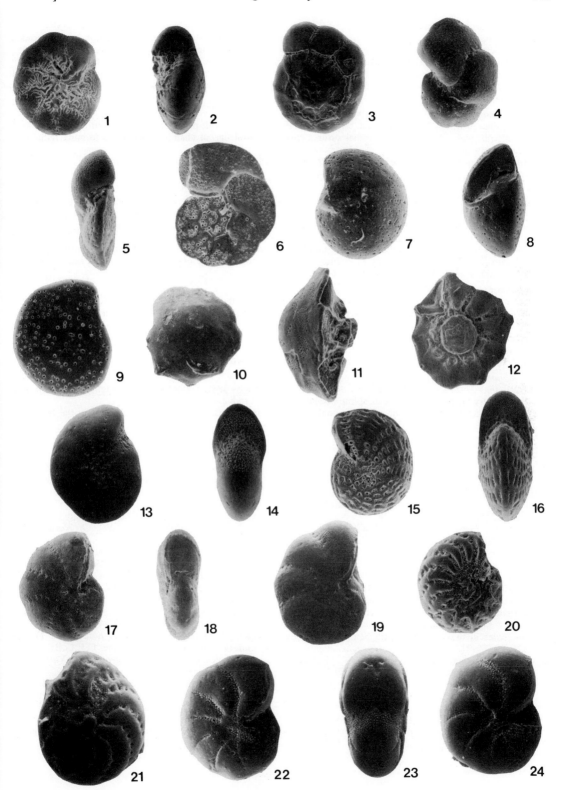

Plate 12.1

Elphidium frigidum Cushman, 1933
Plate 12.1, Figs 17–19 (× 70). Description: about 8–9 chambers in final whorl; rounded periphery, margin usually becoming increasingly lobate; planispiral, flattened umbilical areas often finely granular with granulations extending along radial sutures; indistinct retral processes and single pore openings from canal system along sutures.
Remarks: consistently present in early Pleistocene as small percentages, from Butley Crag onwards; more abundant in cold (glacial) stages. Present-day distribution Arctic, and, if *E. subarcticum* is included in synonymy, also east coast of North America.

Elphidium haagensis van Voorthuysen, 1949
Plate 12.1, Figs 20–21 (× 75). Description: circa 12 chambers in final whorl; periphery rounded, but tendency to keel, margin smooth; planispiral, biconvex; radial sutures curved, central line of chambers thickened forming radial ribs; umbilical boss or bosses.
Remarks: not known later than the Antian stage of the early Pleistocene of the North Sea basin.

Elphidium pseudolessonii Ten Dam and Reinhold, 1941
Plate 12.1, Figs 15–16 (× 70). Description: circa 18 chambers in the final whorl; periphery acute, margin smooth occasionally becoming slightly lobate; planispiral, biconvex; retral processes strong and numerous, usually extending across full width of chambers, umbilical tubercles.
Remarks: Similar to modern *E. crispum/macellum*, it differs in having less well-developed umbilical bosses, less strong and consistent development of a peripheral keel, and more variable development of retral processes, with anterior edges of chambers sometimes slightly inflated. Occurs commonly in the early Pleistocene from the Butley Crag onwards. At first accompanied by *E. crispum*, but from Thurnian stage onwards usually alone; last known from the Pre-Pastonian stage, but may possibly occur later.

Pararotalia serrata (Ten Dam and Reinhold)
Plate 12.1, Figs 10–12 (× 70). = *Rotalia serrata* Ten Dam and Reinhold, 1941. Description: circa nine chambers in the final whorl; periphery acute with margin showing tendency to, but not actual development of radial spines; biconvex with single strong umbilical boss on umbilical side.
Remarks: representative of a much larger number of Lusitanian, Pliocene-relict species found in the Polymorphinid–textulariid-containing assemblages of the Red Crag Formation. Resembles the modern Mediterranean *Rotalia calcar* but in *P. serrata* the spines are never more than incipiently developed.

Protelphidium orbiculare (H. B. Brady)
Plate 12.1, Figs 22–24 (× 75). = *Nonionina orbicularis* H. B. Brady, 1881. Description: circa eight chambers in final whorl; periphery broadly rounded; planispiral, strongly biconvex; radial sutures incised.
Remarks: particularly characteristic of transitional temperate (interglacial)/cold (glacial) conditions in the North Sea basin. Present-day distribution Arctic and colder temperate waters.

REFERENCES

Adams, T. D. and Haynes, J. 1965. Foraminifera in Holocene Marsh cycles, at Borth, Cardiganshire (Wales). *Palaeontology*, **8**, 27–38.

Beck, R. B., Funnell, B. M. and Lord, A. R. 1972. Correlation of Lower Pleistocene Crag at depth in Suffolk. *Geol. Mag.*, **109**, 137–139.

Bristow, C. R. and Gregory, D. M. 1983. Notes on the high-level (?marine), Late-Anglian Woolpit Beds, Suffolk. *Trans. Suffolk Nat. Soc.*, **18**, 310–317.

Cameron, T. D. J., Bonny, A. P., Gregory, D. M. and Harland, R. 1984. Lower Pleistocene dinoflagellate cyst, foraminiferal and pollen assemblages in four boreholes in the southern North Sea. *Geol. Mag.*, **121**, pp. 85–97.

Coles, B. P. L. and Funnell, B. M. 1981. Holocene palaeoenvironments of Broadland, England. In *Holocene marine sedimentation in the North Sea Basin*. Internat. Assoc. Sediment. Spec. Publ. No. 5, 123–131.

Culver, S. J. and Banner, F. T. 1978. Foraminiferal assemblages as Flandrian palaeoenvironmental indicators. *Palaeogeog., palaeoclimatol., Palaeoecol.*, **24**, 53–72.

Fisher, M. J., Funnell, B. M. and West, R. G. 1969. Foraminifera and pollen from a marine interglacial deposit in the Western North Sea. *Proc. Yorks. geol. Soc.*, **37**, 311–320.

Funnell, B. M. 1961. The Palaeogene and Early Pleistocene of Norfolk. *Trans. Norfolk Norwich Nat. Soc.*, **19**, 340–364.

Funnell, B. M. 1980. Palaeoenvironmental analysis of the Dobb's Plantation Section, Crostwick (and comparison with type localities of the Norwich and Weybourne Crag). *Bull. geol. Soc. Norfolk*, **31**, 1–10.

Funnell, B. M. 1983a. The Crag at Bulcamp, Suffolk. *Bull geol. Soc. Norfolk*, **33**, 33–44.

Funnell, B. M. 1983b. Preliminary note on the Foraminifera and stratigraphy of the C.E.G.B. Sizewell boreholes L and S. *Bull. geol. Soc. Norfolk*, **33**, 54–62.

Funnel, B. M. 1987. Late Pliocene and Early Pleistocene stages of East Anglia and the adjacent North Sea. *Quaternary Newsletter*, **52**, 1–11.

Funnell, B. M. and Booth, S. K. 1983. Debenham and Stadbroke, two Crag boreholes in Suffolk compared. *Bull. geol. Soc. Norfolk*, **33**, 45–53.

Funnell, B. M. and West, R. G. 1962. The Early

Pleistocene of Easton Bavents, Suffolk. *Q. Jl. Geol. Soc. Lond.*, **117**, 125–141.

Funnell, B. M. and West, R. G. 1977. Preglacial Pleistocene deposits of East Anglia. In Shotton, F. W. (ed.) *British Quaternary Studies — Recent Advances*. Oxford University Press, pp. 247–265.

Funnell, B. M., Norton, P. E. P. and West, R. G. 1979. The crag at Bramerton, near Norwich, Norfolk. *Phil. Trans. Roy. Soc.*, **B287**, 489–534.

Gregory, D. M. and Bridge, V. A. 1979. On the Quaternary foraminiferal species *Elphidium ustulatum*, Todd, 1957: its stratigraphic and palaeocological implications. *J. Foramin. Res.*, **9**, 70–75.

Gregory, D. M. and Harland, R. 1978. The late Quaternary climatostratigraphy of IGS Borehole SLN 75/33 and its application to the palaeoceanography of the north-central North Sea. *Scott. J. Geol.*, **14**, 147–155.

Harland, R., Gregory, D. M. Hughes, M. J. and Wilkinson, I. P. 1978. A late Quaternary bioclimatostratigraphy for marine sediments in the north central part of the North Sea. *Boreas*, **7**, 91–96.

Haynes, J. R., Kitely, R. J., Whatley, R. C. and Wilks, P. J. 1977. Microfaunas, microfloras and the environmental stratigraphy of the Late Glacial and Holocene in Cardigan Bay. *Geol. J.*, **12**, 129–158.

Hughes, M. J., Gregory, D. M., Harland, R. and Wilkinson, I. P. 1977. Late Quaternary foraminifera and dinoflagellate cysts from boreholes in the UK sector of the North Sea between 56° and 58° N. In Holmes, R. Quaternary deposits of the central North Sea, 5. *Rep. Inst. Geol. Sci.*, **77/14**, 36–46.

Jansen, J. H. F., Doppert, J. W. C., Hoogendoorn-Toering, K., Jong, D. E. J. and Spaink, G. 1979. Late Pleistocene and Holocene deposits in the Witch and Fladen Ground area, northern North Sea. *Neth. J. Sea Res.*, **13**, 1–39.

Jones, T. R. 1865. Microzoa of the deposits of the Nar Valley. *Geol. Mag.*, **2**, 306–307.

Jones, R. T., Parker, W. K., Brady, H. B., Burrows and Holland, 1866–1897. A Monograph of the Foraminifera of the Crag. *Mon. Palaeontogr. Soc.*

King, C. 1983. Cainozoic micropalaeontological biostratigraphy of the North Sea. *Rep. Inst. Geol. Sci.*, **83/7**, 1–40.

Lees, B. J. 1975. Foraminiferida from Holocene sediments in Start Bay, *Jl. Geol. Soc. Lond.*, **131**, 37–49.

Lord, A. R. 1980. Interpretation of the Late glacial marine environment of North-West Europe by means of Foraminifera. In Lowe, J. J., Gray, J. M. and Robinson, E. (eds) *The Lateglacial environment of the British Isles and possible correlations with North-West Europe*, 103–114. Pergamon, Oxford.

Lord, A. R. and Robinson, J. E. 1978. Marine Ostracoda from the Quaternary Nar Valley Clay, west Norfolk. *Bull. geol. Soc. Norfolk*, **30**, 113–118.

Macfadyen, W. A. 1932. Foraminifera from some Late Pliocene and Glacial Deposits of East Anglia. *Geol. Mag.*, **69**, 481–497.

Macfadyen, W. A. 1933. The Foraminifera of the Fenland Clays at St. Germans, near King's Lynn. *Geol. Mag.*, **70**, 182–191.

Macfadyen, W. A. 1942. A post-glacial microfauna from Swansea Docks. *Geol. Mag.*, **79**, 133–146.

Macfadyen, W. A. 1952. Foraminifera and other microfauna from Post-glacial clays from the middle Bure valley, Norfolk. *Roy. Geogr. Soc. Res. Mem.*, **2**, 59–62.

Macfadyen, W. A. 1955. Appendices 1 and 2, Foraminifera. In Godwin, H. Studies of the post-glacial history of British vegetation XIII. The Meare Pool region of the Somerset Levels. *Phil. Trans. R. Soc.*, **B239**, 185–190.

Mitchell, G. F., Penny, L. F., Shotton, F. W. and West, R. G. 1973. A correlation of Quaternary deposits in the British Isles. *Geol. Soc., Lond., Special Report*, **4**, 1–99.

Murray J. W. and Hawkins, A. B. 1976. Sediment transport in the Severn Estuary during the past 8000–9000 years. *Jl. Geol. Soc. Lond.*, **132**, 385–398.

Peacock, J. D., Graham, D. K., Robinson, J. E. and Wilkinson, I. P. 1977. Evolution and chronology of late glacial marine environments at Lochgilphead, Scotland. In Gray, J. M. and Lowe, J. J. (eds) *Studies in the Scottish Lateglacial Environment*, pp. 89–100, Pergamon, Oxford.

Peacock, J. D., Graham, D. K. and Wilkinson, I. P. 1978. Late-Glacial and post-Glacial marine environments at Ardyne, Scotland, and their significance in the interpretation of history of the Clyde Sea Area. *Rep. Inst. Geol. Sci.*, **78/77**, 1–25.

Ten Dam and Reinhold, T. 1941. Die stratigraphische Gliederung des Neiderlandischen Plio-Pleistozäns nach Foraminiferen. *Meded. geol. Sticht. C-V*, **1**, 1–66.

Van Voorthuysen, J. H. 1949. Foraminifera of the Icenian of the Netherlands. *Verh. geol. mijnb. Genoot. Ned. Kolon. geol. ser.*, **15**, 63–68.

Van Voorthuysen, J. H. 1955. In Baden-Powell, D. F. W. Report on the marine fauna of the Clacton Channels. *Q. Jl. geol. Soc. Lond.*, **111**, 301–305.

West, R. G., Funnell, B. M. and Norton, P. E. P. 1980. An Early Pleistocene cold marine episode in the North Sea: pollen and faunal assemblages at Covehithe, Suffolk, England. *Boreas*, **9**, 1–10.

West, R. G., Devoy, R. J. V., Funnell, B. M. and Robinson, J. E. 1984. Pleistocene deposits at Earnley, Bracklesham Bay, Sussex. *Phil. Trans. Roy. Soc.*, **B306**, 137–157.

13

An outline of faunal changes through the Phanerozoic

J. W. Murray

Within the small area of the British Isles there is a fairly complete Phanerozoic succession, much of which is represented by fossiliferous marine strata. Rocks older than Carboniferous have so far yielded only sparse foraminiferal faunas but diverse assemblages are known from the Carboniferous, Jurassic, Cretaceous, Palaeogene, Neogene and Quaternary. The succession of faunas is described below. However, although the changes observed are partly due to evolution, they are also the consequence of changing environments. The assemblages described all come from shelf seas, or marginal marine lagoons and estuaries.

Our knowledge of pre-Carboniferous faunas is still patchy. The oldest Cambrian form is *Platysolenites*, a flattened siliceous agglutinated tube, which occurs in the Lower Cambrian of Wales, the East European Platform and Baltic Shield, Newfoundland and California. Primitive, agglutinated, long-ranging taxa are characteristic of the Ordovician but the calcareous ?parathuramminacean *Saccamminopsis* is recorded from the Middle and Upper Ordovician of Scotland and North Germany. Simple agglutinated textulariids of the superfamilies Ammodiscacea, Hyperamminacea, Astrorhizacea and Saccamminacea are present in the Silurian. The Devonian record is mainly from carbonate sediments and a fairly diverse fauna of early fusulininids has now been recorded (Chapter 2).

The shallow shelf seas of the Carboniferous provided habitats favourable for benthic foraminifera and the faunas are rich and diverse. The majority of genera are of the suborder Fusulinina, especially superfamily Endothyracea, but there are also representatives of the superfamilies Parathuramminacea (e.g. *Saccamminopsis*) and Fusulinacea (e.g. *Millerella*). The suborder Miliolina makes its first appearance with *Eosigmoilina* and *Nodosigmoilina*. All these forms are confined to limestones or calcareous mudstones (Chapter 3) believed to have been deposited in water of near normal marine salinity. They are absent from rocks believed to have been deposited in hypersaline environments. In the late Carboniferous simple agglutinated forms are found in the marine bands.

In much of Britain deposits of Permian age are of continental origin but in northern England normal marine and very slightly hypersaline deposits have yielded foraminiferal faunas (Chapter 4). There are simple tubular Miliolina: *Agathammina*, *Orthovertella* and *Calcitornella*; simple Textulariina: *Ammobaculites*, *Ammodiscus*, *Hyperammina*; and the Lagenina are represented by the nodosariaceans *Dentalina*, *Geinitzina*, *Nodosaria* and *Frondicularia*. The Miliolina alone occur in the slightly hypersaline and patch reef deposits.

The marine assemblages include representatives of all the groups listed above. These Permian assemblages are of very low diversity as compared with the Carboniferous and none of the Carboniferous genera are present in them. However, the Permian assemblages do include early representatives of the Nodosariacea which were to become so important during the early part of the Mesozoic.

Most of the British Triassic succession is non-marine. However, in the Late Triassic the first marine incursion brought with it an essentially Jurassic microfauna quite different from that observed in the Permian. The earliest occurrences are of *Glomospira*, *Eoguttulina* and *Dentalina*. Exclusively agglutinated assemblages composed of *Glomospirella*, *Ammodiscus*, *Trochammina*, *Reophax* and *Bathysiphon* are thought to represent dysaerobic marine conditions (Chaper 5). Shallow water limestones are characterised by *Eoguttulina* with rare nodosariacean taxa. Near the Triassic/Jurassic boundary the nodosariacean *Lingulina* makes its first appearance.

Jurassic sediments range from moderately deep basinal deposits to shallow shelf sands, silts and carbonates to marginal marine and continental deposits. At the present time sampling for foraminifera has been concentrated on the clays. Some of the sands have been looked at but few of the carbonates have received attention. So our knowledge of Jurassic faunas is not truly representative.

The early Jurassic transgression which drowned much of the British Isles brought with it a fauna which diversified into the newly available niches (Chapter 6). The common genera are, Lituolacea: *Textularia*, *Haplophragmoides*, *Ammobaculites*, *Trochammina* and *Triplasia*. Miliolina: *Ophthalmidium* and *Nubecularia*. Nodosariacea: *Nodosaria*, *Dentalina*, *Frondicularia*, *Lingulina*, *Lenticulina*, *Marginulina*, *Vaginulina*, *Vaginulinopsis*, *Marginulinopsis*, *Saracenaria*, *Citharina*, *Eoguttulina*, and *Tristix*. Spirillinacea: *Spirillina*. Cassidulinacea: *Involutina* and *Paalzowella*. Robertinacea: *Epistomina*, *Conorboides* and *Reinholdella*.

The middle Jurassic shallow water carbonate sediments, unlike their modern counterparts, are not characterised by Miliolina-rich assemblages. Instead they have a nodosariacean fauna composed of *Lenticulina*, *Citharina* and *Nodosaria*.

Representatives of the Nodosariacea occur in assemblages from most environments. An abundance of Miliolacea (*Ophthalmidium*, *Nubecularia*, *Nubeculariella*) and Spirillinacea (*Spirillina*) is thought to indicate well oxygenated shallow, normal marine conditions. Assemblages of small agglutinated genera (*Trochammina*, *Haplophragmoides*, *Ammobaculites*) may be indicative of brackish water and perhaps less oxygenated bottom conditions. Larger Lituolacea (*Ammobaculites coprolithiformis*, *Triplasia* spp.) are found in deeper marine waters. Among the Robertinacea, *Epistomina* is from normal marine conditions but *Reinholdella*, when occurring alone, may indicate waters of low oxygen content (Chapter 6).

All the major faunal changes in the Jurassic can be related to changes in the environments of deposition. In each stage there is an introduction of new species. Even though many of the forms are relatively long-ranging, the faunas of each stage are fairly distinct.

The lower Cretaceous faunas of the Ryazanian to Aptian are essentially Jurassic in character. *Lenticulina muensteri* continues through to the basal Albian. The following genera, which were present in the Jurassic, are found: *Ammobaculites, Textularia, Trochammina, Nodobacularia, Citharina, Saracenaria, Lenticulina, Marginulopsis, Frondicularia, Planularia, Vaginulina, Lingulina, Epistomina,* and *Conorboides.* New arrivals in the Hauterivian include Lituolacea: *Tritaxia,* Miliolina: *Wellmanella* and Nonionacea: *Gavelinella.* During the Barremian and Aptian, many of the Nodosariacea die out and after this time they never again dominate the benthic assemblages.

There are major additions to the fauna in the Albian, Lituolacea: *Arenobulimina, Dorothia, Flourensina.* Nonionacea: *Lingulogavelinella.* These herald the incoming of the Upper Cretaceous fauna.

The other major event recorded in the Lower Cretaceous succession is the appearance, in the Barremian, of the first planktonic foraminifera (*Hedbergella infracretacea*). This is joined in the Albian by *Hedbergella delrioensis, H. planispira, Guembelitria cenomana, H. moremani* and *Globigerinelloides bentonensis* (Chapter 7).

The Upper Cretaceous, typified by chalk deposition, is characterised by diverse benthic assemblages dominated by Textulariina and Rotaliina.

New genera include the following:

Cenomanian: Lituolacea, *Pseudotextulariella*
Turonian: Discorbacea, *Valvulineria,* Nonionacea, *Globorotalites*
Coniacian: Buliminacea, *Eouvigerina,* Nonionacea, *Stensioeina*
Santonian: Orbitoidacea, *Cibicides*
Campanian: Cassidulinacea, *Loxostomum,* Nodosariacea, *Neoflabellina.*

Evolutionary relationships have been proposed for a number of benthic groups including *Bolivinoides* (Barr, 1966), *Arenobulimina* (Carter and Hart, 1977; Price, 1977), *Flourensina* (Carter and Hart, 1977) and *Gavelinella* (Price, 1977).

The late Cretaceous planktonic fauna underwent both diversification and increase in number of individuals. The *Hedbergella* and *Guembelitria* groups continued on from the Albian. In the Cenomanian *Rotalipora* appeared and became extinct, and *Praeglobotruncana* and *Whiteinella* made their appearance. The first *Globotruncana* are found in the Coniacian, *Rugoglobigerina* at the top of the Campanian and *Abathomphalus* in the Maastrichtian. The late Cretaceous chalks are richer in planktonic foraminifer than any other part of the land based stratigraphic succession in Britain.

Restricted and/or deeper water (*Rhabdammina*-biofacies) faunas with non-calcareous agglutinated genera are developed in parts of the North Sea, from Hauterivian to Middle Albian, Turonian to Maastrichtian (Chapter 8) and continue into the Palaeogene (Chapter 9).

At the end of the Cretaceous the whole of Britain became emergent and underwent erosion. The oldest Palaeogene deposits (Thanet Formation) are of late Palaeocene age. The Palaeogene successions are of variable facies and many of them are shallow water shelf, marginal marine or continental. Deeper water (mid/outer shelf) deposits are found in the London Clay of the London basin and in the North Sea (Chapter 9).

The foraminiferal faunas are very different from those of the Cretaceous. Among the new genera are: Buliminacea: *Bulimina, Uvigerina, Turrilina, Praeglobobulimina.* Discorbacea: *Asterigerina, Cancris, Discorbis.* Rotaliacea: *Elphidium, Protelphidium, Pararotalia, Nummulites.* Cassidulinacea: *Caucasina.* Nonionacea: *Alabamina, Anomalinoides.* Miliolina: *Orbitolites, Fasciolites.*

Because of the rapid facies variations many of the extinctions and new appearances are due to environmental causes.

The Palaeogene planktonic faunas are generally sparse. The genera represented include *Globigerina, Globorotalia, Pseudohas-*

tigerina, *Guembelitria* and *Globigerinatheka*. Most of the Cretaceous genera were extinct by the Palaeogene but reworked tests are common at certain Palaeogene levels even in non-marine deposits (Chapter 10).

The Neogene is represented by the Jones Formation of the English Channel, the Crags of East Anglia, and the St. Erth Beds and clays of the Hordaland and Nordland Groups. The benthic assemblages are quite similar to modern day faunas not only at generic level but also at specific level. Extant forms include *Florilus grateloupi*, *Cibicides dutempli*, *C. lobatulus*, *Elphidium crispum*, *Bulimina alazanensis* and *Textularia sagittula*. The planktonic faunas of the Jones Formation are moderately diverse but they are rare in the rest of the deposits (Chapter 11).

The Quaternary assemblages of the warmer intervals are closely similar to the modern ones and are dominated by shallow water forms (Chapter 12). During the cooler intervals the genus *Elphidiella* was common together with cold-water species of *Elphidium* and *Protelphidium*.

The principal changes are listed below (- - - = change in fauna).

Quaternary Neogene	benthic fauna of modern aspect
	- - - - - - - - - - - - - - - - - - -
Palaeogene	common Miliolina, Discorbacea, Buliminacea, Rotaliacea, Cassidulinacea, Nonionacea. Planktonic forms generally rare.
	- - - - - - - - - - - - - - - - - - -
Cretaceous (Albian-Maastrichtian)	common Lituolacea, Nodosariacea, Buliminacea, Discorbacea, Cassidulinacea, Nonionacea. Planktonic forms are common.
	- - - - - - - - - - - - - - - - - - -
(Ryazanian-Aptian)	common Lituolacea, Nodosariacea, Nonionacea. Incoming of planktonic fauna.
Jurassic	common Lituolacea, Nodosariacea, Spirillinacea, Cassidulinacea, Robertinacea. Subordinate Miliolina.
Triassic	start of 'Jurassic' faunas.
	- - - - - - - - - - - - - - - - - - -
Permian	common Ammodiscacea, Lituolacea, Miliolina, Nodosariacea
	- - - - - - - - - - - - - - - - - - -
Carboniferous	common Parathurammi-nacea, Endothyracea, subordinate Fusulinacea.
	- - - - - - - - - - - - - - - - - - -
Pre-Carboniferous	mainly Ammodiscacea.

From this it can be seen that the really major events in the foraminiferal faunas of Britain took place at the end of the Carboniferous, Permian, early Cretaceous, end Cretaceous and end Palaeogene.

REFERENCES

Barr, F. T. 1966. The foraminiferal genus *Bolivinoides* from the Upper Cretaceous of the British Isles, *Palaeontology*, **9**, 220–243.

Carter, D. J. and Hart, M. B. 1977. Aspects of mid-Cretaceous stratigraphical palaeontology. *Bull. Brit. Mus. (Nat. Hist.), Geol Ser.*, **29**, 1–135.

Price, R. J. 1977. The evolutionary interpretation of the Foraminiferida *Arenobulimina*, *Gavelinella* and *Hedbergella* in the Albian of North-West Europe. *Palaeontology*, **20**, 503–527.

General Index

Aalenian, 190, 193
Aalenian-Bajocian, 125
Aalenian to Callovian, 189
Abadehian Stage, 91
Agglutinating foraminiferid zones, NSA12-NSA1, 443–445
albani Zone, 249
Albian, 277, 375
Alpine Rhaetian, 98
Alston Block, 42
Ampthill Clay, 247
Amundsen Formation, 133
Anglian Stage, 564
angulata Zone, 147
Anisian, 99, 100
'anoxic events', 282
Aptian, 277, 375
Arctic, 131
Arnsbergian, 41
Arundian, 39
Asbian, 40
Ashgillian, 22
Assemblage Zone 1 (*macrocephalus-calloviense* Zones), 202
Assemblage Zone 2 (*calloviense-coronatum* Zones), 202
Assemblage Zone 3 (*coronatum-lamberti* Zones), 202
Atherfield Clay, 278

Bajocian, 125, 189, 195
Bakevellia Sea, 87, 88
Balder Formation, 445
Baltic Shield, 22, 273
Bankfield East beds, 38
Barremian, 277
Bathonian, 123, 189, 200
Baventian Stage, 564

baylei Zone, 248
Bearreraig Sandstone Formation, 195
Beatrice Field, 203
Benthonic Zones, NSB17-NSB1, 429–441
bifrons Zone, 154
biofacies, 422
Biozones, 426
Black Rock Limestone, 37
Blue Lias, 143
Boreal, 133
'Boreal' realm, 278
Boulonnais, 250
Brent Group, 125, 190
Brigantian, 41
Bristol area, 39
British Isles, 13
British Triassic subdivisions, 100
Brora, 245
Brora Argillaceous Formation, 203
bucklandi Zone, 147
Burdigalian, 541
Butley Crag, 564
Byfield, 129

Cadeby Formation, 91, 92
Calcareous agglutinant (CA), 423
California, 22
Callovian, 125, 189, 202
Cambrian, 5, 20, 21
Campanian, 277
Canadian Arctic, 450
Carboniferous, 5
Cardigan Basin, 160
Carnian, 98, 100
Caradoc, 22
cautisnigrae Zone, 247

General Index

Celtic Sea, 98, 273
Celtic Sea Basin, 237, 241, 250
Cenomanian, 277
Cenozoic, 6, 418
Central Graben, 155, 251, 378, 420
Chadian, 37
Chalk, 273
Chalk Group, 378, 384
Channel Tunnel Site, 273
Chillesford Crag, 566
Claymore, 241
Cleveland Basin, 160
Clitheroe, 38
Clitheroe Formation, 38
Cloughton Formation, 197
Cockburn Formation, 539
Cockle Pits Borehole, 107
concavum Zone, 199
Concretionary Limestone, 92
Coniacian, 277
Cook and Drake Formations, 154
Corallian, 125, 238
Coralline Crag, 537, 543
Cotham Member, 98, 103
Cotswolds, 126, 189
Courceyan, 37
'Crag Foraminifera', 563
Craven Basin, 37
Cretaceous, 6, 273, 372
Cretaceous Stages, 378
Cromer Knoll Group, 279, 376
Culverhole Point, 107
cymodoce Zone, 248

Deltaic Series, 195
decipiens Zone, 247
Devensian, 566
Devon, 279
Devonian, 5, 20, 23
Dibdale Stream Section, 107
Dinantian, 32
dinoflagellates, 420
discites-laeviuscula Zone, 189, 197
Dorset, 97, 105, 129, 152, 189, 237, 279
Dover, 279
Dublin Basin, 37

Early Cretaceous, 250, 375, 380
Early Eocene, 420
Early Jurassic, 125
Early Miocene, 426, 446
Early Palaeocene (Danian), 420
Early Pliocene, 426
Early Turonian, 278, 375
East Anglia, 563
East European Platform, 22
East Greenland, 450
East Midlands Shelf, 160
Ekofisk Formation, 445
Embsay Limestone, 38
English Channel, 273, 490
Eocene, 419, 447, 490
Eoguttulina liassica Assemblage, 114

European Jurassic, 126
Exogyra Bed, 249

Faeroes Basin, 156
Faeroes Trough, 378, 425, 449, 539
falciferum Zone, 154
Fammenian, 20
Fastnet, 98, 109
Fastnet Basin, 189
Fastnet–Celtic Sea Basins, 125
Faunule A (*parkinsoni-zigzag* Zones), 200
Faunule B (*tenuiplicatus-progracilis* Zones), 200
Faunules B_2–C_3 (*subcontractus*-basal *aspidoides* Zones), 201
Faunule C_4 (*aspidoides* Zone), 201
Faunule D (*discus* Zone), 201
fittoni Zone, 248
Flandrian, 566
Folkestone, 279
Ford Formation, 92
Frasnian, 20
Frome Clay, 189

Gastrioceras cumbriense Marine band, 42
Gault Clay, 275, 278
German Dogger, 195
Glomospira/Glomospirella Assemblage, 114
Gower, 37
Great Estuarine Group, 190
Gully Oolite, 38

Hampshire Basin, 490
Hauterivian, 277, 375
Haw Bank Limestone, 37
Haw Crag Limestone, 38
Headon Formation, 493
Heather Formation, 203, 241
Hebrides Basins, 160
Heligoland, 383
Hettangian, 129, 132
Hettangian-Pliensbachian, 129
Hettangian-Sinemurian, 125
Hetton Beck Limestone, 38
Hobbyhorse Bay, 39
Holkerian, 39
Holocene, 564
Hordaland Group, 445
Hoxnian, 566
Hoxnian Nar Valley Clay, 563
hudlestoni Zone, 247

ibex Zone, 152
Inferior Oolite, 125
Inner Moray Firth, 125, 143, 203
Inner sublittoral biofacies, 422
Ipswichian, 566
Ireland, 39
Isle of Wight, 279, 490

jamesoni Zone, 152
JF 1 Zone, 115, 158
Jones Formation, 539
Jura Mountains, 190

Jurassic, 125
Jurassic-Cretaceous boundary, 278
Jurassic sea level change, 126

Kimmeridge Clay Formation, 202, 250, 379
Kimmeridgian, 125, 237
Kinderscoutian, 42

Labrador Sea, 450
Ladinian, 98, 100
lamberti Zone, 189
Langport Member, 98
Late Cenomanian anoxic event, 278
Late Cimmerian, 278
Late Cretaceous, 375, 378, 380, 420
Late Miocene, 426, 446
Late Neogene, 419
Late Palaeocene, 420
Late Palaeozoic, 32
Late Pliocene, 426
Late Ryazanian, 379
Later Eocene, 419
Lavernock, 106
Lenham Beds, 537
Lias Group, 125, 132
Lias Paper Shale, 103
liascius Zone, 146
Lilstock Formation, 98
Lincolnshire Limestone Formation, 125, 193
Lista Formation, 445
Little Sole Formation, 539
Llandeilo, 22
Llandovery, 22
Llandovery-Wenlock, 20
London Basin, 490
London Clay Formation, 490
Lower Cretaceous, 275
Lower Greensand, 278
Lower Inferior Oolite, 189
Lower Jurassic, 129
Ludlow, 22

Maastrichtian, 277
Magnus field, 203
Manchester Marls, 91, 92
margaritatus Zone, 152
mariae Zone, 251
Maureen Formation, 445
Middle-Lake Eocene, 426
Mediterranean, 131
Mendip High-Radstock Shelf, 160
Mendips, 37
Mendips-Glamorgan, 146
Mercia Mudstone Group, 97, 105
Mesozoic syn-rift, 420
Middle Bathonian-Callovian, 203
Middle Jurassic, 125, 189
Middle Miocene, 426, 446
mid-Toarcian, 125
Millbrook, 241
Milton Keynes Boreholes, 189
Miocene, 537, 541
Mochras Borehole, 130, 153

Moray Firth, 251
Moray Firth Basin, 383
morrisi-aspidoides Zones, 189
Much Wenlock Limestone Formation, 26
murchisonae Zone, 193
Mutterflöz, 92

Namurian, 41
Nannofossil, 420
Neogene, 7, 537
Newfoundland, 22
Non-calcaerous agglutinants (NCA), 381, 423
Norian, 100, 105
Norian-Rhaetian, 98, 99
Normandy, 250
North Celtic Sea Basin, 189
North Atlantic, 420
North Atlantic-Arctic rift, 420
Northern Ireland, 279
North Sea, 130, 153, 241, 372, 490, 564
North Sea Basin, 155, 273, 418, 448, 537
Northumberland, 32
Norwegian Sea, 448
Norwegian sector, 376, 418
Norwich Crag Formation, 564

Olduvai, 564
Oligocene, 447
okusensis Zone, 248
opalinum Zone, 193
opalinum-murchisonae Zones, 189
Ordovician, 22
Outer Moray Firth Basin, 195, 375
Outer sublittoral-epibathyal biofacies, 422
Oxford Clay, 189, 237
Oxfordian, 125
oxygen isotope analysis, 545
oxygen minimum zone, 279
oxynotum Zone, 151

Palaeocene, 448, 490
Palaeoecology, 133, 422, 494, 543
Palaeogene, 6, 490
Paper Shale unit, 107, 139
parkinsoni Zone, 196
pectinatus Zone, 248
Penarth Group, 97, 98, 103
Pendleian, 41
Permian, 5, 87, 90
Permian eastern England, 90
Permian north Germany, 90
Permian Zechstein Sea, 87
Phanerozoic, 7
Pinhay Bay, 107
Piper, 241
Planktonic Zones NSP16-NSP1, 441-443
planorbis Zone, 132
Platt Lane Borehole, 107
Pleistocene, 426, 446, 563
Pliensbachian, 129, 132
Pliocene, 446, 537, 541, 563
Polygnathus asymmetricus Zone, 23
Portland Group, 241

Index of Genera and Species

Portlandian, 238
Portland Sand, 249
Portland Stone, 249
Post-Glacial, 563
Praetiglian (pollen) Stage, 564
Pre-Carboniferous, 20
Pre-*planorbis* Beds, 103, 132

Quaternary, 7, 418, 563

Raisby Formation, 92
raricostatum Zone, 151
Rattray Formation, 195
Ravenstonedale, 38
Reading Formation, 490
Recent foraminifera, 13
Red Chalk, 275
Red Crag, 537
Red Crag Formation, 564
Redonian, 539
Reuverian 'c' (pollen) stage, 564
Rhabdammina biofacies, 422
Rhaetian, 97, 100
Rockall-Faeroes Rift, 420
rotunda Zone, 248
Rupelian transgression, 426
Ryazanian, 277

Santonian, 277
Scapa Sand, 378
scitulus Zone, 247
Scotland, 32, 279
Scottish Rhaetian, 98
Scythian, 100
Scythian-Rhaetian, 98
Sealab Trial Borehole, 540
sea level cycles, 425
Sele Formation, 445
Selsey Formation, 505
semicostatum Zone, 149
Seine Estuary, 250
Shetland Group, 378
Shetland Platform, 420
Silurian, 22
Sinemurian, 132
Skye, 247
Solent Group, 493
South Devon, 23
South Viking Graben, 195, 378
South Wales-North Somerset, 160
South West Approaches Basin, 273, 543
Speeton Clay, 275, 370
Spilsby Sandstone, 376
spinatum Zone, 152
Staffin Shale, 202
Statfjord Field, 130
St. Audrie's Slip, 105, 106, 109
St. Erth Beds, 539, 540
Stone Gill Beds, 38
subfurcatum-garantiana Zones, 199
Svalbard, 450
S.W. Province, 37
Swinden No. 1 borehole, 37

Tarbert and Heather Formations, 189
tenuicostatum Zone, 154
Tethyan, 133, 278
Thanetian, 505
'Thanetian ash marker', 445
Throckley Borehole, 42
Toarcian, 132
Tolcis quarry, 107
Torquay, 23
Triassic, 51, 97
Triassic/Jurassic boundary, 103, 107, 130, 161
'Triml;ey Sands', 537
turneri Zone, 151
Turonian, 277, 282

UKB Zones, 1–22, 308–314
UKP Zones, 1–17, 294–308
Ulster Basin, 160
Upper Albian, 279
Upper Bringewood Formation, 26
Upper Cretaceous, 275
Upper Fuller's Earth Clay, 189
Upper Greensand, 278
Upper Jurassic, 237
Upper Llandeilo, 22
Uppermost Triassic, 125
'Upper Rotliegend', 92

Valanginian, 277
Viking Graben, 151, 202, 375, 420
Viséan, 39
Vlieland Sandstone, 376

Warboys, 241
Watchet, 106, 109
Weald Basin, 193, 490
Wedmore, 97
Welsh Basin, 23
Wenlock, 22
Wessex Basin, 160
Wessex-English Channel basin, 195
Westbury Formation, 97, 98, 103
Western Approaches, 98, 109, 155, 539, 541
West Sole Group, 195
Weymouth, 247
Whitecliff Bay, 505
Wilkesley Borehole, 107
Winterbourne Kingston Borehole, 105
Withycombe Farm Borehole, 21
Worlaby Boreholes, 189

Yeadonian, 42
Yoredale facies, 32
Yorkshire Lias, 129

Zechstein cycles, 89, 91, 92
Zechstein foraminifera, 87, 88
Zechstein rocks, 89
Zones FCN 21-1, 400–408
Zones FCS 231, 386–400

Index of Genera and Species

abbreviata, *Uvigerinella*, 485
Abathomphalus mayaroensis, 322
aberystwythi, *Asterigerina*, 514
aculeata, *Eouvigerina*, 330
acuminata, *Citharina*, 326
acuminata, *Uvigerina*, 482
acuta, *Geinitzina*, 94
acuta, *Hyperammina*, 94
acutidorsatum, *Reticulophragmium*, 458
acutus, *Anomalinoides*, 512
adamsi, *Bolivinopsis*, 508
advena, *Arenobulimina*, 316
advena, *Plectofrondicularia*, 466
aequale, *Haplophragmium*, 318
affine, *Melonis*, 318
Agathammina, 91, 110
    miloides, 94, 110
    pusilla, 94, 110
agglutinans, *Ammobaculites*, 218, 230, 252
aksuatica, *Bulimina*, 468
Alabamina obtusa, 512
    tangentialis, 467
alaskensis, *Lingulina*, 99
alazanensis, *Bulimina*, 550
albiumbilicatum, *Elphidium*, 475
alleni, *Cibicidoides*, 518
Alliatina excentrica, 548
alsatica, *Turrilina*, 482
alta, *Falsogaudryinella*, 410
althoffi, *Triplasia*, 220
Almaena osnabrugensis, 467
Ammarchaediscus spirillinoides, 62
Ammobaculites, 87, 91, 99, 140
    agglutinans, 218, 252
    cf. agglutinans, 230
    coprolithiformis, 218, 252

    eiselei, 92
    cf. eiselei, 116
    fontinensis, 218
    irregulariformis, 408
    reophacoides, 316
    sp. 1, 106, 116
Ammodiscus, 21, 23, 24, 92, 99, 137
    auriculus, 105, 110, 116
    cretaceus, 452
    roessleri, 94
    siliceus, 164
    sp.A, 452
    sp.B, 452
Ammolagena clavata, 452
Ammomarginulina macrospira, 452
Ammonia beccarii, 566
Ammovertella cellensis, 316
Amphitremoida, 22
ampla, *Plectogyranopsis*, 76
amplectens, *Reticulophragmium*, 458
ampulla, *Globulina*, 529
andreaei, *Turrilina*, 528
anglica, *Arenobulimina*, 316
anglica, *Brizalina*, 514
anglica, *Clavulina*, 454, 508
cf. angulata, *Irregularina*, 23, 28
Angulogavelinella bettenstaedti, 412
angulosa, *Trifarina*, 482
angustissima, *Planularia*, 259
angustiumbilicatus, *Gyroidinoides*, 520
'*Annulina*' *metensis*, 116
Anomalinoides
    acutus, 512
    aragonensis, 512
    aff. A. 512
    nobilis, 512

Index of Genera and Species

rubiginosus, 467
antiqua, Bolivina, 467
antiqua, Quinqueloculina, 320
antiqua, Wellmanella, 322
antiquissimus, Platysolenites, 21
antonium, Elphidium, 475
appenninica, Rotalipora, 360
applinae, Nonion, 522
aprica, Whiteinella, 364
aquitanicus, Nummulites, 532
aragonensis, Saracenella, 184
araucana, Rosalina, 528
Archaediscus itinerarius, 81
  karreri, 76
  krestovnikovi, 63
  moelleri, 82
  reditus, 70
  sp., 66
  stilus, 74
  stilus eurus, 78
Archaelagena cf. *borealia*, 23, 28
Archaeochitosa lobosa, 22
Archaeoglobigerina cretacea, 322
Archaespira firmata, 52
  inaequalis, 52
  reitlingerae, 52
arcuatus, Astacolus, 462
Arenobulimina advena, 316
  anglica, 316
  chapmani, 316
  macfadyeni, 316
  sabulosa, 316
'*Arenobulimina*' sp. A, 452
arguta, Vaginulina, 364
Aschemonella, 89
'*Astacolus dorbignyi*', 99
Astacolus arcuatus, 462
  gladius, 462
  platypleura, 514
  schloenbachi, 412
  semireticulata, 170
  speciosus, 170, 210
Asterigerina aberystwythi, 514
  bartoniana, 514
  guerichi, 297
  guerichi guerichi, 467
  guerichi staeschei, 467
  staeschei, 550
Asteroarchaediscus gregorii, 82
  occlusus, 78
  sp., 82
  ?sp., 78
Atelikamara, 21, 23
  incomposita, 24
Astrononion perfossus, 550
atlantica, Neogloboquadrina, 558, 564
audouini, Pararotalia, 524
auriculus, Ammodiscus, 105, 110, 116
auriculus, Cancris, 470
avonensis, Rhodesinella, 64, 70

Baituganella sp., 52
balthica, Hyalinea, 478

baltica, Gavelinella, 334
baltica, Whiteinella, 364
barnardi, Vaginulina, 260
barremiana, Gavelinella, 334
barri, Globigerinatheka, 530
barroisi, Pleurostomella, 356
aff. *barsae, Glomospiranella*, 64
cf. *barsae, Glomospiranella*, 57
bartensteini, Conorotalites, 412
bartletti, Cribroelphidium, 474
bartoniana, Asterigerina, 514
Bathysiphon, 21, 22, 24
  eocenicus, 454
  nodosariaformis, 454
  ? sp. A, 454
Bathysiphon spp., 105
batjesi, Uvigerina, 484, 528
beaumontianus, Cibicides, 326
beccariformis, Stensioeina, 482
beccarii, Ammonia, 566
beierana, Planularia, 228, 259
beisseli, Eponides, 332
bentonensis, Globigerinelloides, 340
Berthelinella involuta, 170
bettenstaedti, Angulogavelinella, 412
bettenstaedti, Textularia, 411
bicarinata, Quinqueloculina, 510
bifurcata, Rhabdammina, 24
bigoti, Nubeculinella, 254
biloba, Draffinia, 64
Biseriella bristolensis, 58
  parava, 74, 82
Bisphaera cf. *elegans*, 23, 28
bituba, Colonammina, 24
Bogushella ziganensis, 71
Bolivina antiqua, 467
  cookei, 467
  decurrens, 322
  dilatata, 467
  hebes, 550
  incrassata, 322
Bolivinoides culverensis, 322
  decoratus, 324
  draco, 324
  laevigatus, 324
  miliaris, 324
  paleocenicus, 324
  peterssoni, 324
  pustulatus, 324
  sidestrandensis, 326
  strigillatus, 326
Bolivinopsis adamsi, 508
cf. *borealia, Archaelagena*, 23, 28
borealis, Lingulina, 99
boueanus, Florilus, 476
bowmani, Endothyra, 56
bradyana, Howchinia, 76, 78
bradyana, Martinottiella, 456
bradyi, Cassidulinoides, 471
bradyi, Globigerina, 554
bradyi, Trifarina, 482
Bradyina cribrostomata, 82
  rotula, 76

brassfieldensis, Stomasphaera, 23, 26
breviscula, Mediocris, 63
brevispira, Turrilina, 482, 528
brielensis, Gavelinella, 334
bristolensis, Biseriella, 58
bristolensis, Hedbergella, 344
brizaeformis, Frondicularia, 172
Brizalina anglica, 514
    cookei, 515
    liasica, 187
    oligocaenica, 515
brizoides, Frondicularia, 99
broedermanni, Globorotalia, 530
Brunsia pseudopulchra, 58
    sp., 64
    spirillinoides, 52
Brunsiarchaediscus sp., 62
Buccella aff. *frigida*, 550
    frigida, 467
    propingua, 515
    tenerrima, 468
budensis, Frondicularia, 463
Bulimina aksuatica, 468
    alazanensis, 550
    dingdenensis, 468
    elongata, 468
    gibba, 468
    marginata, 468
    midwayensis, 470
    sp. A., 470
    thanetensis, 515
    trigonalis, 470
    truncata, 470
bulimoides, Rotaliatina, 480
bulloides, Globotruncana, 342
aff. *buskensis, Pseudoammodiscus*, 78
Buzasina cf. *pacifica*, 454

Calcitornella spp., 91
Cancris auriculus, 470
    sp. A., 470
    subconicus, 470, 516
canningensis, Trochammina, 166, 220, 234
cf. *cannula, Paratikhinella*, 23, 28
cantii, Cibicides, 516
canui, Haplophragmoides, 230, 252
canui, Pararotalia, 480, 524
capitosa, Tritaxia, 411
caracolla, Hoeglundina, 346
carapitana, Cassidulina, 471
carinata, Quinqueloculina, 510
carinatum, Ophthalmidium, 221
carpenteri, Hoeglundina, 348
carseyae, Praebulimina, 356
carteri, Saccammina, 22
Cassidulina carapitana, 471
    laevigata, 471
    pliocarinata, 471
Cassidulinoides bradyi, 471
Cassigerinella chipolensis, 554
cassivelauni, Cibicides, 516
Caucasina coprolithoides, 516
caudata, Pelosina, 411

cava, Psammosphaera, 23, 24
cavernula, Ichtyolaria?, 95
celata, Sigmoilopsis, 462
cellensis, Ammovertella, 316
cenomana, Guembelitria, 344
cenomana, Plectina, 318
cenomanica, Gavelinella, 334
Ceratammina, 23
    cornucopia, 24
Ceratobulimina aff. *contraria*, 551
cernua, Lingulina, 120
chapmani, Arenobulimina, 316
chapmani, Hoeglundina, 348
chapmani, Textularia, 320
chapmani, Verneuilinoides, 411
charoides, Glomospira, 455
chascanona, Globigerina, 530
Chernyshinella glomiformis, 52
chilostoma, Karreriella, 456
Chilostomellina fimbriata, 471
chipolensis, Cassigerinella, 554
Cibicides alleni, 518
    beaumontianus, 326
    cantii, 516
    cassivelauni, 516
    (*Cibicidina*) *cunobelini*, 516
    (*Cibicidina*) *mariae*, 516
    (*Cibicidina*) *succedens*, 518
    grossus, 471
    lobatulus, 516, 566
    tenellus, 518
    ribbingi, 326
    cf. *simplex*, 516
    subhaidingerii, 566
    ungerianus, 551
    cf. *ungerianus*, 518
    westi, 518
Cibicidoides alleni, 518
    dayi, 471
    dutemplei peelensis, 472
    eocaenus, 472
    limbatosuturalis, 472
    mexicanus, 472
    pachyderma, 472
    pygmeus, 518
    succedens, 474
    truncanus, 474
    velascoensis, 474
    (?) *voltziana*, 326
ciperoensis angulisuturalis, Globigerina, 554
ciperoensis ciperoensis, Globigerina, 554
ciryi inflata, Lingulogavelinella, 414
Citharina acuminata, 326
    clathrata, 210
    clathrata eypensa, 229
    cf. *C discors*, 326
    colliezi, 170, 210
    flabellata, 222
    harpa, 328
    pseudostriatula, 328
    serratocostata, 256
    sparsicostata, 328
cf. *C. discors, Citharina*, 326

Index of Genera and Species

Citharinella laffittei, 328
    moelleri, 222
    nikitini, 222
    pinnaeformis, 328
    sp. A., 222
clathrata, Citharina, 210
clathrata eypensa, Citharina, 229
clathrata, Vaginulina/Citharina, 184
clavata, Ammolagena, 452
clavaticamerata, Protentella, 557
Clavulina anglica, 454, 508
    cocoaensis, 455
'*Clavulina*' *gaultina*, 408
clementiana, Gavelinella, 336
Climacammina postprisca, 82
cocoaensis, Clavulina, 455
colliezi, Citharina, 170, 210
Colonammina, 21
    bituba, 24
    conea, 24
communis, Martinottiella, 456
complanata, Spiroplectinata, 411
compressa, Vissariotaxis, 71
compressum, Ophthalmidium, 254
concentricum, Spirophthalmidium, 221
cocinna, Eponides, 332
concavata, Dicarinella, 330
conciliata, Valvulinella, 78
concinna, Lugtonia, 56
cf. *concinnus, Planoarchaediscus*, 62
cf. *conciliata, Valvulinella*, 66
conea, Colonammina, 24
conica s.l., *Tetrataxis*, 63
Conicosprillina cf. *trochoides*, 210
Conilites dinantii, 58
Conorboides lamplughi, 328
    valendisensis, 328
Conorotalites bartensteini, 412
    intercedens, 414
contorta, Quinqueloculina, 510
aff. *contraria, Ceratobulimina*, 551
contusa, Globotruncana, 342
conversa, Karrerulina, 456
convexa, Endothyra, 60
cookei, Bolivina, 467
cookei, Brizalina, 515
cooperensis, Sorosphaerella, 26
coprolithiformis, Ammobaculites, 218, 252
coprolithoides, Caucasina, 516
cf. *cordata, Parathurammina* (S.), 23, 28
cordieriana, Osangularia, 354
cornua, Nodosarchaediscus, 74
cornucopia, Ceratammina, 24
Cornuspira liasina, 221
    echenbergensis, 254
coronata, Marginotruncata, 352
costata, Marginulina, 259
costata, Planulina, 480
crassa, Endothyranopsis, 60, 68, 82
crebra, Reinholdella, 229
crepidularis, planularia, 356
cretacea, Archaeoglobigerina, 322
cretaceus, Ammodiscus, 452

cretosa, Pseudotextulariella, 318
cribriformis, Koskinotextularia, 78
Cribroelphidium bartletti, 474
    vulgare, 474
Cribospira mira, 66
    pansa, 70
    sp., 82
cribostomata, Bradyina, 82
Cribostomum lecomptei, 72
Cribostomoides scitulus, 455
cristata, Gavelinella, 336
cristata, Pseudouvigerina, 358
cubensis, Plectina, 456
culverensis, Bolivinoides, 322
cunobelina, Cibicides (Cibicidina), 516
curiosa, Pseudoglomospira, 58
curryi, Pararotalia, 524
cushmani, Rotalipora, 360
cf. *cushmani, Haplophragmoides*, 208
cf. *cushmani, Parathurammina (P.)*, 23, 28
curva, Hyperammina, 24
Cyclogyra kinkelini, 94
Cystammina sp. A., 455

Dainella cf. *elegantula*, 57
    holkeriana, 64
?*Dainella fleronensis*, 56
danica, Endothyra, 54
danvillensis, Globorotalia, 530
danvillensis var., *gyroidinoides,*
    Gyroidinoides, 520
Darjella monilis, 58
dayi, Cibicidoides, 471
decorata, Vaginulinopsis, 466
decorata subsp., A, *Vaginulinopsis*, 466
decoratus, Bolivinoides, 324
decrescens, Textularia, 459
decurrens, Bolivina, 322
dehiscens, Globoquadrina, 556
delepinei, Endothyra, 56
delrioensis, Hedbergella, 344
delrioensis, Praeglobotruncana, 356
Dentalina, 22
    filiformis, 222
    intorta, 222
    langi, 171
    matutina, 171
    mucronata, 222
    permiana, 94
    pseudocommunis, 105, 118, 222, 256
    tenuistriata, 171
    terquemi, 171
    varians haeusleri, 171
denticulatacarinata, Vaginulinopsis, 186
depressa, Trochammina, 320
deslongchampsi, Palmula, 99, 183, 224
Dicarinella concavata, 330
    hagni, 330
    primitiva, 330
dictyodes, Lenticulina, 226
dilatata, Bolivina, 467
dinantii, Conilites, 58
dingdenensis, Bulimina, 468

dingdeni, Florilus, 552
Diplosphaerina cf. *isphaeramensis*, 23, 28
discoidea, Tournayella, 58
Discorbis propinqua, 518
gr. *discreta, Psammosiphonella*, 458
diversa, Eotextularia, 63
dividens, Gaudryina, 410
'*dorbignyi, Astacolus*', 99
dorbignyi, Lenticulina, 176, 212
Dorothia filiformis, 316
    retusa, 408
    trochoides, 408
dorsetensis, Massilina, 221
draco, Bolivinoides, 324
Draffania biloba, 64
dreheri, Reinholdella, 216
dumontana, Frondicularia, 463
dutemplei peelensis, Cibicidoides, 472
dytica, Nevillella, 66

Earlandia elegans, 57
    minor, 57
    pulchra, 82
    sp., 81
    vulgaris, 52
Eblanaia michoti, 54
ectypa costata, Lenticulina, 258
ectypa, Lenticulina, 228, 256
Eggerellina mariae, 318
Ehrenbergina healyi, 551
    serrata, 474
eichenbergensis, Cornuspira, 254
eichenbergi, Lenticulata, 348
eiseli, Ammobaculites, 92
cf. *eiseli, Ammobaculites*, 116
elegans, Earlandia, 57
cf. *elegans, Bisphaera*, 23, 28
cf. *elegantula, Dainella*, 57
eleyi, Loxostomum, 352
elongata, Bulimina, 468
elongata, Lugtonia, 84
elongata, Stegnammina, 26
Elphidiella hannai, 474, 566
Elphidium albiumbilicatum, 475
    antoninum, 475
    excavatum, 475
    frigidum, 568
    groenlandicum, 475
    haagensis, 568
    hiltermanni, 520
    inflatum, 475, 551
    latidorsatum, 520
    oregonense, 475
    pseudolessonii, 568
    subarcticum, 476
    subnodosum, 476
    ungeri, 476
    '*ustulatum*', 476
Endostaffella fucoides, 76
    sp., 74
Endothyra, 32
    bowmani, 56
    convexa, 60

    danica, 54
    delepinei, 56
    excellens, 68, 82
    cf. *freyri*, 54
    laxa, 54
    maxima, 66
    nebulosa, 52
    obsoleta, 80
    cf. *pandorae*, 82
    ex. gr. *phrissa*, 64, 68
    prisca, 60
    cf. *prokirgisana*, 57
    sp., 52, 57, 63, 72, 84
    ex. gr. *spira*, 68
    tumida, 54
Endothyranopsis crassa, 60, 68
    pechorica, 68
    sphaerica, 66, 74
eocaena, Uvigerina, 484
eocaenus, Cibicidoides, 472
eocenicus, Bathysiphon, 454
Enodosaria, 21
    cf. *evalensis*, 23, 28
Eoguttulina, 98
    liassica, 97, 103, 105, 106, 118, 129, 171
Eoparastaffella simplex, 60
Eosigmoilina robertsoni, 54
Eostaffella mosquensis, 72
    parastruvei, 66
    cf. *parastruvei*, 60
    sp., 80, 84
Eotextularia diversa, 63
Eouvigerina aculeata, 330
Eovolutina cf. *magna*, 23, 28
epigona, Rzehakina, 459
Epistomaria rimosa, 520
    separans, 520
Epistomina hechti, 330
    mosquensis, 260
    cf. *nuda*, 222
    ornata, 260, 330
    parastelligera, 262
    regularis, 224
    reticulata, 262
    spinulifera, 332
    stellicostata, 224
    stelligera, 224
    tenuicostata, 262
Epistominella oveyi, 520
    vitrea, 520
Eponides beisseli, 332
    concinna, 332
erectum inconstans, Haplophragmium, 410
cf. *esnaensis, Globorotalia*, 530
esseyana, Lingulina, 214
eugenii, Planularia, 228
 cf. *evlanensis, Eonodosaria*, 23, 28
exarata, Vaginulinopsis, 186
excavata, Millerella, 64
excavatum, Elphidium, 475
excellens, Endothyra, 68, 82
excentrica, Alliatina, 548
excilis, Vissariotaxis, 70

Index of Genera and Species

exelikta, Plectogyranopsis, 58
exgaleata, Lenticulina, 226
exsculpta exsculpta, Stensioeina, 362
exsculpta gracilis, Stensioeina, 362
exsertus, Ammodiscus, 24

Falsogaudryinella alta, 410
    moesiana, 410
    sp.1, 410
    sp.x, 410
fallax, Karreria, 479
farinosa, Hopkinsina, 552
farinosa, Uvigerina, 484
Faujasina subrotunda, 552
Favusella washitensis, 332
feifeli, Paalzowella, 216, 220, 255
filiformis, Dentalina, 222
filiformis, Dorothia, 316
fimbriata, Chilostomellina, 471
firmata, Archaesphaera, 52
flabellata, Citharina, 222
fleronensis, ?Dainella, 56
florealis, Nuttallinella, 414
Florilus boueanus, 476
    dingdeni, 552
Flourensina intermedia, 318
fluens, Trifarina, 482
foeda, Marginulinopsis, 352
fontinensis, Ammobaculites, 218
fornicata, Globotruncana, 342
Forshica, cf. *parvula*, 60
    sp., 76
Forschiella prisca, 60, 81
fragilis, Urbanella, 63
fragraria, Marginulinopsis, 463
franconica, Frondicularia, 224, 256
cf. *freyri, Endothyra*, 54
frigida, Buccella, 467
aff. *frigida, Buccella*, 550
frigidum, Elphidium, 568
Frondicularia brizaeformis, 172
    brizoides, 99
    budensis, 463
    dumontana, 463
    franconica, 224, 256
    hastata, 332
    irregularis, 212
    lignaria, 212, 224
    nikitini, 256
    nympha, 224
    oblonga, 463
    oolithica, 212
    pseudosulcata, 256
    rhaetica, 99
    terquemi, 99
    terquemi bicostata, 172
    terquemi muelensis, 174
    terquemi subsp. A, 174
    terquemi subsp. B, 174
    terquemi sulcata, 174
    terquemi terquemi, 175
    terquemi plexus, 172
fucoides, Eostaffella, 76

Fursenkoina schreibersiana, 476
fusulinaformis, Saccamminopsis, 80
cf. *fusulinaformis, Saccamminopsis*, 22
Fusulinia, 28

cf. *gallowayi, Nanicella*, 23, 28
Gaudryina dividens, 410
    hiltermanni, 455, 510
    sherlocki, 132, 252
    ? sp., 218
Gaudryinella sherlocki, 318
gaultina, 'Clavulina', 408
Gavelinella baltica, 334
    barremiana, 334
    brielensis, 334
    cenomanica, 334
    clementiana, 336
    cristata, 336
    intermedia, 336
    lorneiana, 336
    monterelensis, 338
    pertusa, 338
    rudis, 414
    sigmoicosta, 338
    stelligera, 338
    thalmanni, 340
    usakensis, 340
Gavelinopsis tourainensis, 414
geinitzi, Nodosaria, 95
Geinitzina, 87
    acuta, 94
germanica, Robertina, 526
germanica, Uvigerina, 482, 528
gibba, Bulimina, 468
gibba myristiformis, Globulina, 478
gibba rugosa, Globulina, 478
Glabratella? sp. A, 476
gladius, Astacolus, 462
Glandulina laevigata, 528
Globigerina, 249
    bradyi, 554
    chascanona, 530
    ciperoensis angulisuturalis, 554
    ciperoensis ciperoensis, 554
    patagonica, 530
    triloculinoides, 530
    woodi, 554
Globigerinatheka barri, 530
Globigerinella praesiphonifera, 556
Globigerinelloides bentonensis, 330
'Globigerinelloides' gyroidinaeformis, 412
globigeriniformis, Trochammina, 234
Globigerinoides ruber, 556
    sacculifer, 556
    trilobus, 556
Globobulimina socialis, 476
Globoquadrina dehiscens, 556
Globorotalia broedermanni, 530
    danvillensis, 530
    cf. *esnaensis*, 530
    inflata, 558
    obesa, 556
    perclara, 530

peripheroronda, 557
praehirsuta, 558
praescitula, 557
pseudocontinuosa, 557
puncticulata, 558
reissi, 532
saginata, 557
semivera, 557
subbotinae, 532
tosaensis, 558
wilcoxensis, 532
zealandica, 557
Globorotalites hiltermanni, 340
  micheliniana, 342
globosa, Lingulogavelinella, 350
Globotruncana bulloides, 342
  contusa, 342
  fornicata, 342
  linneiana, 342
  plummerae, 344
Globotruncanella havanensis, 344
globulata, Nodosaria, 216
Globulina ampulla, 529
  gibba myristiformis, 478
  gibba rugosa, 478
  inaequalis, 529
Glomodiscus sp. 62
glomoformis, Chernyshinella, 52
Glomospira charoides, 455
  perplexa, 116
  subparvula, 98, 105, 110, 116
Glomospiranella aff. *barsae*, 64
  cf. *barsae*, 57
  sp., 81
Glomospirella sp. 1, 105, 118
  sp. A., 455
glomospiroides, Lituotubella, 80
?*gordiformis, Orthovertella*, 94
gracillima, Valvulineria, 414
gracilis caledoniae, Mikhailovella, 72
gracilissima, Marginulinopsis, 354
granosum, Nonion, 480
granulata granulata, Stensioeina, 362
granulata humilis, Stensioeina, 414
granulata kelleri, Stensioeina, 414
granulata levis, Stensioeina, 414
granulata polonica, Stensioeina, 362
gravata, Pseudolituotuba, 60
greenhornensis, Rotalipora, 360
gregorii, Asteroarchaediscus, 82
gregorii, Nodosperodiscus, 78
grilli, Grillina, 99
Grillina grilli, 99
groenlandicum, Elphidium, 475
grossus, Cibicides, 471
Guembelitria cenomana, 344
  triseriata, 532
guerichi guerichi, Asterigerina, 467
guerichi staeschei, Asterigerina, 467
guttata, Lenticulina, 348
gutticostata, Lenticulina, 463
Guttulina pera, 224
  problema, 529

Gyroidina soldanii girardana, 478
  soldanii mamillata, 478
gyroidinaeformis, '*Globigerinelloides*', 412
Gyroidinoides angustiumbilicatus, 520
  danvillensis var. *gyroidinoides*, 520
  octacamerata, 522
  aff. *parvus*, 552

haagensis, Elphidium, 568
hagni, Dicarinella, 330
hannai, Elphidiella, 474, 566
Hanzawaia producta, 522
Haplophragmium aequale, 318
  inconstans erectum, 410
  cf. *pokrovkaensis*, 232
Haplophragmoides, 162
  cf. *cushmani*, 208
  canui, 230, 252
  cf. *H. (Cribrostomoides) dolininae*, 230
  sp. cf. *Haplophragmoides* sp. 143, 232
  infracalloviensis, 230
  kirki, 455
  lincolnensis, 164
  sp. 1, 164
  walteri, 455
harpa, Citharina, 328
harpa, Vaginulina, 229
hastata, Frondicularia, 332
hauteriviana, Lagena, 348
havanensis, Globotruncanella, 344
Haynesina orbiculare, 478
healyi, Ehrenbergina, 551
hebes, Bolivina, 550
hebesta, Stegnammina, 26
hechti, Epistomina, 330
Hedbergella brittonensis, 344
  delrioensis, 344
  infracretacea, 346
  planispira, 346
  simplex, 346
heiermanni, Lenticulina, 350
helenae, Islandiella, 478
helvetica, Praeglobotruncana, 358
helvetica, Reophax, 105, 118
Hemisphaerammina, 21, 22, 23
  thola, 23, 24
?*Hemisphaerammina* sp., 22
Heronallenia lingulata, 552
Heterohelix moremani, 346
Heterolepa pegwellensis, 522
hiltermanni, Elphidium, 520
hiltermanni, Gaudryina, 455, 510
hiltermanni, Globorotalites, 340
Hippocrepina, 22
hispida, Lagena, 348
Hoeglundina caracolla, 346
  carpenteri, 348
  chapmani, 348
hofkeri, Protelphidium, 480, 526
holkeriana, Dainella, 64
Hopkinsina farinosa, 552
hortensis, Nodosaria, 216, 228
Howchinia bradyana, 76, 78

Index of Genera and Species

humboldti, *Sabellovoluta*, 459
humilis praecursoria, *Vaginulinopsis*, 364
Hyalinea balthica, 478
Hyperammina, 21, 22, 23, 92
    acuta, 94
    curva, 24

Ichtyolaria? cavernula, 95
impressa, *Quinqueloculina*, 510
inaequalis, *Archaesphaera*, 52
inaequalis, *Globulina*, 529
inaequistriata, *Planularia*, 183
incertus, *Neoarchaediscus*, 82
incomposita, *Atelikamara*, 24
inconstans erectum, *Haplophragmium*, 410
incrassata, *Bolivina*, 322
inermis, *Pararotalia*, 524
infima, *Spirillina*, 129, 216
inflata, *Globorotalia*, 558
inflatum, *Elphidium*, 475, 551
infracalloviensis, *Haplophragmoides*, 230
infracretacea, *Hedbergella*, 346
ingens, *Nodosaria*, 228
intercedens, *Conorotalites*, 414
intermedia, *Flourensina*, 318
intermedia, *Gavelinella*, 336
intermittens, *Nodosaria*, 464
intorta, *Dentalina*, 222
involuta, *Berthelinella*, 170
Involutina liassica, 167
irregulariformis, *Ammobaculites*, 408
Irregularina cf. *angulata*, 23, 28
irregularis, *Frondicularia*, 212
irregularis, *Rhabdammina*, 24
irregularis, *Trochamminoides*, 460
islandica, *Islandiella*, 478
Islandiella helenae, 478
    islandica, 478
cf. isphaeramensis, *Diplosphaerina*, 23, 28
issleri, *Nodosaria*, 182
aff. issleri, *Nodosaria*, 228
itinerarius, *Archaediscus*, 81

jacksonensis, *Uvigerina*, 484
jactata, *Brunsia*, 74
?*Janischewskina minuscularia*, 81
Janischewskina operculata, 76
jankoi, *Uvigerinammina*, 411
jarzevae, *Lingulogavelinella*, 352
jonesi, *Textularia*, 89
juleana, *Quinqueloculina*, 462, 510
jurassica, *Miliammina*, 218
jurassica, *Textularia*, 252
jurensis, *Thurammina*, 164

karreri, *Archaediscus*, 76
Karreria fallax, 479
Karreriella chilostoma, 456
    siphonella, 456
    sp. A, 456
Karrerulina conversa, 456
karsteni, *Neoponides*, 479
kelleri, *Reussella*, 358

kinkelini, *Cyclogyra*, 94
kirki, *Haplophragmoides*, 455
kisella, *Tournayella*, 58
konickii, *Nodosaria*, 464
Koskinobigenerina sp., 74
Koskinotextularia cribriformis, 78
    sp., 71
krainica, *Septabrunsiina*, 54
krestornikovi, *Archaediscus*, 63
kummi, *Marssonella*, 410

Laagensis, *Elphidium*, 568
labradoricum, *Nonion*, 480
laffittei, *Citharinella*, 328
Lagena hauteriviana, 348
    hispida, 348
    sp. A, 212
laxa, *Endothyra*, 54
laeve, *Nonion*, 522
laevigata, *Cassidulina*, 471
laevigata, *Glandulina*, 528
laevigatus, *Bolivinoides*, 322
laevigatus, *Nummulites*, 532
laevis, *Praebulimina*, 356
Lagenammina, 21, 23, 140
    stilla, 24
lammersi, *Loxostomoides*, 479
lamplughi, *Conorboides*, 328
langi, *Dentalina*, 171
latejugata, *Nodosaria*, 464, 522
latidorsatum, *Elphidium*, 520
Latiendothyranopsis menneri, 62
latissima, *Valvulinella*, 81
lecomptei, *Cribrostomum*, 72
legumen, *Vaginulina*, 229
lenticula, *Valvulineria*, 364
Lenticulina dictyodes, 226
    dorbignyi, 176, 212
    ectypa, 228, 256
    ectypa costata, 258
    eichenbergi, 348
    exgaleata, 226
    guttata, 348
    guttiostata, 463
    heiermanni, 350
    cf. limata, 226
    major, 226, 258
    muensteri, 258, 350
    muensteri acutiangulata, 176
    muensteri muensteri, 178
    ex. gr. muensteri, 214
    nodosa, 412
    oachaensis wisselmanni, 350
    platypleura, 463
    quenstedti, 176, 226, 258
    saxonica, 350
    subalata, 226, 258
    tricarinella, 226, 258
    varians, 176, 214
    varians subsp. D., 176
    volubilis, 226
'*Lenticulina*' schloenbachi, 350
    schreiteri, 350

liasicum, Ophthalmidium, 167
liasina, Cornuspira, 221
liasina, Tristix, 184
liassica, Brizalina, 187
liassica, Eoguttulina, 97, 103, 105, 106, 118, 129, 171
liassica, Involutina, 167
lignaria, Frondicularia, 212, 224
cf. *limata, Lenticulina*, 226
limbatosuturalis, Cibicidoides, 472
lincolnensis, Haplophragmoides, 164
lingulata, Heronallenia, 552
Lingulina, 87
    alaskensis, 99
    borealis, 99
    cernua, 120
    esseyana, 214
    longiscata, 226
    longiscata longiscata, 214
    cf. *longiscata alpha*, 214
    tenera, 99
    tenera, Bornemann plexus, 178
    tenera collenoti, 103, 106, 120, 178
    tenera occidentalis, 178
    tenera var. *octocosta*, 120
    tenera plexus, 97
    tenera pupa, 178
    tenera subprismatica, 179
    tenera substriata, 179
    tenera tenera, 103, 106, 120, 179
    tenera tenuistriata, 179
    testudinaria, 180
    tenera subsp. A, 180
'*Lingulina tenera*', 99
Lingulogavelinella ciryi inflata, 414
    globosa, 350
    jarzerae, 352
    sp. cf. *L. vombensis*, 352
linneiana, Globotruncana, 342
aff. *lipinae, Palaeotextularia*, 76
listi, Vaginulina, 186
lithuanica, Miliospirella, 254
Lituotubella glomospiroides, 80
    magna, 68
    aff. *magna*, 74
lobosa, Archaeochitosa, 22
lobatulus, Cibicides, 516, 566
cf. *lobedevae, Septatournayella*, 54
Loeblichia paraammonoides, 78
Loftusia, 16
cf. *longiscata alpha, Lingulina*, 214
longiscata, Lingulina, 226
longiscata longiscata, Lingulina, 214
aff. *longiseptata, Palaeotextularia*, 66
lorneiana, Gavelinella, 336
Loxostomoides lammersi, 479
Loxostomum eleyi, 352
    sinuosom, 479
ludwigi, Quinqueloculina, 510
luenebergensis, Marginulinopsis, 464
Lugtonia concinna, 56
    elongata, 84

macfadyeni, Arenobulimina, 316

macfadyeni, Ophthalmidium, 167
macfadyeni, Reinholdella, 186
mackeei, Septabrunsiina, 80
macrospira, Ammomarginulina, 452
magna, Lituotubella, 68
aff. *magna, Lituotubella*, 74
cf. *magna, Eovolutina*, 23, 28
major, Lenticulina, 226, 258
malloryi, Storthosphaera, 26
Marginotruncana coronata, 352
    pseudolinneiana, 352
    sigali, 352
margarita, Reinholdella, 186
marginata, Bulimina, 468
Marginulina costata, 259
    prima incisa, 180
    prima insignis, 180
    prima interrupta, 180
    prima praerugosa, 182
    prima prima, 182
    prima rugosa, 182
    prima spinata, 182
    cf. *scapa*, 216
    undulata, 259
Marginulinopsis foeda, 352
    fragraria, 463
    gracilissima, 354
    luenebergensis, 464
    pedum ackneriana, 464
    wetherelli, 522
mariae, Cibicides (Cibicidina), 516
mariae, Eggerellina, 318
Marssonella kummi, 410
    ozawai, 318
    subtrochus, 410
Martinottiella bradyana, 456
    communis, 456
Massilina dorsetensis, 221
    secans, 462
Matanzia varians, 456
matutina, Dentalina, 171
mauritii, Verneuilinoides, 166
maxima, Endothyra, 66
mayaroensis, Abathomphalus, 322
Mediocris breviscula, 63
    mediocris, 66, 71, 72
mediocris, Mediocris, 66, 71, 72
medocarinata, Vaginulina, 364
mellina, Palaeospiroplectammina, 56
Melonis affine, 522
    pompilioides, 479
menneri, Latiendothyranopsis, 62
metensis, 'Annulina', 116
metensis, Nodosaria, 120, 182, 216
mexicana grammensis, Valvulinaria, 485
mexicanus, Cibicidoides, 472
micheliniana, Globorotalites, 342
michoti, Eblanaia, 54
midwayensis, Bulimina, 470
midwayensis, Stilostomella, 466
Mikhailovella gracilis caledoniae, 72
Miliammina jurassica, 218
miliaris, Bolivinoides, 322

Index of Genera and Species

milioloides, Agathammina, 94, 110
Miliospirella lithuanica, 254
Millerella excavata, 64
minima, Omphalotis, 63
minor, Earlandia, 57
minuscularia,? Janischewskina, 81
minuta, Reophax, 411
mira, Cribrospira, 66
miranda matura, Urbanella, 80
cf. *mitchelli, Spinoendothyra,* 56
moelleri, Archaediscus, 82
moelleri, Citharinella, 222
moesiana, Falsogaudryinella, 410
monilis, Darjella, 58
Monspeliensina pseudotepida, 479, 552
monterelensis, Gavelinella, 338
moremani, Heterohelix, 346
moremani, Saccammina, 24
mosquensis, Eostaffella, 72
mosquensis, Epistomina, 260
Mstiniella, sp. 68
mucronata, Dentalina, 222
muensteri acutiangulata, Lenticulina, 176
muensteri, Lenticulina, 258, 350
ex. gr. *muensteri, Lenticulina,* 214
muensteri muensteri, Lenticulina, 178
muensteri, Verneuilina, 320
muralis, Uvigerina, 528
mutabilis, Nodosaria, 216

Nanicella, 21
    cf. *gallowayi,* 23, 28
navarroana, Osangularia, 354
nebulosa, Endothyra, 52
Neoarchaediscus incertus, 82
Neobulimina sp. 2, 187
neocomiensis, Verneuilinoides, 320
Neoflabellina praereticulata, 354
    reticulata, 354
    rugosa, 354
Neogloboquadrina altantica, 558, 564
    pachyderma, 558
Neoponides karsteni, 479
Nevillella dytica, 66
    tetraloculi, 64
nibelis, Quasiendothyra, 64, 66, 72
nikitini, Citharinella, 222
nikitini, Frondicularia, 256
nobilis, Anomalinoides, 512
Nodasperodiscus gregorii, 78
    sp., 78
    stellatus, 76
Nodobacularia nodulosa, 320
nodosa, Lenticulina, 412
nodosariaformis, Bathysiphon, 454
Nodosarchaediscus cornua, 74
    sp., 64, 68
Nodosaria geinitzi, 95
    globulata, 216
    hortensis, 216, 228
    ingens, 228
    intermittens, 464
    issleri, 182

    aff. *issleri,* 228
    koninckii, 464
    latejugata, 522
    metensis, 216
    mutabilis, 216
    obscura, 216
    opalina, 228
    pectinata, 228
    regularis, 183
    soluta, 464
    sp. A, 466
    torsicostata, 464
Nodosinella digitata, 87, 89
nodulosa, Nodobacularia, 320
nodulosa, Trocholina, 255
Nonion applinae, 522
    granosum, 480
    labradoricum, 480
    laeve, 522
    parvulum, 524
    reculverensis, 524
    sp. A, 480
northamptonensis, Ophthalmidium, 169
Nubeculinella bigoti, 254
    tibia, 167
cf. *nuda, Epistomina,* 222
Nummulites aquitanicus, 532
    laevigatus, 532
    planulatus, 532
    prestwichianus, 534
    rectus, 534
    variolarius, 534
Nuttallinella florealis, 414
nympha, Frondicularia, 224

obesa, Globorotalia, 556
obliqua, Palmula, 224
oblonga, Frondicularia, 463
obscura liassica, Nodosaria, 216
obsoleta, Endothyra, 80
obtusa, Alabamina, 512
obtusa, Praebulimina, 356
occlusus, Asterorarchaediscus, 78
octocamerata, Gyroidinoides, 522
oligocaenica, Brizalina, 515
Omphalotis minima, 63
    cf. *omphalotis,* sp. 72
    samarica, 74
    cf. *volynica,* 70
oolithica, Frondicularia, 212
oolithica, Tristix, 229
opalina, Nodosaria, 228
operculata, Janischewskina, 76
Ophthalmidium, 99
    carinatum, 221
    compressum, 254
    liasicum, 167
    macfadyeni, 167
    northamptonensis, 168
    sp. 2, 168
    sp. A, 221
    strumosum, 208, 221, 255
orbiculare, Haynesina, 478

orbiculare, Reticulophragmium, 458
orbiculare, Praeophthalmidium, 168
orbiculare, Protelphidium, 568
ornata, Epistomina, 260, 330
oregonense, Elphidium, 475
Orthovertella? *gordifomris*, 91, 94
Osangularia cordieriana, 354
    navarroana, 354
    plummerae, 480
    schoenbachi, 414
    whitei, 354
osnabrugensis, Almaena, 467
ouachensis wisselmanni, Lenticulina, 350
ovata, Praeglobobulimina, 524
oveyi, Epistominella, 520
oxfordiana, Saracenaria, 229, 260
ozawai, Marssonnella, 318

Paalzowella feifeli, 216, 220, 255
pachyderma, Cibicidoides, 472
pachyderma, Neogloboquadrina, 558
cf. *pacifica, Buzasina*, 454
Palaeospiroplectammina mellina, 56, 58
    syzranica, 71
    tschernyshinensis, 52
Palaeotextularia aff. *lipinae*, 76
    aff. *longiseptata*, 66
paleocenicus, Bolivinoides, 324
'*Palmula desclongchampsi*', 99
Palmula desclongchampsi, 183, 224
    obliqua, 224
    tenuistriata, 183
cf. *pandorae, Endothyra*, 82
pansa, Cribrospira, 70
papillata, Thurammina, 26, 460
paraammonoides, Loeblichia, 78
cf. *paracushmani, Parathurammina* (*P.*), 23, 28
Pararotalia audonuini, 524
    canui, 480, 524
    curryi, 524
    inermis, 524
    serrata, 552, 568
    spinigera, 524
parastelligera, Epistomina, 262
parastruvei, Eostaffella, 66
sp. cf. *parastruvei, Eostaffella*, 60
Parathurammina, spp., 21
Parathurammina (Parathuramminites)
    cf. *cushmani*, 23, 28
    cf., 23, 28
Parathurammina (Salpingothurammina)
    cf. *cordata*, 23, 28
Paratikhinella, 21
    cf. *cannula*, 23, 28
Parkeri, 16
parva, Biseriella, 74, 82
cf. *parvula, Forschia*, 60
parvulum, Nonion, 524
aff. *parvus, Gyroidinoides*, 552
patagonica, Globigerina, 530
pechorica, Endothyranopsis, 68
pectinata, Nodosaria, 228
pedum ackneriana, Marginulinopsis, 464

pegwellensis, Heterolepa, 522
Pelosina caudata, 411
pera, Guttulina, 224
perclara, Globorotalia, 530
peregrina, Uvigerina, 484
perfossus, Astrononion, 550
peripheroronda, Globorotalia, 557
perlata, Svatkina, 482
permiana, Dentalina, 94
cf. *permira, Uslonia*, 23, 28
perplexa, Glomospira, 116
pertusa, Gavelinella, 338
pertusa, Virgulinella, 485, 554
petrolei, Valvulineria, 485
peterssoni, Bolivinoides, 324
ex. gr. *phrissa, Endothyra*, 64, 68
pictonica, '*Placentula*', 168
pinnaeformis, Citharinella, 328
placenta, Reticulophragmium, 458
'*Placentula*' *pictonica*, 168
planiconvexa, Reinholdella? 122, 187
planispira, Hedbergella, 346
Planoarchaediscus cf. *concinnus*, 62
Planoendothyra sp. 70
Planularia angustissima, 259
    beierana, 228, 259
    crepidularis, 356
    eugenii, 228
    inaequistriata, 183
planulatus, Nummulites, 532
Planulina costata, 480
platypleura, Astacolus, 514
platypleura, Lenticulina, 463
Platysolenites antiquissimus, 21
Plectina cenomana, 318
    cubensis, 456
Plectofrondicularia advena, 466
    seminuda, 466
Plectogyranopsis ampla, 76
    exelikta, 58
Pleurostomella barroisi, 356
pliocarinata, Cassidulina, 471
plummerae, Globotruncana, 344
plummerae, Osangularia, 480
plummerae, Textularia, 459
pokornyi, Stensioeina, 414
cf. *pokrovkaensis, Haplophragmium*, 232
pommerana, Stensioeina, 362
pompilioides, Melonis, 479
postprisca, Climacammina, 82
Praebulimina carseyae, 356
    laevis, 356
    obtusa, 356
    reussi, 356
cf. *praeclara, Spinoendothyra*, 56
Praeglobobulimina ovata, 522
Praeglobotruncana delrioensis, 356
    helvetica, 358
    stephani, 358
praehirsuta, Globorotalia, 558
Praeophthalmidium orbiculare, 168
praereticulata, Neoflabellina, 354
praescitula, Globorotalia, 557

Index of Genera and Species

praesiphonifera, Globigerinella, 556
pressula pressula, Textrataxis, 78
prestwichianus, Nummulites, 534
primaeva, Septaglomospiranella, 52
primalis, Pulleniatina, 558
prima incisa, Marginulina, 180
prima insignis, Marginulina, 180
prima interrupta, Marginulina, 180
prima praerugosa, Marginulina, 182
prima prima, Marginulina, 182
prima, Pulsiphonina, 526
prima rugosa, Marginulina, 182
prima spinata, Marginulina, 182
'*prima*', *Vaginulina*, 260
primitiva, Dicarinella, 330
prisca, Forschiella, 60, 81
problema, Guttulina, 529
producta, Hanzawaia, 522
cf. *prokirgisana, Endothyra*, 57
propingua, Buccella, 515
propinqua, Discorbis, 518
Protelphidium hofkeri, 480, 526
    orbiculare, 568
    cf. *P. roemeri*, 526
    sp. 1, 526
    sp. 2, 526
    sp. 4, 526
Protentella clavaticamerata, 557
Psammosiphonella gr. *discreta*, 458
Psammosphaera, 21, 22
    cava, 23, 24
Pseudoammodiscus aff. *bukensis*, 78
    sp. 62
    aff. *volgensis*, 74
pseudocommunis, Dentalina, 105, 118, 222, 256
pseudocontinuosa, Globorotalia, 557
Pseudoendothyra sublimis, 72
Pseudoglomospira curiosa, 58
Pseudohastigerina wilcoxensis, 532
Pseudolamarckina rjasanensis, 229
pseudolessonii, Elphidium, 568
pseudolinneiana, Marginotruncana, 352
Pseudolituotuba gravata, 60
    wilsoni, 72
Pseudolituotubella sp. 63
Pseudonodosaria radiata, 259
    vulgata, 183, 259
pseudopulchra, Brunsia, 58
pseudosulcata, Frondicularia, 256
pseudostriatula, Citharina, 328
Pseudotaxis, sp. 63
pseudotepida, Monspeliensina, 479, 552
Pseudotextulariella cretosa, 318
Pseudouvigerina cristata, 358
pulchra, Earlandia, 82
Pullenia quaternaria, 358
    quinqueloba, 526
Pulleniatina primalis, 558
Pulsiphonina prima, 526
puncticulata, Globorotalia, 558
pusilla, Agathammina, 94, 110
pustulatus, Bolivinoides, 324
pygmaea langenfeldensis, Uvigerina, 484

pygmaea langeri, Uvigerina, 484
pygmeus, Cibicidoides, 518
pyramidata, Tritaxia, 320

Quasiendothyra nibelis, 64, 66, 72
quaternaria, Pullenia, 358
quenstedti, Lenticulina, 176, 226, 258
quinqueloba, Pullenia, 526
quinqueloculina antiqua, 320
    bicarinata, 510
    carinata, 510
    contorta, 510
    impressa, 510
    juleana, 462, 510
    ludwigi, 510
    reicheli, 512
    seminulum, 512
    simplex, 512

radiata, Pseudonodosaria, 259
cf. *ramsbottomi, Spinobrunsiina*, 60
Ramulina spandeli, 358
recta, Spinoendothyra, 54
rectus, Nummulites, 534
reculverensis, Nonion, 522
Recurvoides sublustris, 232
    gr. *walteri*, 458
reditus, Archaediscus, 70
regularis, Epistomina, 224
regularis, Nodosaria, 183
regularis, Sigmomorphina, 480
recticulata, Siphonina, 482
reicheli, Rotalipora, 360
reicheli, Quinqueloculina, 512
riedeli, Vaginulina, 364
Reinholdella, 135
    crebra, 229
    dreheri, 216
    macfadyeni, 186
    margarita, 186
    ? *planiconvexa*, 103, 122, 187
reissi, Globorotalia, 532
reitlingerae, Archaesphaera, 52
reophacoides, Ammobaculites, 316
Reophax, 99
    helvetica, 105, 118
    minuta, 411
    sterkii, 208
reticulata, Epistomina, 262
reticulata, Neoflabellina, 354
reticulata, Siphonina, 482
Recticulophragmium acutidorsatum, 458
    amplectens, 458
    orbiculare, 458
    placenta, 458
    sp. A, 458
retusa, Dorothia, 408
Reussella kelleri, 358
    szajnochae praecursor, 358
    szajnochae szajnochae, 360
reussi, Praebulimina, 356
Rhabdammina, 21, 22, 23
    bifurcata, 24

Index of Genera and Species

irregularis, 24
?*Rhabdammina* sp., 21
rhaetica, *Frondicularia*, 99
Rhodensinella avonensis, 64, 70
ribbingi, *Cibicides*, 326
rimosa, *Epistomaria*, 520
rjasanensis, *Pseudolamarckina*, 229
Robertina germanica, 526
robertsoni, *Eosigmoilina*, 84
cf. *P. roemeri*, *Protelphidium*, 526
roessleri, *Ammodiscus*, 94
rohri, *Truncorotaloides*, 532
Rosalina, *araucana*, 528
Rotaliatina bulimoides, 480
Rotalipora appenninica, 360
    cushmani, 360
    greenhornensis, 360
    reicheli, 360
rotula, *Bradyina*, 76
rotundus, *Uralodiscus*, 62
ruber, *Globigerinoides*, 556
rudis, *Gavelinella*, 414
rubiginosus, *Anomalinoides*, 467
Rugoglobigerina rugosa, 360
rugosa, *Neoflabellina*, 354
rugosa, *Rugoglobigerina*, 360
ruthvenmurrayi, *Trochammina*, 460
Rzehakina epigona, 459

Sabellovoluta humboldti, 459
sabulosa, *Arenobulimina*, 316
Saccammina, 21, 23
    carteri, 22
    moremani, 24
Saccamminopsis cf. *fusulinaformis*, 22, 80
Saccamminiopsis? sp. cf. *syltensis*, 22
    sp. cf. *teschenhagensis*, 22
sacculifer, *Globigerinoides*, 556
saginata, *Globorotalia*, 557
Sagrina selseyensis, 528
samarica, *Omphalotis*, 74
Saracenaria oxfordiana, 229, 260
    sublaevis, 184
    spinosa, 412
    valanginiana, 362
Saracenella aragonensis, 184
    sp. A, 184
saxonica, *Lenticulina*, 350
scalariformis, *Vaginulinopsis*, 364
cf. *scapa*, *Marginulina*, 216
schloenbachi, *Astacolus*, 412
schloenbachi, '*Lenticulina*', 350
schloenbachi, *Osangularia*, 414
schlumbergeri, *Sigmoilopsis*, 462
schreibersiana, *Fursenkoina*, 476
schreiteri?, '*Lenticulina*', 350
scitulus, *Cribrostomoides*, 455
sculpturata, *Siphotextularia*, 459
secans, *Massilina*, 462
sedentata, *Tholosina*, 26
selmensis, *Tappanina*, 528
selseyensis, *Sagrina*, 528
seminuda, *Plectofrondicularia*, 466

seminulum, *Quinqueloculina*, 512
semiornata saprophila, *Uvigerina*, 484
semiornata, *Uvigerina*, 554
semireticulata, *Astacolus*, 170
semivera, *Globorotalia*, 557
separans, *Epistomaria*, 520
Septabrunsiina krainica, 54
    mackeei, 80
    sp., 54
Septaglomospirinella primaeva, 52
Septatournayella cf. *lobedevae*, 54
serrata, *Ehrenbergina*, 474
serrata, *Pararotalia*, 552, 568
serratocostata, *Citharina*, 256
sherlocki, *Gaudryina*, 252
sherlocki, *Gaudryinella*, 318
sidestrandensis, *Bolivinoides*, 326
sigali, *Marginotruncana*, 352
sigmoicosta, *Gavelinella*, 338
Sigmoilopsis celata, 462
    schlumbergeri, 462
Sigmomorphina regularis, 480
siliceus, *Ammodiscus*, 164
similis, *Webbinelloidea*, 26
cf. *simplex*, *Cibicides*, 516
simplex, *Eoparastaffella*, 60
simplex, *Hedbergella*, 346
simplex, *Quinqueloculina*, 512
singularis, *Tritaxia*, 320
sinuosum, *Loxostomum*, 479
siphonella, *Karreriella*, 456
Siphonina reticulata, 482
Siphotextularia sculpturata, 459
socialis, *Globobulimina*, 476
soldanii girardana, *Gyroidina*, 478
soldanii mamillata, *Gyroidina*, 478
soluta, *Nodosaria*, 464
Sorosphaerella, 23
    cooperensis, 26
spandeli, *Ramulina*, 358
sparsicostata, *Citharina*, 328
speciosus, *Astacolus*, 170, 210
spectabilis, *Spiroplectammina*, 459
sphaerica, *Endothyranopsis*, 66, 74
spinicostata, *Uvigerina*, 484
spinigera, *Pararotalia*, 524
spinigera, *Vaginulina*, 466
Spinobrunsiina cf. *ramsbottomi*, 60
Spinoendothyra cf. *mitchelli*, 56
    cf. *praeclara*, 56
    recta, 56
spinosa, *Saracenaria*, 412
spinulifera, *Epistomina*, 332
ex. gr. *spira*, *Endothyra*, 68
Spirillina, *infima*, 129, 216
spirillinoides, *Ammarchaediscus*, 62
spirillinoides, *Brunsia*, 52
Spiroloculina striatula, 462
Spirophthalmidium concentricum, 221
Spiroplectammina spectabilis, 459
    thanetana, 510
Spiroplectinata complanata, 411
Spirosigmoilinella sp. A, 459

squamosa, Trochammina, 105, 118, 234
staechei, Asterigerina, 550
Stegnammina, 21, 23
    elongata, 26
    hebesta, 26
stellatus, Nodasperodiscus, 76
stellicostata, Epistomina, 224
stelligera, Epistomina, 224
stelligera, Gavelinella, 338
Stensioeina beccariiformis, 482
    exsculpta exsculpta, 362
    exsculpta gracilis, 362
    granulata granulata, 362
    granulata humilis, 414
    granulata levis, 414
    granulata kelleri, 414
    granulata polonica, 362
    porkornyi. 414
    pommerana, 362
stephani, Praeglobotruncana, 358
sterkii, Reophax, 208
stilla, Lagenammina, 24
stilus, Archaediscus, 74
stilus eurus, Archaediscus, 78
Stilostomella midwayensis, 466
Stomasphaera, 21
    brassfieldensis, 23, 26
Storthosphaera, 21, 23
    malloryi, 26
striatula, Spiroloculina, 462
strigillatus, Bolivinoides, 326
strumosum, Ophthalmidium, 208, 221, 255
subalata, Lenticulina, 226, 258
subarcticum, Elphidium, 476
subbotinae, Globorotalia, 532
subconicus, Cancris, 470, 516
subeocaenus, Verneuilinoides, 460
subfavosa, Thurammina, 166
subhaidingerii, Cibicides, 566
sublaevis, Saracenaria, 184
sublimis, Pseudoendothyra, 72
sublustris, Recurvoides, 232
subnodosum, Elphidium, 476
subparvula, Glomospira, 98, 105, 110, 116
subquadrata, Warnantella, 82
subrotunda, Faujasina, 552
subtrochus, Marssonella, 410
aff. *subvesicularis, Trochammina*, 460
succedens, Cibides (Cibicidina), 518
succedens, Cibicidoides, 474
Svatkrina perlata, 482
syzranica, Palaeospiroplectammina, 71
szajnochae praecursor, Reussella, 358
szajnochae szajnochae, Reussella, 360

tangentialis, Alabamina, 467
Tappanina selmensis, 528
tenellus, Cibicides, 518
tenera collenoti, Lingulina, 106, 120, 178
tenera, Lingulina, 99
'*tenera, Lingulina*', 99
tenera occidentalis, Lingulina, 178
tenera pupa, Lingulina, 178

tenera subprismatica, Lingulina, 179
tenera subsp. A, *Lingulina*, 180
tenera substriata, Lingulina, 179
tenera tenera, Lingulina, 106, 120, 179
tenera tenuistriata, Lingulina, 179
tenera var. *octocosta, Lingulina*, 120
tenerrima, Buccella, 468
tenuicostata, Epistomina, 262
tenuipustulata, Uvigerina, 484
tenuistriata, Dentalina, 171
tenuistriata, Palmula, 183
terquemi bicostata, Frondicularia, 172
terquemi, Dentalina, 171
terquemi, Frondicularia, 99
terquemi muelensis, Frondicularia, 174
terquemi sulcata, Frondicularia, 174
terquemi subsp. A., *Frondicularia*, 174
terquemi subsp. B., *Frondicularia*, 174
terquemi terquemi, Frondicularia, 175
testudinaria, Lingulina, 180
tetraloculi, Nevillella, 64
Tetrataxis conica, 63
    pressula pressula, 78
    sp., 64, 208, 220
Textularia bettenstaedti, 411
    chapmani, 320
    descrescens, 459
    jonesi, 89
    jurassica, 252
    plummerae, 459
    sp. 1, 411
thalmanni, Gavelinella, 340
thanetana, Spiroplectammina, 510
thanetensis, Bulimina, 515
thola, Hemisphaerammina, 23, 24
Tholosina, 21, 23
    sedentata, 26
    sp., 26
Thurammina, 21, 23
    jurensis, 164
    papillata, 26, 460
    subfavosa, 166
    tubulata, 26
tibia, Nubeculinella, 167
Tolypammina, 21, 22, 23
Tolypammina, sp., 26
    vagans, 460
torsicostata, Nodosaria, 464
tosaensis, Globorotalia, 558
tourainensis, Gavelinopsis, 414
Tournayella discoidea, 52
    kissela, 58
triangularis, Tristix, 260
tricarinella, Lenticulina, 226, 258
Trifarina angulosa, 482
    bradyi, 482
    fluens, 482
trigonalis, Bulimina, 470
trigonula, Triloculina, 512
trilobus, Gloigerinoides, 556
Triloculina trigonula, 512
triloculinoides, Globigerina, 530
Triplasia althoffi, 220

triseriata, Guembelitria, 532
Tristix liasina, 184
   oolithica, 229
   triangularis, 260
Tritaxia capitosa, 411
   pyramidata, 320
   singularis, 320
Trochammina, 98, 140
   canningensis, 166, 220, 234
   depressa, 320
   globigeriniformis, 234
   ruthvenmurrayi, 460
   squamata, 234
   squamosa, 105, 118
   aff. *subvesicularis*, 460
   ?*volupta*, 510
Trochammina, sp. 1, 166
Trochamminoides iregularis, 460
cf. *trochoides, Cornicospirillina*, 210
trochoides, Dorothia, 408
Trocholina nodulosa, 255
truncana, Bulimina, 470
truncanus, Cibicidoides, 474
Truncorotaloides rohri, 532
tryphera, Verneuilinoides, 220, 234
tschernyshinensis, Palaeospiroplectammina, 52
tuberculata, Tuberendothyra, 56
Tuberendothyra tuberculata, 56
tubulata, Thurammina, 26
tumida, Endothyra, 54
Turrilina alsatica, 482
   andreaei, 528
   brevispira, 482, 528
Turritellella, 22

undulata, Marginulina, 259
ungerianus, Cibicides, 551
 cf. *ungerianus, Cibicides*, 518
ungeri, Elphidium, 476
Unicosiphonia zsigmondyi, 552
Uralodiscus rotundus, 62
Urbanella fragilis, 63
   miranda matura, 80
usakensis, Gavelinella, 340
Uslonia cf. *permira*, 23, 28
ustulatum, Elphidium, 476
Uvigerina acuminata, 482
   batjesi, 484, 528
   eocaena, 484
   farinosa, 484
   germanica, 482, 528
   jacksonensis, 484
   muralis, 528
   peregrina, 484
   pygmaea langenfeldensis, 484
   pygmaea langeri, 484
   semiornata, 554
   semiornata saprophila, 484
   spinicostata, 484
   tenuipustulata, 484
   venusta saxonica, 484
   sp. A, 484

Uvigerinella abbreviata, 485
Uvigerinammina jankoi, 411

vagans, Tolypammina, 460
Vaginulina arguta, 364
   barnardi, 260
   citharina clathrata, 184
   harpa, 229
   legumen, 229
   listi, 186
   mediocarinata, 364
   '*prima*', 260
   spinigera, 466
   reideli, 364
Vaginulinopsis
   decorata, 466
   decorata subsp. A, 466
   denticulatacarinata, 186
   exarata, 186
   humilis praecursoria, 364
   scalariformis, 364
   sp. B, 466
valanginiana, Saracenaria, 352
valendisensis, Conorboides, 328
Valvulinella cf. *conciliata*, 66
   conciliata, 78
   latissima, 81
Valvulineria gracillima, 414
   lenticula, 364
   mexicana grammensis, 485
   petrolei, 485
varians haeusleri, Dentalina, 171
varians, Lenticulina, 176, 214
varians, Matanzia, 456
varians subsp. D, *Lenticulina*, 176
variolarius, Nummulites, 534
velascoensis, Cibicidoides, 474
venusta saxonica, Uvigerina, 484
Verneuilina muensteri, 320
Verneuilinoides, 140
   chapmani, 411
   mauritii, 166
   neocomiensis, 320
   tryphera, 220, 234
   subeocaenus, 460
   sp. 1, 140, 166, 234
   sp. 2, 234
Virgulinella pertusa, 485, 554
Viseidiscus sp., 58
Vissariotaxis compressa, 71
   exilis, 70
vitrea, Epistominella, 520
aff. *volgensis, Pseudoammodiscus*, 74
?*voltziana, Cibicidoides*, 326
volubilis, Lenticulina, 226
volupta, Trochammina?. 510
cf. *volyanica, Omphalotis*, 70
sp. cf. *L. vombensis, Lingulogavelinella*, 352
vulgare, Cribroelphidium, 474
vulgaris, Earlandia, 52
vulgata, Pseudonodosaria, 183, 259

walteri, Haplophragmoides, 455
gr. *walteri, Recurvoides*, 458
Warnantella subquadrata, 82
washitensis, Favusella, 332
Webbinelloidea, 21, 23
   similis, 26
Wellmanella antiqua, 322
westi, Cibicides, 518
Wetheredella sp., 23
wetherelli, Marginulinopsis, 522
whitei, Osangularia, 354

Whiteinella aprica, 364
   baltica, 364
wilcoxensis, Globorotalia, 532
wilcoxensis, Pseudohastigerina, 532
wilsoni, Pseudolituotuba, 72
woodi, Globigerina, 554

zealandica, Globorotalia, 557
ziganensis, Bogushella, 71
zsigmondyi, Unicosiphonia, 552